Springer Series in
Surface Sciences

24

Editor: Robert Gomer

Springer Series in **Surface Sciences**

Editors: G. Ertl and R. Gomer   Managing Editor: H. K. V. Lotsch

S. Y. Tong   M. A. Van Hove
K. Takayanagi   X. D. Xie (Eds.)

The Structure of Surfaces III

Proceedings of the 3rd International Conference on the Structure of Surfaces (ICSOS III) Milwaukee, Wisconsin, USA, July 9–12, 1990

With 374 Figures

Springer-Verlag

Berlin Heidelberg New York
London Paris Tokyo
Hong Kong Barcelona
Budapest

Professor Shuk Yin Tong
Department of Physics, University of Wisconsin,
Milwaukee, WI 53201, USA

Michel A. Van Hove, Ph. D.
Lawrence Berkeley Laboratory,
Berkeley, CA 94720, USA

Professor Kunio Takayanagi
Materials Science and Engineering, Tokyo Institute of Technology,
4259 Nagatsuda, Midori-ku, Yokohama 227, Japan

Professor Xide D. Xie
Department of Physics, Fudan University,
Shanghai, People's Rep. of China

Series Editors

Professor Dr. Gerhard Ertl
Fritz-Haber-Institut der Max-Planck-Gesellschaft, Faradayweg 4–6,
W-1000 Berlin 33, Fed. Rep. of Germany

Professor Robert Gomer, Ph. D.
The James Franck Institute, The University of Chicago, 5640 Ellis Avenue,
Chicago, IL 60637, USA

Managing Editor: Dr. Helmut K. V. Lotsch
Springer-Verlag, Tiergartenstrasse 17,
W-6900 Heidelberg, Fed. Rep. of Germany

ISBN 3-540-54171-3 Springer-Verlag Berlin Heidelberg New York
ISBN 0-387-54171-3 Springer-Verlag New York Berlin Heidelberg

Typesetting: Camera ready by authors
54/3140-543210 – Printed on acid-free paper

Preface

This book collects together selected papers presented at the Third International Conference on the Structure of Surfaces (ICSOS-III). The conference was held at the University of Wisconsin, Milwaukee, Wisconsin, USA on July 9-12, 1990. The International Organizing Committee members were:

M. A. Van Hove (Chairman)	X. D. Xie (Vice-Chairman)
S. Y. Tong (Treasurer)	D. L. Adams
R. J. Behm	A. M. Bradshaw
M. J. Cardillo	J. E. Demuth
J. Eckert	G. Ertl
S. Ino	J. B. Pendry
M. N. Read	J. R. Smith
J. Stohr	R. Tromp
J. F. van der Veen	D. P. Woodruff
K. Yagi	W. S. Yang

The ICSOS meetings serve to assess the status of surface structure determination and the relationship between surface or interface structures and physical or chemical properties of interest. The contributions to this book cover: theoretical and experimental structural techniques; structural aspects of metal and semiconductor surfaces, including relaxations and reconstructions, as well as adsorbates and epitaxial layers; phase transitions in two dimensions, roughening and surface melting; defects, disorder and surface morphology.

Milwaukee, Berkeley, Tokyo, Shanghai
September 1990

S. Y. Tong
M. A. Van Hove
K. Takayanagi
X. D. Xie

V

Acknowledgements

We wish to acknowledge the many organizations and individuals whose contributions made possible the Third International Conference on the Structure of Surfaces and these Proceedings. We express our gratitude to our host institution: the University of Wisconsin-Milwaukee; and our sponsors: IUPAP (International Union of Pure and Applied Physics), University of Wisconsin-Milwaukee, Lawrence Berkeley Laboratory and Marquette University. We also thank our exhibitors: Springer-Verlag and Kluwer Academic Publishers.

Particular thanks to all the individuals who contributed much to the smooth running of both the conference and the preparation of the proceedings, especially Laurie Girard, the conference secretary, and the members of the Local Organizing Committee: S.Y. Tong, C.R. Aita, T.L. Barr, D.K. Saldin, B.P. Tonner, W.T. Tysoe, M.B. Webb, W.E. Brower, C.G. Chen. An important element was of course the contribution from the International Advisory Committee members: M. Aono, J.C. Bertolini, D.J. Chadi, G. Comsa, M. De Crescenzi, F. Garcia Moliner, V.A. Grazhulis, D. Haneman, A. Kawazu, A.A. Lucas, T.E. Madey, S. Nakamura, A.G. Naumovetz, F. Netzer, D. Norman, J. Nørskov, P.R. Norton, M. Prutton, I.K. Robinson, G. Rovida, G.A. Somorjai, D.S. Wang.

Contents

Part II   Clean Metals

II.1 Elemental Metals

II.2 Alloys

Part III   Adsorption on Metals

III.1 Metallic Adsorption and Growth

Part IV   **Clean Semiconductors**

IV.1 Elemental Semiconductors

Part V Adsorbates on Semiconductors

V.1 Metallic Adsorption

Part VI   Oxides

Introduction

In recent years, our understanding of atomic-scale surface structure and properties has advanced considerably. Thanks to a variety of theoretical and analytical tools, surface structure is being determined and understood with ever increasing detail and accuracy. For instance, subtle adsorbate-induced relaxations and reconstructions are observed ever more frequently. Dynamical processes, including phase transitions, diffusion, growth, roughening, melting and reactions are also being explored in greater structural detail than in the past. At the same time, the relationship between surface structure and chemical, electronic, vibrational and other properties is rapidly being elucidated in many instances.

Particularly noteworthy developments include: novel forms of electron microscopy, such as forward focusing, holography and various microscopic versions of well-established techniques (e.g. photoemission microscopy); major progress in the theory of LEED; investigations of surface stress; advances in scanning tunneling and atomic-force microscopy; exploration of growth processes on the atomic scale, especially for metallic layers growing on metallic and semiconducting substrates.

These proceedings cover the latest developments in the field, grouped according to the following scheme.

Part I reports developments in theoretical and experimental techniques, including in particular theory of surface structure, forward focusing, and LEED. Progress is also reported in other techniques that use the scattering of electrons and atoms. New applications of scanning tunneling microscopy are of particular interest.

Part II addresses clean metal surfaces, their reconstructions and relaxations. This includes alloy surfaces, as well as kinetic effects and defect structures.

Part III, adsorbates on metal surfaces are described: metallic adsorption and growth play a prominent role here, while non-metallic adsorbates receive special attention due to their structural effect on the substrate, and due to induced phase transitions.

Part IV focuses on clean semiconductor surfaces and inlcudes both elemental and compound materials. Properties of vicinal surfaces and reconstructions receive the most attention here.

Adsorption on semiconductors, covered in Part V, concerns overwhelmingly the question of metal layer growth, where considerable progress emerges.

Finally, Part VI addresses oxide surfaces, as well as their interfaces, which are studied by a wide variety of techniques.

Part I

Techniques

Microscopic Origins of Stress on Semiconductor Surfaces

R.D. Meade and D. Vanderbilt

Lyman Laboratory of Physics, Harvard University, Cambridge, MA 02138, USA

Abstract. First-principles calculations of stress have been performed on substitutional 1×1 Si(111)and 2×1 Si(100)surfaces, using a variety of column-III, IV, and V adsorbates. Trends in surface stresses are understood in terms of three contributing factors: the relative atomic size of the adsorbate and substrate atoms, the chemical nature of the adsorbate species, and the bonding topology of the surface reconstruction.

1. Introduction

Despite the mounting importance of the role of stress in understanding surface phenomena[1]-[6], there has been little work on the microscopic origins of this important quantity. In this paper, we report the results of a series of state-of-the-art LDA pseudopotential calculations of surface stress on 1×1 substitutional surfaces of Si(111) and the 2×1 substitutional surfaces of Si(100). The method of calculation and some of the results have been discussed in Ref.[7]. By examining the trends of the stress as function of the row and column of the adsorbate species, we have identified three principal sources of surface stress. First, there may be an atomic size mismatch between the adsorbate and substrate; although this is most often cited as the origin of surface stress, we find that it is frequently not the dominant effect. A more important origin of surface stress is the chemical nature of the adsorbate, which will effect the hybridization of the surface atoms. Finally, we find that an unusual bonding topology of a surface reconstruction will also induce a significant surface stress.

We also discuss the recent experiments of Martinez *et al.*[8] and Schell-Sorokin *et al.*[9]. They found that a difference of surface stress on two faces of a thin wafer will induce the wafer to bend. They exploited this effect to determine differences of surface stress, and these measurements compare favorably with our calculations.

2. Size Effect

The most obvious source of surface stress is an atomic size effect, corresponding to the intuitive notion that as the atomic size of the adsorbate increases, the compressive stress of the surface increases. For example, a comparison of the 1×1 Si(111) and 1×1 Ge:Si(111) surfaces in Table 1 reveals a greater compressive (more negative) stress for the latter. This increased compressive stress is a result of the larger size of the Ge atom, which has a covalent radius 4%

Springer Series in Surface Sciences, Vol. 24   **The Structure of Surfaces III**
Editors: S.Y. Tong · M.A. Van Hove · K. Takayanagi · X.D. Xie
© Springer-Verlag Berlin, Heidelberg 1991

Table 1. Stresses in eV/(1×1 cell) and bond angles of 1×1 substitutional surfaces on Si(111). σ_{ii} is the diagonal component of the stress tensor, which must be isotropic by symmetry. A bond angle of 120° (109.4°) corresponds to an sp^2 (sp^3) hybridization.

	B:Si(111)		
σ_{ii}	4.87		
θ	119.0°		
	Al:Si(111)	Si(111)	
σ_{ii}	-6.45	-0.54	
θ	117.9°	114.0°	
	Ga:Si(111)	Ge:Si(111)	As:Si(111)
σ_{ii}	-4.45	-1.12	2.27
θ	119.3°	111.5°	104.5

larger that of Si[10]. Similarly, the large tensile stress of the 1×1 B:Si(111) (see Table 1) is an extreme example of the atomic size effect. The ideal B-Si bond length, as derived from covalent radii, is significantly less the minimum bond length consistent with the C_{3v} symmetry, inducing a strong tensile stress. (The decrease in the magnitude of the compressive stress between the 1×1 Al:Si(111) and the 1×1 Ga:Si(111) surfaces is somewhat counterintuitive. However, gallium has a number of anomalous properties, e.g. its atomic size is the same as that of aluminum, because it is the first Group III element with a filled d-shell[11]).

3. Chemical Effect

Whereas atomic size effects were sufficient to understand the changes in surface stress as a function of the row of the adsorbate or substrate species, it does not explain the trends in the stress as the column of the adsorbate species is varied. Arsenic, for instance, is the same size as silicon, but 1×1 As:Si(111) is under a strong tensile stress, whereas 1×1 Si(111) is under mild compressive stress. In order to understand these trends, we must consider the *chemical* nature of the adsorbate species. For the Group III elements, such as Ga, the adsorbate contributes only three electrons to the bonds with its neighbors. In this case, the surface band is empty, and so the surface energy is lowered by sp^2 hybridizing the Ga atom. Note, from Table 1, that the bond angles of the Group III adsorbates are all near 120°. Because increasing the bond angles necessarily decreases the bond lengths (see Fig. 1), the relaxed bond lengths of Ga and Al are less than the ideal bond lengths, and so the surface is under a compressive stress. Each arsenic adsorbate, on the other hand[7], prefers to hybridize with bond angles less than those of a perfect tetrahedron. With decreasing bond angles, the bond lengths increase, inducing a tensile stress.

Thus the surface stresses and relaxations for different adsorbates within a row of the periodic table may be understood in terms of the *chemical* nature

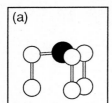 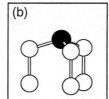 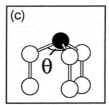

Figure 1. Relaxed positions of three 1×1 Si(111) substitutional surfaces. The top layer (dark spheres) represent Ga, Ge and As in (a), (b), and (c) respectively. From left to right, bond angles decrease and surface stresses increase.

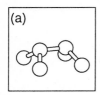 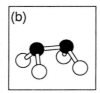

Figure 2. A side view of the relaxed dimers of the (a) 2×1 Si(100) and (b) 2×1 As:Si(100) dimers.

of the chemisorbed species. Note that the surface relaxations *cannot* be understood in terms of a size effect alone, as the calculated bond lengths obey the *opposite* trend from that predicted by covalent radii.

4. Topological Effect

In addition to size and chemical effects, surface stress may also result from any unusual bonding configurations present at a surface. For instance, we have found that the tensile stress associated with the 2×1 Si(100) is a consequence of the relaxation mechanism of the surface cell[7]. Our calculations indicate that the surface is under a strong tensile stress parallel to the dimer bond, and a weaker compressive stress perpendicular to the dimer bond (see Table 2), consistent with the results of Payne *et al.*[12]. We can understand the topological source of this stress in a simple manner. The two surface atoms which form a bond at the (100) surface would normally be second neighbors. When the surface reconstructs, they move closer together, breaking the symmetry of the ideal surface by dimerizing. The tensile surface stress simply reflects the fact that decreasing the lattice constant parallel to the dimer bond moves these atoms closer together, strengthening the dimer bond and lowering the surface energy. The compressive stress perpendicular to the bond, though harder to understand, may result from hybridization effects analogous to the 1×1 Si(111)

Table 2. Stresses of 2×1 Si(100) dimer surfaces, in eV/(1×1 cell).

Structure	$\sigma_\parallel$	$\sigma_\perp$
2×1 Si(100)	1.56	-0.88
2×1 As:Si(100)	2.41	2.35

surface. The anisotropy of the surface stress is quite important, since it leads to the formation of stress domains [2].

A number of other cases in which the bonding topology of a surface has a dominant influence on the surface stress have been discussed previously. These include the tensile stress associated with adatom covered Si(111) and Ge(111) surfaces[7], the π-bonded chain structure[13] of cleaved Si(111), which displays a strong tensile stress primarily along the chain direction[14]; and the dimers and surface stacking faults which are present in the 7×7 Si(111) reconstruction, both of which introduce a tensile stress[14].

5. Combinations of Sources

In general, the observed surface stress may result from a combination of the factors discussed above, which is exemplified by the chemisorbed adatom surfaces on Si(111) [7], and the chemisorbed 2×1 Si(100)surfaces. Both the 2×1 Si(100) and the 2×1 As:Si(100) surfaces have $\sigma_{\parallel} > \sigma_{\perp}$, although the anisotropy of the stress tensor is small in the latter case. The degree of tensile stress, however, is modulated by the chemical nature of the adsorbate (see Table 2). Like the 1×1 As:Si(111) surface, the arsenic adsorbates on the 2×1 As:Si(100) surface have lone pairs, they relax by decreasing the bond angles, which increases the bond lengths, leading to a tensile stress.

6. Measurement of Surface Stress

In this section, we discuss the recent experiments of Martinez et al.[8]and Schell-Sorokin et al.[9], who measured differences of surface stresses. The experimental method consists of adsorbing an element on just one side of a thin wafer of silicon. Differing stresses on the two faces will induce the wafer to bend, lowering the surface energy at the cost of distorting the bulk. The details of these experiments can be found elsewhere[8],[9]. The results of the experiments are given in Table 3. The agreement between theory and experiment is quite good, especially on the Si(111) surface.

Table 3. Comparison of surface stress as predicted by calculations and measured by experiment. For the (111) surfaces, $\bar{\sigma}$ is the diagonal element of the stress tensor. In the second and third columns, $\Delta\bar{\sigma}$ is the difference in stress relative to the 7×7 Si(111) surface. For the (100) surfaces $\bar{\sigma}$ is the average of the stress parallel and perpendicular to the dimer rows, and $\Delta\bar{\sigma}$ is the difference in $\bar{\sigma}$ relative to the 2×1 Si(100) surface. All stresses in eV/(1×1 cell).

Structure	$\bar{\sigma}$ Theory	$\Delta\bar{\sigma}$ Theory	$\Delta\bar{\sigma}$ Exp.
7×7 Si(111)	~ 2.6		
$\sqrt{3} \times \sqrt{3}$ Ga:Si(111)	1.4	-1.2	-1.0
1×1 As:Si(111)	2.3	-0.3	-0.5
2×1 Si(100)	0.4		
2×1 As:Si(100)	2.4	2.0	1.2

Although Martinez *et al.* and Schell-Sorokin *et al.* have succeeded in measuring differences in surface stress, there are no *absolute* measurements of stress at this time. One such measurement would be important, because it would serve as a reference against which to determine other surface stresses. Such a determination could be made, in principle, by measuring the lattice constant of wafers of different thickness. In practice, this procedure may be difficult to perform.

Thanks to R.E. Martinez and J.A. Golovchenko for making data available prior to publication. Support for this work was provided by the Harvard Materials Research Laboratory under U.S. National Science Foundation (NSF) Grant No. DMR-89-20498. Supercomputer time was provided by the Harvard MRL and by the John von Neumann Center. Finally, D.V. and R.D.M. wish to acknowledge the additional support of the Alfred P. Sloan Foundation and Harvard University respectively.

References

[1] F.K. Men, W.E. Packard, and M.B. Webb, Phys. Rev. Lett.**61**, 2469(1988).
[2] O.L. Alerhand, D. Vanderbilt, R.D. Meade,and J.D. Joannopoulos, Phys. Rev. Lett **61** , 1973 (1988); Erratum: **62** , 116 (1989).
[3] R.S. Becker, T. Klitsner and J.S. Vickers, J. of Microscopy, **152** , 157 (1988).
[4] R. Headrick, I.K. Robinson, Vlieg, C. Feldman, Phys. Rev. Lett **63** , 1253 (1989).
[5] P. Bedrossian, R.D. Meade, K. Mortensen, D.M. Chen, D. Vanderbilt, J. Golovchenko, Phys. Rev. Lett **63** , 1257 (1989).
[6] I.W. Lyo, E. Kaxiras, P. Avouris, Phys. Rev. Lett **63** , 1261 (1989).
[7] R.D. Meade and D. Vanderbilt, Phys. Rev. Lett **63** , 1404, (1989).
[8] R.E. Martinez, W.M. Augustyniak, and J.A. Golovchenko, Phys. Rev. Lett **64** , 1035 (1990).
[9] A. J. Schell-Sorokin and R.M. Tromp, Phys. Rev. Lett **64** , 1039 (1990).
[10] Because the electronegativity of these elements are close in value, we are free to use the covalent radii to estimate bond lengths without correcting for ionic effects.
[11] N.N. Greenwood and A. Earnshaw, *Chemistry of the Elements* (Pergamon, Oxford, 1984).
[12] M.C. Payne, N. Roberts, R.J. Needs, M. Needels and J.D. Joannopoulos, Surf. Sci., **211**, p.1 (1989).
[13] K.C. Pandey, Phys. Rev. Lett. **47** , 1913 (1981), and **49** , 223 (1985).
[14] D. Vanderbilt, Phys. Rev. Lett. **59** , 1456 (1987).

Energy Density Calculations Applied to GaAs Polar Surfaces

N. Chetty and R.M. Martin

Department of Physics and Materials Research Laboratory,
University of Illinois at Urbana-Champaign, 1110 W. Green St.,
Urbana, IL 61801, USA

Abstract. Density Functional Theory has been successfully applied in solid state calculations where it has been customary to extract just a single number, namely the total energy. Within this formalism, we show that we can in addition compute an energy density which is useful despite its inherent non-uniqueness [1]. We present the basic theoretical formulation of the energy density and apply this concept to the study of GaAs polar surfaces. We demonstrate that the energy density is computationally efficient, reproduces known results for (100) surfaces, and enables the calculation of the formation enthalpy of a single isolated polar (111) or ($\overline{1}\overline{1}\overline{1}$) surface which is not possible using conventional total energy methods.

1. Introduction

An energy density function $\mathcal{E}(\mathbf{r})$ is a local function in real space which is defined such that, for appropriate cases, integrals

$$\int_V d\mathbf{r}\, \mathcal{E}(\mathbf{r}) \rightarrow E \tag{1}$$

can be identified as the total energy E of the system enclosed in volume V. If V completely contains an isolated system, then the integral is always well defined. However, if V is only a part of the volume of a system, then the integral (and hence $\mathcal{E}(\mathbf{r})$) is not in general well defined [1]. In the present work we show that, nevertheless, there are cases where integrals over volumes, such as surface regions, are well defined and useful.

The difficulties in constructing such a local function arise from interactions between particles at different positions, and the quantum kinetic energy which make any energy density function non-unique. Any function which integrates to zero can be added to $\mathcal{E}(\mathbf{r})$. An example of such gauge transformations is the well known symmetric and non-symmetric forms of the kinetic energy density. We note that the charge density is also a local function in real-space but it differs from the energy density in the all important fact that it is a physical observable and the solution of the Poisson equation [2] is a *unique* charge density which is not subject to gauge transformations.

2. Theoretical Formulation

The basis for our theoretical development is the integral expression for the total energy in density functional theory [2] (see e.g. Ref. [3]) of a many electron system interacting in an external potential. We use *ab initio* pseudopotentials [4] to describe the crystalline solids of interest here, and the local density approximation (LDA) [2, 5] is assumed.

There are two obvious choices for the kinetic energy density, viz. the symmetric $\mathcal{T}_s$ and the asymmetric $\mathcal{T}_a$ forms [6]. The symmetric form may be argued to be more fundamental since it is the one that directly enters the variation principle from which the Schrödinger equation is derived; however, the integrals over surface regions are entirely equivalent. The exchange-correlation energy density $\mathcal{X}(\mathbf{r})$ is uniquely defined by the charge density within the LDA.

The interaction Hamiltonian can be transformed to a more useful form by first defining a fictitious ionic charge density [7], which is positive definite and which is used to re-write the potential energy terms in a way that is equivalent to the Ewald transformation [8]. The final contributions to the energy density are

$$
\begin{aligned}
\mathcal{E}(\mathbf{r}) \quad \rightarrow \quad & \mathcal{T}(\mathbf{r}) + \mathcal{X}(\mathbf{r}) + \mathcal{M}(\mathbf{r}) + \mathcal{L}(\mathbf{r}) \\
& + \sum_{\mathbf{R}_\mu} \mathcal{N}^\mu \, \delta(\mathbf{r} - \mathbf{R}_\mu) + \sum_{\mathbf{R}_\mu} \mathcal{S}^\mu \, \delta(\mathbf{r} - \mathbf{R}_\mu) \,.
\end{aligned} \tag{2}
$$

$\mathcal{M}(\mathbf{r})$ is the Maxwell energy density [9, 10] which depends explicitly on the long range electric fields in the system. $\mathcal{L}(\mathbf{r})$ incorporates the non-Coulombic part of local pseudopotential interaction and it is a sum of short range contributions centered on each of the lattice sites $\mathbf{R}_\mu$. The non-local pseudopotential interaction does not facilitate the construction of a local energy density. This interaction is, however, only operative in the core region and its integral over this region is well defined. We can therefore associate a nonlocal energy $\mathcal{N}^\mu$ at each of the individual lattice sites. The final constant contribution to the total energy is the self-interaction energy $\mathcal{S}^\mu$ of the pseudo ionic cores. This contribution to the total energy is essential to cancel the effects of the fictitious pseudo ionic charges included in the Maxwell term.

The physical quantity that we are interested in is the total energy associated with definite regions in space. If $\mathcal{E}(\mathbf{r})$ is to be a useful construct, then one must in addition consider those regions in space for which the integrated result is gauge independent. For periodic systems there is always a smallest unit , namely the primative unit cell of volume Ω such as the Wigner Seitz (WS) unit cell, for which all gauge terms must integrate to zero. We will use this fact to devise a means to determine gauge independent results. We begin by defining a macroscopic average of the energy density by convolving $\mathcal{E}(\mathbf{r})$, which has microscopic variations that are gauge dependent, with a function $\omega(\mathbf{r})$ which, apart from being real, non-zero in some neighbourhood of $\mathbf{r}$ and having unit normalization, is arbitrary in form [10, 11]

$$\bar{\mathcal{E}}(\mathbf{r}) = \int d\mathbf{r}'\, \omega(\mathbf{r} - \mathbf{r}')\, \mathcal{E}(\mathbf{r}')\,. \tag{3}$$

$\bar{\mathcal{E}}(\mathbf{r})$ is termed the cell averaged energy density, and its constancy defines the bulk region in the crystal with a resolution that is characteristic of the filter function $\omega(\mathbf{r})$. Any deviation from its constant bulk value will be attributed to the perturbation in question. Then the simplest such filter function may be chosen as $\omega(\mathbf{r}) = 1/\Omega$ inside the WS cell and 0 outside. Associated with each point $\mathbf{r}$ is an average of the energy density function over a Wigner Seitz cell for which the gauge terms exactly integrate to zero. It follows that $\bar{\mathcal{E}}(\mathbf{r})$ is gauge independent in the bulk crystal, and integrals over regions such as surfaces can be used to determine physically realizable quantities [12].

3. GaAs Polar Surfaces

The energy density is a computationally more efficient means of studying the (100) surfaces than traditional total energy methods. To demonstrate this, we investigated the simplest non-stoichiometric GaAs (100) surface, namely the 1×1 ideal surface and we compared the results of the surface energies of formation of this ideal surface using the two methods.

Total energy calculations require that the supercell be symmetric, ie. both surfaces in the supercell must be the same [13, 14]. To determine the results for both Ga terminated and As terminated surfaces, two separate calculations must therefore be done, namely one with both Ga

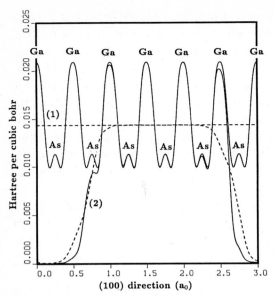

Figure 1 The planar averaged energy density (solid) and the corresponding 1-dimensional macroscopic average (dashed) for the (1) bulk, and the (2) asymmetric GaAs (100) (1 × 1) surface cell.

terminated surfaces and the other with both As terminated surfaces. By using the local energy density function the information for both surfaces can be extracted directly from a single asymmetric unit cell that has both Ga terminated and As terminated surfaces.

The present calculations were done using a supercell with a 6 Ry cutoff and 3 special points in the surface Brillouin zone. Although these 3 points are not sufficient for an accurate calculation, they are sufficient to demonstrate the use of the energy density. The asymmetric unit cell is comprised of 8 atomic layers and 4 vacuum layers, and it has both Ga and As terminations. The planar averaged energy densities and the corresponding one-dimensional macroscopic averaged quantities are plotted in Figure 1 for the asymmetric cell, and comparison is made with the results for the 12 atom bulk calculation. The figure shows clearly that the slab is sufficiently large that the energy density relaxes to its constant bulk value only a few atomic layers away from the surfaces.

The most important difficulty with the use of the asymmetric cell is the incorrect occupation of the gap surface states. Using the simple symmetry arguments by Appelbaum *et al.*[13] it appears that there is a $\frac{3}{4}$ filled dangling bond band at the Ga terminated surface and a $\frac{1}{4}$ filled bridge bond state at the As terminated surface. Charge transfer between these fractionally occupied states will hamper the self-consistency process, and the final lowest energy state in general corresponds to charged surfaces. The charge depletion from the Ga terminated surface is calculated to be $0.06 \pm 0.01e$.

The energy density was integrated for the Ga terminated surface in the asymmetric cell, and the results were compared with that for the Ga symmetric cell. Our results show that the energy differences are less than 0.1 eV per surface atom and that this difference is fairly insensitive to the limits of integration, the only condition being that the limits extend from a (100) plane of symmetry in the bulk to a plane in the vacuum where the energy density has decayed to zero.

The main conclusion from this investigation is that despite the charged surfaces resulting from the incorrect occupation of the surface states in the asymmetric cell, the surface energy of formation for the Ga terminated surface can still be determined to within 0.1 eV. To arrive at the final result for the surface energy of formation for this surface, the chemical potential for adding extra Ga atoms must be included [14, 15] because of the non-stoichiometric nature of this surface; we have, however, chosen not to pursue this here. We note that the same conclusions were reached regarding the As terminated (100) surface.

We turn now to the asymmetric (111) orientation where total energy calculations are unable to extricate the separate energies of formation of the (111) and the ($\overline{111}$) surfaces inherent in the supercell. In their study of the (111) GaAs surfaces, Kaxiras et al. [15] imposed mirror symmetry by using fictitious atoms for the two central layers which were modelled by fractionally charged atoms so chosen to prevent charge transfer between the central region and the surface. The energies of formation for the different reconstructed models that they considered were referred to this ideal fictitious system.

The absolute formation enthalpy for these surfaces can be computed by using an energy density, and the results for the ideal such surface will place the work of Kaxiras et al. on an absolute energy scale. This will enable the direct comparison between other surface orientations in the crystal which is crucial in the study of faceting [14, 16].

We studied the ideal 1×1 GaAs (111) surface using a 12 Ry cutoff and 13 special points, and we found the charge transfer from the Ga terminated surface to be $0.06 \pm 0.01e$. We conclude that the discrepancy in the energy of formation is not greater than 0.1 eV. Assuming equilibrium with the bulk, total energy calculations give the sum of the energy of formation for both surfaces as 2.7 eV. For this non-stoichiometric ideal surface there is $\frac{1}{4}$ extra surface atom whose source we have assumed to be the free atomic gaseous state. This corresponds to a choice of reference that is different for the two surfaces and our preliminary results for the energy of formation for the Ga terminated surface and As terminated surface are 0.4 eV and -0.2 eV respectively. For realistic experimental situations, the chemical potentials must be included [14, 16] to arrive at the final energies of formation for the individual surfaces. Under As rich conditions, the formation enthalpy for the As terminated surface is 0.6 eV and for the Ga terminated surface is 2.1 eV. Under conditions of excess Ga, the results are 1.1 ev and 1.6 eV respectively.

References

[1] Forces and Stresses in Molecules by R.P. Feynman, thesis, Massachusetts Institute of Technology (1939) *unpublished*.

[2] P. Hohenberg and W. Kohn, Phys. Rev. **136**, 3864 (1964), W. Kohn and L.J. Sham, Phys. Rev. **140** (4A), A1133 (1965), Theory of the Inhomogenous Electron Gas ed. S. Lundqvist and N.H. March, Plenum Press, New York (1983).

[3] J. Ihm, A. Zunger and M.L. Cohen, J. Phys. C**12**, 3792 (1979).

[4] D.R. Hamann, M. Schlüter and C. Chiang, Phys. Rev. Lett. **43** (20), 1494 (1979); G.B. Bachelet, D.R. Hamann and M. Schlüter, Phys. Rev Bbf 26 (8), 4199, (1982), D. Vanderbilt, Phys. Rev. B**32** (12), 8412 (1985).

[5] D.M. Ceperley and B.J. Alder, Phys. Rev. Lett. **45**, 566 (1980), J. Perdew and A. Zunger, Phys. Rev. B**23**, 5048 (1981).

[6] J.C. Slater, Phys. Rev. **51**, 846 (1937).

[7] G.B. Bachelet, H.S. Greenside, G.A. Baraff and M. Schlüter, Phys. Rev. B**24** (8), 4745 (1981).

[8] R.A. Coldwell-Horsfall and A.A. Maradudin, J. Math. Phys. **1**, 395 (1960).

[9] O.H. Nielsen and R.M. Martin, Phys. Rev. B**32** (6), 3792 (1985).

[10] Classical Electrodynamics by J.D. Jackson, John Wiley and Sons, New York (1975).

[11] A. Baldereschi, S. Baroni, R. Resta, Phys. Rev. Lett. **61** (6), 734 (1988); R. Resta, *unpublished notes*; R. Resta, *private communication*.

[12] N. Chetty, PhD thesis, University of Illinois (1990) *unpublished*.

[13] J.A. Appelbaum, G.A. Baraff and D.R. Hamann, Phys. Rev. B**14** (4), 1623 (1976).

[14] G.-X. Qian, R.M. Martin and D.J. Chadi, J. Vac. Sci. Technol. B**5** (4), 933 (1987).

[15] E. Kaxiras, Y. Bar-Yam, J.D. Joannopoulos and K.C. Pandey, Phys. Rev. B**33**, 4406 (1986); Phys. Rev. Lett. **57**, 106 (1986); Phys. Rev. B**35**, 9625 (1987); Phys. Rev. B**35** 9636 (1987);E. Kaxiras, K.C. Pandey, Y. Bar-Yam and J.D. Joannopoulos, Phys. Rev. Lett. **56**, 2819 (1986).

[16] D.K. Biegelsen, R.D. Bringons, J.E. Northrup and L.E. Swartz, *to be published*

First Principles Calculation of Lattice Relaxation at Low Index Surfaces of Ag and Cu

K.P. Bohnen[1], Th. Rodach[1], and K.M. Ho[2]

[1]Kernforschungszentrum Karlsruhe, INFP, Postfach 3640
  W-7500 Karlsruhe, Fed. Rep. of Germany
[2]Ames Laboratory – USDOE and Department of Physics,
  Ames, IA 50011, USA

Abstract. Using first principles total energy calculations, the lattice relaxation on low index surfaces of Ag and Cu are determined. While the (100) and (111)-surfaces show only small effects, the (110) surfaces exhibit pronounced multilayer relaxation which is typical for all fcc (110) surfaces. The results are in excellent agreement with experimental LEED and ion scattering studies.

For a long time, mostly experimental results from LEED and ion scattering gave information about the atomic arrangements of atoms at surfaces [1]. However, over the past few years it has also become possible to determine the structure of surfaces by first principles total energy calculations. While most of the studies so far have been done for semiconductor surfaces, now also a number of metal surfaces have been investigated successfully [2]. These investigations have established that first principles total energy calculations using local density functional formalism (LDA) [3] can determine surface structures in close agreement with the experimentally observed geometries.

   Most of the studies for metals have been carried out using the pseudopotential method [4]. This procedure allows for the elimination of core electrons which are not involved in the bonding. For determination of the equilibrium structure in complicated cases, the calculation of forces using the Hellman-Fegmann theorem [5] has been very helpful. These pseudopotential studies, however, have been restricted so far to systems with fairly delocalized electrons such that an expansion of the charge density in terms of a plane wave basis is rapidly convergent. With modern supercomputers, this limitation could be overcome for many systems where localized d-electrons play an important role. The noble metals Au, Ag and Cu are typical examples where d-electrons play an important role. While the 5d-electrons of Au are fairly delocalized, the 3d-electrons of Cu are strongly localized, thus excluding the successful study of Cu-surfaces within the pseudopotential method so far. We present here a systematic study of the three low index faces of Ag and Cu. These studies have been made possible by some recent progress in the development of a mixed basis representation for the charge density [6]. This formulation allows for the efficient treatment of systems with strongly localized electrons. Details of the surface calculations, construction of the pseudopotential, bulk properties and extensive convergence tests will be given elsewhere [7]. Results of our studies are summarized in Table 1 together with other theoretical studies and available experimental information.

   The first principles results are in excellent agreement with available experimental information. The (100) and (111) surfaces show only small effects while the (110) surfaces exhibit pronounced multilayer relaxations as seen for other fcc

Springer Series in Surface Sciences, Vol. 24   **The Structure of Surfaces III**
Editors: S.Y. Tong · M.A. Van Hove · K. Takayanagi · X.D. Xie
© Springer-Verlag Berlin, Heidelberg 1991

Table 1.
Relaxations Given in Percentage of Interlayer Spacing

		First Principles Total Energy Calculations	Other Theoretical Approaches		Experiments		
Cu(110)	Δd_{12}	-9.3	-8.7 [9]	-4.9 [8]	-8.5 [10]	-5.3 [11]	-7.5 [12]
	Δd_{23}	2.8	1.6	0.2	2.3	3.3	2.5
	Δd_{34}	-1.1	-1.2				
Cu(100)	Δd_{12}	-3.0	-3.8 [9]	-1.44 [8]	-1.2 [13]		
	Δd_{23}	0.1	-0.5	-0.3	0.9		
	Δd_{34}	-0.2	0.0				
Cu(111)	Δd_{12}	-1.3	-2.5 [9]	-1.4 [8]	-0.7 [14]		
	Δd_{23}	-0.6	-0.	0.1			
	Δd_{34}	-0.3	0.				
Ag(100)	Δd_{12}	-1.3	-3.0 [9]	-1.9 [8]			
	Δd_{23}	1.0	0.	0.			
	Δd_{34}	0.8	0.				
Ag(110)	Δd_{12}	-7.0	-6.9 [9]	-5.1 [8]	-5.7 [15]		9.5 [16]
	Δd_{23}	2.8	2.2	0.3	2.2		6.
	Δd_{34}	-0.2	-1.0		-3.5		-3.5
Ag(111)	Δd_{12}	-0.4	-1.9 [9]	-1.3 [8]			
	Δd_{23}	-0.2	0.1	0.			

(110)-surfaces. Our results for Ag (110) confirmed an earlier study [17] using the standard plane wave expansion of the charge density. The small differences are due to the improved treatment of the charge density, however, they are within the error bar of 0.5-1% which is related to kpt-sampling, slab thickness, etc. Besides having studied successfully the structural properties of the low index faces for Ag and Cu, calculations of the surface phonon spectrum have been carried out. Results will be published elsewhere [18].

Acknowledgment

Ames Laboratory is operated for the U.S. Department of Energy by Iowa State University under Contract No. W-7405-Eng-82. This work was supported by the Director for Energy Research and the Office of Basic Energy Sciences including a grant of computer time on the Cray Computers at the Lawrence Livermore Laboratory and by NATO collaborative research Grant No. RG(86/0516).

References

1. See for example: The Structure of Surfaces II, editors J. F. van der Veen and M. A. Van Hove, Springer Series in Surface Sciences (Springer, NY, 1988).
2. K. M. Ho, K. P. Bohnen, *Phys. Rev. B* **32**, 3446 (1985); K. M. Ho, K. P. Bohnen, *Europhys. Lett.* **4**, 345 (1987); C. L. Fu, S. Ohnishi, E. Wimmer, A. J. Freeman, *Phys. Rev. Lett.* **53**, 675 (1984).
3. P. Hohenberg, W. Kohn, Phys. Rev. B 136, 864 (1964); W. Kohn, L. J. Sham, *Phys. Rev. A* **140**, 1133 (1965).
4. W. F. Pickett, *Computer Physics Reports* **9**, 115 (1989).
5. H. Hellman, Einführung in die Quantentheorie (Deuticke, Leipzig 1937); R. P. Fegnman, *Phys. Rev.* **56**, 340 (1939).
6. C. Elsässer, N. Takeuchi, K. M. Ho, C. T. Chan, P. Braun, M. Fähnle, J. *Phys. Condens. Matter* **2**, 4371 (1990).
7. T. Rodach, K. P. Bohnen, K. M. Ho, to be published.
8. S. M. Foiles, M. J. Baskes, M. S. Daw, *Phys. Rev. B* **33**, 7983 (1986).
9. T. Ning, Q. Yu, Y. Ye, *Surf. Sci.* **206**, L857 (1988).
10. D. L. Adams, H. B. Nielsen, J. N. Andersen, *Surf. Sci.* **128**, 294 (1983).
11. J. Stensgard, R. Feidenhans'l, J. E. Sorensen, *Surf. Sci.* **128**, 281 (1983).
12. M. Copel, T. Gustafsson, W. R. Graham, S. M. Yalisave, *Phys. Rev. B* **33**, 8110 (1986).
13. D. M. Lind, F. B. Dunning, G. K. Walters, H. L. Davis, *Phys. Rev. B* **35**, 9037 (1987).
14. S. A. Lindgren, L. Wolldin, J. Rundgren, P. Westrin, *Phys. Rev. B* **29**, 576 (1984).
15. H. L. Davis, J. R. Noonan, *Surf. Sci.* **126**, 245 (1983).
16. E. Holub-Krappe, K. Horn, J.W.M. Frenken, R. L. Kraus, J. F. van der Veen, *Surf. Sci.* **188**, 335 (1987); and references therein.
17. C. Fu, K. M. Ho, *Phys. Rev. Lett.* **63**, 1617 (1989).
18. T. Rodach, K. P. Bohnen, K. M. Ho, to be published.

Surface Diffusion of Al(110)

P. Stoltze and *J.K. Nørskov*

Laboratory of Applied Physics, Building 307, Technical University of Denmark, DK-2800 Lyngby, Denmark

The self-diffusion of an Al(110) surface at temperatures up to the melting point is studied by molecular dynamics simulations using the effective medium theory to calculate the interatomic forces. The simulations show a diffusion in the surface which is strongly enhanced over that in the bulk. The diffusion is moderately anisotropic, predominantly parallel to the close-packed rows, and strongly dependent on the layer number. At temperatures where pre-melting is observed in the simulations the diffusion is dominated by adatoms 'skating' on top of the first layer.

Many metal surfaces show a gradual loss of order well below the melting point. This phenomenon, which is known as surface pre-melting, may be of vital importance for an understanding of the melting process and is therefore attracting a large amount of attention. The structure of pre-melted or disordered surfaces have been characterized by both ion scattering experiments[1] and by different diffraction techniques[2, 3, 4]. Most recently, the dynamics of atoms at the disordered Pb(110) surface have been studied experimentally by He-scattering[5]. Anomalously large diffusion rates were deduced from the experiments, and the rates were found to increase with increasing temperature. An interesting finding of this work was that the He scattering data could not be fitted by a simple diffusion model on a lattice.

In the present paper we extend our previous simulations of the pre-melting of Al(110)[6] to include the surface diffusion at the pre-melted surface. We find in accordance with the experiments that the pre-melting is accompanied by a large increase in the diffusion rate. Moreover we find additional support for our contention[6] that the initial stage of pre-melting is strongly coupled to the formation of a layer of adatoms on top of the original surface. The adatom layer shows by far the largest diffusivity.

Previous simulations of the surface diffusion at high temperatures have been based on pair potentials to describe the interactions between the atoms[7, 8]. While this may be adequate for the study of the surface of crystals of noble gases, this is not the case for metals. In the present study we have used the Effective Medium Theory to calculate the interactions. Effective Medium Theory is an attempt to capture the essentials of the many body inter-atomic interactions in metallic systems using relatively few approximations, while still keeping the computational effort involved in practical applications at a low level[10]. Molecular dynamics simulations based on this potential give a detailed description of the pre-melting of Al(110) in close agreement with experiment[6, 9].

In the simulations we propagated the system microcanonically for 25 ps using the Verlet algorithm. During this period 300 configurations were collected for analysis. Before the collection of data for this study, the systems had been propagated for at least 500 ps using alternating periods of stochastic temperature control and data collection. The system consists of 512 atoms initially distributed in 16 layers, 32 atoms in each layer. Four layers at one face of the system are kept static, and periodic boundary conditions are applied parallel to the surface. Simulations have been made at 93 K intervals from 465 K to 1030 K (the melting point in the simulations presented here is around 1000 K in reasonable agreement with the experimental value of 933 K. For more details see Ref. [6]).

Atoms may move from one layer to another in addition to diffusion parallel to the surface. As the layer-structure of the crystal remains to some extent even in the premelted state, we may distinguish between bulk and surface by sorting the atoms out on layers and calculating averages for each layer. As atoms may move from one layer to another repeatedly during the calculation, the assignment of layer numbers must be repeated for every configuration.

The diffusivity, D_n, for an atom n is defined through the Einstein formula

$$D_n = \frac{(\vec{r}_n(t) - \vec{r}_n(0))^2}{2dt} \tag{1}$$

where $\vec{r}_n(t)$ is the position of the atom at time t and d is the spatial dimension of $\vec{r}_n(t)$. The implementation of this formula in a least squares calculation of the diffusivity for an atom, n, is straightforward [11]. The least squares procedure takes into account the proper statistical significance of the relatively few observations over long time intervals compared to a much larger number of observations over short time intervals. To generalize to the calculation of the diffusivity for a layer, we make an ensemble average of the D_n's using a weight function $w_l(t)$ selecting exactly those time intervals where the atom remains in layer l without interruptions. We have checked that $< (r(t + \Delta t) - r(t))^2 >$ vs Δt remains linear for values of Δt much larger than a typical vibrational period.

Figure 1 shows the calculated diffusivities for layer number 0 (adatoms), 1 and 2 as a function of temperature. Two important conclusions can be made based on the results of Fig. 1. i) The diffusivity increases dramatically towards the surface, and ii) The diffusivity is anisotropic at low temperatures. It is larger along the closepacked rows (i.e the $[\bar{1}10]$) than across the closepacked rows (i.e in the [010] direction).

Diffusion events are always found to be related to defects, just as in the bulk metal. In the layers 1 and 2 (and the deeper ones) diffusion proceeds via vacancies, and in layer 0 the adatoms are relatively free to move along the surface. For a single adatom on a perfect (110) surface the activation energy for diffusion is only 0.1 eV for diffusion along the close packed rows and a

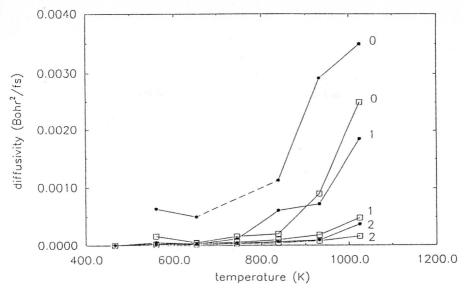

Figure 1: Diffusivities for layers along the x (solid circle) and y (open squares) as a function of temperature. For the points connected by a dashed line, the accuracy of the calculated diffusivity is poor. The poor accuracy is found in a temperature region where only few, but very mobile adatoms exist. In this region we have in some of the simulations observed atoms to move over quite long distances. This was only observed in a few of the simulations but led to high values for D, when it occurred.

little higher perpendicular to them. The increase in diffusivity towards the surface is therefore a simple consequence of the increase in the concentration of vacancies. We have previously shown that the pre-melting of this surface is closely related to the formation of vacancy-adatom pairs and the large diffusivity at temperatures close to the melting point is therefore also closely related to the pre-melting behavior.

Animations of the data generated in the simulation have given a valuable insight in to the mechanism of the diffusion process. At the surface, atoms from reasonably complete layers occasionally jump into the adatom layer. When more than just a single adatom is present, the adatoms tend to cluster into short (2-4 atoms) closepacked rows. This clustering lowers the diffusivity dramatically compared to isolated adatoms. Essentially the atoms not found at the terminals of the row will be vibrating, but not diffusing during the lifetime of the row.

A close inspection of the trajectory of a diffusing atom show that the atom does not move by one or a few brief jumps, but rather through a relatively slow, sliding motion. A few, representative results are shown in Figure 2. It is not surprising that it was not possible to fit the experimental results for Pb(110) using a simple diffusion model on a lattice.

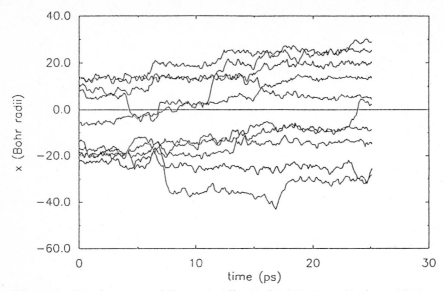

Figure 2: Development of the x-coordinate for 10 atoms in the surface region. This illustrates that the diffusion process does not resemble diffusion on a lattice. A preliminary analysis of the dynamics of the atomic motions indicate that the movement of the atoms proceeds in slow, sliding fashion and that the diffusive motion may not have a characteristic length scale

For some temperatures we were not able to find a diffusivity for layer 0 that was reproducible from one simulation to the next. These temperatures are therefore not included in Fig. 1. The problems seem to be related to a large sensitivity to events, where $|\vec{r}(t_2) - \vec{r}(t_1)|$ is large. Such events were observed occasionally and lead to relatively large calculated diffusivities when they occurred. As such events were also rare and did not occur in every simulation for a given temperature, the reproducibility of the diffusivity becomes poor. More work is needed to elucidate this effect, which may well be inherent in the diffusion process.

Acknowledgements: Financial support from the Danish Research Councils through the *Center for Surface Reactivity* is gratefully acknowledged.

References

[1] J. W. M. Frenken and J. F. van der Veen, Phys. Rev. Lett. **54**, 134 (1985); J. W. M. Frenken, P. M. J. Marée and J. F. van der Veen, Phys. Rev. B **34**, 7506 (1986); B. Pluis, A. W. Denier van der Gon, J. W. M. Frenken and J. F. van der Veen, Phys. Rev. Lett. **59**, 2678 (1987); B. Pluis, J. W. M. Frenken and J. F. van der Veen, Phys. Scr. **T19**, 382 (1987).

[2] P. von Blankenhagen, W. Schommers and V. Vogele, J. Vac. Sci. Technol. A **5**, 649 (1987)

[3] K. C. Prince, U. Breuer and H. P. Bonzel, Phys. Rev. Lett. **60**, 1146 (1988).

[4] P. H. Fuoss, L. J. Norton and S. Brennan, Phys. Rev. Lett. **60**, 2046 (1988).

[5] J. W. M. Frenken, J. P. Toennies and Ch. Wöll, Phys. Rev. Lett. **60**, 1727 (1988).

[6] P. Stoltze, J. K. Nørskov, and U. Landman, Phys. Rev. Lett. **61**, 440 (1988); P. Stoltze, J. Chem. Phys., **92**, 6306 (1990).

[7] J. Q. Broughton and G. H. Gilmer, J. Chem. Phys. **79**, 5095, 5105, 5119 (1983);

[8] V. Rosato, G. Ciccotti and V. Pontikis, Phys. Rev. B **33**, 1860 (1986); V. Pontikis and P. Sindzingre, Phys. Scr. **T19**, 375 (1987).

[9] A. W. Denier van der Gon *et. al.* To be published.

[10] K. W. Jacobsen, J. K. Nørskov, and M. J. Puska, Phys. Rev. B **35**, 7423 (1987); K. W. Jacobsen, Comments Condens. Matter Phys. **14**, 129 (1988).

[11] W. H. Press, B. P. Flannery, S. A. Teukolsky, W. T. Vetterling, *Numerical Recipes*, section 14.4. Cambridge University Press (1986).

Theory of Submonolayer Alkali-Metal Chemisorption

*Dingsheng Wang[1], Kailai Chen[1], Shiwu Gao[1], Ruqian Wu[1],
and Ning Wang[2]*

[1]Laboratory for Surface Physics, Institute of Physics,
  Academia Sinica, P.R. of China
[2]Beijing Vacuum Electronic Devices Research Institute, Beijing, P.R. of China

Abstract. The jellium–slab model combines the flexibility of varying alkali–metal coverage in the submonolayer range and the accuracy of the state–of–the–art band method for the description of the abundant surface states on transition metal and semiconductor surfaces. Chemisorption bonds with strong covalent character are formed by the hybridization of the local substrate states and the alkali–metal s states. This has been shown for a number of transition metal substrates and GaAs (110). The adsorption induced increase of electron density is rather small in the metallic substrates due to strong screening, but is prominent on the surface atomic layer of semiconductor surfaces. This explains why the metallization of the alkali–metal adsorption layer begins at higher coverage on the semiconducting GaAs (110) surface than on the transition metals.

1. Historical Background

As early as the late 1920s, Langmuir and his coworkers discovered that during cesiation of tungsten the work function decreases drastically, and gave an elegant classical explanation. The lowering of work function has been attributed to the formation of a dipole layer between the ionized alkali-metal atoms and the metal substrate. To explain the depolarizing effect it has been considered that the dipole moment p of an adsorbed atom could be decreased from the initial value by the field of neighboring dipoles. Assuming the work function lowering equals the Coulombic contribution of this dipole layer,

$$\Delta\phi \sim -4p(N_a)N_a \sim -p(0)N_a/(1 + \alpha k N_a^{3/2}) \quad ,$$

where N_a is the coverage in terms of atoms per unit area, α is the susceptibility of the adsorbate, and k a constant. This expression not only gives the initial decrease of work function, but also explains the work function minimum.[1]

Springer Series in Surface Sciences, Vol. 24   **The Structure of Surfaces III**
Editors: S.Y. Tong · M.A. Van Hove · K. Takayanagi · X.D. Xie
© Springer-Verlag Berlin, Heidelberg 1991

However, development of surface characterization and theoretical treatment has been successful in revealing more details, which cast doubts on this classical ionic picture. Lang and Williams[2] have studied the adsorption of an isolated atom on jellium surface, which is used to simulate the metal substrate (atom–jellium model). At the other extreme, by use of a very accurate band method, namely, the full potential, linearized augmented planewave method (FLAPW), Wimmer et al.[3] have treated the hypothetical c(2x2) Cs/W(001) adsorption (coverage slightly larger than 1 ML). Two points different from the classical ionic picture clearly emerge: (a) that instead of a pure ionic bonding, the alkali–metal atom hybridizes appreciably with the substrate states, the adsorption electron distributes mainly outside the jellium surface, and the bonding bears prominent covalent feature; (b) in the substrate region, the adsorption induced change of electron density is negligibly small due to strong screening.

Attempts to give a theoretical treatment, aiming at an understanding of coverage dependence encounter serious difficulties. Because the coverage of interest is in the submonolayer region,[1] in an empirical approach no fit to experimental data could be meaningfully done for that large atomic separation. On the other hand, in *ab initio* cluster or band approaches, the low coverage means a huge cluster or unit cell beyond the reach of available computing power. One approach uses Anderson–Newn's model and considers the adsorbate level E_a not only depending on intra–atomic Coulomb repulsion (U), but also on the electrostatic image force and the Coulombic potential of neighboring atoms such that

$$E_a = -I + U < n_a > + e^2/2d_a + (< n_a > -1)V \quad,$$

where the electrostatic potential V is summed over other adsorption atoms,

$$V = e^2 \sum_i' [r_i^{-1} - (r_i^2 + 4d_a^2)^{-1/2}] \quad,$$

and is thus coverage dependent. This is a quantum mechanical elaboration of the earlier point charge approach. In order to achieve a better fit to experiments, Muscat and Batra[4] have suggested that the distance between adsorbate atom and substrate surface, d_a, also varies with the coverage over a wide range, which looks too artificial and is not well justified.

Another approach, inspired by the success of the jellium model for simple metals, uses two jellium layers to represent both the overlayer and the substrate[5]. Though this jellium-jellium model is the simplest, it exhibits similar bonding features to the atom-jellium calculation, and in addition, a qualitatively good dependence of work function vs coverage is obtained.

2. Jellium–slab Model and Slab–jellium Model

Although theoretical discussions have continued for more than fifty years, the theory of alkali–metal chemisorption in the submonolayer range is still far from established. The fundamental questions for a better theoretical treatment are, to our belief, (a) how to take the local states of the substrate and the coverage dependence into proper consideration; (b) how to describe the interaction between alkali–metal adsorbates and determine whether the Coulombic interaction dominates.

In the last five years, two groups have attacked this problem in different but complementary ways. The jellium–slab model has been used by the present authors,[6] which consists of a jellium layer and an atomic slab (Fig.1). With a suitable band method, this model can take the d states of transition metal substrates or the dangling bonds of semiconductor surfaces into account properly, and at the same time address the continuous variation with coverage. The thickness of the jellium layer, D, is set equal to the spacing of the most densely packed lattice plane of bulk metals, independent of coverage in the whole submonolayer range.

This model has been used by Louie and Cohen[7] and Manghi et al.[8] in the investigation of monolayer adsorption of Al/Si and Cs/GaAs, respectively. Present authors have extended this model to the submonolayer coverage,[6] varied the jellium thickness to simulate different alkali-metals[9], and treated various transition and noble metal substrates.[10] Though alkali–metals and their surfaces can be accessed by the jellium model,[11] its use at low coverage has been questioned by a number of authors, because the atomic properties of the overlayer are omitted. How-

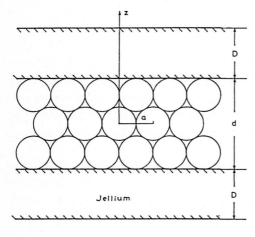

FIG.1 A jellium-slab model system simulating alkali-metal covered substrate atomic slab.

ever, our results show its success in revealing the role of the substrate in the submonolayer range too.

Another Japanese group[12] has attacked the same problem by a slab–jellium model, in which a single layer of atomic slab is put on the jellium with a given density simulating the substrate. The coverage is changed by varying the atomic distance in the slab. Obviously, it is convenient in clarifying the detail of the interaction between adsorbates. An important conclusion of the slab–jellium calculation is that the Coulombic contribution to this interatomic interaction is negligibly small, contrary to all previous point charge descriptions. Dr. Ishida will cover this study in his invited talk at the Osaka satellite meeting of this conference.

Both models have avoided the breaking of 2-dimensional translational symmetry during the continuous variation of coverage, and so could take advantage of the state–of–the–art band calculation. The present paper will concentrate on the results given by the jellium–slab model, emphasizing the role of the surface states of either the transition metal or semiconductor surfaces at submonolayer coverage.

3. Adsorption of Alkali–metals on Transition Metal Surfaces

This system has been thoroughly studied by use of the jellium–slab model, for varying jellium thickness simulating different alkali–metals from Na to Cs on W(001) substrate,[9] and for different 5d substrates.[10] As a comparison, a study has also been carried out for a 3d metal (Ni) and some noble metals (Au and Cu).

3.1 Interaction and Charge Transfer

The energy shift and broadening of adsorbate levels, as predicted in the Anderson–Newn's model, appear in the jellium model as a decrease of the parabolic free–electron band minimum with increasing jellium density. The jellium–slab model is the first to give the coverage dependence of levels of the substrate states from a realistic electronic structure calculation. Typical results are shown in Fig.2 for Cs (jellium, D=8.1au) on a W(001) slab. Two features are worth noting, namely:

(a) Among all occupied states the $\overline{\Gamma}$ d_{z^2} state shows the strongest interaction with the jellium state. According to present calculation, the level of $\overline{\Gamma}$ d_{z^2} state decreases from −0.2eV for the clean surface, to −1.0eV for 1/2 ML coverage, but remains constant from 1/2 ML to 1 ML (Cs atomic density $4.4 \times 10^{14}cm^{-2}$, or jellium density $1.52 \times 10^{-3}au^{-3}$), in accordance with the accurate calculation[3] and the experimental measurement[13]. This agreement is a good proof of the validity of the present jellium–slab model;

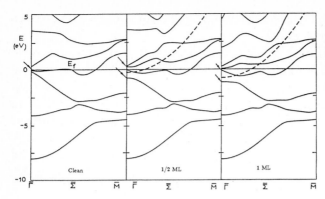

FIG.2 Energy band of Cs (jellium, D=8.1au) on W(001). Dashed lines show the parabolic free-electron band in the jellium region. Arrows show the surface state with d_{z^2} character.

(b) Not the whole d_{z^2} band, but only those states near the center part of the Brillouin zone (about 1/3 from $\overline{\Gamma}$ to $\overline{M}$) show a prominent downward shift. At the boundary of the Brillouin zone the states are almost not affected by the presence of the jellium overlayer. This is easy to understand from two–band perturbation, and is essential to the submonolayer behavior of alkali–metal adsorption.

Accordingly, as shown in Fig. 3a, the distribution of the adsorption electron of the first 1/2 monolayer presents strong d_{z^2} character. It shows that a strong s (jellium) – d (substrate) covalent bonding is formed. However, in the substrate region including the surface layer, the change of electron density is much less pronounced due to metallic screening. Further adsorption from 1/2 ML to 1 ML shows a different picture (Fig.3b). The electron distributes rather uniformly over the whole jellium region either on the plane or along the z direction, and in the substrate region an Friedel oscillation is clearly seen. The planar averaged change is shown in Fig.4. The strong polarization in the jellium region due to the covalent bonding is shown for the first 1/2 monolayer. When $N_a >$ 1/2 ML, there is not any weakening of this bond and depolarization only comes from the uniform metallic distribution (curve c).

From the above analysis, we have seen that at the first 1/2 monolayer the adsorption of alkali–metal on transition metal substrate is covalent due to hybridization of the surface d and alkali–metal s states. Beyond 1/2 ML, the adsorption layer reassumes its metallic property, as has been known for years from experiments.[1] In the whole submonolayer range, the electron density inside the metal substrate does not show any appreciable increase except a minor redistribution due to the screening. This is consistent with the atom–jellium and slab–jellium calculations, and also with the experimental fact that the core level of W atoms doesn't shift during cesiation.[14]

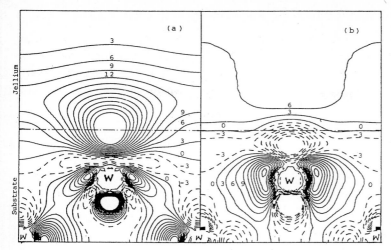

FIG.3 Contour plot showing the difference of the valence charge density of Cs (jellium, D=8.1au) on W(001), (a) $\rho_{1/2ML} - \rho_{clean}$; and (b) $\rho_{1ML} - \rho_{1/2ML}$, in units of $10^{-4}au^{-3}$.

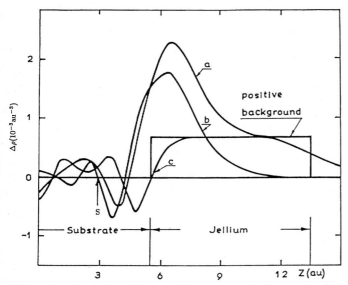

FIG.4 Adsorption induced change of planar averaged electron density for Cs on W(001) by a jellium-slab model calculation. Curve a, $\rho_{1ML} - \rho_{clean}$; b, $\rho_{1/2ML} - \rho_{clean}$; and c, $\rho_{1ML} - \rho_{1/2ML}$.

29

3.2 Induced Dipole Moment and Work Function

The adsorption induced dipole moment is considered to be the origin of the change of work function. Figure 5a shows the coverage dependence of the dipole moment for different jellium thicknesses. Below a certain critical coverage, N_m, the moment keeps constant independent of the

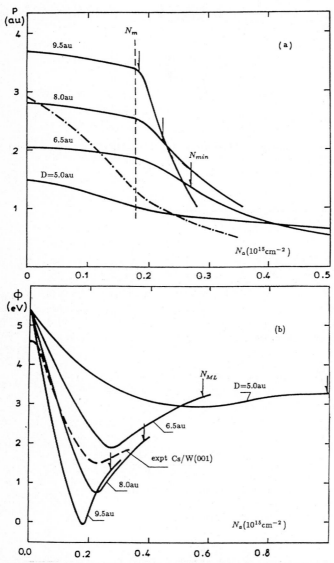

FIG.5 Induced dipole moment (a) and change of work function (b) vs coverage for W(001) covered by jellium with different thicknesses. The dot-dashed line is the result of the jellium-jellium model[5] (overlayer D=8.1au).

coverage and the bonding is covalent, Beyond N_m it starts to decrease, which marks the initiation of metallization. N_m, independent of the jellium thickness as shown in Fig.5a, is $0.18 \times 10^{15} \text{cm}^{-2}$ (about 1/2 ML) for W(001) surface.

Above analysis reveals the essential role of the substrate surface state in determining the adsorption properties, which has been neglected by either jellium–jellium or slab–jellium model study. The dot-dashed curve in Fig.5a depicts the result of the jellium–jellium calculation, which only shows the rapidly decreasing part. In fact it is similar to the result of the jellium-slab calculation for noble metal substrates (Au and Cu) where the d band is almost fully occupied and doesn't contribute to adsorption bonding. The coverage dependence of the work function has been given for W(001) and other substrates [9,10]. Comparison has been made with experiments for the slope of initial decrease of work function, $\phi'(0)$, and especially the coverage of the work function minimum, N_{min}. The jellium–slab model has succeeded in giving the correct N_{min}, for example, experimental data of Cs/W(001) compare well with corresponding curve (Fig.5b), while other calculations neglecting the the effect of local surface states in the covalent polarizing bonding usually give a much lower value of N_{min}. The increase of work function at higher coverage when $N_a > N_{min}$ is connected with the metallization of the overlayer as suggested from the decrease of the dipole moment. The absolute value of work function minimum ϕ_{min} obtained by the present calculation is lower than the experimental one because the core polarization of adsorbed alkali–metal atoms, opposite to the polarization of the valence electrons,[3] is omitted in this jellium–slab approximation.

3.3 Characterizing the Metal Substrates

From the above picture, the submonolayer adsorption of alkali–metal on metal substrates is better characterized by two parameters. The first is N_m representing the maximum number of adsorption electrons held by the localized covalent adsorption bonds. The second parameter is the average position of this portion of the electrons, $< Z_e >$, measured from the substrate boundary toward the jellium side. For example, the induced dipole moment at low coverage is related to $< Z_e >$ by

$$p(0) = \frac{1}{2}D - < Z_e > \quad .$$

Both parameters are expected to be substrate dependent because they are determined by the energy and spatial distribution of the relevant surface states.

A calculation has been made for Ta, W, Ir, Pt and Au.[10] N_m is larger for Ta and W because the empty d_{z^2} state has its lobe extending far into the jellium region and interacts most strongly with the jellium electrons. So $< Z_e >$ is large for Ta and W too. At the end of the $5d$ transition series, the empty d band near Fermi level has a strong $d_{x^2-y^2}$ feature which lies mostly in the surface plane and is less extended toward the jellium region, so the interaction is weaker and gives rise to a smaller N_m and $< Z_e >$. This idea has been used in the analysis of the work function change in alkali–metal chemisorption and co-adsorption systems.[15]

4. Adsorption of Alkali–metal on GaAs Surface

A typical example is Cs on the nonpolar GaAs (110) surface. It is non-reactive and thus represents a simple case. Either from the coverage dependence of the work function,[16] which shows a monotonic decrease in the whole submonolayer range, or from the development of free–electron plasma loss, the intersection of the Fermi level with the bonding energy cutoff, and the position of both n– and p–type semiconductors,[17] it is concluded that, unlike the adsorption on transition metal substrates, up to 1 ML the alkali–metal layer on GaAs (110) is not metallic, and only above 1 ML (deposition at low temperature) does the metallicity start to develop. Encouraged by the success in studying the alkali–metal on transition metal adsorption, we have recently extended this study to GaAs surfaces, trying to elucidate the difference of their adsorption behavior at submonolayer range, which cannot be understood either by the depolarization of point charge Coulombic interaction, or by other methods neglecting the existence of localized surface states, such as the jellium–jellium or slab–jellium model.

A model calculation is carried out for an unrelaxed GaAs (110) surface. Figure 6a gives two surface bands (S_3 and S_4) inside the band gap, similar to previous calculations for an ideal GaAs (110) surface.[18] First 1/2 monolayer of jellium coverage shifts As dangling bond at $\overline{\Gamma}$ downward about 0.8eV (Fig.6b). The interaction is stronger near the center of the Brillouin zone and forms the adsorption bond. Figure 7a shows the charge transfer during the first 1/2 monolayer adsorption and the crucial role played by the As dangling bond. A large number of adsorption electrons are distributed inside the jellium region with prominent localization. The outer peak of the planar averaged $\rho_{1/2ML} - \rho_{clean}$ (Fig.8, curve b) shows this polarization of jellium electrons, corresponding to curve b of Fig.4 for a W(001) substrate. So, below 1/2 ML the general feature of this adsorption bonding is similar to that of W(001) surface, and the bonding possesses covalent features too. Here the As dangling p bond plays

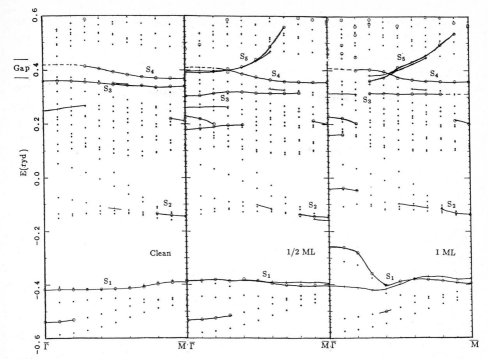

FIG.6 Energy band of Cs (jellium, D=8.1au) on GaAs (110).

the role of the d_{z^2} state of the W(001) substrate. However, it is worth pointing out that due to the reduced screening in the semiconductor, the adsorption induced change is much stronger around the surface As atoms as is evidenced by the contour plot (Fig.7a) and the two lower peaks in curve b of Fig.8, which do not occur for the W(001) substrate (cf. curve b of Fig.4).

Most interesting is the result above 1/2 ML. From 1/2 ML to 1 ML, the dominating Cs surface band S_5 decreases, reaches at $\overline{\Gamma}$ point below the valence band maximum, and becomes partly occupied (Fig.6c). The change of electron distribution from 1/2 ML to 1 ML is concentrated mainly in the jellium region (Fig.7b) with appreciable localization on top of the As–Ga parallel bond. Along the z direction, its distribution (curve c of Fig.8) shows that polarization toward the interface still exists, though much weaker than the first 1/2 monolayer. This is different from the case of metal substrates, and explains the experimental findings that till 1 ML the cesiated GaAs (110) is still not metallic. The essential point leading to this difference is that though in both cases there are surface states (d_{z^2} for W, and As dangling bond and As–Ga parallel bond for GaAs) interacting with the jellium overlayer, in the GaAs case, due to reduced screening,

33

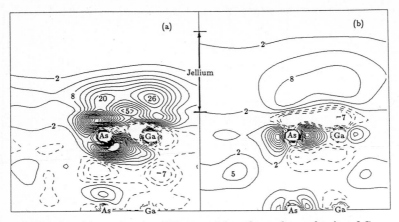

FIG.7 Contour plot showing the difference of the valence charge density of Cs (jellium, D=8.1au) on GaAs (110), (a) $\rho_{1/2ML} - \rho_{clean}$; and (b) $\rho_{1ML} - \rho_{1/2ML}$, in units of 10^{-4}au^{-3}.

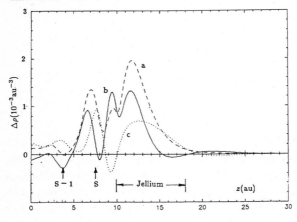

FIG.8 Adsorption induced change of planar averaged electron density for Cs on GaAs (110) by a jellium-slab model calculation. Curve a, $\rho_{1ML} - \rho_{clean}$; b, $\rho_{1/2ML} - \rho_{clean}$; and c, $\rho_{1ML} - \rho_{1/2ML}$.

the adsorption bond goes much deeper into the substrate region than in the metal substrates. Besides being concentrated in the jellium interface region due to covalent bonding, the adsorption electron also occupies the space around the surface atom. For the first 1/2 monolayer (curve b of Fig.8), the electron inside the substrate accounts for about 45% of the total adsorption electrons. That explains why the metallization of a Cs overlayer on GaAs (110) begins at a coverage about twice as high as that on the W(001) surface. The fact that adsorption induced bonding states penetrate into semiconductor substrates also agrees with the well–known idea of metal induced gap state (MIGS) for Fermi level pinning.

The calculated coverage dependence of work function is shown in Fig.9. Though the absolute value of calculated work function $\phi(0)$=7.4eV

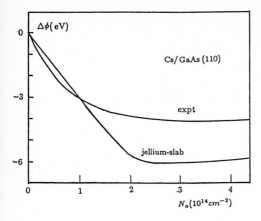

FIG.9 Change of work function vs coverage according to jellium-slab model for GaAs (110) covered by Cs (jellium, D=8.1au)

is larger than the experimental one, its coverage dependence shows the correct relation: it decreases rapidly below 1/2 ML and remains nearly constant from 1/2 ML to 1 ML, in agreement with the experiment.[16] The jellium–slab calculation for relaxed GaAs (110) surface of 1 ML coverage has shown some change in electronic structure,[8] but it does not alter the above conclusion about the difference with respect to the metal substrates.

5. Summary

Though cesiation of the metallic W(001) surface and the semiconductor GaAs (110) surface exhibit quite different behavior at submonolayer range, they are considered in a unified way by the jellium–slab model. The interaction between the empty surface bands of the substrates and alkali–metal s states gives rise to the covalent bonding. It is localized on top of the bonding atoms and concentrated mainly at the jellium interface region. The basic difference between metal and semiconductor substrates lies in the fact that the metal has a large Fermi momentum and short screening length, but the latter not. On the semiconductor surface the adsorption induced electron distribution is also quite prominent on the substrate atoms, and this causes the alkali–metal overlayer to become metallized at much higher coverage.

At low coverage, however, the jellium is better used only to describe the alkali–metal overlayer because the alkali–metal atoms have less ionization energy. In this case, the hybridization near the Fermi level only involves those overlayer states near the adsorbate band minimum, so it doesn't care what the exact adsorbate band structure is, and the problem can be approximated by a jellium model. For other elements with high ionization energy, more adsorbate states will be involved, the effect will depend on the atomic properties and structure of the overlayer and this is beyond the reach of the jellium approximation.

The adsorption bonding features of different transition metals have been summarized in two parameters, namely, N_m and $< Z_e >$. Both are determined by the energy and spatial distribution of the most highly interacting surface state. The relation with the d band filling of the substrate atoms is given for $5d$ metals.

It will be very interesting to extend the present investigation to other semiconductor surfaces, and to the problem of co-adsorption of alkali-metal and other elements (e.g. oxygen), which is of both theoretical and practical importance.

Acknowledgment

This work is supported by the National Science Foundation of China under Grant No. 6881018.

References

1. For a review, see Z. Sidorski, Appl. Phys., A33 (1984) 213
2. N. D. Lang and A. R. Williams, Phys. Rev. Lett., 37 (1976) 212.
3. E. W. Wimmer, and A. J. Freeman, J. R. Hiskes, and A. M. Karo, Phys. Rev., B28 (1983) 3074
4. P. Muscat and I. P. Batra, Phys. Rev., B34 (1986) 2889.
5. N. D. Lang, Phys. Rev., B4 (1971) 4234.
6. N. Wang, K. L. Chen, and D. S. Wang, Phys. Rev. Lett., 56 (1986) 2759
7. S. G. Louie and M. L. Cohen, Phys. Rev., B13 (1976) 2461
8. F. Manghi, C. Calandra, C. M. Bertoni, and E. Molinari, Surf. Sci., 136 (1984) 629
9. R. Q. Wu, K. L. Chen, D. S. Wang, and N. Wang, Phys. Rev., B38 (1988) 3180
10. R. Q. Wu and D. S. Wang, Phys. Rev., B41 (1990) June
11. N. D. Lang, 'Density functional approach to the electronic structure of metal surfaces and metal-adsorbate systems', in 'Theory of Inhomogeneous Electron Gas', ed. S. Lundqvist and N.H. March (Plenum Press, 1983)
12. H. Ishida and K. Terakura, Phys. Rev., B36 (1987) 4510
13. P. Soukiassian, R. Riwan, J. Lecante, E. Wimmer, S. R. Chubb, and A. J. Freeman, Phys. Rev., B31 (1985) 4911
14. D. M. Riffe, G. K. Wertheim, and P. H. Citrin, Phys. Rev. Lett., 64 (1990) 571

15. Yabo Xu et al. Acta Phys. Sinica (to be published)

16. M. Prietsch, M. Domke, C. Laubschat, T. Mandel, C. Xue, and G. Kaindl, Z. Phys., B74 (1989) 21

17. R. Cao, K. Miyano, T. Kendelewicz, I. Lindau, and W. E. Spicer, Phys. Rev., B39 (1989) 12655

18. D. J. Chadi, Phys. Rev., B18 (1978) 1800

19. J. Tersoff, Phys. Rev. Lett. 52 (1984) 465

Downward Funneling Model of Low-Temperature Epitaxial Growth: A Hybrid Molecular Dynamics–Monte Carlo Study

D.E. Sanders[1,*] *and J.W. Evans*[2,†]

[1]Department of Chemistry, Iowa State University, Ames, IA 50011, USA
[2]Department of Physics and Ames Laboratory, Iowa State University,
 Ames, IA 50011, USA

Abstract. Near layer-by-layer growth of thin metal films on fcc(100) metal substrates has been observed experimentally as persistent diffracted intensity oscillations at temperatures as low as 80 K. Three growth mechanisms have been proposed: (i) cyclical nucleation, growth, and coalescence of 2D islands in each layer mediated by thermal diffusion, if assumed still operative at ~ 80 K; (ii) as for (i) except that "transient mobility" of hot atoms immediately after deposition is assumed to play the role of thermal diffusion; (iii) no thermal or transient mobility between four-fold hollow (4fh) sites is involved, but only "downward funneling" during deposition to 4fh sites. Utilizing accurate CEM interaction potentials typically allows us to rule out (i) after determination of Arrhenius rates for intersite diffusion, and to rule out (ii) from MD studies of single atom deposition dynamics. The latter generally support the downward funneling picture. Thus the film/substrate is sufficiently rigid to support funneling, but not so rigid as to support transient mobility. The MD studies also provide input for MC simulations which confirm the surprisingly layer-by-layer character of film growth.

I. Introduction

Recent experimental observations of near layer-by-layer growth in thin metal films on fcc(100) substrates at low temperatures (~ 80 K) have prompted reassessment of the key factors which determine structure during epitaxial growth. The traditional picture is that (near) layer-by-layer growth involves cyclical nucleation, growth and coalescence of 2D islands in each layer mediated by thermal diffusion [1,2]. In this picture, diffusion is necessary to allow atoms deposited on-top of islands to migrate to the edge thereby being incorporated in the layer beneath. At low temperatures (~ 80 K), it has been postulated that thermal diffusion is effectively inoperative and its role is played by "transient mobility" of "hot" adatoms immediately following deposition - mobility envisioned to result from the inability to instantaneously dissipate energy released upon formation of the atom-surface bond [2]. Furthermore it has been argued that such motion must be ballistic in nature for atoms to be able to reach the edge of islands [2]. In contrast to the above, we have recently presented a quite different picture for low-temperature near layer-by-layer growth which involves no motion between the four-fold hollow (4fh) adsorption sites, but relies only on "downward funneling" to such adsorption sites during deposition [3,4].

   In this contribution, we demonstrate unequivocally that thermal diffusion is insignificant at around 80 K in several metal-on-metal systems. We also show that transient mobility is effectively absent, and provide a detailed analysis of the downward funneling characteristics of the deposition dynamics. Finally we describe an <u>exact mapping</u> of the full dynamical deposition process at T = 0K — for physically reasonable low deposition

Springer Series in Surface Sciences, Vol. 24   **The Structure of Surfaces III**
Editors: S.Y. Tong · M.A. Van Hove · K. Takayanagi · X.D. Xie
© Springer-Verlag Berlin, Heidelberg 1991

rates — onto a much simpler stochastic process. Statistical information on film structure is then efficiently obtained from Monte-Carlo (MC) simulation of the latter. This avoids the problem of artificially high deposition rates required in full Molecular Dynamics (MD) simulations.

II. Microscopic Picture of Low-Temperature Epitaxial Growth

Here we investigate thermal diffusion, transient mobility, and downward funneling deposition dynamics during epitaxial growth at low-temperatures in several metal-on-metal systems. We adopt many-body corrected effective medium (CEM) potentials whose accuracy for metallic systems has been confirmed previously [5], in contrast to Lennard-Jones (12,6) potentials [6]. It is also important to provide a sophisticated treatment of energy dissipation into the bulk. Our MD simulations with CEM potentials couple $\geq$ 250 "target" substrate atoms in three movable layers with Langevin coupling to the rest of the bulk [7].

Thermal Diffusion Rates: The influence of diffusion on film structure is determined by the magnitude of the intersite diffusion rate relative to the deposition rate (per site). Studies of diffusion on fcc(100) surface (with Lennard-Jones potentials) [8] show that dynamical corrections to the Transition State Theory (TST) rates are small at low temperatures, that these TST rates can be reasonably approximated by the simple Arrhenius form, $h = \nu_0 \exp(-\beta E_{act})$, from 1D harmonic TST. Here ν_0 and E_{act} should be taken as the vibrational frequency and activation energy barrier for an atom on a relaxed substrate. Accurate CEM calculations yield ν_0 = 3.1, 1.6, 3.2 x 10^{13} sec^{-1}, and E_{act} = 12.5, 15.0, 14.3 kcal/mol for Cu/Cu(100), Pt/Pd(100), and Cu/Ag(100), respectively. Thus for these systems at ~ 80 K, h is negligible compared with the deposition rate of $O(10^{-2})$ per site per second, and so thermal diffusion is insignificant.

Transient Mobility: The propensity for transient mobility clearly increases with the rigidity of the substrate/film. Thus it is important that the potential accurately predicts the bulk modulus and to allow a sufficient number of movable substrate layers in the simulation [9]. Of course, an accurate (e.g., Langevin) treatment of energy dissipation to the bulk is also necessary.

Here we search for the occurrence of transient mobility via MD studies of single atom deposition dynamics on various substrate+film micro-configurations at 80 K. Vertical impingement is assumed unless otherwise stated. We have noted previously that for deposition on a clean substrate, the atom always settles into the 4fh - site on which it impinges [3]. Thus transient mobility does not provide a mechanism for incorporation at island edges of atoms deposited at some distance from such edges. As a further test of the absence of a transient mobility mechanism for smoothing films, we consider deposition of an atom on top of a quartet of four atoms placed on a clean substrate (Fig. 1a). Again the atom always settles into the second layer 4fh site created by this quartet, rather than hopping down to a first layer four-fold hollow site (as required for film smoothing).

We do not absolutely rule out transient motion between neighboring 4fh sites for film micro-configurations which are locally more rigid (see the next subsection). However any such limited motion certainly will not have the film smoothing effect postulated in Ref. 2.

Downward Funneling Dynamics: Here we again utilize MD studies of single atom deposition dynamics on various substrate+film micro-configurations to assess the validity of the downward funneling picture. Clearly for downward funneling to occur, any protrusions, e.g., micropyramids, in the growing film must be sufficiently rigid to funnel the impinging atom without restructuring. For convenience we shall provide a detailed analysis only for T = OK, where the deposition process is deterministic, i.e., the

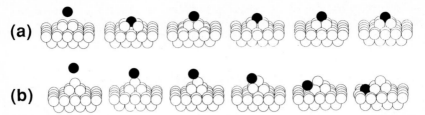

Fig. 1. 0.25 eV-Cu atom (black) impinging on a Cu-film held at 80 K.

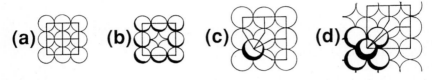

Fig. 2. Some capture zones for "slimy" hard-sphere dynamics.

asymptotic lateral coordinates, $x_{||}$, of the impinging atom (and its energy) determine into which 4fh site it will eventually settle. Thus the $x_{||}$-space can be uniquely partitioned into "capture zones" for the various 4fh adsorption sites. Film structure is completely determined by the relative "capture areas" of these zones. Dependence of this partition and the associated capture areas on the initial kinetic energy is weak since most of the kinetic energy upon impact comes from the atom-substrate bond energy. Thus below we only consider initial energies of 0.25 eV.

For a detailed analysis of downward funneling dynamics, it is instructive to compare against the following simple "reference" model: atoms are regarded as "slimy" hard-spheres; impinging gas atoms, on contact with substrate/film atoms, slide downhill until reaching a 4fh adsorption site. For a clean substrate the capture zones are trivially the squares shown in Fig. 2a. We set their area A_o to unity. The capture zone for a "hole" in a perfect monolayer is a square of area 4 (Fig. 2b). An impinging atom aimed at a single adsorbate atoms funnels to either a diagonally or laterally adjacent 4fh site (Fig. 2c). The associated capture areas A_D and A_L (respectively) clearly satisfy $4A_D + 4A_L = 9$. In fact capture areas for substrate 4fh sites with any surrounding configuration of first layer atoms are trivially determined by $A_D = 7/6$ for this model [3,4]. Fig. 2d shows some capture zones near a 5-atom pyramid. Note that some atoms impinging within the "lateral" zones shown contact two bottom pyramid atoms. These end up in one of two 4fh sites with equal probability.

We now examine the actual funneling dynamics for Cu/Cu(100) at T = 0K. For a clean substrate, the capture zones remain as in the above model (Fig. 2a). The capture zone for a hole is slightly distorted from Fig. 2b (Fig. 3a). For funneling off a single adsorbate, the capture zones of Fig. 2c are significantly distorted (Fig. 3b). In general, they become topologically disconnected. This feature reflects the propensity for atoms impinging near a zone boundary (e.g., near a bridge site) to "bounce" to a non-adjacent 4fh site. The dynamics for funneling off a 4+1 atom pyramid is correspondingly more complicated, however the simple downward funneling picture basically still applies (Fig. 1b). For larger free standing pyramids, the simple funneling picture begins to break down. For a 9 + 4 + 1 atom pyramid, on rare occasions the impinging atom "reconstructs" the pyramid replacing a second layer atom, which is then pushed out to a nearby three-fold hollow site. This reflects the stronger coupling of the "localized atomic impact"

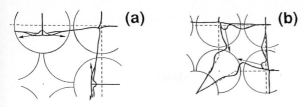

Fig. 3.   Capture zones (a) near a hole, and (b) near an isolated adsorbate,
for 0.25 eV-Cu/Cu(100) at 0 K.

to the vibrational excitation of the pyramid. This coupling is enhanced by
the reduced frequency of the softest vibrational mode for this larger
pyramid [10]. For a 9 + 4 + 1 atom (and larger) pyramids, one occasionally
sees the impinging atom hop to a site neighboring that first reached at the
pyramids base. Deposition dynamics on large pyramids, corresponding locally
to impingement at 45° on (111) faces, can be quite "erratic". However we
emphasize that the large protrusions, and thus the non-funneling dynamics,
described in this paragraph are statistically rare, and thus do not
significantly affect film structure.

III.   Film Structure for Downward Funneling Deposition Dynamics

Once the capture areas for various configurations are catalogued, we can
<u>exactly map</u> the full dynamical deposition process at T = 0K onto a simple
stochastic growth process where 4fh sites are filled at rates proportional
to their capture areas. This assumes no atoms are knocked out during
deposition. Generation of an extensive set of capture areas and
incorporating them into a MC algorithm would be unwieldly. However we shall
see below that our funneling models produce surprisingly smooth films, so
most rough local configurations and associated capture areas are
statistically irrelevant. One expects that it is most important to input
correct relative values A_0, A_p, A_L, ... of capture areas for funneling of
atoms in the same layer as the adsorption site. However even dependence of
structure on these values seems weak [4]. Thus here we make only a crude
attempt to incorporate these areas using a one-parameter "discretized"
downward funneling model, which also approximately treats funneling off
higher layer atoms [3,4]. We choose δ = 0 in that model [3] (A_0 = 1, A_L =
5/4, A_p = 1, ...) to roughly fit the Cu/Cu(100) dynamics. Film smoothness
is probed here through the time or θ-dependence of the distribution, θ_j, of
coverages deposited layer j > 0 (j = 0 denotes the top substrate layer, θ_0 =
1). From Table 1 one sees that layer coverage distribution quickly
approaches a quasi-steady state profile, $\theta_j \sim F(j - \bar{j})$, where $\bar{j} = \sum_{j=1}^{\infty} \theta_j = \theta$.
Thus, for example, θ_3, θ_4 and θ_5, when θ = 4, are almost equal to $\theta_4^{j=1}$, $\theta_5^{j=1}$, and
θ_6, when θ = 5, respectively. Further examination of Table 1 shows that the
spreading of the interface is negligible for 2 $\leq$ θ $\leq$ 0(10). In this range
no more than two to three layers are partially filled at any time, and one
expects that e.g., free standing 9 + 4 + 1 atom pyramids effectively do not
occur. The θ_j-data in Table 1 is also incorporated in Fig. 4 below.

   Consider now substrate temperatures T > 0K for which thermal diffusion
is still effectively inoperative. Here the deposition dynamics is no longer
deterministic. As T increases above zero, the boundaries of the capture
zones become fuzzy, and so the zones themselves become <u>fuzzy sets</u> [11].
Points, $x_{||}$, well inside the zone will presumably have near full membership
in that zone, but points near the boundaries will have only partial

Table 1. Evolution of the layer coverage distribution, θ_j, with increasing θ.

	θ_0	θ_1	θ_2	θ_3	θ_4	θ_5	θ_6	θ_7	θ_8	θ_9	θ_{10}	θ_{11}
$\theta=0$	1	0	0	0	0	0	0	0	0	0	0	0
$\theta=1/2$	1	0.490	0.010	0	0	0	0	0	0	0	0	0
$\theta=1$	1	0.845	0.154	0.001	0	0	0	0	0	0	0	0
$\theta=3/2$	1	0.977	0.498	0.026	0	0	0	0	0	0	0	0
$\theta=2$	1	0.999	0.822	0.178	0.002	0	0	0	0	0	0	0
$\theta=5/2$	1	1	0.967	0.499	0.034	0	0	0	0	0	0	0
$\theta=3$	1	1	0.997	0.811	0.189	0.003	0	0	0	0	0	0
$\theta=7/2$	1	1	1	0.961	0.499	0.040	0	0	0	0	0	0
$\theta=4$	1	1	1	0.996	0.803	0.197	0.004	0	0	0	0	0
$\theta=9/2$	1	1	1	1	0.956	0.500	0.044	0	0	0	0	0
$\theta=5$	1	1	1	1	0.995	0.798	0.202	0.005	0	0	0	0
$\theta=11/2$	1	1	1	1	1	0.953	0.499	0.047	0	0	0	0
$\theta=6$	1	1	1	1	1	0.994	0.794	0.206	0.006	0	0	0
$\theta=13/2$	1	1	1	1	1	1	0.950	0.499	0.050	0	0	0
$\theta=7$	1	1	1	1	1	1	0.994	0.790	0.209	0.007	0	0
$\theta=15/2$	1	1	1	1	1	1	1	0.948	0.500	0.052	0	0
$\theta=8$	1	1	1	1	1	1	1	0.993	0.788	0.212	0.007	0
$\theta=17/2$	1	1	1	1	1	1	1	1	0.945	0.500	0.055	0.001
$\theta=9$	1	1	1	1	1	1	1	1	0.992	0.785	0.215	0.008
$\theta=19/2$	1	1	1	1	1	1	1	1	1	0.944	0.500	0.057

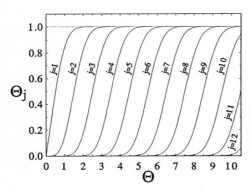

Fig. 4. Evolution of the layer coverage distribution, θ_j, with increasing θ.

membership. The dynamical deposition process can be exactly mapped onto a stochastic process where now capture areas are simply taken as the integrals of membership functions. Film structure should differ little from T = OK.

*D.E.S. was supported by an IBM predoctoral fellowship and NSF grants CHE-8609832 and CHE-8921099 to A.E. DePristo.
†Ames Laboratory is operated for the USDOE under contract No. W-7405-Eng-82. J.W.E. was supported by the Division of Chemical Sciences, Office of Basic Energy Sciences.

References

1. J. A. Venables, G. D. T. Spiller and M. Handbuchen, Rep. Prog. Phys. 47, 399 (1984).
2. W. F. Egelhoff and I. Jacob, Phys. Rev. Lett. 62, 921 (1989).
3. J. W. Evans, D. E. Sanders, P. A. Thiel, and A. E. DePristo, Phys. Rev. B 41, 5410 (1990).
4. J. W. Evans, Vacuum, 41, 479 (1990); Phys. Rev. B (submitted).
5. J. D. Kress, M. D. Stave, and A. E. DePristo, J. Chem. Phys. 93, 1556 (1989); T. Raeker and A. DePristo, Phys. Rev. B. 39, 9967 (1989).
6. M. Schneider, A. Rahman, and I. K. Schuller, Phys. Rev. Lett. 55, 604 (1985).
7. A. E. DePristo and H. Metiu, J. Chem. Phys. 90, 1229 (1989).
8. A. F. Voter and J. D. Doll, Ann. Rev. Phys. Chem. 38, 413 (1987).
9. D. E. Sanders and A. E. DePristo, in preparation.
10. A pyramid of Hookean springs of height j attached to a rigid fcc(100) plane has $j(j+\frac{1}{2})(j+1)$ vibrational modes. The softest correspond to lateral oscillations of the entire pyramid, and would naturally be excited by impingement on a face. They have frequencies in the ratio 1 : 0.78 : 0.63 : 0.52 : ... for j = 1, 2, 3, 4,... respectively.
11. A. Kandel, "Fuzzy Mathematical Techniques with Applications" (Addison-Wesley, Reading, 1986).

Mapping of Crystal Growth onto the 6-Vertex Model

C. Garrod[1], *M. Kotrla*[2], *A.C. Levi*[2], *and M. Touzani*[3]

[1]University of California, Department of Physics, Davis, CA 95616, USA
[2]SISSA, Strada Costiera 11, Miramare, I-34014 Trieste, Italy
[3]Ecole Normale Supérieure Takaddoum, B.P. 5118, Rabat, Morocco

Two models of crystal growth are developed and solved using the 6-vertex model of statistical mechanics, onto which two different mappings are considered. One mapping establishes a direct correspondence between surface steps and the lines occurring in the line representation of the 6-vertex model; the other mapping is that of van Beijeren. These two mappings correspond to two different physical situations: a vicinal surface near the (100) surface of a simple cubic crystal, and the (100) surface of a bcc crystal (together with its vicinal surfaces) respectively. In the latter case, nucleation phenomena are explored by studying finite-size systems with both Monte Carlo and (for very small systems) analytic methods.

Most models of 3-dimensional crystal growth are complicated, beyond any hope of an exact solution. They are usually studied either using simulations or in some simple approximation. Here we report on an application of the 6-vertex model [1,2] to the problem of crystal growth, which yields some exact results. Two stochastic models have been developed and solved. They correspond to two different physical situations:

model I - surface of a simple cubic crystal with low Miller indices,

model II - (001)-surface of a bcc-crystal together with its vicinal surfaces. In each case a different mapping of the surface onto the 6-vertex model is used.

The 6-vertex model is the two-dimensional version of the *ice model*, introduced by Pauling and Slater[3] in order to study the residual entropy of ice at $T = 0$ as well as the ferroelectric phase transitions. In a lattice where each atom has four nearest neighbours (e.g. in 2D, a square lattice; in 3D, a tetrahedral arrangement), arrows are drawn along the bonds to represent the electric dipoles. Charge neutrality causes the *ice rule*: "two arrows in, two arrows out" to hold at every lattice point. Only six arrow configurations satisfy the ice rule: in 2D they are the well known six vertices, shown in Fig.1.

Springer Series in Surface Sciences, Vol. 24   **The Structure of Surfaces III**
Editors: S.Y. Tong · M.A. Van Hove · K. Takayanagi · X.D. Xie
© Springer-Verlag Berlin, Heidelberg 1991

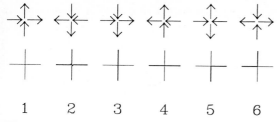

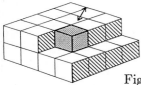

Fig.1 - The six vertices and their line representation.

Fig.2 - Growth-evaporation process for model I.

In model I [4] only one kind of growth-evaporation process (Fig.2) is considered. The model describes a vicinal surface, whose level is assumed to increase down and to the right. The mapping is performed by letting the steps correspond directly to the lines occurring in the *line representation* of the 6-vertex model. In such a mapping, vertex 2 turns out to be absent, so that a 5-vertex model is obtained. The 5-vertex model is an interesting mathematical problem in itself. It is possible to obtain an exact solution for the rate of growth using dimer statistics [5,6], when the equilibrium probability distribution of configurations is assumed. Under more general conditions exact solutions are obtained for small-size versions of the model, and larger sizes can easily be treated by simulations [4]. Model I appears very satisfactory as a description of situations where growth takes place solely by the motion of steps, causing the higher terraces to expand at the expense of the lower ones. Due to its simplicity it has, however, the defect that it predicts a zero growth rate for the surfaces with one or two Miller indices equal to zero.

To remove this unrealistic effect we have to include also the other processes important for the growth of the surfaces near the principal planes. In the case of the sc lattice this would lead, however, to a rather complicated model, containing at least 19 different vertices. The situations is simpler in the case of a bcc crystal [7] where the same processes are possible. One can then use the more conventional mapping of the surface of a body-centred solid-on-solid model onto the 6-vertex model, proposed by van Beijeren [8]. This leads to model II. The rates for the different processes depend on the change of the energy of the 6-vertex model (given by the change of the number of bonds to n.n.n. in the bcc-lattice). The dependence on the temperature is introduced here using Glauber kinetics. The temperature is compared with the roughening temperature T_R of the 6-vertex model $k_B T_R = \frac{\epsilon}{\ln 2}$, where ϵ is the vertex energy for vertices 1,2,3,4 (i.e. the energy of a broken bond). Model II is more complicated than model I, so that it is unlikely that the exact solution in the thermodynamic limit can be obtained; on the other hand it is more rich, so that it permits to study the transition between the different modes of growth.

To obtain exact results finite-size systems have been studied for both models. The number of configurations increases very rapidly (exponentially) with size. The configurations, however, separate naturally into distinct classes, characterized by the values taken by the conserved quantities (related to the mean Miller indices, which do not change in the growth process). The classes, in turn, are partitioned into translation equivalence subclasses. The time evolution within a class is described by a master equation for the probabilities $P_m(t)$ that the crystal configuration, as a function of time, belongs to different subclasses within a given class. Solving the master equation the rate of growth is calculated.

This programme can be implemented only for small-size versions of the models. For larger sizes simulations have been devised. They are founded on the ergodicity of the models and do not require any classification of the configurations into subclasses. In each simulation step the processes possible in the given configuration together with their multiplicities are known. One random number is used to decide which process will take place and a second random number is generated to select one of the possible active sites. Then the configuration is changed and multiplicities are recalculated. The rates of growth obtained from the simulations agree with the exact results when the latter are available.

The dependence of the rate of growth rate per site on the size shows that for model II the finite size effects are small for $N \geq 32$. Hence, the majority of our calculations have been done for $N = 32$. The necessary number of steps for $N = 32$ is at least 500000. For small disequilibrium or low temperature usually many more steps are needed.

In the case of model II for the (001)-surface the two expected modes of growth have been found. For high temperature and disequilibrium $\Delta\mu$ the dependence of the rate of growth G on disequilibrium follows the Wilson-Frenkel law[9] $G \sim (e^{\beta \Delta\mu} - 1)$ (Fig.3). For low temperature and disequilibrium the nucleation phenomena are important and the predicted Becker-Döring law[9] for the rate of growth by nucleation $G \sim e^{-\beta E^2/\Delta\mu}$ is approximatively fulfilled (E being proportional to the step energy per unit length). A crossover between nucleation and Wilson-Frenkel type behaviour corresponds to the transition between layer-by-layer (Frank-van der Merwe) and

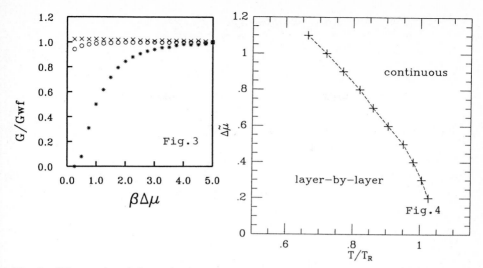

Fig.3 - The ratio of the calculated growth rate G to the Wilson-Frenkel law normalized to the value at $\beta\Delta\mu = 5$.

Fig.4 - Growth phase diagram showing the generalized roughening transition separating the nucleation regime from the Wilson-Frenkel (3-dimensional growth) regime ($\Delta\tilde{\mu} = \Delta\mu/(kT_R)$).

continuous (Volmer-Weber) growth, i.e. to the roughening transition. This can be represented as a transition curve [10] of the kinetic phase diagram in the $(T, \Delta\mu)$-plane (Fig.4). The points in the figure are simply obtained by finding the temperature where the growth rate G is maximal for each value of $\Delta\mu$ (the fact that a part of the curve lies to the right of T_R is due to this simplistic procedure).

References

[1] E.H.Lieb and F.Y.Wu, in *Phase transitions and critical phenomena* (C.Domb and M.S.Green, editors), Vol.**1**, Academic Press, London, 1972

[2] R.J.Baxter, *Exactly solvable models in statistical physics*, Academic Press, London, 1982

[3] L.Pauling, J.Am.Chem.Soc. **57**(1935)2680; J.C.Slater, J.Chem.Phys. **9**(1941)16

[4] C.Garrod, A.C.Levi and M.Touzani, to appear in Sol.State Comm.

[5] F.Y.Wu, Phys.Rev.**168**(1968)539

[6] C.Garrod, to be published.

[7] M.Kotrla and A.C. Levi,to be published.

[8] H.van Beijeren, Phys.Rev.Letters **38**(1977)993

[9] For a general review see J.D.Weeks and G.H.Gilmer, Adv.Chem.Phys. **40** (1979)157

[10] P.Bennema and G.H.Gilmer, in *Crystal growth, an introduction* (P.Hartman, editor), North-Holland, Amsterdam 1973.

Characterizing the Evolution
of Non-equilibrium Structure During Adsorption

J.W. Evans

Ames Laboratory* and Departments of Physics and Mathematics,
Iowa State University, Ames, IA 50011, USA

Abstract. We review a diverse array of stochastic concepts and models which can provide a sophisticated description of the evolution of structure during chemisorption or layer-by-layer epitaxy. We consider far-from-equilibrium processes involving quasi-irreversible birth, growth and coalescence of 2D islands. The structure of individual growing islands is influenced by whether atom incorporation is governed by local rules or by diffusion, but is probably dominated by restructuring. Their characteristic (correlation) length scales non-trivially with the ratio of birth to growth, or deposition to hopping rates. The pair correlations, determining the diffracted intensity, scale with this characteristic length, but crossover to fast asymptotic decay. Concepts from stochastic geometry and correlated percolation theory provide a framework for characterizing the ramified structure of clusters of coalesced islands. Islands or domains of a single phase must percolate, and of two-phases "nearly" percolate, in contrast to multi-phase islands where map-coloring ideas elucidate structure.

I. Introduction

We consider far-from-equilibrium chemisorption and layer-by-layer epitaxial growth processes involving quasi-irreversible birth, growth and coalescence of two-dimensional (2D) islands.

Chemisorption often occurs via a mobile precursor, and is essentially irreversible, with chemisorbed atoms having little mobility [1-3]. It is instructive to consider the two limiting regimes: (i) an equilibrated precursor, where precursor adsorption/desorption and diffusion rates dominate the irreversible chemisorption rates. Here incorporation at islands perimeters occurs with enhanced rates because of the corresponding enhancement of the precursor density (due to attractive interactions). Such incorporation rates depend only on the local environment [2]; (ii) an unequilibrated precursor where incorporation at island perimeters is diffusion mediated, with contributions from intrinsic (extrinsic) precursor atoms above empty (island-covered) regions of the substrate [1,3,4].

A direct adsorption model is more common for epitaxial growth [5,6]. The conventional picture is of thermal-diffusion mediated, quasi-equilibrated "classical" 2D nucleation and growth [5,6]. Here we focus on the low-temperature regime where adsorption is irreversible, diffusion of isolated atoms is significant, but 2D island dissolution is rare, so island growth is effectively irreversible [7]. In both chemisorption and epitaxial growth

---

* Ames Laboratory is operated for the USDOE by ISU under contract No. W-7405-Eng-82. This work was supported by the Division of Chemical Sciences, Office of Basic Energy Sciences.

models, restructuring of the 2D islands is presumably significant, and produces compact islands.

In Section II, we discuss adlayer structure on the characteristic length scale (reflecting island size), and adsorption kinetics. We determine this length scale, and discuss the structure of the growing islands, and the nature of spatial pair correlations. The latter, which have a distinctly non-equilibrium form, are important as they determine diffracted intensity behavior (in the kinematic approximation). The large scale structure, e.g., associated with ramified clusters of coalesced islands, is discussed in Section III. We identify appropriate measures of cluster (or domain) size. Structure is characterized separately for the cases of one-, two-, and multi-phase islands/domains. We close with some comments on the relationship between structure and diffracted intensity in Section IV.

II. Structure on the Characteristic Length Scale; Kinetics

The processes considered here are assumed to involve competition between continuous birth and growth (followed by possible coalescence) of 2D islands with compact structure. The coverage, in monolayers, will be denoted by θ. We now consider several fundamental questions:

The Characteristic (or Correction) Length, ξ_c, which is associated with the linear dimension of individual (precoalesced) islands. Consider first processes involving competition between random birth of islands, at rate d per empty site per unit time (a Poisson process), and stochastic growth of compact islands which expand at some "average" rate of v lattice vectors per unit time. Then a dimensional analysis shows that $\xi_c \sim (v/d)^{1/3}$ [2,8-11]. If the growth rate v equals aL, where L is some island dimension and 'a' is the acceleration rate, the above result with L $\sim \xi_c$ yields $\xi_c \sim (a/d)^{1/2}$.

Second consider processes which involve competition between random deposition at rate d, and diffusion at rate h per atom per unit time leading to irreversible aggregation. Clearly the spatial statistics of this model are determined entirely by the ratio h/d. Of particular interest here is the island density $C \sim (h/d)^{-x}$ and the associated mean lattice area per island $A = C^{-1} \sim (h/d)^x$. It has been argued based on random walk theory that $x = 1/3$ [12]. One expects ξ_c to scale like $A^{1/2}$ at least for moderate θ.

The Structure of Growing Islands (before coalescence). There is clearly a competition between the evolution of "kinetic structure" determined by the growth rules, and restructuring or "relaxation". For the moment, ignore the latter effect. If incorporation at island perimeters is governed by local rules, then growth is in the Eden universality class [11]. However it is important to note that any variation in incorporation rates for different perimeter sites could significantly affect the structure of "small" islands. In some cases, the well known circular Eden cluster structure may never be achieved before coalescence [2,9,10]. If island incorporation is diffusion mediated, then growth from diffusion of atoms exterior to the island produces Mullins-Sekerka or Diffusion Limited Aggregation (DLA) type shape instabilities [11]. Growth from diffusion of atoms on top of islands to the edge produces anti-DLA type shape stabilization [4]. In practice, however, we expect that restructuring will dominate these kinetic growth aspects of island structure.

Kinetics. Again consider processes involving random birth at rate d and stochastic growth of compact islands with some characteristic shape and perhaps variable rate of expansion. Here the kinetics can be elucidated by consideration of semideterministic lattice models where island birth still occurs randomly, but growth is modeled as a deterministic rather than stochastic process [9]. Specifically, assume that islands have a fixed shape with boundary expanding to incorporate sites at a prescribed rate of exactly $m(\phi) \, v(\delta t)$ lattice vectors per unit time in direction ϕ. Here $m(\phi)$

50

determines shape where any convenient normalization can be chosen, and δt denotes the time since birth. These quantities determine $\tau(\ell)$, the time for an island nucleated at site ℓ to travel to site 0. Clearly the probability that 0 is empty at time t, depends only on sites in the "domain of influence" satisfying $\tau(\ell) \leq t$, and is given by

$$1 - \theta = \prod_{\ell:\tau(\ell)\leq t} \exp[-d(t - \tau(\ell))] , \tag{1}$$

since ℓ must be empty at time $t - \tau(\ell)$ for 0 to be empty at time t.

In the large-ξ_c regime, (1) reduces to the Avrami form [9]

$$1 - \theta \sim \exp[-\frac{d}{2\alpha} (\int_0^{2\pi} d\phi \, m(\phi)^2) \int_0^t dt' (\int_0^{t'} ds \, v(s))^2] , \tag{2}$$

where α is the area per site in units of (lattice vector)^2. It is instructive to note that the scaling of the characteristic time, t_c, can be determined from (1) or (2), or more directly using $\xi_c \sim \int_0^{t_c} v(s) \, ds$.

Clearly diffusion mediated processes, with deposition on top of islands as well as in empty regions, have zero order kinetics with d/dt $\theta \sim$ d.

Form of the Spatial Pair-Correlations, $C(\delta\ell)$, where $\delta\ell$ denotes the separation. Consider any adsorption process starting with a clean substrate at t = 0. Here the $C(\delta\ell)$ must exhibit superexponential asymptotic decay, as $|\delta\ell| \to \infty$, at any finite time. This is in marked contrast to equilibrium behavior. A rigorous general proof follows from the theory of Interacting Particle Systems [13], and exact explicit 1D examples are available for irreversible cooperative filling [14], and the Glauber (adsorption-desorption) model [15]. It is important to note that the crossover to the fast asymptotic decay occurs only for large separations, certainly greater than ξ_c. One anticipates [2] that for $|\delta\ell| = O(\xi_c)$, the correlations roughly scale like $C(\delta\ell/\xi_c)$. This scaling primarily determines the diffracted intensity behavior, with the full width at ½-maximum (FWHM) of the diffuse profile scaling like ξ_c^{-1} (at fixed θ).

For processes involving random birth and stochastic growth of compact islands with some characteristic shape and expansion rate, the corresponding semideterministic models (described above) are particularly useful for elucidating the form of $\hat{C}$. Here correlations are determined by the overlap in "domains of influence" of the two sites separated by $\delta\ell$, analogous to the determination of θ by (1) [16]. Clearly $C(\delta\ell)$ will be identically zero for $|\delta\ell| > (m(\phi) + m(-\phi)) \int_0^t ds \, v(s)$, where ϕ is the direction of $\delta\ell$ (in contrast the original model involving stochastic island growth). From these $C(\delta\ell)$, one can then determine the scaled form of the diffracted intensity.

III. Large Scale Structure

For islands which have only one or two phases, significant coalescence of individual growing islands will occur producing ramified clusters of islands (Fig. 1). These will be termed filled regions (for one phase) or domains or clusters. A sophisticated characterization of this structure is provided here. We start by identifying appropriate measures of domain or cluster size. One measure of linear size is obtained by taking 1D slices through the 2D pattern in a fixed direction, and measuring the "chord" lengths intersecting the clusters (of a specific phase) [2,10,17,19]. Other standard "percolation theoretic" measures [10,18,19] include the number of atoms (which we term the "size"), and the radius of gyration. All these distributions and their averages depend on the specific definition of clusters or domains (i.e., on the choice of connectivity rule).

First <u>one-phase islands</u> are considered. Clearly percolation (coalescence or linkage of islands to span the substrate) will occur for some coverage $\theta_p < 1$. (See Fig. 1.) The average cluster size and radius of gyration diverge as $\theta \rightarrow \theta_p$, but the average chord length increases smoothly through θ_p to diverge only when $\theta = 1$. In general the introduction of clustering will reduce θ_p from the random percolation value of 0.59. However as the "strength" of the clustering and thus ξ_c increases, θ_p will eventually "turn around" (for compact islands) often attaining values near 0.7, familiar from standard continuum percolation problems [10].

Since the correlation length, ξ_c, is finite in these models (even the range of the correlations is finite for semideterministic or Avrami models), the percolation transition is the random percolation universality class [10]. This implies that the fractal dimension of the spanning filled cluster at the percolation threshold is determined by random percolation critical exponents [18] as 91/48. Many relevant concepts here are addressed in the subdiscipline of stochastic geometry termed "random area patterns", e.g., chord lengths, covariance (correlations), clumping (percolation), and contact distributions [17].

Typically the statistics of empty regions at coverage 1 - θ is <u>not</u> equivalent to that of filled regions at coverage θ. Separate consideration of the former is appropriate, particularly at high coverages where only small isolated empty regions remain. If, by definition, empty and filled regions cannot cross, then they cannot simultaneously percolate. However in the continuum limit $\xi_c \rightarrow \infty$, either one or the other must percolate [10].

For <u>two-phase islands</u>, e.g., c(2x2) ordering, where the phases are statistically equivalent (i.e., no long-range order) percolation is not possible--assuming domains of different phase cannot cross. However the "saturation state" where there are no empty regions left may be "close to percolation", and the average domain size and radius of gyration may increase very quickly near saturation (in contrast to the average chord length). The concept of a "ghost percolation" threshold slightly "beyond" saturation has been introduced to describe this behavior [19]. Note that the saturation state does approach percolation in the continuum limit, $\xi_c \rightarrow \infty$, where domains have the fractal dimension 91/48. (See Fig. 1.)

For <u>multi-phase islands</u>, it is instructive to consider an auxiliary adsorption model where no two distinct islands are allowed to coalesce upon impingement. The saturation state of this model clearly generates a random partition of the plane (also called random cell or area pattern, tiling, and in one special case random tessellation or mosaic). If one then assigns a "color" for each phase, the original model is recovered by randomly coloring the cells, and removing "domain boundaries" between touching cells of the same color. Clearly the amount of coalescence or linkage is quickly reduced

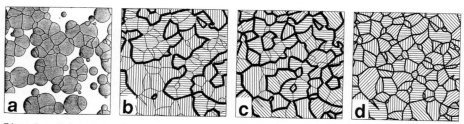

Fig. 1. Schematic derived from Avrami random area patterns of: (a) single phase islands above percolation; (b) 2-phase and (c) 4-phase domains with a random phase assignment, (d) 4-phase domains with a cartographer's (four-color problem) phase assignment, at saturation.

with increasing number of phases or colors, c. In fact one knows from the four-color problem [20], that for $c \geq 4$, there exists a coloring of the saturation state with <u>no</u> coalescence. (See Fig. 1.)

Finally we note that the problem of assessing the (limited) extent of coalescence here can be viewed as a percolation problem on a random lattice. The latter is constructed by placing a "site" in each cell, and bonds between "sites" in adjacent cells [19,21]. The average cluster size at occupancy $1/c$ quantifies the amount of coalescence.

IV. Structure Versus Diffracted Intensity

The simplest analysis determines the diffracted intensity as a sum of incoherent contributions from individual islands, at least near superlattice beams [22,23]. However this scheme is clearly inappropriate when island coalescence is significant and individual islands are not discernable [2,24]. In such cases, quasi-1D geometric or Markovian terrace width distribution models [22] and their extensions [25] are more appropriate. At least they indicate that the FWHM of the diffuse profile is related to a quasi-1D "terrace width" measure of island size, which we identify here more generally as a chord length measure of the type described above.

Our focus here is on characterizing the variation in the FWHM and island size measures with increasing θ, and the relationship between them. (As noted previously the FWHM scales like ξ_c^{-1} at fixed θ.) Consider the simple 1D Markov model for one-phase islands. Let D denote the density of empty-filled pairs, and M_e [M_f] = $(1-\theta)/D$ [θ/D] denote the average number of sites in empty [filled] stretches. These are the average chord lengths (and also average island sizes in 1D). One can show that the FWHM is proportional to $1/M_f + 1/M_e$. Thus as θ increases (from 0 to 1), the FWHM first decreases with increasing $M_f \ll M_e$, and then increases with decreasing $M_e \ll M_f$. Analysis of the corresponding 1D filling model involving competition between birth and growth of islands is much more complicated, even for large ξ_c. However from separate pair-correlation (cf. Sec. II) and island size [9] analyses, one finds that essentially the same asymptotic relationships hold for low and high θ. One anticipates that this relationship between the FWHM and filled/empty regions chord lengths apply at least approximately in 2D.

For one- or two-phase islands where there is significant coalescence, it is clear that the diffuse profile behavior does not correlate strongly with that of the average island size or radius of gyration. The former is insensitive to percolation transitions in these models, in contrast to the latter. For multi-phase islands all size measures should behave similarly and correlate reasonably with diffuse profile behavior.

References

1. E. S. Hood, B. H. Toby, and W. H. Weinberg, Phys. Rev. Lett. <u>55</u>, 2437 (1985).
2. J. W. Evans, R. S. Nord, and J. A. Rabaey, Phys. Rev. B <u>37</u>, 8598 (1988)
3. O. M. Becker and A. Ben-Shaul, Phys. Rev. Lett. <u>61</u>, 2859 (1988).
4. J. W. Evans, Phys. Rev. A <u>40</u>, 2868 (1989).
5. J. A. Venables, G. D. T. Spiller, and M. Handbuchen, Rep. Prog. Phys. <u>47</u>, 399 (1984).
6. J. D. Weeks and G. H. Gilmer, Adv. Chem. Phys. <u>40</u>, 157 (1989).
7. J. W. Evans, D. K. Flynn, and P. A. Thiel, Ultramic. <u>31</u>, 80 (1989).
8. J. D. Axe and Y. Yamada, Phys. Rev. B <u>34</u>, 1599 (1984).
9. J. W. Evans, J. Bartz and D. Sanders, Phys. Rev. A <u>34</u>, 1434 (1986), see also S. Ohta, T. Ohta, and K. Kawasaki, Physica A <u>140</u>, 478 (1987).

10. D. E. Sanders and J. W. Evans, Phys. Rev. A $\underline{38}$, 4186 (1988), J. W. Evans, J. Phys. A, $\underline{23}$, L197 (1990).
11. "On Growth and Form", edited by H. E. Stanley and N. Ostrowsky (Martinus Nijhoff, Dordrecht, 1986).
12. Y.-W. Mo, Ph.D. thesis, University of Wisconsin-Madison (1990).
13. T. M. Liggett, "Interacting Particle Systems" (Springer-Verlag, Berlin, 1985), Ch. I.
14. J. W. Evans, D. Hoffman and D. Burgess, J. Math. Phys. $\underline{25}$, 3051 (1984).
15. J. W. Evans and J. H. Luscombe, unpublished.
16. Again, as $\xi_c \to \infty$, these correlations reduce to the form of Ref. 12.
17. D. Stoyan, W. S. Kendall and J. Mecke, "Stochastic Geometry and its Applications" (Wiley, New York, 1987).
18. D. Stauffer, Phys. Rep. $\underline{54}$, 1 (1979).
19. J. W. Evans and D. E. Sanders, Phys. Rev. B $\underline{39}$, 1587 (1989); J. W. Evans, Surf. Sci. $\underline{215}$, 319 (1989).
20. T. C. Saaty and P. C. Kainen, "The Four-Color Problem" (Dover, New York, 1986).
21. J. F. McCarthy, J. Phys. A $\underline{20}$, 3465 (1987).
22. M. G. Lagally, G.-C. Wang, and T.-M. Lu, CRC Crit. Rev. Solid State Mater. Sci. $\underline{7}$, 233 (1978).
23. T.-M. Lu, L.-H. Zhao, M. G. Lagally, G.-C. Wang, and J. E. Houston, Surf. Sci. $\underline{122}$, 519 (1982); E. D. Williams and W. H. Weinberg, Surf. Sci. $\underline{109}$, 574 (1981).
24. J. W. Evans and R. S. Nord, Phys. Rev. B $\underline{85}$, 6004 (1987).
25. J. M. Pimbley and T.-M. Lu, J. Appl. Phys. $\underline{57}$, 1121 (1985); $\underline{57}$, 4583 (1985); $\underline{59}$, 2439 (1986).
26. J. M. Pimbley, T.-M. Lu and G.-C. Wang, J. Vac. Sci. Technol. A$\underline{4}$, 1357 (1986).

Dependence of Cluster Diffusivity upon Cluster Structure

H.C. Kang, P.A. Thiel, and J.W. Evans

Departments of Chemistry and Physics and Ames Laboratory,
Iowa State University, Ames, IA 50011, USA

We show that the diffusion coefficient of an adsorbed cluster can be decomposed
into two factors, one depending upon a weighted length of the 'active' perimeter
and the other depending upon a subtle correlation between hops taken by cluster
atoms during diffusion. Both of these factors are structure dependent. Monte-
Carlo simulations are performed for clusters on a triangular lattice.

1. Introduction

An important component of many surface phenomena is the diffusion of adsorbed
atoms, either singly or in clusters. Experimental studies of cluster diffusion
include numerous observations of diffusion of small metal clusters on clean metal
surfaces (1-8). There have also been many numerical and theoretical
investigations of cluster diffusion (9-20). In these studies general
mathematical frameworks for the description of cluster diffusion have been
established. In this paper we present an analysis of cluster diffusion which, in
particular, provides insight into the dependence of the diffusion coefficient
upon the size and the structure of the cluster. We also show that cluster
diffusion can be considered as a correlated walk of the cluster center of mass
on a lattice. Monte-Carlo simulations are performed for clusters on a triangular
lattice.

2. Definition of Model and Analysis

Here a cluster is defined as a group of particles any two of which are connected
directly or indirectly, through other particles in the cluster, by nearest-
neighbor bonds. The cluster diffuses as a result of its particles executing
independent hops to vacant nearest-neighbor sites, with the constraint that hops
which cause the cluster to break apart are not allowed. Molecular dynamics
simulations of cluster diffusion have been performed with and without this 'no-
dissociation' constraint (9). Only at sufficiently low temperatures do the two
cases give the same results. Although similar considerations could be
incorporated into Monte-Carlo simulations of cluster diffusion, we restrict
ourselves to the case of· 'no-dissociation' so that the effects of the cluster
size and structure can be disentangled from the effects of cluster lifetime.
Although there exist systems where diffusion and dissociation occur at
approximately the same temperatures (21), typically the activation energy for a
dissociative hop is larger than the activation energy for a non-dissociative hop
(13,14). In such cases there is a well-defined range of low temperatures where
our model applies. It should also be noted that for some systems exhibiting
surface reconstructions concerted motion (i.e., simultaneous jumps) of several
atoms are observed (22). Since this probably results from the peculiar energetics
exhibited by a reconstructive system, we do not incorporate such concerted motion
into our model.

    The hop rate for particles in the cluster is $hw(\alpha,\delta b)$, where h is the rate at
which hops are attempted, $w(\alpha,\delta b)$ is the probability of success of a hop and is
taken to be $\alpha^{\delta b/2}/(\alpha^{\delta b/2}+\alpha^{-\delta b/2})$, and δb is the change in the number of nearest-
neighbor bonds b in the cluster if the hop were successful. The parameter α can

be considered to be the Boltzmann factor for the lateral interaction between two nearest-neighbor particles (with $\alpha > 1$ corresponding to attractive lateral interactions) even though the 'no-dissociation' constraint would be somewhat non-physical with this interpretation of α. For each cluster size there can be many cluster configurations each consisting of different arrangements of the particles and different orientations of the cluster. It is clear that the probability of occurrence of a cluster configuration in the equilibrium distribution is proportional to α^b. When α is equal to one, the probability of occurrence for each cluster configuration is equal. Thus, the clusters are random animals. When α is larger than one (attractive lateral interactions), the clusters correspond to the correlated animals that result from cooperative lattice filling (23). In this case the clusters with the largest number of nearest-neighbor bonds occur most frequently, i.e., compact clusters occur more frequently than ramified ones.

For all values of α, however, the equilibrium ensemble averaged perimeter $<<t>>$ of the clusters, in the limit of infinite number of particles N in the cluster, scales as $<<t>> \sim N$ (24). For compact clusters, which have a larger weight than ramified clusters when α is larger than one, the perimeter scales as $N^{1/2}$. Thus, we would expect a crossover from $<<t>> \sim N^{1/2}$ for small N to $<<t>> \sim N$ for large N. The crossover occurs at progressively larger N when the α is increased. There are several measures of the cluster perimeter. Conventional ones include the number of vacant sites t_v which are nearest-neighbors of particles in the cluster and the number of particles t_f which have at least one vacant nearest-neighbor vacant site. As shown below, it is more appropriate here to consider a weighted 'active' perimeter t_α by $t_\alpha = \Sigma w(\alpha, \delta b)$, where the sum is over all possible hops for that cluster configuration. Note that t_1 is simply half the number of hops that can possibly occur.

We can define the diffusion coefficient using

$$D = \lim_{\tau \to \infty} \tau^{-1} <R_{cm} \cdot R_{cm}>_\tau, \tag{1}$$

where R_{cm} is the position of the center of mass of the cluster, and $< >_\tau$ is the average taken over many walks at time τ. To elucidate the behavior of D, it is instructive to consider the analogous quantity

$$\lim_{n_h \to \infty} n_h^{-1} <R_{cm} \cdot R_{cm}>_{n_h} \equiv C(\alpha, N)/N^2, \tag{2}$$

in which the number of successful hops n_h plays the role of time. The motivation for this definition of the 'correlation factor' $C(\alpha, N)$ is as follows. If the center of mass were undergoing a pure random walk, then $C(\alpha, N)$ would simply equal one, noting that after each particle jumps, the center of mass moves a distance $1/N$ (25). Thus, deviations of $C(\alpha, N)$ from unity measure the correlation in its walk.

The process by which the cluster diffuses on the lattice is ergodic so that, for $\tau \to \infty$, we obtain $<n_h>_\tau \sim h\tau <<t_\alpha>>$. Therefore, in the limit of $\tau \to \infty$, we can write

$$<R_{cm} \cdot R_{cm}>_\tau = C(\alpha, N) <n_h>_\tau /N^2$$

$$= hC(\alpha, N) \ \tau <<t_\alpha>>/N^2, \tag{3}$$

from which, using Equation (1), we obtain

$$D = hC(\alpha, N) \ <<t_\alpha>>/N^2. \tag{4}$$

Equation (4) shows that the diffusion coefficient for a cluster can be decomposed into a factor depending upon the weighted 'active' perimeter and a factor depending upon the correlation between hops. We present below the results for $<<t_\alpha>>$ and $C(\alpha, N)$ from simulations of clusters on a triangular lattice.

3. Simulation Results

In Fig. 1, we plot log $<<t_\alpha>>$ as a function of log N. The value of α for each plot is indicated in the figure. When α is equal to one, it is clear that $<<t_\alpha>>\sim N$ even for cluster size N as small as 10. However, as the value of α increases, the asymptotic regime, in which $<<t_\alpha>>\sim N$, is reached at progressively larger values of N. For the case in which $\alpha=5$, the asymptotic regime is not reached even at N=100, and the scaling $<<t_\alpha>>\sim N^{1/2}$, expected for compact clusters, is observed for the cluster size range that we simulated.

In Fig. 2, we plot the mean-square displacement as a function of time for $\alpha=5.0$. The cluster size for each plot is indicated in the figure. It can be seen that the correlation factor $C(\alpha,N)$ is less than one, indicating (negative) correlation between hops. This means that the correlation between hops causes the center of mass of the cluster to diffuse slower than the 'correpsonding' random walker. That is, the cluster center of mass undergoes a correlated walk. The results for $\alpha=1.0$ and $\alpha=2.5$ are similar. For the case in which $\alpha=1$ and $N=3$, i.e., the random animal trimers, the correlation factor is exactly one because there is no correlation between hops (25). (All dimers, regardless of the value of α, have a correlation factor of one.)

In Fig. 3 log $C(\alpha,N)$ is plotted as a function of log N. The confidence limits are approximately ±0.05 for each of the data points. Since the range of cluster

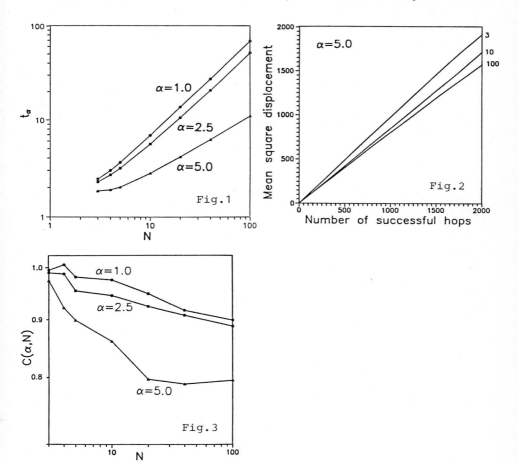

Fig.1

Fig.2

Fig.3

57

sizes simulated is rather limited, it is not sufficient to clearly establish whether a scaling relation exists between $C(\alpha,N)$ and N. However, assuming that $C(\alpha,N)\sim N^{-\epsilon}$ is valid, we obtain a value of approximately 0.03 for ϵ in the cases in which $\alpha=1.0$ and $\alpha=2.5$. The results for $\alpha=5.0$, in which case a crossover in cluster structure occurs, suggest that a higher effective exponent may be observed for small N and large α.

4. Discussion

Using Equation (4) and the simulation results it is easy to understand how a scaling relation $D\sim N^{-y}$ can arise. For metal clusters on clean metal surfaces, in which case α is relatively high (approximately 20 for Rh on Rh(100) at a temperature of 2000 K), our analysis predicts an exponent of $y=3/2+\epsilon$ for cluster sizes up to, at least, 100. It should be noted that this holds regardless of the geometry of the lattice on which the cluster diffuses, so long as the mechanism of diffusion consists of particles moving along the perimeter of the cluster. In the limit of large N, the exponent obtained would be $y=1+\epsilon$. Real clusters for which this exponent holds probably do not exist as ramified clusters would be quite unstable to dissociate. Therefore, for experimentally observable clusters the diffusion coefficient is more likely to behave as $D\sim N^{-3/2-\epsilon}$ than as $D\sim N^{-1-\epsilon}$. Molecular dynamics simulations of Rh clusters of size up to 75 atoms show an exponent of $y\approx1.76$ (13,14). In comparing this with our result of $y\approx1.53$, it should be noted that there are many differences between molecular dynamics simulations and the Monte-Carlo simulations used here, such as the absence of dynamical correlations in the Monte-Carlo simulations.

5. Conclusions

We have shown that the diffusion coefficient of a cluster is a product of $<<t_\alpha>>$ which depends on the weighted length of the 'active' perimeter and $C(\alpha,N)$ which depends on the correlation between hops. The analysis that we have presented provides a basis whereby the dependence of the diffusion coefficient of a cluster upon its structure and size can be readily understood. In particular, if a scaling relation $D\sim N^{-y}$ is valid, it is possible to understand, using Equation (6), how the structure of the cluster and the correlation between its hops determines the value of the exponent y. Clusters for which the diffusion coefficient is experimentally accessible would probably have a value of $3/2+\epsilon$ for the exponent y. The simulations show the interesting result that, in general, there is correlation between the hops taken by a diffusing cluster so that cluster diffusion results in a correlated walk by the cluster center of mass.

Acknowledgments: Primary support for this work is derived from a Camille and Henry Dreyfus Foundation Teacher-Scholarship. In addition, computing facilities are provided by the Ames Laboratory. Ames Laboratory is operated for the U.S. Department of Energy by Iowa State University under contract No. W-7405-Eng-82. J.W. Evans is supported by the Division of Chemical Sciences, Office of Basic Energy Sciences.

References

1. G. Ehrlich and F.D. Hudda, J. Chem. Phys. **44**, 1039 (1966).
2. D.W. Bassett and M.J. Parsley, Nature (London) **221**, 1046 (1969).
3. G. Ehrlich, CRC Crit. Rev. Solid Stat. Sci. **4**, 205 (1974).
4. W.R. Graham and G. Ehrlich, J. Phys. F **4**, L212 (1974).
5. T. Sakata and S. Nakamura, Surf. Sci. **51**, 313 (1975).
6. T.T Tsong, P. Cowan and G. Kellogg, Thin Solid Films **25**, 97 (1975).
7. D.W. Bassett, J. Phys. C **9**, 2491 (1976).
8. K. Stolt, W.R. Graham and G. Ehrlich, J. Chem. Phys. **65**, 3206 (1976).
9. J.C. Tully, G.H. Glimer and M. Shugard, J. Chem. Phys. **71**, 1630 (1979).

10. S.H. Garofalini, T. Halichioglu and G.M. Pound, J. Vac. Sci. Tech. **19**, 717 (1981).
11. S.H. Garofalini, T. Halichioglu and G.M. Pound, Surf. Sci. **114**, 161 (1982).
12. S.M. Levine and S.H. Garofalini, Surf. Sci. **163**, 59 (1985).
13. A.F. Voter, Phys. Rev. B **34**, 6819 (1986).
14. A.F. Voter, Ann. Rev. Phys. Chem. **38**, 413 (1987).
15. K. Kitihara, H. Metiu, J. Ross and R. Silbey, J. Chem. Phys. **65**, 2871 (1976).
16. U. Landman and M.F. Schlesinger, Phys. Rev. B **16**, 3389 (1977).
17. S. Efrima and H. Metiu, J. Chem. Phys. **69**, 2286 (1978).
18. D.A. Reed and G. Ehrlich, J. Chem. Phys. **64**, 4616 (1976).
19. J.D. Wrigley, D.A. Reed and G. Ehrlich, J. Chem. Phys. **67**, 781 (1977).
20. U.M. Titulaer and J.M. Deutch, J. Chem. Phys. **77**, 472 (1982).
21. P. Cowan and T.T. Tsong, Phys. Lett. **53A**, 383 (1975).
22. T.T. Tsong and Q. Gao, Surf. Sci. Lett. **182**, L257 (1987).
23. D.E. Sanders and J.W. Evans, Phys. Rev. A **38**, 4186 (1988).
24. C. Domb and E. Stoll, J. Phys. A **10**, 1141 (1977).
25. H.C. Kang, P.A. Thiel and J.W. Evans, J. Chem. Phys (in press).

Interaction of Heavy Rare-Gas Atoms with Metal Surfaces: A Model Based on the Effective Medium Theory

M. Karimi[1], D. Ila[1], I. Dalins[2], and G. Vidali[3]

[1]Alabama A&M University, Physics Department, Normal, AL 35762, USA
[2]EM-22, Marshall Space Flight Center, AL 35812, USA
[3]Syracuse University, Physics Department, Syracuse, NY 13244, USA

Abstract. We have calculated the interaction potentials between Ne, Ar, Kr, and Xe atoms and surfaces of Cu, Ag, and Au. The repulsive part of the potential is constructed using the Effective Medium Theory while the attractive part is evaluated from the superposition of damped dipole–dipole and damped dipole–quadrupole interactions. We find that our potentials, without any fitting parameter, give a deeper well depth than other calculations in the case of Ar but a shallower one in the case of Ne, Kr, and Xe. We discuss the sensitivity of the potential to the parameters of the short (α_0) and long (C_6) range parts and to the charge density of the crystal.

1. Introduction

Recently, several groups have directed their attention to the application of the Effective Medium Theory (EMT) and related density functional approaches to the interaction of atoms with surfaces [1–5]. Part of this interest is spurred by the fact that some of these approaches are less empirical than the ones based on pair potentials, and they take into account many–body interactions which are very important in some cases [3]. Although these methods are more accurate than pair potentials, computationally they are not much more demanding in simulation applications.

In order to assess the applicability and accuracy of some of the models mentioned above, we decided to evaluate the interaction potential of rare–gas atoms adsorbed on surfaces of noble metals.

Our interest in the study of adsorption of rare gases on metal surfaces is due to the following reasons: First, rare–gas atoms have a closed electronic shell and therefore the repulsive part of the atom–surface potential is simply proportional to the charge density of the surface probed by the adatom [1–3]. Second, the superposition of atomic charge densities seems to be a reasonable approximation of the actual charge densities of many metal surfaces [5,6]. Third, these systems have been studied by a variety of techniques, including atom beam scattering, neutron scattering and calorimetry.

Besides the immediate need of describing experimental results, these potentials find applications in other areas. For example, the interpretation of diffraction patterns of helium beams from an overlayer of adatoms (such as Xe or Kr) requires an accurate description of the overlayer–adatom/surface interaction [7]. Other areas in which a good knowledge of the interaction forces is required are molecular dynamics [8] and MonteCarlo [9] simulations, as well as in statistical mechanics calculations of overlayer properties [10].

2. The Model

In recent publications [1], we applied EMT to calculate the interaction of H_2 and light rare–gas atoms with surfaces of insulators, semimetals, and metals. Here

Springer Series in Surface Sciences, Vol. 24 **The Structure of Surfaces III**
Editors: S.Y. Tong · M.A. Van Hove · K. Takayanagi · X.D. Xie
© Springer-Verlag Berlin, Heidelberg 1991

we extend EMT to heavier atoms and discuss the applicability of the method to these systems. Work is in progress to implement these potentials in simulations of phase tranformations within the overlayer.

The model used to obtain the interaction potential has been described in detail elsewhere [1–3]. Hence, we briefly outline its main features. The interaction between a rare–gas atom and a metal surface can be written as:

$$V(\underline{r}) = V_A(\underline{r}) + V_R(\underline{r}) \tag{1}$$

where V_A and V_R are the attractive and repulsive parts of the potential, respectively. V_R is calculated using EMT:

$$V_A(\underline{r}) = \alpha_{eff.} \, \bar{\rho}(\underline{r})$$

$$\alpha_{eff.} = \alpha_0 - \alpha_{at}$$

$$\alpha_{at} = \int_\Omega \phi_a(\underline{r}' - \underline{r}) \, d\underline{r}'$$

$$\bar{\rho}(\underline{r}) = \frac{1}{\alpha_{at}} \int_\Omega \phi_a(\underline{r}' - \underline{r}) \rho(\underline{r}') dr'$$

where ρ is the charge density of the surface, $\alpha_0 = 99.2, 247, 474$, and 755 eV–Å^3 [11,12] for Ne, Ar, Kr, and Xe, respectively; ϕ_a is the electrostatic potential of the adatom, Ω is a sphere of radius $R_c = 2.5$ Å, and $\bar{\rho}$ is the average of ρ. All the integrals can be evaluated analytically using an expression of the charge density of the surface which depends on the z coordinate only, $\rho(Z)$. If the charge density $\rho_a(r)$ of each atom of the crystal is given by: $\rho_a(r) = A \exp(-\lambda r)$, then from Ref. [13] we find that A and λ are equal to 0.79 Å^{-3}, 2.34 Å^{-1}; 1.19 Å^{-3}; 2.34 Å^{-1}; 1.51 Å^{-3}, 2.40 Å^{-1} for Cu, Ag, and Au atoms, respectively.

V_A is obtained using superposition of anisotropic damped dipole–dipole and damped dipole–quadrupole terms and has been evaluated previously [1,2]. The laterally averaged part and the Fourier components of V_A are [1,2,14]:

$$V_{00} = \frac{2\pi}{A_s} \beta_{00} \left[-C_6 \frac{\xi(4,Z/d)}{4 \, d^4} - C_8 \frac{\xi(6,Z/d)}{6 \, d^6} + H_0(Z) \right]$$

$$V_{GA} = \frac{2\pi}{A_s} \beta_G \left[-\frac{C_6}{2} \left(\frac{G}{2Z}\right)^2 K_2(GZ) - C_8 \left(\frac{G}{2Z}\right)^3 K_3(GZ) + H_G(Z) \right]$$

$$C_6 = \frac{6}{\pi n_0}, \quad C_8 = \frac{15}{\pi n_0} C_5$$

where β_{00} is number of atoms per surface unit cell, A_s is the area of the surface unit cell, d is the interplanar spacing, C_6 and C_8 are dispersion coefficients of the rare–gas atoms, n_0 is number density of atoms in the solid, C_3 and C_5 are dispersion coefficients of the adatom with the surface [15], $\xi(n,Z/d)$ is the Riemann zeta function, K_2 and K_3 are modified Bessel functions, H_G is a damping function [1,16], $\beta_G = \exp(i\underline{G} \cdot \underline{\delta}_i)$ where $\underline{\delta}_i$ is the position of the i–th atom in the unit cell.

3. Results and Discussion

In Table I we report the most important results along with the parameters used in our calculations. From $V_{00}(Z)$ we calculated the eigenenergies E_n and

TABLE I

Comparison between results of Ref. [4] and calculations of this paper. ξ is the corrugation of the surface as seen by an atom of energy 15 meV (from four–fold hollow to saddle site), E_0 is ground state energy of the potential, and D is the well depth.

Systems	$\alpha_0(a)$ (eV-Å³)	$\langle Z \rangle$ (Å)	D(b) (meV)	$C_3(c)$ (meV-Å³)	$C_5(c)$ (meV-Å⁵)	ξ (Å)	E_0 (meV)
Ne–Cu	99.2	4.25	14.4 (10.9) 12.4e	488 [581]	364	0.22	8.07
Ar–Cu	247	3.0	54.6 (67.5) 124d	1500 [1320]	1751	0.55	53.30
Kr–Cu	474 [369.7]	3.11	78.5 (39.6) 119f	1730	2712	0.36	77.18
Xe–Cu	755 [580.6]	3.23	107.7 (53.4) 196d	2430	4373	0.31	106.56
Ne–Ag	99.2	4.05	13.7 (11.6)	501 [556]	398	0.36	9.04
Ar–Ag	247	3.13	50. (78.3) 72f	1620 [1228]	2565	0.78	48.61
Kr–Ag	474 [391.5]	3.43	71.6 (44.9) 107f	2260	3944	0.46	67.98
Xe–Ag	755 [596.4]	3.37	109.6 (62.6) 211f	3380	6351	0.46	108.38
Ne–Au	99.2	3.41	24. (14.6)	554 [736.8]	442	0.45	22.40
Ar–Au	247	2.87	85.2 (92.9)	1770 [1681]	2814	0.44	83.23
Kr–Au	474 [365.9]	2.97	120 (52.8)	245	4317	0.71	118.59
Xe–Au	755 [566.6]	3.04	167 (69.8)	353	6938	0.64	165.39

a) values are taken from Refs. [11,12].
b) are from Ref. [4] (top row); values in parentheses are from this model with no adjustable parameters.
c) are from Ref. [15].
d) are from Ref. [18].
e) are from Ref. [20] for the (110) face.
f) are from Ref. [21] for the (111) face.
[ ] are fitted values in order to obtain well depths of Ref. [4].

eigenfunctions $\psi_n(Z)$. From $\psi_n(Z)$ we evaluated the expectation value $<Z>$ for each adatom on the metal surfaces.

We constructed our potentials first by using the values of C_6 and α_0 available in the literature. However, both C_6 and α_0 are not well known. For lighter rare-gas atoms there exists a few independent calculations of α_0. On the other hand, for heavier rare gas atoms only one determination of α_0 exists. We decided to consider C_6 as the free parameter for Ne and Ar but α_0 for Kr and Xe in such a way to reproduce the well depths of the pseudopotential calculation of Ref. [4a]. We did this rather than fitting our only free parameter to experimental data in order to have a comparison with a homogeneous set of results; in fact, experimental data show a wide range of values and are not available for all systems studied here [17]. Overall, comparably smaller changes are necessary for C_6 than for α_0 to obtain the results of Ref. [4], pointing at the necessity of obtaining better values for α for Kr and Xe. In fact, a 10% or more correction on C_6 has been observed before [1,5b]. However, the well depths for Xe are unreasonably small and we think this is due to α_0. Notice that the experimental values for Ar and Xe on Cu(001) [18] are much higher than this and Ref. [4] models. The reason for this discrepancy is still not clear. We also reported experimental values for the well depth for other faces (i.e., the (110) and (111)). Our calculations were done for a (100) surface, since we believe that the approximation of summing charge densities should hold better here than for other faces; for a (110) face the well depths change by a few percent.

We have also calculated and reported in Table I, the "corrugation" of the surface, i.e. the maximum variation of distance of closest approach of a 15 meV atom along the $<110>$ direction of (001) face (that is, from the four-fold hollow to the bridge site). The corrugation of the surface estimated using the present model is larger than the value inferred from experiments or other models [1,2]. According to the explanation of Ref. [19], it is the repulsive part of the potential which is mostly responsible for probing the corrugation of the surface. Due to larger sizes of these atoms as compared with He, one expects that they will probe a smaller corrugation than He. This is not necessarily the case because the stronger polarizability of a bigger atom might draw it closer to the surface.

In conclusion, our EMT-based applied previously to lighter rare gas atoms and H_2 [1,2] has been extended to the interactions of heavier rare-gas atoms with metal surfaces. Our model predicts a shallower well depth for Ne, Kr, Xe and deeper well depth for Ar without any fitting parameter.

EMT, as expected [1,2,6], overestimates the corrugation of the surface. Because EMT is sensitive to C_6, α_0, and surface charge density, more accurate values of these parameters will help us assess where the model needs improvement. Our analysis suggests that α_0 is the parameter that causes the largest disagreement in this model's results as compared with other calculations or experimental data. The wide range of well depths obtained experimentally [17] or with models is perplexing, and we tried to point out where more work is necessary.

ACKNOWLEDGMENTS

We thank Professor M.W. Cole for helpful discussions and the Alabama Supercomputer Network for Cray time and particularly Dr. D.S. Retallek for technical assistance. This work was supported in part by NASA grant #NAG8–127.

References

1. M. Karimi and G. Vidali, Phys. Rev. B39, 3854 (1989); Surf. Sci. 208, L73 (1989); Phys. Rev. B38, 7759 (1988); 36, 7576 (1987).
2. F. Toigo and M.W. Cole, Phys. Rev. B32, 6989 (1985); A. Frigo, F. Toigo, M.W. Cole, and F.O. Goodman, Phys. Rev. B33, 4184 (1986).
3. K.W. Jacobsen, J.K. Norskov, M.J. Puska, Phys. Rev. B35, 7423 (1987); J.K. Norskov, J. Chem. Phys. 90, 7461 (1989).
4. A. Chizmeshya and E. Zaremba, Surf. Sci. 220, 443 (1989); C.Y. Fong, L.H. Yang and I.P. Batra, Phys. Rev. B (in press).
5. D. Eichenauer, U. Harten, J.P. Toennies and V. Celli, J. Chem. Phys. 86, 3693 (1987); V. Celli, D. Eichenauer, A. Kaufhold, J.P. Toennies, J. Chem. Phys. 83, 2504 (1983).
6. I.P. Batra, Surf. Sci. 148, 1, (1984).
7. R.A. Aziz, U. Buck, H. Jonsson, J.C. Ruiz–Suarez, B. Schmidt, G. Scoles, M.J. Salaman, and J. Xu, J. Chem. Phys. 91, 6477 (1989).
8. P. Stoltze, J.K. Norskov and U. Landman, Phys. Rev. Lett. 61, 440 (1988).
9. M.S. Daw and S.M. Foiles, Phys. Rev. B35, 2128 (1987).
10. J. Unguris, L.W. Bruch, M.B. Webb, and J.M. Phillips, Surf. Sci. 114, 219 (1982).
11. M.W. Cole and F. Toigo, Phys. Rev. B31, 727 (1985).
12. P. Nordlander, S. Holloway, and J.K. Norskov, Surf. Sci. 136, 59 (1984).
13. F. Herman and Skillman, Atomic Structure Calculations (Prentice Hall, 1963).
14. W.A. Steele, Surf. Sci. 36, 317 (1973).
15. G. Vidali and M.W. Cole, Surf. Sci. 110, 10 (1981); X. Jiang, F. Toigo, and M.W. Cole, Surf. Sci. 145, 281 (1984).
16. K.T. Tang and J.P. Toennies, J. Chem. Phys. 80, 3726 (1984).
17. G. Vidali, G. Ihm, H–Y Kim and M.W. Cole, Surf. Sci. Rep., submitted.
18. J. Lapujoulade, Y. Lejay, and G. Armand, Surf. Sci. 95, 107 (1980); A. Glachant and V. Bardi, Surf. Sci. 87 (1979).
19. Y. Takada and W. Kohn, Phys. Rev. Lett. 54, 470 (1985).
20. B. Salanon, J. Physique 45, 1373 (1984).
21. J. Unguris, L.W. Bruch, E.R. Moog, and M.B. Webb, Surf. Sci. 109, 522 (1981) and 115, 219 (1982).

Angular Diffraction Patterns of Photoelectrons and Auger Electrons from Single-Crystal Cu(111)

X.-D. Wang, Y. Chen, Z.-L. Han, S.Y. Tong, and B.P. Tonner

Department of Physics and Laboratory for Surface Studies,
University of Wisconsin–Milwaukee, Milwaukee, WI 53211, USA

Abstract. Complete 2π steradian photoelectron and Auger electron diffraction patterns from single-crystal Cu(111) have been measured. Iso-intensity contours are constructed to display the diffraction pattern as a standard stereographic projection of momentum space. In this form, the patterns show the dominance of forward scattering effects along low-index crystallographic directions, as well as some deviations from the simple forward-scattering model which are only apparent from two-dimensional displays. Detailed simulations of these diffraction patterns using multiple-scattering techniques are used to form a quantitative theory of the angular anisotropy of emission electrons from single crystals.

1 Introduction

The characteristics of angle-resolved X-ray photoelectron scattering and Auger electron scattering (ARXPS and ARAES) angular distributions have been extensively used for qualitative determinations of atomic structure, particularly in application to epitaxial films [1, 2]. A number of theoretical models have been proposed to explain the angular anisotropies in a quantitative context, in order to increase the precision with which ARXPS/ARAES can be used for determining the structures of unknown surface and thin-film samples [3, 4, 5].

Most of the work in this area to date has been confined to either single azimuthal or polar angle scans through the diffraction hemisphere, or in isolated cases a small section of the 2π steradian diffraction pattern[1]. These limitations, which apply to both the available data and theoretical calculations, are due in part to the difficulty in accumulating such a large set of data and multiple-scattering (MS) intensities. In this work, however, we have deliberately surmounted this barrier. Using apparatus automated to accumulate emission-electron diffraction patterns, we have measured complete 2π steradian ARXPS/ARAES patterns from Cu(001), Cu(011), and Cu(111). Further, full MS calculation of the ARXPS patterns from these substrates has also been achieved. We present here the results from the Cu(111) surface.

The fundamental contributions to the ARXPS/ARAES angular distributions can be broadly classed into 'initial-state' and 'final-state' effects. Initial state effects contribute to differences between diffraction patterns measured for Auger electrons as opposed to photoelectrons, for example. Initial state ef-

Springer Series in Surface Sciences, Vol. 24   **The Structure of Surfaces III**
Editors: S.Y. Tong · M.A. Van Hove · K. Takayanagi · X.D. Xie
© Springer-Verlag Berlin, Heidelberg 1991

fects have been found to be particularly important in describing the angular distribution of low-energy Auger electron emission [3].

We consider here experiments using core-level AES/XPS at relatively high electron kinetic energy ($\sim 500 - 1500 eV$). For these energies, the most important contribution to the angular anisotropies is due to final-state scattering. This results in strong similarities between the diffraction patterns from Auger electrons and primary photoelectrons, a fact which has been empirically documented [1].

Our choice of Cu(111) substrates for this study was influenced by the simplicity of the structure of this surface, and the ease with which well-defined, clean surfaces can be prepared. This results in an excellent system for testing quantitative comparisons between theory and experiment. Figure 1 is a summary of the experimental measurements of Cu(111) LVV Auger anisotropy and 2p photoemission anisotropy, along with a full multiple-scattering calculation for the photoemission case.

2   Methods

The photoelectron and Auger electron diffraction patterns were accumulated using an apparatus whose components have been described in prior publications [6, 7]. X-ray radiation from an $Al_{K\alpha} - Mg_{K\alpha}$ conventional twin-anode was used as the excitation source. The sample normal is rotated in the plane containing the incident photon beam and angle-resolved electron detector axis. An azimuthal angle Φ is selected, and the polar angle Θ is scanned in steps of 0.2^o by a spectrometer-controlled stepping motor which drives the sample manipulator. At each emission direction (Θ, Φ), a set of electron kinetic-energy scans are performed, integrated, and stored. The kinetic-energy windows are selected to collect emission from 'primary' core-level electrons and Auger electrons, and a portion of the inelastically-scattered 'background' intensity.

The experimental angular dependence of the electron intensity, $I(\Theta, \Phi)$, can be written in the form:

$$I(\Theta, \Phi) = [1 + \chi(\Theta, \Phi)]I_o(\Theta, \Phi). \tag{1}$$

In this form, I_o represents the intensity distribution that would be measured from an amorphous sample of the same material, using the same apparatus. It contains slowly varying angular structure due to the scattering geometry and instrument function, as well as any atomic-scattering factors. We approximate the form of this distribution by measuring a portion of the inelastic background, which is assumed to be proportional to I_o.

Angular anisotropies $\chi(\Theta, \Phi)$ were measured at azimuthal angles Φ every 5° from the $[11\bar{2}] - 5^o$ to the $[2\bar{1}\bar{1}] + 5^o$ directions (a total of 15 individual azimuths). The azimuthal symmetry of the anisotropy was established experimentally by comparing χ for angles around the $[11\bar{2}]$ and $[2\bar{1}\bar{1}]$ symmetry axes. The data for one 60° sector of the hemisphere of emission angles was reflected around the mirror-plane and 3-fold replicated, and fit to a Cartesian mesh of 200×200 control points, from which iso-intensity contours were calculated.

Multiple-scattering calculations were performed for the case of 3p photoemission from Cu(111) with a kinetic energy of 1400 eV. The renormalised-forward-scattering method (RFS) was used for an 8-layer slab. Atomic photoemission matrix elements were calculated to include both s and d-wave final states in the scattering calculation. Inelastic scattering was modeled by a constant imaginary part of the inner potential with a magnitude of 4 eV (corresponding to a mean-free-path of $\sim 6\text{Å}$). Calculations for phase shifts with $l_{max} = 10, 12$, and 14 were performed. No significant changes were seen between $l_{max} = 12$ and $l_{max} = 14$; the results shown here are from the calculation with 14 phase shifts.

3 Hemispherical diffraction patterns

Much of the work to date in the use of photoemission diffraction for structure determination has relied upon either azimuthal scans at fixed polar angle, or polar-angle scans at fixed azimuthal angle. An important observation that is made obvious by our full-hemisphere angular distributions is that the general structure of the diffraction pattern can only be understood from the full-angle display.

The shape of the anisotropies for Auger and photoelectron emission from Cu(111) are shown in Figure 1. Polar-angle scans correspond to radial sections from the origin to the edge of the figure (note that the display is a stereographic projection, radius is proportional to $\sin(\Theta)$). Azimuthal scans correspond to circles in these plots.

Some general observations can be made directly from the shape of the iso-intensity contours. The three-fold symmetry of both Auger and photoelectron contours shows that the anisotropy samples at least two atomic layers. The relevant crystallographic directions of the substrate are indicated in the figures. The clear presentation of three-fold symmetry, while not unexpected for this substrate, is not always apparent in Auger anisotropies measured at low kinetic energy.

A second observation is that there is a qualitative agreement between low-index crystallographic directions, and local intensity maxima in the contour plots for both cases. This trend has been observed in all cases of single-crystal anisotropies studied to date (Cu(111), Cu(100), Cu(110), Ag(100), Fe(100), Ir(111)) using this technique. The explanation of anisotropy maxima in terms of forward-scattering along low-index directions has been used to qualitatively interpret single-scan azimuthal or polar-angle distributions. The full-contour plot shows in an elegant way that the emission intensity is not necessarily confined to approximately conical regions surrounding these low-index directions. Instead, the character of the anisotropy is more accurately described as channels of scattering which intersect at low-index directions.

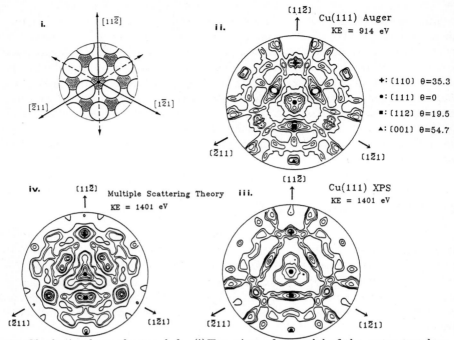

1. Clockwise from the top-left: (i)Top view of a model of the outer two layers of an ideal Cu(111) crystal surface, showing the crystal in the same orientation as the electron angular distributions. (ii)Iso- intensity contour of Auger electrons at 914eV in stereographic projection. Contour spacing is linear in emission intensity. Local intensity maxima occur at low-index crystallographic directions [110], [111], [112], and [001]. (iii)Iso-intensity contour of 3p photoelectrons. A narrowing of the angular width of the major channeling directions is apparent due to the increased final-state kinetic energy, compared to the Auger case. (iv)Multiple-scattering calculation of the 3p photoemission angular distribution. The major peaks and channels of the experimental contours are reproduced.

This, of course, is very similar to the description of the angular anisotropy of quasi-elastic (Kikuchi) scattering. This is not an accident, but rather a further confirmation of the dominance of final-state scattering on the shape of the anisotropy function at kilovolt electron energies. The similarity between Auger emission anisotropy and Kikuchi scattering has been shown experimentally for Ni(100) by Heinz et al.[8]

4   Layer-by-layer analysis

The multiple-scattering calculations provide information about the relative contributions of each layer of the slab to the total anisotropy which is measured from the Cu(111) substrate. An electron originates as a primary photoelectron or Auger electron in a specific layer of the substrate. All multiple-scattering paths of this outgoing electron interfere coherently to produce an angular distribution from the atoms in this specific layer. Atoms in other layers will in general have different angular distributions due to the difference in the number of layers between the emitting layer and the surface. Angular distributions from individual layers are added incoherently to produce the total anisotropy measured in the experiment. Given the good agreement between theory and experiment for the complete single-crystal substrate, we can use the layer-by-layer decomposition of the anisotropy from the theoretical calculation to interpret the origin of specific features in the substrate contour plots.

Figure 2 shows a model of a cut through the mirror plane of an ideal Cu(111) crystal, showing atoms in the plane perpendicular to the surface. The solid curve is a sketch of a possible anisotropy distribution, which has intensity maxima only at the illustrated low-index crystallographic directions.

The measured anisotropy in this plane is shown in Figure 3, along with the theoretical calculation. The mirror-plane geometry contains the major intensity-maxima from the contour plots (see Fig. 1), and corresponds to polar plots in the $[11\bar{2}]$ (right-hand side of Fig. 3) and $[\bar{1}\bar{1}2]$ directions (left-hand side of Fig. 3). The good agreement between theory and experiment can be seen in the comparison shown at the top of Fig. 3.

The layer-by-layer contribution to the total photoemission anisotropy is shown for the first five layers. The emission from the surface layer (layer number 1) is essentially featureless as expected, since only back-scattering events can cause a modulation in the anisotropy, and these events are weak at 1400eV (note that the trivial geometric variations in intensity have been removed from both theory and experiment). The layer-2 emission shows the presence of the

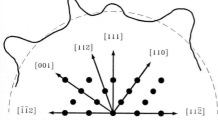

2. Model of atom positions in the mirror plane of the Cu(111) surface, perpendicular to the surface plane. The low-index crystallographic directions which contribute to intensity maxima in photoemission anisotropy are shown. The orientation of this model is the same as the experiment shown in Figure 3.

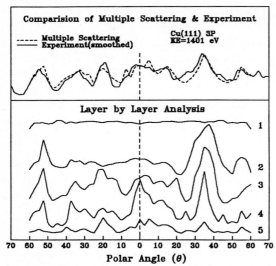

3. Experimental and theoretical photoemission polar-angle distributions in the Cu(111) mirror plane. The $[\bar{1}\bar{1}2]$ direction is shown to the left of the vertical dashed line, the $[11\bar{2}]$ direction is shown to the right of the dashed line. The theoretical emission from individual layers is shown in the bottom half of the graph, with layer 1 being the surface layer. The comparison between experiment and theory for the total substrate emission is shown at the top.

forward-scattering peaks along the [110] direction (35.3° in $[11\bar{2}]$ azimuth) and along the [001] direction (54.7° in $[\bar{1}\bar{1}2]$ azimuth). These directions are illustrated in Fig. 2. The structure in the [112] direction (see Fig. 2) appears as the peak near 20° in layer 3 of Fig. 3, as expected in the simple forward-scattering model. Similarly, the normal-emission forward-scattering peak, which requires 4 layers, can be seen in the layer 4 theory. This peak is weak, however, in both experiment and theory for the complete substrate due to obscuration from emission from the other layers. The rapid decrease in intensity of the forward-scattering enhancements shown in layer 5 is a clear illustration that multiple-scattering restricts the sensitivity of forward-scattering peaks to only the outer atomic planes.

Acknowledgements

This work was supported by the National Science Foundation through grant DMR-88-05171 and DMR-88-05938.

References

[1] For a review of applications of photoelectron diffraction techniques, see C.S. Fadley, in *Synchrotron Radiation Research: Advances in Surface Science*, R.Z. Bachrach, ed., (Plenum Press, New York, 1990).

[2] W.F. Egelhoff, Jr., Critical Reviews in Solid State and Materials Sciences, **16**, 213 (1990).

[3] H.L. Davis and T. Kaplan, Solid State Commun. **19**, 595 (1976); D.M. Zehner, J.R. Noonan, and L.H. Jenkins, Phys. Lett. **62A**, 267 (1977).

[4] H.C. Poon and S.Y. Tong, Phys. Rev. B **30**, 6211 (1984), S.Y. Tong, H.C. Poon, and D.R. S?????, Phys. Rev. B **32**, 2096 (1985).

[5] M.L. Xu, J.J. Barton, and M.A. Van Hove, Phys. Rev. B **39**, 8275 (1989).

[6] Y.C. Chou, M. Robrecht, and B.P. Tonner, Rev. Sci. Instrum.

[7] H. Li and B.P. Tonner, Phys. Rev. B **37**, 3959 (1988)

[8] H. Hilferink, E. Lang, and K. Heinz, Surf. Sci. **93**, 398 (1980).

Surface Termination of Binary Epitaxial Compounds by High Angular Resolution X-Ray Photoelectron Diffraction

S.A. Chambers

Boeing High Technology Center, P.O. Box 3999, MS 9Z-80,
Seattle, WA 98124, USA

Abstract. We demonstrate the utility of x-ray photolectron diffraction at high photoelectron kinetic energy and high angular resolution in determining the surface termination of binary epitaxial compounds. Identification of which atomic species occupies the surface layer is straightforward when the two sublattices possess different symmetries. In this case, x-ray photoelectron diffraction angular distributions in conjunction with low-energy electron diffraction patterns permit an unambiguous determination of the occupation of the sublattices and, therefore, the surface termination. This determination is, however, much more difficult when the sublattices possess the same symmetry. We illustrate these concepts with Ga-stabilized GaAs(001) and epitaxial NiAl on GaAs(001).

The ability to grow binary intermetallic compounds of very high structural quality on III-V semiconductors is of considerable interest for both fundamental and technological reasons [1]. The task is made difficult by the fact that most metals react with III-V semiconductors, thereby disturbing the template for epitaxy. The interplay between interface reactivity and epitaxy must be well understood in order to maximize the structural quality of such epifilms. It is highly desirable to have experimental probes that can elucidate structural details of ultrathin binary epifilms. In this paper, we illustrate how x-ray photoelectron diffraction (XPD) can be used to determine one of the important structural characteristics of these systems.

In order to perform such experiments, it is desirable to combine molecular beam epitaxial (MBE) growth capability with XPD so that epifilms can be grown and characterized without exposure to air. Such a system has been developed and was used in the present work. Details can be found elsewhere [2].

Ga-terminated GaAs(001)-c(8x2) surfaces were prepared by performing a standard liquid degrease, dilute acid etch, and heating in air to ~200C prior to insertion into the ultrahigh vacuum (UHV) preparation chamber associated with the experimental system. After a mild outgassing at ~250C, specimens were rapidly flashed to ~600C in UHV to desorb the surface oxides. The resulting surfaces were Ga-terminated, free of contaminants as judged by XPS, and atomically ordered, exhibiting a (1x1) low-energy electron diffraction (LEED) pattern with weak c(8x2) reconstruction streaks. NiAl epifilms were then grown at ~250C by MBE. Quartz crystal oscillators were used to accurately monitor the two fluxes. The resulting overlayers were composed of $50 \pm 2\%$ Ni and $50 \pm 2\%$ Al, and exhibited a (1x1) LEED pattern.

Springer Series in Surface Sciences, Vol. 24   **The Structure of Surfaces III**
Editors: S.Y. Tong · M.A. Van Hove · K. Takayanagi · X.D. Xie
© Springer-Verlag Berlin, Heidelberg 1991

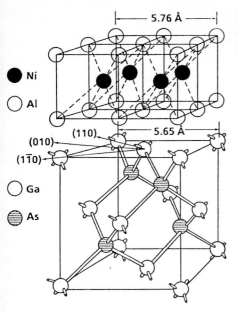

Figure 1. Crystal structures of NiAl and GaAs in which the {100} family of planes are aligned parallel.

Ni
Al

(110)
(010)
(1̄10)

Ga
As

5.76 Å
5.65 Å

In fig. 1 we show the crystal structures of NiAl and GaAs in such a way that the lattice match between the two materials is obvious. NiAl possesses a CsCl crystal structure with a lattice constant which is very close to half that of GaAs (lattice mismatch ~2%). Furthermore, both materials consist of alternating layers of their respective two elements along the [001] direction. Therefore, it is reasonable to expect that the epitaxial growth of one material on the template of the other will lead to a unique surface termination (i.e either all Ni or all Al in the surface layer of NiAl). It is this particular surface property that we have attempted to determine in the present work.

We show in fig. 2 measured and theoretical (single scattering) Ga and As 3d azimuthal angle distributions at a polar angle (θ) of 45^0 and intermediate angular resolution ($\Delta\theta_{1/2} = \Delta\phi_{1/2} = 3.3^0$) for Ga-terminated GaAs(001). The Ga and As 3d angular distributions are different from one another and lack mirror symmetry about $\phi=90^0$ because of the different symmetries of the two sublattices. One sublattice can be transformed into the other by performing a 90^0 rotation about the surface normal. Thus, a 90^0 phase shift transforms the Ga (As) 3d angular distribution into the As (Ga) 3d angular distribution. Furthermore, agreement between theory and experiment is clearly superior when Ga termination is assumed. Thus, it is a simple matter to determine the surface termination of a binary material by XPD, provided the symmetries of the two sublattices are different.

In fig. 3 we show high-angular-resolution ($\Delta\theta_{1/2} = \Delta\phi_{1/2} = 1.7^0$) data at $\theta=15.7^0$ for Al 2p and Ni 3p photoemission from 25 ML NiAl/GaAs(001) grown at 250C. The angular distributions of Al 2p and Ni 3p intensities are very similar for this and other polar angles investigated, in contrast to the results for GaAs(001) [3].

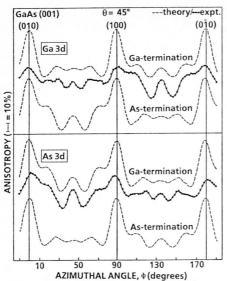

Figure 2. Measured and calculated Ga and As 3d azimuthal angle intensity distributions for Ga-terminated GaAs(001) at a polar angle (θ) of 45° relative to the plane of the surface.

This observation stems from the fact that the two sublattices in the CsCl structure possess the same symmetry. Therefore, the positions of near-neighbor atoms are the same for both Al and Ni atoms in all layers except the surface layer. Inasmuch as XPD probes the immediate structural environment of the emitter, the angular distributions for the two elements are very nearly the same. Nevertheless, grazing-emission azimuthal scans (such as those shown in fig. 3) do show some subtle but key differences by virtue of the fact that atoms in the surface layer possess a different coordination than those in subsurface layers. Azimuthal scans at higher polar angles to which subsurface emissions contribute more strongly are essentially completely insensitive to the composition of the surface layer.

In order to illustrate how subtle differences in the raw data can be used to determine the composition of the surface layer, we also show in fig. 3 angular distributions calculated by means of single scattering theory [4] in which three different surface terminations have been assumed--100% Ni, 50% Ni/50% Al, and 100% Al. Inspection of the figure reveals that superior agreement (as judged by R-factor analysis) between theory and experiment is reached for the Al 2p (Ni 3p) data when Al (Ni) termination is assumed. Intuitively, this result suggests that the surface layer is composed of both kinds of atoms. Emission from the surface layer is accentuated since the data were obtained at a small exit angle relative to the surface plane. Ni 3p and Al 2p photoelectrons generated in the top layer then probe the short-range structural environment of the surface layer and dominate their respective angular distributions. This result suggests that good agreement with

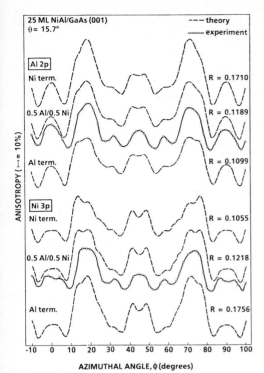

Figure 3. Measured and calculated Al 2p and Ni 3p azimuthal angle intensity distributions at θ=15.7° for 25 ML NiAl/GaAs(001) grown at ~250C.

theory for *both* Al 2p and Ni 3p data might be achieved by modeling a mixed surface termination. We have carried out a complete set of calculations in which the R factors were computed as a function of composition in the surface layer for both Ni 3p and Al 2p photoemission intensities [3]. R factors for both kinds of emission which are within 10-15% of those obtained for Ni 3p (Al 2p) when Ni (Al) termination is exclusively assumed are achieved when a mixed termination involving 50% Ni and 50% Al is modeled. In addition to generating the best overall R-factor agreement, a 50/50 mixed termination results in a better reproduction of diffraction peak asymmetries than do exclusive Al or Ni terminations. The asymmetries associated with features at φ=10-20° and φ=70-80° in the experimental data are very well reproduced by the mixed-termination model for both Al 2p and Ni 3p photoemission. However, agreement is not as good when exclusively Ni or Al termination is assumed. Significantly, these asymmetries vanish when lower angular resolution is employed [5].

The conclusion about a mixed surface termination is corroborated by Al 2p to Ni 3p intensity ratio measurements down to grazing emission (θ→0°) in the (010) azimuthal plane. This ratio, which is expected to rise (fall) as θ→0° for Al (Ni) termination is observed to remain very near unity for all angles. Single scattering cluster and simple inelastic damping calculations confirm that this ratio should remain near unity for a 50/50 mixed termination, as observed [3].

In summary, we have used high-angular-resolution x-ray photoelectron diffraction to investigate the surface termination of single-crystal binary compounds with alternating layer sturctures. We have shown that determining which kind of atom is situated in the surface layer is rather straightforward when the two atoms occupy sublattices of different symmetry. In this case, the angular distributions of photoemission intensities from the two atoms are substantially different, and are well reproduced by single scattering calculations which model the surface termination correctly. In contrast, the determination is much more difficult when the two atoms occupy sublattices of the same symmetry. In this case, the angular distributions are virtually identical unless they are obtained at or near grazing emission. Subtle but key differences exist near grazing emission which yield useful information about the surface termination when used in conjunction with theoretical scattering calculations. The use of high angular resolution results in peak asymmetries that are not present at lower angular resolution.

References

[1] T. Sands, J.P. Harbison, N. Tabatabaie, W.K. Chan, H.L. Gilchrist, S.A. Schwarz, C.L. Schwartz, L.T. Florez, and V.G. Keramidas, in *Proceedings of the 1988 Fall Meeting of the Materials Research Society, Symposium W: Advances in Materials, Processing and Devices in III-V Compound Semiconductors*, Eds. D.K. Sadana, L. Eastman and R. Dupuis (1989).
[2] S.A. Chambers, J. Vac. Sci. Technol **B 7**, 737 (1989).
[3] S.A. Chambers, Phys. Rev **B**, to appear (1990).
[4] C.S. Fadley, in *Progress in Surface Science*, edited by S.G. Davison (Pergamon, New York, 1984), pp. 327-365.
[5] S.A. Chambers, J. Vac. Sci. Technol. **A 8**, 2062 (1990).

Multiple-Scattering Effects in Auger Electron Diffraction and Photoelectron Diffraction: Theory and Applications

A.P. Kaduwela[1], D.J. Friedman[1], Y.J. Kim[1], T.T. Tran[1], G.S. Herman[1], C.S. Fadley[1], J.J. Rehr[2], J. Osterwalder[3], H.A. Aebischer[3], and A. Stuck[3]

[1]University of Hawaii, Department of Chemistry, Honolulu, HI 96822, USA
[2]University of Washington, Department of Physics, Seattle, WA 98195, USA
[3]University of Fribourg, Institute of Physics, CH-1700 Fribourg, Switzerland

Abstract. We apply a new separable-Green's function matrix method due to Rehr and Albers to Auger electron- and photoelectron-diffraction. This method permits building up successive orders of scattering and judging the approach to convergence in a convenient and time-saving way. The systems studied are linear chains of Cu, Al, Ge, CO adsorbed in the tilted α_3 state on Fe(001), and $(\sqrt{3} \times \sqrt{3})R30°$ Ag on Si(111).

1. Introduction

Single-scattering cluster (SSC) theories, by now often with spherical-wave effects included, have been widely used in simulating experimental Auger electron- and photoelectron-diffraction patterns [1]. However the possible effects of multiple-scattering (MS) on such patterns have also been discussed [1-7]. As a result, several cases where a single-scattering (SS) approach is not fully adequate in explaining experimental results have been pointed out [3-5]. In this communication we present numerical results of a new approach to MS due to Rehr and Albers [8] which is conceptually simple and computationally efficient. It is based on a separable approximation to the scattering Green's function and allows building up successive orders of scattering and judging the approach to convergence in a convenient and time-saving way. Scattering events up to 10th order have been included in some cases to insure convergence. The differences between the MS treatments of Barton and Shirley [3] and Rehr and Albers are discussed elsewhere [8]. These methods differ primarily in how the scattering matrices are defined and calculated. They are similar in that they both use scattering matrices to uncouple the nested sums over angular momentum indices that plague LEED theories. The method of Barton and Shirley [4] is based on a low energy Taylor expansion. The Rehr-Albers [8] approach is based on a separable approximation for the propagator and has the correct behavior in both high and low energy limits at low order which gives the theory improved convergence properties. (i.e. smaller matrices are needed to achieve convergence). In particular, the Rehr-Albers method [8] reduces to the point scattering approximation [9] at lowest order.

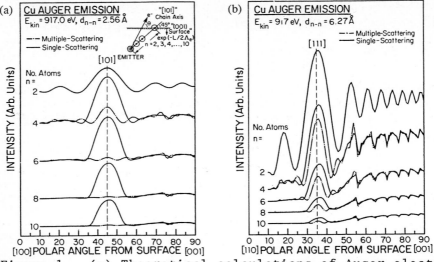

Figure 1. (a) Theoretical calculations of Auger electron diffraction from linear Cu [101] chains at 917.0 eV. The emitter is at one end of the chain, as shown in the inset. (b) As in (a), but for Cu [111] chains.

2. Results and discussion

2.1. Linear Chains

We begin with results on linear chains of Cu atoms placed at 45° with respect to a fictitious "(001) surface" as shown in Fig. 1a. The surface is used only to determine that region of space over which an exponential decay of intensity due to inelastic scattering is included. These chains represent the nearest-neighbor [101] direction in the fcc Cu crystal, with a single emitting atom at the bottom of the chain. Cu LMM Auger emission at 917.0 eV is simulated. Some results are shown in Fig. 1a. It is evident that the longer the chain is the greater are the MS effects: the forward-scattering peak height systematically diminishes as the number of atoms in the chain increases. It is also interesting to notice that the MS peak widths are consistently narrower than those of corresponding SS peaks, becoming systematically narrower in FWHM as the number of atoms in the chain is increased. For a 2-atom chain, MS effects are negligible, a result which is applicable to emission from an oriented diatomic molecule such one that to be discussed later. The 4- to 10-atom cases are applicable to substrate emission or to grazing-angle emission from adsorbate/substrate systems. In such cases, one expects that intensity along linear or nearly-linear chains of atoms with small interatomic distances will be significantly reduced. Such intensity reductions have been termed 'defocusing' in the first discussion by Tong and co-workers [2].

78

At about six atoms the MS intensity in the forward-scattering direction is reduced to the background level. After five atoms the SS intensity begins to reduce due to both interference effects and inelastic attenuation. But even at ten atoms, SS shows a pronounced forward-scattering peak. A similar set of calculations were reported earlier by Xu and Van Hove [6] for 2-, 3- and 5-atom Cu [101] chains at 917 eV. The good agreement between these two different approaches to MS is very encouraging.

We now investigate the dependence of these MS effects in chains for different crystallographic directions and materials. In Fig. 1b we first look at the [111] direction in the fcc Cu lattice at 917.0 eV, which has a nearest-neighbor distance of $d_{n-n} = 6.27$ Å compared to 2.56 Å for the [101] direction considered previously. In this case, the chain is placed at 35.3° with respect to a fictitious (001) surface and hence the forward-scattering peak is at that angle with respect to the surface. The intensities fall more rapidly due to inelastic attenuation, because the interatomic distance is about 2.5 times larger than that for Cu [101] chains. Also, the SS and MS results are much closer to one another because of this increased distance and a concomitant reduction of defocusing effects; thus the importance of this type of MS effect depends strongly on the direction of observation. Both SS and MS peaks are very small after about eight atoms and hence major contributions to the photoelectron intensity are coming from the top eight layers in each case. As an example for a different material, we next show in Fig. 2a results for linear [101] Al chains at 1336.0 eV, which correspond to Mg Kα excited Al 2s

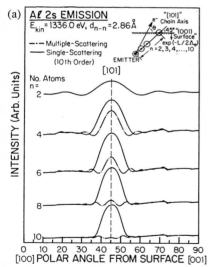

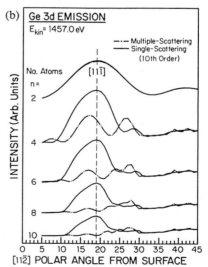

Figure 2. (a) Theoretical calculations of Al 2s photoelectron diffraction from a linear Al [101] chain at 1336.0 eV. The emitter is at one end of the chain, as shown in the inset. (b) As in (a), but for Ge 3d photoelectron diffraction from a linear Ge [11-1] chain at 1457.0 eV.

emission. Defocusing effects are again obvious, since the SS curves show a pronounced peak for an emitter as deep as 10 layers in the solid, but the MS intensity is lost in the background when the emitter is about eight layers deep. Ge [11-1] chains are another interesting case related to a recent study by our group on a surface phase transition of Ge(111) [10]. In this case, the behavior of a forward-scattering peak at 19° with respect to the Ge(111) surface was monitored as a function of temperature. The Ge [11-1] chains on which we have carried out MS calculations represent this direction. As shown in Fig. 2b, even though the SS peak has contributions from more than ten layers, the MS signal is coming from the top eight layers only.

2.2. CO(α_3) on Fe(001)

The CO/Fe(001) system has recently been studied by Saiki et al. [11] using x-ray photoelectron diffraction. From previous studies, it was known that, for the α_3 state of CO, the C-O bond is tilted away from the Fe(001) surface normal by about 45±10°; however, the tilt angle was not very precisely known and there was no information available on the azimuthal preference of this

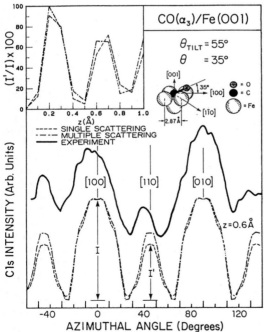

Figure 3. Azimuthal dependence of the C 1s photoelectron intensity from CO(α_3)/Fe(001). The ratio of the intensities I' and I is also shown in the inset as a function of the distance z of the C atom above the first Fe layer. The geometry of the cluster used in the calculation is also shown.

tilting. By comparing experimental polar and azimuthal C 1s data with spherical-wave SSC results for a 7-atom CO/Fe cluster, it was possible to determine both of these structural parameters more precisely [11]: the tilt is 55±2° from the surface normal and it is preferentially along <100> azimuths. Single-scattering calculations have also been performed on much larger CO/Fe clusters but these results are very similar to those for the $COFe_5$ cluster [12]. In order to further estimate the vertical height of the C atom with respect to the first Fe plane on the assumed 4-fold hollow site, Saiki et al. [11] used the intensity ratio I'/I of the peaks at 0° and 45° as defined in Fig. 3. Comparison of experiment and SS theory then yielded possible C vertical heights of about 0.3 or 0.6 Å.

In order to assess the influence of MS effects on this analysis, we have performed azimuthal MS calculations for the 7-atom $COFe_5$ cluster. In Fig. 3, we now compare SS and MS results, including the ratio I'/I as a function of the C vertical distance. The SS and MS curves are very similar. This indicates that MS effects are not important in using such high-energy x-ray photoelectron diffraction to determine the structural parameters for such diatomic adsorbates. This is consistent with our previous observation that there are no significant MS contributions in two-atom clusters.

2.3. ($\sqrt{3}x\sqrt{3}$)R30° Ag on Si(111)

This system has been studied using almost every technique in surface science, including Ag 3d x-ray photoelectron diffraction [13]. The most recent study by photoelectron diffraction [13-b] indicated that the Ag cannot be more than 0.5 A below the surface, and furthermore from an R-factor analysis of azimuthal Ag 3d results that the structure consists of two closely related types of Ag honeycomb domains that grow on the second Si layer, with the top Si layer missing. This geometry is shown in Fig. 4a. For this geometry, linear Ag-Si-Si-Si chains can be seen, e.g. along the arrow at $\phi = 160°$. Especially at low photoelectron emission angles relative to the surface, these chains could cause defocusing near the chain axis and hence the resulting MS intensity patterns could be different from their SS counterparts. We have thus simulated azimuthal intensity patters for three take-off angles (4.1°, 10.0° and 14.7°) with respect to the surface as a function of the percentage of domain 1. The Ag photoemitter is placed 0.2 Å below the surface and the compression parameter s is 0.86 Å, with both parameters being found via a SS R-factor analysis [13-b]. The corresponding R-factors are shown in Fig. 4b. The total R-factor is a sum of individual R-factors for the three theta values involved, with each R value being weighted by the experimental anisotropy at that polar angle. The SS total R-factor, which is smaller than the MS total R-factor, indicates a mixture of 40% of domain 1 to 60% of domain 2. The ratio indicated by the MS total R-factor is nearly 50:50. The experiment-theory comparison in Fig. 4c is for 40% of domain 1. The overall fit to the experiment changes very little when using

(a)

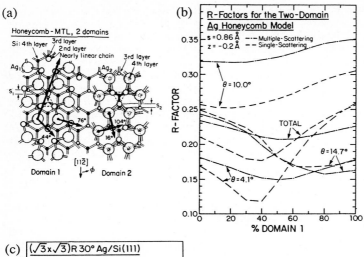

(b)

Figure 4.   (a) 2-domain missing-top layer model proposed in ref. 12(b) for the ($\sqrt{3} \times \sqrt{3}$)R30° Ag on Si(111).   The nearly linear Ag-Si-Si-Si chains along which enhanced MS effects could occur are shown on domain 1.   (b) R-factors for Ag 3d emission, as a function of the percentage of domain 1.   The total R-factor is a sum of the individual R-factors for three polar angle values weighted by experimental anisotropies.   (c) Azimuthal dependence of Ag 3d SS and MS intensity from Ag/Si(111).   The theory curves are for 40% domain 1 and 60% domain 2.

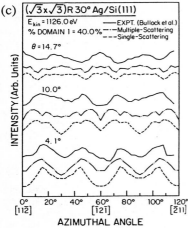

SS or MS, even though the total SS R-factor is slightly lower than the MS R-factor.   The remaining small discrepancies between experiment and theory could be due to: the use of too small a cluster (here 22 atoms for SS and MS), the need to more accurately allow for the true final-state angular momenta involved (here approximated as being from an s-to-p emission event) and for vibrational attenuation of diffraction effects, slight errors in the calibration of the experimental polar angle scale, and/or need for further structural refinement.

3. Conclusions

As shown in cases of linear chains of atoms, it is clear that the amount of defocusing and peak narrowing depends on the interatomic distance as well as the nature of the material under investigation. Such data should permit estimating the strength of MS in experimental systems as well as the effective emission depths along different directions. The diatomic adsorbate $CO(\alpha_3)/Fe(001)$ shows no significant MS effects, which is fully consistent with our previous calculations on two-atom clusters. In the more complicated case of $(\sqrt{3}x\sqrt{3})R30°$ Ag on Si(111) there are short and nearly linear chains of Si atoms, but the defocusing is not large enough to cause significant differences between the structures predicted by SS and MS.

Acknowledgments

Work supported by the Office of Naval Research under contract N00014-87-K-0512 and Grant N00014-90-J-1457 and by the National Science Foundation under Grant CHE83-20200. Calculations were performed at the San Diego Supercomputer Center.

References

[1]. C.S. Fadley, Prog. in Surf. Sci. 16 (1984) 275; C.S. Fadley, Physica Scripta T 17 (1987) 39; C.S. Fadley, Synchrotron Radiation Research: Advances in Surface Science, Ed. R.Z. Bachrach (Plenum, NY, 1990).
[2]. S.Y. Tong, H.C. Poon, and D.R. Snider, Phys. Rev. B32 (1985) 2096.
[3]. J.J. Barton and D.A. Shirley, Phys. Rev. B32 (1985) 1892; J.J. Barton and D.A. Shirley, Phys. Rev. B32 (1985) 1906; J.J. Barton, Ph.D. Thesis, University of California at Berkeley (1985); J.J. Barton, S.W. Robey, and D.A. Shirley, Phys. Rev. B 34 (1986) 778.
[4]. W.E. Egelhoff, Phys. Rev. Lett. 59 (1987) 559.
[5]. J.-C. Tang, J. Vac. Sci. Technol. 5 (1987) 658.
[6]. M.-L. Xu and M. Van Hove, Surf. Sci. 207 (1989) 215.
[7]. A.P. Kaduwela, G.S. Herman, J. Osterwalder, C.S. Fadley, J.J. Rehr, and J. Muestre de Leon, Physica B 158 (1989) 564; A.P. Kaduwela, G.S. Herman, D.J. Friedman, C.S. Fadley, and J.J. Rehr, Physica Scripta, 41 (1990) 948.
[8]. J.J. Rehr and E.A. Albers, Phys. Rev. B 41 (1990) 8139.
[9]. P.A. Lee and J.B. Pendry, Phys. Rev. B11 (1975) 2795.
[10]. T.T. Tran, D.J. Friedman, Y.J. Kim, G.A. Rizzi, and C.S. Fadley, to appear in these proceedings.

[11]. R.S. Saiki, G.S. Herman, M. Yamada, J. Osterwalder, and C.S. Fadley Phys. Rev. Lett. 63 (1989) 283.

[12]. G.S. Herman and C.S. Fadley, unpublished results.

[13]. (a) S. Kono, K. Higashiyama, and T. Sagawa, Surf. Sci. 165 (1986) 21.

(b) E.L. Bullock, G.S. Herman, M. Yamada, D.J. Friedman, and C.S. Fadley, Phys. Rev. B41 (1990) 1703.

Inelastic Photoelectron Diffraction

G.S. Herman, A.P. Kaduwela, T.T. Tran, Y.J. Kim, S. Lewis,
and C.S. Fadley

Department of Chemistry, University of Hawaii, Honolulu, HI 96822, USA

Abstract. Photoelectron- and Auger electron- diffraction studies to date have involved the study of the elastically scattered or "no-loss" features in spectra. In this work, we consider the diffraction patterns associated with inelastically-scattered electrons well removed from the elastic peak. These inelastic photoelectron diffraction (IPD) patterns, associated with plasmon satellites, are found to be very similar in peak positions to those of the elastically-scattered electrons. However, the intensities along low-index directions in both W and Ge crystals are found to be significantly lower in the IPD results. These differences are explained for the model case of Al in terms of defocussing due to multiple scattering effects.

1. Introduction

The structures of surfaces are presently being studied by photoelectron- and Auger electron- diffraction (PD and AED, respectively) in several different ways [1,2,3]. Common to each of these methods is the study of only the quasielastically-scattered or "no-loss" photoelectrons. In scanned-angle PD or AED, the emission direction above a surface is varied, and intensity modulations are produced by scattering and interference effects. Forward scattering peaks are the strongest features at high kinetic energies E_{kin} greater than 500eV, with these peaks being produced by scattering from near-neighbor atoms along low-index directions in a crystal. In this study, inelastic photoelectron diffraction (IPD) patterns resulting from inelastically-scattered electrons are found for W and Ge to be significantly different in their behavior along such low-index directions. We also present a theoretical model for explaining such differences, and apply it to the only other IPD study of which we are aware: a very recent investigation of Al by Osterwalder et al. [4].

2. Experiment

We have measured scanned-angle elastic and inelastic PD from W(110) and Ge(111) surfaces. These data were taken as azimuthal scans at a constant polar angle θ defined with respect to the surface. The measurements for W were performed on a Vacuum Generators ESCALAB5 spectrometer, and those for Ge on a Hewlett Packard 5950A spectrometer, with both instruments being modified for automated angle scanning as described elsewhere [1]. Al Kα radiation was used for excitation in

Springer Series in Surface Sciences, Vol. 24   **The Structure of Surfaces III**
Editors: S.Y. Tong · M.A. Van Hove · K. Takayanagi · X.D. Xie
© Springer-Verlag Berlin, Heidelberg 1991

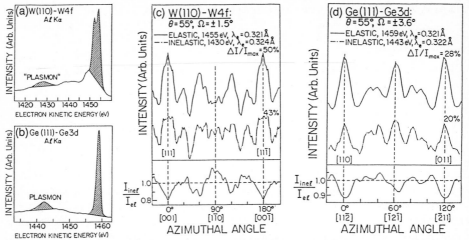

Figure 1   X-ray photoelectron spectra showing elastic and inelastic peaks studied: (a) W 4f region for W(110), and (b) Ge 3d region for Ge(111). Azimuthal scans of the elastic intensity and its associated plasmon loss intensity, together with the normalized ratio of the plasmon loss intensity to the elastic intensity: (c) for W 4f at a polar angle of 55°, and (d) for Ge 3d at a polar angle of 55°.

both systems.   For the VG system, the angular acceptance cone was precisely limited to ±1.5° by the use of tube arrays [5], while that of the HP system was ±3.6°.

The photoelectron peak intensities that were measured as a function of azimuth are shown in Fig. 1(a) and 1(b).   Fig. 1(a) shows both the elastic W 4f and its inelastic "plasmon" loss satellite at 25eV lower kinetic energy [6].   Fig. 1(b) contains analogous curves for the elastic Ge 3d and its inelastic plasmon loss feature at 16eV lower kinetic energy [6a].   A simple linear background was subtracted from each of the peaks and the area was recorded as a function of azimuth.

3. Results and Discussion

In the upper panel of Fig. 1(c) are azimuthal scans of the W 4f elastic and inelastic intensities with a polar angle of $\theta = 55°$.   The degree of mirror symmetry across the $[1\bar{1}0]$ azimuth indicates the high reproducibility of the fine structure observed.   The two strongest features in both diffraction patterns are due to forward scattering along $[111]$ and $[11\bar{1}]$ emission directions, as noted.   The other features found between these peaks are due principally to higher order diffraction features, although the maxima near the azimuthal angles of 55° and 125° probably contain a strong contribution from forward scattering that is separated by only 3.5° in θ from the $[131]$ and $[13\bar{1}]$ emission directions.   The IPD curve in Fig. 1(c) reproduces the fine structure in the elastic curve, but some peaks are greatly diminished in intensity.   In particular, the forward scattering peaks along the above mentioned emission directions are significantly reduced in relative intensity and for <111> become much flatter over their maxima.

In order to quantitatively measure the differences between the inelastic and elastic curves, we have in the lower panel of Fig. 1(c) taken a ratio between the two normalized diffraction patterns. The largest changes in relative intensity in this format are seen to be reductions of ~20% in the IPD curve along the forward scattering directions.

Ge was studied as well to determine whether the characteristics noted for W were general in occurrence. In the top panel of Fig. 1(d) are azimuthal scans of the Ge 3d elastic and inelastic intensities with $\theta = 55°$. Again, the two curves show very similar fine structure, but different relative intensities for the major peaks. The strongest features in both of these curves are due to second-nearest-neighbor forward scattering along the [110] and [011] emission directions, as noted. The strong peak at $\phi = 60°$ is also close to a [131] forward scattering direction. The largest differences between the inelastic and elastic diffraction patterns are found along the [110] and [011] directions for which a reduction in intensity of 13% is observed. Analogous azimuthal scans for both W and Ge at different polar angles and for different elastic and inelastic peak pairs (4f and 4p in particular) gave very similar results.

Another experimental IPD study has recently been performed by Osterwalder et al. [4] on Al(001). In this work, similar azimuthal scans at $\theta = 45°$ were performed on the Al 2s photoelectron peak and three of its loss peaks associated with the creation of n = 1, 2, and 3 bulk plasmons. In agreement with our results, it was found that the elastic and inelastic diffraction curves had very similar fine structure, and that the forward scattering relative intensity along the nearest-neighbor <110> emission directions was reduced in the elastic curves. This reduction became more pronounced as the number of plasmon losses increased, until for n = 3, the forward scattering peaks were effectively reduced to zero relative to an average "background" intensity away from the low-index directions. The resulting experimental plasmon relative intensities in the <110> forward scattering directions [4] are presented in Table 1; they have been normalized to the corresponding elastic peak.

We now attempt to understand this selective reduction of the inelastic peaks along forward scattering directions in a more quantitative way, using the data of ref. 4 as a first test case because of the ready availability of input parameters for Al. Based upon prior studies of inelastic scattering in simple metals [6-8], the inelastic processes can be divided into two classes: 1) short-range scattering associated with electron-hole pair production and large momentum transfer, and 2) long-range scattering associated with plasmon production and small momentum transfer. The total mean-free-path λ_T for photoelectrons in this picture can be written as $\lambda_T^{-1} = \lambda_{SR}^{-1} + \lambda_B^{-1} + \lambda_S^{-1}$, where λ_{SR} is the short-range mean-free-path, and λ_B and λ_S are the mean-free-paths for <u>extrinsic</u> plasmon production of bulk and surface plasmons, respectively. For electrons propagating at more than 1-2 layers below the surface (> about 3Å below for Al [6,7]), the λ_S^{-1} term becomes negligible. The depth-dependent bulk plasmon excitation probability $P_{nB}(l)$ including both extrinsic <u>and</u> intrinsic creation is then given in this model by a Poisson distribution [6-8]: $P_{nB}(l) = \exp[-Q_B(l)] \cdot Q_B(l)^n/n!$, where l is the distance from the surface along the electron path, n is the number of plasmons excited, and $Q_B(l)$ is the position-dependent bulk plasmon excitation probability [7,8]. The total intensity observed for the nth plasmon peak is then given by: $I_{nB}=\int_0^\infty P_{nB}(l)\exp[-l/\lambda_{SR}]dl$, where the exponential in λ_{SR} allows phenomenologically for the attenuation of plasmon loss intensities due to short-range scattering.

A further important addition that we make to this model concerns the effect of multiple <u>elastic</u> scattering along high-density linear chains of atoms. It has been shown previously that such multiple scattering (MS) effects along chains of atoms can cause an effective "defocussing" of the electron flux, thus dramatically reducing the forward scattering intensity from that expected on the basis of a simple single elastic scattering model [9-11]. These effects, which can be much larger than simple inelastic attenuation, should be present in the elastic scattering of both elastic and inelastic electrons if they are observed parallel to such near-neighbor chains. Although a full-cluster MS calculation with allowance for length-dependent plasmon creation, short-range inelastic scattering, and multilayer emission would be the correct way to allow for such effects, this would be very time consuming. We can make a good approximation to the results from such a calculation by extending in a simple way the model of the last paragraph. Along near-neighbor rows of atoms, we assume that the length dependence of the combined effects of multiple elastic scattering and short-range inelastic scattering can be approximately described by the intensities $I_i(\lambda_{SR}, MS)$ resulting from simple MS calculations for emission parallel to a linear chain with an emitting atom at a length l_i inward along the electron path length (cf. inset of Fig. 2). The total intensity for the nth plasmon loss peak can be estimated from: $I_{nB}=\Sigma(i) P_{nB}(l_i) I_i(\lambda_{SR},MS)$, where the sum is over chains of

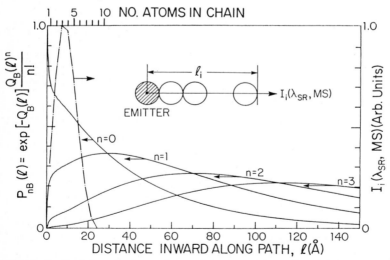

Figure 2 Path-length-dependent plasmon excitation probabilities $P_{nB}(l)$ for Al are shown together with the multiple scattering intensities $I_i(\lambda_{SR},MS)$ for Al chains containing different numbers of atoms. The chain geometry is shown in the inset.

Table 1. Experimental and theoretical values for the relative intensity of the nth plasmon loss peak.

n	Al2s[4] I_{nB}(Expt.)	Al2p I_{nB}(MS)
0	1.00	1.00
1	0.50	0.54
2	~0.24	0.16
3	~0.0	0.04

$1,2,3,\cdots$ atoms in length up to a limit at which I_i effectively goes to zero due to defocussing. The previous exponential attenuation of electrons due to short-range scattering is thus replaced by an effective intensity that includes this effect, as well as both single- and multiple- scattering effects along chains of atoms.

In Fig. 2, we summarize these calculations. The depth-dependent plasmon excitation probabilities $P_{nB}(l_i)$, calculated using values of $Q_B(l)$ from Sunjic et al. [7] for Al 2p photoelectrons, are shown as the curves labelled as n = 0(no loss), 1, 2, and 3. Overlaid on this is a plot of the multiple scattering intensities $I_i(\lambda_{SR},MS)$ for Al chains of different lengths. It can be seen that, as the chain length increases, the factor I_{nB}(MS) strongly suppresses the higher-order plasmon intensities, $I_i(\lambda_{SR},MS)$ down to nearly zero for emitters at depths of ≥ 8 atoms. Our final values for I_{nB}(MS) are given in Table 1 along with the values of Osterwalder et al. [4]. There is excellent agreement between theory and the experiment.

In conclusion, we find that for W and Ge, in agreement with the results of ref. 4, inelastic photoelectron diffraction patterns are very similar in peak position and fine structure to the elastic patterns associated with them, but that the inelastic patterns are systematically reduced in relative intensity along the forward scattering directions of low-index chains of atoms. These reductions of the inelastic features can be explained by defocussing effects due to multiple elastic scattering. In addition, these IPD results indicate that measuring the ratio I(inelastic)/I(elastic) as a function of direction should provide a convenient way to locate the low-index directions above a surface.

We gratefully acknowledge the support of the Office of Naval Research under Contract No. N00014-90-J-1457. We also thank J. Osterwalder for helpful comments and for making results available to us before publication.

References

1. C.S. Fadley, Prog. in Surf. Sci. 16, 275 (1984).
2. J.J. Barton, S.W. Robey, and D.A. Shirley, Phys. Rev. B34, 778 (1986).
3. C.S. Fadley, Physica Scripta T17, 39 (1987).
4. J. Osterwalder, T. Greber, S. Hufner, and L. Schlapbach, Phys. Rev. B 41, 12495 (1990).

5. R.C. White, C.S. Fadley, and R. Trehan, J. Electron Spectrosc. Relat. Phenom. 41, 95 (1986).
6. (a) H. Raether, Springer Tracts in Modern Physics, Vol. 88 (Springer, Berlin, 1980). (b) N.R. Avery, Surf. Sci. 111, 358 (1981).
7. M. Sunjic, D. Sokcevic, and A. Lucas, J. Electron Spectrosc. Relat. Phenom. 5, 963 (1974). Q_B values obtained from Fig. 5.
8. R.J. Baird, C.S. Fadley, S.M. Goldberg, P.J. Fiebelman, and M. Sunjic, Surf. Sci. 72, 495 (1978).
9. S.Y. Tong, H.C. Poon, and D.R. Snider, Phys. Rev. B32, 2096 (1985).
10. M.-L. Xu, and M. Van Hove, Surf. Sci. 207, 215 (1989).
11. A.P. Kaduwela, G.S. Herman, D.J. Friedman, C.S. Fadley, and J.J. Rehr, Physica Scripta, to appear.

Electron Diffraction from Al(001)

J. Osterwalder, T. Greber, S. Hüfner*, H.A. Aebischer, and L. Schlapbach

Institut de Physique, Université de Fribourg, CH-1700 Fribourg,
Switzerland
*Permanent address: Fachbereich Physik, Universität des Saarlandes,
 W-6600 Saarbrücken, Fed. Rep. of Germany

Abstract. Azimuthal diffraction patterns of 1s, 2s and valence-band photoelectrons, of plasmon-loss peaks and inelastic-background intensities and medium-energy electrons are presented for a polar angle of 45° off the surface normal of a clean Al(001) crystal. All the data have been measured with high angular resolution, and in the high-energy regime above 1keV all diffraction curves are found to be very similar, including all of the fine structure, as long as similar kinetic energies are involved. Different behaviour of plasmon-loss and inelastic-background intensities is essentially seen only in forward-scattering features and can be associated with different depths of origin of such electrons. This rather complete data set should provide a testing ground for electron scattering theories.

1. Introduction

The angular intensity modulations associated with diffraction of core-level photoelectrons and Auger electrons from single-crystal surfaces is widely used for characterizing structural properties of surfaces [1,2]. The measuring process for photoelectron diffraction (PD) and Auger-electron diffraction (AED) commonly consists in recording energy spectra containing the peak of interest, removal of some suitable background, and integration of the relevant peak intensity. This procedure is then repeated at various sample orientations to produce the characteristic angular diffraction patterns.

When considering an overall x-ray-excited photoelectron spectrum extending from valence-band emission down to the low-energy secondary tail, one may wonder if spectral features other than direct photoelectron or Auger peaks show angular dependences as well, and if so, whether one can extract additional information from such anisotropies. We have chosen an Al crystal for this study because of the relatively simple electronic structure of aluminium, and because Al shows very pronounced discrete loss features due to plasmon excitations. We have recorded many of the spectral features appearing in MgKα- and SiKα-excited photoelectron spectra in an angle-scanned fashion, such as valence-band photoelectrons, 2p, 2s and 1s core-level photoelectrons, many of the associated plasmon-loss peaks, low-energy Auger electrons and inelastic background intensities. Some of these results have recently been published [3-5] and are only briefly reviewed in this paper, together with some additional data to give a rather complete picture of the angular dependence of various spectral features. We furthermore compare PD angular patterns to medium-energy electron diffraction (MEED) patterns.

2. Experimental Results and Discussion

The experimental procedures have been described in detail elsewhere [4]. We present in Fig.1 a summary of our experimental data on quasielastically-scattered electrons from Al(001), measured at a polar angle θ of 45° off normal. Azimuthal scans are shown over one symmetry-equivalent quadrant of this fourfold-symmetric surface, and overall anisotropies $A=(I_{max}-I_{min})/I_{max}$ are also indicated. The <011> directions included in these scans at $\phi=0°$ and 90° are of particular interest, because forward scattering along high-density nearest-neighbour atomic rows produces strong intensity maxima, which are clearly exposed in all curves in Fig.1 except for the bottom one. Theoretical analyses indicate that multiple-scattering (MS) effects are particularly important along these directions [6]. We will now discuss in more detail the curves for different excitation processes.

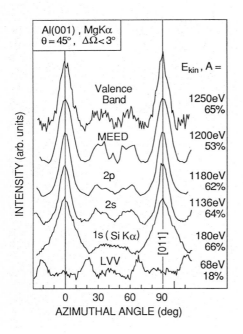

Fig.1 Comparison of azimuthal diffraction curves from Al(001) at a polar angle θ of 45° off normal, obtained with various primary excitations: MgKα–excited valence-band and 2p and 2s core-level emission, SiKα-excited 1s core-level emission, x-ray excited LVV Auger emission and medium-energy electron diffraction (MEED). The kinetic energies and the overall anisotropies in these curves are also indicated. All the data have been measured at a high angular resolution defined by a full-cone opening angle $\Delta\Omega$ of less than 3°.

2.1. Core-Level and Valence-Band Photoelectron Diffraction

The MgKα-excited 2p and 2s core levels show almost identical diffraction patterns both at $\theta=45°$ and 55° (not shown), with essentially identical anisotropies. The only difference in the patterns is the importance of <011> forward-scattering peaks relative to the off-symmetry diffraction features between $\phi=20°$ and 70°, with a local maximum at 27°. The relative anisotropy ratio as expressed by $\Delta A=(I(27°)-I_{min})/(I(0°)-I_{min})$ is 16% for 2p and 21% for 2s photoemission. The kinetic energies of 1180eV and 1136eV correspond to electron wave lengths of 0.357Å and 0.363Å, respectively, which are thus different by less than 2%. MS calculations along linear chains of Al atoms indicate that even such small wave length differences can yield modulations of up to 5% in the intensity along the chain [7]. The different angular momentum character of the photoelectron waves associated with s and p emission may also be important [8].

In Fig.1 we also show a diffraction curve for SiKα-excited 1s emission, with a kinetic energy of only 180eV. Again, the <011> maxima are clearly seen. They are here considerably broader and with a peculiar triangular shape, which is well reproduced in single-scattering-cluster (SSC) calculations [4]. The region between $\phi=20°$ and $70°$ has a distinctly different appearance in this low-energy scan, indicating the higher-order or true diffractive character of these features.

In valence-band emission from aluminium, a nearly-free electron metal, it is surprising to see how similar the associated diffraction pattern (Fig.1) is to those from the core states, both in terms of shape *and* overall anisotropy. In a recent publication [3] we point out that this similarity is evidence for hole-state localization in the final state of high-energy valence-band photoemission.

2.2 Low-Energy Auger Electron Diffraction

It is evident from Fig.1 that at a low kinetic energy of only 68eV, the LVV AED pattern is very different from all other curves. The most striking difference is the absence of the characteristic <011> forward-scattering maxima, which are here replaced by a region of rather low intensity. Polar intensity scans comprizing the [011] direction, and further azimuthal scans (not shown here), show that the forward-scattering peak has transformed into a rather broad, volcano-shaped structure. SSC calculations indicate that first-order interference effects within the top one or two layers are at the origin of such seemingly more complex diffraction patterns. At 68eV one is close to the minimum of the electron inelastic mean free path (MFP), which explains this strong surface sensitivity.

2.3 Medium-Energy Electron Diffraction

Our spectrometer configuration permits to measure medium-energy electrons scattered at a fixed angle of $90°$ while the crystal is being rotated about polar or azimuthal axes. At $\theta=45°$ one is therefore observing the specular beam. The MEED pattern shown in Fig.1 for electrons of 1200eV kinetic energy is rather similar to the PD patterns of similar energy, including the positions of all of the fine structure. Some minor variations in relative peak intensities do occur in the MEED data. The relative anisotropy ΔA of first-order features and <011> forward-scattering peaks described in Section 2.1 is 28% in this case. Also, at the high angular resolution used ($\Delta\Omega<3°$), one can detect additional fine features which are not there in the PD data, as e.g. the local maximum at $\phi=45°$ in the $\theta=45°$ scan (Fig.1). Moreover, the anisotropies are significantly different from those of PD data. Still, the overall similarity of all of these high-energy data is remarkable, and Chambers et al. [9] have already pointed out that structural information can be conveniently obtained from such MEED data, with much faster data acquisition times than with PD. A more detailed comparison of the two diffraction processes can be found elsewhere [5,9].

2.4 Plasmon-Loss Photoelectron Diffraction

The high plasmon creation rate in Al leads to very pronounced discrete loss features in electron spectra from Al. These loss peaks represent a unique opportunity to systematically study the interplay between elastic and inelastic processes leading to observed diffraction patterns. In Fig.2 we present such diffraction patterns for the 2s main line (n=0) and the first three associated plasmon-loss peaks (n=1-3). The remarkable result from this experiment is that the <011> forward-scattering maxima are suppressed with increasing plasmon number, whereas the off-symmetry features between ϕ=20° and 70° remain more or less unchanged. This observation represents one of the most striking manifestations of MS effects in electron diffraction: The more plasmon losses an electron has suffered, the longer it has, on the average, travelled through the crystal. Now, it is a well-known fact that electrons scattering along a dense linear chain of atoms are first focussed into the forward direction by the first 2-4 atoms and then defocussed out of the chain by subsequent atoms [6,10]. Electrons with three plasmon losses originate on the average deep enough along <011> chains to exhibit complete defocussing in their PD pattern. It is therefore not a different scattering process that leads to deviations from the main-line PD pattern, but the different spacial distribution of photoemitters probed.

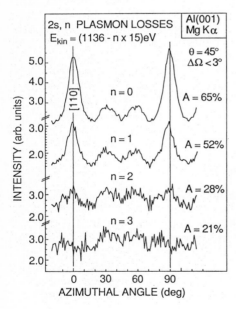

Fig.2 Azimuthal PD curves at a polar angle θ of 45° off the surface normal of Al(001) measured with MgKα radiation, for the 2s core level (n=0) and the first three associated plasmon-loss peaks (n=1-3). The plasmon excitation energy is 15eV. Overall anisotropies A (see text) are also indicated.

2.5 Inelastic Background Diffraction

A few years ago it has been stated in an experimental AED study on Cu(001) by Egelhoff [11] that all anisotropy disappears if one goes a few tens of eV away from the elastic Auger peak. This we do not find confirmed in our experiments on Al(001). We observe anisotropies of almost 10% in azimuthal scans at θ=45°, even 300eV below the 2s core-level peak [5], with patterns showing the correct symmetry of the crystal.

Very similar to the discrete plasmon-loss peaks, these remote inelastically-scattered electrons show a pronounced reduction of forward-scattering intensity along <011> producing sharp minima along these same directions (see Ref. 5).

3. Conclusions

We think that the most remarkable finding from this experimental data set is the fact that *all* spectral features in a full x-ray photoelectron spectrum from Al(001) show anisotropies of at least 5%, with patterns exhibiting the proper fourfold symmetry of this surface. At energies between 1250eV and 1136eV, with a maximum wave length difference of 3%, all *elastic* lines show rather similar patterns, independent of the primary excitation process. *Inelastic* features show significantly different patterns. These differences are primarily associated with <011> forward-scattering directions, and they become more pronounced when more loss processes are involved. Comparison with MS calculations along atomic chains permits to relate these differences with different depth distributions of the primary excitation process.

Acknowledgements. The authors have benefitted from stimulating discussions with A. Stuck. This work has been supported by the Swiss National Science Foundation.

References

[1] C. S. Fadley, in "Synchrotron Radiation Research: Advances in Surface and Interface Science", R. Z. Bachrach, Ed. (Plenum Press, New York, 1990) to appear.

[2] W. F. Egelhoff, Jr., CRC Crit. Rev. Solid State Mater. Sci. **16**, 213 (1990).

[3] J. Osterwalder, T. Greber, S. Hüfner, and L. Schlapbach, Phys. Rev. Lett. **64**, 2683 (1990).

[4] J. Osterwalder, T. Greber, S. Hüfner, and L. Schlapbach, Phys. Rev. B**41**, 12495 (1990).

[5] S. Hüfner, J. Osterwalder, T. Greber, and L. Schlapbach, to appear in Phys. Rev. B.

[6] M. L. Xu, J. J. Barton, and M. A. Van Hove, Phys. Rev. B**39**, 8275 (1989); M. L. Xu and M. A. Van Hove, Surf. Sci. **207**, 215 (1989).

[7] H. A. Aebischer, unpublished.

[8] D. J. Friedman and C. S. Fadley, Phys. Scr., to appear.

[9] S. A. Chambers, I. M. Vitomirov, S. B. Anderson, and J. H. Weaver, Phys. Rev. B**35**, 2490 (1987); S. A. Chambers, I. M. Vitomirov, and J. H. Weaver, Phys. Rev. B**36**, 3007 (1987).

[10] H. A. Aebischer, T. Greber, J. Osterwalder, A. P. Kaduwela, G. S. Herman, D. J. Friedman, and C. S. Fadley, to appear in Surf. Sci.

[11] W. F. Egelhoff, Jr., Phys. Rev. B**30**, 1052 (1984).

Photoelectron Diffraction Patterns by Display-Type Spherical Mirror Analyzer

H. Daimon*, Y. Tezuka, N. Kanada, A. Otaka, S.K. Lee, S. Ino, H. Namba, and H. Kuroda

Faculty of Science, University of Tokyo, Bunkyo-ku, Tokyo 113, Japan
*Present address: Faculty of Engineering Science,
  Osaka University, Toyonaka, Osaka 560, Japan

Abstract   A two-dimensional photoelectron diffraction pattern of bulk Si 2p core emission from Si(111)7 × 7 surface, which has been obtained by a display-type spherical mirror analyzer, was analyzed with a single scattering cluster calculation.   The pattern of Si 2p core emissions from Si(111) surface does not have the apparent three-fold symmetry of the bulk crystal but has nearly six-fold symmetry.   The peak positions in the pattern were well reproduced by the calculation and they are explained well considering the directions of strong scatterers from the emitter.   Although the final state anisotropy is strongly expected in the calculation considering the polarization of the incident synchrotron radiation, the observed data suggests that there is little anisotropy in the final state.

1. Introduction

The study of the two-dimensional angular distribution pattern of photoelectrons or Auger-electrons may be a powerful tool to analyze the atomic structure of bulk crystals or their surfaces by utilizing the phenomenon of photoelectron (or Auger electron) diffraction.   The experiments of two-dimensional angular distribution patterns are, however, very time-consuming and have scarcely been made[1-3].   The analysis so far showed the validity of single (or multiple) scattering cluster calculations[4], and in this case the wave from the emitter atom is enhanced behind the scatterer atoms.   The recent analysis of Auger pattern by Frank et al.[3], however, is opposite to this model.   They assumed the phenomenon of "silhouettes", a model in which the wave behind the scatterer decreases as a shadow.

We have developed a new two-dimensional display-type spherical mirror analyzer[5-7] using a hemispherical grid and electrode.   It has been effectively used for surface studies to observe one-dimensional diffraction[8] or ESDIAD(electron stimulated desorption ion angular distribution)[9].   In the present study, the two-dimensional photoelectron diffraction patterns were obtained and examined theoretically.

2. Experiment

The experimental details have been reported previously[10], and are described here briefly.   The experiment was done at BL-7A in Photon Factory, a 2.5 GeV positron synchrotron radiation facility.   Photons of 10 to 1000 eV from an ultra-high-vacuum

Springer Series in Surface Sciences, Vol. 24   **The Structure of Surfaces III**
Editors: S.Y. Tong · M.A. Van Hove · K. Takayanagi · X.D. Xie
© Springer-Verlag Berlin, Heidelberg 1991

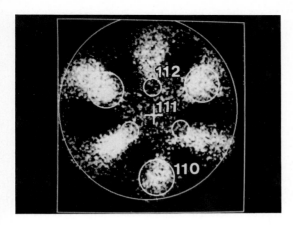

Fig. 1 Bulk photoelectron diffraction pattern of Si 2p core from clean Si(111) 7 × 7 surface at a kinetic energy of 500 eV. This picture was obtained from the original patern by subtracting the pattern which was obtained by the same process with a 30˚-rotated sample. The largest circle shows the size of the screen of the analyzer. The cross shows the surface normal direction [111]. The medium circles show the direction of [110], [101], and [011]. The small circles show those of [112], [121], and [211].

plane-grating grazing-incidence monochromator[11] were focused on the sample as a spot of 1 mm diameter. Because the monochromator is a vertical dispersion type, the electric vector of these photons is linearly polarized about 85 - 90% in a horizontal plane.

The new two-dimensional display-type spherical mirror analyzer was used for the measurement of angular distributions of photoelectrons as well as their kinetic energy distributions. This analyzer displays the angular distribution pattern of the charged particles of one kinetic energy without distortion. The acceptance cone of the analyzer is variable, and was 96.6˚ in the present work.

Figure 1 shows the bulk photoelectron diffraction pattern of Si 2p core from clean Si(111)7 × 7 surface at the kinetic energy of 270 eV. This picture was obtained from the original pattern by subtracting the pattern which was obtained by the same process with a 30˚-rotated sample. This picture is only the positive part of the difference pattern. This method of background subtraction is not standard but was necessary because the non-uniformity of the background is not the same as the photoelectron diffraction pattern. This procedure produces a six-fold symmetrical pattern when the original pattern is six-fold, but does not produce a six-fold pattern when the original one is three-fold.

Almost six-fold symmetrical peaks are clearly observed in Fig. 1. The largest circle shows the size of the screen of the analyzer, which corresponds to ± 48.3˚. The cross shows the surface normal direction [111]. The medium circles show the directions of [110]. [101], and [011]. The small circles show the directions of [112], [121], and [211]. Although the

Si(111) surface has three-fold symmetry, this figure shows that the pattern is nearly six-fold symmetric.

The pattern at the kinetic energy of 500 eV was shown in Fig. 1 of the previous paper[10]. Although the kinetic energies are very different, the two patterns are very similar.

3. Calculations

The observed nearly six-fold symmetry of the pattern from the Si(111) surface is somewhat strange. Although the directions of the peaks are near the crystallographic axes, the axes are not arranged in a hexagon and are not six-fold symmetric. Hence, we examined whether the observed pattern is reproduced by a usual cluster calculation.

The calculational method used here[12] was similar to the usual cluster calculations[4], including about 100 atoms in five Si layers. The Debye-Waller factor, absorption etc. have been included in the usual manner. The spherical nature of the emitting wave has also been included[13]. The effects of multiple scattering and the anisotropy of the final state wave have been considered. In the following calculation, only single scattering has been included, because the effect of multiple scattering was small.

Figure 2(a), (b), (c), (d), and (e) show the calculated patterns for the photoelectrons emitted from first, second, third, fourth, and fifth layer Si atoms, respectively. In these calculations, the final state wave was assumed to be an

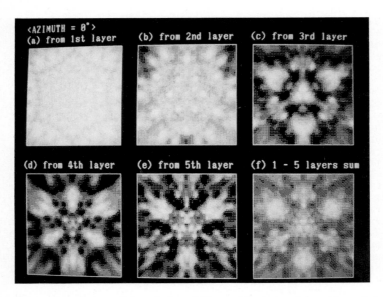

Fig. 2 (a), (b), (c), (d), and (e) show the calculated patterns from first, second, third, fourth, and fifth layer Si atoms, respectively. (f) is the sum of (a), (b), (c), (d), and (e), which is the expected pattern from the Si(111) surface. The angular range in the calculation is ± 48.3˚.

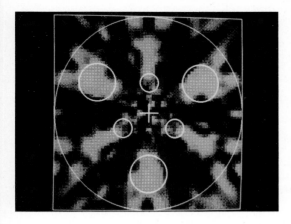

Fig. 3   A simulated pattern of the experiment,   which was
generated from Fig. 2(f) by subtracting the 30° -rotated calculated pattern.

isotropic s wave, because the inclusion of the anisotropy in
the final state wave worsens the agreement between the theory
and experiment, as shown later.   Figure 2(f) is the sum of (a),
(b), (c), (d), and (e), which is the expected pattern from the
Si(111) surface.   This simple summation postulates that the
incident light hardly decreases in this short length, 6.27 Å.
Figure 3 is a simulated pattern for the experiment, which was
generated from Fig. 2(f) by subtracting the 30° -rotated calculated   pattern.   The   angular   range   in the calculation is
± 48.3° in the horizontal and vertical axes, and the inscribed
circle corresponds to the screen of the detector.   The small
circles and cross indicate the same directions as in Fig. 1.
The simulated pattern of Fig. 3 agrees well with the observed
one, Fig. 1.
    Figure 4 shows the unit cell of the diamond structure of Si.
The observed peak positions can be explained simply by considering the directions of the strong scatterers from the emitter.
Along the [110] direction, there are two directions of the
combination of the emitter and scatterer.   One is that from O
to C, and the other is that from D to C etc., which is the
[110] direction.   The expected peak position is almost   the
[110] direction.   The contribution from combinations such as O
and C is negligible because the number of combinations is one
third of those along [110] direction and the distance between
the emitter and scatterer is longer than them.
    There are also two directions near the [112] axis.   One
corresponds to that from O to A ([112] direction), and the
other corresponds to that from D to A.   Their intensities are
expected to be not very different because the numbers of combinations are the same and the distances from the emitter to the
scatterer are not very different in the two directions.   Hence,
the expected peak position is almost the midpoint between the
two directions, which corresponds well to the observed data.
These peaks behind the strong scatterer have usually been
observed in photoelectron diffraction[4] or electron diffraction patterns[8].

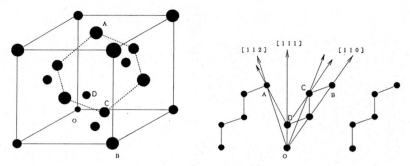

Fig. 4   The diamond structure of Si.   The bottom figure is a cross section in the plane including the [110] and [112] directions.

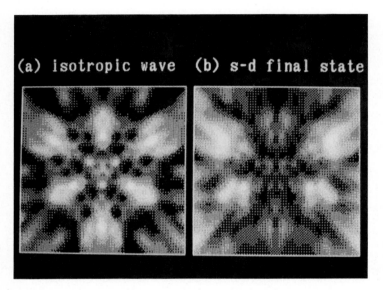

Fig. 5   The effect of anisotropy of final wave.   (a) was calculated assuming an isotropic final wave (s wave) from the fourth layer Si atoms, which is identical with Fig. 2(d).   (b) was calculated including the s-d mixing of the final wave.

Figure 5 shows the effect of anisotropy of the final wave. Fig. 5(a) is a calculated pattern assuming an isotropic final wave (s wave) from the fourth layer Si atoms, and is identical with Fig. 2(d).   Fig. 5(b) was calculated including the s-d mixing of the final wave.   The mixing ratio and the phase difference between them used here were those calculated by Goldberg et al.[14].   The two spots at the top and the bottom of the figure almost disappeared in Fig. 5(b).   This effect is explained by the fact that electrons are not likely to be ejected in the direction perpendicular to the electric vector of the incident light.   The intensities of these top and bottom spots seem weaker than other spots in Fig. 1, although they are

100

not as weak as in Fig. 5(b). This difference between the theory and the experiment is not clarified yet. The theoretical s-d mixing ratio used here might be not accurate, and these experiments may offer the experimental value of the s-d mixing ratio. In other cases, the interatomic multiple scattering may more important than that expected here.

4. Conclusions

The two-dimensional photoelectron diffraction pattern from Si(111) surface has been analyzed. The observed pattern of Si 2p core emissions from the Si(111) has nearly six-fold symmetry. The peak positions in the pattern were well reproduced by the conventional calculation and can be explained simply by considering the directions of strong scatterers from the emitter. Hence, the phenomenon of "silhouettes" could not be observed. Although a final state anisotropy is strongly expected in the calculation due to the polarization of the incident synchrotron radiation, the observed data suggest that there is little anisotropy in the final state.

References

[1] S. Kanayama, M. Owari, E. Nakamura, and Y. Nihei, Rev. Sci. Instrum. 60 (1989) 2231.
[2] R.J. Baird, C.S. Fadley, and L.F. Wagner, Phys. Rev. B 15 (1977) 666.
[3] D.G. Frank, N. Batina, T. Golden, F. Lu, and A.T. Hubbard, Science 247 (1990) 182.
[4] C.S. Fadley, in "Synchrotron Radiation Research: Advances in Surface Science" (R.Z. Bachrach, eds.) Plenum Press, New York, (1990); and refs. therein.
[5] H. Daimon, Rev. Sci. Instrum. 59 (1988) 545.
[6] H. Daimon, and S. Ino, J. Vacuum Soc. Japan 31 (1988) 954.
[7] H. Daimon, and S. Ino, Rev. Sci. Instrum. 61 (1990) 57.
[8] H. Daimon, and S. Ino, Surf. Sci. 222 (1989) 274.
[9] H. Daimon, and S. Ino, Proceedings of the 7-th Int. Conf. Solid Surfaces (1989, Cologne); Vacuum 41 (1990) 215.
[10] H. Daimon, Y. Tezuka, A. Otaka, N. Kanada, S.K. Lee, S. Ino, H. Namba, and H. Kuroda, Proceedings of the 26th. Int. YAMADA Conf. on "Surfaces as New Materials" (1990, Osaka), and will be published in Surf. Sci.
[11] H. Namba, H. Daimon, Y. Idei, N. Kosugi, H. Kuroda, M. Taniguchi, S. Suga, Y. Murata, K. Ueyama, and T. Miyahara, Rev. Sci. Instrum. 60 (1989) 1909.
[12] H. Daimon, and S. Ino, to be published.
[13] H. Daimon, H. Ito, S. Shin, and Y. Murata, J. Phys. Soc. Japan, 53 (1984) 3488.
[14] S.M. Goldberg, C.S. Fadley, and S. Kono, J. Electron Spectroscopy and Related Phenomena, 21 (1981) 285.

Energy-Angle Multidetection-Type Electron Spectrometer for X-Ray Photoelectron Diffraction Studies

S. Kanayama[1], *S. Teramoto*[2], *M. Owari*[2], and *Y. Nihei*[2]

[1]Mitsui Toatsu Chemicals, Inc., 1190 Kasama-cho,
  Sakae-ku, Yokohama 247, Japan
[2]Institute of Industrial Science, University of Tokyo,
  7-22-1 Roppongi, Minato-ku, Tokyo 106, Japan

Abstract. In order to obtain an X-ray photoelectron diffraction (XPED) pattern with high speed and high resolution, a two-dimensional multidetection-type electron spectrometer was constructed by use of a 180° deflection toroidal analyzer. This analyzer has a rectangular detection plane which consists of energy and polar angle distribution sides, and two-dimensional multidetection is easily achieved. Energy and polar angle resolutions of the spectrometer are high enough for chemical state and emission angle analysis in XPED measurements.

1. Introduction

X-ray photoelectron diffraction (XPED) has recently come to play an important part in the analyses of solid surfaces[1,2]. But in a conventional point-to-point collection-type analyzer system[3], it takes a long time to obtain an XPED pattern. In order to measure energy and emission angle distributions of charged particles rapidly, several types of simultaneous detection analyzers have been constructed. Among these analyzers, a retarding field analyzer and a 135° deflection toroidal analyzer seem more practical than the others.

The retarding field analyzer[4] enables very effective measurements by imaging all electrons emitted into a cone of large solid angle. But they have problems of poor energy resolution, low signal-to-noise ratio, and distortion of the electron image. When applying the analyzer to XPED measurements which require high energy resolution and deal with electrons of very low density, only low-resolution data can be acquired by rather complicated data handling[5].

The 135° deflection toroidal analyzer[6] enables simultaneous measurements of the polar angle distribution of electrons with high energy resolution. But with this analyzer, two-dimensional multidetection for practical use is difficult to realize, because the sector shape of the detection plane makes energy multidetection difficult. Since the toroidal analyzer requires a small source spot, two-dimensional multidetection is desirable when using X-rays as the excitation source. For such reasons, the 135° deflection analyzer is not always efficient enough when utilized for XPED measurements.

The electron spectrometer presented here is constructed by combining a two-dimensional position sensitive detector (PSD) with a 180° deflection toroidal analyzer. Because this analyzer gives a rectangular detection plane defined by energy and polar angle dispersion sides, simultaneous measurement of energy as well as angle is easily achieved. In the following, details of the spectrometer and spectra for performance test are presented.

Springer Series in Surface Sciences, Vol. 24 **The Structure of Surfaces III**
Editors: S.Y. Tong · M.A. Van Hove · K. Takayanagi · X.D. Xie
© Springer-Verlag Berlin, Heidelberg 1991

2. Instrument

Figure 1 shows a schematic view of the analyzer system, which is composed of a sectorial input lens system, a 180° deflection toroidal analyzer, a transfer lens, and a two-dimensional PSD. This system covers polar angle θ of -15° < θ < 95° for simultaneous detection. The azimuthal angle can be varied by the rotation of the sample around its normal. Electrons emitted from the sample within a narrow azimuthal range are focused and retarded in the input lens system, and go into the analyzer, where electrons are dispersed according to their kinetic energy, while preserving their polar angle. Then the electrons pass through the transfer lens and go into the PSD. On the detection plane, the energy and the polar angle distributions appear along the directions shown in Fig. 1.

The sectorial input lens system is preceded by a circular slit aperture of 1 mm width, whose center is coincident with that of the sample surface. The lens system has three elements, and each element consists of a pair of sector-shaped parallel plates. The toroidal analyzer is made of aluminum metal plated with gold. The radial radii of the inner and the outer toroids and the mean trajectory are 140 mm, 180 mm and 160 mm, respectively. No slit aperture is placed at the entrance or the exit of the analyzer. A transfer lens system, which is composed of a pair of sectorial plates and held at the retarding potential, follows the analyzer to shield the field around it. For the detection of electrons, a commercial two-dimensional PSD system, which consists of 40 mmϕ five-stage MCP and a resistive anode, is utilized. Signals from the PSD are processed in a position computer. The electrons arriving at different points of the PSD are counted independently. The pressure of the analyzer chamber is kept lower than 1×10^{-9} Torr. The magnetic field is shielded by double mu-metal shields outside the vacuum chamber. All the metallic parts in the vacuum chamber were demagnetized before assembling them.

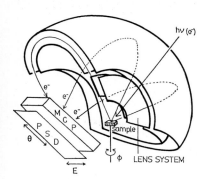

Fig. 1.
Schematic view of the 180° deflection toroidal analyzer.

3. Results and discussion

3.1 Focusing property

The reason why the angle of 135° is adopted in the earlier toroidal analyzer[6] is that the focusing point appears near this angle. In the newly constructed toroidal analyzer, the first focusing point appears in the analyzer and the second one is used for detection. Owing to the deflection angle of 180°, this analyzer has a rectangular detection plane, so that energy-angle multidetection is easily achieved. But the focusing properties of this analyzer system for different polar angles are not quite the same, and the polar angle distribution of the detection plane becomes non-linear. This is because theoretical focusing points of the 180° deflection analyzer for different polar angle lie on a curved surface, while the electrons are detected with a planar surface. Though the linearity can easily be corrected in the data processing, the difference of focusing property is important. For the purpose of correcting this aberration, the input lens system was, in its early design, designed so that different lens action is produced for different polar angle.

In order to estimate the influence of the aberration on spectra, energy spectra of elastically scattered electrons at different polar angles were compared with each other, but no significant difference in the peak FWHM was found. The reason for this is probably that the difference of the length of electron orbits for different polar angles is very small compared with the full length of the orbits, and that only electrons in a narrow azimuthal range are allowed to enter the analyzer. The result shows that the aberration is negligible and does not matter in practical measurements.

3.2 Angular resolution

The azimuthal angle resolution ($\Delta\phi$) can be estimated from the diameter of the source spot (d), slit width (s), and the distance (l) between the sample and the slit. When d = 1.0 mm, an azimuthal angle resolution of $\Delta\phi$ = ±0.7° can be acquired from s = 1.0 mm and l = 42 mm.

In order to estimate polar angle resolution, an aperture for polar angle restriction was placed between the input lens system and the analyzer, and an energy-angle image of elastically scattered electrons was measured. The aperture has openings of 1.4 mm to 1.8 mm widths cut on every 15° step. The primary electrons were accelerated to 177 eV and the source spot size is about 1 mm. Figure 2 shows the section view of the two-dimensional image along the polar angle side. The peak FWHMs are 1.4° to 1.8°, while the theoretical values, which were calculated from the dimension of the spectrometer and the aperture, are 0.9° to 1.2°.

As an additional test of polar angle resolution, LEED measurements are performed for MoS_2. The FWHM of the peaks observed in the LEED profile at one azimuthal angle was about 1.4°. This shows that the polar angle resolution is better than ±0.7°, which is good enough for XPED measurements.

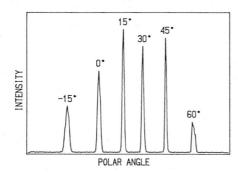

Fig. 2.
Polar angle cross section pattern measured with an angular restricting aperture. The peak FWHMs are 1.4° to 1.8°, while the calculated values are 0.9° to 1.2°.

3.3 Energy resolution

In order to evaluate the energy resolution of the 180° deflection toroidal analyzer, energy spectra of elastically scattered electrons and X-ray excited photoelectrons were measured.

In Fig. 3, curves (1) and (2) are the energy spectra of elastically scattered electrons measured with the toroidal analyzer under E_{pass} = 200 eV and E_{pass} = 100 eV, where both the spectra are the cross sections of the energy-angle two-dimensional image, while the curve (3) is obtained with a usual hemispherical analyzer under E_{pass} = 52 eV. Taking the difference in the pass energy into account, we can conclude that the energy resolution of the toroidal analyzer is as high as that of the hemispherical one and good enough for chemical state analysis.

As an additional test to evaluate the energy resolution, an X-ray photoelectron spectrum of silver was measured. Figure 4 shows the Ag 3d region of the spectrum over the polar angle between -1° and 1° measured with a Mg target X-ray tube. Because sufficient photoelectron intensity was not acquired when the X-ray was defined to 1 mmϕ by an aperture in the X-ray tube, the irradiation area was expanded to 6 mmϕ. The pass energy of the analyzer was 50 eV.

The resolution of the spectrum in Fig.4 is a little lower than that of a spectrum acquired with a usual photoelectron spectrometer, for it was sacrificed considerably by the large source spot. But this resolution is much better than that of a retarding field analyzer and chemical state analysis is possible for peaks which show large chemical shift.

The best way to measure X-ray photoelectrons with higher intensity and resolution is, of course, to make use of a focusing type X-ray source. But even if it is not available, it will be possible to get better resolution with sufficient intensity, by using an aperture of improved geometry.

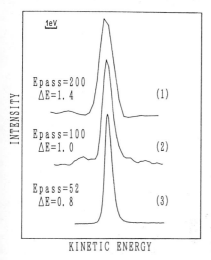

Fig. 3.
Energy spectra of elastically scattered electrons. Curves (1) and (2) were measured with the 180° deflection toroidal analyzer under pass energy of 200 eV and 100 eV, respectively. The curve (3) was measured with usual hemispherical analyzer under E_{pass} = 52 eV.

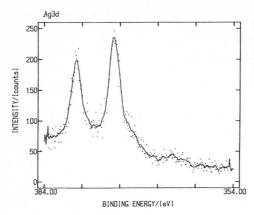

Fig. 4.
Ag 3d photoelectron spectrum of silver measured with 50eV pass energy and 6 mmφ X-ray source.

4. Conclusion

For the purpose of high speed and high resolution measurements of X-ray photoelectron diffraction patterns, an energy-angle two-dimensional multidetection-type electron spectrometer was constructed by combining a commercial PSD system with a 180° deflection toroidal analyzer. This spectrometer has merits of rapid data acquisition by two-dimensional multidetection, high energy resolution which is enough for chemical state analysis in XPS and XPED, high angular resolution enough for XPED measurements, and simple data handling. Besides high speed XPED, high sensitivity XPS by integrating the spectra over all polar angles and rapid depth profiling in XPS are realized. This spectrometer is very effective especially when a focusing-type X-ray source is used.

References

[1] Y. Nihei, M. Owari, M. Kudo, and H. Kamada, Jpn. J. Appl. Phys. 20, L420 (1981).
[2] M. Owari, M. Kudo, Y. Nihei, and H. Kamada, J. Electron Spectrosc. Relat. Phenom. 22, 131 (1981).
[3] M. Owari, M. Kudo, Y. Nihei, and H. Kamada, Jpn. J. Appl. Phys. 24, L394 (1985).
[4] S. P. Weeks, J. E. Rowe, S. B. Christman, and E. E. Chaban, Rev. Sci. Instrum. 50, 1249 (1979).
[5] S. Kanayama, M. Owari, E. Nakamura, and Y. Nihei, Rev. Sci. Instrum. 60, 2231 (1989).
[6] H. A. Engelhardt, W. Back, and D. Menzel, Rev. Sci. Instrum. 52, 835 (1981).

3D Images of Surface Structure from Photoelectron Holography

J.J. Barton and L.J. Terminello

IBM T.J. Watson Research Center, P.O. Box 218,
Yorktown Heights, NY 10598, USA

Abstract: Photoelectron holography combines the image reconstruction theory of inline holography with the electron scattering theory of photoelectron diffraction to open an avenue to fully three dimensional imaging of surface adsorption sites. Image degradation from electron multiple scattering and holographic twin images can be reduced by an analytically exact Fourier filtering process. Experimental work has produced holograms and I will discuss the progress to date on solving the practical problems which remain in the multidimensional data analysis leading toward three dimensional images.

1   Introduction

We have been working on experimental implementation of a new technique for direct visualization of surface structures from measurements of core-level photoelectron partial cross-sections. By recognizing the phenomenon of photoelectron diffraction as a holographic technique, theoretical work has shown[1] that three dimensional surface site images (3 space dimensions) can be extracted from two dimensional photoelectron diffraction patterns (2 angle dimensions). In this paper we review the fundamental physics behind photoelectron holography, outline our proposed experimental procedure, and report our progress to date on experimental realization. Our discussion purposefully avoids detailed mathematical discussion in favor of basic principles to show that the key ingredients are remarkably simple and follow from our understanding of electron scattering.

2   Fundamentals of Photoelectron Holography

Our experiment is simply angle resolved photoemission. Photons strike our sample and photoelectrons are ejected. With some apparatus we must measure the energy and the two angles describing the path of the electron, and of course detect it. The map of the photoelectron current versus the two angles for a fixed energy is the raw hologram. The more appropriate view of the measurement is that we are determining the photoemission partial cross section as a function of its wavevector $\mathbf{k}$. The two angles fix k_x and k_y, while the energy measurement fixes the vector length k and hence k_z. The first section describes the theory of photoelectron holography and in the second we outline our experimental procedure.

2.1 Photoelectron Waves

We begin with a qualitative description of the photoemission process. This description is not precise because we shall use macroscopic ideas to visualize microscopic quantum events. Specifically, we shall use a time-independent wave model. The wave mechanical model for the electron, must be interpreted as usual as the "probability amplitude" for the electron propagation. We square the amplitude and compare it to the average arrival rate of photoelectrons. Moreover we shall ignore electron indistinguishability for ease of description.

Photon Coherence Not Relevant. First we must recognize that the average atom on the surface does not participate in photoemission. That is, of the 10^{16} surface atoms, only a small fraction absorb a photon and emit an electron. The atoms which do participate are widely separated in space and their absorption events are separated in time: the individual absorption events are completely uncorrelated. Thus the coherence of the photon beam in space and in time plays no role in photoelectron holography.

Space Coherence . If we zero in on an atom which has absorbed a photon we can begin to understand what *is* important for holography. Photon absorption is very fast: we suddenly find one of the atom's electrons with high energy, well above the binding energy for electrons in the solid and thus far above the binding energy for that electron in the atom. This energy places the electron in a propagating state whose initial shape is determined by the electron's original orbital. If that orbital is a deep core level, the electron wave will emerge from near the atom center and its shape reflect the nearly spherically symmetric core potential environment: we get an emerging "spherical wave". Thus the photoelectron wave is coherent across spherical surfaces centered on the emitting atom to the extent that the original orbital is a point.

Time Coherence. From the moment of photoabsorption until the photoelecton's core hole is filled, the photoelectron wave emanates from the atom center. The wave "pulse" contains a range of energies centered about $E_k = h\nu - E_b$ where $h\nu$ is the incoming photon energy and E_b is the binding energy of the original orbital. The longer the core-hole lifetime, the narrower is the range of energies in the pulse; conversely the "natural" or lifetime broadening of the observed core photopeak width directly tells us the length of the photopeak pulse train. Wide pulses allow the wave to bounce and still interfere with itself: long core hole lifetimes mean photoelectron waves have longitudinal or time coherence.

2.2 Photoelectron Interference

Interference occurs when this photoelectron wave emerges from the core of the emitter and travels to the detector by two or more paths of different length. For the simplest case of an atom on top of a surface, the main path is direct emission away from the surface and into the detector, but secondary paths exist via elastic scattering from surface atoms. To the extent that the scattering path length differences can be extracted from the measured interference extrema, we can determine atomic positions from the photoemission measurements. In conventional photoelectron diffraction analysis this extraction is done by trial and error fits to the extrema; here we directly image the positions from the interference pattern.

2.3 Photoelectron Holography

Once we have a wave and multiple paths we have interference; once we have interference we need only these fundamental ingredients for holography:

Coherent Waves. Holography is an interference phenomenon through and through: it only works with waves which are coherent.

Identifiable Reference Wave. Holography works by interfering a strong reference wave with waves scattered from objects: we need to know what the reference wave is.

Weak Object Waves. Holographic reconstruction only succeeds if the square of the object wave amplitude can be neglected when compared to the interference between reference and object waves.

Photoelectron waves satisfy these requirements in most cases, with the unscattered, direct photoemission serving as the reference wave and scattered wavelets serving as objects. The core orbital excitation provides spatial coherence: parts of the electron wave headed directly to the detector and parts headed toward the scattering atoms are coherent. The long core hole lifetime provides time coherence: even though scattering in the solid requires time, the pulse is long enough for the front part of the pulse to scatter and return to the detector to interfere with the back part of the pulse. A reference wave can be identified: we use the atomic photoemission of the zero order perturbation theory. And the object waves are weak compared to the reference wave.

Where these ingredients fail we expect photoelectron holography to fail. Short hole lifetimes and broad initial states characteristic of valence orbital photoemission precludes holographic work with these states. Ranges of emission angles where atomic photoemission gives low direct photoemission (final state nodes) will not produce holographic information. But in other cases the ingredients for holography exist in photoelectron diffraction.

To demonstrate the power of the holographic model, electron scattering computations have been used to simulate a photoelectron hologram.[1] The accuracy and details of this theory in no way influence the validity of the holographic model for photoelectron interference, but the theory does provide an important base for analytic study and numerical simulation. Thus we briefly review its critical components.

Geometrical Optics Does Not Apply. There are physical problems in which the wave propagation can be computed by geometrical "optics", using an analogy to photon propagation by ray-tracing and computing wave amplitudes by assigning optical properties to the electrostatic potentials encountered by the wave. This gives a particle interpretation to the wave equation solutions as the geometrical rays follow the trajectories of kinematical particles. Ray tracing is only adequate in cases where the optical properties are constant over the dimension of several wavelengths. Photoelectron scattering never qualifies for geometrical optics treatment: at 150eV the electron wavelength is 0.1nm and only decreases to 0.03nm at 1500eV. Thus over the whole of the useful energy range, diffractive optical equations must be used. This is true, independent of any other approximations or in fact of the particular Hamiltonian, classical or quantum, used to model electron motions.

Perturbation Theory Works. While a wave mechanical treatment is dictated by the scales involved, the resulting model is not very complex and can comfortably compete with kinematic or geometrical optics treatments for simplicity. Key ingredients for a physical model are:

Weak Scattering. Core-level photoemission is dominated by atomic cross-section.

Atomistic Model. The solid can be adequately modeled with spherically symmetric ion cores in a sea of valence electrons.

Long Core Hole Lifetimes. Time dependence can be ignored and the photoelectron pulse can be treated as infinitely long (stationary state).

Potential Scattering. Exchange, correlation, and inelastic scattering can all be represented by a suitable electrostatic potential.

Curved Waves. Our photoemitter and all scattered waves are spherical and within a few wavelengths of their origin they encounter other ion cores whose potentials are about a wavelength in diameter: we must compute with spherical waves. This only leads to replacing classic plane wave scattering factors with curved wave factors.

Multiple Scattering. Electron scattering is strong enough to require that multiple scattering be considered at all energies of interest for structural studies.

The physical reasoning and detailed justification for these approximations are quite involved but lengthy study has shown them to be appropriate even if debate remains on the accuracy of each for detailed numerical calculations.[2]

Holographic Reconstruction.   Once the connection to holography has been established, the considerable apparatus of optical holography can be applied to provide direct three-dimensional image reconstruction of photoelectron holograms. Given that we construct the hologram with electron waves scattering from atoms, reconstruction cannot be done with electron waves and we need very large magnification to image atoms. Thus we forgo direct physical reconstruction in favor of numerical simulation with large magnification. In concept, the hologram acts like a transmission diffraction grating, forming an image by interference near the center of the sphere. Numerically, the procedure boils down to multiplying the hologram by a two-dimensional phase function and Fourier transforming, repeating the process with a different phase function for each plane perpendicular to the holographic axis. As shown in [1], this procedure does indeed invert the photoelectron scattering equations and produce bright spots of intensity around atomic centers in full three dimensions. This paper also shows the results from applying the holographic reconstruction to a simulated hologram, verifying the analytic inversion result.

Connection to Laser-Based Optical Holography   To help understand the nature of photoelectron holography, it is useful to compare it to more conventional laser based holography. Table 1   compares the individual components. Note in particular that the photon acts only as an energy source – its coherence is not relevant – and Auger excitation could also be used to create holograms. Moreover, Saldin and de Andres[7] have shown that diffuse LEED can also be interpreted with this holographic model.

For optical holography the wavelength of light is so short compared to the size of the object that extremely high resolution film and very stable recording systems are required. In photoelectron holography, the wavelength is comparable to the object so we need wide aperture, not high angular resolution, and the stability of the recording system is fixed by atomic vibrations.

Table 1: Comparing Optical and Photoelectron Holographies

Laser-based holography	Photoelectron holography
Flash lamp or discharge	Photon
Lasing medium	Photoemitting atom core
Plane Wave + Beam splitter	Spherical wave
Reference wave	Direct wave
Object waves	Scattered waves
Hologram	Two-angle diffraction pattern

Single Particle Detection of Holograms. One slightly confusing aspect of the analogy between typical optical and photoelectron holographies concerns the detection of the hologram. In laser based systems especially, the light intensity is very high and holograms are usually recorded on film. We do not have extremely intense photoelectron sources and hence we are driven to collect the photoelectron holograms one electron at a time. This difference is not fundamental and in fact optical holograms are also sums of a great many single photon events.

Each photoelectron emerges from one particular atom on the surface: the recorded hologram is the sum of a great many individual photoemission events. For the hologram to be useful for structural studies, the vast majority of these events must originate in one or a few identically arranged atoms. We call this a requirement for short-range "orientational order", but of course few systems can be imagined which have such order without also being ordered over a long range.

2.4 Filtering Out Multiple Scattering

The critical difference between optical holography such as that employing lasers and the photoelectron holography we propose for surface structure work remains: optical holography rarely confronts the problem of multiple scattering, particularly in the regime where geometrical optics fails. We must face this problem without the aid of optical analogs.

Multiple Scattering Events Are Imaged At Atoms. Before proceeding, we should put this problem in perspective. Multiple scattering so thoroughly dominates the theoretical treatment of LEED and even photoelectron diffraction, that we might be tempted to think that our simple holographic analogy will only apply to a crude single scattering model for photoelectrons and that correct treatment of multiple scattering will lead to complexities too difficult to master. Such is not the case: every wavelet component in the full wave mechanical solution is centered close to an atom and hence all orders of multiple scattering will be imaged by holographic reconstruction near a true atom position. Conventionally, the zero order atomic final state interacts with the scattering potentials (atoms) to create a first order "single scattering" wave. This wave interacts again to form a second order "double scattering" wave, and so on. However, since the atoms do not overlap, the scattering waves can be decomposed into wavelets centered about each atom. These wavelets can be treated independently: we can rearrange the final wave summed over scattering order, to express it as a sum of wavelets from each atom, where each of these wavelets has contributions from first order, many re-scattered second-order wavelets and so on. Each wavelet component is shifted slightly from the atom center due to the potential phase shift, so all components are imaged near though not at the atom center.

112

The Impact of Multiple Scattering On Resolution. This is not to say that multiple scattering has no impact on photoelectron holography. Unfortunately, the impact is more than a small, correctable inaccuracy in the atom positions. Recall that the three dimensional image is formed by interference of waves from all across the hologram. Thus the individual multiple scattering wavelets interfere, rather than simply add. Depending upon the relative phase and strength of each component the final image spot may have any range of shapes. Moreover, at longer photoelectron wavelengths (lower energies), the correspondingly lower resolution allows image spots from two different atoms to overlap. This quickly confuses the surface site image, eliminating the advantages of photoelectron holography. Specifically, while experimentally the most appropriate energy range for holographic measurements is 100-300eV, good image resolution requires 400-500eV photoelectrons.

A Solution. The very origin of the resolution problem holds the key to its solution. The relative phase of the scattering wavelets is dominated not by the potential phase shift term which shifts it off the atom center, but by the propagation delay phase differences. As we trace along a wave from the photoemitter to a nearby atom, we pass a certain number of wave crests, given by the distance divided by the wavelength. The remainder from this division gives the wave's phase at the atom, for single scattering. Multiple scattering paths travel longer distances and encounter other potentials and hence give rise to different wave phases at the atom. The key to separating these longer paths from the single scattering path lies in the wavelength dependence of the phase. As we tune the wavelength, the remainder from the above divisions also changes. Only one of these phases depends solely on the radial distance from the photoemitting origin: the single scattering wave phase. By phase-locking onto this signal we can filter out multiple scattering.

Wavefront Reconstruction Allows Filtering of Multiple Scattering. In the next section we shall describe how this phase locking can be accomplished in practice. For the remainder of this section we shall try to understand why photoelectron holography alone among electron scattering techniques affords the possibility for filtering out multiple scattering. The difference lies in the holographic interpretation of photoelectron interference. Holographic reconstruction is often referred to as "wavefront reconstruction".[6] What is meant by this phrase is that the original wavefronts from the object are reconstructed by interference when the hologram is illuminated with the reference wave. Thus when we apply the Fourier transformation to our photoelectron hologram we retrieve, in the complex image, numerical imitations of the original scattered wavelets. In the region of any atom numerous wavelets combine, each with their own phase, to give the full wavefield of a scattering atom. As the wavelength changes, the phase of each wavelet cycles. Only one wavelet cycles with the phase of the single scattering wave. By multiplying the complete reconstructed wave with a wave carrying the single scattering phase, repeating this for several energies and adding, we cause the multiple scattering components to cancel at different energies, leaving only the single scattered wave.

Filtering is a Fully 3D Fourier Transform. Alternatively we can say that we are simply performing the correct multi-dimensional Fourier transform to invert the equations of photoelectron diffraction. The dominant phase of a single scattering wavelet starting from the emitter at atom a and scattering from atom b is

$$e^{-i\mathbf{K}\cdot\mathbf{r}_{ab}}e^{ikr_{ab}} \tag{1}$$

where $\mathbf{K}$ is the wavevector from emitter to hologram point with length k, $\mathbf{r}_{ab}$ runs from emitter to scatterer, and r_{ab} is the distance from emitter to scatterer. For a double scattering wavelet which passes through atom c on its way to b the phase is

$$e^{-i\mathbf{K}\cdot\mathbf{r}_{ab}}e^{ikr_{ac}}e^{ikr_{cb}} \quad . \tag{2}$$

Obviously if we multiply by

$$e^{i\mathbf{K}\cdot\mathbf{r}_{ab}}e^{-ikr_{ab}} \tag{3}$$

and integrate over $\mathbf{K}$ then the single scattering term will tend to integrate to a constant while the double scattering term will oscillate in the integrand and tend to be eliminated.

3 Experimental Procedure and Progress

Experimentally, photoelectron holography is straight-forward, but unfortunately it requires unusual technology not often already available from other experiments. In this section we describe our progress in assembling that technology for practical and routine measurements.

A Photoelectron Hologram Apparatus. Based upon our theoretical development, we expect that we shall need to measure core-level photoemission angular distributions, over a wide range of angles, for a series of electron energies. This can be accomplished most readily with a display-type electron energy analyzer[5][4] located on a synchrontron beam line with an energy scanning photon monochromator. After reconstruction to correct serious transmission inhomogenities, the display-type ellipsoidal analyzer designed by Eastman[5] on the IBM U8 beamline at Brookhaven National Laboratory has been fitted with a high-brightness phosphor screen and a SIT video camera attached to a digital integrating frame buffer card in an IBM PC AT computer. This system accepts a 80° full angle cone, scans photon energies from 25 to 800eV, and accumulates the hologram digitally.

Extracting A Hologram Function $\chi(\mathbf{K})$ from Photoemission. The ellipsoidal analyzer resolves energy by passing the electrons between a low pass elliptical and high pass spherical grid. The difference in the geometrical fig-

114

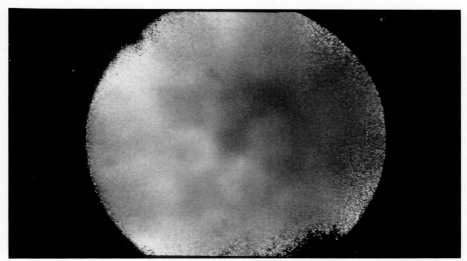

Figure 1: Experimental Cl 2p photoelectron hologram from c(2x2)Cl/Cu(100).

ure of these surfaces leads to a transmission difference across the frame; dust particles on the spherical grid scatter electrons leading to circular spots. Both effects can be eliminated by normalizing the measured photoemisson intensity with a signal which is expected to be uniform at the source but which suffers the same analyzer transmission. We model the analyzing process as follows. The measured intensity I' is a view through an analyzer with transmission T at a source containing the raw hologram intensity I on top of an inelastic electron background B and with dark count rate D:

$$I'(\mathbf{K}) = T(\mathbf{K}) * (I(\mathbf{K}) + B(\mathbf{K})) + D \qquad (4)$$

We estimate the transmission function by measuring the background under the photopeak after we move the photon energy to push the photoelectron energy out of the analyzer acceptance. The intensity $I = I_0 + \chi' I_0$ can be expressed in terms of a zero baseline hologram function χ' which oscillates about the atomic partial cross-section I_0. A raw hologram measured using 104eVCl 2p photoelectrons from c(2x2)Cl/Cu(001) normalized for analyzer transmission is shown in Fig. 1. Note that this image contains about 150,000 data points.

Angle Distortions Must Be Removed. The normalized photoemission cross-section cannot be Fourier transformed with digital algorithms unless the $\mathbf{k}$ coordinates of the pixel corners are linearly spaced. The emission angles from the ellipsoidal analyzer are projected onto a slanted microchannel plate resulting in a slightly distorted projection of $\mathbf{k}$ onto a plane. This distortion can be removed by so-called warping algorithms[3]. The central difficulty in this work is measurement of the experimental angle distortion to verify the

115

theoretical predictions in a ideal analyzer. We are presently developing physical masks for this purpose.

Filtering Out Multiple Scattering. Schematically, the procedure for completing the photoelectron holography experiment is shown in Fig. 2. We must repeat the photoelectron measurement at perhaps 10 electron energies spread over 25-30eV. Each hologram is Fourier transformed and the resulting complex image is phased by the single scattering wave. Then the images are summed and squared to yield the final image. Note that the raw holograms at each energy need only be measured 1/10 as long a we might for a single acceptable hologram, since the final Fourier sum adds the signal power from all 10 measurements.

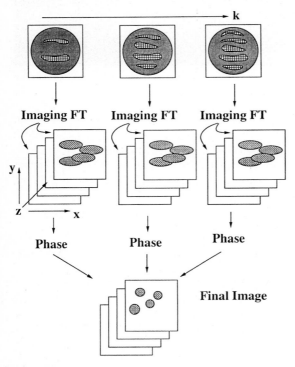

Figure 2: Procedure for filtering out multiple electron scattering. Holograms at several energies are reconstructed into a series of three dimensional complex spaces. Each space is multiplied by a complex phase function and added, then squared to yield the final image. The procedure also removes the twin image.

Experimental Prognosis. Although the experimental measurements of photoelectron holograms are complicated, the major problems only plague the initial setup. Display analyzers, synchrotron radiation beamlines, three-dimensional data analysis algorithms, and angle calibrations require time to develop but once conquered they do not affect routine measurements. Each individual hologram is measured with no "moving parts": neither the electron nor photon energy need be scanned and with a display analyzer the angles are fixed by the detector physical geometry once and for all. Thus many time-dependent fluctuations like photon beam intensity have little influence on the measurements. As only the electron's final kinetic energy is needed for the analysis, the photon energy need not be carefully calibrated and the much simpler electron energy analyzer voltage at the top of the photoelectron peak can be converted to electron kinetic energy. Since the photoemitting atom acts like the beamsplitter of optical holography, macroscopic vibrations are not relevant: to reduce vibrational smearing we need to cool our sample. Moreover high precision sample positioning is not important: a rotation of the sample simply rotates the image in the reconstruction. Experimental photoelectron holography looks promising and we are pursuing a full three dimensional imaging capability for surface structures.

References

[1] J. J. Barton. *Phys. Rev. Lett.*, 61:1356, 1988.

[2] J. J. Barton, S. W. Robey, and D. A. Shirley. *Phys. Rev. B*, 34:3807, 1986.

[3] E. Catmull and A. R. Smith. *SIGGRAPH 80 Proceedings*, 14:279–285, 1980.

[4] H. Daimon. *Rev. Sci. Instrum.*, 59:545, 1988.

[5] D. E. Eastman, J. J. Donelon, N. C. Hien, and F. J. Himpsel. *Nucl. Instrum. and Meth.*, 172:327, 1980.

[6] D. Gabor. *Nature*, 161:777, 1948.

[7] D. K. Saldin and P.L. de Andres. *Phys. Rev. Lett.*, 64:1270, 1990.

New Developments in the Theory of LEED

P.J. Rous

The Blackett Laboratory, Imperial College of Science, Technology and Medicine, Prince Consort Rd, London SW7 2BZ, UK

Abstract. The last decade has seen a number of significant developments in the theory of low-energy electron diffraction which have opened up exciting opportunities for surface crystallography in the 1990s. By reviewing the current status of surface crystallography I identify the theoretical limitations which have confined LEED analysis to relatively simple adsorbates on low Miller index substrates. Two recently developed solutions to these difficulties are presented; directed search strategies with Tensor LEED and a new real-space method for the calculation of LEED IV spectra from high Miller index surfaces.

1 The Current Status of Surface Crystallography

Over the past decade, surface crystallography has continued to play a predominant rôle in the study of the chemistry and physics of surfaces. The beginnings of quantitative surface crystallography can be traced to the late 1960s when the first reports of LEED structure determinations appeared in the literature. Since that time the number of scientific papers published each year reporting surface structure determinations has risen steadily to close to 200 in 1989; a grand total of almost 2000 papers [1].

Figure 1 shows the growth in the number of reported structure determinations, classified by technique. The 1980s saw a significant increase in the number of determinations by SEXAFS, LEIS, MEIS, HEIS, X-ray diffraction, angle resolved photoemission and ab-initio theoretical determinations by total energy minimisation. These techniques now make up over 50% of all reported structural studies; the remainder being by Low-energy electron diffraction. Despite the preponderance of alternative methods, the number of determinations by LEED has remained stable over the last 5 years and the continued popularity of LEED means that it is appropriate to consider the future prospects for LEED. In this paper I shall examine some of the remaining obstacles to the future progress by LEED surface structure determination which have their source in the present theory of LEED [3]. Perhaps suprisingly, the deficiencies of LEED theory are most readily appreciated by looking to the past and examining those surface structures which have been solved over the last 20 years.

Springer Series in Surface Sciences, Vol. 24   **The Structure of Surfaces III**
Editors: S.Y. Tong · M.A. Van Hove · K. Takayanagi · X.D. Xie
© Springer-Verlag Berlin, Heidelberg 1991

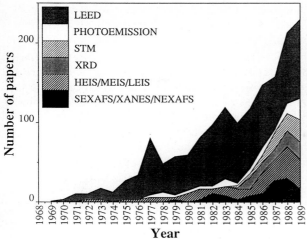

Figure 1. A yearly analysis of the number of published papers reporting a surface structure determination, classified by technique. Of these reports approximately 500 can be regarded as complete in the sense that they report more than the odd bond length [2].

1.1 LEED and Complex Surfaces

Whilst an a-priori LEED structure determination of a simple surface such as Cu(100) can be performed on a microcomputer [4], an organic molecule adsorbed on a catalyst substrate or a complex reconstructed semiconductor surface requires many hours of supercomputer time. Despite this fact, a number of molecular adsorption geometries have been determined over the past few years, and our knowledge of such systems now extends to approximately 19 structures compared to only 2 at the beginning of the last decade. If such determinations are to become routine we need to overcome two theoretical limitations presently suffered by LEED, both of which are consequences of the complexity arising from the presence of many independent atoms in the surface unit cell [5].

The first limitation is the time taken to perform a single calculation of LEED intensity spectra for a given surface structure which scales as the cube of the number of atoms in the surface unit cell, N^3, or at best N^2. Most of the development of LEED theory has concentrated upon ways of reducing the computational resources needed for a single LEED calculation through the use of reliable approximations to full multiple scattering such as Renormalised Forward Scattering, Reverse Scattering Perturbation, Quasidynamical LEED and the Beam Set Neglect method [5]. More recently Huang et al, [6] have delivered a *tour-de-force* analysis of the Si(111)-(7x7) using a new and highly symmetrised LEED code.

The second, and in many ways more serious, limitation arises from the conventional trial-and-error method of determining a surface structure by LEED since the time taken to perform this type of structure search scales exponentially with the number of structural parameters being sought. For this reason, the prospects for reliably determining the best fit structure when we are varying as many as 10-20 parameters are clearly limited; most likely significant regions of such a large volume of parameter space will remain unexplored.

These limitations seem especially acute given the current interest in adsorbate induced restructuring [7] and asymmetric adsorption sites [8]. Even simple adsorption systems seem, upon further examination, to be more complex than we might have imagined at first, perhaps displaying asymmetric adsorption sites or substrate buckling [9]. Indeed, it is probably fair to say that the majority of structural analyses performed during the last two decades are incomplete in the sense that what now seems to be important structural refinements could not be considered. It is for this reason that the future of LEED lies in the development of novel theoretical techniques which can reliably retrieve many structural parameters. At the present time it would appear that the best way of achieving this goal is the use of either directed search methods or attempting the inversion of LEED IV spectra to obtain the atomic positions directly. Both of these approaches will be discussed in section 2.

1.2 Stepped Surfaces

Table 1 shows the number of LEED structure determinations of monatomic fcc, bcc and hcp surfaces classified according to the Miller indices . Whilst the structures of low Miller index surfaces have been extensively explored; relatively little is known about the structure of higher Miller index or stepped surfaces.

In the case of clean stepped surfaces, those LEED experiments which have been able to determine atomic positions with any precision are small enough in number to be listed as: for Fe the (311), (210), (211) and (310) surfaces [10], for Al (311), (331) and (210) [11], Cu(311) [12], Ni(311) [13]. This sparsity of structural information might seem surprising given the fundamental importance of stepped surfaces to a variety of surface chemical and physical processes. Step sites are known to provide new active sites for bond breaking and formation [14]. The step induced suppression or enhancement of surface diffusion has an important influence on the reaction pathway followed by intermediates [14] and governs the microscopic mechanism of nucleation and growth [15].

Low-energy electron diffraction *patterns* have provided information on the qualitative nature of steps such as the step height and terrace width distribu-

Table 1. Number of LEED structure determinations of clean and adsorbate covered monatomic fcc, bcc and hcp surfaces. Repeat analyses are not included.

	Clean Surfaces									
	111	100	110	311	331	210	211	310	0001	1010
fcc	9	9	9	3	1	1	0	0	-	-
bcc	6	6	3	1	0	1	1	1	-	-
hcp	-	-	-	-	-	-	-	-	7	3
	With Adsorbates									
fcc	32	41	16	1	0	0	1	0	-	-
bcc	3	14	5	1	1	0	0	0	-	-
hcp	-	-	-	-	-	-	-	-	8	0

tion [16]. Field-ion microscopy (FIM) [17] and scanning tunneling microscopy (STM) [18] have now delivered graphical confirmation of these results. Indeed it is STM which has shown the prevalence of wide-terrace steps on many surfaces that appear step-free to LEED due to the limited instrumental response of most LEED diffractometers.

Various kinds of theory have been applied to clean stepped surfaces, but few have attempted to optimize atomic positions. One semi-empirical study that did obtain atomic relaxations has shown the presence of complex multilayer relaxations in which atomic displacements occur both perpendicular and parallel to the step terraces [19]. Such displacements have been verified experimentally in a limited number of less stepped surfaces [10].

Structural analysis of adsorbates on stepped surfaces has been performed in only a very few cases. Examples from alternative techniques include an angle-resolved photoelectron diffraction study of atomic oxygen on Cu(410) in which the in-step and terrace locations of the oxygen atoms were determined [20]. An ESDIAD study of CO on stepped Pd surfaces [21] has shown that the CO molecules tilt away from the terrace plane toward the macroscopic surface normal although the adsorption site could not be determined. In certain cases, high-resolution electron energy loss spectroscopy (HREELS) can elucidate the mode of bonding of an adsorbate to a stepped surface [22] whilst a few theoretical studies of adsorption at steps have been performed using both empirical or semiempirical methods [23].

The apparent sparsity of structure determinations beyond the low Miller index surfaces can be traced directly to the lack of a LEED theory capable of treating such surfaces. The reason why this is so and the possible ways of overcoming this obstacle will be discussed in section 3.

2   Directed Searches and Complex Surfaces

The use of a directed search through the parameter space of structural variables has been suggested since the beginning of quantitative LEED structure determinations [24]. Recently there has been a resurgence of interest in applying such searches to conventional full dynamical LEED calculations [25]. Our approach has been to combine Tensor LEED with a numerical optimisation algorithm to perform a highly efficient and automatic search for the best fit surface structure [26].

Tensor LEED is a perturbative approach for the calculation of LEED intensities [27]. The fundamental concept of Tensor LEED is that of the reference structure: a particular surface structure which we guess to be as close as possible to the actual surface structure. This surface is distorted by moving some of the atoms to new positions corresponding to a given trial structure. Provided these displacements are small then the changes in the IV spectra induced by this distortion can be calculated by a highly efficient perturbation theory.

To first order, the difference between the amplitudes of a LEED beam scattered from the reference and trial surface, δA, can be written as an expression which is linear in the atomic displacements which generate the trial structure. Thus if we move N atoms through δr_{ij} (i=1..N, j=1,2,3):

$$\delta A = \sum_{i=1}^{N} \sum_{j=1}^{3} T_{ij} \delta r_{ij} \qquad (1)$$

The quantity T is the tensor which depends only upon the scattering properties of the reference surface and is calculated by performing what is essentially a full dynamical calculation for this surface. Once T is known then the diffracted intensities for many trial surfaces can be evaluated extremely efficiently by summing eqn(1) for the appropriate set of atomic displacements.

This linear version of tensor LEED is limited to atomic displacements of less than 0.1Å, beyond which eqn(1) becomes a poor approximation. In this case we can appeal to a more sophisticated version of our theory, one which allows displacements of up to 0.4Å [27]. In either case, the computational effort per trial structure can be reduced by a factor of 50 for a simple surface such as Cu(100) to 10,000 for a p(2x2) overlayer system [27] compared to an equivalent full dynamical calculation. The other important aspect of Tensor LEED is that the time taken to evaluate a set of IV spectra is independent of the presence or lack of symmetry within any given trial surface. Therefore we can consider highly asymmetric systems, such as off-center adsorption sites, with no loss of efficiency. This particular feature of Tensor LEED is especially desirable if we are to use an automated structure search since we cannot predict in advance that the path to be taken through parameter space by the optimisation procedure will pass through only symmetrical trial surfaces.

Our search strategy is implemented as two separate computer programs. The first program performs the reference structure calculation and stores the tensors to a disk file. The second program utilises those tensors to calculate LEED IV spectra for a sequence of trial structures using the tensor LEED approximation. As the IV spectra of each trial structure are calculated they are immediately compared to the experimental spectra by an in-situ R-factor calculation. A steepest descent optimisation algorithm chooses the next trial structure from the result of the R-factor comparison and this procedure is repeated until an R-factor minimum is found.

Of course this method does not guarantee that the trial structure at the termination of the search is the actual surface structure since this structure may correspond to a local rather than the global R-factor minimum. Considerable care must be taken to circumvent this possibility by repeating the entire optimisation procedure starting with different widely-spaced reference structures which span the parameter space of physically reasonable surface structures. However, since only one full dynamical calculation is needed for each reference structure the overall computational savings are retained.

We have applied this search strategy to a number of surface structure determinations. For Pd(111) and Pd(111)-($\sqrt{3} \times \sqrt{3}$)$R30°$-CO we obtain the same optimised atomic positions as previous full-dynamical, trial-and-error analyses [26]. More recently we have determined the adsorption geometry and substrate restructuring of Mo(100)-c(2x2)-S and C by varying no fewer than 15 structural parameters simultaneously [26].

In addition to determining atomic positions we can also generalise the optimisation procedure to search for non-structural parameters such as the inner potential and surface Debye temperature. This is especially important in the case of the inner potential, variations of which are strongly correlated with the values of all structural parameters.

2.1   Direct Methods

Another alternative method of structural solution lies in avoiding the structure search altogether and developing a direct method of inverting LEED spectra to yield the surface structure.

The relative simplicity of the analysis of X-ray diffraction data has led many researchers to seek a direct method of obtaining atomic positions from a set of IV spectra. Early attempts, employing conventional X-ray methods such a Patterson functions, were found unreliable and were abandoned in the early 1970s in favour of trial and error comparison to full dynamical calculations [28]. However, recently a new direct method has been proposed by Pendry′ et al and successfully applied to a few simple systems [29].

The fundamental idea is not to attempt an ab-initio inversion of a set of LEED IV spectra but to try to determine the *difference* between the actual surface structure and a "best guess" reference surface for which the IV spectra can be calculated. The direct method allows us to take the difference between the measured IV spectra and those evaluated for the reference structure and use it to directly determine the way in which the actual surface structure deviates from that of the reference surface. In essence the direct method is used to lead the structural solution from the reference surface to the actual surface structure. This is the analogue of the heavy atom method in X-ray diffraction in which the "heavy atom" is the reference surface.

This is possible because the Tensor LEED theory tells us how the difference between the calculated IV spectra of the reference surface and those measured experimentally is related to the difference in the position of the atoms in the reference and actual surface. Symbolically:

$$\Delta \mathbf{I} = \mathbf{M} \Delta \mathbf{r} \qquad (2)$$

where $\Delta \mathbf{I}$ is the column vector constructed from intensity differences for each LEED beam and incident electron energy and $\Delta \mathbf{r}$ is the column vector made up of products of *moments*, $< \delta r^n >$, for each displaced atom. Provided that the differences in the atomic coordinates are less that 0.4Å we have recently shown that the coefficient matrix $\mathbf{M}$ can be calculated exactly by Tensor LEED theory [30].

Initial applications of this method to Ni(100)-c(2x2)-O and Rh(110)-H [29] have successfully inverted eqn(2) to retrieve the *static* positions (i.e. the first moment) of adsorbate and substrate atoms, when the reference surface has atomic positions not more than about 0.1Å from the actual atomic coordinates.

The ability to determine the higher moments of the atomic positions would open up the possibility of determining the spatial probability distribution of each surface atom.

3   A LEED Theory for Stepped Surfaces

A characteristic feature of all high Miller index surfaces is the large area of the surface unit cell and the consequent reduction of the spacing between atomic planes. These features present three problems to conventional LEED theory, an unfortunate consequence of the use of a plane wave basis to describe the scattering of the LEED electron within the surface. Firstly, the number of beams or plane waves required for a convergent LEED calculation scales as the area of the surface unit cell. Secondly, as the interplanar spacing shrinks the number of evanescent beams which must be taken into account increases exponentially [5]. This leads to a disproportionate increase in the computational effort expended in such a calculation which scales as the *cube* of the number of beams. High Miller index surfaces also have low symmetry; the best one can hope for is one mirror plane. Therefore the efficient symmetrised LEED codes [5] which work so well for high symmetry, low Miller index systems, are largely redundant when applied to stepped surfaces. The final problem is that even the most robust LEED algorithm for constructing a surface, layer doubling, often diverges for planes spaced less than 1Å apart, a consequence of the ability of such closely spaced layers to support strong interplanar shape resonances.

Thus new theories are needed. Several approaches have already been proposed. One is based on bundles of chains of atoms parallel to the steps [31]. Two alternative methods employ the spherical wave basis to combine together a few atomic planes and then proceed to treat the surface as a stack of these "composite" layers separated by an effectively larger interlayer spacing [32]. These techniques improve convergence with respect to the actual interplanar spacing at the expense of the computational effort required to build the composite layers from their constituent atomic planes.

Recently we have proposed a new theory [33] which dispenses with plane waves and instead uses an entirely angular momentum basis. Then, in principle, the computational effort required becomes independent of both the area of the unit cell and the interplanar spacing. To do this we have developed a novel algorithm for building a surface from its constituent atomic planes, much in the spirit of the layer doubling algorithm [5] but in a spherical rather than plane wave basis. This algorithm is based upon the concept of removal invariance which exploits the fact that the bulk termination of any crystal exhibits a universal property, a consequence of the semi-infinite periodicity perpendicular to the surface. In simplest terms it implies that the reflectivity of the crystal is unchanged (within a trivial phase factor) if one atomic plane is peeled away from the surface.

We can use the removal invariance principle to construct a self-consistent equation for the reflectivity of the surface. Consider a surface made up of a

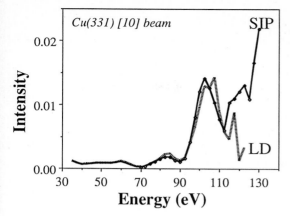

Figure 2. Comparision of the IV spectrum of the (10) beam of Cu(331) calculated using conventional layer-doubling (LD) and our real space algorithm based upon semi-infinite periodicity (SIP). The layer doubling calculation fails to converge above about 100eV.

stack of atomic planes each of which has a reflection matrix $\mathbf{r}$ and a transmission matrix $\mathbf{t}$. The full reflectivity of the surface is not changed if we remove the topmost atomic plane, so that

$$\mathbf{R} = \mathbf{r} + \mathbf{t}\mathbf{R}\,(1 - \mathbf{r}\mathbf{R})^{-1}\,\mathbf{t} \tag{3}$$

where, for clarity, we have dropped the phase factors arising from free propagation between the planes. Therefore the reflectivity $\mathbf{R}$ is a solution of the following quadratic matrix equation,

$$-\mathbf{R}\mathbf{t}^{-1}\mathbf{r}\mathbf{R} + \left(\mathbf{r}\mathbf{t}^{-1}\mathbf{r}\mathbf{R} + \mathbf{R}\mathbf{t}^{-1} - \mathbf{t}\mathbf{R}\right) - \mathbf{r}\mathbf{t}^{-1} = 0 \tag{4}$$

This non-linear matrix equation can be solved iteratively by a modified Newton-Raphson method, starting with an initial guess for the bulk reflectivity $\mathbf{R}_o$ evaluated for a thin slab of 2 or 3 atomic planes.

It is important to note that in eqns(3) and (4) we have made no assumptions about the particular basis set within which the reflectivity matrix is expressed. Thus by choosing a spherical wave basis we can avoid the conventional scaling and convergence problems associated with the usual plane wave basis set. Finally we note that, although we began by considering the case of equidistant planes corresponding to the bulk termination of the crystal, any structural distortion of the selvedge is amenable to treatment by a fast perturbative approach such as Tensor LEED [26].

In figure 2 we show the results of our new real-space approach compared with a conventional layer-doubling calculation for the (10) beam of Cu(331). Note that the layer doubling calculation diverges noticeably at 105 and 115eV; this is a consequence of the small interplanar spacing of 0.83Å for this surface. Figure 2 clearly demonstrates the utility of our new approach [33].

Acknowledgements

I would like to acknowledge the major contribution to the research reported in this paper made by my coworkers: M.A. Van Hove, J.B. Pendry, G.A. Somorjai, X.-G Zhang, and A. Gonis.

References

1. I am indebted to Dr. M.A. Van Hove for supplying me with his structural database from which these statistics were compiled.

2. J.M. MacLaren, J.B. Pendry, P.J. Rous, D.K. Saldin, G.A. Somorjai, M.A. Van Hove and D.D. Vvedensky, *Surface Crytallographic Information Service - A Handbook of Surface Structures*, (Reidel, Dordrecht, 1987); F. Jona and P.M. Marcus, in "The Structure of Surfaces II", eds. J.F. van der Veen and M.A. Van Hove, (Springer-Verlag, Heidelberg, 1988), p. 90.

3. For an experimental review: K. Heinz, Progr. Surface Sci. **27** 239 (1988)

4. N Bickel and K. Heinz, Surface Sci. **163** 435 (1985).

5. J.B. Pendry, *Low Energy Electron Diffraction*, (Academic Press, London, 1974); M.A. Van Hove, W.H. Weinberg and C.-M. Chan, *Low-Energy Electron Diffraction: Experiment, Theory and Surface Structure Determination*, (Springer-Verlag, Heidelberg), 1986.

6. H. Huang, S.Y. Tong, W.E. Packard and M.B. Webb, Phys. Lett. **A130** 166 (1988).

7. M.A. Van Hove and G.A. Somorjai, Progr. Surface Sci. **30** 201 (1989).

8. Proceedings of the ESF workshop: *Reconstructive and Asymmetric Adsorption on fcc(100) Metal Surfaces*; April 1989, Erlangen, F.R.G.

9. W. Oed, H. Lindner, U. Starke, K. Heinz, K. Müller and J.B. Pendry, Surface Sci. **224** 179 (1989).

10. J. Sokolov, F. Jona and P.M. Marcus, Phys. Rev. **B29** 5402 (1984); *ibid* **B31** 1929 (1985); *ibid* **B33** 1397 (1986).

11. J.R. Noonan, H.L. Davis and W. Erley, Surface Sci. **152/153** 142 (1985); D.L. Adams and C.S. Sørenson, Surface Sci. **166** 495 (1986); D.L. Adams, V. Jensen, X.F. Sun and J.H. Vollensen, Phys. Rev. **B38** 7913 (1988).

12. R.W. Streater, W.T. Moore, P.R. Watson, D.C. Frost and K.A.R. Mitchell, Surface Sci. **72** 744 (1978).

13. D.L. Adams, W.T. Moore and K.A.R. Mitchell, Surface Sci. **149** 407 (1985).

14. G.A. Somorjai, *Chemistry in Two Dimensions: Surfaces*, (Cornell, Ithaca and London, 1981).

15. E.G. Bauer, B.W. Dodson, D.J. Ehrlich, L.C. Feldman, C.P. Flynn, M.W. Geis, J.P. Harbison, R.J. Matyi, P.S. Peercy, P.M. Petroff, J.M. Phillips, G.B. Stringfellow, A. Zangwill, J. Mater. Res. **5** 852 (1990).

16. B. Lang, R.W. Joyner and G.A. Somorjai, Surface Sci. **30** 454 (1972). M.A. Van Hove and G.A. Somorjai, Surface Sci. **92** 489 (1980). G.-C. Wang

and M.G. Lagally, Surface Sci. **81** 69 (1979). M. Henzler, Surface Sci. **22** 12 (1970).

17. E.W. Müller, Z. Phys. **136** 131 (1951); E.W. Müller and T.T. Tsong, *Field Ion Microscopy* (American Elsevier, New York) 1969; G. Ehrlich, Surface Sci. **63** 422 (1977).

18. D.F. Ogletree, C. Ocal, B. Marchon, G.A. Somorjai and M. Salmeron, J. Vac, Sci. Technol., (in print).

19. G. Allan, Surface Sci. **85** 37 (1979); P.M. Marcus, P. Jiang and F. Jona, in "The Structure of Surfaces II", eds. J.F. van der Veen and M.A. Van Hove, (Springer-Verlag, Heidelberg, 1988), p. 100.

20. K.A. Thompson and C.S. Fadley, Surface Sci. **146** 281 (1984).

21. T.E. Madey, J.T. Yates Jr., A.M. Bradshaw and F.M. Hoffmann, Surface Sci. **89** 370 (1979)

22. S. Lehwald and H. Ibach, Surface Sci. **89** 425 (1979).

23. H. Kobayashi, S. Yoshida, H. Kato, K. Fukui and K. Tarama, Surface Sci. **79** 189 (1979); H. Kobayashi, S. Yoshida and M. Yamaguchi, Surface Sci. **107** 321 (1981); J.P. Jardin, M.C. Desjonqueres and D. Spanjaard, J. Phys. **C18** 5759 (1985).

24. D.L. Adams , *Proceedings of the First International Seminar on Surface Structure Determination By LEED*, Erlangen 1985.

25. P.G. Cowell, M. Prutton, S.P. Tear, Surface Sci **177** L915ʹ (1986); P.J. Rous and J.B. Pendry, in *'The Structure of Surfaces II'*, eds. J.F. van der Veen and M.A. Van Hove, (Springer-Verlag, Heidelberg, 1988); G. Kleine, W. Moritz, D.L. Adams and G. Ertl, Surface Sci. **219** L637 (1988). See also W. Moritz in these proceedings.

26. P.J. Rous, M.A. Van Hove and G.A. Somorjai, Surface Sci. **226** 15 (1990); P.J. Rous, D. Jentz, D.G. Kelly, R.Q. Hwang, M.A. Van Hove and G.A. Somorjai, these proceedings.

27. P.J. Rous, J.B. Pendry, D.K. Saldin, K. Heinz, K. Müller and N. Bickel, Phys. Rev. Lett. **57** 2951 (1986) ; P.J. Rous and J.B. Pendry, Comput. Phys. Commun. **54** (1989) 137; *ibid* **54** 157 (1989); P.J. Rous and J.B. Pendry, Surface Sci **219** 355 (1989); *ibid* **219** 373 (1989).;

28. D.P. Woodruff in *The Chemical Physics of Solid Surfaces and Heterogeneous Catalysis* eds. D.A. King and D.P. Woodruff (Elsevier, Amsterdam, Oxford, New York, 1981).

29. J.B. Pendry, K. Heinz and W. Oed, Phys. Rev. Lett. **61** 2953 (1988). K. Heinz, W. Oed and J.B. Pendry, in these proceedings.

30. P.J. Rous, W. Oed, K. Heinz and K. Müller, to be published.

31. P.J. Rous and J.B. Pendry, Surface Sci. **173** 1 (1986).

32. D.W. Jepsen, Phys. Rev. **B22** 5701 (1980); P. Pinkava, S. Crampin and J.B. Pendry, Surface Sci. (in print)

33. P.J. Rous, X.-G. Zhang, J.M. MacLaren, A. Gonis, M.A. Van Hove and G.A. Somorjai, Surface Sci. (in print); see also X.-G. Zhang, P.J. Rous, J.M. MacLaren, A. Gonis, M.A. Van Hove and G.A. Somorjai, these proceedings.

Progress in Automatic Structure Refinement with LEED

W. Moritz[1], H. Over[2], G. Kleinle[2], and G. Ertl[2]

[1]Institut für Kristallographie und Mineralogie, Universität München,
  Theresienstr. 41, W-8000 München 2, Fed. Rep. of Germany
[2]Fritz-Haber-Institut der Max-Planck-Gesellschaft,
  Faradayweg 4–6, W-1000 Berlin 33, Fed. Rep. of Germany

It is demonstrated that conventional least squares optimisation techniques can be successfully used for automatic structure refinement with LEED. Examples are given
for two adsorbate systems, H/Ni(110)-(1x2) and
O/Ni(110)-(2x1) where rapid convergence is reached in a
simultaneous optimisation of all structural parameters
within the top three layers.

1.Introduction

The application of LEED as a standard technique for
surface structure determination relies critically on its
convenient use and its capability to solve complex structures. An automatic structure refinement technique, a
clear recipe to localise the best fit model and a standa-
rdised criterion to judge the quality of the result
would be most useful to make the method applicable by
the non-specialist. Clearly, this situation has not yet
been reached and, compared to X-ray diffraction, the
method is still limited. Recent developments, however,
show that considerable improvements are possible. The
computational effort, which is still a limiting factor,
can be reduced to a large extent [1-3]. Further improvements can be made introducing optimisation techniques
into the structure analysis.

Several procedures have been proposed for an automatic structure refinement in which a minimisation of the
conventionally used R-factors is obtained by gradient
methods or search procedures [4-6]. The first approach
had been proposed by Powell and de Carvalho [4], who
applied a search procedure with an independent optimisation of each parameter. The method is generally applicable and does not require the calculation of derivatives.
A more sophisticated method which allows the simulta-

Springer Series in Surface Sciences, Vol. 24   **The Structure of Surfaces III**
Editors: S.Y. Tong · M.A. Van Hove · K. Takayanagi · X.D. Xie
© Springer-Verlag Berlin, Heidelberg 1991

neous refinement of all parameters has been recently proposed by Rous, Van Hove and Somorjai [6]. This method is a gradient method which, in the formulation used there, also requires R-factor calculations only and no derivatives. A different approach, somewhat related to optimisation methods is the direct method proposed by Pendry, Heinz and Oed [5] in which the deviation from a reference structure is determined in one step. The procedure can be iterated if the linear expansion from the reference structure is not sufficient. The combination with Tensor LEED techniques [1] makes this method very efficient.

We propose here an alternative approach which closely resembles the methods conventionally applied in x-ray crystallography [7,8]. Both methods are diffraction techniques and the structure is usually determined by fitting model calculations to experimental data. The only difference in the two techniques is given by the way the experimental and theoretical data are compared, and, of course, in X-ray diffraction, further methods are applicable which cannot be applied with LEED. Using X-rays a set of diffracted intensities is measured at constant wavelength, while with LEED full spectra are measured and fitted to theoretical curves. The fit is usually done by adjusting the position of maxima and minima in the spectra.

There is a-priori no obvious reason why the simple comparison of relative intensities, which works well in the case of X-rays, should not work in the case of LEED. It should be noted that the position of maxima and minima in the spectra is not directly used. Measuring the distance between experimental and calculated spectra by the linear or quadratic deviation has the advantage that well developed optimisation techniques can be used. With the conventional R-factors, such as defined by Zanazzi and Jona [9] and Pendry [10] the fit-function becomes fairly complicated. A simpler R-factor therefore seems to be advantageous and it has indeed been shown that an R-factor based on the mean square deviation is well applicable and leads to reliable results. A full description of this method and the optimisation procedure has been published recently [7,8]. We therefore give here a short review of the novel structure refinement technique and discuss some calculational improvements in detail.

2. Optimisation procedure

The R-factors used in X-ray crystallography are either the linear or the mean square deviation between observed and calculated intensities at a fixed energy. The analogous approach in LEED has been proposed previously as the I(g) method [11]. The shortcomings of that method are that a superstructure producing only weak extra spots, such as a hydrogen superstructure, cannot be well determined because the I(E) spectrum of the superstructure spot is not properly weighted. This problem can be overcome by evaluating each beam with a separate weight factor. The R-factor is defined for discrete energies and is given by

$$R_{DE} = \sum_g W_g \frac{\sum_i |J_{i,g}^{ex} - c_g J_{i,g}^{th}|}{\sum_i J_{i,g}^{ex}} \; ; \qquad (1)$$

$$c_g = \sum_i J_{i,g}^{ex} / \sum_i J_{i,g}^{th} \; ; \qquad W_g = n_g / \sum_g n_g . \qquad (2)$$

n_g is the number of points per beam g. This R-factor can be compared with the usual R-factor in x-ray diffraction. The quantity which is actually minimised in the optimisation procedure is, however, the mean square deviation R_2. R_{DE} is used for comparison with x-ray diffraction where the unweighted linear deviation is a standard R-factor.

To save computing time it is advantageous to choose a step width on the energy scale which is as large as possible. We have extensively tested which step width can be chosen without losing precision of the result. A step width up to 15 - 20 eV corresponding to about 10 -15 data points per spectrum seems to be completely sufficient [9,10]. The upper limit of the step width appears to be set by the requirement that enough points per beam remain.

Having defined the fit-function an automatic minimisation procedure can be introduced. A very efficient method has been developed by Marquardt [12]. His method combines the advantages of the gradient method and of the expansion method. In the gradient method the steepest decent of the R-factor in the parameter space is determined from its partial derivatives with respect to all variable parameters. The method works well far away from the mimimum but converges only slowly near the minimum where the derivatives become very small. The expansion method, on the other hand, works well near the minimum and may lead to serious errors far away from the minimum.

In the expansion method a linear approximation of the intensity function is used

$$I^{th}(\underline{p}_0+\Delta\underline{p}) = I^{th}(\underline{p}_0) + \sum_{j=1}^{k} \left(\frac{\partial I^{th}(\underline{p}_0)}{\partial p_j}\right) \Delta p_j + \ldots \tag{3}$$

where $\underline{p}_0 = (p_1,\ldots p_k)$ denotes the set of structural para- meters and I^{ex}, I^{th} are the normalized intensities. Eq. (3) is inserted into the minimum condition

$$\partial R/\partial p_j = 0, \quad (j=1,\ldots k) \tag{4}$$

leading to a set of linear equations

$$\sum_{i=1}^{n} \left\{I_i^{ex} - I_i^{th}(\underline{p}_0) - \sum_{j=1}^{k} \frac{\partial I_i^{th}(\underline{p}_0)}{\partial p_j} \Delta p_j\right\} \cdot \frac{\partial I_i^{th}(\underline{p}_0)}{\partial p_m} = 0 \tag{5}$$

which is solved by matrix inversion

$$\Delta p = \beta \cdot \alpha^{-1}, \tag{6}$$

$$\beta_m = \sum_{i=1}^{n} (I_i^{ex} - I_i^{th}(\underline{p}_0)) \frac{\partial I^{th}(\underline{p}_0)}{\partial p_m}), \tag{7}$$

$$\alpha_{jm} = \sum_{i=1}^{n} \left(\frac{\partial I^{th}(\underline{p}_0)}{\partial p_i} \cdot \frac{\partial I^{th}(\underline{p}_0)}{\partial p_m}\right) . \tag{8}$$

The method of Marquardt replaces α_{jm} by

$$\alpha'_{jm} = \alpha_{jm} (1+\delta_{jm} \cdot \lambda) \tag{9}$$

δ_{jm} is the Kronecker symbol. If λ is large then the dia- gonal terms dominate and the result is similiar to the gradient method, if λ is small the expansion method is recovered. λ is dynamically adjusted by

$$\lambda = c \sum_{i=1}^{k} \left(\frac{\beta_i^2}{\alpha_{ii}}\right) \left(\frac{1}{R_2}\right) \qquad (10)$$

The speed of the optimisation can be influenced by the parameter c.

3. Calculation of derivatives

The calculational effort in the procedure described above increases linearly with the number of free parameters because the numerical calculation of each derivative requires an additional full dynamical calculation per parameter at all energies. This calculation could be done very efficiently by applying the Tensor LEED technique [1]. However, an approximate calculation of derivatives seemed to be easier to implement in the existing program and turns out to be quite efficient. A linear approximation in calculating derivatives can be used.

The layer scattering matrices are approximated by:

$$M_{gg'} = M_{gg'}(\underline{p}_0) + \sum_j \frac{\partial M_{gg'}(\underline{p}_0)}{\partial p_j} \Delta p_j + \ldots \qquad (11)$$

where $\partial M_{gg'}/\partial p_j$ can be obtained from a linear expansion of the inverse of the propagator matrix. The definition of the propagator matrix and the layer scattering matrices is given in ref. 13. In a linear expansion

$$\left(1 - X(\underline{p}_0 + \Delta \underline{p})\right)^{-1} = \left(1 - X(\underline{p}_0)\right)^{-1} + \left(1 - X(\underline{p}_0)\right)^{-1} \cdot$$
$$\left(X(\underline{p}_0 + \Delta \underline{p}) - X(\underline{p}_0)\right) \cdot \left(1 - X(\underline{p}_0)\right)^{-1} \qquad (12)$$

and only matrix multiplications are required. The inverse of $\left(1 - X(\underline{p}_0)\right)$ has been already calculated for the reference structure and can be reused again. Next an approximate calculation of the lattice sum is desirable. The sum over scattered waves from all atoms in the layer within a limiting radius of about 10 - 15 interatomic distances requires a good part of the computing time and it is worth while considering simplifications. The lattice sum of scattering paths between planes v and v' is given by [13]

$$F_{\ell,m}^{\upsilon\upsilon'} = \sum_{P} i^{\ell} h_{\ell}(|k| \cdot |P+p_{\upsilon}-p_{\upsilon'}|) \cdot$$

$$Y_{\ell m}(\Omega_{P+p_{\upsilon}-p_{\upsilon'}}) e^{-ik(P+p_{\upsilon}-p_{\upsilon'})} \qquad (13)$$

The sperical harmonics $Y_{\ell m}(\Omega_{P+p_V-p_{V'}})$ do not change much by increasing p_{υ} by Δp_{υ} and for the Hankel functions the asymptotic behaviour at large P can be used to calculate the derivatives for values $P > P_{min}$. P_{min} can be set to 3-4 interatomic distances without losing precision.

$$\frac{\partial h_1(z)}{\partial z} \approx i h_{\ell}(z) + e^{i(z-\ell\frac{\pi}{2} + \frac{\pi}{4})} \qquad (14)$$

The lattice sum is therefore split up into two parts where only one part containing the near neighborhood of an atom needs to be recalculated. The minimum distance P_{min} can be chosen to about three interatomic distances. The calculational effort for the lattice sum is reduced by a factor of 15 by this approximation. The comparison of an approximate calculation of derivatives with the full dynamic calculation is shown in fig. 1.

The approximate calculation of derivatives turns out to be completely sufficient for all types of parameters.

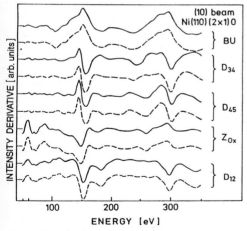

Fig.1. Comparison of approximate (dashed line) and full dynamical (solid line) calculation of derivatives. The structure parameters are displayed in fig. 2.

The increase of the speed of the calculation depends, however, on the number of variable parameters and the number of phase shifts used. In the examples shown below with 6 and 8 variable parameters a factor of 2.5 for the whole calculation is gained. Further improvements are possible. A linear expansion similar to that described in eq. (12) can be used for the matrices to be inverted in the layer doubling method.

4. Application to H/Ni(110)-(1x2) and O/Ni(110)-(2x1)

To illustrate the capability of the method the results of two adsorbate systems will be presented. Full structural results have been published recently [7,14,15], we present here only the results of the fit-procedure using the bulk values of Ni with a slight buckling in the third layer as start parameters.

The structure of the H/Ni(110)-(1x2) is shown in fig. 2; the hydrogen atoms are ignored in the calculation. The final structural parameters as well as the result of two calculations with different start parameters are given in table 1. The choice of the bulk structure as start parameter is not possible in this case. The bulk structure does not produce superstructure beams and the derivatives with respect to the superstructure parameters LS and BU vanish. The derivative vanishes because at a highly symmetry point two choices of the derivative are symmetrically equivalent. It is therefore necessary to shift at least one atom off its bulk lattice position.

A second example is O/Ni(110)-(2x1). The structural model is shown in fig. 3. Here 6 independent parameters within the top three layers had to be refined, assuming

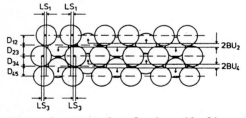

Fig. 2. Model of the (1x2) structure of H/Ni(110).
        Hydrogen atoms are not shown.

Table 1. Result of a simultaneous fit of 8 structural parameters. The final structure was reached after 10 iterations, $R_{DE} = 0.3$

Parameter	Start Value	Final Value [Å]
D_{12}	1.246	1.223
D_{23}	1.246	1.331
D_{34}	1.246	1.272
D_{45}	1.246	1.220
LS_1	0	0.30
LS_3	0	0.12
BU_2	0.1	0.25
BU_4	0	0.02

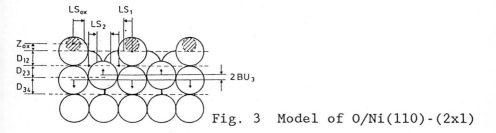

Fig. 3 Model of O/Ni(110)-(2x1)

Table 2. Result of a simultaneous fit of 6 parameters, the minimum R-factor was reached after 7 iterations, $R_{DE} = 0.24$

Parameter	Start Value	Final Value [Å]
Z_{ox}	0.3	0.224
D_{12}	1.246	1.293
D_{23}	1.246	1.246.
LS_2	0	0.011
BU_4	0	0.052

that oxygen sits in the symmetric position. Results are shown in table 2. 15 eV steps were used in the energy range between 40 and 340 eV, corresponding to 106 data points in 8 I/V spectra.

In the above calculations oxygen was fixed in the symmetric site. A detailed study showed that a slight preference was found for an asysmmetric site [14]. The search for an asymmetric site was stimulated by a HREELS study [16] which showed an additional oxygen mode not compatible with the assumption of a symmetric oxygen position. This asymmetry has been neglected here, because it is felt that from the LEED data at normal incidence alone this asymmetry cannot be definitely concluded [14]. The influence of a lateral shift of the oxygen atom on the other structural parameters can be neglected.

5. Discussion

A central point in any optimisation scheme is of course the radius of convergence within which a minimum will be localised. That the minimum may be a local mimimum has been already pointed out. Local minima can be avoided only by choosing start structures on a wide grid in the parameter space. The radius of convergence therefore determines the grid size which must be applied to exclude local minima. This same problem occurs, by the way, in the conventional R-factor analysis. The average distance between local minima in the R-factor hyperface may be estimated to be about 0.5 Å. This results from the simple consideration that with an average wavelength of 1.0 Å an interference maximum of backward and forward scattering between two atoms occurs again after a shift of 0.5 Å. It follows that roughly a deviation of 0.2 - 0.3 Å can be tolerated in the start parameters. This is a rough estimate, of course, and applies to parameters parallel to $\Delta \underline{k}$. Smaller distances between local minima may also occur due to domain averaging [6].

To check the radius of convergence we performed several runs with different start values for two parameters in the O/Ni(110)-(2x1) structure, Z_{ox} and D_{12}, keeping all other parameters at their optimum value. The results are shown in table 3.

It may be concluded that the radius of convergence is about 0.2 Å in agreement with the estimate considered above. In test runs with a simultaneous fit of all parameters the same radius of convergence was found which indicates that the parameters are only weakly correlated.

The convergence depends also on the parameter c in eq. 10, this has been chosen as 0.2 in the above examples. A larger value decreases the speed of the calcula-

Table 3 Check of the radius of convergence for two parameters.

	start	final	R_{DE}	No. of iterations
Z_{ox}	0.5	0.83	0.43	6
Z_{ox}	0.4	0.22	0.243	4
D_{12}	1.5	1.58	0.73	1
D_{12}	1.4	1.31	0.242	3

tion and has been found sometimes to increase the radius of convergence because the large steps in the beginning of the iteration process are damped.

Acknowledgement

Part of this work was financially supported by the Deutsche Forschungsgemeinschaft, SFB 338.

Literature

[1] P.J. Rous, J.B. Pendry, D.K.Saldin, K. Heinz, K. Müller and N.Bickel, Phys. Rev. Letters 57 (1986) 2951.
[2] W. Moritz, J. Phys. C17 (1984) 353.
[3] J.B. Pendry and K. Heinz, Surf. Sci., in press.
[4] P.G. Powell and V.E. de Carvalho, Surf. Sci. 187 (1987) 175
[5] J.B. Pendry, K. Heinz and W. Oed, Proceedings of the 11th Intern Vac. Cong. (Colonge, 1989)
[6] P. Rous, M.A. Van Hove and G.A. Somorjai, Surf. Sci. 226 (1990) 15, and this volume.
[7] G. Kleinle, W. Moritz, D.L. Adams, and G. Ertl, Surf. Sci. 219 (1989) L637.
[8] G. Kleinle, W. Moritz and G. Ertl, Surf. Sci. in press.
[9] E. Zanazzi and F. Jona, Surf. Sci. 62 (1977) 61.
[10] J.B. Pendry, J. Phys. C 13 (1980) 937.
[11] a) L.J. Clarke, Surface Crystallography - An Introduction to Low Energy Electron Diffraction (Wiley Chichester, 1985).
     b) L.J. Clarke, Vacuum 29 (1979) 405.

[12] D.W. Marquardt, J. Soc. Indust. Appl. Math. 11 (1963) 431

[13] J.B. Pendry, Low Energy Electron Diffraction, Academic press, London and New York 1974.

[14] G. Kleinle, R.J. Behm, F. Jona, W. Moritz, J. Wintterlin and G. Ertl, Surf. Sci. 225 (1990) 171.

[15] E. Kleinle, V. Penka, R.J. Behm, G. Ertl and W. Moritz, Phys. Rev. Letters 58 (1987) 148.

[16] B. Voigtländer, S. Lehwald and H. Ibach, Surf. Sci. 225 (1990) 162

Application of Direct Methods in LEED: Multilayer Relaxation, Adsorption Sites and Adsorbate Induced Reconstruction

K. Heinz[1], W. Oed[1], and J.B. Pendry[2]

[1]Lehrstuhl für Festkörperphysik, University of Erlangen-Nürnberg,
   Staudtstr. 7, W-8520 Erlangen, Fed. Rep. of Germany
[2]The Blackett Laboratory, Imperial College, London SW7 2BZ, UK

Abstract. The inversion of linear Tensor LEED is used to retrieve structural data directly from measured LEED spectra, i.e. without applying the usual trial-and-error method. The structure results as a change with respect to a certain reference structure. The latter should be near the true structure but can be of much less complexity. We show that the method allows the retrieval of multilayer relaxation, adsorption sites and adsorbate induced reconstruction. Also, the moments of vibrational amplitudes can be determined.

1. An Approach to Complex Structures Using LEED

It is well known that the amount of computational work necessary to solve surface structures increases tremendously with the complexity of the structure. Not only does the calculation of LEED spectra for one single structural model become more time consuming - the increase of the number of structural models to be introduced in the trial-and-error procedure is even more serious. Therefore, ways out of this dilemma are needed and are indeed under way. On the one hand there are automatic search procedures including the method of steepest descent applied to the R-factor hypersurface /1,2/. Also, recently a least square method was proposed using intensity derivatives with respect to structural parameters /3/. On the other hand efforts were made to allow for more rapid calculations in order to speed up the variations of structural parameters whereby the most powerful method in this respect seems to be Tensor LEED /4,5/. Additionally, it was found that for the theory-experiment comparison via R-factors a much less dense energy grid than usually taken can be applied because of the partial redundancy of the intensity data with respect to structural information /3,6/.

In this paper we propose a different method which basically is an inversion of the Tensor LEED method. Experimental data are compared to spectra computed for a certain reference structure and their differences are directly converted to structural differences between the reference and the real structure.

Springer Series in Surface Sciences, Vol. 24   **The Structure of Surfaces III**
Editors: S.Y. Tong · M.A. Van Hove · K. Takayanagi · X.D. Xie
© Springer-Verlag Berlin, Heidelberg 1991

2. Inversion of Tensor LEED and Error Minimization

The main idea of Tensor LEED comes from the fact that intensity spectra change smoothly with a smooth variation of structural data. So, deviations from the amplitude A_0 of a given reference structure according to a structural change described by atomic displacements δR_i of certain atoms can be calculated by perturbation theory, which, in the linear version, yields the amplitude change $\delta A = A_0 + {}_i\Sigma T_i\, \delta R_i$. The power of Tensor LEED is that the Tensor T must be calculated only once for the reference structure, so that structural modifications subsequently can be calculated with high speed. Also, the reference structure may be rather simple, e.g. non-reconstructed, though the atomic displacements may produce rather complex structures.

In the experiment intensities rather than amplitudes are collected. The intensity of the modified structure is $I = |A_0 + \delta A|^2 \simeq |A_0|^2 + {}_i\Sigma M_i\, \delta R_i$ with $M_i = 2\mathrm{Re}(A_0 T_i^*)$ and the second order term neglected as justified by experience for $\delta R_i \leq 0.1$ Å /7/. Obviously, in order to determine unknown displacements at least m data points I_k must be known and the reference calculation must be done for m different beams and/or energies to produce amplitudes A_{0k}. Then $I_k = |A_{0k}|^2 + {}_i\Sigma M_{ik}\, \delta R_i$ (k=1,...m) results which by inversion produces directly the deviations from the reference structure

$$\delta R_i = {}_k\Sigma (M_{ik})^{-1}\, (I_k - |A_{0k}|^2) \qquad (i = 1,...m).$$

Using experimental data for I_k could lead to inaccurate results because of unavoidable experimental errors. Assuming the errors to be statistical the correct structure should produce a minimum deviation between experimental and calculated intensities, i.e. a minimum of $E = {}_k\Sigma (I_k^{exp} - I_k(\delta R_i))^2$ (i = 1,..m) whereby more than m data points may enter the sum. The minimum develops for vanishing derivatives $\partial E/\partial \delta R_i = 0$ (i = 1,...m) which can be calculated analytically using the equation $I_k = |A_{0k} + {}_i\Sigma T_{ik}\, \delta R_i|^2$. By neglecting the quadratic terms this leads to the linear system of equations

$$_k\Sigma (I_k^{exp} - |A_{ok}|^2 - 2\Sigma M_{ik}\delta R_i)\, M_{jk} = 0 \quad (j = 1,...m).$$

The solution of this system can be taken as starting point for solving the non-linear system where the quadratic terms are also considered. Also, E can be taken as an error estimation, its minimization with respect to the inner potential determines this quantity. It should be noticed that the above direct procedure works accurately for $\delta R_i < 0.1$ Å as shown in the applications below. For larger deviations from the reference structure a new reference calculation can be done as demonstrated in one of the examples, too. Of course this increases the amount of computational work and proper convergence of the procedure must be checked.

3. Applications

3.1. Determination of Multilayer Relaxation
The direct method is almost ideal for the determination of layer relaxations /7/ because they usually are only in the percentage region of bulk distances. Results for some surfaces are given in Table 1 in comparison to traditional calculations /8,9/. Evidently, the results agree within the usual limits of error. The reference was the bulk like terminated crystal in each case. The advantage of the method is more obvious for more open surfaces i.e. for small layer distances where neither RFS nor the layer doubling procedure converge and time consuming calculations for each trial structure are necessary.

<u>Table 1:</u> Multilayer relaxations determined by direct and conventional (in brackets) analyses

Surface	Rh(110)	Rh(110)-2H	W(100)1x1
$\Delta d_{12}/d_0$ (%)	-5.7 (-6.9)	-0.8 (-1.3)	-6.3 (-8.2)
$\Delta d_{23}/d_0$ (%)	+1.6 (+1.9)	+0.9 (+0.2)	+1.9 (±0.0)

3.2 Determination of Adsorption Sites
Our present computer codes do not allow for superstructures as reference. However, at least for weak scattering adsorbates a 1x1 adlayer can be taken with a suitable array of adatoms removed again using Tensor LEED. This seems to work for negligible intralayer scattering in the adlayer and for not too many atoms to be removed. We give results in the following section.

3.3 Determination of Adsorbate Induced Reconstruction
Oxygen and sulfur adsorb in c(2x2) /10,12/ and p(2x2) /11,13/ overlayers on Ni(100) inducing relaxations and buckling in the substrate (see model for c(2x2) in fig. 1). The reference was the bulk-like substrate with oxygen (sulfur) in hollow sites at height 0.90 Å (1.30 Å). In case of c(2x2)O a change of the

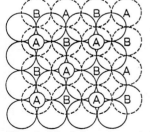 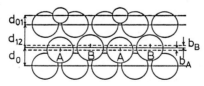

Fig. 1: Model and model parameters for the c(2x2)O,S/Ni(100) structure. For the p(2x2) structure the centered atoms have to be removed.

<u>Table 2:</u> Adsorbated induced substrate reconstructions for O and S on Ni(100). Model parameters according to fig. 1 are given (b_C corresponds to the uncovered site in the p(2x2) phase) as determined by the direct and conventional (in brackets) method. The substrate reference was always the unrelaxed crystal ($d_{12} = 1.76$ Å)

Structure	p(2x2)O	c(2x2)O	c(2x2)S
Reference	$d_{01} = 0.90$ Å	$d_{01} = 0.90$ Å	$d_{01} = 1.30$ Å
d_{01} (Å)	0.83 (0.80)	0.73 (0.77)	1.28 (1.30)
d_{12} (Å)	1.81 (1.80)	1.86 (1.86)	1.78 (1.79)
b_A (Å)	-0.20 (-0.05)	-0.02 (-0.02)	-0.02 (-0.03)
b_B (Å)	+0.13 (-0.02)	+0.01 (+0.02)	+0.01 (-0.02)
b_C (Å)	-0.13 (+0.05)	---	---

oxygen height of larger than 0.1 Å had to be expected by comparison to the conventional analysis /10/ and therefore a second reference calculation was made after the first step of the analysis. Table 2 presents the results which again are of convincing accuracy except for the p(2x2)O phase. To simulate this phase too many atoms had to be removed from the initial 1x1 overlayer entailing restricted accuracy.

3.4 Determination of Thermal Vibration Moments

Obviously, the direct method can be further developed to determine moments of thermal vibrations. The expansion $A = A_0 +_i\Sigma T_i \, \delta R_i$ is only of first order and e.g. for a single displacement can be extended to $I = |A_0|^2 +_n\Sigma M_n \, (\delta R)^n$. If δR is a vibrational rather than a static amplitude, experimentally the average $<I> = |A_0|^2 + _n\Sigma M_n <(\delta R)^n>$ results and so the moments $<(\delta R)^n>$ can be determined. A first application was reported recently for O/Rh(100) /14/.

4. Discussion

The direct method such as the inversion of the linear version of tensor LEED allows the proper determination of structural data as long as the reference structure guessed is not too far away from the real structure. But even if it is, a second, nearer reference structure can be calculated. The method saves all the more computer time the more complex the real structure is (a factor of 10 compared to the conventional method for O,S/Ni(100)). As reconstructions frequently happen upon adsorption and as the displacements are small in many cases the direct method is a valuable tool to retrieve the correct structure. In any case its application gives an idea about the possible reconstruction process, i.e. about which atoms move upon adsorption and in which direction. In any case this gives valuable hints even if a final refinement by the conventional method remains necessary. Nevertheless the precision of the method has proved to be in the range of the conventional analyses at least for most of the examples given. Increased accuracy can be expected by using computer codes allowing reference structure calculations also for superstructures. Additionally we want to emphasize that the method can be further speeded up by calculating intensities only at a restricted number of energies as proposed earlier /3,6/. We also think that the potential of the method is not yet exhausted as can be seen from the possibility of tackling the problem of thermal vibrations.

Acknowledgements:
The authors K.H. and W.O. are indebted to Deutsche Forschungsgemeinschaft (DFG) and Höchstleistungsrechenzentrum (HLRZ, Jülich).

References:
/1/ P.G.Powell and V.E.de Carvallio, Surface Sci.187(1987)175

/2/ P.J.Rous, M.A.Van Hove and G.A.Somorjai, Surface Sci.226(1990)15

/3/ G.Kleinle, W.Moritz and G.Ertl, Surface Sci., to be published

/4/ P.J.Rous, J.B.Pendry, D.K.Saldin, K.Heinz, K.Müller and N.Bickel, Phys. Rev. Lett. 57 (1986) 2951

/5/ P.J.Rous and J.B.Pendry, Surface Sci.219 (1989)335 and 373

/6/ G.Kleinle, W.Moritz, D.L.Adams and G.Ertl, Surface Sci.219(1989)L637

/7/ J.B.Pendry, K.Heinz and W.Oed, Phys.Rev.Lett.61(1988)2953

/8/ W.Nichtl, N.Bickel, L.Hammer, K.Heinz and K.Müller, Surface Sci.188 (1987)L729

/9/ J.B.Pendry, K.Heinz, W.Oed, H.Landskron, K.Müller and G.Schmidtlein, Surface Sci.193(1988)L1

/10/ W.Oed, H.Lindner, U.Starke, K.Heinz, K.Müller and J.B.Pendry, Surface Sci.224(1989)179

/11/ W.Oed, H.Lindner, U.Starke, K.Heinz, K.Müller, D.K.Saldin, P.de Andres and J.B.Pendry, Surface Sci.225(1990)242

/12/ U.Starke, F.Bothe, W.Oed and K.Heinz, Surface Sci., in press

/13/ W.Oed, U.Starke, F.Bothe and K.Heinz, Surface Sci., in press

/14/ J.B.Pendry and K.Heinz, Surface Sci.230(1990)137

Real-Space Multiple Scattering Theory Calculations of LEED Intensities for Stepped Surfaces

X.-G. Zhang[1], *P.J. Rous*[1], *J.M. MacLaren*[2], *A. Gonis*[3], *M.A. Van Hove*[1], and *G.A. Somorjai*[1,4]

[1]Center for Advanced Materials, Materials and Chemical Sciences Division,
 Lawrence Berkeley Laboratory, Berkeley, CA 94720, USA
[2]Theory Division, Los Alamos National Laboratory,
 Los Alamos, NM 87545, USA
[3]Division of Chemistry and Materials Sciences, Lawrence Livermore
 National Laboratory, Livermore, CA 94550, USA
[4]Department of Chemistry, University of California, Berkeley, CA 94720, USA

We use a newly developed real-space multiple scattering theory (RS-MST) to calculate low-energy electron diffraction (LEED) intensities from stepped surfaces. In this calculation the electron wavefunctions are expanded in terms of an angular momentum basis, utilizing the property of removal invariance of systems with semi-infinite periodicity. This strongly reduces the dependence of the calculation on the interlayer spacing and thus opens up the possibility of treating more open surfaces. This includes in particular stepped surfaces, to which conventional methods cannot be applied. Applications of the formalism to various stepped surfaces are presented. In particular, the results for Cu(311) and (331) surfaces obtained from both the layer doubling and RS-MST methods are compared. In addition, numerical techniques which can improve the convergence as well as the speed of the RS-MST approach are discussed.

1. Introduction

It is known that stepped surfaces are important in studying the properties of many materials and understanding processes in many technologies. For example, steps can provide active sites for bond breaking and bond formation in heterogeneous catalysis. They can also affect the mechanical properties of solids through the pinning of dislocations and crack propagation. In order to understand these phenomena, it is often necessary to obtain accurate information about atomic positions on the steps of a surface.

Quantitative determination of the structures of clean stepped surfaces has been difficult. Conventional theoretical techniques for the study of LEED spectra cannot be readily applied to stepped surfaces. This is due to the small interplanar spacing and large two-dimensional (2-d) unit cell of a stepped surface, which causes difficulties with the convergence of the plane wave expansion used in a conventional LEED theory.

However, a recently developed technique [1-3] has greatly alleviated this problem. This technique is based on a real-space formulation of multiple scat-

Springer Series in Surface Sciences, Vol. 24   **The Structure of Surfaces III**
Editors: S.Y. Tong · M.A. Van Hove · K. Takayanagi · X.D. Xie
© Springer-Verlag Berlin, Heidelberg 1991

tering theory (RS-MST) [1] and dispenses completely with plane waves (except of course in the trivial propagation through vacuum from the electron gun to the surface and back to the detector). Within the surface, it only uses spherical waves and, as discussed in Ref.[3], strongly reduces the scaling problems of the plane wave basis used by the earlier methods [4,5]. RS-MST is based on the principle of removal invariance which holds for semi-infinite periodic lattices: removing a layer from the free end of such a lattice does not change the electronic states (except for a trivial phase factor), because the resulting surface is identical to the original one, being merely displaced with respect to that by a single layer. This removal invariance provides a self-consistency condition for the electronic states, which can be solved numerically. Detailed discussion of this new approach and RS-MST can be found in references [1-3]. In this paper we will focus on the application of the method to several stepped surfaces of Cu.

2. Theory

We first consider the ideal bulk termination of the crystal. In general, however, the selvedge will undergo some form of restructuring, exhibiting planar relaxation perpendicular and/or parallel to the surface and perhaps reconstruction of the first few atomic layers, and may also involve adsorbates. Such deviations from the ideal semi-infinite bulk lattice can be readily incorporated and will be discussed later in this section.

The bulk termination of an ideal crystal exhibits a universal property, a consequence of the semi–infinite periodicity perpendicular to the surface. In simplest terms it implies that the reflectivity of the crystal is unchanged if an atomic plane is peeled away from the surface. More generally, we can say that the full scattering t-matrix of the system is invariant with respect to the removal of any finite number of layers from the surface. This *removal invariance* property is the foundation of our new approach to LEED theory. It allows us to derive an equation within a purely angular momentum basis to determine the t-matrix corresponding to a semi-infinite periodic system, which in turn can be used to evaluate the reflectivity of the entire surface.

Instead of using a layer doubling approach as in the plane–wave representation, in the angular momentum representation we use a self-consistency condition to determine the reflectivity of the half solid. To do this we use the property of removal invariance in the presence of semi-infinite periodicity [1]. We imagine replacing the half solid with a single *renormalized* layer which is constructed in such a way as to possess the scattering properties of the entire half solid. This renormalized layer is described by a t-matrix, T, in angular momentum representation. According to the property of removal invariance, a system consisting of a *bare* atomic layer, represented by the monolayer t-matrix, τ_{lay}, and a renormalized layer, T, can also be represented by the t-matrix, T. This constitutes a self-consistency condition from which T can be determined. In practice, a more rapidly convergent procedure is to consider a half-solid being represented by a stack of N layers, properly renormalized to reproduce all

145

the scattering properties of the half-solid. The self-consistency condition can be constructed by adding one more layer, represented by η_{ay}, and demanding that the resulting system can still represented by $\mathcal{T}$. We call such a procedure, in which N "dressed" layers represent the half-solid, an N-layer calculation, or $(1, N)$ mode.

It can be shown that the solution of the self-consistent equation, as formulated in Ref.[3], takes exact account of all multiple scattering paths within any $N+1$ nearest neighbor layers. All multiple scattering terms between layers further apart are represented by the products of Green's functions and translation operators, which involve internal summations over angular momentum states that are truncated at a finite l. This consideration provides a basis upon which one can estimate the rate of convergence in terms of the angular momentum truncation and make a proper choice of the value of N in a calculation. Specifically, N should scale roughly by Nd =const., where d is the interplanar spacing of the system. Therefore, the size of the basis set for a RS-MST calculation scales as $1/d$, compared with that of a plane wave basis set which scales as $1/d^3$ [3].

The above discussion applies to an ideally truncated unrelaxed surface. A more realistic model of a surface usually includes several relaxed layers and sometimes adsorbates. The use of the angular momentum representation provides an easy multiple scattering approach for the treatment of surface relaxations and adsorbates. Once the t-matrix, $\mathcal{T}$, of the unrelaxed substrate is known, one can use multiple scattering theory to construct the t-matrix, $\mathcal{T}_{relax}$, for the relaxed surface by combining the t-matrix, $\tau_{overlay}$, of an overlayer with $\mathcal{T}$ of the substrate. This process can be used repeatedly for relaxations and adsorbates involving more than one layer.

3. Results

In this section we present LEED I–V spectra for Cu fcc surfaces obtained with the new RS-MST for several surface structures. First we compare the I-V spectra for unrelaxed surfaces with that obtained by the layer doubling technique. In all calculations the atomic Cu phase shifts are obtained from solving the Schrödinger equation for the potential provided by Moruzzi, Janak and Williams [6]. We found that six atomic phase shifts ($l_{max} = 5$) were sufficient for energies below 80eV, and $l_{max} = 6$ was required from 80eV to 200eV. We used a three-layer calculation ($N = 3$ in the self-consistent equation, see Ref.[3]) for the (311) surface, a four-layer calculation ($N = 4$) for the (331) surface, and a six-layer mode ($N = 6$) for the (321) surface when $E < 55$eV, a seven-layer mode ($N = 7$) when $E \geq 55$eV. All calculations are for normal incidence.

Figure 1 shows I–V curves for (10) and ($\bar{1}0$) LEED beams calculated for a Cu(311) surface using both a conventional layer doubling technique and the real-space multiple scattering theory method. This system has an interplanar

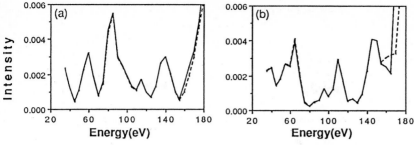

Fig. 1. The comparison of the I-V curves obtained by the RS-MST method (solid curves) and the layer doubling method (dashed curves) for Cu (311) unrelaxed surface, (a), (10) beam, and (b), ($\bar{1}$0) beam.

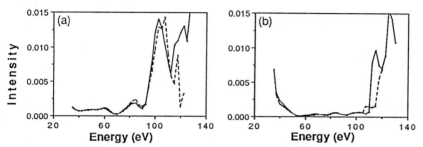

Fig. 2. Similar comparison as in Fig. 1 for Cu (331) unrelaxed surface, (a), (10) beam, and (b), ($\bar{1}$0) beam.

spacing of about 1.09Å. The comparison between the RS-MST results (solid curves) and those obtained from the layer doubling calculation (dashed curves) are in good agreement. We found that both the numerical stability and the converged solution of the self-consistent equation are sensitive to the accuracy of the Green functions (structure constants) at high energies. In Fig. 2, we present similar I-V curves for the Cu(331) surface, which has an interplanar spacing of 0.83Å. In this case, the layer doubling method (dashed curves) begins to show signs of failure. We see that the I-V curves obtained by layer doubling show sharp spikes at about $E \approx 120$eV, indicating convergence problems.

Finally, in Fig. 3 we show the I-V curves of Cu (321) surface, with an interplanar spacing 0.48Å, obtained by RS-MST, and for three different top layer positions relative to the bulk truncation, -5%, 0, 5%, respectively. We included in the calculation the temperature effects corresponding to $T = 300$K. The rather small change in the I-V curves with respect to the surface relaxation reflects that a 5% change in d represents a much smaller change in the lattice position when d is small.

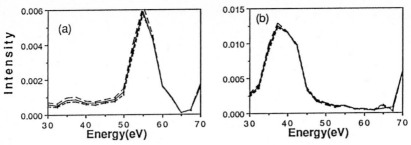

Fig. 3. I-V curves obtained by RS-MST for Cu (321) surface with three different top layer positions, -5% (broken lines), 0% (solid lines), 5% (dashed lines), relative to the bulk truncation, (a), (20) beam, and (b), ($\bar{1}\bar{1}$) beam.

4. Discussion

We have applied the RS-MST method for the calculation of LEED I–V curves of stepped surfaces. The technique is based on equations expressed entirely in terms of matrices in the angular momentum representation. The technique, unlike conventional LEED theory, is equally applicable to both low and high Miller index surfaces since the increase of the matrix size associated with decreasing interplanar spacing is much more modest than in conventional LEED calculations. It has clear advantages in treating high-Miller index surfaces. The new method has been tested on Cu (311) and (331) stepped surfaces where excellent agreement with the layer doubling method was obtained, where the latter converges. The I–V curves for Cu (321) surface were calculated for the first time with the new method. This calculation demonstrates the ability of the RS-MST method to deal with high Miller-index surfaces and to treat surface relaxations with relative ease.

Acknowledgement

The work reported here was supported in part by the US Department of Energy under a *Grand Challenge* program. The project was partially supported by the Director, Office of Energy Research, Office of Basic Energy Sciences, Materials Sciences Division, US Department of Energy, under contract No. DE-AC03-76SF0098, and the Lawrence Livermore National Laboratory under Contract No. W-7405-ENG-48.

5. References

[1] X.-G. Zhang and A. Gonis, Phys. Rev. Letters 62, 1161 (1989).

[2] J.M. MacLaren, X.-G. Zhang and A. Gonis, Phys. Rev. B40, 9955 (1989).

[3] X.-G. Zhang, P.J. Rous, J.M. MacLaren, A. Gonis, M.A. Van Hove and G.A. Somorjai, Surf. Sci., in press.

[4] J.B. Pendry, **Low Energy Electron Diffraction**, (Academic Press, London, 1974).

[5] M.A. Van Hove, W.H. Weinberg and C.-M. Chan, **Low-Energy Electron Diffraction: Experiment, Theory and Surface Structure Determination** (Springer-Verlag, Heidelberg, 1986).

[6] V.L. Moruzzi, J.F. Janak and A.R. Williams, **Calculated Electronic Properties of Metals** (Pergamon, New York, 1978).

A Simple Scheme for LEED Intensity Calculations

R.F. Lin[1,*], *P.M. Marcus*[2], *and F. Jona*[1]

[1]College of Engineering and Applied Science, State University of New York,
  Stony Brook, NY 11794, USA
[2]IBM Research Center, Yorktown Heights, NY 10598, USA

Abstract. We examine the possibility of improving the quasidynamical approximation by introducing some intraplanar scattering. An exact equation is given relating the reflection matrix of a semi-infinite truncated bulk crystal to the reflection and transmission matrices of a single layer, and applied to a quasidynamic calculation of the beam intensity-energy curves of Cu(001). The solution by iteration is shown to converge quickly and to reproduce well the peak positions of a full-dynamical calculation. A cluster formulation is given which should provide better single-layer matrices to improve on the quasidynamic calculation.

1. Introduction

The lengthy character of LEED calculations when complex structures or extended energy ranges are studied sustains interest in approximate methods of calculation of reasonable accuracy. Of the many methods proposed, the quasidynamical (QD) approximation is one of the most useful and successful[1]. Since the QD method treats interlayer scattering accurately, but neglects intralayer multiple scattering, a natural way to try to improve the QD method is to introduce some scattering within the layer. We formulate a way to make this introduction by using a cluster of atoms in the layer around the origin atom and finding linear equations for the spherical wave amplitudes at each atom of the cluster that describe multiple scatterings between cluster atoms to all orders. Formulas are given which use these amplitudes to obtain beam-reflection matrices $\mathscr{R}$ and beam-transmission matrices $\mathscr{T}$ for the single layer, which are better than kinematic and can be introduced into the exact matrix equation relating the reflection matrix for the semi-infinite crystal $\mathscr{R}^S$ to $\mathscr{R}$ and $\mathscr{T}$. We give that equation for the truncated bulk Bravais crystal, and solve it here for Cu(001) using just the kinematic $\mathscr{R}$ and $\mathscr{T}$ matrices. This quasidynamic solution is shown to compare well with the full-dynamical (FD) solution. We then give linear equations for the spherical-wave amplitudes at each atom produced by a plane wave incident on a cluster of atoms. The equations take account of all multiple scatterings between atoms of the cluster.

---

*On leave from the Department of Physics, Fudan University, Shanghai, The People's Republic of China.

Springer Series in Surface Sciences, Vol. 24   **The Structure of Surfaces III**
Editors: S.Y. Tong · M.A. Van Hove · K. Takayanagi · X.D. Xie
© Springer-Verlag Berlin, Heidelberg 1991

Simple expressions for the $\mathscr{R}$ and $\mathscr{T}$ matrices of a single layer, which use the solutions of the cluster equations to go beyond the kinematic formulas, are given. These expressions can then be put into the equation for $\mathscr{R}^S$. It is not yet clear whether this procedure based on the cluster equations is more efficient than the standard layer-KKR method. However a pairing approximation is suggested which would be more efficient if it proves reasonably accurate. All waves propagate in an attenuating medium with a complex constant potential.

2. The Exact Interlayer Reflection Equation and the QD Approximation

We start with the layer matrices $\mathscr{R}$ and $\mathscr{T}$ whose components $\mathscr{R}_{h'h}$, $\mathscr{T}_{h'h}$ give scattering amplitudes from incident plane wave $\exp(i\mathbf{k}_h^+ \cdot \mathbf{r})$ into reflected and transmitted plane waves $\mathscr{R}_{h'h} \exp(i\mathbf{k}_{h'}^- \cdot \mathbf{r})$, $\mathscr{T}_{h'h} \exp(i\mathbf{k}_{h'}^+ \cdot \mathbf{r})$. The $\mathbf{k}_h^\pm$ are wave numbers of the beam set with the same parallel components of $\mathbf{k}$ ($+$ or $-$ means toward $+z$ or $-z$) and the same energy E, propagating in an attenuating medium[2] with complex potential with respect to vacuum $V_0 = V_{0r} + iV_{0i}$. We relate $\mathscr{R}$ and $\mathscr{T}$ to $\mathscr{R}^S$, the reflection matrix of the semi-infinite crystal by

$$\mathscr{R}^S = \mathscr{R} + \mathscr{T} \, \mathscr{R}_1^S (1 - \mathscr{R} \, \mathscr{R}_1^S)^{-1} \, \mathscr{T} \tag{1}$$

$$\mathscr{R}_1^S \equiv \mathscr{P}_{-d}^- \mathscr{R}^S \, \mathscr{P}_d^+ \tag{2}$$

where $\mathscr{P}_d^\pm$ are diagonal phase matrices, elements $\exp(i\mathbf{k}_h^\pm \cdot \mathbf{d}) \, \delta_{h'h}$, and $\mathbf{d}$ is the translation vector between successive layers. Equation (1) is a special case of general multiple-scattering t-matrix relations[3] and is easily derived using the idea of "removal invariance"[4] for reflection from a semi-infinite stack of identical layers, and the same stack with one layer removed. We consider only the simplest case of a truncated bulk Bravais crystal and assume symmetry of left and right reflections and transmissions at a layer.

The QD approximation uses the kinematic[5] formulas (3) and (4) for $\mathscr{R}$ and $\mathscr{T}$ and solves (1) for $\mathscr{R}^S$.

$$\mathscr{R}_{h'h} = (2\pi i \, / \, \mathscr{A} k_{h'z}) f(\mathbf{k}_h^+ \cdot \mathbf{k}_{h'}^-) \tag{3}$$

$$\mathscr{T}_{h'h} = (2\pi i \, / \, \mathscr{A} k_{h'z}) f(\mathbf{k}_h^+ \cdot \mathbf{k}_{h'}^+) + \delta_{h'h} \tag{4}$$

where $f(\mathbf{k} \cdot \mathbf{r})$ is the form factor for single-atom scattering of a plane wave $\exp(i\mathbf{k} \cdot \mathbf{r})$, which gives the amplitude of the asymptotic outgoing scattered spherical wave $f(\mathbf{k} \cdot \mathbf{r}) \exp(i k r)/r$ in direction $\mathbf{r}$; $\mathscr{A}$ is the area of the unit mesh; $f(\mathbf{k} \cdot \mathbf{r})$ is a simple sum over the phase shifts of the scattering atom[5]. Equation (1) can be solved readily by iteration, starting from $\mathscr{R}^S = \mathscr{R}$. Figure 1 shows the development of the beam intensities I(E) after 2, 4, and 15 iterations to compare with the FD result for Cu(001) at normal incidence for the 00 beam using 8 phase shifts, 29 beams and $V_0 = (-10 - 3.5i)$ eV. We see the rapid development of peak structure by multiple scattering between layers, and the close correspondence of that peak structure to the FD result, although relative amplitudes differ. Note the correspondence is better at higher E, say above 100eV.

Closer correspondence at higher E is also shown in Fig. 2, where the sum S_1 is plotted against E; S_h, which expresses flux conservation for a beam $\mathbf{k}_h$, is given by (5), and is unity for $\mathscr{R}$ and $\mathscr{T}$ matrices of a layer of atoms all in one plane, when the matrices satisfy flux conservation.

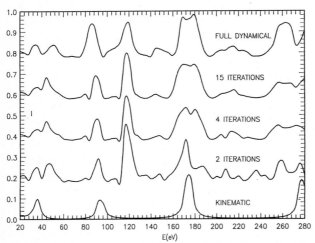

Fig. 1 I(E) curves for 00 beam of Cu(001) at $\theta = 0°$ from successive iterates of the interlayer scattering Eq.(1) starting from the kinematic curve (lowest), after 2, 4, 15 iterations and the FD curve (top).

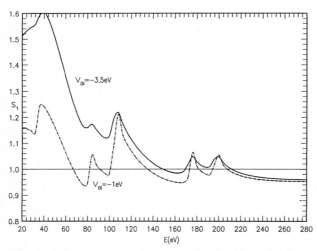

Fig. 2 Flux conservation sum for incident 00 beam on Cu(001) at $\theta = 0°$ in QD approximation using 29 beams. The upper (solid) curve is for $V_{0i} = -3.5eV$ and the lower (dashed) curve is for $V_{0i} = -1.0eV$. Differences from unity show the extent of the failure of kinematic $\mathscr{R}$ and $\mathscr{T}$ matrices of a layer to satisfy flux conservation.

$$S_h = \sum_{h'} (Re(k_{h'z})/Re(k_{hz}))[\,|\,\mathcal{R}_{h'h}\,|^2 + |\,\mathcal{T}_{h'h}\,|^2\,],\tag{5}$$

where $Re(X)$ means the real part of X. In (5) k_{hz} is the z or perpendicular component of the wave number of the incident wave $\mathbf{k}_h$, and $k_{h'z}$ the same for the outgoing waves $\mathbf{k}_{h'}$; $h = 1$ is the 00 beam. Note that above about 120eV the conservation condition is satisfied to better than 10%, indicating greater accuracy for the QD I(E) of the 00 beam above 120eV.

3. Formulation of the Cluster Amplitude Equations

Let a given atom in a layer be the origin atom of a cluster of N atoms, $n = 1$ to N, which will modify the scattering properties of the origin atom ($n = 1$) for an incident plane wave $\mathbf{k}_h$. In the cluster calculation the plane wave will strike only atom 1, and atoms 2 to N will act as passive reflectors of the spherical waves scattered from atom 1. We do not neglect the plane wave striking atoms 2 to N, but we take account of incidence on these atoms only when they are in turn origin atoms of a cluster. Then self-consistent equations for the spherical wave amplitudes on the atoms of the cluster can be formulated with the aid of the expansion (6) which expresses outgoing spherical waves $\mathcal{H}_{L_1}(\mathbf{r} - \mathbf{R}_{n'})$ around atom n' as a sum of incoming spherical waves $\mathcal{I}_{L_2}(\mathbf{r} - \mathbf{R}_n)$ around atom n,

$$\mathcal{H}_{L_1}(\mathbf{r} - \mathbf{R}_{n'}) = \sum_{L_2} D_{L_1 L_2}(\mathbf{R}_{nn'})\,\mathcal{I}_{L_2}(\mathbf{r} - \mathbf{R}_n)\tag{6}$$

$$\mathcal{H}_L(\mathbf{r}) \equiv i^{l+1} Y_{lm}(\hat{\mathbf{r}})\, h_l^{(1)}(kr)\tag{7}$$

$$\mathcal{I}_L(\mathbf{r}) \equiv i^l Y_{lm}(\hat{\mathbf{r}})\, j_l(kr)\tag{8}$$

In (7) and (8) $Y_L(\hat{\mathbf{r}}) \equiv Y_{lm}(\hat{\mathbf{r}})$ is a spherical harmonic of the direction of the unit vector $\hat{\mathbf{r}}$, and $\mathbf{R}_{nn'} \equiv \mathbf{R}_n - \mathbf{R}_{n'}$; $j_l(kr)$, $h_l^{(1)}(kr)$ are spherical Bessel functions and k is the complex magnitude of the wave vectors $\mathbf{k}_h$ at energy E in the constant potential V_0. The propagator $D_{L_1 L_2}(\mathbf{R}_{nn'})$ is given by

$$D_{L_1 L_2}(\mathbf{R}_{nn'}) = 4\pi(-)^{m_1} \sum_{L_3} C_{\bar{L}_1 L_2 L_3}\,\mathcal{H}_{L_3}(\mathbf{R}_{nn'})\tag{9}$$

where $C_{L_1 L_2 L_3}$ is a Gaunt coefficient[6]. Then the equations for the B's are

$$B_L^{(nh)} - t_l \sum_{n', n' \neq n} \sum_{L_1} D_{L_1 L}(\mathbf{R}_{nn'}) B_{L_1}^{(n'h)} = t_l A_L^{(nh)} \delta_{n1}\tag{10}$$

$$A_L^{(nh)} = 4\pi(-)^m Y_{\bar{L}}(\hat{\mathbf{k}}_h)\tag{11}$$

where n = 1 to N, t_l is the t-matrix element for scattering from the spherical atoms, and the inhomogeneous term comes from the expansion

$$\exp(i\mathbf{k}\cdot\mathbf{r}) = \exp(i\mathbf{k}\cdot\mathbf{R}_n)4\pi \sum_L (-)^m Y_{\bar{L}}(\hat{\mathbf{k}})\,\mathcal{I}_L(\mathbf{r} - \mathbf{R}_n).\tag{12}$$

A simple refinement of (10) takes approximate account of scattering from the atoms of the layer not in the cluster, designated by n''. The incident plane wave on atom 1 scatters spherical waves to the atoms n'', which scatter them back to the atoms of the cluster. The resulting inhomogeneous terms in the nth equation of (10) are given by

$$t_1 \sum_{n''} \sum_{L_2} t_{l_2} \sum_{L_1} t_{l_1} A_{L_1}^{(1h)} D_{L_1 L_2}(\mathbf{R}_{n''1}) D_{L_2 L}(\mathbf{R}_{nn''}) \tag{13}$$

where $A_{L_1}^{(lh)}$ is given by (11).

The solution of the $N(l_{max}+1)^2$ equations (10), where l_{max} is the index of the largest significant phase shift, gives a set of outgoing wave amplitudes $B_L^{(nh)}$ on the atoms of the cluster. If we superpose clusters with origin atoms at all atoms of the layer with the phase factor $\exp(i\mathbf{k}_h^+ \cdot \mathbf{R}_n)$ given in (12), the sum over the layer is analytically converted into beams $\mathbf{k}_{h'}^+$ with amplitudes[7]

$$\mathcal{R}_{h'h} = \frac{2\pi i}{\mathcal{A} k k_{h'z}} \sum_{L} Y_L(\hat{\mathbf{k}}_{h'}^-) \sum_{n'=1}^{N} B_L^{(n'h)} \exp(-i\mathbf{k}_{h'}^- \cdot \mathbf{R}_{n'}) \tag{14}$$

$$\mathcal{T}_{h'h} = \delta_{h'h} + \frac{2\pi i}{\mathcal{A} k k_{h'z}} \sum_{L} Y_L(\hat{\mathbf{k}}_{h'}^+) \sum_{n'=1}^{N} B_L^{(n'h)} \exp(-i\mathbf{k}_{h'}^+ \cdot \mathbf{R}_{n'}). \tag{15}$$

Then (14) and (15) provide the modified $\mathcal{R}$ and $\mathcal{T}$ matrices to use in (1). Note that we do not expand the outgoing spherical waves from atoms $n = 2$ to N around atom 1. Such an expansion generates an effective t-matrix for the entire cluster, no longer diagonal, which is formally correct and sometimes used [8]. However these expansions introduce an infinite number of L values in the components of the effective t-matrix, and we lose the important simplification that scattering from an atom at given E generates only a finite number of scattered spherical waves. Hence in (14) and (15) we retain the spherical wave amplitudes on the cluster atoms as a distributed source.

We must note, however, that if the superposition of clusters over the layer is carried out directly on the equations (10), the spherical wave amplitudes on the atoms of the cluster differ only by a phase factor, and the equations collapse into $(l_{max}+1)^2$ equations with coefficients that are structure constants with only a finite number of terms N. The equations are then equivalent to the standard layer-KKR equations with a cutoff in the structure-constant sums[9]. If we must solve more equations than the usual FD procedure to improve the $\mathcal{R}$ and $\mathcal{T}$ matrices over the kinematic forms, we have not gained in efficiency. The value of the cluster equations (10) must come from approximations that reduce the calculation for complex structures. One possibility is to solve (10) for $N = 2$ with the origin atom paired with each neighbor with which significant multiple scattering occurs. The amplitudes on the origin atom would then be averaged over all pairs, and the complete coherent set of amplitudes used in (14) and (15).

The authors thank Y.S. Li for programming assistance and gratefully acknowledge partial support of this work by the National Science Foundation with grant DMR8709021.

References.

1. See, for example, N. Bickel and K. Heinz, Surf. Sci. **163**, 435 (1985) and earlier references therein.

2. The notation $k_{i}^{\pm}$ for beam wavevectors in an attenuating medium is defined in *Low-Energy Electron Diffraction* by M.A. Van Hove, W.H. Weinberg, and C.-M. Chan, (Springer 1986), p. 98 (where it is called $k_{g}^{\pm}$).

3. Ref. 2, Chapt. 5 relates the t matrices of an atom and a crystal.

4. X.-G. Zhang and A. Gonis, Phys. Rev. Lett. **62**, 1161 (1989).

5. See Ref. 2, pp. 103, 127 for kinematic formulas.

6. The compact notation for spherical waves, the derivation of the propagator D, and the Gaunt coefficients are given by D.W. Jepsen, F. Jona and P.M. Marcus, Phys. Rev. B5, 3933 (1972) in the derivation of the layer-KKR equations. More detailed mathematical discussion is given by P.M. Marcus in *Proc. Conf. on Determination of Surface Structure by LEED*, ed. by P.M. Marcus and F. Jona, (Plenum 1984).

7. The transformation from spherical wave sums over a plane lattice to plane wave sums over beams is given in the two publications in Ref. 6.

8. As in S. Andersson and J.B. Pendry, J. Phys. C13, 3547 (1980) and in the paper by P.M. Marcus in Ref. 6.

9. J.N. Andersen and D. Adams in the *Proc. of the International Seminar on Surface Structure Determination by LEED and Other Methods*, ed. by K. Müller and K. Heinz, (Erlangen 1985), p.99, report that limitation of intralayer scattering to near neighbors gave good results, better than QD.

A Video-LEED System Based on a Personal Computer and Frame-Grabber

D.L. Adams, S.P. Andersen, and J. Buchhardt

Institute of Physics, Aarhus University, DK-8000 Aarhus C, Denmark

Abstract. A video-LEED system for rapid measurement of LEED patterns and intensities has been constructed from widely-available commercial components. The design and operation of the system is described.

1. Introduction

A system for rapid and automatic measurement of LEED intensities via digitization of a video image of the LEED pattern on the fluorescent screen of a LEED optics was first described by Heilmann et al. in 1976 /1/ and the usefulness of the method was convincingly demonstrated in some of the first applications /2/. More recent developments of the Erlangen system have been reviewed by Heinz /3/. An alternative implementation of the method has been described by Jona et al ./4/.

The video-LEED system described here has the same general features as systems described more recently by Anderegg and Thiel /5/ and by Ogletree et al. /6/, but is based on a PC/AT computer and adaptors compatible with the AT bus rather than a PDP-11 computer and Q-bus adaptors. The present system shares with these latter systems the facilities of control by high-level language programs, and of digitization of the full video image, so that the complete information content of the LEED pattern is available.

2. System Components and Configuration

The main components of the system are: a reverse-view LEED optics and electronics (Omicron); a JAI 733 SIT video-camera (Jørgen Andersen Ingeniør a/s); and a PC/AT computer (IBM), with DT2851 frame-grabber and DT2801 A/D (and D/A) adaptor cards (Data Translation) installed in the computer. The functions of the adaptor cards can be controlled by Fortran programs running in the PC/AT which access machine-language subroutines in subroutine libraries DT-IRIS and PCLAB (Data Translation).

The LEED optics contains a miniature electron gun (12mm diameter), 4-retarding grids, and a phosphor/tin oxide-coated glass fluorescent sceen, which enables the LEED pattern to be viewed from the rear of the optics. The electron gun and connections block about 2% of the LEED pattern.

Springer Series in Surface Sciences, Vol. 24 **The Structure of Surfaces III**
Editors: S.Y. Tong · M.A. Van Hove · K. Takayanagi · X.D. Xie
© Springer-Verlag Berlin, Heidelberg 1991

The electron beam voltage is programmed and read via the 12-bit A/D converter, which is interfaced to the LEED electronics via home-made 1:1 isolation amplifiers. The output filter stage of the electron beam voltage supply has been modified to reduce the settling time from $\sim$100ms to $\sim$40ms. The electron beam voltage can be set with a precision of $\pm$0.1V in the range 0 - 500V using the 0 - 5V range of the A/D converter.

The video camera contains an RCA 4804/H silicon intensifier target (SIT) phototube with maximum sensitivity of $6{\times}10^{-5}$ lux faceplate illumination and a dynamic range of about 300. The response of the phototube falls off by $\sim$20% from the center of the active area to the edge. Thus accurate relative measurements of the intensities in different parts of the LEED pattern require that the response of the tube be calibrated. This is carried this out by recording a digital image of a 99.5% uniform light source (Labsphere Uniform Light Source 600). The calibration has proved to remain valid during two years of use. The camera was supplied with manual control of black level and gain of the video amplifier. The gain is switch selectable between two adjustable levels. The camera is usually used with a Cosmicar 25mm f/1.4 lens with 1.5mm extension ring at a distance of $\sim$33cm from the focal point of the LEED optics, so that the LEED pattern just fills the image field of the camera. An aperture of f/8 is used under typical operating conditions. The camera is mounted on a simple home-made x-z goniometer to allow accurate positioning of the camera with respect to the axis of the LEED optics.

The video image of the LEED pattern is digitized in real time (40ms per full frame) using the frame grabber, which includes a 10MHz a/d converter and two 250Kbyte frame buffers. The pattern is an inscribed circle in the video frame, which is digitized with resolution of 512 x 512 x 8-bit pixels, giving an angular resolution of $\sim$0.2° per pixel. The frame grabber can continuously transmit digitized images to a video monitor via a frame buffer. Alternatively, a single image frame, or the average of several successive frames, can be acquired to a frame buffer and displayed. The contents of one of the onboard frame buffers can be processed simultaneously with acquisition of an image to the second buffer. The frame buffers are memory-mapped to the extended memory space of the PC/AT and can be accessed by a program running in the AT. Thus the individual pixel intensities in an image frame and, in particular, the intensities in a window in the frame centered on an arbitrary LEED beam spot, can be measured. Input and output look-up tables can be modified, and different bits can be masked, allowing the superposition of real-time images and graphical overlays. The effective pixel aspect ratio of the combination of video camera and frame-grabber is 1:1.5.

3. Data Acquisition and Processing

The system functions which have been implemented include: 1) calibration of the magnification of the optical system and frame-grabber, by determination of the positions of LEED spots in the digitized image for a crystal of known 2-D surface structure. This calibration forms the basis for calculation (RMS deviation ~4 pixels) of the position of an hk beam in the digitized image. 2) Rapid alignment of the crystal (to ±0.5°) by superposition of a real-time video image of the LEED pattern with an image calculated for a given electron-beam/crystal geometry. 3) Simultaneous (to within the 40ms digitization time) measurement of intensity-energy spectra for an arbitrary number of LEED beams in an arbitrary energy range on a reproducible, arbitrarily large, discrete energy grid. 4) Measurement of k-space intensity-contours at fixed electron energy.

The algorithm for measurement of intensity-energy spectra is shown schematically in Fig. 1. Prior to the measurement, various starting conditions including choice of beams to be measured, energy range and grid,

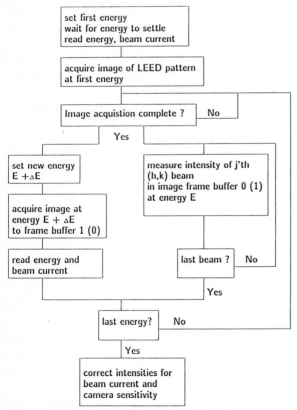

Fig. 1. Flowchart of the algorithm for measurement of intensity-energy spectra

and camera gain can be set interactively by the operator and the values written to a data file. Choice of beams is carried out at the highest energy of interest with the help of a real-time image of the LEED pattern. Superposition of this pattern with a stored image of the region of the pattern blocked by the electron gun enables automatic deselection of beams which fall into this dead area to be performed.

In the loop over energies, processing of the image acquired at the current energy is carried out in parallel with the setting of the next energy and the acquisition of the next image. Thus in most cases of interest the ~80ms per energy required for these two operations does not contribute to the total time for the measurement of a set of intensity-energy spectra. In the loop over beams, the position of a beam is calculated and the pixel intensities in a window (typically of size 15 x 29) centered on the beam are transferred from the frame buffer to the AT's memory. The position of maximum intensity is determined (and saved) and a new window centered on this position is defined and the pixel intensities in this window are transferred to the AT. Finally the spot intensity is determined by summing a central part of the window (typically 15 x 23) and corrected for the background, which is determined by summing the top and bottom borders of the spot window. The time required for determination of a beam's intensity is ~40ms, but this can be reduced to ~20ms if the spot positions for all beams at all energies are known from a previous run. It is emphasized that the beam-tracking algorithm is completely stable. The calculation of each beam's position at each energy relies only on the calibration described in function 1) above.

At the end of the loop over energies the beam intensities are normalized for the variation of beam current with energy and are corrected for the variation in camera sensitivity with spot position. The intensity-energy spectra and spot positions are then optionally written to files and the spectra are plotted on the AT's monitor.

A particularly useful consequence of the near-simultaneous measurement of beam intensities follows if possible adsorbed contaminants do not affect the symmetry of the intensities. In this event, the process of adjusting the crystal/electron beam geometry to set normal incidence, by minimizing χ^2 for the comparison of intensity-energy spectra of symmetry-equivalent beams, is independent of the surface condition. A comparison of measured spectra for three pairs of symmetry-equivalent beams from Al(111) is shown in Fig. 2.

The analysis of intensity-energy spectra measured with the present system for clean and hydrogen-covered W(310) is described elsewhere in these proceedings /7/.

Finally, an example of a contour plot of LEED pattern intensities is shown in Fig. 3. The pattern is for the Al(111)−($\sqrt{3}$x$\sqrt{3}$)R30°−K structure.

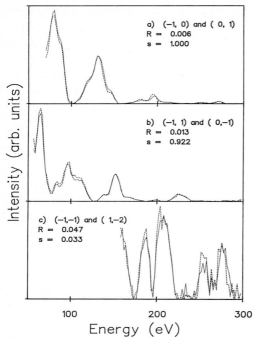

Fig. 2. (a - c) Intensity-energy spectra for 3 pairs of symmetry-equivalent beams from clean Al(111). The R values /7/ for the comparisons of the pairs of beams are noted on the plots, as are the factors s by which the plotted intensities must be multiplied to bring all the beams on to the same intensity scale.

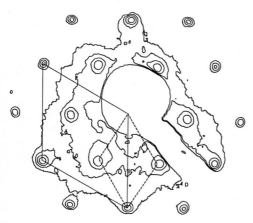

Fig. 3. Contour plot of LEED pattern for Al(111)$-(\sqrt{3}\text{x}\sqrt{3})$R30°$-$K at 123eV and 130K. The substrate and overlayer unit cells are indicated by full and broken lines respectively. The area blocked by the electron gun and connections is evident in the figure.

160

Acknowledgements

We gratefully acknowledge useful discussions with Professor Klaus Heinz on the choice of video camera. This work has been supported by the Danish Natural Science Research Council and the Center for Surface Reactivity.

References

1. P. Heilmann, E. Lang, K. Heinz and K. Müller, Appl. Phys. 9 247 (1976).
2. K. Müller, E. Lang, L. Hammer, W. Grimm, P. Heilmann and K. Heinz, in *Determination of Surface Structure by LEED*, eds. P. M. Marcus and F. Jona (Plenum, New York, 1984).
3. K. Heinz, Progress in Surface Science 27 239 (1988).
4. F. Jona, J. A. Strozier and P. M. Marcus, in *The Structure of Surfaces*, eds. M. A. Van Hove and S. Y. Tong (Springer, Berlin, 1985).
5. J. W. Anderegg and P. A. Thiel, J. Vac. Sci. Technol A4 1367 (1986).
6. D. F. Ogletree, G. A. Somorjai and J. E. Katz, Rev. Sci. Instrum. 57 3012 (1986).
7. D. L. Adams and S. P. Andersen, these proceedings.

One-Beam RHEED for Surface Structure Analysis

A. Ichimiya

Department of Applied Physics, School of Engineering,
Nagoya University, Chikusa-ku, Nagoya 464-01, Japan

Abstract. Surface normal components of atomic positions are determined by RHEED intensity rocking curve analysis at one–beam conditions at which the azimuthal angle of the incident beam direction is set at several degrees off from a certain crystallographic direction in order to avoid simultaneous reflections. The rocking curves obtained by one–beam calculations are examined by many–beam dynamical calculations at the off angle incidences. From 37–beam calculations the one–beam condition for Si(111) surface is realized near 7° off the $[11\bar{2}]$ direction. For Si(001) and Ag(001) surfaces the one–beam condition is obtained near 22° off the [110] direction by 43–beam calculations.

1. Introduction

For surface structure determination by reflection high energy electron diffraction(RHEED), intensity rocking curves are analyzed by RHEED dynamical calculations[1-6]. In ordinary cases of the RHEED intensity analysis[7-12], a trial and error approach with enormous RHEED dynamical calculations is imposed for the structure determination as well as the method of LEED dynamical analysis. High energy electrons are scattered dominantly in the forward direction by atoms. Therefore dynamic diffraction mainly occurs in a forward direction. Using this feature, it is possible to choose an orientation of the incident beam at which electrons are diffracted mainly by lattice planes parallel to the surface. Here this diffraction condition is named the one–beam condition[13], because the main diffraction beam is simply the specular one. At this condition a rocking curve of the specular reflection intensity is a function of surface normal components of atomic positions, but scarcely depends on lateral components of them. Therefore surface normal components of atomic positions of surface layers are determined by dynamical calculation analysis of a one–beam rocking curve with short computation times. We can analyze easily structure changes by adsorption, epitaxial growth and phase transition processes by the one–beam RHEED analysis[10-15]. Recently

Springer Series in Surface Sciences, Vol. 24   **The Structure of Surfaces III**
Editors: S.Y. Tong · M.A. Van Hove · K. Takayanagi · X.D. Xie
© Springer-Verlag Berlin, Heidelberg 1991

Zhao *et al.*[5] have shown that many–beam effects from higher order Laue zones are significant on RHEED intensity rocking curves. Therefore we examined the reliability of the one–beam RHEED method by many–beam dynamical calculations, in the cases of Si(111), Si(001) and Ag(001) surfaces.

2. One–beam dynamical calculation

The wave function of fast electrons at the one–beam condition is given by

$$\psi(\mathbf{r}) = \varphi(z) \exp(i\mathbf{K_t} \cdot \mathbf{r_t}),\tag{1}$$

where z and $\mathbf{r}_t$ are surface normal and parallel components of the position vector $\mathbf{r}$, and $\mathbf{K}_t$ the surface parallel component of the wave vector. Using eq.(1) the Schrödinger equation for the electrons becomes

$$-\frac{\hbar^2}{2m}\frac{d^2}{dz^2}\varphi(z) - eV(z)\varphi(z) = E_z\varphi(z),\tag{2}$$

where $V(z)$ is a complex mean potential including an imaginary potential at z, E_z the surface normal component of the kinetic energy of incident electrons, and m, $\hbar$ and e are the ordinary physical constants. According to the multi-slice RHEED dynamical theory[3] the transfer matrix $\mathbf{P}_j$ at the j-th slice is

$$\mathbf{P_j} = \begin{pmatrix} \tau_j & \rho_j \\ \rho_j & \tau_j \end{pmatrix} \begin{pmatrix} \exp(-i\gamma_j \cdot \Delta z_j) & 0 \\ 0 & \exp(i\gamma_j \cdot \Delta z_j) \end{pmatrix} \begin{pmatrix} X_j & Y_j \\ Y_j & X_j \end{pmatrix},\tag{3}$$

where

$$\tau_j = \Gamma + \gamma_j, \qquad \rho_j = \Gamma - \gamma_j,$$

$$X_j = \frac{\Gamma + \gamma_j}{4\Gamma\gamma_j}, \qquad Y_j = -\frac{\Gamma - \gamma_j}{4\Gamma\gamma_j},$$

$$\Gamma = \sqrt{(2m/\hbar^2)E_z}, \qquad \gamma = \sqrt{\Gamma^2 + (2me/\hbar^2)V(z_j)}$$

and

$$\Delta z_j = z_j - z_{j-1}.$$

RHEED intensity is obtained easily from products of the transfer matrices as described in the previous paper[3].

3. Many–beam effects on one–beam rocking curves

The incident beam directions at the one–beam condition depend on surface indices and incident electron energies. For the Si(111) surface the

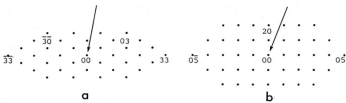

Fig. 1 (*a*) A set of reciprocal rods of Si(111) surface. (*b*) A set of reciproc rods of Si(001) and Ag(001) surfaces.

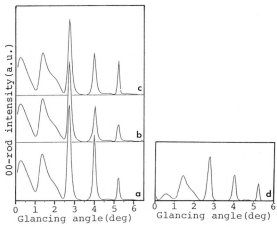

Fig. 2 Rocking curves from Si(111)7×7 surface at one–beam conditions. (*a*) One–beam rocking curve (*b*) Many–beam rocking curve at 7.2° o [11$\bar{2}$] (*c*) Many–beam rocking curve at 7.5° off [11$\bar{2}$] (*d*) Experimen tal rocking curve at 7.5° off [11$\bar{2}$]

incident beam direction of 10keV electrons was experimentally set at near 7° off the [11$\bar{2}$] direction[10-13]. Therefore rocking curves at azimuthal angles near 7° off [11$\bar{2}$] for Si(111) were calculated with 37 beams shown in fig. 1*a* in order to examine effects of simultaneous reflections on the one–beam rocking curve. For the Si(001) surface, on the other hand, the one–beam condition was obtained at an incident beam direction set at about 22° off the [110] direction for 10keV electrons[14]. In this case the rocking curves were calculated with 43 beams shown in fig. 1*b*. The dynamical calculations were carried out with real and imaginary poten- tials calculated from Doyle and Turner's table[16] and a formula proposed earlier[13]. Fig. 2 shows rocking curves obtained by the one–beam and 37–beam calculations(called one–beam and many–beam rocking curves respectively) at one–beam condition from Si(111)7×7 surfaces with the Dimer-Adatom-Stacking fault(DAS) structure[17]. In the calculations the surface normal components of atomic positions used were those obtained previously[13] and the lateral components were those obtained by Tong et al[18]. In these calculations, however, the rocking curve of the specular beam hardly depends on the lateral components. The many–beam rocking curves are very similar to the one–beam one: The peak and shoulder positions of the many–beam curves are the same as those of the one–beam curve, and there are no extra peaks appearing in the curves b and c. Peak and shoulder heights of the many–beam curves are not significantly different from those of the one–beam one. In most of the cases, we analyze a surface structure using peak and shoulder positions. Therefore peak height difference at a glancing angle of about 4° is not so significant. This result confirms the reliability of one–beam analysis for surface structures such as Si(111)"1×1" at least[12]. For the Si(111)7×7 case there are many fractional order spots. Since experimental one–beam rocking curves at several one–beam conditions[13,19] are not significantly different, it is expected that dynamic diffraction between fractional and integral order beams is negligibly weak at one–beam conditions. The curve at 7.5° off [11$\bar{2}$] is in very good agreement with an experimental curve shown in fig. 2d, obtained at the same azimuthal angle.

For the Si(001) surface the many–beam effect is very significant at off axis incident directions. From experimental observations for the Si(001) surface, the one–beam condition was realized at an angular region near 22° off the [110] direction. Fig. 3a shows rocking curves obtained by one–beam and 43–beam calculations at one–beam conditions. In these cases, weak disturbance by many–beam dynamical effect is detected in the rocking curves at a glancing angle of about 3°. Such weak disturbance seems to give no significant effect for one–beam analysis if we examine carefully a many–beam calculation at the final stage of an analysis. For the Ag(001) surface the one–beam condition was obtained at a very narrow angular region near 22° off the [110] direction for 20 keV electrons as shown in fig. 3b. However the global profiles of many–beam rocking curves are very similar to that of the one–beam one. Therefore careful analysis at the one–beam condition leads to reliable results for surface structure. As mentioned above, with careful use of the one–beam method, surface structure analysis is easily carried out with short computation times.

In RHEED experiments the one–beam conditions are determined empirically as strong oblique Kikuchi lines that do not cross on the specular spot in a range of glancing angles of the incident beam for rocking curve measurements. For Si(111) and Si(001) surfaces, it has been confirmed that many–beam effects are not significant at the one–beam conditions

165

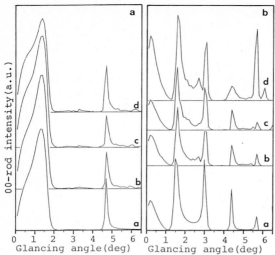

Fig. 3 (*a*) Rocking curves from Si(001) at 10keV electrons at one–beam conditions: curve *a* is a one–beam rocking curve, curve *b* is a many-beam rocking curve at 21.5° off, curve *c* is at 22° off and curve *c* is at 22.8° off the [110] direction. (*b*) Rocking curve from Ag(001 at 20keV: curves from *a* to *d* are at the same condition as those o Si(001).

which are determined empirically from Kikuchi pattern configurations in RHEED.

References

[1] N. Masud and J.B. Pendry : J. Phys. C9, 1833(1976)

[2] P.A. Maksym and J.L. Beeby : Surf. Sci. 110, 423(1981)

[3] A. Ichimiya : Jpn. J. Appl. Phys. 22, 176(1983); 26, 1365(1985)

[4] L.M. Peng and J.M. Cowley : Acta Cryst. A42, 545(1986)

[5] T.C. Zhao, H.C. Poon and S.Y. Tong : Phys. Rev. B38, 1172(1988)

[6] G. Meyer-Ehmsen : Surf. Sci. 219, 177(1989)

[7] J.V. Ashby, N. Norton and P.M. Maksym : Surf. Sci. 77, 131(1978)

[8] A. Ichimiya and S. Mizuno : Surf. Sci. 191, L765(1987)

[9] M.G. Knibb and P.A. Maksym : Appl. Phys. A46, 25(1988)

[10] A. Ichimiya, S. Kohmoto, T. Fujii and Y. Horio : Appl. Surf. Sci. 41/42, 82(1989)

[11] S. Kohmoto, S. Mizuno and A. Ichimiya : Appl. Surf. Sci. 41/42, 107(1989)

[12] S. Kohmoto and A. Ichimiya : Surf. Sci. 223, 400(1989)
[13] A. Ichimiya : Surf. Sci. 192, L893(1987)
[14] T. Makita, S. Kohmoto and A. Ichimiya : Surf. Sci. in press
[15] H. Nakahara and A.Ichimiya : Surf. Sci. in press
[16] P.A. Doyle and P.S. Turner : Acta Cryst. A24, 390(1968)
[17] K. Takayanagi, Y.Tanishiro, M. Takahashi and S. Takahashi : Surf. Sci. 164, 367(1985)
[18] S.Y. Tong, H. Huang, C.M. Wei, W.E. Packard, F.K. Men, G. Glanden and B. Webb : J Vacuum Sci. Techol. A6, 615(1988)
[19] H. Nakahara and A. Ichimiya : Proceedings of First International Conference on Epitaxial Growth(Budapest) in press

Direct Observation of Surface Structures by a High Resolution UHV-SEM

*S. Ino, A. Endo, and H. Daimon**

Department of Physics, Faculty of Science, University of Tokyo, Bunkyo-ku, Tokyo 113, Japan
* Present address: Faculty of Engineering Science Osaka University, Toyonaka Osaka 560 Japan

Abstract: We have constructed a high resolution UHV-SEM(scanning electron microscopy)apparatus which can obtain 1.7×10^{-10} Torr through baking all the parts related to the ultra-high vacuum. In the SEM mode, a resolution of about 5Å has been obtained for an image taken from deposited Au particles on a carbon film. In the STEM (scanning transmission electron microscopy) mode, a lattice image(1.43Å) of Au(220) has been clearly observed.

Using this UHV-SEM, we found that the domain contrast of the two dimensional surface structures which are formed by metal deposition less than one monolayer on a Si(111) surface can be observed clearly in detail. For the Si(111) surface on which about 1/4 monolayer of Au was deposited, three kinds of areas were recognized. They are of bright contrast(B), corresponding to the 5x2-Au structure, dark contrast(D) corresponding to the 7x7 structure and middle contrast(M). The last has never been reported in the LEED and RHEED studies. For a Ag deposited Si(111) surface, similar contrasts B, D and M were observed but their external forms of boundaries were very different from those of the Au-deposited Si(111) surface.

1. Introduction

Osakabe et al.[1] have shown that REM(reflection electron microscopy) is very useful to observe the surface images, especially for the observation of steps, imperfections, adsorption and desorption processes at clean surfaces. In this method, however, detailed observations have been difficult because of the image shortening effect due to the grazing incidence of the electron beam.

Before this REM experiment we have shown that RHEED is very powerful for surface structure studies just like LEED[2] and has been successfully used thereafter[3]. During such RHEED experiments, we have found that TRAXS(total reflection angle X-ray spectroscopy) has high sensitivity for surface elemental analysis and named it as RHEED-TRAXS[4,5]. Applying this method we have been studying the adsorption process, the desorption process, and the growth process of metals on Si and Ge surfaces[6,7].

If a surface is hit by a very fine electron beam whose diameter is comparable to an atomic one, it may be possible to detect the X-rays emitted from a single atom or from a few atoms adsorbed on the surface. Following this idea we planned to make a high resolution UHV-SEM about seven years ago[8] immediately after the discovery of the TRAXS method and recently

Springer Series in Surface Sciences, Vol. 24   **The Structure of Surfaces III**
Editors: S.Y. Tong · M.A. Van Hove · K. Takayanagi · X.D. Xie
© Springer-Verlag Berlin, Heidelberg 1991

it has been nearly perfected[8,9]. Section 2 describes this instrument.

Using this UHV-SEM, we tried to see the two dimensional surface structures and succeeded in observing their domain structures clearly. In section 3, some SEM images taken from Si(111) surface on which Au and Ag were deposited are shown with some explanations.

2. Construction of an UHV-SEM

For the construction of the UHV-SEM, we aimed to attain the following six conditions[8], (1) a fine electron beam comparable to the atomic size, (2) an electron beam current of 1×10^{-10}A, (3) high stability for electron beam position, (4) low drift for electron beam position, (5) UHV of 1×10^{-10} Torr, (6) X-ray detection under the TRAXS condition.

Figure 1(a) shows a section of the UHV-SEM newly constructed. To obtain an electron beam as fine as possible, we chose a higher acceleration voltage than in the usual SEM in which the acceleration voltages are generally less than 30keV. In this SEM, the acceleration voltage is variable from 40keV to 100keV.

Spherical aberration and chromatic aberration coefficients at 100keV are 0.6mm and 1.2mm, respectively. When it was operated at 100keV, the lattice image of Au(220) with the STEM mode has been clearly observed. Then the resolution of 1.43Å was obtained. In the SEM mode in which secondary electrons are detected, a resolution of 5-7A was obtained for an image of gold particles deposited on a carbon film.

For the vacuum seal of the microscope column, only metal O rings and Cu-gaskets were used and all the parts related to ultra-high vacuum can be baked out at about 160℃ . Thus this SEM can obtain 1.7×10^{-10} Torr at present. The main residual gases were H_2, H_2O, CO, CO_2 and their partial pressures are 4.1, 6.2, 1.6, 1.3×10^{-11} Torr, respectively.

As shown in Fig.1(a), an X-ray detector was mounted on a plane which is the extended plane of the sample surface. That is, when the electron beam is incident perpendicularly upon the sample surface, the X-rays excited nearly perpendicularly to the electron beam can be detected under the TRAXS condition. The X-rays emitted from the sample surface propagate into air through a thin Be window and are detected by using a conventional Si(Li) type detector.

This UHV-SEM has been developed as a joint research project of Tokyo University and JEOL[8,10]. At present the above six conditions are nearly attained. They are, however, not enough and we are further improving.

Figure 1(b) shows a section which is perpendicular to the electron beam and through the sample. A sample preparation chamber which can also attain ultra-high vacuum was constructed and was combined with the microscope. This chamber is equipped with a RHEED system, a sample holder which has a sample heating system, an evaporator etc. as shown in Fig.1(c). Thus, we can easily prepare a necessary sample surface by observing the RHEED patterns. The sample was then transferred into the SEM column and the SEM images are observed in detail.

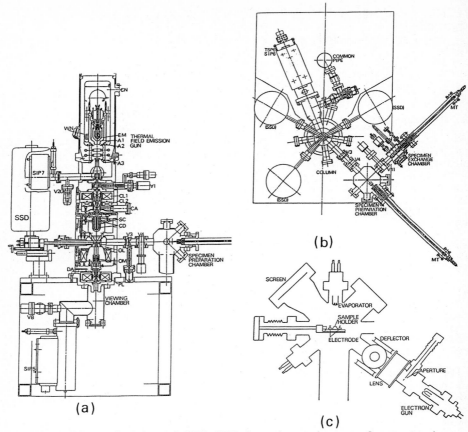

(a)

(b)

(c)

Fig.1.　Newly developed UHV-SEM for observing surface atomic arrangement by applying electron beam TRAXS. (a) A section of the UHV-SEM through the axis parallel to the electron beam. (b) A section of the UHV-SEM through the sample plane which is perpendicular to the electron beam. (c) Sample preparation UHV chamber in which sample cleaning, metal depositions and RHEED observation etc. are possible and the sample can be transferred into the main SEM column.

3. SEM images of Si(111)-Au,Ag surfaces

When 0.2-0.3 monolayer of Au was deposited at 600°C on a clean Si(111) surface showing the 7x7 structure, a RHEED pattern with the 7x7 and 5x2-Au structure spots was observed[3].　After this sample was formed and investigated by RHEED in detail in the sample preparation chamber, it was transferred into the SEM column.

　　Figure 2 shows a typical SEM image taken from such a sample. The acceleration voltage was 80keV.　The electron beam is incident perpendicular to the sample surface.　The secondary electrons were spiraled up by the strong magnetic field of a magnetic lens and gathered and detected by an electron multi-

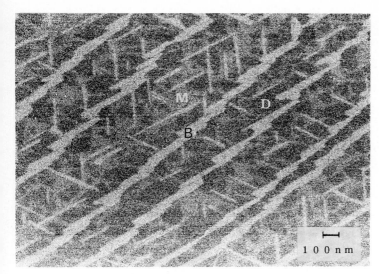

Fig.2. SEM image taken from a Si(111)-Au surface showing the 7x7 and the 5x2-Au structures. About 1/4 monolayer of Au was deposited at 600℃ .

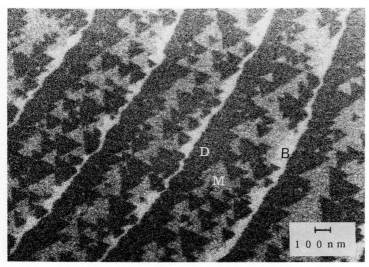

Fig.3. SEM image taken from a Si(111)-Ag surface showing the 7x7 and the √3x√3-Ag structures. About 1/2 monolayer of Ag was deposited at 450℃ .

plier. In Fig.2, three kinds of contrast B, D and M are observed. The contrasts B and D correspond to the 5x2-Au and the 7x7 structures, respectively. The domains with the contrast B appear mainly along the surface steps, but are also formed on the terraces. These are three kinds of elongated domains whose directions make a fixed angle 120° to each other. The origin of the contrast M is not clear at present.

171

When a small amount of Ag, less than 1ML of coverage, was deposited on a clean Si(111) surface held at 450℃ , 7x7 and $\sqrt{3}$ x$\sqrt{3}$-Ag structures were observed simultaneously in a RHEED pattern[11]. Figure 3 shows an example of the SEM image taken from such a surface. Three kinds of domains B, D and M are clearly observed just like in the case of Au. The dark area D corresponds to the 7x7 structure. The boundary of the domain D on the terrace makes regular triangles. The bright areas B may correspond to the $\sqrt{3}$x$\sqrt{3}$-Ag structure, formed along the surface steps and have irregular external forms. The structure of the area M is not clear at present. More detailed results will be published soon[12].

4. Conclusion

We have constructed a high resolution UHV-SEM. It has a sample preparation chamber in which RHEED observation is possible and the sample can be transferred into the SEM column under UHV conditions. By applying it to surface structure observation of the Si(111) surface on which less than 1ML of Au and Ag were deposited, we succeeded for the first time in observing three kinds of domains B, D and M. The domain D corresponds to the 7x7 structure. The domain B corresponds to the 5x2-Au and the $\sqrt{3}$x$\sqrt{3}$-Ag structures for Au and Ag depositions, respectively. For the domains M induced by Au and Ag, we could not give a definite explanation. Such structures have never reported in the LEED and RHEED studies[13,11].

Because the electron beam is incident perpendicularly upon the sample surface, very high resolution is obtained compared with REM. The contrast of the REM images arises by Bragg reflections whereas that of the SEM arises mainly by the differences of atoms. It is worth mentioning that the limit of the detection of the contrast by this SEM reaches about 1/10 monolayer.

Thus, this UHV-SEM is very powerful for surface structure study and applicable for many kinds of studies. We believe that this new method may shed new light on material sciences such as semiconductor, superconductor, and metal sciences.

References

1. N. Osakabe, Y. Tanishiro, K. Yagi and G. Honjo:Surf. Sci. 98, 393 (1980), ibid. 102, 424 (1981), ibid. 109, 353 (1981).
2. S. Ino:Japan. J. Appl. Phys. 16, 891 (1977).
3. S. Ino:Proc.NATO Advanced Research Workshop,June15-19(19-87),Netherlands,"RHEED and Reflection Electron Imaging of Surfaces "NATO ASI Series B 188,3(1988)(Plenum,New York).
4. S. Ino, T. Ichikawa and S. Okada:Japan. J. Appl. Phys. 19,1451 (1980).
5. S. Hasegawa, S. Ino, Y. Yamamoto and H. Daimon:Japan. J. Appl. Phys. 24, L387 (1985).
6. S. Hasegawa, H. Daimon and S. Ino:Surf. Sci.186,138(1987).
7. S. Ino, S. Hasegawa, H. Matsumoto and H. Daimon:Proc. 2nd Int. Con. on the Structure of surfaces(ICSOS II), Amsterdam, The Netherlands, June 22-25, p334 (1987).

8. S. Ino:The Report for "Grant-in-Aid for specially
   Promoted Research" supported by the Ministry of Education,
   Science and Culture of Japan, 0-290 (1988).
9. S. Ino:7th Int. Workshop on Future Electron Devices-
   Superlattice and Quantum Functional Devices-, Oct 2-4,
   Toba Hotel International, Toba, Mie-ken, Japan, p55 (1989).
10. T. Tomita, Y. Kokubo, Y. Harada, H. Daimon and S. Ino:to
    be published at the XIIth Int. Con. for Electron
    Microscopy, Seattle 1990.
11. Y. Gotoh and S. Ino:J. Appl. Phys. 17, 2097(1978).
12. A. Endo, H. Daimon and S. Ino:to be published at the
    XIIth Int. Con.for Electron Microscopy, Seattle 1990.
13. K. Spiegel:Surf. Sci.27,125(1967).

High Resolution Electron Energy Loss Spectroscopy as a New Tool to Determine the Structure of a Surface Reconstructed System

J. Szeftel, J. Colin de Verdière, G.S. Dong, and D. Muller

CEN/Saclay, DPhG/SPAS, F-91191 Gif-sur-Yvette Cedex, France

It is shown by taking advantage of vibrational EELS results that Ni(100)/p(2 x 2)0 is actually reconstructed Ni(100)/c(2 x 2). The reconstructed structure is identified.

1. LEED AND SEXAFS VERSUS VIBRATIONAL EELS IN THE INVESTIGATION OF THE SURFACE STRUCTURE

The determination of the surface structure of a crystal by means of diffraction techniques is a difficult task because the strong multiple scattering undergone by the incoming beam precludes any straightforward structural assignment based on the diffraction pattern. Nevertheless the diffraction pattern provides the size and shape of the two-dimensional (2D) unit-cell and the space-group. However there are in general several structural models consistent with the diffraction pattern and still differing by the number of atoms per 2D unit-cell, and closely related to that, the occurrence of surface reconstruction. Further information can be obtained by calculating the diffraction signal and comparing with the measured one. However, since such calculations involve numerous ill-known parameters describing the particle-matter interaction and the surface geometry, the inferred structural assignments turned out to be wrong in several examples. SEXAFS is a useful complement to diffraction techniques. It enables one primarily to measure the interatomic distances between adsorbate and substrate atoms. But if two distinct structures differ too little, given the SEXAFS sensitivity, by their respective interatomic distances, the SEXAFS measurements do not permit a clear-cut structural identification.

A method taking advantage of the 2D vibrational data to ensure a structural identification will be presented here and exemplified in the case of Ni(100)/0. Adsorbing oxygen on Ni(100) gives rise first to a p(2 x 2) structure which turns into a c(2 x 2) at higher exposure. The well-accepted structures on the basis of diffraction and SEXAFS results /1, 2/ are recalled in fig. 1. Our EELS vibrational results confirm the c(2 x 2) structure and invalidate the p(2 x 2). They show that the p(2 x 2) is actually a reconstructed c(2 x 2).

Springer Series in Surface Sciences, Vol. 24   **The Structure of Surfaces III**
Editors: S.Y. Tong · M.A. Van Hove · K. Takayanagi · X.D. Xie
© Springer-Verlag Berlin, Heidelberg 1991

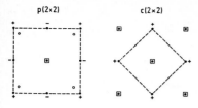

p(2×2)   c(2×2)

Fig. 1 : Commonly accepted structural models for Ni(100)/p(2 x 2)
and Ni(100)/c(2 x 2). The open
rectangles stand for the oxygen
atoms, and the open and black
circles for Ni atoms in the first
and second layers, respectively.
The dashed lines indicate the
respective unit-cells. The + and -
signs represent a static buckling
normal to the (100) surface.

2. THE EELS VIBRATIONAL RESULTS

Our discussion relies on the high-frequency vibrational modes
[3] observed in Ni(100)/0. The corresponding frequencies are
derived by looking at the EEL spectra in fig. 2. Each spectrum
is labelled by the observed LEED pattern. All these modes are
dipole-active and have thus been observed in specular
reflection at the Brillouin zone centre Γ. The modes at
230 cm^{-1} and 420 cm^{-1} in the p(2 x 2) and at 310 cm^{-1} in the
c(2 x 2) had been previously [4] observed. The feature at
230 cm^{-1} in the p(2 x 2) has been shown to consist of a
doublet [5] of dipole-active modes while the loss at 420 cm^{-1}

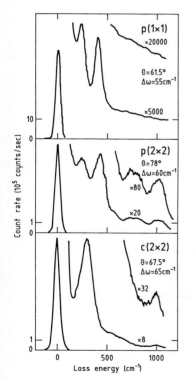

Fig. 2 : HREEL spectra
recorded in Ni(100)/0
at three different O-
coverages. The correspon-
ding observed LEED pattern
is indicated in each
case. Δω indicates the
energy resolution and
θ the incident angle.

is also a doublet which however comprises a single dipole-active component. As for the p(1 x 1) spectrum, the presence of a low O-coverage is witnessed by the two features at 230 and 420 cm^{-1}, characteristic of the p(2 x 2).

3. WORKING OUT THE STRUCTURAL CONCLUSIONS

Two dipole active modes are expected for the c(2 x 2) structure in fig. 1 consistently with the EELS measurements of fig. 2 : i) given its C_{4v} symmetry, the adsorbate can contribute only one such mode per c(2 x 2) unit cell. We assign the loss at 1000 cm^{-1} to this mode ; ii) the second dipole active mode originates from the edge point M of the (1 x 1) Brillouin zone in the (010) direction and is carried to Γ by the c(2 x 2) overlayer induced Brillouin zone folding. Due to symmetry, it can be attributed only to the S_2 resonance [6] at 220 cm^{-1} on clean Ni(100).

Three dipole active modes are expected for the p(2 x 2) structure of fig. 1. This includes the two modes discussed hereabove for the c(2 x 2) case. The oxygen mode keeps practically the same frequency of 1020 cm^{-1} whereas the mode at 310 cm^{-1} undergoes a subsequent shift up to 420 cm^{-1}. In addition the p(2 x 2) structure induces a further folding of the edge point X in the (011) direction onto Γ. This gives rise to an extra dipole active mode which, for symmetry reasons, can only stem from the S_6 phonon observed [7] at 252 cm^{-1} on Ni(100). We assign this mode to the resonance at 230 cm^{-1}. However the origin of the fourth dipole active loss at 770 cm^{-1} cannot be explained within the p(2 x 2) structural model of fig. 1. Accordingly a single dispersion curve is expected for the p(2 x 2) structure of fig. 1 in each case where a doublet is actually observed in fig. 2. This statement of fact indicates that there are rather two oxygen atoms per (2 x 2) unit-cell, instead of one as in the p(2 x 2) model of fig. 1. As this corresponds to the coverage of the c(2 x 2), we are led to look for the reconstruction models of the c(2 x 2) in fig. 1 which give rise to a (2 x 2) structure of the C_{4v} group, as evidenced by the LEED pattern.

By proceeding as in ref. 8 it can be shown that there are only two reconstruction models of the c(2 x 2) which give rise to a (2 x 2) structure of the C_{4v} symmetry. They are represented in fig. 3. They are obtained by freezing, respectively, the A_1 and B_2 phonon at X in the c(2 x 2) Brillouin zone, which produces the static distortion field, characteristic of each (2 x 2) reconstructed structure. Although both $A_1(X)$ and $B_2(X)$ structural models have the same number of adsorbate and substrate atoms per (2 x 2) unit-cell and consequently the same number of vibrational modes, the adsorbate modes are not mapped onto the representations of the C_{4v} group in the same manner. The $A_1(X)$ model has exactly twice more dipole active modes than the $B_2(X)$ one owing to the following structural property : in the $A_1(X)$ model, there are two types of adsorbate site, located at the centre of either a large or a small square, made up of four Ni atoms in the first layer ; by contrast in the $B_2(X)$ model there is only one O-site, located at the centre of a single type of rectangle of

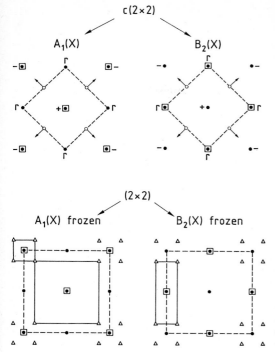

Fig. 3 The upper figure represents the vibrational pattern of the A_1 and B_2 phonons at X in the c(2 x 2). The lower figure gives the (2 x 2) reconstruction models deduced by freezing $A_1(X)$ and $B_2(X)$. The open rectangles, open and black circles, and the dashed lines have the same meaning as in fig. 1. The + and - signs and the arrows in the upper figure denote a motion normal and parallel to the (100) surface respectively while atoms marked r remain at rest. In the lower figure the open triangles indicate the Ni atoms in the first Ni plane and the solid lines designate the rectangle and squares, at the centre of which the adsorbate is sitting.

four Ni atoms in the first layer (fig. 3). Since the dipole active modes arise pairwise at 1020 and 770 cm^{-1} on one hand, and 230 and 180 cm^{-1} on the other hand, we conclude that the structure which so far was believed to be a p(2 x 2) is actually a A_1(X)-like reconstructed c(2 x 2) system.

It must be noticed that the (1020 cm^{-1}, 770 cm^{-1}) doublet consists of two dipole active components whereas the unresolved doublet at 420 cm^{-1} contains only a single one. This shows that the non-dipole active feature [4] around 400 cm^{-1} at Γ originates from a phonon of B_2 symmetry at X in the c(2 x 2), while the loss at 770 cm^{-1} results from a A_1 phonon at X. For, if there would have been a B_2(X) reconstruction instead of a A_1(X) one, a doublet of dipole active modes rather than a singlet would have arisen in the reconstructed (2 x 2) around 400 cm^{-1}, while the loss at 770 cm^{-1} would not have been dipole active.

4. CONCLUSION

The approach, presented here, uses the 2D vibrational data obtained by EELS to secure the structural identification of a reconstructed system. It takes advantage of two basic properties of the 2D phonon spectrum : i) as a phonon dispersion curve is attached to each vibrational degree of freedom of any atom, the number of measured dispersion curves yields in particular the number of adsorbate atoms inside the 2D unit-cell, that is the absolute coverage rate ; ii) even though two distinct structures have the same 2D unit-cell, space group and number of adsorbate and substrate atoms, they still differ by the symmetry properties of the 2D phonons. This has been taken advantage of owing to the EELS selection rules.

This approach applied to Ni(100)/0 has enabled us to propose a reconstruction model which reconciles all LEED, SEXAFS and EELS results obtained in Ni(100)/p(2 x 2)0.

R E F E R E N C E S

1. S.R. Chubb, P.M. Marcus, K. Heinz, K. Müller, Phys. Rev. B, 41, 5417 (1990)
2. L. Wenzel, D. Arvanitis, W. Daum, H.H. Rotermund, J. Stöhr, K. Baberschke, H. Ibach, Phys. Rev. B, 36, 7689 (1987)
3. J. Colin de Verdière, J. Szeftel, P. Soukiassian, at press in Phys. Rev. B
4. J. Szeftel, S. Lehwald, Surf. Sci., 143, 11 (1984)
5. S. Andersson, P.A. Karlsson, M. Persson, Phys. Rev. Lett., 51, 1876 (1983)
6. M. Rocca, S. Lehwald, H. Ibach, T.S. Rahman, Surf. Sci., 138, L123 (1984)
7. M.L. Xu, B. Hall, S.Y. Tong, M. Rocca, H. Ibach, S. Lehwald, J.E. Black, Phys. Rev. Lett., 54, 1171 (1985)
8. J. Szeftel, F. Mila, J. Phys. C, 21, L1131 (1988) ; J. Szeftel, F. Mila, A. Khater, Surf. Sci., 216, 125 (1989)

Auger Photoelectron Coincidence Spectroscopy of Cobalt

S.M. Thurgate

School of Mathematical and Physical Science, Murdoch University,
Murdoch, WA 6150, Australia

Abstract. Auger photoelectron coincidence spectroscopy (APECS) involves measuring an Auger line in coincidence with the corresponding photo-electron line of an X-ray excited spectrum from a surface. Such spectra are free of many of the complicating effects found in conventional spectra [1]. We have measured the $L_{23}VV$ of cobalt in coincidence with the $2 P_{1/2}$ and $2 P_{3/2}$ photoelectrons. The loss of intensity of the L_2VV component is fully accounted for by redistribution of part of the intensity to the L_3VV line via a Coster-Kronig process. The broadening of the L_3VV on the low energy side due to electron-hole scattering processes is clearly seen. The L_3VV component in coincidence with the $2 P_{3/2}$ is narrower than the L_3VV component in coincidence with the $2 P_{1/2}$, probably due to the longer life-time of the final 3 hole state.

1. Introduction

Auger photoelectron coincidence spectroscopy (APECS) is a technique that simplifies Auger and photoelectron spectra. In an APECS measurement, the Auger electron is measured in coincidence with the ejected photoelectron that caused the initial hole. Either the Auger peak [1] or the photo-electron peak [2] can be measured. As it is usually possible to identify the origin of the photoelectron, one can uniquely identify and separate parts of the Auger line. This is of particular advantage if there is more than one Auger process taking place, such as with multi-element systems with overlapping lines, or systems rich with satellites. The coincidence technique discriminates against inelastically scattered electrons, hence the data appear on a flatter background with the low energy broadening removed. The resultant spectra are considerably simplified compared to singles spectra and are hence more amenable to comparison with theory.

The major difficulty with this technique has been low count rates, although Jensen et al.[2] have been able to get good rates by using synchrotron radiation. The rates are low as only those events for which both the photo-electron and Auger electron are collected can contribute to the signal. These events are superimposed on a background from random events. In order to separate the random signal from the true signal, it is necessary to have very good timing resolution from both spectrometers.

We have recently completed construction of a pair of analysers suitable for measuring APECS spectra. These differ from conventional analysers in that they make use of an idea first suggested by Völkel and Sandner [3] to improve the timing resolution of deflecting analysers. The analysers achieve very good timing resolution (~ 1.6 nS [4]) without a consequent degradation of energy resolution.

We have used these analysers to measure the L_3 VV Auger spectra of cobalt. The coincidence data shows that the L_3 VV spectra has a component which appears in coincidence with the $2\ P_{1/2}$ photoelectron. This is due to the $L_2 L_3 V$ Coster-Kronig process. This process is responsible for the apparent reduction in intensity of the L_2 VV component in the singles spectrum.

2. Experimental

Both the photoelectron and Auger electron analysers were 127° cylindrical deflecting analysers, with a mean radius of 61.5 mm. They used micro-channel plates (MCPs) inclined at an angle to the electron path at the output slit to detect the electrons, as shown in figure 1. Following the prescription set out by Völkel and Sandner [3], the angle of tilt was chosen to correct for the time of flight differences in different traject-ories through the analysers. A full description of the analysers and electronics can be found in reference [4].

All measurements were made in a UHV system (pressure ~1×10^{-9} Torr). The target was fabricated from a piece of 99.999% cobalt foil. It was formed into a square of dimensions 1×1 mm.

The target was cleaned by ion bombardment. After initial cleaning, it was bombarded for 20 minutes each 24 hours. The analysers and multichannel analyser were controlled by computer.

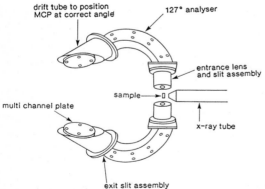

FIG. 1. Layout of the electron analysers with respect to the X-ray tube and sample. Note the position of the MCP (Multi-Channel Plate) detector.

In order to reduce the effect of drift and sample contamination, the Auger analyser was scanned in energy for a number of complete cycles in each 24 hour run.

3. Results

The L_{23} VV - 2 $P_{3/2}$ and L_{23} VV - 2 $P_{1/2}$ spectra, together with the corresponding conventional singles spectra are shown in figure 2 and 3. These data are very similar in overall features to those collected for Cu L_{23} VV [1]. However, they do not show the sharp, atomic-like features found in the Cu spectra. This is due to the smaller hole-hole interaction in Co compared to that in Cu. In Co the hole-hole interaction is estimated to be 1.2 eV [5] while twice the valence band has a width of 6.7 eV. Hence, the two hole

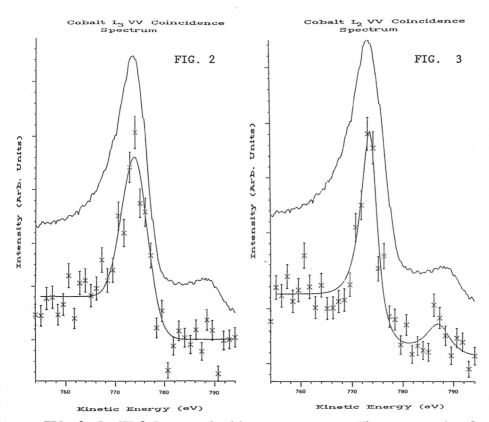

FIG. 2. L_{23} VV-2 $P_{3/2}$ coincidence spectrum. The conventional L_{23} VV spectra is also shown for comparison.

FIG. 3. L_{23} VV-2 $P_{1/2}$ coincidence spectrum. The peak appearing at 774 eV is due to the Coster-Kronig process $L_2 L_3 V$ followed by L_3 VV. The shift and narrowing of this peak compared to the L_3 VV is due to the effect of the valence band hole.

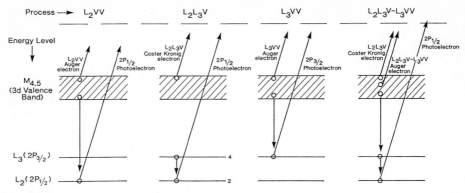

FIG. 4. Processes responsible for the L_{23}VV Auger spectrum in Cobalt.

final state is short-lived as the holes can distribute their energy to other 3d valence band electrons. In the case of Cu, the hole-hole interaction energy is 8.0 eV while twice the valence band has a width of 5.3 eV [1]. Hence, the final state is long-lived as the two holes cannot readily redistribute their energy. The resulting spectra is hence characteristic of Auger spectra from isolated atoms [1]. Figure 4 illustrates the processes responsible for the L_{23}VV Auger spectra. The Auger data shown in figures 2 and 3 have been collected by looking at the coincidence signal with either the 2 $P_{1/2}$ photoelectron or the 2 $P_{3/2}$ photoelectron. The ratio of the L_3VV peak to the L_2VV peak in the singles spectrum is approximately 7.4±1:1. This does not reflect the relative multiplicity of the 2 $P_{3/2}$ to 2 $P_{1/2}$ states as the L_2L_3V Coster-Kronig process causes loss of intensity from the L_2VV line. In this process, rather than an electron from the valence band filling the hole, it is filled from an electron in the 2 $P_{3/2}$ level with consequent emission of an electron from the valence band. The remaining 2 $P_{3/2}$ hole is filled by an electron from the valence band together with emission of an Auger electron. This second process differs from the mainline process as it occurs in the presence of the initial valence band hole.

In figure 3, the ratio of the L_3VV peak to the L_2VV, both in coincidence with the 2 $P_{1/2}$ is approximately 3.8±1:1. Hence, the ratio of the L_3VV to the sum of the L_2VV plus L_2L_3V - L_3VV intensities is approximately 2:1, as would be expected.

The data of figure 2 shows that the spectra are narrower than the corresponding singles spectra. In the L_3VV spectra this reduction in width can be fully attributed to the discrimination of the coincidence experiment against inelastically scattered electrons. This results in a narrowing of the peak on the low energy side and a shifting of the peak centroid towards higher energies. The effect can be simulated, as shown in figure 5. In this figure, a model of the Auger emission is shown in panel (A). This is convoluted with the two different response functions in

182

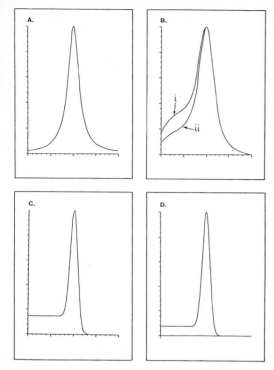

FIG. 5. Effect of change in the response function of the
analyser. The coincidence experiment discriminates against
secondaries, hence changing the response function. The model of
the Auger emission is shown in (A), the two response functions are
shown in (C) and (D) and the convolution of these is shown in (B).
Curve (i) is the convolution of (A) and (C) while curve (ii) is
the convolution of (A) and (D). Note the reduction in the
sensitivity to inelastic electrons reduces the low energy step but
does not eliminate it.

panels (C) and (D). These functions have similar half-widths but
different low energy tails. The tails are meant to model the
effect of inelastic scattering. It is necessary to include these
in the instrument response function as every point in the Auger
line will produce emission that has its own inelastic tail. Thus
the total response function is the sum of the analyser instrum-ent
response function plus the inelastic tail [6]. The convolution of
these two response functions with the model Auger emission is
shown in panel (B). Curve (i) shows the convolution with the
function in panel (C) and curve (ii) shows the convolution with
the curve in panel (D).

  As can be seen from the simulations, the step on the low energy
side of the peak is reduced, but not eliminated, indicating that
electrons that have undergone some inelastic scattering remain in

coincidence with the photoelectron. The sensitivity to scattered electrons we estimate to be reduced by a factor of approximately half in going to the coincidence experiment.

The estimated FWHM of the L_3VV-$2P_{3/2}$ component in coincidence is 6.5 eV. The L_2L_3V-L_3VV-$2P_{1/2}$ component shown in figure 3 has an estimated FWHM of 4^3 eV. The peak centroid is also shifted by .5 eV with respect to the L_3VV-$2P_{3/2}$ component. The narrowing of this line is clear evidence of the partial localization of the initial valence band hole created by the L_2L_3V process and points to a finite value of U_{eff}. This is quite clear as the only difference between the two processes is in the initial valence band hole. It is further evidenced by the .5 eV shift. The inclusion of this somewhat shifted and narrowed component points to the difficulty of measur-ing the valence band width by deconvolution of the Auger lineshape [5]. The reason why the L_2L_3V-L_3VV-$2P_{1/2}$ component should be narrower than the L_3VV-$2P_{3/2}$ is not clear. The localization of the spectator hole may increase the final state lifetime, resulting in a lower overall width. Clearly, a full-blown calculation of the lineshape, including appropriate band structure calculations, is necessary to answer this question unequivocally.

4. Conclusion

The $L_{23}VV$ coincidence spectra of cobalt show a number of interesting facets. The peaks are narrowed due to the reduction in sensitivty to inelastically scattered electrons. The Coster-Kronig component of the L_3VV peak can be removed. The L_2L_3V-L_3VV-$2P_{1/2}$ peak is narrowed by the presence of the spectator hole in the valence band. Hence, all the data is essentially simpler than the corresponding conventional Auger data. However, its full interpretation requires a complete treatment of the Auger process, including band structure calculations.

5. Acknowledgements

I gratefully acknowledge useful and continuing discussions of this work with Andris Stelbovics. The work was supported by the Australian Research Council.

6. References

[1]H.W. Haak, G.A. Sawatzky, L. Ungier, J.K. Gimzewski and T.D. Thomas, Rev. Sci. Inst. 55, 696 (1984).
H.W. Haak, *Auger Photoelectron Coincidence Spectroscopy - A Study of Correlation Effects in Solids* (Ph.D. Thesis), University of Gröningen, The Netherlands (1983).
[2]E. Jensen, R. Bartynski, S. Hulbert, E. Johnson and R. Garrett, Phys. Rev. Lett. 62, 71 (1989).
[3]M. Völkel and J. Sandner, J. Phys. E. 16, 456 (1983).
[4]S.M. Thurgate, B.D. Todd, B. Lohmann and A. Stelbovics, Rev. Sci. Instru. 61, 3733 (1990).
[5]D.M. Zehner, J.R. Noonan and H.H. Madden, J. Vac. Sci. Technol, 20, 859 (1982).
[6]H.H. Madden, Surf. Sci. 126, 80 (1983).

Exact Theory of One-Phonon Amplitudes in Diffractive Atom–Surface Scattering

M.E. Flatté

Department of Physics, University of California, Santa Barbara, CA 93106, USA

Abstract I derive an expression for the probability of creating or annihilating one long-wavelength surface or bulk phonon during a scattering event in a system with diffraction scattering. This expression depends on the *bulk* elastic constants but is independent of force constant changes near the surface and of the details of the atom-target potential. If the inelastic scattering is weak (as defined in the text), the expression is exact.

In a system where the only elastic scattering is specular, the bulk elastic constants of the target and the quantum numbers of the decoupled atom-target system entirely determine the probability for the weak inelastic creation (or annihilation) of a long-wavelength phonon [1]. A heuristic explanation follows of why an expression with similar qualities should also exist for systems with non-specular elastic scattering. In particular, the expression is independent of (1) deviations of force constants from bulk values near the surface of a target uncoupled to atoms and (2) the functional form of the atom-target interaction potential $V(\vec{r}; \{\vec{u}_i\})$, where $\vec{r}$ is the position of the colliding atom and $\vec{u}_i$ the displacement of the i'th target atom.

Displacements in the surface region associated with a phonon of small wave-vector $\vec{q}$ are quasi-rigid for $(qa)^{-1}$ layers perpendicular to the surface, where a is the lattice constant. If $qa \ll 1$, force constant changes in a few layers near the surface are irrelevant to the displacements associated with this phonon.

The functional form of V is irrelevant because to lowest order in qa, the first-order expression for the probability is proportional to

$$\left[\int_{\mathbf{V}} \langle \Psi_o^{(1)} | \vec{r} \rangle \frac{\partial V(\vec{r})}{\partial z} \langle \vec{r} | \Psi_o^{(2)} \rangle d\vec{r}\right]^2 . \tag{1}$$

Here $V(\vec{r}) = V(\vec{r}; \{\vec{0}\})$ and the $|\Psi_o\rangle$'s are wave functions of the atom-rigid target system with the same crystal momentum parallel to the surface ($\vec{K}$) and energy (E). When there is only elastic scattering, the case with a rigid target, for a given $\vec{K}$ and E there is a discrete set of asymptotic perpendicular momenta $k_{\vec{G}}$ the atom may have. These channels, incoming if $k_{\vec{G}} < 0$, outgoing if $k_{\vec{G}} > 0$, are labelled by surface reciprocal lattice vectors $\vec{G}$. The atom involved in an inelastic scattering process where $qa \ll 1$ will emerge near one of these channels. The probability that the atom emerges near channel $\vec{G}_o$ is calculated by choosing the amplitude of all incoming channels except $\vec{0}$ to be 0 for $|\Psi_o^{(1)}\rangle$, and the amplitude of all outgoing channels except $\vec{G}_o$ to be 0 for $|\Psi_o^{(2)}\rangle$. When both wave functions are normalized to unit incoming flux in channel $\vec{0}$, the integral represents the total force applied in transferring the atom from incoming channel $\vec{0}$ to outgoing channel $\vec{G}_o$. This is rigorously given by the change in perpendicular momentum per unit time and area, $i.e.$, $-\hbar(k_{\vec{0}} + k_{\vec{G}})$.

The results are exact in the limit where the following conditions are satisfied:

$$\frac{\hbar}{Mcb} \ll 1 \quad (2a) \qquad \frac{m}{M}\left(\frac{V_o}{\hbar\omega_D}\right)^2 \ll 1 \quad (2b) \qquad E_a \lesssim \hbar\omega_D \quad (2c)$$

$$k_B T \lesssim \hbar\omega_D \quad (2d) \qquad qa \ll 1 \quad (2e),$$

where c is a characteristic velocity of a phonon in the target, b is the range of the atom-target potential and V_o is its depth. m is the mass of the colliding atom and E_a is its incident energy. M is the mass of a target atom, ω_D is the Debye frequency, and T is the temperature of the target.

The condition (2a) allows one to stop the expansion of V after the first order term

$$V(\vec{r}; \{\vec{u}_i\}) = V(\vec{r}) + \vec{u}_i \cdot \nabla_{\vec{u}_i} V(\vec{r}; \{\vec{0}\}) \tag{3}$$

Condition (2b) ensures that in calculating the inelastic scattering, the last term in (3) can be treated to lowest order.

Derivation of the Probability for Production of Long-Wavelength Phonons

This derivation requires the use of a variational principle developed for multichannel scattering in nuclear collisions [2]. The following assumptions simplify this presentation: (1) the atom only interacts with the uppermost layer of target atoms and (2) the units are chosen so that the target's volume and the area of its interacting surface are 1. The results remain unchanged if these assumptions are dropped.

To first order in $\{u_i\}$, the Hamiltonian consists of an unperturbed Hamiltonian H_o, which ignores phonon coupling to the colliding atom, and H_1, which couples the atom linearly to the phonons.

$$H = H_{particle} + H_{target} + V(\vec{r}, \{\vec{u}_i\}) = H_o + H_1 + \mathcal{O}(u^2) \quad (4)$$

where

$$H_o = \frac{p^2}{2m} + \sum_{\vec{q},\nu} \hbar\omega_{\vec{q},\nu} n_{\vec{q},\nu} + V(\vec{r}) \quad (5a) \qquad H_1 = \sum_{i,\alpha} \frac{\partial V}{\partial u_{i,\alpha}}(\vec{r}) u_{i,\alpha}. \quad (5b)$$

Here α denotes a cartesian coordinate and $n_{\vec{q},\nu} = a_{\vec{q},\nu}^{\dagger} a_{\vec{q},\nu}$ is the occupation of phonon mode $\vec{q}, \nu$, where ν is the phonon's branch.

Let $|\Psi^{(1)}\rangle$ and $|\Psi^{(2)}\rangle$ be two wave functions for the combined target-atom system with the same E and $\vec{K}$. Their asymptotic form is

$$\langle \vec{r}|\Psi^{(\mu)}\rangle = \sum_{\{n\},\vec{G}} |\{n\}\rangle e^{i(\vec{K}_{\{n\}}+\vec{G})\cdot\vec{R}} \left[A^{(\mu)}_{\vec{G},\{n\}} e^{-ik_{\vec{G},\{n\}}z} + B^{(\mu)}_{\vec{G},\{n\}} e^{ik_{\vec{G},\{n\}}z} \right] \quad (6)$$

where the sum is carried out over all surface reciprocal lattice vectors $\vec{G}$ and phonon occupations $\{n\}$, $\vec{R}$ is the atom's displacement parallel to the surface, and

$$\vec{K}_{\{n\}} = \vec{K} - \sum_{\vec{q},\nu} n_{\vec{q},\nu}\vec{q}, \qquad k^2_{\vec{G},\{n\}} + (\vec{K}_{\{n\}}+\vec{G})^2 = \frac{2m}{\hbar^2}(E - \sum_{\vec{q},\nu} \hbar\omega_{\vec{q},\nu} n_{\vec{q},\nu}). \quad (7)$$

Choose two trial functions $|\Psi_t^{(1)}\rangle$ and $|\Psi_t^{(2)}\rangle$, such that $A^{(2)}_{\vec{G},\{n\},t} = A^{(2)}_{\vec{G},\{n\}}$ and $B^{(1)}_{\vec{G},\{n\},t} = B^{(1)}_{\vec{G},\{n\}}$. Then the equation

$$\int_V \langle \Psi_t^{(1)}|\vec{r}\rangle (H-E)\langle\vec{r}|\Psi_t^{(2)}\rangle d\vec{r} = \frac{-i\hbar^2}{m} \sum_{\vec{G},\{n\}} k_{\vec{G},\{n\}} B^{(1)*}_{\vec{G},\{n\}} \delta B^{(2)}_{\vec{G},\{n\}}, \quad (8)$$

where $\delta B^{(2)}_{\vec{G},\{n\}} = B^{(2)}_{\vec{G},\{n\},t} - B^{(2)}_{\vec{G},\{n\}}$, is correct to first order in the errors of the wave functions.

This variational principle is of value because the integral can be evaluated analytically to lowest order in qa for trial functions which are eigenfunctions of H_o. These wavefunctions have the form [3]

$$\langle \vec{r}|\Psi_{\vec{K},E,\{n\}}\rangle = |\{n\}\rangle \sum_{\vec{G}} e^{i(\vec{K}_{\{n\}}+\vec{G})\cdot\vec{r}} \phi_{\vec{G},\vec{K},E,\{n\}}(z) \quad (9)$$

The set of all $\phi_{\vec{G}}(z)$ for a given $\vec{K}$, E and $\{n\}$ satisfy the following coupled equations, where $E_a = E - \sum_{\vec{q},\nu} \hbar\omega_{\vec{q},\nu} n_{\vec{q},\nu}$ and $\sum_{\vec{H}} e^{i\vec{H}\cdot\vec{r}} V_{\vec{H}}(z) = V(\vec{r})$:

$$\left[-\frac{\hbar^2 \partial_z^2}{2m} - \left(E_a - \frac{\hbar^2 (\vec{K}_{\{n\}} + \vec{G})^2}{2m}\right)\right] \phi_{\vec{G}}(z) = -\sum_{\vec{G}'} V_{\vec{G}-\vec{G}'}(z) \phi_{\vec{G}'}(z).$$

(10)

Let $|\Psi^{(2)}_{\{n_2\},o}\rangle$ be an eigenfunction of H_o with energy E, crystal momentum $\vec{K}$ and $|\{n_2\}\rangle = |\ldots n_{\vec{q},\nu}\ldots\rangle$ for which $A_{\vec{0}} = (m/\hbar k_{\vec{0}})^{1/2}$ and all other A's are 0. Let $|\Psi^{(1)}_{\{n_1\},o}\rangle$ be an eigenfunction of H_o with eigenvalues E, $\vec{K}$ and $|\{n_1\}\rangle = |\ldots n_{\vec{q},\nu} + 1\ldots\rangle$ for which $A_{\vec{0}} = (m/\hbar k_{\vec{0}})^{1/2}$ and $B_{\vec{G}_o}$ is the only nonzero B.

For these trial functions, the integral is $\int_V \langle \Psi^{(1)}_{\{n_1\},o}|\vec{r}\rangle H_1 \langle \vec{r}|\Psi^{(2)}_{\{n_2\},o}\rangle d\vec{r}$. Expanding the $\vec{u}_i$ in normal modes and rewriting yields

$$H_1 = \sum_{\vec{q},\nu,\alpha} e^{-i\vec{q}\cdot\vec{R}} u_{\vec{q},\nu,\alpha} \sum_i \frac{\partial V}{\partial u_{i,\alpha}}(\vec{r}) e^{-i\vec{q}\cdot(\vec{R}_i - \vec{R})} + \text{c.c.}$$

(11)

The second exponential can be ignored if the wavelength of the phonon is much larger than the range of the coupling potential, similar to a and b, since unless $q(R_i - R) \ll 1$ $(\partial V/\partial u_i)$ will vanish. Furthermore, $\sum_i \partial V/\partial u_{i,\alpha}(\vec{r}) = -\partial V(\vec{r})/\partial r_\alpha$ since $V(\vec{r}; \{u_i\})$ depends only on $\vec{r} - \vec{u}_i$. Finally, defining $\vec{u}_{\vec{q},\nu} = \vec{U}_{\vec{q},\nu}(a^\dagger_{\vec{q},\nu} + a_{-\vec{q},\nu})$ yields

$$H_1 = -\sum_{\vec{q},\nu,\alpha} e^{-i\vec{q}\cdot\vec{R}} \frac{\partial V}{\partial r_\alpha}(\vec{r}) U_{\vec{q},\nu,\alpha}(a^\dagger_{\vec{q},\nu} + a_{-\vec{q},\nu}) + \text{c.c.}$$

(12)

Transforming the resulting integral with (5a) and Green's Theorem yields

$$(n_{\vec{q},\nu} + 1)^{1/2} U_{\vec{q},\nu,\alpha} \sum_{\vec{G},\vec{G}',\alpha} \int_S dS \left[e^{-i\vec{G}\cdot\vec{R}} \phi^{(1)*}_{\vec{G},\{n_1\}}(z) \frac{\partial}{\partial r_\alpha} \hat{n} \cdot \nabla \phi^{(2)}_{\vec{G}',\{n_2\}}(z) \right.$$
$$\left. - \hat{n} \cdot \nabla \left(e^{-i\vec{G}\cdot\vec{R}} \phi^{(1)*}_{\vec{G},\{n_1\}}(z) \right) \frac{\partial}{\partial r_\alpha} \phi^{(2)}_{\vec{G}',\{n_2\}}(z) \right]$$

(13)

where $\hat{n}$ is the unit normal to the surface S which bounds V. Because of the small difference in atomic energy and parallel crystal momentum between $|\Psi^{(1)}_{\{n_1\},o}\rangle$ and $|\Psi^{(1)}_{\{n_2\},o}\rangle$, $\phi^{(1)*}_{\vec{G},\{n_1\}}$ may be replaced by $\phi^{(1)*}_{\vec{G},\{n_2\}}$. The vanishing of the resulting integrals for $n_x(\partial/\partial x)$ and $n_y(\partial/\partial y)$ and evaluation of the remainder yields

$$\frac{\hbar^2 U_{\vec{q},\nu,\alpha}(n_{\vec{q},\nu} + 1)^{1/2}}{2m} \sum_{\vec{G}} \left[\phi^{(1)*}_{\vec{G}} \frac{\partial^2 \phi^{(2)}_{\vec{G}}}{\partial r_\alpha \partial z} - \frac{\partial \phi^{(1)*}_{\vec{G}}}{\partial z} \frac{\partial \phi^{(2)}_{\vec{G}}}{\partial r_\alpha} \right]_{z=\infty}$$

(14)

Since $k_{\vec{G}_o}(B^{(1)*}B^{(2)})_{\vec{G}_o,\{n_2\},o} = k_{\vec{0}}(A^{(1)*}A^{(2)})_{\vec{0},\{n_2\},o}$, (14) is zero if α denotes x or y, and when α denotes z it is $-(n_{\vec{q},\nu}+1)^{1/2}U_{\vec{q},\nu,z}\hbar(k_{\vec{0}}+k_{\vec{G}_o})$.
Define $\delta B^{(1)}_{\vec{G}_o,\{n_2\}} = -\delta B^{(2)*}_{\vec{G}_o,\{n_2\}}(B^{(1)}/B^{(2)*})_{\vec{G}_o,\{n_2\},o}$.

The differential probability of scattering an atom into final angle Ω with energy E_a, per unit incident perpendicular flux is

$$\frac{d^2P}{d\Omega dE_a} = (\pi\hbar^2 k_{\vec{G}_o}/m)|\delta B^{(1)}_{\vec{G}_o,\{n_2\}}/A^{(1)}_{\vec{0},\{n_2\}}|^2\rho(E_a)g(\hbar\omega_{\vec{q},\nu},\vec{Q}) \qquad (15)$$

$$= (k_{\vec{G}_o}k/4\pi^2)(k_{\vec{0}}+k_{\vec{G}_o})^2U^2_{\vec{q},\nu,z}(n_{\vec{q},\nu}+1)|B^{(1)}_{\vec{G}_o,\{n_2\},o}/A^{(1)}_{\vec{0},\{n_2\}}|^2g(\hbar\omega_{\vec{q},\nu};\vec{Q})$$

where k is the atom's final momentum, $\rho(E_a)$ is the density per unit energy of atomic states with momentum in final angle Ω, and $g(\hbar\omega_{\vec{q},\nu};\vec{Q})$ is the density per unit energy of phonons with frequency $\omega = \omega_{\vec{q},\nu}$ and surface wavevector $\vec{Q}$.

For a target which is isotropic in the long wavelength limit, $U_{\vec{q},\nu,z}$ and $g(\hbar\omega_{\vec{q},\nu};\vec{Q})$ are well known, depend only on bulk elastic constants, density, ω and Q, and are given in simple form in Ref. 1. For non-isotropic targets, $U_{\vec{q},\nu,z}$ and $g(\hbar\omega_{\vec{q},\nu};\vec{Q})$ depend in a more complicated way on the bulk elastic constants, density, Miller indices of the surface, ω and $\vec{Q}$.

Acknowledgments
This work is supported by the National Science Foundation through a graduate fellowship and Grant No. DMR87-03434 and by the U.S. Office of Naval Research through Grant No. N00014-89-J-1530.

References

[1] M. E. Flatté and W. Kohn, *Phys. Rev. B* **43**, 7422, (1991).

[2] W. Kohn, *Phys. Rev.* **74**, 1763, (1948).

[3] N. Cabrera, V. Celli, F. O. Goodman and R. Manson, *Surface Science* **19**, 67, (1970).

Metastable Spin-Polarized He as a Probe of Surface Magnetism on NiO(100)

A. Swan[1], *W. Franzen*[1], *M. El-Batanouny*[1] *and K.M. Martini*[2]

[1]Physics Department, Boston University, Boston, MA 02215, USA
[2]Physics and Astronomy Department, University of Massachusetts, Amherst, MA 01003, USA

Abstract We discuss an extremely surface sensitive non-destructive method for determining the surface magnetic order of antiferromagnetic (AF) materials by using spin-polarized metastable He scattering. Information about the AF order is obtained from angle resolved measurements of surviving metastable atoms. This new technique makes use of the local nature of the interaction to enhance the survivability of scattered metastables from one of the two sublattices. Application to the NiO(111) surface will be discussed.

1. Introduction

In recent years there has been an explosion of new experimental techniques exploring magnetic properties on surfaces[1-3]. The development of both spin-polarized electron-sources and spin-sensitive detectors together with improved vacuum-techniques has made it feasible to investigate surface magnetism. The techniques of making thin films and epitaxially grown layered materials has become very refined, and it is possible to synthesize quasi-two dimensional magnetic layers. Important technical applications in computer industry such as modern magnetic recording and memory devices have also contributed to the interest in magnetic surfaces. Most methods used for surface magnetic measurements such as spin-polarized angle-resolved photoemission[2], spin-polarized inverse photoemission, spin-polarized low energy electron energy diffraction, electron capture spectroscopy[1] and surface magneto-optic Kerr effect studies[1] rely on difference-spectra which give information about majority and minority spins as well as the exchange splitting in ferromagnetic materials. However, these techniques are of no use for anti-ferromagnetic surfaces since they have no net macroscopic spin. The information about the properties of anti-ferromagnetic (AF) surfaces still remains sparse. There are very few experimental techniques that can distinguish between AF ordering and non-magnetic surfaces. Angle-resolved photoelectron emission measurements on Cr [1] have shown a band structure with symmetry consistent with anti-ferromagnetic ordering. Sinkovic *et al.*[4] have reported a new technique which uses spin-polarized photoelectron diffraction to measure local magnetic ordering of the near surface, of both AF and ferromagnetic order. Moreover, LEED can, in principle, be used to determine AF structure by displaying extra fractional or

Springer Series in Surface Sciences, Vol. 24   **The Structure of Surfaces III**
Editors: S.Y. Tong · M.A. Van Hove · K. Takayanagi · X.D. Xie

incommensurate order peaks due to exchange scattering from the AF super-lattice unit cell. However, the intensity of these spots is calculated to be only a few percent of the integral order spots[5]. Palmberg *et al.*[6] used this technique to study the (100) surface of NiO and did indeed find weak intensity half-order spots which they attributed to an AF spin arrangement. This result is very interesting but somewhat inconclusive since later investigators[7] where not able to reproduce these results, indicating that the observed extra spots may be due to some reconstruction or imperfection on the sample surface.

In this paper we discuss the use of elastic spin-polarized metastable helium-atom (He*) scattering. This is a technique which promises to be especially well suited for obtaining new information about AF order on transition metal oxide (TMO) surfaces. Unlike some of the spectroscopies mentioned above, the metastable He* beam which is elastically scattered depends on the local nature of interaction with the spin of the surface[8]. Thus this technique has the potential to reveal new information about surfaces and overlayers of AF insulators. The TMO's from MnO to NiO have presented interesting yet elusive systems for the solid state community ever since de Boer and Verwey in 1937[9] reported that these oxides are insulators. Recently, these oxides have drawn more attention as a result of the discovery of high temperature superconductors, since it is believed that the layers of AF CuO in these materials are intimately linked to the superconducting mechanism. The TMO surfaces are therefore of great value as model systems for studies of electronic and magnetic structure. Since elastic scattering of metastable atoms is a relatively new technique, it is worthwhile to discuss briefly the physical mechanisms involved.

2. Interaction of Metastable He-atoms with Solid Surfaces

A few groups have studied the scattering of metastable He-atoms with kinetic energies in the range 25-60 meV from solid surfaces[10-12]. Rather than measuring elastically scattered He*, they focused on the process of electron emission from clean and adsorbate covered surfaces to study the surface/adsorbate local density of state. These measurements have also shown that upon impact with a surface, there are two possible channels for decay of the metastables;
1) resonance ionization + auger neutralization (RI+AN), and
2) auger deexcitation (AD).
The former process dominates overwhelmingly when the excited 2s electron of the incoming He* is degenerate with an unoccupied local surface density of state (LSDOS). In this case, as a He* 2^3S atom approaches a solid surface, it is first resonantly ionized by the tunneling of the excited He* 2s-electron into the unfilled state above the Fermi level of the solid surface, as illustrated by the arrow in fig. 1a. The resulting He$^+$ ion continues to travel towards the surface where it is neutralized by a surface electron which fills the empty 1s hole, thus deexciting the He-atom. The released energy is absorbed by a second surface electron which may escape into vacuum if the energy is sufficiently large. This two-electron process, termed auger neutralization, leads to an energy distri-

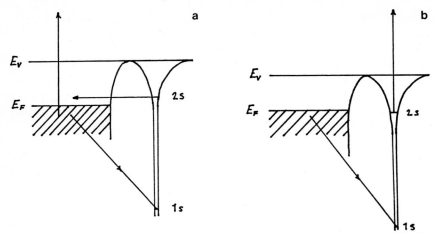

Fig 1. The two possible decay processes for He*. 1a.Resonance ionization and auger neutralization (RI+AN). As the He* atom approaches the surface, the He* 2s electron tunnels into an unfilled state above the Fermi level. Closer to the solid, a surface electron fills the empty He 1s level and another surface electron is ejected. 1b. Auger Deexcitation (AD). The He* 2s level lies below the Fermi level and RI is inhibited. A surface electron fills the empty He* 1s level and the He* 2s electron is ejected.

bution of ejected electrons reflecting a self-convolution of the LSDOS. Such a behavior is characteristic of most clean transition metal surfaces.

The second process, AD, prevails when the 2s excited electron level falls into band gap, or is located below the Fermi level as in fig. 1b. In this case tunneling is inhibited and the deexcitation takes place through the AD process. The AD process leads to the filling of the empty He* 1s level by a surface electron and the ejection of the electron in the He* 2s level. The resulting electron energy distribution represents the one-electron character of the process. This mechanism is usually operative in some metal oxides and non-metallic surfaces as well as in alkali metal overlayers and surfaces. Measurements of survival probabilities (SP) on a range of non-magnetic crystals show SP from 10^{-3} to 10^{-6} for the two processes[13].

From the discussion above it is clear that the decay of the metastable state is determined solely by the presence of a local surface spin density of states[8] that would favour such an event: either to allow the electron in the He* 2s orbit to tunnel into the surface, or to provide a surface electron with opposite spin that can fill the empty He* 1s orbit. In some AF insulators, such as the transition metal oxides mentioned above, the 2s orbital energy, -4.7eV, lies in the insulating gap. Resonance ionization is then inhibited and the only deexcitation channel available would be AD. On an AF surface, we argue that the AD process would be significantly reduced due to the local character of the interaction and spin-selection rules. This can be seen as follows; since the He*-

atoms have such low kinetic energy (60meV), they scatter from the very tail of the electron distribution on the surface and have thus only a significant overlap with the outermost electrons. If at a particular site on the surface the spin of the outermost electron is aligned with the incoming spin orientation of the He*-atom, then the empty 1s state can not be filled; spin-flip events are exluded since the encounter-time is of the order of 10^{-14}s[13]. From these arguments we can see that SP can be enhanced considerably by eliminating RI+AN and supressing the AD process. An AF surface which is properly aligned with the polarization of the metastable beam is then expected to have one sublattice that would favor decay and another that would inhibit it. The latter establishes an elastic diffractive scattering channel for the surviving metastables which reflects the magnetic structure.

3. Experimental Set-Up

The experimental set-up consists of a scattering-chamber with surface cleaning and diagnostic devices, and a differentially pumped beamline as shown in fig. 2. The first chamber on the beamline is the beam generator where a conventional nozzle-skimmer system forms a mono-energetic He-beam with a kinetic energy of ~60meV. The second chamber houses the He*-exciter. The excitation into the He* 2^3S state is achieved by electron impact in a colinear geometry where electrons travel in opposite direction of the He-beam[14]. The excitation of the He atom is done at electron energies about 0.5eV above

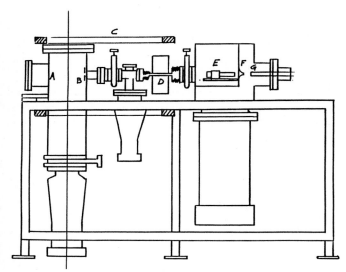

Fig 2. Schematic of experimental set-up. A. Scattering chamber, B. Movable orifice, C. Helmholtz coils, D. Movable magnet with entrance orifice, E. He*2^3S exciter, F. Skimmer, G. Nozzle.

the excitation threshhold(19.81eV) to optimize the cross section for He*2^3S, yet avoiding the excitation of He*2^1S. The spin-polarization is achieved in the third chamber. The polarizer consists of a Stern-Gerlach magnet. The spin-orientation selection is achieved by means of a set of orifices placed between the exciter and the scattering chamber. In the scattering chamber, the sample is mounted on a manipulator which allows polar- and azimuthal rotation as well as xyz-translation. The chamber is also equipped with an ion-sputter gun and a LEED/Auger spectrometer for structure and composition characterization of the sample surface. The angle-resolving detector, which has a rotation of 360°, consists of a channeltron biased for the detection of elastically scattered He*-atoms.

4. Application of Metastable Spin-polarized He 2^3S Scattering to NiO(100)

We have chosen NiO to be the first material to test this method since it is chemically stable, the (100) surface does not reconstruct and the Neel temperature is conveniently well above room temperature (523 K).

The NiO crystal is arranged in a rocksalt structure with the Ni-atom surrounded by a cage of 6 oxygen-atoms and 12 nearest neighbor Ni-atoms as in a FCC crystal. Atomic Ni has a $3d^84s^2$ configuration, but Ni in NiO is in a $3d^8$ configuration where the two s-electrons are transfered to fill the oxygen 2p orbitals. The remaining eight 3-d electrons are split in energy by mainly two interactions; exchange interaction which splits the spin-up and spin-down electrons and the crystal field which lifts the degeneracy of the e_g and t_{2g} bands. These energy splittings give rise to the insulating behaviour of NiO. The microscopic details of this system as well as other TMO's are not fully understood and there are two very different schools of thought concerning the electronic structure of NiO[15,16].

The electronic structure measured by photo emission spectroscopy and bremsstrahlung isochromat spectroscopy [17] is shown in fig.3. From the figure we can see that the He* 2s electron level lies clearly in the band gap and thus resonance ionization is inhibited. The prominent peak above the gap is attributed to the empty $e_g \downarrow$, while the topmost peak below the gap is due to the occupied $t_{2g} \downarrow$ state. The remaining occupied d-states with up-spin seem to lie at least 2 eV below $t_{2g} \downarrow$ peak. Consequently, we expect that the He* will overlap mainly with the $t_{2g} \downarrow$ electron wavefunctions.

The magnetic ordering of the NiO-bulk[18] is known to be of type II AF ordering, which has the spins aligned within a (111)-plane but where consecutive (111) planes are anti-ferromagnetically ordered. If the AF ordering on the (100) surface is a continuation of the bulk, we would expect the presence of two Ni spin-sublattices. In this case our technique would lead to a diffraction from the Ni sublattice which has the top t_{2g} electron spin aligned with the spin of the He*.

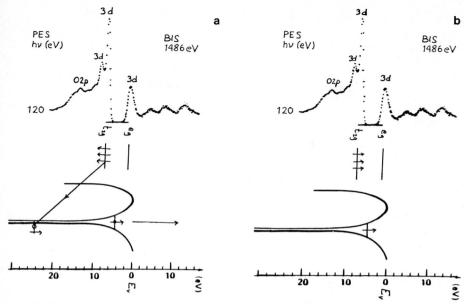

Fig 3. Schematic energy diagram of NiO and He*2^3S. This fig illustrates how the He* 2s level lies in the band gap and thus inhibits RI. Fig 3a shows how He* with opposite spin alignment to the minority t_{2g} level undergoes AD. Fig. 3b illustrates how also the AD process is supressed on Ni sites where the electron spin of the minority t_{2g} level is aligned with the H* polarization.

However, the magnetic structure of the surface might be quite different from the bulk, because of the change of symmetry at the surface. In a similar system Chrzan and Falicov[19] even predicted an incommensurate AF surface spin ordering as a possible ground state. The lack of experimental results on these systems, given their significance, prompts us to pursue this investigation.

We have completed the sample preparation stage and verified the structure and surface stoichiometry using LEED and Auger. We are currently setting up the He* beam alignment. We expect to complete the measurements shortly and extend these measurements to MnO(100)

5. Acknowledgement

This work is supported by US DOE Contract DE-FG02-85ER45222

References

1. *Magnetic Properties of Low-Dimensional Systems*, Ed: L.M. Falicov and J.L. Moran-Lopez, (Springer, 1986).
2. *Polarized Electrons in Surface Physics*, R. Feder, (World Scientific, 1986).

3. *Polarized Electrons at Surfaces*, J. Kirschner, (Springer, 1985)
4. B. Sinkovic and C.s. Fadley, Phys. Rev. B, **31**, 4554, (1985), and references therein.
5. R.E. De Wames, Phys. Stat. Sol., **39**, 437, (1970).
6. P.W. Palmberg, R.E. De Wames, and L.A. Vredevoe, Phys. Rev. Lett., **21**, 682, (1968).
7. M. Prutton, J. Phys. C,**8**, p.2401, (1975), B. Sinkovic, private communication.
8. J. Lee, C. Hanrahan, J. Arias, F. Bozso, R.M. Martin, and H. Metiu, Phys. Rev. Lett., **54**, 1440, (1985).
9. J.H. de Boer and E.J.W. Verwey, Proc. Phys. Soc., **49** (extra), 59, 59, (1937).
10. H. Conrad, G. Ertl, J. Kuppers, W. Sessleman, b. Woratschek, and H. Haberland, Surf. Sci., **117**, 98, (1982).
11. W. Sesselmann, B. Woratschek, J. Kuppers, G. Ertl, and H. Haberland, Phys. Rev. B., **35**, 1547, (1987).
12. M. Onellion, M.W. Hart, F.B. Dunning, and G.K. Walters, Phys. Rev. Lett., **52**, 380, (1984).
13. H. Haberland in*Inelastic Ion-Surface Collisions*, Ed. N.H. Tolk, J.C. Tully, W. Heiland, and C.W. White, (Academic Press, 1977). p. 14.
14. Anna Swan, W. Franzen, M. El-Batanouny and K.M. Martini, to be published.
15. A. Fujimori and F. Minami, Phys. Rev. B, **29**, 5225, (1984).
16. K. Terakura, T. Oguchi, and A.R. Williams, Phys. Rev. B, **30**, 4734, (1984).
17. G.A. Sawatzky and J.W. Allen, Phys. Rev. Lett., **53**, 2339, (1984).
18. C.G. Shull, W.A. Strauser, and E.O. Wollan, Phys. Rev., **83**, 333, (1951).
19. D.C. Chrzan and L.M. Falicov, Phys Rev. B, **39**, 3159, (1989).

Scattering of O_2 from the Ag(110) Surface

X.Y. Shen, K.Q. Zhang, and Y.L. Tan

Department of Electronic Engineering, Tsinguha University,
Beijing 100084, P.R. of China

The scattering process of an oxygen molecule from the silver (110) surface has
been investigated with different theoretical approaches. A modified semi-empirical
LEPS(London-Eyring-Polanyi-Sato) interaction potential is adopted to describe the gas-
surface system. Quasiclassical trajectory calculations are performed for the system
to examine the microscopic mechanism of the scattering process. In order to study the
quantum phenomenon of selective adsorption,which is basically a resonance scattering
associated with a bound state embedded in the continuum, quantum mechanical bound
state calculations are carried out. Good agreement is found between the quantum and
the classical results on the incident energies of the selectively adsorbed states
above the potential barrier where the tunneling effect is not important. We conclude
that it is necessary to include the vibrational degree of freedom of the projectile
molecule for describing the scattering process.

1. Introduction

Recently, considerable experimental research on gas-surface interaction has been
undertaken [2,22-24]. In conjunction with the recent expansion of experimental work,
there has been a growing theoretical effort directed toward elucidation of the
detailed dynamics of gas-surface interaction [1-3]. One of the primary objectives
of gas-surface studies is to determine the nature of the interaction potential which
exists between a wide variety of incident atomic or molecular particles and different
classes of crystalline surfaces [17]. Knowledge of the gas-surface potential is
fundamental to any study of gas-surface scattering since it determines the dynamics
and energetics , i.e., the time evolution of these phenomena on the molecular level.

The semi-empirical model potentials have been most widely used in the dynamic studies of gas-surface interaction [11-15]. In the present paper, a modified LEPS
(London-Eyring-Polanyi-Sato) potential is adopted for the oxygen molecule-silver
surface system. In the semi-empirical approach, a few experimental data have been
incorporated into the potential surface [10].

With the assumption of classical mechanics, classical trajectory methods provide
exact solutions for model scattering problems. Modern advances in semiclassical
theories offer the possibility of accurate inclusion of quantum effects within a
classical framework, at great savings of computational effort compared to quantum

theories [11-16].   We have recently carried out a quasi-classical trajectory calculation to study the scattering processes of hydrogen molecules on copper surfaces
[4,5]. In the present work, quasi-classical trajectory calculations are performed
to the oxygen molecule-silver surface system to examine the dynamic processes. How
-ever, the classical approaches have limited applications, because the quantum
effects are important as manifested in tunneling, zero point motion, difffraction
and interference, etc.

     As the quantum theory becomes better developed, and is able to interpret more
and more of the forthcoming experimental data, we expect a gradual increase of the
use of quantum mechanics to describe gas-surface scattering phenomena. It has become
clear that more information on gas-surface systems is obtainable from experimental
data needing a quantum mechanical interpretation than from data for which a
classical interpretation is adequate [18-21]. We have recently performed a quasi-
bound state calculation and a close-coupling calculation of the hydrogen molecule-
copper surface collision processes [6,7]. In the present work, we have carried out
a quantum mechanical quasi-bound state calculation for the oxygen molecule-silver
surface system. A reasonably good agreement is found between the quantum and
classical results on the incident energies of the selectively adsorbed states
above the potential barrier where the tunneling effect is not important.

2.   The Potential Energy Surface (PES)

     We have applied the LEPS formulation to the oxygen molecule-silver surface
system [8, 10].   The LEPS approach to molecular electronic energy calculation is
basically an approximate valence bond treatment derived for the determination of the
potential energy of a system of three one-electron atoms. The solid surface is
approximated by a periodic static background potential. In other words, the surface
is assumed to be rigid but corrugated. The individual atom-atom and atom-surface
potentials are assumed to be known and represented by Morse potentials so that the
correct asymptotic behavior is obtained for all final arrangements. The Sato
parameters are adjustable and can be used to construct a family of related potential
surfaces with varying positions and heights of potential barriers, wells, and etc.
Another attractive feature of the LEPS formulation is that the potential is easy to
evaluate numerically. Actually the interaction of the diatomic molecule with the
solid surface is most simply thought of as an interaction of three particles: two
atoms and the solid. Many-body effects involving the solid atoms are automatically
included since the potential involves two atoms interacting separately with the
complete solid.   To account for the structure of the solid in the atom surface
interactions we require that the dissociation energy, the equilibrium distance for
the gas atom above the surface be functions of x and y, where the x-y plane is parallel to the plane of the solid surface. Thus the corrugation of the surface can be
varied by adjusting the parameters [8].

198

3. The Quasi-Classical Trajectory (QCT) Calculation

The quasi-classical trajectory method involves the selection of the initial conditions of the gas molecule in discrete internal energy states corresponding to the quantum states of the projectile. In the present work, the initial internal energies of the oxygen molecule are determined by the semi-classical theory of bound states [4, 8]. The initial positions and momenta are randomly sampled by the Monte-Carlo procedure. As soon as the integration of the trajectory begins, the quantum restriction is relaxed so that the time evolution of the system is solely governed by classical mechanics.

Thousands of trajectories have been integrated on the oxygen molecule-silver(110) surface system with different incident translation energies and different beam angles of the projectile molecule. Elastic and inelastic scattering, diffraction, dissociative adsorption, and selective adsorption have all been observed. Due to the limited space, we shall show only a few representative trajectories for the selective adsorption, the others will be discussed elsewhere.

The representative trajectories of the selective adsorptions are shown in Figure 1, which is for a normal incidence projectile molecule with initial translational energy of 0.4eV. $z1,z2$ are the vertical coordinates of the two nuclei of the oxygen molecule. It is seen that the molecule stays on the surface for quite a long time. It is about 13,000 atomic time units, or roughly 12 vibration periods of the oxygen molecule. The molecule only vibrates strongly and remains undissociated. This can be seen from the trajectory of internuclear separations of the molecule. The energy transfer in the scattering process is carefully examined. In Figure 2, the molecule-

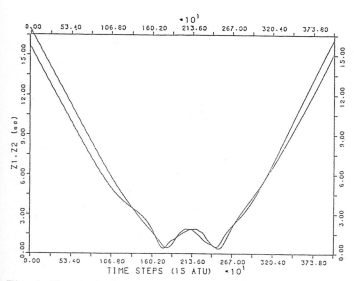

Fig. 1. The vertical coordinates, $z1,z2$ trajectories.

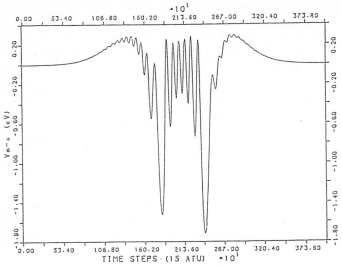

Fig. 2. Molecule-surface interaction potential energy versus the integration steps. Each time step is given by 15 atomic time units.

Table 1. The translational energies (in eV) at which the selective adsorptions are observed.

0.380	0.382	0.392	0.398	0.430	0.460
0.478	0.480	0.482	0.500	0.502	0.504
0.506	0.508	0.524	0.526	0.528	0.530
0.532	0.534	0.536	0.554	0.556	0.558
0.560	0.562	0.564	0.566	0.568	0.570
0.588	0.590	0.592	0.594	0.596	0.598
0.600	0.602	0.604	0.606	0.608	0.626
0.628	0.630	0.632	0.634	0.636	0.638
0.640	0.646	0.648	0.650	0.652	0.654
0.656	0.658	0.660	0.662	0.664	0.666

surface potential energy, which is equal to the total potential energy minus the diatomic potential energy of the molecule, is plotted against the integration steps. The molecule-surface potential is virtually zero when the molecule is far away from the surface. As the molecule gets close to the surface and stretches its internuclear separation, the molecule-surface potential becomes quite attractive.

200

This is why the molecule is adsorbed onto the surface. As stated in our early work [9], in the case of a rigid surface, there is no energy exchange between the gas molecule and the phononless surface; however, the molecule can still transfer energy from internal states into translational motion relative to the surface, and vice versa. Therefore, this type of selective adsorption involves energy transfer between the translational and the internal energies of the molecule and is significant in the molecular projectile case.

It is seen that only at certain incidence energies, is the molecule adsorbed on the surface for a period of time while remaining undissociated, i.e. the selective adsorption is energy-dependent. In Table 1, the incident translation energies of the molecule, at which the selective adsorptions have been observed in the QCT calculations, are listed.

As known, selective adsorption is basically a quantum phenomenon. For this reason, we now proceed to do the quantum mechanical analysis.

4. The Quasi-Bound State Study

The selective adsorption is basically a resonance scattering associated with a bound state embedded in the continuum. This is also called a compound state or quasi-bound state which is the eigenstate of the total Hamiltonian of the system. The internal excitation of the resonance in the sense of the Feshbach formulation is manifested via a compound state/bound state embedded in the continuum.

Molecule-surface selective adsorption is known to be mediated through surface phonons, diffraction, rotational or vibrational degrees of freedom of the molecule. For the time being, we are interested in the rotational and vibrational mediations. Rotationally or vibrationally mediated selective adsorptions arise from the excitations of the internal states of the projectile and the scattering exhibits resonance behavior.

In this section the compound states embedded in the continuum are computed by diagonalizing the total Hamiltonian of the oxygen molecule-silver (110) surface system [4,6]. The same LEPS potential mentioned in the previous section is employed. After the diagonalization of the Hamiltonian, 360 eigenstates that correspond to the selectively adsorbed resonance states have been obtained. We have tested the convergence of the eigenstates by increasing and decreasing the number of basis states. It turns out that for the lowest lying 70-80 states among the 360 states, reasonable convergence seems to be achieved. The energy levels of the lowest lying 20 states and the corresponding translational energies are listed in Table 2. The energy levels are not equally spaced. Comparing the classical and the quantum results of the translational energies, the discrepancy is large in the very low energy region where tunneling effect is significant. This region is generally around and below the potential barrier. For this reason, a larger discrepancy is expected in the energy region below 0.2 eV. It should be noted from the manner the classical

Table 2. The energy levels for the lowest lying 20 states (in eV).

I	Eigenvalues	Tr. Energies	Differences
1	-5.03749418	.07769231	-.207229E-01
2	-5.01677131	.09841512	-.220528E-01
3	-4.99471855	.12046759	-.237012E-01
4	-4.97101736	.14416876	-.257564E-01
5	-4.94526100	.16992521	-.239258E-01
6	-4.92133522	.19385099	-.714598E-01
7	-4.84987545	.26531070	-.202594E-01
8	-4.82961606	.28557020	-.217290E-01
9	-4.80788707	.30729902	-.234370E-01
10	-4.78445005	.33073604	-.255857E-01
11	-4.75886440	.35632205	-.237131E-01
12	-4.73515129	.38003486	-.139427E-02
13	-4.73375701	.38142926	-.675640E-01
14	-4.66619300	.44899333	-.422096E-02
15	-4.66197204	.45321408	-.156007E-01
16	-4.64637136	.46881497	-.214238E-01
17	-4.62494754	.49023861	-.102000E-01
18	-4.61474752	.50043857	-.129862E-01
19	-4.60176134	.51342505	-.252676E-01
20	-4.57649374	.53869277	-.158787E-03

energies were obtained that nearby classical translation energies may form a group that correspond to one quantum resonance level. Considering this remark, we find that the agreement between the classical and the quantum results is reasonably good.

In order to investigate the nature of the resonance scattering of the internal excitations we calculated the internal excitation probabilities in the sense of the Feshbach formulation where the molecular projectile is initially in the rotational and vibrational ground state. The results indicate that for incident energies below 0.2 eV the rotational mediations are dominant in the resonance scattering, whereas the resonance states that arise from the vibrational mediations frequently appear above that energy. Therefore we conclude that it is necessary to include the vibrational degree of freedom of the projectile molecule for describing the scattering process.

The classical and quantum methods have been applied to the oxygen molecule-silver surface system to investigate the mechanism of the scattering and to examine the energy transfer process. For comparison with experimental measurements, the use of more realistic potential is essential. Computations in this direction are in progress.

References

1. J. C. Tully, Annu.Rev.Phys.Chem. 31, 331(1981).
2. J. A. Barker and D. J. Auerbach, Surf. Sci. Rep. 4, 1(1984).
3. R. B. Gerber, Chem. Rev. 87, 29(1987).
4. X. Shen, Ph.D.thesis, University of California at Riverside, (1986).
5. N. L. Liu, B. H. Choi and X. Shen, Surf. Sci. 198, 79(1988).
6. B. H. Choi, N. L. Liu and X. Shen, Surf. Sci. 198, 99(1988).
7. B. H. Choi , N. L. Liu and X. Shen , in Condensed Matter Physics, edited by
   R.L.Orbach (Springer Verlag, New York, 1986), p.82.
8. Y. L. Tan, Ph.D. thesis, Tsinghua University, (1989).
9. X. Shen, J. Vac. Sci. Tech. A7, 2132(1989).
10. J. H. Lin and B. J. Garrison, J. Chem. Phys. 80, 2904(1984).
11. J. H. McCreery and G. Wolken, J. Chem. Phys. 63, 2349(1975).
12. J. H. McCreery and G. Wolken, J. Chem. Phys. 66, 2316(1977).
13. J. H. McCreery and G. Wolken, J. Chem. Phys. 68, 4338(1978).
14. A. Gelb and M. J. Cardillo, Surf. Sci. 64, 197(1977).
15. A. Gelb and M. J. Cardillo, Surf. Sci. 75, 199(1978).
16. J. M. Bowman and S. Park, J. Chem. Phys. 76, 1168(1982).
17. H. Hoinkes, Rev. Mod. Phys. 52, 933(1980).
18. B. H. Choi and R. T. Poe, J. Chem. Phys. 83, 1330(1985); 83,1344(1985).
19. R. B. Gerber, L. H. Beard and D. J. Kouri, J. Chem. Phys. 74, 4709(1981).
20. D. J. Kouri and R. B. Gerber, Israel J. Chem. 22, 321(1982).
21. G. Drolshagen, A. Kaufhold and J. P. Toennies, J. Chem. Phys. 83(2), 827(1985).
22. M. J. Cardillo, Annu. Rev. Phys. Chem. 32, 331(1981).
23. Ph. Avouris, D. Schmeisser and J. E. Demuth, Phys. Rev. Lett. 48(3), 199(1982).
24. Chien-Fan Yu, Charles S. Hogg, James P. Cowin, K. Birgitta Whaley, J. C. Light
    and Steven J. Siebener, Israel J. Chem. 22, 305(1982).

Imaging Adsorbates on Metals by Scanning Tunneling Microscopy

S. Chiang, D.D. Chambliss, V.M. Hallmark, R.J. Wilson, and Ch. Wöll

IBM Research Division, Almaden Research Center, San Jose, CA 95120, USA

Abstract. Scanning tunneling microscopy has been used to examine several adsorbate systems on metals. Images show the interaction of carbon islands on Pt(111) with Pt step edges. The rotational orientations of ordered naphthalene molecules on Pt(111) are shown and used to infer the molecular binding site. A detailed description of the reconstruction of clean Au(111) is given to explain how Ni islands nucleate in ordered arrays on the Au(111) surface.

1. Introduction

Over the past several years, the scanning tunneling microscope (STM) has proved to be a useful tool for determining surface structure. STM data has been more readily interpretable for controlled experiments on single crystal surfaces in ultrahigh vacuum (UHV). For metallic substrates, the STM has been successfully used to elucidate surface structure for both atomic and molecular adsorbates, as well as nucleation and growth of metallic overlayers on metals. Examples of recent work include the imaging of the systems of oxygen chemisorbed on Cu(110) [1,2] and Cu(100) [3] surfaces, ordered arrays of benzene and CO molecules on Rh(111) surfaces [4,5,6], observation of the internal structure of copper phthalocyanine molecules on Cu(100) [7], and nucleation and growth of Ag on Au(111) [8]. Here we show the application of STM to several other adsorption systems on metal substrates. We first describe the imaging of carbon islands and naphthalene molecules on Pt(111)[9]. Then we discuss the reconstruction of clean Au(111) [8,10] and its relationship to the nucleation of Ni islands on the surface.[11]

2. Carbon Islands on Pt(111)

Carbon contamination on metal surfaces has significant implications for surface catalysis. Surface carbon can be either graphitic carbon which acts as a catalyst poison or carbidic carbon which is catalytically active. Growth and stability of carbon islands on platinum have been studied by field emission microscopy and very high resolution Auger electron spectroscopy [12,13]. STM investigations of carbon island structure on low index platinum surfaces should provide useful real space images of the islands to assist in the understanding of such catalytic processes.

Springer Series in Surface Sciences, Vol. 24   **The Structure of Surfaces III**
Editors: S.Y. Tong · M.A. Van Hove · K. Takayanagi · X.D. Xie
© Springer-Verlag Berlin, Heidelberg 1991

The UHV STM used in these experiments has been described previously [14]. All STM images shown here have been taken in the constant current topographic mode, with image acquisition times of ~5 minutes. To reduce the amount of carbon in the bulk, the Pt(111) single crystal was annealed at 600°C in 10^{-6} torr O_2 for many hours. Subsequent cleaning consisted of cycles of Ar ion sputtering and annealing to 800°C for >2 min. Auger spectroscopy of the cleaned Pt surface revealed residual carbon contamination was less than the ~5% detection limit of the instrument. Low energy electron diffraction (LEED) patterns for this cleaned sample exhibited low background intensity and very sharp (111) spots. When higher C coverages remained after an inadequate number of cleaning cycles, LEED observations showed a segmented ring outside the first order spots.

STM observations of such cleaned Pt(111) surfaces reveal impurity islands. The size and number of these islands scaled with the amount of C on the surface, as measured by both Auger spectroscopy and LEED. Such carbon islands usually are located at kinks along monoatomic steps of the Pt(111) surface, as shown in Figure 1, but occasional isolated islands also occur on terraces. The Pt(111) step height in the STM images agrees with the crystallographic value of 2.26Å. The C islands appear to be about half the height of a Pt(111) monoatomic step. Some islands appear to have a pseudo-hexagonal shape. Two types of Pt step edges are observed: smooth, straight steps oriented along the crystallographic directions of the (111) surface, and rough, curved steps which do not follow the crystallographic orientations. In Figure 1, note that the C islands are typically located at junctions between the straight and rough steps, implying that C island growth and step formation are correlated. Surface segregated carbon may diffuse to kinks in the platinum steps or alternatively, kinks on the platinum

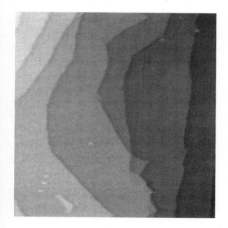

Fig. 1. Top view of STM topographic image of monoatomic steps on bare Pt(111), showing C islands at many of the kinks in the steps. The area of the carbon islands indicates a coverage of ~2%. Note the nearly triangular Pt island on terrace in lower left corner. White indicates elevated areas. Image area is 3000Åx3000Å. Sample bias V_S=+0.22V and tunneling current I=0.2nA.

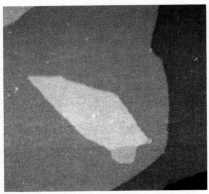

Fig. 2.   Pt island on Pt(111) terrace, with two C islands near edge of Pt one. Image area is 3000Åx3000Å. Vₛ=+0.22V and I=0.2nA.

step edges may be nucleated where carbon islands are located. Experiments on samples annealed at higher temperature so as to cause surface carbon to diffuse into the bulk of the Pt crystal may distinguish between these possibilities. The latter mechanism is supported by the formation of a monoatomic height island of Pt on a Pt terrace, shown in Figure 2, where two carbon islands are located at the edge of the Pt one. Platinum steps also appear to cross one another on the surface in some areas, and are usually accompanied by a C island at the junction. An example of this phenomenon is shown below when we discuss the adsorption of naphthalene on the Pt(111) surface. We have also observed transitions between areas of high step density to areas with terraces ~1000Å in width. The high step density areas typically do not display C islands, but quite large C islands have been found at kinks bordering very large Pt terraces. Our limited attempts to achieve atomic resolution on these carbon islands have so far been unsuccessful.

3.   Naphthalene on Pt(111)

Recent successes in STM imaging of ringed molecules adsorbed flat on surfaces [4,5,6,7] led us to examine the system of naphthalene on Pt(111).Here we will discuss only monolayer coverage naphthalene, which forms an ordered structure when deposited at temperatures between 100 and 200°C. Future publications will discuss our measurements of low coverage and disordered monolayer coverage naphthalene on Pt(111) [9].

Naphthalene was adsorbed onto the cleaned Pt(111) crystal from the gas phase by backfilling the vacuum system with naphthalene from a gas handling line maintained at about 80°C. The naphthalene exposure was $5x10^{-8}$ mbar for 600 sec, with the sample temperature between 100 and 200°C. LEED patterns with (6x3) structure comparable to published data were obtained [15]. Previous work suggested a real space structure

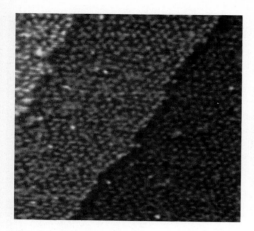

Fig. 3. High contrast image of naphthalene on three different terraces of Pt(111), separated by monoatomic steps. Image shows that naphthalene molecules have 3 distinct molecular orientations, 120° apart. Image area is 300Åx300Å. V_S=+1.01V and I=0.7nA.

with a (6x3) unit cell consisting of two molecules adsorbed with the rings parallel to the surface. The molecules in this model are located at (3x3) Pt lattice points, with molecules in alternating rows having different rotational orientations. The absolute orientation of the molecules with respect to the substrate crystallographic axes was not previously known. The threefold symmetry of the (111) substrate leads to three equivalent domains of this structure.

In the STM images, the naphthalene molecules appear as bi-lobed structures with three different orientations on the surface, 120 degrees apart, as shown in Figure 3. The alternating rows of molecules with parallel orientation expected for the (6x3) unit cell structure are rarely observed, and then only in domains containing six or fewer molecules. By overlaying a Pt lattice mesh onto the naphthalene images, we can form detailed maps of the observed domain structures. The molecular centers fall on the expected (3x3) superlattice, and domain boundaries typically involve shifting this superlattice by one Pt atom. By comparing the orientation of the long axis of the molecules with the domain boundary directions, we can assign an absolute orientation of napthalene parallel to the crystallographic axes of the Pt(111) surface. There are two possible high symmetry adsorption sites which have three-fold symmetry: a three-fold hollow and an on top site. Because of the orientation of the naphthalene long axis along the close packed direction of the platinum lattice, the three-fold hollow site would necessarily be asymmetric. Thus the on top adsorption site seems to be the more likely location for naphthalene molecules on Pt(111).

Naphthalene (3x3) domains are quite small, typically <50Å diameter, and all three possible molecular orientations are equally populated within the domains. Figure 4 shows a highly ordered area of the naphthalene on Pt(111) substrate. The three different rotational orientations of the molecules are very clear. The molecules appear to have

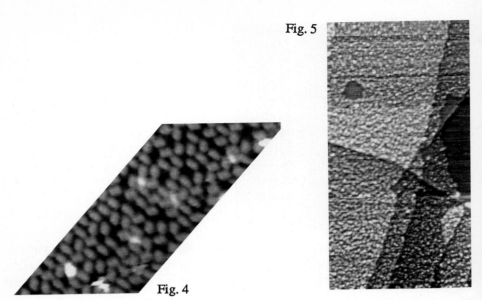

Fig. 4.   Highly ordered image of naphthalene on Pt(111), showing domain size and separation distance of 1 Pt lattice constant between domains. Image resolves two lobes of naphthalene molecules. Note bright "defects", which may be tilted molecules. Image area is 90Åx180Å. V_S=+0.72V and I=2.2nA. Image has been corrected for large amount of piezoelectric creep resulting from movement of tip to new area of sample.

Fig. 5.   Large image, 900Åx1500Å, showing naphthalene on Pt(111). Note that naphthalene does not adsorb on C islands. Some steps appear to cross one another, with C islands at their intersection. V_S=+2.1V and I=0.4nA.

two distinct lobes, but other internal molecular structure seems to be tip dependent and not very reproducible. Also observed in this image are defects which appear to be approximately twice the typical naphthalene height of 1 Å. By the same mapping technique detailed above, we discern that these defects are located at positions where neighboring naphthalene molecules are too near to accommodate another naphthalene in a comparable adsorption site. Because these defects are the same lateral size as a naphthalene molecule and also appear to have a bi-lobed structure, it is reasonable to think that they are indeed naphthalene molecules, possibly tilted to fit into the small space.

In Figure 5, we show a large area scan of naphthalene on Pt(111). This large image, 900 x 1500Å, is composed of 3 smaller ones, measured consecutively from top to bottom. Note that naphthalene does not stick to the carbon islands on the Pt(111) surface. Steps appear to cross one another, with C islands located at their intersections. Such large scans emphasize the power of STM imaging in looking at defects and disorder on surfaces.

4. Reconstruction of Clean Au(111)

We have also observed the reconstruction of clean Au(111). Here we discuss this reconstruction in more detail than our previous work [10], in order to understand the binding sites on this surface for Ni islands, which are discussed in the next section.

The Au(111) crystal was cleaned by Ar ion bombardment (1 keV) and annealed at 600°C for ≥5 min. The reconstruction of the surface appears as a pattern of zigzagging 0.15Å high ridges in STM images that do not resolve individual atoms (Fig. 6(a)). He diffraction [16] and STM [10] experiments have previously determined the atomic structure of the uniaxial ~23×√3 reconstruction of Au(111) to be as follows. Atoms in the contracted surface layer vary in registry from the subsurface layers, with occupation of hollow sites of both face centered cubic (fcc) and hexagonal close packed (hcp) stacking order. The hollow site regions are the lower (i.e., darker) areas in the STM images. Between them are partial surface dislocations in which surface atoms are near bridge sites and thus appear higher in the STM. The pattern of these dislocations is shown in Figure 6(b). The different widths for the hcp and fcc hollow site bands can be attributed to an energy difference between the two sites; the non-bulklike hcp stacking order presumably has the higher energy and corresponds to the narrower bands. On our clean single crystal sample, small islands and terraces may have single domains of the uniaxial reconstruction, but large terraces are covered with a superlattice of alternating uniaxial domains that gives rise to a herringbone ridge pattern. [8]

The STM ridges, however, do not consist of simple alternating straight sections. We can understand the detailed shape of the ridges by looking at the surface crystallography underlying the reconstruction of Au(111). The bridge site ridges in the

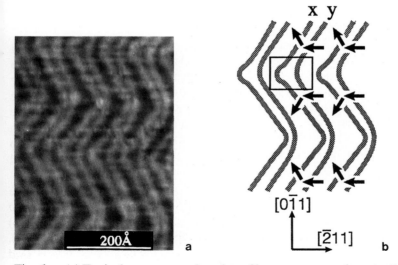

Fig. 6.   (a) Typical reconstructed section of large terrace on clean Au(111). V_S=-2V.
(b) Sketch of herringbone ridges and buckling. Arrows show orientation of bridge sites for surface dislocation segments. The box shows the area depicted in more detail in

209

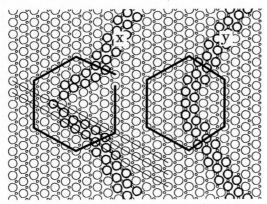

Fig. 7. Possible atomic structure of reconstruction near "elbow". Heavier circles are surface atoms near bridge sites, while fine circles are in hollow sites. Second layer atoms are indicated by very small circles. Burgers circuits of surface lattice (heavy polygons) show that while type x ridge at left has dislocation (i.e., circuit is not closed) type y does not (closed circuit). Fine lines show one visualization as an edge dislocation.

uniaxial domains of the reconstruction are partial surface dislocations with different Burgers vectors. When uniaxial domains are combined into the herringbone pattern these dislocation ridges join in two different ways (Fig. 6(b)). The segments of type x ridges have two different Burgers vectors, and the difference is an edge dislocation of the surface lattice. The type y segments have the same Burgers vector so they can join without any additional dislocations. The surface lattice dislocation on the type x ridge is illustrated in detail in Fig. 7. This difference between the ridges also accounts for systematic distortions of the ridge shape observed in clean Au(111) data like those in Fig. 6(a). At domain walls of one orientation, the narrow (presumably hcp) dark bands tend to bulge, while at oppositely oriented domain walls they tend to pinch off. Fig. 7 incorporates this sort of distortion. The atomic positions at the core of the dislocation have not been measured, and indeed, our STM data suggest that the elbow structure varies from place to place. Regardless of the detailed structure, however, the dislocation implies that there must be at least one site that differs in bond topology from that of the close-packed surface. This chemically distinct site serves to nucleate Ni islands on Au(111), as discussed in the next section.

5. Ni Islands on Au(111)

Nucleation and growth of metal films have been studied for decades using electron microscopy, but with the high resolution of the STM and the cleanliness of a UHV environment we can observe in much finer detail the intrinsic behavior of metals on clean metal sufaces. We have found that the surface dislocations of the herringbone reconstruction of Au(111) act as an ordered array of binding sites for the nucleation of

Fig. 8.   Correlation of Ni island nucleation with Au reconstruction. Completed nucleation and polygonal shape of Ni islands at 0.14ml. V_S=-1.41V; rate 0.1ml/min. A nonlinear gray scale is used to make 0.15Å ridges visible.

Ni islands (Fig. 8). The islands nucleate in well ordered arrays, 73Å apart in rows separated by 140Å. For coverage $0.1 \leq \theta \leq 0.7$, the arrays of Ni islands have a fixed number density, a narrow size distribution, and an average size that can be varied continuously by varying Ni coverage.

After the sample is allowed to cool for at least 60 min. to reach room temperature, it is exposed to Ni flux. Ni of 99.99% purity is evaporated from an alumina coated W basket at a deposition rate of 0.04-0.4 monolayer (ml)/min. The pressure rise is $<10^{-10}$ Torr, and no contamination of the surface with O or C is found by Auger electron spectroscopy (AES). The STM images shown here were obtained with tunnel currents of 0.5-2.0nA. Each image was recorded in ~3 min, at least 1 hour after deposition.

STM data at low Ni coverage confirm that the island nucleation site is the dislocation on the type x ridge. The Ni islands nucleate almost exclusively at the "elbows" formed by the ridges at the boundaries between neighboring uniaxial domains (Fig. 8). As discussed above, the difference in chemical bonding at such an elbow site may facilitate the nucleation of Ni islands.

Our model for Ni island formation is that Ni atoms arrive from the vapor and diffuse freely on the surface until encountering an atomic step, a growing Ni island, or an elbow site. The high bond coordination of sites at the base of a step or island makes it highly likely that an atom would stick there. To produce the compact, nearly hexagonal island shapes seen in Fig. 8, there must also be some motion along the edge, multi-atom rearrangement, or site dependent sticking The atom can also stick at the

211

Fig. 9.   STM image of Ni island arrays on Au(111). In this top view representation of STM topographic data, light regions are higher than dark. Several atomically flat Au terraces are seen, separated by steps of single-atom height. Small light dots on each terrace are monolayer Ni islands. Ni coverage is 0.11±0.03ml deposited at 0.4ml/min. Two orientations of island rows are evident. V_S=+2.0V.

surface lattice dislocation at an elbow. The sticking probability may be smaller here than at the base of a step. Indeed, our measurements of the nucleation behavior at low Ni coverage and the island size distribution at higher coverage suggest a value on the order of 0.01 for sticking at bare elbow sites at room temperature.[11] Once an elbow site is occupied, other atoms are highly likely to stick at the growing island.

The wide area image in Fig. 9 is typical for a 0.1 ml Ni deposit on room temperature Au(111). On the two wide terraces, the island ordering is fairly regular, although the spacing between rows is variable and each row meanders. Island rows are also formed on the narrow terrace near the top center. Three orientations of island arrays are possible, related by 120° rotations; two of these are seen in Fig. 9. The regularity of the island pattern is disrupted near step edges, and on the narrowest terrace segment (bottom center) no islands are observed.

The number of islands in the completed array is fixed by the herringbone spacing at $1 \times 10^{12} cm^{-2}$. Thus a specified average island size can be selected by choosing the appropriate Ni coverage. The regular island spacing causes the size distribution to be narrower than those distributions produced by homogeneous nucleation. At 0.1ml the islands have a mean size of 140 Ni atoms with a standard deviation of 40 atoms. The absence of very large islands and of the large local coverage variations that can occur in diffusion limited aggregation [8] may be important for investigating size dependent properties in the nanometer regime.

This unusual nucleation mechanism may be extended to other metal systems. There are presumably other metals with the tendency to nucleate at the Au(111) elbow sites. Those metals that do not nucleate there (e.g., Ag) [8] could be seeded with 0.05ml Ni exposure to nucleate an island at each elbow. While Au(111) is the only close-packed face of an fcc elemental metal crystal known to reconstruct, it is useful to note that the herringbone reconstruction is essentially a network of misfit dislocations between a contracted surface layer and the bulk. It may be possible to find conditions for metal heteroepitaxy in which the misfit dislocations form an ordered system that can cause similar ordered nucleation on other substrates. In any event, the dependence of this nucleation behavior on deposited species and on temperature can yield important information on the interactions of metal atoms.

Acknowledgments

This work was partially supported by the Office of Naval Research (N00014-89-C-0099). We are grateful to J.P. Toennies for the loan of the Au(111) crystal.

References

1   F. M. Chua, Y. Kuk and P. J. Silverman, *Phys. Rev. Lett.* **63**, 386 (1989).
2   D. J. Coulman, J. Wintterlin, R. J. Behm, and G. Ertl, *Phys. Rev. Lett.* **64**, 1761 (1990).
3   Ch. Wöll, R. J. Wilson, S. Chiang, H. C. Zeng, and K. A. R. Mitchell, to be published.
4   H. Ohtani, R. J. Wilson, S. Chiang, and C. M. Mate, *Phys. Rev. Lett.* **60**, 2398 (1988).
5   S. Chiang, R. J. Wilson, C. M. Mate, and H. Ohtani, *J. Microsc.* **152**, 567 (1988).
6   S. Chiang, R. J. Wilson, C. M. Mate, and H. Ohtani, *Vacuum*, **41**, 118 (1990).
7   P. H. Lippel, R. J. Wilson, M. D. Miller, Ch. Wöll, and S. Chiang, *Phys. Rev. Lett.* **62**, 171 (1989).
8   D.D. Chambliss and R.J. Wilson, to be published.
9   V. M. Hallmark, S. Chiang, and Ch. Wöll, to be published.
10   Ch. Wöll, S. Chiang, R. J. Wilson, and P. H. Lippel, *Phys. Rev. B.*, **39**, 7988 (1989).
11   D. D. Chambliss, R. J. Wilson and S. Chiang, to be published.
12   R. Vanselow and M. Mundschau, *J. de Phys.* **C7**, 117 (1986).
13   M. Mundschau and R. Vanselow, *Surf. Sci.* **160**, 23 (1985); *J. de Phys.* **C7**, 121 (1986).
14   S. Chiang, R. J. Wilson, Ch. Gerber and V. M. Hallmark, *J. Vac. Sci. Techol.* **A6**, 386 (1988)
15   D. Dahlgren and J. C. Hemminger, *Surf. Sci.* **109**, L513 (1981).
16   U. Harten, A. M. Lahee, J. Peter Toennies, and Ch. Wöll, *Phys. Rev. Lett.* **54**, 2619 (1985).

Scanning Tunneling Microscopy of Organic Conductors

A. Kawazu, N. Ara, and M. Yoshimura

Department of Applied Physics, The University of Tokyo,
Hongo, Bunkyo-ku, Tokyo 113, Japan

The surface structures of single crystals of organic (super)conductors and TTF-TCNQ films deposited on a mica substrate were investigated by using scanning tunneling microscopy (STM) at room temperature in air. STM images of crystal surfaces parallel to the conductive layer showed the periodic structures, which are in good agreement with the arrangement of the BEDT-TTF molecule projected onto the observed plane. Images taken perpendicular to the layer revealed the alternating stacking structure of the conductive BEDT-TTF layer and the insulative $Cu(NCS)_2$ layer, as expected from the ESR, ESCA and other measurements. The observed current distribution on the BEDT-TTF molecule was concentrated on the central TTF part, consistent with the *ab initio* calculation. In the study of TTF-TCNQ films, besides the normal region where TTF and TCNQ molecules are arranged as observed in TTF-TCNQ crystal by Sleator and Tycko, we observed domains where TCNQ molecules were tilted in the reverse direction in crystal *ab*-plane with respect to the surface normal.

1. Introduction

Since its invention in 1982 by Binnig et al. [1], scanning tunneling microscopy (STM) has offered novel insight into surface science. Recently, this technique has been widely applied to many kinds of organic conductive, semiconductive and insulative materials such as TTF-TCNQ crystal [2], copper-phthalocyanine molecules on Cu(100) surfaces [3], polyaniline molecules on Si surfaces [4], benzene molecules on Rh(111) surfaces [5], LB films on Si surfaces [6], DNA [7], liquid crystals on graphite [8] and MoS_2[9]. These studies yielded much important information useful to understanding the adsorption mechanism of molecules, electronic structures of molecules and so on. In this paper, we report the study of the surface of organic crystals of (BEDT-TTF) compounds [BEDT-TTF : bis (ethylenedithio) tetrathiafulvalene] and thin films of TTF-TCNQ [TTF: tetrathiafulvalene, TCNQ : tetracyanoquinodimethane] by STM. These materials are conductive and are considered to be appropriate for STM observation.

2. Experimental

Four kinds of single crystals of (BEDT-TTF) salts were prepared by the conventional electrochemical oxidation of (BEDT-TTF). The dimensions of the

Springer Series in Surface Sciences, Vol. 24   **The Structure of Surfaces III**
Editors: S.Y. Tong · M.A. Van Hove · K. Takayanagi · X.D. Xie
© Springer-Verlag Berlin, Heidelberg 1991

crystals obtained are about 1-2 x 2-3 x 0.05-0.1 mm^3 [10]. These surfaces show a metallic feature with high reflectivity. No particular surface treatment was performed except rinsing in deionized water.

TTF-TCNQ crystal was prepared by the mixture of TTF and TCNQ in acetonitrile solution. TTF-TCNQ film was deposited onto the freshly cleaved mica substrate at the deposition rate of 10 nm/s at room temperature in high vacuum (3 x 10^{-4} Pa) by heating the quartz crucible containing TTF-TCNQ crystals. The film was about 100nm in thickness. Epitaxial growth of TTF-TCNQ films on the mica substrate with its crystal ab-plane was confirmed by electron microscopy and diffraction [11].

STM observation was performed in air at room temperature using Pt-Ir tips.

3. Results and Discussion

3.1 κ-(BEDT-TTF)$_2$Cu(NCS)$_2$

κ-(BEDT-TTF)$_2$Cu(NCS)$_2$ is the first organic two-dimensional superconductor whose critical temperature (Tc) exceeds 10 K under ambient pressure [12]. The crystal data determined by X-ray diffraction are shown inTable 1. Figure 1 shows the crystal structure, which consists of alternating stacked layers of (BEDT-TTF) and Cu(NCS)$_2$ molecules along the crystal a^*-axis [13]. According to the ESR, ESCA and thermoelectric power measurements [14], the BEDT-TTF layers become conductive while the Cu(NCS)$_2$ layers become insulative through the electron transfer from the BEDT-TTF layers to the Cu(NCS)$_2$ layers. BEDT-TTF molecules are dimerized in the crystal, and one hole exists on a dimer. As a result, this material demonstrates a two-dimensional conductivity in the conducting layers parallel to the bc-plane.

Table 1. Structural parameters and critical temperature (Tc) of (BEDT-TTF:ET) salts

Compound	κ- ET$_2$Cu(NCS)$_2$	β- ET$_2$I$_3$	ET$_2$KHg(SCN)$_4$	ET$_2$(NH$_4$)Hg(SCN)$_4$
Crystal Type	monoclinic	triclinic	triclinic	triclinic
Space Group	P2$_1$	P$\bar{1}$	P$\bar{1}$	P$\bar{1}$
a [nm]	1.6248	1.5243	1.0082	1.0091
b [nm]	0.8440	0.9070	2.0565	2.0595
c [nm]	1.3124	0.6597	0.9933	0.9963
α [deg]	--.--	109.73	103.70	103.65
β [deg]	110.30	95.56	90.91	90.53
γ [deg]	--.--	94.33	93.06	93.30
Vol. [nm^3]	1.6880	0.8489	1.9970	2.0081
Tc [K]	10.4	1.5, 7	--.--	1.1

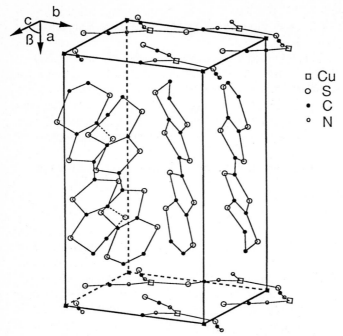

Fig. 1. The crystal structure of κ–(BEDT-TTF)$_2$Cu(NCS)$_2$.

Figure 2(a) shows a current image scanned over the crystal bc-plane. The sample voltage was kept at 29.9 mV during measurement [15-17]. The size of the unit cell drawn was about 0.94 nm x 1.56 nm, which is close to the value of the bc unit cell in the bulk (0.8440 nm x 1.3124 nm). The change of the polarity of the tunneling bias voltage to the tip, Vt (-300 mV < Vt < 300 mV), causes no particular change of the STM image. These results indicate that this material is metallic, and the observed protrusions correspond to (BEDT-TTF) molecules, which are responsible for the two-dimensional conduction.

According to the ab $initio$ calculation on the BEDT-TTF molecules [18], high distribution probability of the highest occupied molecular orbital (HOMO) around the sulfur atoms exists. Figure 2(b) shows the projected arrangement of the upper-half part of BEDT-TTF molecules on the bc-plane. The sulfur atom nearest to the surface in each BEDT-TTF molecule, which is expected to appear as a protrusion in the STM image, is overlapped with the elliptic patterns. The arrangement of the elliptic patterns in the unit cell depicted in fig.2(b) is in good agreement with the protrusions in the unit cell drawn in fig.2(a). The difference in brightness among the four protrusions in a unit cell can be well interpreted referring to the calculated probability distribution of electron density [18].

Figure 3(a) shows a STM image obtained over the crystal ab-plane at the sample voltage of -22.9 mV in the variable current mode [17,18]. Alternating bright

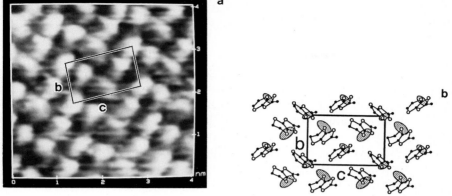

Fig. 2. (a) A 4.1 nm x 4.1 nm current image of the bc-plane of κ–(BEDT-TTF)$_2$Cu(NCS)$_2$. Sample voltage is 29.9 mV to the tip. (b) (BEDT-TTF) molecular arrangement projected onto the bc-plane.

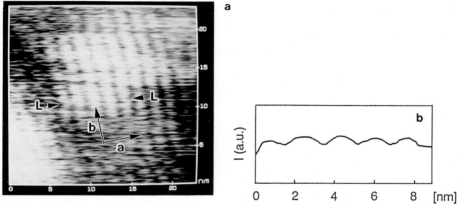

Fig. 3. (a) A 23 nm x 23 nm current image obtained over the ab-plane at the sample voltage of -22.9 mV. (b) The cross section along the line L in fig. 3.

and dark bands along the b-axis can be observed. Figure 3(b) shows a line profile along the line L drawn in Fig.3(a). The experimentally obtained period of this structure was about 1.8 nm and was in good agreement with the lattice constant of $a = 1.6248$ nm. The bright and dark regions in fig.3(a) correspond to the conductive (BEDT-TTF) layers and the insulative Cu(NCS)$_2$ layers, respectively.

The current distribution in the conductive layer is concentrated on the middle region of the bright band as seen in fig.3(b). This is consistent with the results of both the extended Hückel calculation [19] and the *ab initio* calculation [18], which show that the probability density of the HOMO is higher on the central part ; i.e., the density is higher on the TTF of the BEDT-TTF molecule than on either of its end sides. Domain structures were occasionally observed in the large scan image in the *ab*-plane, which were possibly responsible for the electronic conductivity along the *a*-axis together with electron hopping.

3.2 β-(BEDT-TTF)$_2$I$_3$

(BEDT-TTF)$_2$I$_3$ shows many different phases of structure and is very sensitive to the history and environmental conditions such as temperature and pressure [20]. The bulk structure, determined by X-ray diffraction, consists of the conductive BEDT-TTF layers and the insulative triiodide (I$_3$) layers [21]. Two BEDT-TTF molecules and one triiodide were included in the unit cell. The STM image over the *bc*-plane (fig.4(a)) showed the two kinds of rows of molecules along the [011] direction, each of which consisted of two types of protrusions [22]. The size of the surface unit cell shown in fig.4(a) is 1.41 nm x 1.06 nm (*b'* x *c'*, α' = 114 degrees), and is about two

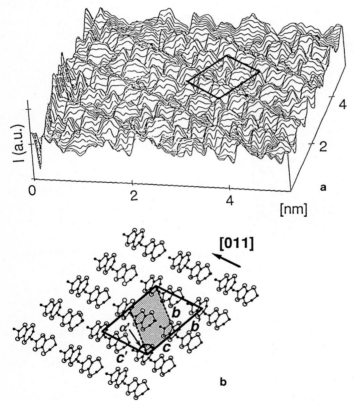

Fig. 4. (a) A 5.2 nm x 5.2 nm current image of β-(BEDT-TTF)$_2$I$_3$ over the *bc*-plane. Two kinds of rows along the [011] direction are visible. (b) BEDT-TTF molecular arrangement of the bulk crystal. The *bc* unit cell is hatched with dotted lines. The size of the observed unit cell (solid line) is two times as large as that of the *bc* unit cell.

times as large as that in the bulk bc-plane (1.319 nm x 0.924 nm, $\alpha' = 112.5$ degrees), as is shown in fig.4(b). The structure can be related to the bulk unit cell in a bc-plane with the matrix of

$$\begin{bmatrix} 0 & -2 \\ 1 & 1 \end{bmatrix}.$$

This reconstruction of the surface can be explained well by considering the instability of anion layers.

3.3 (BEDT-TTF)$_2$KHg(SCN)$_4$

(BEDT-TTF)$_2$KHg(SCN)$_4$ does not have a superconducting phase and maintains a metallic phase to 1.5 K [23]. The crystal structure is shown in fig.5. This compound has the anion layer consisting of a triple sheet of K(NCS)$_4$ and Hg(SCN)$_4$ molecules. As a result, the anion layer of this material is rather thick (about 0.6 nm) and is expected to be rigid and stable compared with the case of β-(BEDT-TTF)$_2$I$_3$. Figure 6(a) shows the projected arrangement of the BEDT-TTF molecules on the crystal ac-plane, where the positions of K and Hg ions in the top anion layer are also marked. Four BEDT-TTF molecules in a unit cell are labeled as A to D. They are different in shape, and the molecules C and D have the K or Hg ion on them. The tunnel currents from the molecules C and D are expected to be modified by the existence of these K and Hg ions.

Figure 7 shows a current image obtained at the sample voltage of 18.3 mV, in which the line profiles along lines L and M are also shown [16,24]. Four types of protrusions with different shapes and brightness can be observed. The observed size of the unit cell is 0.9 nm x 1.2 nm. This is very close to the size of the unit cell in the crystal ac-plane (1.008 nm x 0.993 nm). The protrusions on the line L are much sharper, while the protrusions on the line M are rather distorted. Two different types

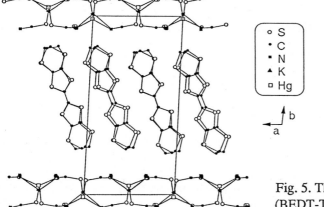

o S
• C
▪ N
▲ K
▫ Hg

Fig. 5. The crystal structure of (BEDT-TTF)$_2$KHg(SCN)$_4$.

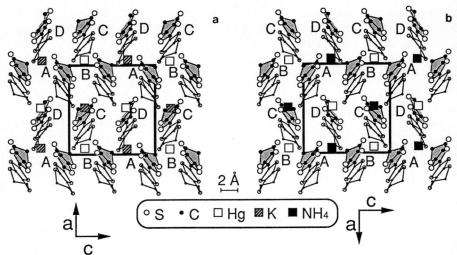

Fig. 6. The arrangement of (BEDT-TTF) molecules projected onto the *ac*-plane in (BEDT-TTF)$_2$XHg(SCN)$_4$ (X = K, NH$_4$) (a) viewed from *b**-axis and (b) from -*b**-axis. The upper part of each (BEDT-TTF) molecule is hatched. The positions of X, Hg ions are also marked.

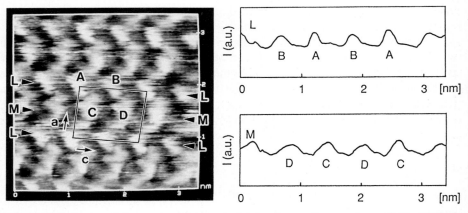

Fig. 7. A 3.5 nm x 3.5 nm current image obtained over the *ac*-plane of the (BEDT-TTF)$_2$KHg(SCN)$_4$. The sample voltage is 18.3 mV. The cross sections along the lines L and M are also shown.

of protrusions, brighter and less bright ones, are lined up alternatively on both lines L and M. As the observed size of unit cell agrees with that of crystal and there are four BEDT-TTF molecules in a unit cell of a (BEDT-TTF)$_2$KHg(SCN)$_4$ crystal, the protrusions in the STM image can be attributed to the BEDT-TTF molecules. We can further assign these four protrusions to the molecules A to D indicated in

220

fig.6(a) by considering the tilt angle to the *ac*-plane and the deformation of each molecule, and referring to the result of the *ab initio* calculation which gives the spatial distribution of charge density of HOMO on sulfur atoms which mainly contribute to the tunnel current [24].

For the protrusions assigned to the BEDT-TTF molecules C and D, the darker parts are observed at the positions of K and Hg ions. These darker parts can be originated from the modification of the tunneling probability of electrons from HOMO due to the effect of the electrostatic potential of K or Hg atom. Strong bias voltage dependence of the STM image was also observed, reflecting the dependence of the probability density of state on the energy of the electrons in BEDT-TTF molecules.

3.4 $(BEDT\text{-}TTF)_2(NH_4)Hg(SCN)_4$

Although the crystal structure of $(BEDT\text{-}TTF)_2(NH_4)Hg(SCN)_4$ is almost the same as that of $(BEDT\text{-}TTF)_2KHg(SCN)_4$, this crystal shows superconductivity below Tc = 1.1K at ambient pressure [25]. It is very interesting to study these crystals with almost the same structure to understand the conduction mechanism in the organic materials.

Figure 8 shows the STM image obtained at the sample voltage of 48.2 mV [26]. The observed image is almost the same as that of the $(BEDT\text{-}TTF)_2KHg(SCN)_4$ surface shown in fig.7. There are four different types of protrusions in a unit cell. The observed size of the unit cell is 1.18 nm x 1.09 nm, and this is in good agreement with that in the crystal *ac*-plane (a = 1.009 nm, c = 0.996 nm). Considering the shape, intensity and the position of protrusions in a unit cell as in the case of the $(BEDT\text{-}TTF)KHg(SCN)_4$ crystal, these bright images can be assigned to the BEDT-TTF molecules A to D drawn in fig.6(b). The darker parts for C and D in fig.8 can also be interpreted considering the existence of NH_4 or Hg ions on the BEDT-TTF molecules. We did not observe such modification of protrusions for the surfaces of κ-$(BEDT\text{-}TTF)_2Cu(NCS)_2$ and β-$(BEDT\text{-}TTF)_2I_3$ *bc*-planes. In these cases, there are no atoms on top of BEDT-TTF molecules and we can observe BEDT-TTF molecules directly through the anion layer.

The precise evaluation of the effect of the topmost anion layer on the tunneling current will be difficult without the electron density calculation taking the existence of both the BEDT-TTF layer and anion layer into consideration. However, this requires an enormous computation time, and it is difficult to put this calculation into practice in the present situation.

Figure 9 shows a wide 50 x 50 nm^2 scan image over the *ac*-plane without any image processing. Individual molecules in alternating line-shaped structures along the *c*-axis are minutely imaged, and neither defect nor step structure is observed. On the contrary, we often observe many defects for the $(BEDT\text{-}TTF)_2I_3$ surface. These remarkable differences in the stability of the surface between these two materials may arise from the different stability of the anion layers.

In the case of a $(BEDT\text{-}TTF)_2(NH_4)Hg(SCN)_4$ crystal, there are 18 atoms in an anion layer of a unit cell. They form a triplet-sheet structure. This quasi-three-

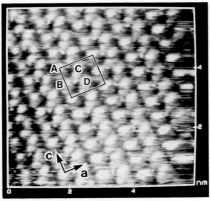 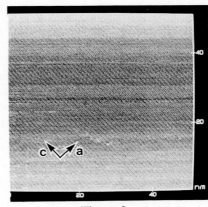

Figure 8   Figure 9

Fig. 8. A 6 nm x 6 nm current image obtained over the ac-plane of the (BEDT-TTF)$_2$(NH$_4$)Hg(SCN)$_4$. The sample voltage is 48.2 mV.

Fig. 9. A wide 50 nm x 50 nm scan image over the ac-plane of the (BEDT-TTF)$_2$(NH$_4$)Hg(SCN)$_4$ obtained at the sample voltage of 50 mV at the constant current of 4.1 nA.

dimensional framework structure can be more stable than that in (BEDT-TTF)$_2$I$_3$ crystal, where only three atoms of I are in a unit cell, they are arranged in a single layer, and each I$_3^+$ ion is weakly bonded to each other or to the BEDT-TTF layer. Thus, in some cases, the sublimation of I atoms from the surface may occur and the BEDT-TTF layer may be exposed to air directly. These results indicate that the anion layer with rigid structure is expected to design stable (BEDT-TTF) salts.

3.5 TTF-TCNQ Thin Films

TTF-TCNQ has a monoclinic structure with lattice parameters of $a = 1.2298$ nm, $b = 0.3819$ nm, $c = 1.8468$ nm and $\beta = 104.46$ degrees [27]. It is known as a one-dimensional conductor and shows high conductivity to the stack of molecules of TTF and TCNQ. The molecular arrangement projected onto ab, ac and bc^*-planes is shown in fig.10.

    Figure 11 shows a current image taken at the sample voltage of 4.9mV [28]. The image is in good agreement with the result of STM observation of TTF-TCNQ crystal by Sleator and Tycko [2]. There are two molecular images (A and B) in the unit cell drawn in fig.11. The image A corresponds to the TCNQ molecule which consists of three bright parts (triplet), two of which arise from N atoms of TCNQ and the other from a C atom. The image B corresponds to the TTF molecules. The experimentally obtained size of the unit cell is 1.2 nm x 0.4 nm and is close to that of the unit cell in the ab-plane of the TTF-TCNQ crystal. This indicates that the TTF-TCNQ film grows on the mica substrate with its ab-plane parallel to the substrate surface, which agrees with the result of TEM and electron diffraction measurements.

222

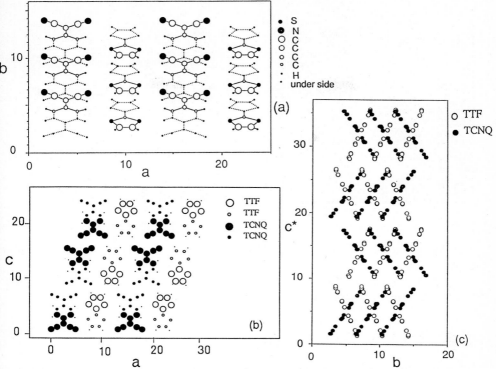

Fig. 10. Molecular arrangement of the TTF-TCNQ crystal projected onto (a) *ab*, (b) *ac* and (c) *bc**-plane.

Figure 12 shows the STM image taken from the other part of the TTF-TCNQ film. Two domains bounded at the line L depicted in fig.12 are observed. The image of TCNQ molecules in the right-hand side of the line L (I) is the same as that observed in fig.11, while rather different molecular images of TCNQ are observed in the left-hand side (II).

Upon comparison of the STM image obtained in region II with that in region I, it can be noted that, in region II : (1) The shapes of triplets are much more symmetric about the line which passes through the center of TCNQ molecules and which is parallel to the *b*-axis. The existence of such symmetric TCNQ molecules indicates that the asymmetric structure of the triplet observed in the normal STM image is originated not from the tip artifact reported by Sleator and Tycko [2], but from the actual electronic structure. If the asymmetry of TCNQ in region I is due to the tip artifact, the same asymmetric patterns should be observed in region II ; (2) The pointed direction of the image of the TCNQ molecule oriented reversely to the direction of the molecules in region I, which was thought to be due to the tilt of TCNQ molecules reversely to the *b*-axis compared with that in region I ; (3) The

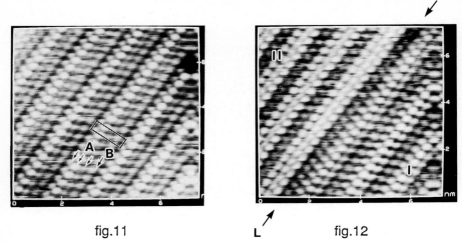

fig.11　　　　　　　L　　　　fig.12

Fig. 11. A 7.5 nm x 7.5 nm current image obtained at the sample voltage of 4.9 mV.

Fig. 12. A 7.0 nm x 7.0 nm current image in the other region of the sample. Two domains (I and II) bounded at L are clearly seen.

protrusion of the center ball in the triplet was darker than that of the outer two balls and was not resolved clearly in the direction parallel to the b-axis. This result shows that the spatial position of TCNQ molecules in the unit cell in region II differs from that in region I. TTF molecules at the boundary were expected to suffer tension from the TCNQ molecules on both sides.

To realize molecular devices in the future, it will be necessary to control the arrangement of molecules precisely. In such a case, STM is indispensable to observe the atomistic structure of thin film, which is clearly shown by the observation of the atomistic change of the arrangement of TCNQ molecules by STM.

4. Summary

The surface structures of the crystals of (BEDT-TTF) salts and thin films of (TTF-TCNQ) were observed by scanning tunneling microscopy. The observed STM images were in good agreement with the arrangement of the (BEDT-TTF) molecules in two-dimensional conductive layers. In the case of $(BEDT-TTF)_2KHg(SCN)_4$ and $(BEDT-TTF)_2(NH_4)Hg(SCN)_4$ surfaces, the modification of protrusions observed at the position where K, NH_4, or Hg existed. Surface reconstruction was observed for the $(BEDT-TTF)_2I_3$ crystal. This reflects the instability of this surface. The existence of conductive and insulative layers in the ab-plane of κ-(BEDT-TTF)$_2$Cu(NCS)$_2$ was also observed directly for the first time.

As for (TTF-TCNQ) films deposited on mica substrate, well-ordered molecular structure coinciding with the bulk structure in the ab-plane was observed. Besides this normal structure, structures characteristic to the thin films were also observed.

These results strongly indicate that the application of STM is essential in understanding the structures or electronic properties of the surface of these organic conductors precisely on an atomistic scale.

Acknowledgement:
The authors are pleased to acknowledge stimulating discussion with Dr. T. Mori, Intstitute for Molecular Science.

References

[1] G. Binnig, H. Rohrer, Ch. Gerber and E. Weibel, Phys. Rev. Lett. 50 (2), 120 (1983).
[2] T. Sleator and R. Tycko, Phys. Rev. Lett. 60, 1418 (1988).
[3] P. H. Lippel, R. J. Wilson, M. D. Miller, Ch. Wöll and S. Chiang, Phys. Rev. Lett. 62, 171 (1989).
[4] D. A. Bonnel and M. Angelopoulos, Synth. Met. 33, 301 (1989).
[5] H. Ohtani, R. J. Wilson, S. Chiang and C. M. Mate, Phys. Rev. Lett. 60, 2398 (1988).
[6] H. Fuchs, W. Schrepp and H. Rohrer, Surf. Sci. 181, 391 (1987).
[7] G. Travaglini, H. Rohrer, M. Amrein and H. Gross, Surf. Sci. 181, 380 (1987).
[8] D. P. E. Smith, J. K. H. Horber, G. Binnig and H. Nejoh, Nature 344, 641 (1990).
[9] M. Hara, Y. Iwakabe, K. Tochigi, H. Sasabe, A. F. Garito and A. Yamada, Nature 344, 228 (1990).
[10] G. Saito, Mol. Cryst. Liq. Cryst. 119, 393 (1985).
[11] K. Yase, O. Okumura, T. Kobayashi and N. Uyeda, Bull. Inst. Chem. Res., Kyoto Univ. 62, 242 (1984).
[12] H. Urayama, H. Yamochi, G. Saito, K. Nozawa, T. Sugano, M. Kinoshita, S. Sato, K. Oshima, A. Kawamoto and J. Tanaka, Chem. Lett. 1988, 55.
[13] H. Urayama, H. Yamochi, G. Saito, S. Sato, A. Kawamoto, J. Tanaka, T. Mori, Y. Maruyama and H. Inokuchi, Chem. Lett. 1988, 463.
[14] H. Urayama, H. Yamochi, G. Saito, T. Sugano, M. Kinoshita, T. Inabe, T. Mori, Y. Maruyama and H. Inokuchi, Chem. Lett. 1988, 1057.
[15] M. Yoshimura, K. Fujita, N. Ara, M. Kageshima, R. Shiota, A. Kawazu, H. Shigekawa and S. Hyodo, J. Vac. Sci. Technol. A8, 488 (1990).
[16] M. Yoshimura, N. Ara, M. Kageshima, R. Shiota, A. Kawazu, H. Shigekawa, H. Mori, M. Oshima, H. Yamochi and G. Saito, *"The Physics and Chemistry of Organic Superconductors"*, edited by G. Saito and S. Kagoshima (Springer-Verlag, Heidelberg, 1990), p.280.
[17] M. Yoshimura, N. Ara, M. Kageshima, R. Shiota, A. Kawazu, H. Shigekawa, H. Mori, M. Oshima, H. Yamochi and G. Saito, Surf. Sci. Soc. of Jpn., 11 353 (1990).
[18] M. Yoshimura, H. Shigekawa, H. Nejoh, G. Saito, Y. Saito and A. Kawazu, Phys. Rev. B, in print.

[19] T. Mori, A. Kobayashi, Y. Sasaki, H. Kobayashi, G. Saito and H. Inokuchi, Bull. Chem. Soc. Jpn. 57, 627 (1984).

[20] T. J. Emge, P. C. W. Leung, M. A. Beno, A. J. Schultz, H. H. Wang, L. M. Sowa and J. M. Williams, Phys. Rev. Lett. 30, 6780 (1984).

[21] T. Mori, A. Kobayashi, Y. Sasaki, H. Kobayashi, G. Saito and H. Inokuchi, Chem. Lett. 1984, 957.

[22] M. Yoshimura, H. Shigekawa, H. Yamochi, G. Saito, Y. Saito and A. Kawazu, to be published.

[23] M. Oshima, H. Mori, G. Saito and K. Oshima, Chem. Lett. 1989, 1159.

[24] M. Yoshimura, N. Ara, M. Kageshima, R. Shiota, A. Kawazu, H. Shigekawa, Y. Saito, M. Oshima, H. Mori, H. Yamochi and G. Saito, Surf. Sci., in print.

[25] H. Mori, S. Tanaka, M. Oshima, G. Saito, T. Mori, Y. Maruyama and H. Inokuchi, Bull. Chem. Soc. Jpn, in press.

[26] A. Kawazu, M. Yoshimura, H. Shigekawa, H. Mori and G. Saito, J. Vac. Sci. Technol., in print.

[27] T. J. Kistenmacher, T. E. Philips and D. O. Lowan, Acta. Crystallogr. Sect. B30, 763 (1974).

[28] N. Ara, M. Yoshimura, H. Shigekawa, A. Kawazu, K. Yase and M. Okada, to be published.

Restructuring and Symmetry of Surfaces: $\sqrt{3} \times \sqrt{3}$-Ag on Si(111)7 × 7

K. Takayanagi

Tokyo Institute of Technology, Materials Science and Engineering,
4259 Nagatsuda, Midori-ku, Yokohama 227, Japan

Abstract. The restructuring of the Si(111)7 × 7 surface upon adsorption of metals is studied by UHV electron microscopy. With a brief review of the electron microscope and diffraction methods to determine density of the Si atoms in the restructured layer, recent results on the $\sqrt{3} \times \sqrt{3}$ structure of Ag are given. The structure consists of a honeycomb-chained-trimer Ag, a honeycomb Si layer, on the missing   top Si layer.

1   Introduction

Clean surfaces often have reconstructions, and they are restructured by adsorption of foreign atoms on them. The restructuring often results in a well-ordered superstructure of the adsorbed layer. It is well known that Si(111)7 × 7 is restructured to form $\sqrt{3} \times \sqrt{3}$ structures upon adsorption of many kinds of metals such as Al, Ag, Au, B, Bi, Ga, In, Pb, Pd, Sb, Sn at temperatures of 300 - 600 °C. Restructuring of these surfaces is well understood in such cases as Al[1] and B[2-4] where only one metal atom is adsorbed in each $\sqrt{3} \times \sqrt{3}$ unit cell. However, most other cases are not understood, because of the lack of convincing results on the coverage of metal, the density of Si included in the restructured layer, thus their atomic position, and of the structure in electronic nature. Among these, the $\sqrt{3} \times \sqrt{3}$- Ag[5,6] is the case that has long been investigated with many conflicting results. It is a key problem to solve the $\sqrt{3} \times \sqrt{3}$-Ag structure. From studies on the restructuring of metal-Si surfaces, we understand the valency of the metal, and the surface states.

Restructuring processes and atomic arrangement of metals (Au, Ag, Cu, Pd) on Si(111)7 × 7 surfaces have been studied by UHV electron microscopy and diffraction. We briefly review the electron microscopic methods of observing the restructuring, and the previous studies on Pd and Cu[7]. Analysis of the $\sqrt{3} \times \sqrt{3}$-Ag structure is presented as a recent topic.

2   UHV Electron Microscopy of Restructuring

2.1   UHV High-Resolution Reflection Electron Microscopy

Figure 1 shows a high-resolution electron microscope(HR-REM) image of a Si(111)7 × 7 surface[8] and the RHEED pattern. The image was obtained by interference between the specular reflection and the 1/7-th order superlattice reflection for the [110] incidence of the electron beam. The fringe image spaced 0.23 nm was interpreted as the image of dimer chains of the 7 × 7 reconstructed surface.

Springer Series in Surface Sciences, Vol. 24   **The Structure of Surfaces III**
Editors: S.Y. Tong · M.A. Van Hove · K. Takayanagi · X.D. Xie
© Springer-Verlag Berlin, Heidelberg 1991

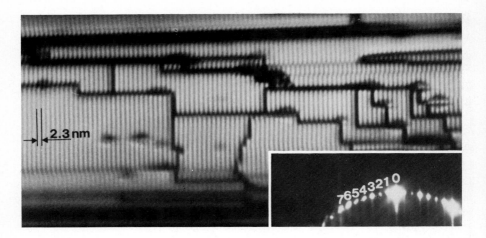

Fig.1 High-resolution reflection electron microscope (HR-REM) image of Si(111) - 7 × 7 surface. Vertical direction parallel to the incidence beam is the [1$\bar{1}$0] axis and lateral direction is the [11$\bar{2}$]. Note the fringe image of the 7 × 7 superlattices of the reconstructed surface.

Positions of the dimer chains are noticed to coincide with the positions of surface steps seen with the dark-bright Fresnel fringe in the figure, as seen in STM images. Thus, REM images, as well as transmission electron microscope (TEM) images, can reveal surface structures at sub-atomic resolution.

2.2 Determination of Si Atom Density in the Restructured Layer

Figure 2 shows an initial deposition process of Au on a Si(111)7 × 7 surface[9]. In Fig.2(a) of a REM image of the clean 7 × 7 surface, the 7 × 7 fringes of 2.3 nm spacing are seen in addition to surface steps and out-of-phase boundaries(OPB) as illustrated. By the deposition the 7 × 7 surface is restructured to 5 × 2 and the 5 × 2 areas are seen with dark contrast in the image of (b). As shown also in the illustration, the 5 × 2 region expands not only over the higher side terrace, but also over the lower side terrace, while the surface step has moved towards the lower side terrace from (a) to (b). This step motion is due to restructuring of the 7 × 7 surface: Excess Si atoms on the upper terrace by the restructuring from the 7 × 7 to 5 × 2 structure migrate to the lower side terrace to grow the step. Provided that the surface region of the 5 × 2 structure expands by s and S over the lower and higher terrace, respectively, the density of the restructured 5 × 2 phase, N is estimated by the following relation.

$$N(s + S) = (N_7 - 2)s + N_7 S \tag{1}$$

where N_7 is the density of Si atoms in the 7 × 7 phase[9].

When we assume that the 7 × 7 phase has the dimer adatom stacking-fault (DAS) structure[10], the density N_7 is 2.08 ML(The unit of density, 1ML is defined as one atom per Si-1 × 1 unit cell: The 7 × 7 unit cell includes 102 Si atoms above the bulk layer). Although vacancies exist on surfaces at elevated temperatures and the excess Si atoms

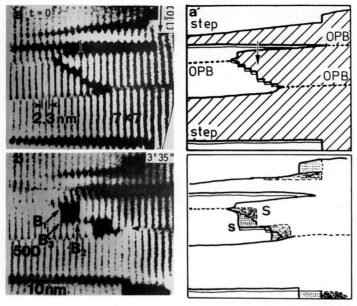

Fig.2 HR-REM images of 5 × 2-Au growing on Si(111)7 × 7, (a) before deposition, and (b) after the deposition. Dark areas in (b) are the 5 × 2 phase.

Table 1 Relation of N and s/S

structure of Si layer	MT	MT+HC	DL	DL+AD
density of Si layer: N	1	1.67	2	2.33
fractional ratio: s/S	1.17	0.26	0.04	-0.11

MT: missing-top, HC: honeycomb, DL: a double layer, AD: adatom.

are not always incorporated into the growth of the adsorbed phase over the lower side terrace, we neglect such facts. Then, by measuring the ratio of s to S from REM images directly, we can determine the density of Si atoms in the restructured layer. In the case of Au-5 × 2, s/S was obtained to be around 1ML, resulting in $N_{5 \times 2} = 1.1$ ML[9]. The relation between N and s/S is given in Table 1.

3   Structures of the $\sqrt{3} \times \sqrt{3}$-Pd, and "5 × 5"-Cu

3.1   Surface Symmetry

The symmetry is closely related with the surface band structure and thus the stability of the surface phases. The 7 × 7 surface (projected structure on the surface plane) has the two dimensional symmetry of p3m1, with the mirror planes perpendicular to the ⟨110⟩

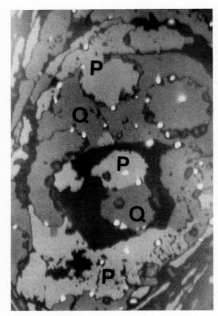

Fig.3

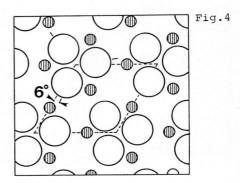

Fig.4

Fig.3 Transmission electron microscope image of $\sqrt{3} \times \sqrt{3}$-Pd on Si(111). Both bright area P and dark area Q have the same structure, but with different orientations.

Fig.4 Structure of $\sqrt{3} \times \sqrt{3}$-Pd on Si(111). Large open circles indicate Pd atoms forming trimers which rotate $6°$. Hatched circles are Si atoms in the top layer (1ML) of the restructured layer (2ML).

axes. Provided that the $\sqrt{3} \times \sqrt{3}$ structure has the same mirror symmetry as the 7×7 surface, they have to have p31m with respect to the unit cell rotated 30° from the Si 1 $\times$ 1 lattice.

In case of Pd on Si(111)7×7, the $\sqrt{3} \times \sqrt{3}$ structure formed at Pd coverage of 1 ML has p3 symmetry[11]. The symmetry has been determined simply from transmission electron microscope images such as Fig.3 where two domains of the $\sqrt{3} \times \sqrt{3}$-Pd structure appear, bright-contrasted region P and dark-contrasted region Q. The contrast never appears if the mirror symmetry exists in the plane normal to the $\langle 110 \rangle$ axes. The lowering of the symmetry is due to the rotation of Pd trimers as much as $6°$ from the symmetric position about the surface-normal as shown schematically in Fig.4: The structure model has been derived from analysis of TED intensities with a reliability factor of 7.6%.

Another example of p3 symmetry is the incommensurate "5 × 5" structure of Cu. Fig.5(a) and (b) are REM images of the same area of the "5 × 5" phase obtained by the specular spot and by a superlattice reflection spot marked as A in the RHEED pattern in (c), respectively. In (a) we see surface steps as dark zigzagging lines, and in (b) we see two domains of bright and dark contrast. The contrast is due to the lack of the mirror symmetry in the Cu layer: The Cu layer forms a hexagonal lattice of almost the closely packed (111) plane of the f.c.c. lattice: The Cu lattice is rotated by 30 +3.5° in the clockwise and counter-clockwise directions about the surface-normal with respect to the Si - 1 × 1 lattice, as shown in Fig.6[12]. The hexagonal lattice is contracted by 4.2 %

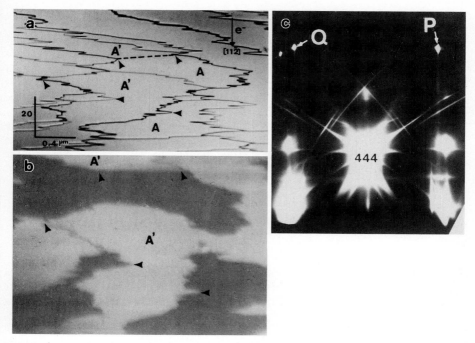

Fig.5 REM images of "5.3 × 5.3"-Cu on Si(111), (a) of the specular reflection, and (b) of the superlattice reflection marked A in the RHEED pattern in (c).

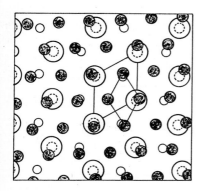

Fig.6 Structure of "5.3 × 5.3"-Cu on Si(111). Solid circles indicate Cu atoms forming hexagonal lattice, and large open circles are Si of the missing-top layer. Small dotted and solid circles are Si atoms in the double layer below the missing-top layer.

from the lattice of bulk Cu to form an incommensurate period of 5.3 × 5.3 with the Si - 1 × 1 lattice, and the density of Cu atoms is 2.45 ML. We refer hereafter to the "5 × 5" structure as the "5.3 × 5.3" structure.

3.2 Density of the Restructured Si Layer

The density of the Si atoms in the restructured layer is interesting since the number of dangling bonds relates closely to the surface electronic state. In the case of the $\sqrt{3} \times \sqrt{3}$-Pd, no apparent movement of the steps is recognized during "in-situ" REM observation of the deposition processes at 300°C, so that the density of the restructured Si layer is estimated to be 2 ML(six Si atoms within the $\sqrt{3} \times \sqrt{3}$ unit cell), in accord with the model derived from TED independently (fig.4).

In the case of the "5.3 × 5.3" Cu, steps move towards the lower side terraces similarly to the case of the 5 × 2-Au. The ratio $s/S = 1.1 - 1.2$, and then the restructured Si lattice has the density of 1 ML. The "5.3 × 5.3" structure, thus, has been concluded to have the missing-top structure shown in Fig.6.

4 Structure of the $\sqrt{3} \times \sqrt{3}$-Ag on Si(111)7 × 7

4.1 Overview of the Proposed Structures

Among many considerations, structure models of three different symmetries have been proposed [5,6,13,14] so far: p31m, p3 (three-fold) and cm (two-fold) symmetries. There are also two conflicting results on the density (coverage) of Ag atoms; 1ML(three Ag atoms per unit cell) or 2/3 ML(two Ag atoms per unit cell). And, related with the density, there are three conflicting arrangements of Ag atoms; trimer models(1ML), honeycomb-chained-trimer models(1ML), and honeycomb models(2/3ML). Furthermore, the density of the restructured Si layer has not been determined. A promising result, from the author's knowledge, is that the surface is semiconducting.

From the analyses[15], the symmetry of $\sqrt{3} \times \sqrt{3}$-Ag has been found to be p31m and the density of the restructured Si layer to be 1.67ML (five Si atoms within each unit cell). The structure model in fig.7 has been derived from intensity analysis of transmission diffraction spots.

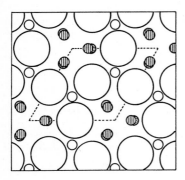

Fig.7 Structure of $\sqrt{3} \times \sqrt{3}$-Ag on Si(111). Large open circles indicate Ag atoms arranged like honeycomb-chained-trimer[5,6]. Small open circles are Si atoms arranged in honeycomb, and hatched circles, in a missing-top layer. Circles with dots are Si atoms below the mising-top layer.

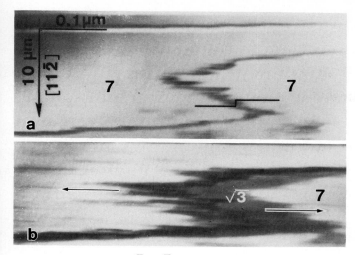

Fig.8 REM images of $\sqrt{3} \times \sqrt{3}$-Ag growing on Si(111), (a) before deposition and (b) after deposition. A zigzagging dark line in (a) is a step, and gray areas in (b) are of the $\sqrt{3} \times \sqrt{3}$ structure. Note the displacement of the step from (a) to (b).

4.2 Restructuring Process

Figure 8(a) and (b) show REM images before and after deposition of Ag on the Si(111)7 × 7 surface at 500 °C. The $\sqrt{3} \times \sqrt{3}$-Ag structure appearing with gray contrast in (b) has been formed on the upper side of the step seen as a dark zigzag line in (a). From (a) to (b) the step has moved towards the lower side terrace (left side), while the step has changed into a zig-zag configuration by the restructuring from the 7 × 7 to the $\sqrt{3} \times \sqrt{3}$ phase. The fractional ratio of the $\sqrt{3} \times \sqrt{3}$ areas expanded over the lower and the upper terraces, s and S, respectively, is about 0.25. Then, the density of the Si atoms in the $\sqrt{3} \times \sqrt{3}$ layer is estimated to be 1.67ML(five atoms per unit cell). A structure model for this density is a double layer composed of a honeycomb layer and a 1 × 1 layer.

It is worth mentioning that the $\sqrt{3} \times \sqrt{3}$ structure nucleates not only at the steps but also on flat region of the terraces.

4.3 Symmetry

No decisive evidence of the lowering of the surface symmetry from p31m has been obtained from REM images[15]. The symmetry still allows us three possible arrangements of Ag, honeycomb, honeycomb-chained-trimer and trimer.

4.4 Intensities of TED Spots

The intensity ordering of the superlattice spots of the $\sqrt{3} \times \sqrt{3}$-Ag is shown schematically in Fig.9, where the area of each circle is proportional to the intensity averaged among 12 or 6 spots of the symmetry-equivalent spots. In model calculations, the density of the Ag layer has been chosen to be 1/3, 2/3, and 1 ML, and that of the restructured Si layers, 1,

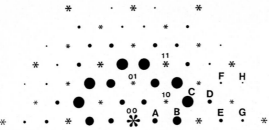

Fig.9 Transmission electron diffraction (TED) intensities for the $\sqrt{3} \times \sqrt{3}$-Ag. Area of each solid circle is proportional to the intensity. Asterisks are fundamental spots.

1.67, and 2ML. For all the combinations of these Ag and Si layers, the reliability factor has been calculated by using amplitudes (not intensities) of the superlattice spots.

Three atomic arrangements which give small reliability factors have been derived for models which consist of a Ag layer of 1ML and of restructured Si layers of 1.67 ML. Similar models for the restructured Si layers of 1 ML give also small reliability factors. One of them has a trimer configuration of Ag atoms. The model seems to be improbable as the true structure, since the distance of Ag-Ag atoms is 70 % to 80 % of the ionic diameter. The second model has a quasi - 1 × 1 layer of Ag which is slightly modulated from the close packed Ag(111) layer. The third model shown in Fig.7, giving reliability factor of around 10%, is a possible structure (see detailed discussion in ref.[15]). If we assume a honeycomb Si layer, the model also satisfies the density of Si layer and the symmetry[15].

4.5   Characteristics of the $\sqrt{3} \times \sqrt{3}$-Ag Structure

In the model structure, Ag atoms form a honeycomb-chained-trimer, and Si atoms form a honeycomb layer on the missing-top layer of Si lattice. From the TED analysis[15], the height of the atomic positions has not been decided, but the lateral positions: Ag-Ag distance is 0.335 nm and Si-Si distance in the missing-top layer is 0.215 nm. The Ag-Ag distance is very close to the models previously derived by X-ray experiments[5,6]. However, the arrangement of the Si atoms in the restructured layer is different from the model derived by X-ray diffraction (model III in ref.[6]). The reliability factors calculated for the X-ray model is around 29%. The X-ray model has a different stacking for the Si atoms from the present model, but the model II[6] rather compatible with the present model.

TED intensities are sensitive to the dynamical diffraction in the bulk Si layer as discussed before[16]. Therefore, dynamical intensity calculation should be done to determine the atomic positions precisely. Dynamical intensities calculated as a function of the thickness of the bulk Si layer[17] do not change their ordering much, which proves that the intensity ordering of the strong intensities are preserved and the kinematical analysis is enough for deriving models. An advantage of the electron diffraction over the X-ray diffraction is high sensitivity of electrons for the detection of light elements.

5 Summary: Structure, Unpaired Electrons, and Symmetry

The $\sqrt{3} \times \sqrt{3}$-Pd, $\sqrt{3} \times \sqrt{3}$-Ag, and "5.3 × 5.3"-Cu have different densities of the restructured Si-layer, as summarized in Table 2. The density of Si dangling bonds (unpaired electrons) per Si - 1 × 1 unit is 1, 1.67 and 3, respectively. Although metal valency and the effect of rehybridization of Si orbits should be taken into account, the number of unpaired electrons per unit cell is estimated to be 0, 2, and 15 for Pd, Ag and Cu. The surfaces of $\sqrt{3} \times \sqrt{3}$-Pd and -Ag , which have even numbers of unpaired electrons, are then expected to become semiconducting.

Table 2 Density of unpaired electrons

Structures	$\sqrt{3} \times \sqrt{3}$-Pd	$\sqrt{3} \times \sqrt{3}$-Ag	"5.3 × 5.3"-Cu
symmetry	p3	p31m	p3
metal (ML)	1	1	2.46
silicon(ML)	2	1.67	1
dangl . bond(ML)	1	1.67	3
unpaired e$^-$	0 (0*)	0.67 (2)	0.54 (15-16)

*; the number of unpaired electrons in a superlattice
otherwise the value indicated is the density in units of ML.

We have analyzed structures of metal-Si surfaces by UHV high-resolution electron microscopy and diffraction, which are useful to detect the density of the restructured Si layer and also atomic positions of composite atoms, particularly of light elements. From the structures derived, we hope that refinements of the models will be made experimentally and theoretically to further our understanding metal/Si surfaces.

Note added in proof: The s/S value is under reinvestigation by reflection electron microscopy because of complexities of the morphology of the domains.

References

1. J.E.Northrup, Phys. Rev. Lett. **53**, 683 (1984).

2. R.L.Headrick, I.K.Robinson, E.Vlieg and L.C.Feldman, Phys. Rev. Lett. **63**, 1253 (1989).

3. P.Bedrossian, R.D.Meade, K.Mortensen, D.M.Chen, J.A.Golovchenko and D.Vanderbilt, Phys. Rev. Lett. **63**, 257 (1989).

4. I.-W.Lyo, E.Kaxiras and Ph. Avouris, Phys. Rev. Lett. **63**, 1261 (1989).

5. T.Takahashi, S.Nakatani, N.Okamoto, T.Ishikawa and S.Kikuta, Jpn. J. Appl. Phys. **27**, L753 (1988).

6. E.Vlieg, A.W.Denier van der Gon, J.F.van der Veen, J.E.Macdonald and C.Norris, Surface Sci. **209**, 100 (1989).

7. K.Takayanagi, Y.Tanishiro, T.Ishitsuka and K.Akiyama, Appl. Surf. Sci. **41/42**, 337 (1989).

8. K.Takayaangi, Y.Tanishiro, K.Kobayashi, K.Yagi, K.Akiyama, Jpn.J.Appl. Phys. **26**, L957(1987).

9. Y.Tanishiro and K.Takayanagi, Ultramicroscopy **31**, 20 (1988).

10. K.Takayanagi, Y.Tanishiro, S.Takahashi and M.Takahashi, Surface Sci. **164**, 367 (1985).

11. K.Akiyama, K.Takayanagi and Y.Tanishiro, Surface Sci. **205**, 177 (1988).

12. T.Ishitsuka, Master Thesis, Tokyo Institute of Technology, and see also ref.[8].

13. M.Copel and R.M.Tromp, Phys. Rev. **B39**, 12688 (1989).

14. S.Kono, T.Abukuwa, N.Nakamura and K.Anno, Jpn. J. Appl. Phys. **28**, L1278 (1989).

15. T.Yoshida, S.Takahashi, Y.Tanishiro and K.Takayanagi, Surface Sci. submitted.

16. K.Takayaangi, Acta Crystallo. bf A46, 83 (1990) and references therein.

17. S.Takahashi, Master Thesis at Tokyo Institute of Technology, unpublished.

Atomic Resolution Imaging of Glycine Molecules on Graphite in Water

W.S. Yang, Wenjun Sun, Jianxun Mou, and Junjue Yan

Department of Physics, Peking University, Beijing 100871, P.R. of China

Abstract. Glycine molecules adsorbed on graphite and covered with glycerol-water solution were imaged with the Scanning Tunneling Microscope. Two of many different 2-D short-range crystalline structures on the samples are reported here. According to our current understanding on these structures, individual C, O, and H atoms of glycine molecule were distinguished. To our knowledge, it is the first time for hydrogen atoms to be resolved with a microscope.

1. Introduction

The Scanning Tunneling Microscope (STM) is capable of resolving surface detail down to the atomic level [1], and could be operated under water and other fluids [2,3]. The possibility of examining biological molecules and substances in air or even in a near-native environment in such detail has been very attractive to scientists. Towards this direction, many excellent works have been published [4-9]. However, to our knowledge, there has not yet been any demonstration of atomic details in organic and biological samples prepared by any method and studied in any environment. In some cases some nearly-atomic resolutions were achieved[4-7], but for under-water works, the situation was even worse [8].

Very recently, we have carried out a series of STM studies on DNA [10], tRNA [11] and proteins(keratin) [12] covered with glycerol-water solution. The best resolution achieved in these works was also nearly-atomic. In the present work we have studied glycine molecules adsorbed on highly oriented pyrolytic graphite (HOPG) and covered with the glycerol-water solution. Here we present atom-resolved images of single glycine molecules in short-range 2-D crystalline aggregates along with our explanations. As far as we know, this is the first time that the atomic resolution is achieved from surfaces of organic or biological substances covered with fluid.

The reason for choosing glycine was: (i) Glycine is the simplest amino acid; (ii) It is very stable [13]; (iii)The structure of its 3-D crystal is known [14,15]; (iv)But many aspects of its properties are still to be studied [13,15].

2. Experimental

The samples were prepared by spreading a drop (4 μl) of dilute glycine water solution (10 μg/μl) onto newly-cleaved HOPG surface and being air-dried, and then covered with a thin layer of 50% glycerol-water solution.

Springer Series in Surface Sciences, Vol. 24   **The Structure of Surfaces III**
Editors: S.Y. Tong · M.A. Van Hove · K. Takayanagi · X.D. Xie
© Springer-Verlag Berlin, Heidelberg 1991

The STM used in this work has been described in earlier works [10,12]. AC-electrochemically etched tungsten tips were used. STM images were collected in the constant-current mode with sample bias voltage $V_b=20$ mV, tunneling current $I_t=2$ nA, and scanning rate of 75 Å /sec. The correspondent gap resistance was 10^7 Ohms. Increase in gap resistance resulted in reduction in image corrugations.

On the samples, patches of glycine adsorbates with or without short-range order were seen. Very nice typical STM images of graphite were observed from the nude areas, thus accurate calibration of the image scaling together with determination of the adsorbing registry of glycine molecules were made from them.

For image processing, smoothing, local-contrast enhancing (high-pass filtering) [10,12], and averaging were used according to the situations.

3. Results and discussion

Fig. 1(a) is a typical raw image, showing some features in short-range order (upper central area) together with some complicated features as well as traces of graphite surface. A zoom of the upper central portion, after high-pass filtering and light smoothing, is shown in Fig. 1(b). A dozen or so similar features are now seen more clearly than in (a). According to the scale, it is plausible to assign a glycine molecule to each of such features.

For image enhancement, averaging is very effective [16]. Fig. 2(a) is an averaged image, based on 8 features in Fig. 1(b). Our understanding of this image is explained

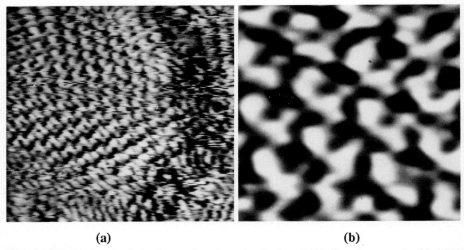

(a)   (b)

Fig. 1. STM images of glycine molecules adsorbed on HOPG and covered with 50% glycerol-water solution, recorded with $V_b=20$ mV and $I_t=2$ nA. **(a).** Unprocessed image of a 60×60Å^2 area, showing short-range ordering of the molecules. **(b).** A zoom of the upper central portion (15×15Å^2) of (a), after local-contrast enhancing and smoothing, showing the 2-D crystalline arrangement of some dozen of the molecules.

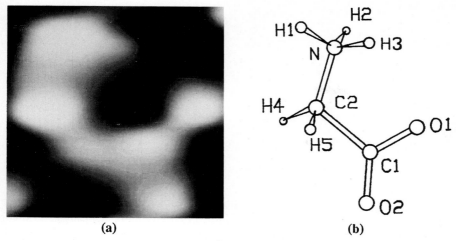

Fig. 2. (a). Averaged image ($3.8 \times 3.8 Å^2$) of a glycine molecule, based on Fig. 1(b). (b). Schematic top view of a glycine molecule lying on a plane, showing the adsorbing geometry of (a).

clearly by Fig. 2(b). Despite that the atomic details of the groups of CH_2 and NH_3 are not seen in Fig. 2(a), the atoms of C_1, O_1, O_2 are unambiguously distinguished. Note that, as indicated in Fig. 2(b), the C_1-O_2 distance is only 1.25 Å . The effectiveness of the explanation implies that in this situation the molecular geometry dominated STM imaging.

Based on the above understanding, Fig. 3(a) is a schematic drawing, which shows the arrangement of the molecules as well as their registry on graphite. A comparison with Fig. 3(b), the molecular arrangement in a layer of a three-dimensional glycine crystal, discloses a strong resemblance. In the crystal case (3D), the arrangement is determined by the van der Waals radii along with the hydrogen bond formation [17]. Therefore, in our two-dimensional (2D) case, the motivation of the arrangement ought to be essentially the same. Besides, from Fig. 3(a), one can see that all the atoms of glycine molecule facing the HOPG have a one-to-one relationship with graphite atoms. This means that there is a strong interaction between the glycine molecules and the HOPG. This interaction ought to have some influence on the adsorption geometry, also.

Indeed, a careful inspection of Fig. 3(a) and (b) can tell that there are some differences between the 2D and 3D cases: (i). The molecular distance is about 20% smaller in the 2D case than in the 3D case; (ii). Their relative orientation are also slightly different in the two cases. In the 2D case, the arrangement is more homogeneous and , therefore, denser. The reasons might be: (i) In the 2D case, the lack of a confinement from the third dimension resulted in more flexibility; (ii) The strong glycine-graphite interaction mentioned above might have deformed the molecules more or less, and also made the van der Waals radii reduced a bit as a result of having less electrons to take

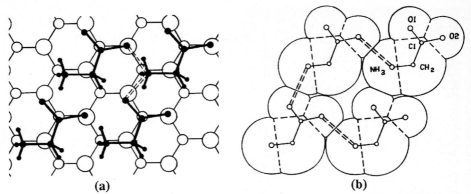

Fig. 3. (a). A schematic drawing showing the arrangement and registry of the glycine molecules adsorbed on HOPG, corresponding to Fig. 1 and Fig. 2. Large and small open circles represent the B- and A- site carbon atoms of graphite, respectively. Thick lines and closed circles represent the glycine molecules. **(b).** The arrangement of molecules in a layer of the crystal glycine [14]. It is seen that the packing of the molecules is determined by the van der Waals radii and hydrogen bonds(double dashed lines). Note the similarity between (a) and (b).

care of the van der Waals forces [17]. In fact, we note that in recent works on liquid crystals a similar change of some 20% has also been detected [7,6].

The molecular arrangement shown in Fig. 3(a) is of $\sqrt{3} \times \sqrt{3}R30°$ relative to the substrate. However, it has been reported that adsorbed molecules on graphite may strongly perturb the surface electronic charge density, giving rise to a $\sqrt{3} \times \sqrt{3}$ structure at a nearly nude graphite surface [18]. Therefore, care has to be taken in dealing with the origin of the $\sqrt{3} \times \sqrt{3}$ structure seen in our work. The physical origin of the $\sqrt{3} \times \sqrt{3}$ structure on the nude graphite area is similar to Friedel oscillations in metals [19], and has been explained in detail by Mizes and Foster [18]. According to their result, in the STM images of the $\sqrt{3} \times \sqrt{3}$ structure, only B-site atoms are seen with their brightness modulated with a $\sqrt{3}$ period, while A-site atoms should never be seen. In other words, in a unit cell of the structure, there are at most three bright spots (with different brightnesses). As a matter of fact, in Fig. 2(a), we see five spots, and the brightest spots are mainly at A sites. Consequently, it is reasonable to suggest that the $\sqrt{3}$ structure seen in Fig. 1(b) was the result of a short-range ordering of the adsorbed glycine molecules.

Images of some other different short-range orders were also collected in our experiment. One of them, after similar image processing as mentioned above, together with our understanding of it are shown in Fig. 4(a), (b) and (c), respectively. We see once more that glycine molecules covered with the glycerol-water solution can be imaged with atomic resolution by the STM and that, in the present case, the STM images were likely dominated by the geometry, instead of the electronic structure.

It should be emphasized that in Fig. 4(a), all of the three hydrogen and one oxygen atoms, which are accessible to the tip, are resolved unambiguously. To our knowledge,

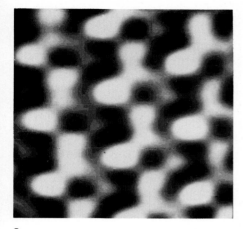

a

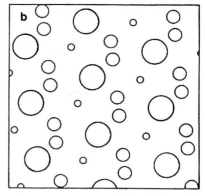

b

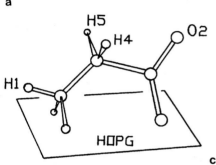

H5
H4
O2
H1
HOPG

c

Fig. 4 (a). Averaged STM image (12 × 12Å^2) of the glycine molecules adsorbed on HOPG and covered with 50% glycerol-water solution, based on an image similar to Fig. 1(b).
(b). Schematic drawing of (a).
(c). Schematic perspective view of a glycine molecule standing on HOPG surface, showing the adsorption geometry of the molecules imaged in (a). H$_1$, (H$_4$, H$_5$), and O$_2$ corresponding to the smallest circles, twined circles, and the large circles in (b), respectively.

this is the first time that individual hydrogen atoms have been imaged with a real-space technique.

4. Conclusion

Glycine molecules adsorbed on graphite and covered with glycerol-water solution can be imaged with atomic resolution by the STM. The imaging is likely dominated by the atomic geometry, thus making the explanation of the images quite straightforward. Individual hydrogen atoms of the molecule were imaged for the first time.

5. Acknowledgment

We should like to thank Cheng Liu for providing the glycine sample.

References

1. G. Binnig, H. Rohrer, Ch. Gerber, and E. Weibel, *Phys. Rev. Lett.* 49, 57(1982).
2. P.K. Hansma, V.B. Elings, O. Marti, C.E. Bracker, *Science* 242, 209(1988).
3. R. Sonnenfeld and P.K. Hansma, *Science* 232, 211(1986).
4. T. Sleator and R. Tycko, *Phys. Rev. Lett.* 60, 1418(1988).
5. H. Ohtani, R.J.Wilson, S.Chiang, and C.M. Mate, *Phys. Rev. Lett.* 60, 2398(1988).
6. J.S.Foster and J.E. Frommer, *Nature* 333, 542(1988).
7. D.P.E. Smith, H. Hörber, Ch. Gerber, and G. Binnig, *Science* 245, 43(1988).
8. S.M. Lindsay, T. Thundat, L. Nagahara, U. Knipping, and R.L. Rill, *Science* 244, 1063(1989).
9. D.D. Dunlap and C. Bustamante, *Nature* 342, 204(1989).
10. Jianxun Mou, Junjue Yan, Wenjun Sun, W.S. Yang, Cheng Liu, and Zhonghe Zhai, *The Chinese Journal of Electron Microscopy* 8(4), 1(1989).
11. Cheng Liu, Zhonghe Zhai, Jianxun Mou, Wenjun Sun, Junjue Yan, and W.S. Yang, *The Bulletin of Chinese Science* 24, 1983(1989).
12. Jianxun Mou et al., to be published.
13. A.R. Slaughter and M.S. Banna, *J. Phys. Chem.* 92, 2165(1988).
14. G. Albrecht and R.B. Corey, *J. Am. Chem. Soc.* 61, 1087(1939).
15. J. Almlöt, Å. Kvick, and J.O. Thomas, *J. Chem. Phys.* 59, 3901(1973).
16. G. Binnig and H. Rohrer, *IBM J. Res. Develp.* 30, 355(1986).
17. L. Pauling, *"The Nature of the Chemical Bond"* (Cornell Univ. Press, Ithaca, 1948), p. 191.
18. H.A. Mizes and J.S. Foster, *Science* 244, 599(1989); T.R. Albrecht, H.A. Mizes, J. Nogami, Sang-il Park, and C.F. Quate, *Appl. Phys. Lett.* 52, 362(1988).
19. J. Friedel, *Philos. Mag. Suppl.* 3, 446(1954); *Nuovo Cimento Suppl.* 7, 287(1958).

Theory of Atomic Force Microscopy on Elastic Surfaces

W. Zhong, G. Overney*, and D. Tománek

Department of Physics and Astronomy and
Center for Fundamental Materials Research, Michigan State University,
East Lansing, MI 48824, USA

We combine *ab initio* Density Functional and Continuum Elasticity theory to determine elastic surface deformations, limits of atomic resolution, and atomic-scale friction in Atomic Force Microscopy (AFM). We apply this formalism to the interaction of a Pd AFM tip with graphite. Our results show that, in the constant–force mode, atomic resolution is marginally possible in a narrow load range which is limited by too small height corrugations for loads below 10^{-8} N (per Pd tip atom) and by irreversible substrate deformations for loads beyond 10^{-8} N. For loads near 10^{-8} N, we determine the microscopic friction coefficient to be $\mu \approx 10^{-2}$.

After being introduced by Binnig, Quate and Gerber in 1986, Atomic Force Microscopy (AFM) has become a very powerful tool in detecting surface structures, especially those of semiconductors and insulators.[1,2] The AFM consists of a very sharp tip suspended on a soft spring. In the constant–force mode, tip deflection due to the interaction with the surface is recorded during the surface scan for a given applied load. Atomic resolution is achieved when the observed surface corrugation is sufficiently large, typically ≥ 0.1 Å.

Calculation of the interaction between an AFM tip and a solid is possible to a high precision within the Density Functional formalism, using the Local Density Approximation (LDA).[3] The predictive strength of these parameter–free *ab initio* calculations is, however, counterbalanced by the practical limitation of applicability to periodic structures with less than $10 - 10^2$ atoms per unit cell. In order to describe large–scale elastic deformations occurring in AFM interacting with elastic surfaces, we develop a new formalism by combining LDA with Continuum Elasticity theory. We apply this theory to describe the interaction between a Pd AFM tip and graphite.

In a first step, we use LDA to determine the Pd–graphite interaction and elastic constants of graphite.[4] In Fig. 1 we show the force acting between graphite and a Pd atom in the on–top and hollow site. We use *ab initio* pseudopotentials of Hamann–Schlüter–Chiang type,[5] a local Gaussian basis, a large energy cutoff of 49 Ry in the Fourier expansion of the charge density ρ, and a fine 47 $\vec{k}$–point mesh sampling the 2–dimensional Brillouin zone. The graphite surface is approximated by a rigid 4 layer slab; the unit cell contains 8 carbon atoms and 2 Pd atoms.[4] The equilibrium structure and elastic constants of the graphite substrate have been obtained from a similar calculation for bulk graphite. The calculated in–plane C–C bond length $a = 1.42$ Å and inter–plane spacing $c = 3.35$ Å

---

* Permanent address: Institut für Physik, Universität Basel, Klingelbergstrasse 82, CH-4056 Basel, Switzerland.

Springer Series in Surface Sciences, Vol. 24   **The Structure of Surfaces III**
Editors: S.Y. Tong · M.A. Van Hove · K. Takayanagi · X.D. Xie
© Springer-Verlag Berlin, Heidelberg 1991

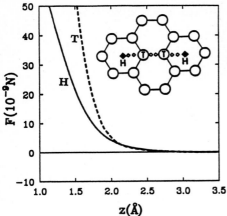

FIG. 1. Force F between a 1–atom Pd AFM tip and graphite, as a function of the tip height z above the surface of hexagonal graphite. The solid and dashed lines correspond to the sixfold hollow (H) and the on–top (T) sites, respectively. The inset shows the adsorption geometry in top view.

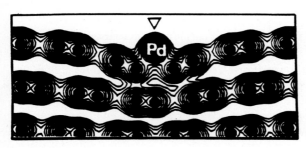

FIG. 2. Total charge density ρ of a 1–atom Pd AFM tip interacting with the elastic surface of graphite near the hollow site, for a load of 5×10^{-9} N. Contours of constant ρ are shown in the xz plane perpendicular to the surface. The position of the AFM tip is indicated by $\bigtriangledown$.

agree very well with the experiment. The forces between Pd and graphite shown in Fig. 1 have been obtained from a polynomial fit of the LDA total energies.

In the next step, we describe the deformation of graphite due to localized forces by Continuum Elasticity Theory. The elastic response of graphite due to external forces (which can be due to the AFM tip or intercalant atoms) are determined by solving a set of coupled differential equations.[6] The distortion of graphite layers, which are approximated by homogeneous plates in this step, is described by elastic constants determined from LDA. Consequently, surface deformations observed by AFM can be used to determine local variations of the elastic constants such as the flexural rigidity which can occur near intercalant impurities or steps.[6]

Once the new equilibrium positions of carbon atoms in the distorted layers have been obtained, the total charge density of the system ρ can be well approximated by a superposition of atomic charge densities obtained from LDA. The resulting charge density of the deformed surface is shown in Fig. 2.

In order to estimate the limits of atomic resolution in AFM, we use the calculated Pd–graphite interaction to determine the equilibrium tip height z at inequivalent surface sites as a function of applied load. We define a load per tip atom $f_{ext} = F_{ext}/N$, where F_{ext} is the total load on the tip and N is the number of tip atoms in contact with the graphite surface. For loads $f_{ext} > 2.0 \times 10^{-9}$N, we find $\Delta z \equiv z_{on-top} - z_{hollow} > 0$. Atomic

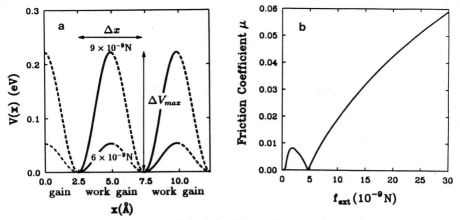

FIG. 3. (a) Potential energy $V(x)$ of the Pd–graphite system as a function of the Pd tip position along the surface x–direction, for external forces $f_{ext} = 6 \times 10^{-9}$ N and 9×10^{-9} N. (b) Microscopic friction coefficient μ as a function of the external force per atom f_{ext} (from Ref. 7). (© American Physical Society)

resolution can be achieved experimentally when Δz is sufficiently large to be detected, i.e. typically $\Delta z \geq 0.1$ Å, which corresponds to $f_{ext} > 10^{-8}$N. On the other hand, our calculation of elastic graphite deformations for $f_{ext} \geq 10^{-8}$N indicates that such large local forces can rupture the graphite surface, in agreement with experiment.[7]

The present results for the Pd–graphite interaction can also be used to describe atomic–scale friction on an ideal substrate.[4] Friction is a dissipative process which is generally dominated by plastic deformations at the interface and dislocation motion within the two bodies in contact. This complexity makes a predictive description of friction a very difficult undertaking. However, progress in surface preparation[7,8] and use of the AFM have made the observation of atomic–scale friction force under near–ideal conditions possible.[8]

We describe the friction process between an AFM tip and graphite by considering a "surface diffusion" of the AFM tip under external load, along a straight trajectory in the surface x direction. The position–dependent part of the potential energy $V(x)$ of the system is shown in Fig. 3(a). The two main components of $V(x)$ are variations of the adsorption bond energy and work against the external force f_{ext} in case of nonzero corrugations Δz. We first consider the energy involved in moving the AFM tip between two neighboring equivalent sites which are separated by Δx. During this process, the system has to cross a potential energy barrier $\Delta V_{max}(F_{ext})$ (shown by a solid line in Fig. 3(a)). Assuming a very slow horizontal motion of the tip, the subsequent energy gain $-\Delta V_{max}(F_{ext})$ (shown by a dashed line in Fig. 3(a)) will be dissipated into heat (occurring as phonons or electron–hole pair excitations).[9] The average friction force is related to this dissipated energy and is given by

$$< F_f > \leq \frac{\Delta V_{max}(F_{ext})}{\Delta x}. \tag{1}$$

The friction coefficient is defined by the ratio of the friction force and the applied load, $\mu = F_f/F_{ext}$. In our calculation, we find the dependence of the Pd–graphite interaction

245

on the tip position x to be very small for loads $f_{ext} \approx 10^{-8}$ N per tip atom. Consequently, the friction coefficient in this load range is very small, $\mu \approx 10^{-2}$. We expect this value to be a realistic estimate of μ for the ideal conditions discussed above. As shown in Fig. 3(b), μ increases with increasing load, in agreement with a recent AFM experiment.[8]

In summary, we developed a theory for the Atomic Force Microscopy (AFM) of deformable surfaces, based on a combination of *ab initio* Density Functional formalism and Continuum Elasticity theory. We applied this theory to graphite and determined quantitatively local distortions in the vicinity of a sharp AFM tip as a function of the applied force. We conclude that AFM should be a unique tool to determine *local* changes of the surface rigidity which can occur near intercalant impurities or steps. Using our formalism, we found that in the constant–force mode, the AFM can marginally achieve atomic resolution for loads (per tip atom) near $f_{ext} \approx 10^{-8}$ N. The load range is limited by a too low corrugation Δz on the lower end and by too large substrate distortions leading to surface rupture on the upper end. We also estimated the friction coefficient between a Pd AFM tip and graphite by considering the energy dissipated along the tip trajectory. We found the friction coefficient to be very small, in the order $\mu \approx 10^{-2}$ for loads near 10^{-8} N, and found μ to increase with increasing applied load.

We thank Professors S. G. Louie, S.D. Mahanti, H. Miyazaki, H.-J. Güntherodt and Mr. Z. Sun for stimulating discussions. This work has been supported by the Office of Naval Research under contract No. N00014-90-J-1396 and, in its initial stage, by the Director, Office of Energy Research, Office of Basic Sciences, Materials Sciences Division of the U.S. Department of Energy under contract No. DE–AC03–76SF00098. CRAY computer time at the National Magnetic Fusion Energy Computer Center and the National Center for Supercomputing Applications and CONVEX computer time have been provided by a grant from the U.S. Department of Energy, the National Science Foundation and Michigan State University, respectively.

[1]G. Binnig, C.F. Quate, and Ch. Gerber, Phys. Rev. Lett. **56**, 930 (1986).

[2]G. Binnig, Ch. Gerber, E. Stoll, T.R. Albrecht, and C.F. Quate, Europhys. Lett. **3**, 1281 (1987).

[3]W. Kohn and L.J. Sham, Phys. Rev. **140**, A1133 (1965).

[4]W. Zhong and D. Tománek, Phys. Rev. Lett. **64**, 3054 (1990).

[5]D.R. Hamann, M. Schlüter and C.Chiang, Phys. Rev. Lett. **43**, 1494 (1979).

[6]D. Tománek, G. Overney, H. Miyazaki, S.D. Mahanti, and H.-J. Güntherodt, Phys. Rev. Lett. **63**, 876 (1989) and ibid. **63**, 1896(E) (1989).

[7]J. Skinner, N. Gane, and D. Tabor, Nat. Phys. Sci. **232**, 195 (1971).

[8]C.M. Mate, G.M. McClelland, R. Erlandsson and S. Chiang, Phys. Rev. Lett. **59**, 1942 (1987).

[9]J.E. Sacco, J.B. Sokoloff, and A. Widom, Phys. Rev. B **20**, 5071 (1979).

Part II

Clean Metals

Microscopic Kinetics of the (1 × 2) Missing Row Reconstruction of the Au(110) Surface*

L.D. Roelofs and E.I. Martir

Physics Department, Haverford College, Haverford, PA 19041, USA

Abstract. Calculations based on the Embedded Atom Method of the microscopic kinetics of the clean Au(110) surface are presented. We have considered elementary 'adatom' moves for both the flat and the reconstructed (1x2) phases of the surface, extracting activation energies and the details of the relaxation of the diffusing and substrate atoms. Full relaxation of underlying atoms must be incorporated to achieve accuracy in the activation energies. It is found that single atom diffusion is anisotropic on both the flat and the reconstructed surface, but that consideration of concerted (2-atom) motions eliminates most of the anisotropy. Activation energies are found to be on the order of 0.25 - 0.35 eV on the defect free surfaces.

1. Introduction

The kinetics of the formation of the (1x2) reconstruction on noble metal (110) surfaces has an interesting history (for a brief sketch see [1]), since the missing row form of the phase requires mass motion. Ref. 1 is an application of the Embedded Atom Method (hereafter EAM) to the energetics of the phase behavior of the Au(110) surface. Issues of kinetics were not explicitly treated, but some light was shed on the development of the reconstruction via consideration of the energy cost of steps on the surface. It was found that steps whose edges lie parallel to the close-packed rows on the (110) surface have extraordinarily low energies, so low indeed that it should be difficult to prepare a surface not having substantial numbers of such steps by ordinary means. If the surface starts in disordered form (1x2) order can develop without long-range movement of individual atoms. Even so, diffusion clearly remains of interest. The EAM approach can also be used to determine the energy barriers that control the kinetics of diffusion like processes on the surface, and in this paper we give   preliminary results from our investigations of simple atomic motion on the flat and the (1x2) surfaces.

Concerted (multi-atom) moves have been shown to be important in diffusion on other surfaces [2], and we report here that such is also the case for Au(110).

2. Method

To determine the basic energetics of diffusion-like moves we used the EAM [3]. A similar approach was used in ref. 1, but there to improve the accuracy sufficiently to

---

* Supported by the National Science Foundation under grant # DMR-8705568

Springer Series in Surface Sciences, Vol. 24   **The Structure of Surfaces III**
Editors: S.Y. Tong · M.A. Van Hove · K. Takayanagi · X.D. Xie
© Springer-Verlag Berlin, Heidelberg 1991

address the subtle energetics of the reconstruction transition a correction was made for the effect of charge density gradients. That degree of accuracy is neither required nor possible to attain for the energy barriers, so gradient corrections, which are costly in terms of computation time, have not been included. It is difficult to estimate the accuracy of the approach, but different sets of EAM functions [4] for Au give results that agree within ±0.02 eV.

All atomic configurations were fully relaxed via the conjugate gradient method using the DYNAMO code developed by Daw and Foiles[5]. (The code works by evaluating the forces acting on each atom and making appropriate adjustments of atomic positions until all forces vanish.) As emphasized by Daw and Foiles [6] for reconstructed noble metal surfaces, failure to allow relaxation of atomic positions leads to errors in the energetics as large as a factor of 2.

All calculations were done with a 'substrate' in the form of a slab of 10 layers of atoms oriented in the (110) direction. The bottom 5 layers were fixed in an fcc structure of lattice constant a = 4.08 Å. (This ensures all unconstrained atoms in the substrate are unaffected by the termination at the bottom of the slab.)

2.1 Application of Constraints

To calculate energy barriers one must control and vary the position of the moving atom while allowing all others to relax in response. To ensure that the lowest energy trajectory is found, it is also necessary to allow relaxation of the moving atom as well in some of its degrees of freedom.

This is illustrated by the simplest diffusion event we considered, a move of a single extra atom (referred to as the 'adatom' even though it is of the same species as the 'substrate' atoms) a distance of $a/\sqrt{2} = 2.885$ Å in the direction of the close-packed rows (the y-direction in our convention). (See the lower right corner of fig. 1.) The relaxed 'adsorption' site is 1.13 Å above a rectangular hollow on the surface. To move to the adjacent adsorption site in the y-direction requires the adatom to pass over a long-bridge site on the surface. To explore the energetics of this move we constrain the adatom to remain in a plane [7] normal to the y-direction and move the constraint plane by 1/10 of the full displacement vector at each stage. After each movement of the plane a full relaxation is done, including that of the adatom within its constraint plane. Five positions of the constraint planes are indicated in fig. 1 on the vector corresponding to this particular move.

In each of the moves discussed below similar constraints were applied to map out the energy variation, giving an upper bound on the energy barrier in each case.

3. Specific Moves

We focus in this report on moves likely to be of importance in well-ordered phases of the surface. Moves with net result of single adatom displacement on the flat surface are shown in fig. 1. These give information about diffusion on the flat, defect-free surface. (Note that the implication of ref. 1 is that such a surface could not be prepared by standard means, but pretreatment with an adsorbate which

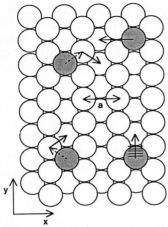

Fig. 1: Flat Au(110) surface with adatoms shaded. Types of diffusion moves considered in the paper are shown: moves parallel to the rows on the lower part of the figure; and those perpendicular to the rows in the upper. In both cases the direct move is on the right and a 2-atom concerted move is the left. For 2-atom moves, the displacement of the constrained atom is denoted by the solid vector; the other atom moves spontaneously (dashed vector) by relaxation. The figure also shows the size of the cell used, but just one 'adatom' was present during any given run

stabilizes the clean surface can produce such an initial condition [8].) The lower part of the figure shows two elementary moves that result in diffusion along the troughs on the surface. The one on the right is a direct move and that on the left a 2-atom concerted move. The constraints for implementing the former have been described in section 2.1. The latter move was accomplished by constraining the substrate atom to planes perpendicular to the solid vector. As the plane was advanced the adatom fell into the spot vacated by the substrate atom as shown with the dashed vector.

The upper portion of the figure shows the perpendicular moves we considered. On the right is a direct move and on the left is a 2-atom concerted move when the same convention is used for the constrained and responding atoms.

In the case of the reconstructed surface we considered similar basic moves: a simple diffusion move along the trough in which the adatom lies; a direct move from one trough to the adjacent one (a distance of 8.16 Å); and a concerted two-atom move of the same net motion. (We could not conceive of a concerted mode for movement along the trough.)

4. Results and Discussion

Table 1 summarizes our results for all the direct and concerted moves we studied. In all cases we have scrutinized all intermediate configurations of atoms to ensure that no anomalous atomic displacements occurred.

Table 1: Energy barriers for diffusion-like moves on Au(110)

Surface	Direction	mode	E_{bar} (eV)
flat	par. to rows	direct	0.27
flat	par. to rows	concerted	0.75
flat	perp. to rows	direct	1.16
flat	perp. to rows	concerted	0.35
(1x2)	par. to rows	direct	0.32
(1x2)	perp. to rows	direct	1.22
(1x2)	perp. to rows	concerted	0.82

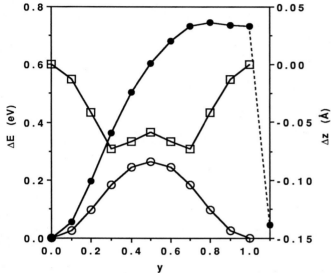

Fig. 2: Energy variation and height of adatom for movement parallel to rows. The energies, denoted ΔE, are plotted as circles: open for the direct move and solid for the concerted move. The height of the atom above the surface during the direct move, Δz, is plotted as the open squares. The variable y is a normalized measure of displacement. y=1 corresponds to the full displacement from one adsorption site to the adjacent one in the y-direction. See text for explanation of y>1 values in the case of the concerted move.

Each energy barrier is deduced from an energy variation with position similar to that shown in fig.2 for the case of movement along a trough on the flat surface. There are various interesting aspects to fig. 2. Note that the concerted move does not lower the energy barrier in this case. For the direct move, the atom does not really 'hop' from one adsorption site to the next. Rather it actually displaces into the sample in order to make the move. (This counterintuitive behavior is due to the fact that in crossing the long bridge the adatom is in an environment of reduced coordination and therefore seeks to move closer to the few neighbors it can find.)

Thirdly, in the case of the concerted move we actually had to push the constrained atom beyond its final position before the original adatom fell into the position originally occupied by the constrained atom (hence the values of y>1). (This does not increase the energy barrier.)

Returning to table 1, we note that if one were to consider direct moves only, diffusion would be highly anisotropic on both surfaces. In both cases, however, concerted moves can be found which decrease the energy barriers for moves perpendicular to the rows. In the case of the flat surface the decrease is nearly enough to restore isotropy (though the prefactors would doubtless be different for one- and two-atom moves). For the reconstructed surface the anisotropy is also decreased, but less so. This is consistent with (but might not be the only explanation for) the observation of better order along the rows than perpendicular to them [9].

In summary, we have determined energy barriers for diffusion-like moves for the Au(110) surface. The values found seem reasonable and their implications are consistent with the observed ordering behavior of this surface. We have also shown that concerted motions are of significance for this surface.

Acknowledgments

It is a pleasure to thank Steven Foiles and Murray Daw for useful conversations and their provision of and assistance with the EAM code.

References

1. L. D. Roelofs, S. M. Foiles, M. S. Daw and M. I. Baskes, Surface Sci.(to appear).
2. D. W. Bassett and P.R. Webber, Surface Sci. 70, 520 (1978); John D. Wrigley and Gert Ehrlich, Phys Rev. Letters 44, 661 (1980); T. T. Tsong and Qiaojun Gao, Surface Sci. 182, L257 (1987); G. Kellogg and P. Feibelman, private communication.
3. M. S. Daw, Phys. Rev. B39, 7441 (1989).
4. The results given below were obtained using a function set that gives an exact fit for the bulk structural properties of Au and in addition gives a good account for some defect energetics including the stacking fault energy.
5. We implemented version 5.2 of DYNAMO, available from Daw and Foiles at Sandia Livermore, on a VAXstation 2000, adding the capability of constraints on particular atoms that also exists in later versions of the code, but not in 5.2.
6. M. S. Daw and S.M. Foiles, Phys Rev. Letters 59, 2756 (1987).
7. The constraint is implemented in the code by setting the force component normal to the plane to zero during minimization.
8. S. Ferrer and H. P. Bonzel, Surface Sci. 119, 234 (1982).
9. See for example, G. Binnig, H. Rohrer, Ch. Gerber and E. Weibel, Surface Sci. 131, L379 (1983).

Surface Structural Analysis of the (1×2) and (1×3) Phases of Pt{110} by Time-of-Flight Scattering and Recoiling Spectrometry

F. Masson and J.W. Rabalais

Department of Chemistry, University of Houston, Houston, TX 77204, USA

1. Introduction

The clean Pt{110} surface is known to reconstruct with (1x2) periodicity [1-4] and, among the different models (buckled surface, pairing rows, saw tooth, missing row) consistent with the (1x2) LEED pattern, the missing row model is generally considered to properly describe this phase. Such a model is supported by LEED [1,2], MEIS [3] and LEIS [4] experiments and is favored by total energy calculations [5]. A picture of the (1x2)-Pt{110} structure is shown in Fig. 1. The atoms are labeled (XYZ) where X is the coordinate along <001>, Y is the coordinate along <$\bar{1}$10> and Z is the layer in which the atoms are located.

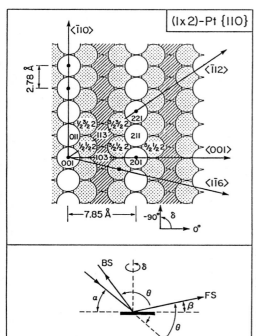

Fig. 1: The missing-row (1x2)-Pt{110} surface. The angular notations used in this work are also indicated (α = incident angle, β = exit angle, θ = scattering angle, δ = azimuthal angle, BS = backscattering, FS = forward scattering).

Quantitative analysis of (1x2)-Pt{110} has been performed by several groups [1-3]. The results indicate that relaxation phenomena are taking place: contraction of the first-to-second layer spacing (>0.20 Å), row-pairing in the second layer (<0.10 Å) and buckling in the third layer.

A (1x3) reconstruction has also been reported [1,3,6] for Pt{110} when the (1x2) phase is submitted to heavy oxygen treatment (>30 min, 900°C, 5×10^{-6} torr). This observation supports theoretical calculations [7], which predict the existence of higher order reconstructions of the type (1xn) with n>2 for fcc{110} metals.

In this work, the technique of time-of-flight scattering and recoiling spectrometry (TOF-SARS) is used to study Pt{110}. It has been shown previously [8-10] that low energy ion scattering can successfully analyze the structure of single crystals.

The objective of this paper is three-fold:

(i)   to apply TOF-SARS to quantitative analysis of clean (1x2)-Pt{110};
(ii) to show that (1x3)-Pt{110} can be obtained by another means than O_2-treatment of the (1x2) phase;
(iii) to demonstrate that TOF-SARS can unambiguously identify the (1x2) and (1x3) periodicities of reconstructed Pt{110} (semi-channeling experiments).

2. Experimental

The TOF-SARS technique and its applications to structural analysis have been described previously [11]. The experimental conditions for this work are: 2 keV Ne^+ primary ion beam; pulse width 50 ns; period = 30 μs; average current density -100-200 pA.mm^{-2}. The base pressure in the analysis chamber was 3×10^{-10} torr. The sample was cleaned by O_2-treatment, Ar^+ sputtering and annealing (1300°C). A fairly sharp (1x2) LEED pattern was obtained.

3. Results and Interpretations

3.1. Study of the (1x2)-Pt{110} Clean Surface

Figs. 2 and 3 are plots of backscattering intensity versus incident angle along several azimuths for two scattering angles, θ = 149° and 95° respectively. In Fig. 2, the peaks are labeled (XYZ)-(X'Y'Z'), indicating that the incident trajectories are focused by atom (XYZ) onto scattering center (X'Y'Z'). In Fig. 3, peaks (A), (B), and (C) are due to incident trajectories being focused at the edge of the shadow cone cast by an atom neighboring the scattering center while peaks (A'), (B'), and (C') are due to scattered trajectories being focused at the edge of a neighboring blocking cone.

The results are in qualitative agreement with those obtained by Niehus [4] and thus support the missing row model. Along certain azimuths some interatomic spacings are known and can therefore be used to calibrate the shadow cone and the blocking cone. The unknown spacings D_{12} (first-to-second layer), D_{23} (second-to-third layer in the trough) and D'_{23} (second-to-third layer below first layer) can then be obtained from the appropriate critical angles (α_c, β_c) along well

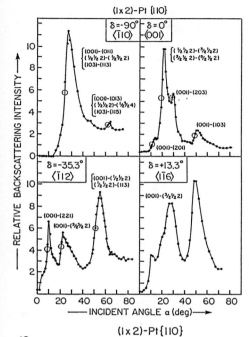

Fig. 2: Scattering intensity versus incident angle scans for (1x2)-Pt{110} at $\theta = 149°$ for different azimuths.

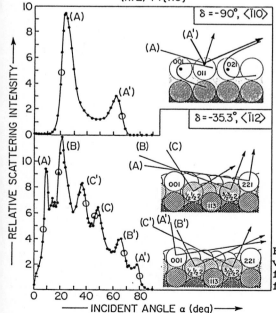

Fig. 3: Scattering intensity versus incident angle scans for (1x2)-Pt{110} at $\theta = 95°$ for different azimuths.

Table 1: Results of the quantitative analysis of (1x2)-Pt{110} by TOF-SARS.

θ	δ	CRITICAL ANGLE α_c, β_c	SPACING
149°	<$\bar{1}$12>	α_c = 20.5°	D_{12} = 1.16 Å
95°	<$\bar{1}$12>	α_c = 19°	D_{12} = 1.14 Å
95°	<$\bar{1}$12>	α_c = 47°	D_{12} = 1.13 Å
95°	<$\bar{1}$12>	β_c = 55°	D_{12} = 1.19 Å
149°	<001>	α_c = 48.5°	D_{23}* = 1.28 Å
149°	<001>	α_c = 29°	D'_{23}*= 1.51 Å

*BASED ON $\overline{D_{12}}$ = 1.15 Å

chosen azimuths. Table 1 summarizes this quantitative analysis and indicates a contraction (0.24 Å) of the first-to-second layer spacing and a buckling of amplitude 0.23 Å in the third layer.

3.2. (1x2)-to-(1x3) Phase Conversion of Pt{110}

After several cycles of high temperature (1300°C) annealing, the (1x2) LEED pattern ultimately evolved into a sharp (1x3) pattern. This phase conversion is due to the diffusion, upon annealing, of bulk impurities to the surface. Direct recoil (DR) spectra [12] induced by 4 keV Kr$^+$ at θ =28°, α = 14°, δ = <001> are shown in Fig. 4. For (1x3)-Pt{110}, a DR

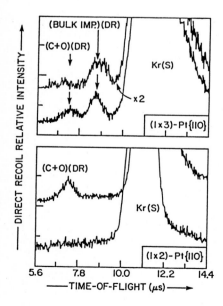

Fig. 4: Direct recoil (DR) spectra, induced by 4 keV Kr$^+$ at θ = 28°, α = 14°, δ = <001>, corresponding to (1x2)- and (1x3)-Pt(110).

structure centered at 9.0 ± 0.3 µs is observed, corresponding to 3p and 4s elements. The resolution of the TOF detection method, especially with a heavy primary ion, is not good enough to clearly identify the impurities. AES and XPS measurements indicate the presence of Ca, K, and P at a coverage < 0.1 ML. Upon O_2-treatment, the Pt{110} surface went back to (1x2) reconstruction and the impurity was no longer observed in the DR spectrum. The carbon and oxygen contamination had no influence on these results.

3.3. TOF-SARS as a Tool to Identify the Periodicity of Pt{110}

Semi-channeling experiments can be used to assign the surface periodicity [10], as an alternative to LEED. If an incident ion (2 keV Ne$^+$ in this work) arrives at glancing incidence ($\alpha = 5°$) at a surface, the trajectories will be channeled between the main crystallographic directions onto the scattering center while this scattering center will be partially shadowed along these main directions. The experiments consist then of measuring the scattering intensity as a function of the azimuthal angle δ. The results are reported in Fig. 5 for (1x2)- and (1x3)-Pt{110}. The two periodicities can be well distinguished by comparing the number of minima in the δ-scan between the <$\bar{1}$11> and <001> directions to the number of crystallographic directions in the same δ window.

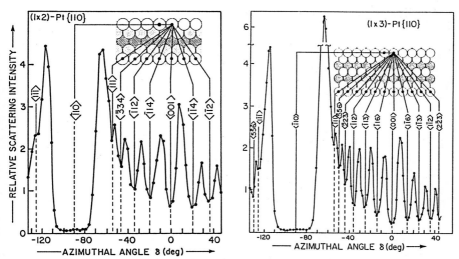

Fig. 5: Scattering intensity versus azimuthal angle scans for (1x2)- and (1x3)-Pt{110} ($\theta = 28°$, $\alpha \sim 5°$).

4. Conclusions

The main conclusions of this study are: (i) TOF-SARS has been successfully used to identify the periodicity of two Pt{110} surfaces; (ii) the clean surface has the (1x2) periodicity whereas diffusion of bulk impurities (Ca,K,P, <0.1 ML) is responsible for conversion to (1x3) periodicity; (iii) a contraction of the 1st-2nd layer spacing (0.24 Å) and a buckling in the 3rd layer are taking place in the (1x2) phase, in agreement with previous results (LEED, MEIS).

5. Acknowledgments

This material is based upon work supported by the National Science Foundation. FM gratefully acknowledges a grant from the Elf-France Corporation.

6. References

1. P.Fery, W.Moritz, and D.Wolf, **Phys.Rev.B**, <u>38</u>, 7275 (1988).
2. E.C.Sowa, M.A.Van Hove, and D.L.Adams, **Surf. Sci.**, <u>199</u>, 174 (1988).
3. P.Fenter and T.Gustafsson, **Phys. Rev. B**, <u>38</u> 10197 (1988).
4. H.Niehus, **Surf. Sci.**, <u>145</u>, 407 (1984).
5. D.Tomanek, H.J.Brocksch, and K.H.Bennemann, **Surf. Sci.**, <u>138</u>, L129 (1984).
6. M.Salmeron and G.A.Somorjai, **Surf. Sci.**, <u>91</u>, 373 (1980).
7. K.-M.Ho and K.P.Bohnen, **Phys. Rev. Lett.**, <u>59</u>, 1833 (1987); M.S.Daw, **Surf. Sci.**, <u>166</u>, L161 (1986); V.Heine and L.D.Marks, **Surf. Sci.**, <u>165</u>, 65 (1986).
8. M.Aono and R.Souda, **Jpn.J. Appl.Phys.**, <u>19</u>, 1249 (1985).
9. S.H.Overbury, **Nucl. Instrum. Methods**, <u>B27</u>, 65 (1987).
10. H.Derks, W.Hetterich, E.Van de Riet, H.Niehus, and W.Heiland, **Nucl. Instrum. Methods**, <u>B48</u>, 315 (1990).
11. H.Bu, M.Shi, and J.W.Rabalais, **Surf. Sci.**, submitted.
12. J.W.Rabalais, **CRC Crit. Rev. Sol. St. Mat. Sci.**, <u>14</u>, 319 (1988).

Analysis of the Reconstructed Ir{110} Surface from Time-of-Flight Scattering and Recoiling Spectrometry

M. Shi, H. Bu, and *J.W. Rabalais*

Department of Chemistry, University of Houston, Houston, TX 77204, USA

1. Introduction

Reconstruction of the {110} surfaces of fcc transition metals into missing-row structures is a well known phenomenon [1–8]. The (1x2) "missing-row" reconstruction has been documented and (1x3) reconstruction has been observed for surfaces with adsorbates for both Pt and Au [2,5]. It has generally been accepted that Ir undergoes a (1x2) reconstruction; the evidence for this was not completely clear [2,3]. In recent work [9] on Ir(110), it was shown that the LEED patterns were neither (1x2) nor (1x3) but could possibly be reconciled by a mixture of the two and that the ion scattering results indicated that the most important structural elements were [1$\bar{1}$0] rows and (1x3) troughs. Using time-of-flight scattering and recoiling spectrometry (TOF-SARS), we have confirmed [10,11] these findings of two-missing rows and extended the model to include adjacent 1st-layer rows, hence a mixed faceted (1x3) and (1x1) structure. A schematic drawing of the Ir{110} surface, showing the faceted (1x3) structure and (1x1) structures, is shown in Fig. 1.

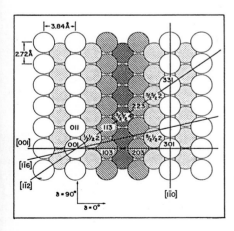

Fig. 1: Reconstructed Ir{110} surface showing coexisting faceted (1x3) and (1x1) structures. Open circles – 1st-layer; Dotted circles – 2nd-layer; Dashed circles – 3rd-layer; Hatched circles – 4th-layer.

2. Experimental Methods

The TOF-SARS technique and applications to structural analysis have been described elsewhere [12,13]. The Ir sample was cleaned by Ar^+ sputtering and O_2 treatment followed by annealing to 1400°C by electron bombardment. Temperature was measured by means of a portable infrared thermometer. Cleanliness was verified by the absence of carbon and oxygen Auger signals and the absence of H, C, and O recoils [13].

3. Experimental Results and Interpretations

3.1. Backscattering Intensity I(BS) Versus Incident Angle α Scans

I(BS) as a function of α was measured along different azimuths δ for θ = 163°. In this <u>shadowing mode</u>, I(BS) is determined by the ability of the incident ions to make almost head-on collisions with Ir atoms. Due to shadowing effects, the trajectories are focused at the edge of the shadow cones. At a critical incident angle $α_c$ where the edge of the cone coincides with a neighboring atom, large enhancements in I(BS) are observed [12]. Selected I(BS) versus α scans [11] are shown in Fig. 2. (1). α Scan Along [1$\bar{1}$0] Azimuth. I(BS) is very low for α>15°, indicating that the surface is well ordered along this azimuth. The peaks at α=28° and α=60° result from atoms emerging from shadow cones of their nearest neighbors in the same layer and in different layers respectively. Scattering along this azimuth provides no information concerning a possible (1xn) reconstruction. (2). α Scan Along [1$\bar{1}$2] Azimuth. Along this azimuth atoms from all layers are aligned in the scattering plane. The 17° peak can result from both (1x3) and (1x1) structures. The shoulder near 50° is identified as deep layer scattering since it

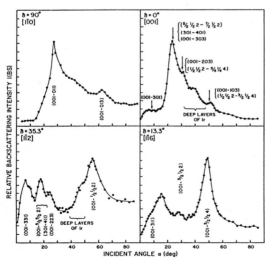

Fig. 2: I(BS) versus incident angle α scans for Ir{110} along four different azimuths δ. The calculated $α_c$ positions for a faceted (1x3) structure are shown. These positions are labeled with the identities of the two atoms involved in the shadowing and scattering.

260

disappears as Θ is reduced from 163° to 115°. (3). α Scan Along [001] Azimuth. The first α_c position at 6° corresponds to 1st-layer interactions such as (001)-(301) occurring at interatomic distances >11 Å. The peak at 39° is attributed to scattering from deeper layers. (4). α Scan Along [1$\bar{1}$6] Azimuth. A very low intensity peak is observed at ≈27°; this is a major peak [10] in the (1x2) missing row Pt{110} structure, i.e. its intensity is comparable to the 45° peak. It arises from 1st-layer (001) atoms shadowing 2nd-layer (3/2 1/2 2) atoms across a trough which has only one missing 1st-layer row. These (3/2 1/2 2) atoms are completely missing in the (1x3) structure. This low intensity peak can result from both traces of (1x2) structures and/or (1x1) structural domains such as (301)-(9/2 1/2 2).

3.2. Forward Scattering I(FS) Versus Scattering Angle Θ Scans

Fig. 3 shows I(FS) versus exit angle β with respect to the surface where β = Θ - α, along selected azimuths δ with fixed α. In this <u>blocking</u> <u>mode</u> for FS at low β, the structural features are determined by focusing of scattered trajectories at the edges of the blocking cones of atoms obstructing their escape along that azimuth. The low value of α = 8° was used

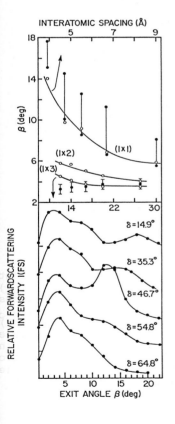

Fig. 3: Lower: I(FS) versus exit angle β scans along five different δ directions with α = 8°. Upper: Experimental and calculated β versus interatomic spacing d for (1x1), (1x2), and (1x3) structures. The experimental points (solid circles) for the low α peak are taken as the peak maximum (along with error bars) while those for the high α peak are drawn as a line from the peak maximum to the half-height position. The calculated points (open circles) were obtained from trajectory simulations of the blocking, assuming two-(1x3), one-(1x2), and zero-(1x1) missing rows. The d values for the (1x2) structure are two times those of the (1x1) structure.

for the scans so that only 1st-layer scattering is obtained
along all azimuths. Thus, the structures are determined by
1st-layer atoms blocking their 1st-layer neighbors. Data is
not shown for $\delta = 0°$ and $90°$ because, along these azimuths,
surface semichanneling [13] effects dominate and the intensity
of the low β peak overwhelms that of the high β peak. This
semichanneling is notably lacking at $\delta = 46.7°$, resulting in
an abnormally high relative intensity for the high β peak.
The two peaks observed in Fig. 3 indicate that there are two
different interatomic spacings in the 1st-atomic layer. The
positions expected for these peaks were estimated from
classical simulations of blocking along the outgoing
trajectory [12], assuming different 1st-layer spacings. The
results are summarized in the upper part of Fig. 3 as plots of
calculated β versus d (i.e. β vs. δ) along with the
experimental data. The experimental and calculated points are
in good agreement for d > 14 Å and > 4 Å for the first- and
second-peaks of Fig. 3, respectively. The poorer agreement at
smaller d is a result of deficiencies in the simulations; the
calculation considers atoms in a plane, which is a poor
approximation as $\delta \rightarrow 0°$ (smaller d) due to surface
semichanneling effects and to the close proximity of off-
planar atoms. The calculated positions for a single missing
row (1x2) do not agree with either of the observed structures
of Fig. 3. Therefore, the data from the I(FS) versus θ scans
support a model of two missing rows and adjacent rows and are
inconsistent with a single missing row model.

4. Estimation of Interatomic Spacings

Scattering peaks from the (1x3) domains can be separated from
those of the (1x1) domains at low θ for the purpose of
quantitative analysis. Details of the estimation of the 1st-
2nd-layer registry and the 1st- through 4th-interlayer
spacings will be presented elsewhere [11]. Analysis of the
(1x1) domains was not attempted because it is a minority
structure and its scattering peaks cannot be separated from
those of the (1x3) domains.

5. Discussion

The TOF-SARS data is consistent with only one of the
reconstruction models that we have considered. This model
consists of primary faceted (1x3) domains and coexisting
secondary (1x1) domains. TOF-SARS is not capable of
determining how many adjacent [1$\bar{1}$0] rows exist together in the
(1x1) domains; it only shows that there are some 1st-layer
spacings that correspond to the 3.84 Å lattice constant along
the [001] direction while others correspond to three times
that number. This shows that, under this annealing condition,
the Ir{110} reconstruction is not complete, i.e. not all of
the unreconstructed (1x1) surface is converted into (1x3)
domains. Thus, there is considerable disorder along the [001]
direction due to the different atomic row spacings. This
disorder is consistent with the streaked LEED patterns.
     Analysis of the (1x3) domains shows that (i) the 2nd-layer
atoms in the facet shift laterally towards the center of the

trough, the estimated shift being -0.12 ± 0.10 Å, (ii) the 1st-2nd layer spacing is contracted from the bulk value, the estimated contraction being 0.11 ± 0.10 Å, and (iii) the 2nd-3rd and 3rd-4th layer spacings are similar to the bulk values.

6. Summary

The TOF-SARS and LEED data agree that after annealing the Ir{110} surface to 1400°C and cooling to room temperature, the surface is reconstructed into major (1x3) domains and minor (1x1) domains. Analysis of the (1x3) structure indicates a lateral shift of the second-layer atoms towards the center of the trough and a contraction of the first-second layer vertical spacing. The second- through fourth-layer spacings appear to be unchanged from the bulk values.

7. Acknowledgment

This material is based on work supported by the National Science Foundation under Grant No. CHE-8814337.

References

1. T. Gustafsson, M. Copel, and P. Fenter, in: The Structure of Surfaces II, Eds. J. F. van der Veen and M. A. Van Hove (Springer, Berlin, 1988) p. 110.
2. C. M. Chang, M. A. Van Hove, W. H. Weinberg and E. D. Williams, **Surface Sci.**, 91, 440 (1980); M. A. Van Hove, W. H. Weinberg, and C. M. Chang, in "Low Energy Electron Diffraction", Vol. 6, Springer Series in Surface Sciences (Springer, Berlin 1986); C.-M. Chang and M. A. Van Hove, **Surface Sci.**, 171, 226 (1986).
3. Q. J. Gao and T. T. Tsong, **J. Vac. Sci. Technol.**, A15, 761 (1987); K. Müller, J. Witt, and O. Schütz, **J. Vac. Sci. Technol.**, A5, 757 (1987); G. L. Kellogg, **J. Vac. Sci. Technol.**, A5, 747 (1987).
4. J. Möller, K. J. Snowdon, w. Heiland, and H. Niehus, **Surface Sci.**, 176, 475 (1986).
5. P. Haberle, P. Fenter, and T. Gustafsson, **Phys. Rev. B**, 39, 5810 (1989); M. Copel, P. Fenter, and T. Gustafsson, **J. Vac. Sci. Technol.**, A5, 742 (1987).
6. I. K. Robinson, **Phys. Rev. Lett.**, 50, 1145 (1983).
7. L. D. Marks, **Phys. Rev. Lett.**, 51, 1000 (1981).
8. G. Binnig, H. Rohrer, Ch. Gerber, and E. Weibel, **Surface Sci.**, 131, L379 (1983).
9. W. Hetterich and W. Heiland, **Surface Sci.**, 210, 129 (1989).
10. H. Bu, M. Shi, F. Masson, and J. W. Rabalais, **Surface Sci.** 230, 1410 (1990).
11. M. Shi, H. Bu, and J. W. Rabalais, **Phys. Rev. B.**, in press.
12. O. Grizzi, M. Shi, H. Bu, J. W. Rabalais, and P. Hochmann, **Phys. Rev. B**, 40, 10,127 (1989).
13. O. Grizzi, M. Shi, H. Bu, and J. W. Rabalais, **Rev. Sci. Instru.** 61, 740 (1990).

Thermally Activated Defects on Crystal Surfaces

B. Salanon, H.J. Ernst, F. Fabre, L. Barbier, and J. Lapujoulade

DPhG/PAS CEN, Saclay, F-91191 Gif-sur-Yvette Cedex, France

Abstract. We show that the creation of defects at the initial stage of roughening far below T_R can explain the loss of coherent intensity in diffraction on Cu(113), Cu(110) and Cu(100). Our energetic model originally developed to account for the broadening of peaks near T_R turns out to give a satisfactory description of low T data as well. The values we obtain for the incoherent scattering cross section σ are comparable to those obtained in other experiments. The values of the kink creation energy J_1 are found to be rather similar on all faces and compare favorably to calculations.

1. Introduction

In recent years numerous experimental studies have been devoted to understanding the thermal behavior of metal faces. Among several possible processes, such as melting or others the roughening transition predicted by Burton, Cabrera and Frank [1] was evidenced on Cu [2] and Ni [3] stepped surfaces. As for (110) surfaces of fcc crystals various phenomena seem to occur. For instance Pb(110) was shown to roughen first [4] and subsequently to undergo surface melting at a higher temperature [5]. Grazing X-ray diffraction experiments have shown that a roughening transition occurs on Ag(110) [6]. The case of Cu(110) is still a matter of controversy, although the anomalous thermal behavior of this surface was demonstrated by several techniques, such as X-ray diffraction [7], thermal helium scattering [8], photoemission [9] or low energy ion scattering [10]. In each case the measured intensities strongly decrease when the crystal temperature exceeds 500 K. This was attributed either to defects [7] or to anharmonicity [8].

Similar strong decreases of diffraction intensities were reported on stepped Cu faces such as Cu(113) and Cu(115) [11]. On the basis of such experimental observations Villain et al [12] could infer the existence of roughening transitions and estimate the critical temperatures T_R. Their analysis was confirmed by subsequent accurate measurements of the shape of diffraction peaks, which is a better means to study roughening than looking at intensities only [13]. The shape analysis method takes advantage of the fact that in the vicinity of the critical temperature T_R

Springer Series in Surface Sciences, Vol. 24   **The Structure of Surfaces III**
Editors: S.Y. Tong · M.A. Van Hove · K. Takayanagi · X.D. Xie
© Springer-Verlag Berlin, Heidelberg 1991

numerous levels (or terraces) appear · on the face in the same time as steps proliferate. This type of disorder is unambiguously evidenced through the interference effects between waves scattered from different terraces. Such effects are largely governed by the phase shift φ: A phase shift $\varphi = 2n\pi$ corresponds to constructive interferences (or in-phase conditions) for which the peak of interest remains narrow, $\varphi = (2n+1)\pi$ corresponds to destructive interferences (or out-of-phase conditions) for which the peak broadens when passing through T_R. Varying the kinematical conditions allows one to vary the phase φ. So in such experiments the oscillation of peak widths (or shapes) with the kinematical conditions implies multilevels. Then following the behavior of the out-of-phase peaks as functions of surface temperature leads to understanding the roughening process even quantitatively.

Surprisingly a careful analysis of Cu(113) diffraction data clearly shows that a decrease of intensity is also seen for in-phase conditions, although it is less pronounced than for out-of-phase conditions. It is clear that if the broadening accounts for the difference between the intensities, the cause for the overall decrease of coherent scattering is to be looked for elsewhere. We will show that the proliferation of steps and pointlike defects gives rise to an incoherent scattering and thus can well explain the observed loss of intensity in the Bragg peaks. Based on the energetic model for roughening of Cu(113) which could successfully explain the broadening of out-of-phase peaks [14] we will show that this very model also accounts for the loss of intensity of in-phase diffraction peaks. The extension of this theory to the cases of Cu(100) and Cu(110) can describe the thermal behavior of these surfaces as well. We find that in all cases the loss of intensity in Bragg peaks can be attributed to the early stages of roughening i.e. the creation of adatoms and vacancies.

2. Models for roughening and scattering applied to Cu(113)

The perfect (113) face is made of (100) terraces separated by (111) steps. We note that this structure is very similar to that of Cu(110) except for the distance between dense rows which is larger for (113). Using Helium diffraction it was found that the width of the specular peak oscillates as a function of incidence angle. As recalled in the introduction this clearly shows that the surface becomes rough at sufficiently high temperature. More specifically, kinks are created on the original steps and domains of different heights appear, separated by lines of kinks that represent domain walls. Following Schulz [15] we call these walls secondary steps. It is the free energy for creating such secondary steps which vanishes at the roughening temperature T_R. The creation of secondary steps determines the long wavelength behavior of the correlation function of heights and is therefore reflected in the low Q shape of the diffraction peaks. In agreement with the Kosterlitz-Thouless theory of roughening we found that at high enough temperature (above $T_R \simeq 700$ K) the shape of the specular peak was a power law function of the momentum transfer Q. This proves that the correlation function of heights is a logarithmic function of distance, which is the signature for thermal roughening. From the quantitative analysis of the peak shapes we could determine the critical temperature T_R. We have described successfully the thermal evolution of the peak shapes as functions of T by a microscopic energetic model similar to the six vertex model, whose main ingredients are the kink creation energy J_1 and the step-step interaction energy J_2 [14]. Apart from small details in the description of the crystallographic structure such a model would also be very appropriate for (100) or (110) faces of FCC metals. Analytical expressions or Monte Carlo (MC) simulations were used in order to decribe the thermal evolution of peak shapes.

As expected, one finds experimentally that the in-phase specular peak remains narrow even at the highest temperatures, since it is insensitive to vertical disorder. Under such conditions the intensity is in principle determined solely by thermal mean square displacements via the Debye-Waller type attenuation. However, it has been demonstrated that point or linelike defects reduce Bragg intensities via incoherent scattering [16]. We suggest that the creation of such defects as the initial stage of roughening can account for the loss of coherent intensity. This assumption is supported by the very fact that an appropriate model for thermal attenuation, such as the Armand Manson procedure including bulk anharmonicity, underestimates the observed loss, as shown in fig.(1). Of course one could get a good agreement by adjusting the mean square displacements of surface atoms over the bulk values as achieved by Armand et al [8]. However there is compelling evidence from the measurements of Zeppenfeld et al. on Cu(110) that point defects are created at high temperatures [8]. In fact the ratio of incoherent to coherent intensity increases with temperature, which is a clear manifestation of the creation of defects at high T.

In order to estimate the loss of coherent intensity due to diffuse scattering from defects, we have used a procedure similar to the one described by Comsa and Poelsema [17]. We have considered mainly the low T data corrected for inelastic effects as indicated hereabove.In this scheme the defects are considered as absorbing (or black) regions the area of which is the total cross section for diffuse scattering from individual defects. Multiple scattering events are included as overlap effects between individual cross sections. In our case MC calculations could provide us with typical configurations for each temperature. In order to keep the number of adjustable parameters as low as possible we have chosen to assign a black square area to every broken bond on the surface . We define a broken bond as the lack of a neighboring atom either in the direction of dense rows or perpendicular to the rows. With this definition an isolated adatom (or vacancy) is surrounded by four broken bonds and its cross section results from the overlap of four squares (see insert

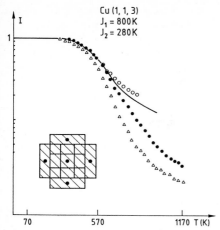

Figure 1 Specular intensity on Cu(113) for θ_i = 59° (in-phase,•)
and θ_i = 68° (antiphase,△) these values were corrected using a
linear Debye Waller factor. (○) corresponds to the in phase data
corrected by using the Armand Manson procedure with bulk
anharmonicity. The full line is the result of the fit with
$\sigma \simeq 70$ Å^2 . The insert shows the configuration of black squares
around an adatom.

fig.1). A secondary step on Cu(113) is covered with black squares
and the cross section is defined by the width of the dark
ribbonlike area following the step. With this procedure the
diffuse scattering from all types of defects on the surface can be
treated readily, the only adjustable parameter being the size of
the elementary black square associated with a broken bond.

We have performed MC calculations with the same energetic model
that could successfully describe the peak shape evolution i.e.
with J_1 = 800 K and J_2 = 280 K. For each configuration generated
by the algorithm the broken bonds were identified and covered by
black areas. The remaining area was calculated and sampled over
10^4 configurations in order for the thermodynamically averaged
intensity to be calculated. As the energetic parameters J_1 and J_2
were in principle determined by the analysis of peak shapes we
have considered that the cross section was the only adjustable
parameter. Fig.(1) shows that a very good agreement for T < 700 K
can be obtained by choosing σ = 6.5 unit cells, which corresponds
to $\sigma \simeq 70$ Å^2 for an isolated adatom. This is consistent with
values obtained for other metals and it should be considered as a
lower bound because vacancies have the same cross section as
adatoms in our calculation.

That means that our energetic model can describe not only the
creation of domains and secondary steps in the course of
roughening at high T but also the creation of point defects at
lower temperature. This is a very good indication that both the

267

creation of point defects at low T and that of secondary steps at somewhat higher T are inherent to the roughening transition.

3. Cu(110) and Cu(100)

The six vertex model was shown to be appropriate for (100) and (110) faces [18]. Since the local geometry is different from that of Cu(113), we have to treat the interaction parameters J_1 and J_2 as adjustable parameters for these surfaces. Following Armand et al [19] we have corrected the measured specular intensity for thermal attenuation using only bulk anharmonicity and ascribed the remaining loss to defects.

On Cu(100) for symmetry reasons one has $J = J_1 = J_2$. Fig.(2) shows the comparison between the experimental data and the results of MC calculations using two different values of the cross section σ for diffuse scattering from defects, the mean value being $\sigma \simeq 80$ Å^2. Again the experimental results are well described within our model for the creation of defects. Incidentally, the results of MC calculations can reasonably well be approximated by the following formula for the coherent intensity I where σ is measured in unit cells :

$$ I \simeq 1 - 2\sigma e^{-\dfrac{4J}{T}} \tag{1} $$

A good fit of the initial decrease is obtained with $J = 1200$ K, which would correspond to an hypothetical roughening temperature

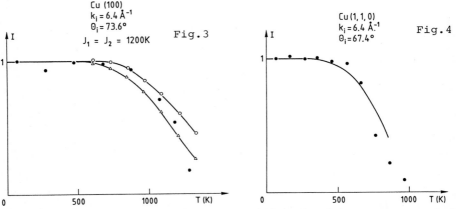

Cu (100)
$k_i = 6.4$ Å^{-1}
$\theta_i = 73.6°$
$J_1 = J_2 = 1200$K

Fig.3

Cu (1,1,0)
$k_i = 6.4$ Å^{-1}
$\theta_i = 67.4°$

Fig.4

Figure 2 Specular intensity on Cu(100), the data were corrected for thermal attenuation using the Armand Manson formalism and are represented as (•). The results of the fit are shown as (△) for $\sigma = 110$ Å^2 and as (○) for $\sigma = 65$ Å^2.

Figure 3 Specular intensity on Cu(110), the data were corrected for thermal attenuation using the Armand Manson formalism and are represented as (•). The results of the fit are shown as a full line for $\sigma \simeq 90$ Å^2.

of the order of $T_R = 1800$ K . That means, that although point defects are created as the initial stage of roughening, the surface cannot develop its fully developed rough state as the crystal melts at lower T .

For Cu(110) the measured intensity was corrected in the same way as for the other surfaces. Here calculating the influence of defects is a little bit more difficult, as we have to fit J_1, J_2 and σ. It is reasonable to estimate $J_1 \simeq 1000$ K , which is intermediate between the values for Cu(113) and Cu(100). Then J_2 and σ can be fitted. The result is shown on fig.(3). It turns out that $J_1 = 1000$ K and $J_2 = 500$ K together with $\sigma \simeq 90$ Å^2 is rather satisfactory. This would predict $T_R \simeq 1000$ K, just slightly above the lower bound estimated by Zeppenfeld et al. [8].

4 Conclusion

We have shown that the creation of defects at the initial stage of roughening far below T_R can explain the loss of coherent intensity in diffraction from Cu(113), Cu(100) and Cu(110). Our energetic model had been shown to account for the broadening of out-of-phase diffraction peaks due to the creation near T_R of steps on flat surfaces or of secondary steps on vicinal surfaces. This very model describes satisfactorily the loss of intensity in diffraction peaks due to incoherent scattering from point and linelike defects far below T_R as well. In fact the creation of point defects far below T_R preceding the creation of steps in the vicinity of T_R is shown to be actually a precursor to roughening itself.

The incoherent scattering cross-section σ of the point defects could be deduced from the low temperature data when corrected for thermal attenuation using bulk anharmonicity only. We found consistently $\sigma \simeq 90$ Å^2 for all copper surfaces, this value is quite comparable to the one obtained from MBE experiments where copper atoms are deposited on a copper surface [20]. Moreover the values of the kink creation energy $J_1 \simeq 1000$ K are found to be similar on all copper surfaces of interest. This value of J_1 has to be compared to the value determined by Loisel from a molecular dynamics calculation for Cu(110) ($J_1 \simeq 1500$ K) [21]. We expect the discrepancy to vanish when including vibrational entropy terms which were so far not considered in our analysis.

References

1 W.K.Burton, N.Cabrera and F.C.Frank, Philos. Trans. Roy. Soc. London A **243**, 299 (1951)
2 F.Fabre, D.Gorse, B.Salanon and J.Lapujoulade, Europhys. Lett. **3**, 737 (1987)
3 E.H.Conrad, R.M.Aten, D.S.Kaufman, L.R.Allen and T.Engel, J. Chem. Phys. **84**, 1015 (1986)
4 H.N.Yang, T.M.Lu and G.C.Wang, Phys. Rev. Lett. **63**, 1621 (1989)

5   J.W.M.Frenken and   J.F.van der   Veen, Phys.   Rev. Lett.   **54** 134 (1985)

6   G.A.Held, J.L.Jordan-Sweet,   P.M.Horn, A.Mak and R.J.Birgeneau, Phys. Rev. Lett. **59** 2075 (1987)

7 S.G.J.Mochrie, Phys. Rev. Lett. **59** 304 (1987)

8   P.Zeppenfeld, K.Kern,   R.David and G.Comsa, Phys. Rev. Lett. **62** 63 (1989)

   G.Armand and P.Zeppenfeld, Phys. Rev. B **40** 5936 (1989)

9   R.S.Williams, P.S.Wehner,   J.Stöhr and   D.A.Shirley, Phys. Rev. Lett. **39** 302 (1977)

10   T.Fauster,   R.Schneider,   H.Dürr,   G.Engelmann and E.Taglauer, Surf. Sci. **189/190** 610 (1987)

11 J.Lapujoulade, J.Perreau and A.Kara, Surface Sci. **129** 59 (1983)

12   J.Villain, D.R.Grempel and J.Lapujoulade, J. Phys.   F **15** 809 (1985)

13 B.Salanon,F.Fabre,D.Gorse and J.Lapujoulade, J.   Vac.   Sci. Technol. A **6(3)** 655 (1988)

14   B.Salanon, F.Fabre, J.Lapujoulade and W.Selke, Phys. Rev. B **38** 7385 (1988)

15 H.J.Schulz, J. Physique **46** 257 (1985)

16   L.K.Verheij, B.Poelsema and G.Comsa, Surface Sci. **162** 858 (1985)

17   B.Poelsema and G.Comsa, in Scattering of Thermal Energy Atoms from Disordered Surfaces, Springer-Verlag (1989)

18 H.van Beijeren, Phys. Rev. Lett. **38** 993 (1977)

19   G.Armand, D.Gorse, J.Lapujoulade and J.R.Manson, Europhys. Lett. **3** 1113 (1987)

20 L.J.Gomez, S.Bourgeal, J.Ibanez and M.Salmeron, Phys. Rev. B **31** 2551 (1985)

21 B.Loisel Thèse Paris (unpublished)

The Surface Barrier Structure of Metals from Barrier-Induced Surface States and Resonances

A.S. Christopoulos and M.N. Read

School of Physics, University of New South Wales,
P.O. Box 1, Kensington, Sydney, Australia 2033

Abstract. We have used a plane-wave scattering method to calculate the band structure of unoccupied states at a metal surface in which the wave function is localised between the top row of atoms and the surface barrier. The method can be applied to any metal surface and calculations have been performed along $\overline{\Gamma}(\overline{\Delta})\overline{X}$ for the W(001) and Cu(001) surfaces. A number of features are found to be very sensitive to the surface barrier and a close fit to experimental inverse and two-photon photoemission data using the present method could lead to the determination of the surface barrier structure.

1. Introduction

The surface barrier is the potential experienced by an electron located in the transition region between the metal substrate potential and the zero potential far from the metal. A number of different approaches to calculating the effective one-electron potential at a metal surface [1-3] have concluded that for distances $|z| > 5$ to 6 a.u. from the centre of the top row of atoms at the surface, the potential should be of the shifted image form

$$U(z) = \frac{1}{2(z-z_o)} \quad \text{(Rydberg units)}$$

where z_o specifies the position of the image plane for each metal surface and the z co-ordinate is directed into the metal. Closer to the surface the potential should weaken and join smoothly to the inner potential of the metal.

Information about the surface barrier potential is contained in very low energy electron diffraction (VLEED) data [4 - 6] and inverse and two-photon photoemission data [7, 8]. The latter two techniques map the electronic band structure of surface states which arise because of the surface barrier potential. Electrons in these normally unoccupied states have wave functions which are localised beyond the first row of atoms at a metal surface with energies lying from the Fermi energy to past the vacuum energy.

2. Theory

We have used a plane-wave scattering method to calculate the band structure of these surface states [9]. The wave function in a region of constant potential between the top row of atoms and the surface barrier is expanded as a sum of forward and backward travelling plane waves

$$\psi(r) = \sum_{v} [a_1(k_v^+) \exp(i\, k_v^+ . r) + b_1(k_v^-) \exp(i\, k_v^- . r)]$$

where $k_v^\pm = (k_{\|x} + 2\pi v_x)\, \hat{x} + (k_{\|y} + 2\pi v_y)\, \hat{y} \pm (\sqrt{E - |k_\| + 2\pi v|^2})\hat{z}$

The x,y directions are parallel to the surface, v extends over an arbitrarily large number of reciprocal net vectors of the surface, $k_\|$ is the component of the electron momentum parallel to the surface, E is the energy in Rydberg atomic units with respect to the constant potential which is the muffin-tin zero in this case.

The reflections of the plane waves at the surface barrier, which is taken here to have no lateral variation, are represented as amplitude reflection coefficients such as $\rho_b(k_v^+ \ k_v^-)$ which form the elements of the matrix S^{II}. The reflections at the substrate are represented as amplitude reflection coefficients such as $\rho_c(k_v^- \ k_v^+)$ and form the elements of matrix M. By summing the multiple scattering contributions to the total back-scattered amplitude b_1 of each plane wave, the probability of finding the electron in this region is a maximum and the electron is trapped in a surface state or surface resonance when

$$|\det[I - S^{II} M]| \quad \text{is a minimum}$$

where I is the identity matrix. Using this method, the phase shift on reflection at the crystal substrate can be calculated exactly for any crystal surface within the muffin-tin potential model.

In an extensive study, Smith and co-workers [10] have established the systematics of the surface barrier states from the Fermi energy to the vacuum energy in bulk band gaps on the principal faces of the noble and near noble fcc metals. This work calculates the phase shift ϕ_c at the crystal substrate from a nearly-free-electron (NFE) band structure approximation fitted to the bulk band gap projected onto the surface. In this approximation, surface resonances which occur outside bulk band gaps are treated approximately if close to band gap edges but cannot be treated well if far from band gap edges. Even for energies inside bulk band gaps ϕ_c will only be accurate if the bulk band gap in this energy range for this metal can be represented in the NFE approximation. Certainly for the transition bcc metals W, Ta and Mo the Fermi and vacuum energies occur within d-bands and NFE models are not useful.

3. Calculations

We have calculated the band structure of unoccupied surface states and resonances which have wave functions localised in the barrier region on the W(001) 1x1 surface along $\bar{\Gamma}(\bar{\Delta})\bar{X}$ by mapping the values of E and $k_\|$ for which $|\det[I - S^{II} M]|$ is a minimum in the plane-wave scattering approach. This surface is not amenable to Smith's approach because there are few bulk band gaps, so that most features are expected to be surface resonances, and those gaps that are present are not NFE like. We have also performed a calculation along $\bar{\Gamma}(\bar{\Delta})\bar{X}$ for Cu(001) which is amenable to both Smith's method and the self-consistent FLAPW slab method. The surface

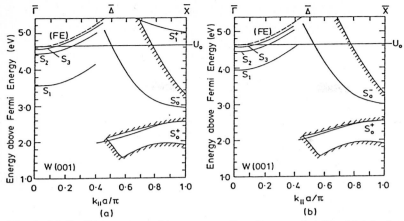

Fig. 1. (a) Surface states and resonances for electrons localised in the barrier region of the W(001) surface calculated from the present plane-wave scattering method for a surface model with $z_0 = -3.5$ a.u. Hatched areas represent the calculated surface projected bulk band edges of states of even symmetry. U_0 is the height of the barrier and the Fermi energy is 9.97 eV above the muffin-tin zero. FE denotes the free electron band structure.

(b) Same as (a) except for a surface barrier with $z_0 = -2.9$ a.u.

barrier is represented as the shifted image form joined onto the bulk potential with a quadratic function which allows variable saturation [12]. The elements of the matrix S^{II} were calculated by integrating the z-dependent Schrödinger equation and those of M were calculated using MacRae's method [13]. The 6% contraction of the surface layer of atoms for the W(001) 1x1 structure would not have a large effect on our calculations here therefore we have not included it at this stage. The potential of Mattheiss was used for W [14] and the self-consistent potential of Snow and Weber was used for Cu [15].

The results for the W(001) surface using a barrier with $z_0 = -3.5$ a.u. are shown in Fig. 1(a). The plane-wave scattering method also allows calculation of the surface projected bulk band gaps for the symmetry of the incident plane waves which is even and these are indicated on the figure. Exactly at $\bar{\Gamma}$ a bulk band gap exists for the entire energy range shown and there is only one propagating plane wave in the barrier region corresponding to the 00 reciprocal net vector. The lower state at $\bar{\Gamma}$ corresponds to this plane wave being incident on the barrier with perpendicular energy 13.3 eV with respect to the muffin-tin zero. Here the phase shift on reflection from the barrier ϕ_b is varying more rapidly with energy than the substrate phase shift ϕ_c and the feature is identified as the n = 1 member of the Rydberg series of barrier-induced surface states. Here we choose the quantum number according to the scheme where it represents the number of extrema in the wave function [16]. The higher energy states are the n = 2 and n = 3 members of the Rydberg series. At $\bar{\Gamma}$ the states are true surface states but they become resonances as they disperse away from $\bar{\Gamma}$ since they are then degenerate with bulk bands. The n = 0 so-called crystal-induced

273

surface state occurs where the substrate reflection phase shift ϕ_c is varying more rapidly with energy than ϕ_b and for the choice of barrier used here this occurs at an energy below that shown. The $n = \infty$ dispersion corresponds to that of a free electron which has just sufficient energy to escape from the barrier region into vacuum. A feature identified as a barrier-induced state has been observed experimentally at about 0.7 eV below the vacuum level U_0 and probably corresponds to the $n = 1$ state of the series [17, 18].

At $\overline{X}$ the states and resonances result from the substrate induced interaction of the propagating plane waves with reciprocal net vectors 00 and $\overline{1}0$. The lowest state and higher resonance state correspond to even and odd combinations of the plane waves incident at the surface barrier with perpendicular energies 8.6 eV and 8.9 eV with respect to the muffin-tin zero. Again ϕ_b is not changing as rapidly with energy as ϕ_c here and the states are identified as the $n = 0$ crystal-induced states. For the upper state at $\overline{X}$ above the vacuum level the even combination of the incident plane waves have perpendicular energy 10.9 eV and ϕ_b is changing more rapidly with energy than ϕ_c so it is the $n = 1$ barrier induced surface state. The other $n = 1$ state at $\overline{X}$ occurs at 7.1 eV and is not shown on the diagram. Away from $\overline{X}$ the crystal-induced surface state S_o^+ and resonance S_o^- disperse in the same way as the bulk band gap edges as expected from the analysis of Pendry et al,.[19]. The barrier-induced states here all show a dispersion which is more free-electron-like.

We show a recalculation of the surface states and resonances on W(001) for a shift of the image plane to $z_0 = -2.9$ a.u. in Fig. 1(b). The $n = 1$ members of the Rydberg series labelled S_1 and S_1^+ are shifted considerably in energy while the $n = 0$ crystal-induced states S_o^+ and S_o^- are little affected.

For our Cu(001) calculation all the features from the self-consistent FLAPW slab method with image potential inclusion [11] and from Smith's method [10] are reproduced and a good fit to the experimental result is obtained [20].

4.   Conclusion

In summary, our present plane-wave scattering method for the calculation of unoccupied surface states and resonances localised in the surface barrier region provides an accurate method for predicting their energy location as a function of $k_\parallel$ for any metal. The crystal substrate reflection is calculated accurately (within the muffin-tin approximation) for all metals and the method calculates the position of resonances as accurately as states unlike the slab calculations and Smith's approximate scheme. Different surface potentials, surface layer contractions and reconstructions can be included. A further advantage is that the potentials are of the same form as those used in LEED and details of the surface barrier obtained from the present technique can be directly carried over into the VLEED calculation. It is anticipated that analysis of the barrier states and resonances measured by inverse and two-photon photoemission techniques using the present method could lead to a definitive determination of the surface barrier on metal surfaces.

The present calculation for W(001) shows that barrier surface states and resonances exist on a bcc surface and for a metal with a far more complicated projected bulk band structure than for the noble and near noble metals studied so far. Two of the features shown here are particularly sensitive to the position of the image plane of the surface barrier and accurate experimental data from inverse and two-photon photoemission corresponding to the S_1 and S_1^+ state would provide information about the surface barrier. A number of highly barrier sensitive surface states also exist at energies along $\bar{X}$ above the vacuum level and experimental investigations in this region are urged.

References

[1] S. Ossicini, C.M. Bertoni and P. Gies, Europhys. Lett. 1, 661 (1986).
[2] N.D. Lang and W. Kohn, Phys. Rev. B 7, 3541 (1973).
[3] J.A. Appelbaum and D.R. Hamann, Phys. Rev. B 6, 1122 (1972).
[4] J.-M. Baribeau, J.-D. Carette, P.J. Jennings, R.O. Jones, Phys. Rev. B 32, 6131 (1985).
[5] J.-M. Baribeau, J. Lopez and J.-C. Le Bossé, J. Phys. C 18, 3083 (1985).
[6] M.N. Read, Phys. Rev. B 32, 2677 (1985).
[7] D. Straub and F.J. Himpsel, Phys. Rev. B 33, 2256 (1986).
[8] K. Giesen, F. Hage, F.J. Himpsel, H.J. Riess and W. Steinmann, Phys. Rev. B 33, 5241 (1986).
[9] M.N. Read and A.S. Christopoulos, to be published.
[10] N.V. Smith and C.T. Chen, Phys. Rev. B 40, 7565 (1989).
[11] S.L. Hulbert, P.D. Johnson, M. Weinert, R.F. Garrett, Phys. Rev. B 33, 760 (1986).
[12] E.G. McRae and M.L. Kane, Surface Sci. 108, 435 (1981).
[13] E.G. McRae, Surface Sci. 25, 491 (1971).
[14] L.F. Mattheiss, Phys. Rev. 139, A1893 (1965).
[15] E.C. Snow and J.T. Weber, Phys. Rev. 157, 570 (1967).
[16] M. Weinert, S.L. Hulbert and P.D. Johnson, Phys. Rev. Lett. 55, 2055 (1985).
[17] W. Drube, D. Straub, F.J. Himpsel, P. Soukiassian, C.L. Fu and A.J. Freeman, Phys. Rev. B 34, 8989 (1986).
[18] I.L. Krainsky, J. Vac. Sci. Tech. A5, 735 (1987).
[19] J.B. Pendry, C.G. Larsson and P.M. Echenique, Surface Sci. 166, 57 (1986).
[20] M.N. Read and A.S. Christopoulos, to be published.

Anomalous Surface Phase Formation on Pt₃Sn⟨110⟩

A.N. Haner[1], P.N. Ross[1], and U. Bardi[2]

[1]Materials and Chemical Sciences Division, Lawrence Berkeley Laboratory, Berkeley, CA 94720, USA
[2]Department of Chemistry, University of Florence, Florence, Italy

Abstract. LEED analysis of the clean annealed surface of a <110> oriented Pt₃Sn single crystal surface indicates the formation of a multilayer surface phase which does not have the Ll₂ bulk structure. LEISS analysis indicates a surface stoichiometry of ca. 1:1 with Sn atoms displaced ca. 1.4Å above the plane of Pt atoms. The surface phase is hypothesized to be a rhombic distortion of the <0001> plane of PtSn, which has a B8₁ (NiAs-type) bulk structure. It is not clear whether the phase forms by precipitation of PtSn due to a slight (0.5%) stoichiometric excess of Sn in the bulk, or due to multilayer reconstruction driven by surface segregation.

1. Introduction

Pt-Sn alloys are of considerable commercial interest as both heterogeneous catalysts for hydrocarbon conversion [1-2] and as electrocatalysts for the direct electro-oxidation of methanol in fuel cells [3-4]. Of particular fundamental interest is the behavior of the highly ordered exothermic alloy Pt₃Sn [5]. Experimental studies of polycrystalline Pt₃Sn [6-8] found surface enrichment in Sn by LEISS, with surface compositions as high as 50-60 at.% Sn. To date we are not aware of any experimental study of the surface structure and composition of a Pt₃Sn single crystal.

The ordered Pt₃Sn alloy has the Cu₃Au(Ll₂) structure with tin atoms on the corners of the face centered cubic unit cell. Other members of the Ll₂ family (Cu₃Au [9-11], Ni₃Al [12-13], and Pt₃Ti [14]) have bulk termination of the <111> and <100> single crystals, with the minority atom positioned slightly upward from the plane of the majority atoms [12-13,15]. If Pt₃Sn single crystal were to behave similarly, one would also expect the <111> and <100> surfaces to be bulk-terminated with a slight displacement of the Sn atoms upward from the plane of Pt atoms. However, because of the surface enrichment of Sn observed

Springer Series in Surface Sciences, Vol. 24 **The Structure of Surfaces III**
Editors: S.Y. Tong · M.A. Van Hove · K. Takayanagi · X.D. Xie
© Springer-Verlag Berlin, Heidelberg 1991

on polycrystalline Pt$_3$Sn, it is not obvious that the Pt$_3$Sn single crystal samples would behave like Cu$_3$Au and Ni$_3$Al. In another paper [16], we report results on the surface structure and composition of <111> and <100> oriented single crystals of Pt$_3$Sn using LEED, AES, and LEISS. We found that the Sn atoms are in p(2x2) and c(2x2) arrays on the <111> and <100> crystals, respectively, but are displaced upward from the plane of Pt atoms. In this paper, we report the first study of the surface structure and composition of the <110> oriented single crystal of Pt$_3$Sn using the same combination of techniques. For this surface, we found an unusual form of surface reconstruction with Sn atoms raised above the plane of Pt atoms. However, unlike the <111> and <100> surfaces, which still have Sn atoms in Ll$_2$ lattice positions, the Sn atoms appear to occupy lattice positions related to the B8$_1$ (NiAs-type) structure of PtSn.

2. Experimental

The apparatus and procedures for analysis of the <110> crystal is reported elsewhere [16]. For LEISS analysis, the angle of incidence of the ion beam with the surface normal could be varied continuously, but the azimuthal orientation could not. For these experiments, the 1 keV ^{20}Ne$^+$ ion beam at glancing incidence was aligned along the <002> direction of the crystal and the scattering angle fixed at 135°.

3. Results

A poorly defined (110)-(3x1) pattern (Fig. 1a) was observed after Ar$^+$ ion bombardment and low temperature annealing. The (3x1) pattern was associated with a low Sn/Pt AES ratio which we attribute to the reconstruction [17] of a nearly-pure Pt (110) surface. Further annealing of this surface at higher temperature produced a diffuse (1x2) pattern (Fig. 1b), a transition through a sharper but streaked (1x2) co-existing with a rhombic pattern shown in Fig. 1c. The pattern is commensurate, having the structure given in matrix notation as $\begin{vmatrix} 1 & 0 \\ 1/2 & 3/2 \end{vmatrix}$ which we will refer to as the "rhombic" structure. One of the surprising aspects of the rhombic pattern is the disappearance of many of the fundamental beams from the (110) substrate, e.g. the (01), (02), (12), etc. (see Fig. 1d). The changes in LEED structure with annealing were accompanied by concomitant

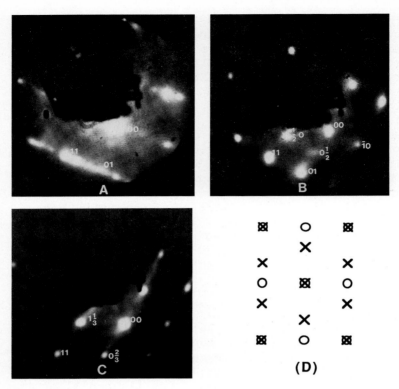

Figure 1. LEED patterns (a-c) at 90 eV: a.) after ion bombardment, b.) after annealing at 500 °C, c.) after annealing at 800 °C, and d.) schematic of (c), where (x) is superimposed on LEED pattern for (110)-1x1 (o).

increases in the Sn/Pt AES ratio (factor of four times higher after complete annealing). The absence of the fundamental beams means the structure associated with the rhombic pattern must be sufficiently thick to attenuate the substrate beams.

The LEISS spectra at normal incidence for all the clean annealed surfaces of all three low index surfaces are shown in Fig. 2. The Sn/Pt intensity ratios on the <110> and <100> surfaces are practically identical, and are about 2 times that for the <111> surface.

The variation of Sn/Pt LEISS intensities with the angle of incidence was similar to the variation we observed on the <100> surface [16]. There was little change between normal incidence and 45°, then rapid change between 45° and 25°, with Sn/Pt ratio increasing by a factor of 8-10 (see Fig. 3a for definition of angles). Below 15°, the Pt

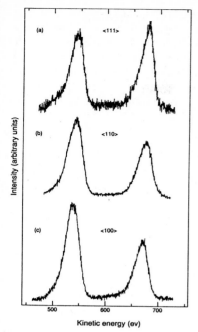

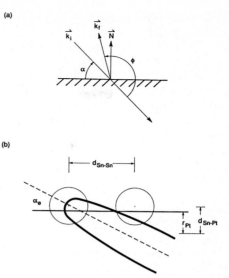

Figure 3. Off-normal incidence scattering geometry and Sn shadow cone geometry. N - surface normal, α - angle of incidence, α_e - angle of incidence for complete shadowing of the Pt plane, φ - scattering angle.

Figure 2. Normal incidence LEIS spectra for all three crystal orientations.

signal was completely attenuated. Such an angular variation is consistent with a structure with Sn atoms above the plane of Pt atoms, shadowing the Pt atoms in the plane below. A shadow cone analysis (Fig. 3b) of these intensity variations together with knowledge of the Sn-Sn spacing (from the LEED pattern) enabled us to determine the distance the Sn atoms were displaced above the Pt atom plane [16]. The separation between the Sn atom and the Pt atom planes for the rhombic structure on the <110> calculated from this analysis is ca. 1.4Å.

4. Discussion

Unlike Pt_3Sn <111> and <100>, where tin surface segregation produced structures having the same symmetry as the bulk, the LEED pattern of the Pt_3Sn <110> crystal surface does not have Ll_2 symmetry. What was observed was a reconstructed surface *region* at least three atomic layers deep, with quasi-hexagonal (actually rhombic) symmetry and not the rectangular symmetry of either (110)-(1x1) or (110)-(2x1). The LEISS analyses indicate the composition of the <110> surface is the same as the <100> surface, i.e. 50% Sn. It is possible that the reconstruction is related to a surface phase of another bulk alloy phase, e.g. the PtSn phase. PtSn has the $B8_1$ or NiAs-type structure [18] and its heat of formation is even more exothermic than that of Pt_3Sn [19]. Bulk termination of this structure normal to <0001> produces a surface with hexagonal symmetry. A rhombic distortion of this plane (ca. 5% contraction in one direction and 10% expansion in the other) would fit reasonably well onto the (110)-1x1 substrate, but this is the only reasonable match of any ordered Pt-Sn phase onto a fcc (110) lattice. The spacing between the Sn atom plane and the Pt atom plane in PtSn [18] agrees well with the interplanar spacing we found here by LEISS.

It is not clear why a new surface phase would form only on the <110> oriented crystal and not on either the <111> or <100> crystals. One obvious factor, which is difficult to eliminate *entirely*, is a shift difference in bulk composition between crystals, particularly a slight, e.g. 0.5%, *excess* of Sn in the <110> crystal. Given the bulk phase diagram of Pt-Sn [20] and the relative free energies of formation of PtSn and Pt_3Sn [19], *any* excess of Sn in the normally Pt_3Sn bulk crystal could be expected to produce *precipitation* of a PtSn phase at the grain boundaries and/or free surfaces of the Pt_3Sn crystal, the total quantity/thickness of the precipitate being proportional to the amount of excess Sn. Analysis of all three crystals by emission spectroscopy (Galbraith Labs.) did not indicate any significant (<0.5%) difference in bulk composition between crystals, all being slightly *deficient* (0.5-0.8%) in Sn. However, we are continuing these analyses by other methods.

If, in fact, the surface phase forms on a Pt_3Sn bulk by reconstruction rather than by precipitation, it would be a unique form of multilayer reconstruction. Presumably the driving force for reconstruction is surface segregation of Sn [5], and the lower free energy of surfaces enriched in Sn. Why multilayer reconstruction is required to achieve the lowest energy state on the <110> surface, and not the <111> or <100> is puzzling.

Acknowledgments

This work was supported by the Assistant Secretary for Conservation and Renewable Energy, Office of Energy Storage and Distribution. Energy Storage Division of the US DOE under Contract No. DE-AC03-76SF00098. We are pleased to acknowledge the invaluable assistance of Michel Van Hove in identifying the structure on the <110> crystal surface.

References

1. F. Dautzenberg, J. Helle, P. Biloen, and W. Sachtler, J. Catal. 63 (1980) 119.
2. Z. Karpinski and J. Clarke, J. Chem. Soc., Faraday Trans. 1 71 (1975) 893.
3. K.J. Cathro, J. Electrochem. Soc. 116 (196) 1608.
4. B. McNicol, R. Short and A. Chapman, J. Electroanal. Chem. 72 (1976) 2735.
5. R. Van Santen and W. Sachtler, J. Catal. 33 (1974) 202.
6. R. Bouwman, L. Toneman and A. Holscher, Surface Sci. 35 (1973) 8.
7. R. Bouwman and P. Biloen, Surface Sci. 41 (1974) 348.
8. P. Biloen, R. Bouwman, R. Van Santen and H. Brongersma, Appl. Surface Sci. 2 (1979) 532.
9. H. Potter and J. Blakely, J. Vac. Sci. Tech. 12 (1975) 635.
10. Y. Fujinaga, Surface Sci. 64 (1977) 751.
11. V.S. Sundaram, B. Ferrell, R.S. Alben, and W.D. Robertson, Phys. Rev. Lett. 31 (1973) 1136.; and V.S. Sundaram, R.S. Alben, and W.D. Robertson, Surface Sci. 46 (1974) 653.
12. D. Sondericker, F. Jona, and P.M. Marcus, Phys. Rev. B 34 (1986) 6770.
13. D. Sondericker, F. Jona, and P.M. Marcus, Phys. Rev. B 33 (1986) 900.
14. U. Bardi and P. Ross, Surface Sci. 146 (1984) L555.
15. T.M. Buck, G.H. Wheatley, and L. Marchut, Phys. Rev. Lett. 51 (1983) 43.
16. A. Haner, P. Ross and U. Bardi, "Surface Composition and Structure of the <111> and <100> Oriented Single Crystals of Pt$_3$Sn," Lawrence Berkeley Laboratory Report 28074, submitted to Appl. Phys. Lett.
17. P. Fery, W. Moritz and D. Wolf, Phys. Rev. B 38 (1988) 7275.
18. R. Wyckoff, "Crystal Structures," Volume 1, Second Ed., Wiley-Interscience, New York and London, 1965, p. 123.
19. R. Ferro, R. Capelli, A. Borsese and S. Delfino, Atti Accad. Naz. Lincei Cl. Sci. Fic., Mat. Nat. 54 (1973) 634.
20. M. Hansen and K. Anderko, "Constitution of Binary Alloys, Second Edition," Plenum Publishing, New York 1989, p. 1141.

Surface Relaxation of $Fe_{72}Cr_{28}(110)$

D.J. O'Connor and C. Xu*

Department of Physics, University of Newcastle, NSW 2308, Australia
*Current address: 1152 Davey Laboratory, The Pennsylvania State University,
  University Park, PA 16802, USA

INTRODUCTION

It is now recognised that the surface composition of an alloy may not reflect that of the bulk material and that the surface makeup plays an important role in determining its physical and chemical properties. The driving force for the preferential segregation of one element to the surface is the energetic benefit of a lower surface free energy and the composition profile near the surface of an alloy changes with a length characteristic of the interatomic spacing. While there exist a number of techniques which can probe the surface composition to such a depth resolution few can give structural information as well. In an analysis of the $Fe_{72}Cr_{28}(110)$ surface it has been concluded that the segregation of Cr results in a surface Cr composition of 70% and that the Cr clusters into islands which experience a different interlayer relaxation to that of the remaining Fe rich surface regions. Much of this is based on structural measurements performed with Medium Energy Ion Scattering as direct determinations of composition are hampered by the similarity of mass between the constituents.

To measure the structure of the surface region Low Energy Electron Diffraction (LEED) and Medium Energy Ion Scattering (MEIS) have been used as probes with a depth resolution of the order of the interatomic layer spacing. LEED can only be used effectively if there is long range order in the surface while MEIS only probes the degree of local order. Both techniques rely on the determination of surface structure by the comparison of a computer simulation of a proposed structure with experiments. To make this comparison an R factor analysis is used which is a measure of the quality of agreement between the measurements and the simulation. The R factor is monitored as one or more structural parameters are varied to establish the set of structural parameters which give best agreement between simulation and experiment. To ensure a unique set of structure parameters a series of measurements under different geometries are performed and compared with the computer models. For single element targets, surface structure analysis involves a contour plot of the R factor as a function of the first and second layer relaxation. Usually a single clear minimum in the contour will be found which corresponds to the optimum structure. In general the progression from single element targets to alloys involves increasing complexity, however, that is offset by the increased information which can be obtained from the separate measurements for each element. In a recent study of a PtNi alloy [1], MEIS analysis provided layer-by-layer composition analysis, interlayer spacings and thermal vibration enhancements for both constituents. Such an analysis is more complicated for iron-chromium alloys as the component atomic masses are so similar that they would not be energetically distinguishable using H or He as projectiles.

Stainless steel (involving only Fe and Cr), as a typical alloy crystal, has been studied by various techniques [2-4] and it has been shown

Springer Series in Surface Sciences, Vol. 24   **The Structure of Surfaces III**
Editors: S.Y. Tong · M.A. Van Hove · K. Takayanagi · X.D. Xie
© Springer-Verlag Berlin, Heidelberg 1991

that the lower chromium steels have properties similar to iron, while stainless steels (Cr>12%) behave similar to pure chromium [6]. Chromium surface enrichment in these alloys is expected from bonding breaking theories as the segregation of Cr will minimize the surface energy [7]. Analysis of $Fe_{72}Cr_{28}$ by XPS [8] has indicated that there is significant chromium enrichment in the surface region with a concentration of 65% of Cr on the top atomic layer. In addition iron-chromium alloys have been confirmed to be weakly clustering alloys, which means that the number of like atom nearest neighbor bonds is larger than the number which would occur if the atoms randomly filled the lattice. The size of the cluster has been calculated for a 65% chromium rich surface and it was predicted that more than 50% of the Cr clusters contain at least 30 chromium atoms [8].

The limitations of techniques used previously have prevented the investigation of the surface reconstruction or relaxation of the iron-chromium alloy and only have concentrated on the analysis of segregation at the surface. In an attempt to measure the surface relaxation for this alloy an analysis of a clean $Fe_{72}Cr_{28}(110)$ surface has been made using MEIS. As the use of channeling and blocking of medium energy ions for surface structure analysis has been reviewed elsewhere [9,10] only a brief description will be included here.

The use of MEIS allows the determination of the relative positions of surface layer atoms by the use of channelling and blocking. The surface thermal vibration amplitude and surface structure can be simultaneously obtained from the surface blocking curve which is the angular scan of the surface peak about one low index direction while the incident ion beam is parallel to another low index direction. This dependence on reconstruction and thermal vibration arises because the position of the surface blocking minimum reflects the surface atom positions while the width of the blocking dip reflects both structure and thermal vibration amplitude. If relaxation exists at the surface, the position of the minimum will be changed from that of an ideal surface.

EXPERIMENTAL

All experiments have been performed in a stainless steel UHV chamber (base pressure less than 1×10^{-9} mB) which is coupled to a 100 keV accelerator via a differentially pumped beamline. Backscattered ions are analyzed with a compact electrostatic energy analyzer which is mounted on a goniometer with two rotational degrees of freedom adjusted by stepper motor drives and is computer controlled. The bulk composition of the alloy sample has been confirmed by electron microprobe analysis. In the MEIS experiment, the alloy sample was mounted on a three axis goniometer, which allows for two independent rotations and three translations. As the mass of Fe and Cr are so similar one would ideally need an energy analyzer with a resolution of better than 0.25% in order to resolve the separate elemental contributions when using protons as projectiles. Even with such an analyzer the two contributions would not be resolved as the natural line width after scattering (which has contributions from thermal vibration, kinematic broadening and inelastic energy loss straggling) exceeds the energy difference caused by the mass difference of the two elements. For this reason no attempt has been made to separate the Fe and Cr contributions in any simulations used in this analysis.

RESULTS

In order to measure the surface relaxations, three different scattering geometries were used to measure the interlayer spacings for the first three atomic planes. From an analysis of the results from these three geometries, the surface structure was determined and the surface

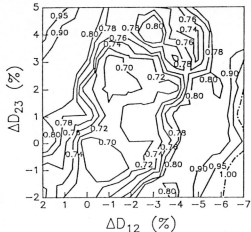

Figure 1   The R factor contour plot as a function of the first two layers
            comparing experimental data from the (001) scattering plane and
            the computer simulation.

composition was inferred. The (001), (110) and (112) scattering planes
were used as the first two were equally sensitive to the first and second
interlayer scattering geometries while the (112) was more sensitive to the
first interlayer spacing. Figure one is a typical R factor contour plot
for this surface using the (110) scattering plane geometry and similar
results were obtained in the other two geometries[11]. R is defined [12]
as

$$R = 100 * \left[1/N \sum_{i=1}^{N} ((Y_i^c - wY_i^e) / wY_i^e)^2 \right]^{\frac{1}{2}}$$

where Y_i^c are the calculated results, Y_i^e are the experimental data and w is
a weighting factor normally allowed to range from 0.95 to 1.05 while
minimising R to account for calibration errors.

Contour plots of the R factor as a function of displacements of the first
and second atomic layers are used to identify the condition of best fit and
in most cases a single clear minimum is observed. In the case of this
surface, in all three geometries, two minima are found. The three possible
explanations for the observation of two minima in the R factor contour are;

a) For an ordered alloy surface (eg NiAl, Ni_3Al, Cu_3Au) it could be a
periodic buckling of the surface layer. This is not the case for FeCr as
it does not show any evidence of an ordered alloy.

b) For a disordered alloy each surface atom experiences a relaxation
depending on its identity. This is not expected to apply here as Cr will
tend to cluster and hence form islands of Cr rich surface regions.

c) For a clustering alloy like FeCr the Cr forms islands on the surface and
these regions may experience a different relaxation distribution to the
remaining surface region. This is the preferred structure to explain the
results for this $Fe_{72}Cr_{28}(110)$ surface.

To test this model a surface of two independent structures has been
simulated simultaneously. These structures involve parts of the surface

which are either Cr or Fe rich and they appear (from the R factor contour) to experience different relaxations between the first and second interlayers. In this model there are six free parameters to determine, four of which are structural parameters associated with the relaxations of the first and second layers in two independent regions of the surface and the remaining two are the surface composition and the thermal vibration amplitude (assumed to be the same for Fe and Cr). To let all six parameters remain free and endeavour to find the optimum parameter set which fits the data set poses two problems in the amount of CPU time involved and the method of graphically representing 6 dimensional data. A simpler approach is to fix the difference between the two sets of relaxation parameters (estimated from the relative positions of the two R factor minima in figure one) and while varying the relaxation parameters for one region automatically alter the second set to maintain the difference. The Monte Carlo simulation Y_i^{cal} used in the determination of the R factor is a weighted sum of two independent simulations giving $(A\ Y_{iA}^{cal} + B\ Y_{iB}^{cal})$. The surface composition is taken to be the sum of the concentrations A and B of the two elements in the different regions on the surface which must sum to 100% and in this analysis the surface thermal vibration amplitude is kept at $\rho = 0.115\text{Å}$ (corresponding to a Debye temperature of 260K) which was determined from a previous study of the Fe(110) surface [11].

The R factor analysis was performed as an iterative procedure. The R factors were calculated as a function of the first two layer relaxation of one of the surface regions (with the relaxation of the other changing automatically to maintain the relative shifts) for a range of concentrations of the two components. Once the parameters for the best agreement between the experimental data and computer simulations were obtained, the relative shifts between different parts of the surface were varied independently. This iteration was repeated until the best fit was obtained with all of the surface structure parameters. The final results reveal that the two regions of the surface, with compositions of 30% (A) and 70% (B) respectively, undergo different relaxations. The final results are presented in a concise form as a composite R factor plot which has been

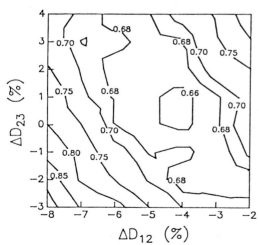

Figure 2   The R factor contour plot for the simulation of the two
            independent regions with one region contributing 70% (Cr) to
            the surface structure and different structures for each region.

derived from the sum of the three geometries. The best fit should be correspond to a single minimum R factor in the contour plot and a significantly improved R factor value however, as the R factors obtained are approaching the noise limit it was necessary to determine the best fit by contour shape alone. The optimum contour shape is a single minimum with no or minimal lobe structure which was found for a 30%, 70% combination of structures and a visible degradation is observed for a 10% departure from this composition. The previous XPS study has identified Cr as the principal surface element so that has been associated with the 70% structural component while Fe is associated with the 30% component in agreement with previous measurements of the surface composition.

$Fe_{72}Cr_{28}(110)$ Structure

	Composition	ΔD_{12}	ΔD_{23}
Fe	$30\pm10\%$	$-3.5\pm1\%$	$-1\pm1\%$
Cr	$70\pm10\%$	$-2.5\pm1\%$	$2\pm1\%$

CONCLUSIONS

Medium Energy Ion Scattering in combination with channeling and blocking has been used for alloy surface structural analysis of $Fe_{72}Cr_{28}(110)$. Despite the inability of separating the Fe and Cr contributions to the scattered ion signal it has been possible to determine the surface composition and structure from structural information.

REFERENCES

[1] S. Deckers, F.H.P.M. Habraken, W.F. van der Weg, A. van der Gon, B. Pluis, J.F. van der Veen, J.W. Gues and R. Baudoing, accepted by Phys Rev.
[2] P.A. Dowben, M. Grunze and D. Wright, Surf. Sci. 134 (1983) L524
[3] C.Leygraf, S.Ekelund and G.Schon, Scand. J. Met. 2 (1973) 313
[4] R.Schubert, J.Vacuum Sci.Technol. 11 (1974) 903
[5] C.Leygraf, G.Hultquist, S.Ekelund and J.C.Eriksson, Surface Sci. 146 (1974) 157
[6] J.E.Holliday and R.P.Frankenthal, J.Electronchem.Soc. 119 (1972) 1190
[7] I.Mirebeau, M.Hennion and G.Parette, Phys. Rev. Lett. 53 (1984) 687
[8] P.A.Dowben, A.Miller, H.J.Ruppender and M.Grunze, Surface Sci. 193 (1988) 336
[9] W.C.Turkenburg, W.Soszka, F.W.Saris, H.H.Kersten and B.G.Colenbrander, Nucl Instr Methods 132 (1976) 587
[10] J.F.Van der Veen, Surface Sci Rep 5 (1985) 119
[11] C. Xu and D.J. O'Connor, submitted to Nucl Instr Methods in Phys Res.
[12] D.L. Adams, H.B. Nielsen, L.N. Anderson, I. Stensgaard, R. Feidenhans'l and J.E. Sorenson, Phys. Rev. Letts. 49 (1982) 669

The Structure of CuPd(110)(2 × 1):
A Case of Underlayer Ordering

M. Lindroos[1], C.J. Barnes[2], M. Bowker[3], and D.A. King[2]

[1]The Department of Physics, Tampere University of Technology,
 P.O. Box 527, SF-33101 Tampere, Finland
[2]The Department of Chemistry, The University of Cambridge,
 Lensfield Road, Cambridge CB2 1EP, UK
[3]The Leverhulme Centre for Innovative Catalysis, The Department of Chemistry,
 The University of Liverpool, P.O. Box 147, Liverpool L69 3BX, UK

The surface structure of a $Cu_{0.85}Pd_{0.15}(110)$-(2x1) alloy has been characterised primarily by quantitative low energy electron diffraction I(E) analysis. We show that the superstructure is due to an ordered CuPd (2x1) underlayer below a Cu-rich p(1x1) termination. Ordering does not extend further into the bulk for annealing conditions required to form a well ordered p(2x1) surface structure This has been established by varying the number and position of ordered CuPd layers in the selvedge. A first interlayer spacing of 1.24±0.04 Å corresponding to a contraction of -4.7±3% has been determined, with changes in deeper interlayer spacings being outside the uncertainty of the analysis. A small rippling in the (2x1) CuPd layer has also been detected.
Preliminary investigation of deviation from the ideal bulk composition in the outermost and third layers has been carried out within the average T-matrix approximation (A.T.A) and compared to conclusions of He^+ low energy ion scattering and angle-resolved XPS measurements for the same sample.

1. Introduction

A new area of application of the technique of low energy electron diffraction (LEED) is studies of the surface geometry and composition of ordered and substitutionally disordered bulk alloys as well as surface two-dimensional alloy phases. Studies of stoichiometric ordered alloy phases such as Ni_3Al, NiAl [1] and Cu_3Au [2] have concentrated on extraction of structural parameters such as surface termination and buckling amplitudes in mixed surface layers as well as the usual changes in interlayer spacing of the outermost layers. In contrast, extensive studies of the low index faces of substitutionally disordered Pt_xNi_{1-x} alloys have suggested that the average T-matrix approximation adequately describes the electron scattering properties of LEED electrons [3]. This suggestion has been confirmed by a theoretical study of S. Crampin and P. Rous [4]. A third and perhaps the most complex alloy surface consists of an ordered two-dimensional surface layer(s) with a substitutionally disordered bulk above or below. To our knowledge only one such alloy surface has been studied to date by LEED I(E) analysis, namely $Cu_{0.84}Al_{0.14}(111)$ [5]. In this case chemical ordering occurs within the outermost atomic layer in a $\sqrt{3}\times\sqrt{3}R30°$ superstructure.

In this study we report results of the structural study of a $Cu_{0.85}Pd_{0.15}(110)$ alloy surface, prepared in a bulk disordered state. Annealing the sample in vacuum to temperatures of 700K, required to remove sputter damage due to argon ion bombardment, led to the irreversible formation of a sharp p(2x1) LEED pattern. Interest in study of this particular system was motivated by our interest to utilise the sample in chemisorption and catalytic studies of CuPd alloy surfaces [6]. Furthermore, our continued interest in the surface structure of clean and alkali metal induced surface phase transformations of the (110) surfaces of Cu, Ag, Ni and Pd and their bimetallic combinations further stimulated the work reported here [7].

2. Results

The LEED I(E) measurements were made in an ion and titanium sublimation pumped ultra-high-vacuum chamber at a base pressure of 5×10^{-11} Torr. LEED patterns were displayed on a standard Varian 4-grid optics and spot intensities measured with a commercial video-LEED analyser consisting of a video camera interfaced to a micro-computer. All calculations were performed with the symmetrised codes of Van-Hove and Tong [8] and theory-experiment agreement tested mainly with the Pendry R-factor [9]. Substitutional disorder in mixed CuPd layers was modelled within the A.T.A approximation. A full account of this work will be published elsewhere containing full details of our experimental procedures such as sample cleaning and preparation, data acquisition and of our theoretical modelling.

Structural parameters varied within the analysis (in addition to the real part of the inner potential) were initially the number and position of the ordered (2x1) CuPd slabs and the first interlayer spacing (dz_{12}). Parameters varied which proved less crucial to the level of theory-experiment agreement included the second interlayer spacing (dz_{23}), rumpling within the composite CuPd slabs(s) due to the 8% size mismatch between Cu and Pd and the composition within the substitutionally disordered surface layers.

In order to solve the first major structural question, namely the origin of the (2x1) superstructure, models involving a) a mixed CuPd outermost atomic layer, b) a mixed (2x1) CuPd second atomic layer and c) the possibility that ordering extends deeper into the selvedge were tested. In this initial search, dz_{12}, dz_{23}, and the concentration in the outermost A.T.A layers were varied. The range of parameter space investigated was as follows: interlayer spacings were allowed to vary up-to 0.25 Å either side of the equilibrium value initially in steps of 0.05 Å (steps of 0.02 Å were utilised in the final structural optimisation). Buckling in the ordered (2x1) CuPd slabs was varied between -0.15 and +0.15 Å in steps of 0.01 Å in the composite buckled layers. Finally, the concentration within the substitutionally disordered layers was varied between 0 and 50 at% palladium in steps of 5at%.

Calculations showed that models involving an outermost ordered (2x1) CuPd layer may be discarded. The possibility of more than one (2x1) ordered layers in the selvedge may also be ruled out, producing Pendry R-factors >25% higher than the favoured model. The favoured geometric structure, yielding a Pendry R-factor of 0.30 consists of an ordered CuPd (2x1) second layer. This model was refined allowing variation of the outermost interlayer spacings, the surface buckling in the (2x1) CuPd layer and the Pd concentration in the outermost atomic layers as well as the sample lattice constant. The lattice constant was varied ±0.1 Å in steps of 0.01 Å with respect to the calculated Vergards law value, yielding a minimum R-factor of 0.28 at $a_{CuPd} = 3.68$ Å.

Figure 1 illustrates the sensitivity of the analysis with respect to variation dz_{12},s and the palladium concentration in the first and third atomic layers (c_1 and c_3). The first two interlayer spacings were determined to be 1.24±0.04 Å and 1.29±0.04 Å [10]. A second layer buckling of 0.07±0.04 Å with the Cu atoms outermost in the composite layer was obtained. Finally, the first and third layer palladium concentrations are 30±15at% and 0+15at% respectively. Deeper layer spacings and compositions were equal to that of the bulk within the sensitivity of the analysis.

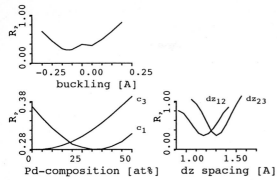

Figure 1. Sensitivity curves for the Pendry R-factor to a) first and second interlayer spacings, b) second layer buckling and c) first and third layer palladium concentration. The minima correspond to the optimal geometry, all other parameters are fixed.

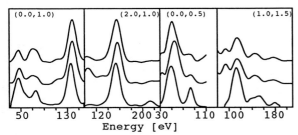

Figure 2. Theory-experiment agreement for two integral and two half order beams. The lower curve in each window illustrates experimental data, the middle the theoretical spectra corresponding to the optimal agreement ($c_1 = 30$ at%) and the uppermost curve the theoretical spectra for an outermost layer of pure Cu ($c_1 = 0$ at%).

Figure 2 illustrates the optimal theory-experiment agreement for two integral and two half order beams (the full analysis was based on a data set including six integral and five half order spectra). Also shown in figure 2 are theoretical spectra for the optimal geometry for a pure Cu outermost layer.

We now briefly summarise results of a number of other surface sensitive techniques commonly utilised for surface compositional analysis of alloy surfaces. Figure 3 illustrates a set of He^+ low energy ion scattering (LEISS) spectra taken at a primary beam energy of 1 keV, with the incoming beam perpendicular to the close packed [110] rows. A clear maximum in the weak Pd signal is seen at an emission angle of 45°, droping to zero at both grazing incidence and emission (the scattering angle is fixed at 90° as shown in the inset of figure 3).

The maximum at 45° would, of course, be consistent with models involving a mixed CuPd second layer of composition 50:50 at% and a richer, but not necessarily pure Cu layer above. However, the rapid loss of palladium signal at angles >10° either side of the maximum and subsequent drop to zero suggest to us that the surface layer is highly Cu rich and possibly terminated by a pure Cu outermost layer. The LEISS results are supported by angle-resolved XPS

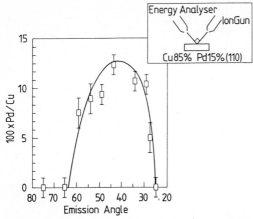

Figure 3. Dependence of the He+ LEISS Pd/Cu intensity ratio as a function of angle of emission of the primary beam ($E_p \sim 1$ keV). The scattering geometry is indicated in the inset of the figure. The sample was held at 650 K throughout the measurements to minimise sputtering damage during the measurement.

measurements along with model calculations based on a continuum model of the selvedge for a wide range of surface compositional profiles (not shown here for brevity). A satisfactory fit is obtained for an outermost layer of pure Cu with a (2x1) CuPd layer below.

3. Discussion

This study has demonstrated two important points. Firstly, to our knowledge for the first time we have observed the existence of a surface alloy phase above a disordered bulk which has a mixed sub-surface ordered phase. Bimetallic surfaces studied to date with sequential stacking of layers such as Cu_3Au always terminate in a mixed rather than pure A or B layer [2]. In the case of ordered overlayer formation in the bulk disordered system $Cu_{0.84}Al_{0.16}(111)$ alloy the $\sqrt{3}x\sqrt{3}R30°$ phase has also been attributed to formation of an ordered Cu_2Al outermost atomic layer [5]. Systems with constituent atoms of different metallic radii are able to lower their surface energy within a mixed outermost layer in the form of buckling due to the lack of nearest neighbours on the vacuum side of the interface more effectively than would be the case in sub-surface layers. This must be balanced against the increase in surface energy associated with the unsaturated bonds of elements with a higher heat of vaporisation (e.g Pd in CuPd). Whether or not the surface structure described here is a true equilibrium structure or simply a metastable state remains open at present.

Regarding the top layer composition, our present LEED analysis yields a palladium outermost layer composition of 30±15at% i.e. a Cu-rich outermost layer, however showing a net palladium enrichment with respect to the bulk composition. This result contrasts with LEISS and angle-resolved XPS studies which indicate strong palladium depletion in the outermost layer. This apparent discrepancy is being further investigated at present by LEISS and via further refinement of the LEED analysis utilising a larger normal and off-normal data base in an attempt to reduce the uncertainty in the top layer composition.

The relatively large error bars presently quoted of ±15at% are much greater than those quoted in recent studies of PtNi disordered alloys where values of $< \pm5$at% have been reported [3]. This is primarily due to the method adopted to estimate random error. The quoted values in this study have been based on the variance of the Pendry-R factor, lower values being obtainable by utilising a larger data set (lower RR) or increasing the level of theory-experiment agreement. In contrast in the PtNi studies the quoted composition is obtained from an average of values obtained for a range of metric-distance R-factors [11]. The error is then quoted as the standard deviation about this average value. Using this particular prescription and the R-factors of Pendry [9], Andersen et al. [12] and Zanazzi-Jona [13] yield a spread of around ±5at%, similar to the aforementioned PtNi results.

To summarise, the (2x1) overlayer on the equilibriated surface of a bulk disordered $Cu_{0.85}Pd_{0.15}(110)$ alloy has been demonstrated to arise due to chemical ordering of Cu and Pd in a mixed CuPd second layer, with a Cu rich layer residing below and above in a "surface sandwich" type structure. A layer-wise A.T.A scheme has been utilised to model substitutional disorder in all layers other than the ordered CuPd (2x1) slab in an attempt to determine the surface segregation properties of this system. Our preliminary results indicate a first and third layer palladium of 30 ± 15at% and $0+15$at%, the first layer composition determined by LEED being higher than other surface sensitive techniques which favour a Cu-terminated surface.

Acknowledgements

C.J.B and M.L. would like to acknowledge Neste Foundation and Academy of Finland for financial support.

References

[1] H.L. Davis and J.R. Noonan, in The structure of Surfaces II, Editors J.F. Van der Veen and M.A. Van Hove, Springer 1988 p.152 and references therein. D. Sondericker, F. Jona and P.M. Marcus, Phys. Rev. **B34**(1986)6770, Phys. Rev. **B34**(1986)6775 and references therein.
[2] T.M. Buck and E.G. McRae, Surface Modifications and coatings, edited by R.D. Sisson, Am. Soc. Metals, New York, p. 337,1986.
[3] Y. Gauthier, R. Baudoing and J. Jupille, Phys. Rev. **B40**(1989)1500.
[4] S. Crampin and P.J. Rous, Submitted to Surface Science. [5] D.F. Ogletree, M.A. Van Hove, G.A. Somorjai and R.J. Baird, Bull. Am. Phys. Soc. **29**(1984)222; Surface Science **165**(1986)345.
[6] D.J. Holmes, D.A. King and C.J. Barnes, Surface Science **227**(1990)179.
[7] C.J. Barnes, M. Lindroos, D.J. Holmes and D.A. King, in Physics and Chemistry of Alkali Metal Adsorption, Material Science Monographs,57, Editors H.P. Bonzel, A.M. Bradshaw and G. Ertl, Elsevier 1989 p.129.
[8] M.A. Van Hove and S.Y. Tong, Surface Chrystallography by LEED, Springer, Berlin, 1979. An improved version of the programs was obtained by private communication with M.A. Van Hove.
[9] J.B. Pendry, J. Phys. C **13**(1980)937.
[10] The first interlayer spacing is defined as the distance between the outermost atomic layer and the upwardly buckled second layer atoms and the second layer spacing between downward buckled second layer atoms and the third atomic plane.

[11] J. Philip and J. Rundgren, In Determination of Surface Structure by LEED, edited by P.M. Marcus and F. Jona, Plenum, New York, 1984.
[12] J.N. Andersen, H.B. Nielsen, L. Petersen and D.L. Adams, J. Phys. C **17**(1984)173.
[13] E. Zanazzi and F. Jona, Surface Science **63**(1977)61.

Effect of Segregation on Angular Distributions of Cu and Pt from a CuPt Alloy Sputtered by Low-Energy Ar Ions

L.-P. Zheng[1], R.-S. Li[2], M.-Y. Li,[1] and W.-Z. Shen[1]

[1]Shanghai Institute of Nuclear Research, Academia Sinica,
  P.O. Box 800-204, Shanghai 201800, P.R. of China
[2]Institute of Metal Research, Academia Sinica,
  Shenyang 110015, P.R. of China

Based on experimental observation of a Bombardment-Induced Gibbsian Segregation (BIGS) of Cu in a sputtered $Cu_{0.81}Pt_{0.19}$ alloy under low-energy Ar ion bombardment, a modified dynamic Monte Carlo program TCIS including a BIGS process was used to calculate angular distributions of individual elements from a sputtered $Cu_{0.5}Pt_{0.5}$ alloy in the Ar ion energy range between 0.2 and 1.5 keV. Calculations show that for each investigated incident energy the angular distribution of Pt is more forward-pointed than that of Cu, and, as the incident energy increases, the angular distributions of Cu and Pt are transformed from under-cosine to cosine-like and even over-cosine shapes.

1. Introduction

It is now recognized that under ion bombardment at high doses, an altered layer with depth of the order of the sputter depth [1, 2] exists at the alloy surface due to preferential sputtering or Bombardment-Induced Gibbsian Segregation (BIGS). The effect of the altered layer at the surface on sputtering angular distributions of alloys has created considerable excitement in sputter investigations in recent years. The existence of concentration gradients at the CuPt surface was found by *Andersen* et al. [3] for Ar ion bombardment with energies greater than 20 keV by means of measurements of angular distributions of composition, which demonstrated that Pt was preferentially ejected in the near-normal direction. Below 20 keV, their measurement results did not confirm the presence of segregation layers of Cu on the sputtered CuPt alloy. Recently, using different energy Auger line combinations, *Ri-Sheng Li* et al. [4] have observed bombardment-induced segregation of Cu in the CuPt alloy at Ar ion energies from 0.2 to 2 keV. *Li*'s experiments may imply the preferred ejection of Pt in the normal direction at very low energies. For element targets, it is well established that, as the projectile energy increases from close to the sputtering threshold to keVs, the angular distribution changes from under-cosine to cosine-like or even over-cosine shapes. However, for alloy targets, little information has previously been published concerning the shape change of angular distributions of individual elements in the above projectile energy range.

Springer Series in Surface Sciences, Vol. 24   **The Structure of Surfaces III**
Editors: S.Y. Tong · M.A. Van Hove · K. Takayanagi · X.D. Xie
© Springer-Verlag Berlin, Heidelberg 1991

In this work, we try to calculate the angular distributions of individual elements from a sputtered $Cu_{0.5}Pt_{0.5}$ alloy in the Ar ion energy range between 0.2 and 1.5 keV at normal incidence with the help of a modified dynamic TCIS Monte Carlo program including the BIGS process [5] in order to further research on the effect of concentration gradients with the order of the sputter depth on sputtering angular distributions of alloys.

2. Simulation Description

The TCIS program series [5, 6] can be used to calculate not only surface composition profiles with depth [5] but also angular distributions [7] for sputtered targets. A modified dynamic TCIS program is employed in the present study to stimulate collision cascade, BIGS and BED (Bombardment-Enhanced Diffusion) processes.

A brief description of BIGS is as follows: After each instantaneous collision cascade, the surface concentration is built up by atomic jumps from the second layer into the first layer at the surface. At the instantaneous state (i), the number of segregating atoms jumping from the second layer into the first layer is equal to $KN_i(1 - C_i)$. Here, N_i is the number of pseudovacancies, which is equal to the difference between the total number of target atoms in the first layer in the initial state and in the instantaneous state (the pseudovancancies can be assumed to exist in the first layer due to sputtering and atomic recoil implantation into other layers, etc.); K is an appropriate constant, and C_i is the concentration of segregating atoms in the first layer. In addition, it is assumed that the region of BED extends in depth to about the range of the incident ions and ceases abruptly at the first layer. In the program, BIGS is assumed to occur at the surface between the first and second layers, so the sharp variance of the composition is between these layers.

In this simulation, we note that *Brongersma* et al. [8] have found that the topmost surface composition assessed with ISS (Ion Scattering Spectroscopy) is the same as the bulk in the $Cu_{0.20}Pt_{0.80}$ alloy sputtered by 1 keV Ne ions when the target temperature is below 100°C. We also note *Sigmund*'s idea [9] that, if the true preferential sputtering has been ignored, then at high dose, i.e., assuming that the composition of the sputtered flux is equal to the bulk composition, the composition of the segregating atoms at the topmost surface is equal to that in the bulk (if sputter depth < segregation layer). The above observations can be adopted to examine simulated surface composition profiles, which significantly influence sputtering angular distributions of alloys.

3. Results and Discussion

Our calculations show that at high enough doses of incident Ar ions, steady composition gradients appear at the sputtered $Cu_{0.5}Pt_{0.5}$ surface in the energy range

Table 1. Steady-state compositions at the sputtered $Cu_{0.5}Pt_{0.5}$ surface as a function of Ar ion energies

Ar ion energy [keV]	Steady-state compositions [at.%]			
	Cu element		Pt element	
	First layer	Second layer	First layer	Second layer
0.2	50	38	50	62
0.7	50	34	50	66
1.1	50	29	50	71
1.5	50	25	50	75

0.2–1.5 keV. In these steady surface composition profiles, the Cu concentration is enriched in the first layer (50 at. %), depleted in the second layer and increases slowly to the bulk composition. Corresponding to the Cu concentration, the Pt concentration in the second layer is much greater than that in the first layer. The calculated concentration data of Cu and Pt are shown in Table 1.

It should be mentioned here that, for all incident energy–ion–target combinations, each layer thickness is chosen to be about 2.5 Å (mean atomic spacings [2]). It is found in our calculations that, for all the above systems, the mean depths of origin of sputtered atoms for Cu and Pt are either below or close to 2.5 Å, i.e., segregation layer thicknesses of Cu. As explained before, not only the calculated composition data and the tendency of the composition variation with depth may be realistic, but also the chosen segregation layer thicknesses of Cu agree with *Sigmund*'s conclusion [9] about the relation between the sputter depth, the segregation layer and the surface concentration.

In Fig. 1, the shape transformation of the Pt angular distribution is always ahead of that of the Cu angular distribution as the energy of Ar ion bombardment of the $Cu_{0.5}Pt_{0.5}$ alloy increases. Double under-cosine distributions emerge at quite low incident energies, but the Pt distribution is more outward peaked than the Cu distribution (Fig. 1a). The Pt distribution shows a cosine-like shape when the Cu distribution still has the under-cosine shape (Fig. 1b). The Pt distribution already shows an over-cosine shape before the Cu distribution changes into the cosine-like shape (Fig. 1c).

It is well known that most sputtered atoms are kicked out of the target from the first two layers at the surface. For alloy sputtering, *Sigmund* et al. have pointed out [1] that if a concentration gradient exists within the sputter-escape depth, the angular distribution of the element(s) in which the surface is depleted will be the more forward-pointed, the differences in the angular distributions being due to the surface layer reducing the fraction of the atoms originating deeper in the target that exit at oblique angles. From a knowledge of the sputter depth and *Sigmund*'s consideration, one can deduce that if the concentration gradient exists within the sputter-escape depth the angular distribution of the element

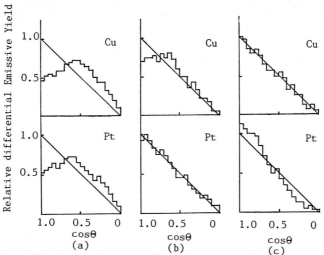

Fig. 1. Stable angular distributions of Cu and Pt from the sputtered $Cu_{0.5}Pt_{0.5}$ alloy under Ar ion bombardment with 0.2 (**a**), 0.7 **b**) and 1.1 keV (**c**). θ is the polar emission angle

whose concentration in the first layer is much less than that in the second layer will be the more forward-pointed. Table 1 shows a large composition difference between the first and second layers at the sputtered $Cu_{0.5}Pt_{0.5}$ surface for each investigated energy; the Pt concentration in the first layer is much less than that in the second layer whereas the Cu concentration in the first layer is much greater than that in the second layer. This is why the angular distribution of Pt is always more forward-pointed than that of Cu. It is easily seen from Fig. 1 that Pt is preferentially ejected in the normal direction at very low energies.

Andersen et al. have considered [3] the significant influence of surface contamination on the experimental $Cu_{0.5}Pt_{0.5}$ angular distributions due to the combination of low sputter yields an available beam current under Ar ion bombardment below 10 keV. Indeed, such an impurity layer on the $Cu_{0.5}Pt_{0.5}$ surface may strongly perturb the detection of the Cu segregation layer by the measurement of sputtering angular distributions. If the surface contamination could be dispelled during the Ar ion bombardment, the experimental angular distributions of Cu and Pt from the sputtered $Cu_{0.5}Pt_{0.5}$ surface might be similar to those calculated by us for the above combinations.

References

1   P. Sigmund, A. Oliva, G. Falcone: Nucl. Instrum. Methods **194**, 541 (1982)
2   R. Kelly, A. Oliva: Nucl. Instrum. Methods B **13**, 283 (1986)
3   H.H. Andersen, B. Stenum, T. Sorensen, H.J. Whitlow: Nucl. Instrum. Methods **209/210**, 487 (1983)
4   R.S. Li, C.F. Li, W.L. Zhang: Appl. Phys. A **50**, 169 (1990)
5   L.P. Zheng, R.S. Li, M.G. Li, F.Z. Cui: Chin. Phys. Lett. **7**, 140 (1990)
6   F.Z. Cui, H.D. Li, J.P. Zhang: Nucl. Instrum. Methods B **21**, 478 (1987)
7   L.P. Zheng, F.Z. Cui: Vacuum **39**, 353 (1989)
8   H.H. Brongersma, M.J. Sparnaay, T.M. Buck: Surf. Sci. **71**, 657 (1978)
9   P. Sigmund: Nucl. Instrum. Methods B **27**, 1 (1987)

Studies of Surface and Interface Structure of Ni₃Al Alloy by AP-FIM

D.G. Ren

Institute of Metal Research, Academia Sinica, Shenyang 110015, P.R. of China

Abstract   The field ion image of Ni_3Al alloy exhibits ordered crystal structure. The probe hole of AP is aimed along <100> at direction of Ni_3Al crystal. The AP depth profile shows that the composition of one atomic layer changes essentially from pure nickel to mixed nickel plus aluminum. The field ion image with the sample of B-doped 0.52 at% displays that the extent of ordering is reduced by adding boron. The results of AP also show that the interface of NiO appears in this sample.

1. Introduction

Polycrystalline Ni_3Al is an intermetallic compound, having the $L1_2$ ordered crystal structure. The Ni_3Al alloy is of interest for elevated temperature structure materials, the study of microstructure of Ni_3Al was currently undertaken in order to improve mechanical properties. The surface structure of ordered alloys has been generally studied by low energy electron diffraction or low energy ion scattering [1]. The compositions of the first few layers at ordering surfaces can be determined by LEIS. But most materials in engineering alloys are polycrystalline, they are made of many small single crystals or two phases. Therefore the research on interfacial phenomena is currently an active field in the materials science. The studies of surface and interface structure of Ni_3Al alloy is of great interest for understanding the microstructure of alloys. In material science, the observation of surface structure and the determination of chemical compositions from surface to bulk are very important problems. The field ion microscope is capable of producing images of the surface of a specimen in which each distinct point on the image is an individual atom. So FIM has provided interesting information about the atomic arrangement of surface layer of alloy. The atom probe (AP) is a time-of-flight mass spectrometer that is used to analyze the specimen chemically with single atom sensitivity for all elements and so the AP-FIM is one of the most powerful instrument for study of microstructure of materials [2]. This paper reports a study of the surface and the interface structure of Ni_3Al alloy by AP-FIM.

2. Experimental procedure and results

The samples used in this work have nearly the stoichiometry of Ni_3Al with 0.52 at% boron and without boron. The samples were subjected to a homogenizing heat treatment at 1200°C for 4 h followed by furnace cooling to 1100°C for 1 h then by air cooling to room temperature. The specimens for FIM were prepared by electropolishing from thin strips of 0.4 mm. The experiments were

Springer Series in Surface Sciences, Vol. 24   **The Structure of Surfaces III**
Editors: S.Y. Tong · M.A. Van Hove · K. Takayanagi · X.D. Xie
© Springer-Verlag Berlin, Heidelberg 1991

carried out in an AP-FIM with a straight flight tube and controlled by a computer. Field ion images are produced by image gas atoms that are ionized close to the specimen because of the high positive voltage (Vdc). The cold finger of the microscope was cooled continuously with liquid nitrogen. The background pressure of the FIM is 3×10^{-9} torr. The image gas is neon at a pressure of 5×10^{-5} torr. In the field evaporation, the surface atoms are removed by applying a higher voltage (Vdc) and a pulse voltage (Vp). The area of analysis can be made as small as a few atoms in diameter by adjusting the effective size of the probe aperture.

Figure 1a shows the field ion image which is the sample without boron. This image displays essentially the ordered f.c.c. crystal structure ($L1_2$) with the centre of (100) face. The mass spectra of this sample exhibited only Ni^{+2} and Al^{+2} elements peaks (Fig.2a). The plane index on the image explained by the stereographic projection of the f.c.c. lattice of (001) orientation (Fig.3). But the contrast of the whole image is slightly different, because the extent of close arrangement of surface atoms Ni and Al was not uniform (Fig.4). The close packed planes (111) of image in Ni_3Al alloy appear as dim regions because a close packed plane with high coordination sites leads to low field regions and the bright region for the less close packed planes (110) with high field regions (Fig.1a).

The probe hole is aimed in the <100> direction of the Ni_3Al crystal. The AP depth profile shows that the composition of one atomic layer changes essentially from pure nickel to mixed nickel plus aluminum (Fig.5). But the result of AP analysis has a deviation compared with ordered face centered cubic structure (Fig.5, insert), it exhibits a slight enrichment of Ni or Al in the mixed layer. A few Al atoms were also mixed in the pure nickel layer. The sample of Ni_3Al without boron is a Ni-rich hypostoichiometric alloy (22.89 at% Al). The atomic defects with antisite defects and vacancies in the Ni_3Al were believed to lead the deviations of atomic sites in the lattice. However those may be related to the conditions of solidification and heat treatment of the sample.

Figure 1b shows the field ion image of B-doped 0.52 at% sample. The extent of ordering is reduced by the addition of boron. The ordered region only appears in the planes of low index. Besides Ni^{+2}, Al^{+2}, B^{+1} and B^{+2} peaks, the mass spectra of

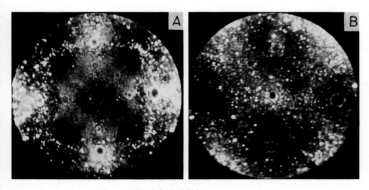

Fig.1 The field ion images of Ni_3Al alloys:(a)   without boron (b)   B- doped 0.52 at%.

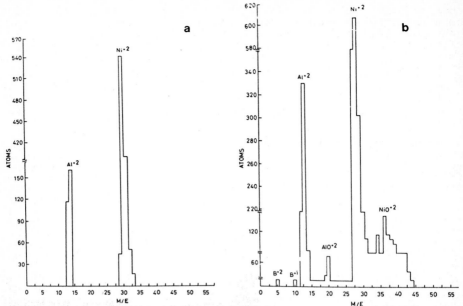

Fig.2 The AP mass spectrum of Ni₃Al alloys:(a) without boron
(b)   B- doped 0.52 at%.

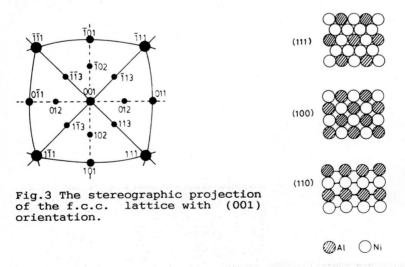

Fig.3 The stereographic projection
of the f.c.c. lattice with (001)
orientation.

Fig.4 The arrangement of surface
atoms Ni and Al in Ni₃Al alloy with
(111),(100) & (110) crystal planes.

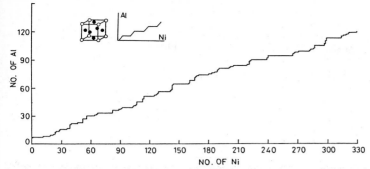

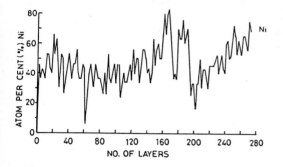

Fig.5 Ladder diagram of Al atoms versus Ni atoms on
the pole in <100> direction of B-undoped Ni₃Al alloy.

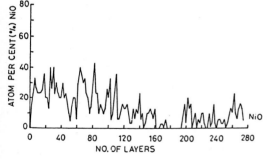

Fig.6 Concentration depth
profiles on Ni and NiO in
B-doped 0.52 at% alloy.

this sample display the peaks of NiO^{+2} and AlO^{+2} (Fig.2b). AP
depth profiles show that the distribution of Al atoms with depth
is essentially homogeneous but that of Ni atoms exhibits some
fluctuation which was mainly induced by an interface of NiO in
the Ni₃Al (Fig.6). The results of AP analysis also show that the
AlO is very little in the interface of NiO. The results show that
oxygen diffused into the B-doped Ni₃Al alloy during the heat
treatment process of the sample. But it is not remarkable for the
sample without boron. In addition, thin layers of solid solution
with various ratios of Ni/Al were found in the B-doped sample
(Fig.7). The thickness of these thin layers is about 4 nm. It is
shown that the extent of ordering in the B-doped sample was
decreased due to the addition of boron to induce a thin layer of

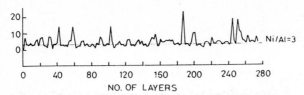

Fig.7 Concentration depth profiles of Ni/Al in B-doped 0.52 at%
Ni₃Al alloy.

Ni-rich solid solution and an interface of NiO in the Ni₃Al
crystal of long range ordering.

References

1. T.M. Buck, Chemistry and Physics of Solid Surfaces, IV
435(1982).
2. M.K. Miller, International Materials Reviews, 32, 221(1987).

Simulation of the $Fe_{80}B_{20}$ Metallic Glass Surface

W. Kowbel[1], P. Tlomak[2], and W.E. Brower, Jr.[3]

[1]Department of Mechanical Engineering, Auburn University,
  Auburn, AL 36849, USA
[2]Department of Mechanical Engineering and Energy Processes,
  Southern Illinois University, Carbondale, IL 62901, USA
[3]Department of Mechanical and Industrial Engineering, Marquette University,
  Milwaukee, WI 53233, USA

Abstract. A model of the surface of $Fe_{80}B_{20}$ metallic glass is presented. The bulk simulation is similar to Gaskell's procedure. The roughness of the surface is described in terms of surface coordination numbers (CN). The amorphous surface is displayed via computer graphics. The roughness of the surface is also described by a fractal dimension, D, which was found to be 2.3 (the value for the Brownian island). This value of D fits several experimental observations on rapidly solidified $Fe_{80}B_{20}$ metallic glass. First, the density is observed to be close to that of the crystalline phases, indicating a non-porous surface structure. A porous structure would be indicated by a D near 3. Second, unusual catalytic selectivity is observed on the $Fe_{80}B_{20}$ glass, possibly due to the absence of smooth terraces on the metallic glass surface. A surface containing mostly terraces would have a fractal dimension nearer to 2 than the model generated value of 2.3.

1. Introduction

1.1 Bulk Model Generation

The computer generation procedure in this work is similar to Gaskell's bulk model for P_4Si glass [1]. The NM_6 trigonal prism was chosen to form the basic structural unit of the bulk glass structure (Fig. 1a). The six iron atoms are at the corners of the trigonal prism and the boron atom is put at the center of the prism. The next units are related to the original prism by clockwise rotations through 215° around the axis in the same plane. Repeated rotations around the axis in the same plane produce the edge-sharing arrangement of trigonal prisms.

The result of starting with a trigonal prism and generating the bulk structure by repetitive rotations is shown in Fig. 1b, a 78-atom wooden ball model. A bulk

---

Figure 1
Ball model of the short range order and bulk simulation procedure. (a) This
work for boron in an undistorted iron prism; (b) Kowbel and Brower [4] for a
bulk simulation for a distorted prism showing the locations of metalloid atoms; the dashed line represents the smooth, unrelaxed boundary condition.

structure is built up in our computer model by adding successive layers of trigonal prisms to the original trigonal
prism until 860 total atoms have been added. This procedure is continued until a more dense packaged arrangement
of trigonal prisms is reached which is a maximum within
the algorithm of the model. The generating procedure maintains the NM_6 short range order, while packing the trigonal prisms.

At this stage the model is not adequately physical
and needs a bulk relaxation procedure. This goal was accomplished minimizing the internal energy. The cohesive
energy of such a system was assumed to be adequately represented by a sum of pairwise interactions [2]. For the
present work a 6-12 Lennard-Jones potential was chosen.
The interatomic potential

$$V_{ij} = 4\varepsilon_{ij} \ [R_{ij}^{12} - R_{ij}^{6}],$$

where

$$R_{ij} = 2^{-1/6} \ (R_{ij}^{0}/r_{ij}),$$

r_{ij} is the distance between atoms i and j and R_{ij}^{0} is the
equilibrium interatomic distance. The quantity ε_{ij} is the
binding energy between atoms i and j which is assumed to
be equal to the cohesive energy ε. The explicit form of
this potential for our system is given by the following
equations where values of ε_{ij} and R_{ij} for Fe and B are
taken from Boudreaux [2]:

304

$$V_{Fe-Fe} = 0.94 \left(\frac{2^{-1/6} \times 2.60}{r}\right)^{12} - \left(\frac{2^{-1/6} \times 2.60}{r}\right)^{6} ;$$

$$V_{Fe-B} = 1.88 \left(\frac{2^{-1/6} \times 2.18}{r}\right)^{12} - \left(\frac{2^{-1/6} \times 2.18}{r}\right)^{6} ;$$

$$V_{B-B} = 0.002 \left(\frac{2^{-1/6} \times 3.68}{r}\right)^{12} - \left(\frac{2^{-1/6} \times 3.68}{r}\right)^{6} .$$

Compared to Boudreaux' potential terms [2], the above potentials do not include terms due to truncation and in our case the total energy, E_t, was calculated taking into account all atoms in the model

$$E_t = \frac{1}{2} \sum_{i,j}^{N} V_{ij} .$$

The energy minimization was accomplished by steepest descent method [3]. Each atom is simultaneously moving along the direction of the net force on that atom [3]. The atoms are moved a distance proportional to the magnitude of the force on them. After each such move, the forces are recalculated and the atoms moved again. The force on atom i is given by $F_i = -\Delta V$. The atoms are then moved according to $x_{i,n} = h_i \nabla_i V_n$. The constant h was chosen for each step separately. The values of h in this study were 0.1 to 1.0. The new set of coordinates in the nth interaction $X_{i,h} = X_{i,0} + \Delta x_{i,h}$. The new value of energy is calculated in new coordinates. To assure that the minimum of the total energy is reached the average force

$$F = N^{-1} \left(\sum_{k=1}^{N} F_k^2\right)^{1/2}$$

was introduced, where N is the total number of atoms in the model. The procedure is stopped when $F < (\varepsilon/R_0)$ which assures that the final energy E_t is approached within 1% of the minimum value.

1.2 Surface Model Generation

The smoothest possible unrelaxed surface was generated in the following way. The same procedure was used in a previous study of the Pd_4Si metallic glass surface [4]. The boundary condition in the Z direction is imposed such that

no atom is allowed to have the Z-component greater than a
fixed number (Fig. 1b). This boundary condition is the
equivalent of a fracture surface at the dashed line in
Fig. 1b. This generation procedure allows composition to
be maintained up to the prism layer below the top surface.
In other words, up to the surface, the structure consists
of trigonal prisms which generate the bulk composition
$Fe_{80}B_{20}$ utilizing a total of 860 atoms.

The equilibrium surface segregation was finally
reached by utilizing minimization of binding energy as the
driving force for segregation, while neglecting the effect
of entropy. The surface was assumed to be segregated and
relaxed when the forces acting on the surface atoms are
negligible ($E_t - E_{t\ min}/E_{t\ min} < 1\%$) and the effect of
entropy in this approach was again assumed to be negligible, similar to a previous surface model of Pd_4Si glass
[4].

1.3   Catalytic Properties

Metallic glass catalysts have been observed to have high
activity [5,6] and unusual selectivity [7,8,9] in a number
of hydrocarbon reactions. In each case, the glassy catalysts were the most selective of hydrogenation over isomer-
ization. Yoshida has reported about 60% selectivity of
ZrNi glass for ethane in the Fischer-Tropsch reaction [10].
The results given below show similar selectivity for ethane for an $Fe_{80}B_{20}$ glass.

2.   Experimental Procedure

The metallic glasses specimens used in this work were produced by the Allied Chemical Corporation (Metglas 2605)
and were used as received.

The Fischer-Tropsch reaction was run in a Pressure
Differential Scanning Calorimeter (PDSC) utilized as a
flow reactor [12]. Use of the PDSC allowed in situ crystallization of the metallic glass and the reevaluation of
the catalytic selectivity and activity of the same catalyst by re-running the Fischer-Tropsch reaction over the
crystallized glass. Small quantities (5-10 mg) of various
catalysts-glasses, crystallized glasses, pure metal foils,
and commercial catalysts were evaluated to allow the comparison between $Fe_{80}B_{20}$ metallic glasses and various forms
of crystalline iron catalysts.

3. Results and Discussion

Based upon this model, probability distributions of the number of nearest neighbors, CN for the atoms located on the surface were calculated at the first coordination shell. Three types of distributions were analyzed, each of which describes the vicinity of the iron atoms on the surface. The Fe-Fe distribution was calculated at a distance of 0.259 nm from the original atom, corresponding to the calculated position of the first coordination shell. Similarly, the Fe-B and B-B distributions were calculated at 0.225 nm and 0.365 nm, respectively. Average coordination numbers, CN, were 4.90 for Fe-Fe neighbors, 2.33 for Fe-B neighbors, and 6.72 for both Fe and B neighbors for the relaxed surface for the smooth boundary condition.

By using a surface graphics system the simulated relaxed surface is shown in Fig. 2. The picture illustrates the roughness of the surface in terms of predominantly low coordinated sites. The roughness of the surface was described by the use of the fractal dimension, D. For the surface depicted in Fig. 2, D = 2.3, as compared to 2.30 for a Brownian island [11]. D was determined by how many initial squares can be found in the rectangles obtained

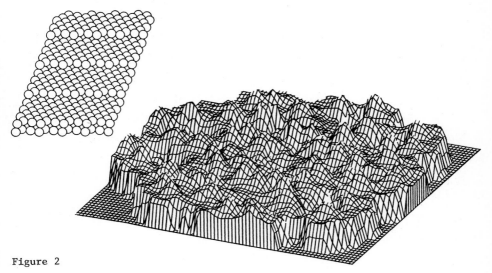

Figure 2

The simulated surface of the relaxed $Fe_{80}B_{20}$ metallic glass displayed via graphics system as compared to a stepped (10,8,7) single crystal surface. Balls represented a schematic metal atom in a (111) single crystal surface are 2 Å in diameter. Smooth surface boundary condition. 2 trans-section lines per iron atom.

from the surface simulation. In our case we assume that a rectangle becomes a square when $a_2 \geq 1.5 (a_1)$, which leads to an error of 4% in D, or D = 2.3 $\pm$ 0.09.

The $\overline{CN}$ of 6.72 for the $Fe_{80}B_{20}$ glassy surface is significantly greater than that generated previously for Pd_4Si glass of 5.94 [4]. This difference appears to be due to the ability of the much smaller boron metalloid atoms to occupy an essentially interstitial position in the undistorted Fe-B trigonal prisms, Fig. 1a. The larger silicon metalloid atoms distort the Pd_4Si trigonal prisms, hinder efficient packing of the prisms, and slightly roughen the Pd_4Si surface as compared to the $Fe_{80}B_{20}$ surface. Both the $Fe_{80}B_{20}$ and the Pd_4Si surfaces are significantly rougher than a stepped and terraced fcc (10,8,7) plane, whose $\overline{CN}$ = 8.75 [4].

The fractal dimension of 2.3 for the $Fe_{80}B_{20}$ glassy surface was similar to that for the Pd_4Si glassy surface, and indicates an atomically rough, but not porous surface. The surface simulation, Fig. 2, indicates a rough surface in the plane of the surface which is free of terraces on the scale of an adsorbate molecule.

Such a difference in surface structure from crystalline surfaces should result in a shift in catalytic selectivity. Comparisons of Fischer-Tropsch reaction results are given in Table 1 for several reaction products at a

Table 1 Fischer-Tropsch Reaction Results (Average of N Runs) at 320°C.

CATALYST		FISCHER-TROPSCH REACTION RESULTS[1]						
Alloy	Form	Nominal Phases	N, # Runs	# Methane	# Ethane	# Propane	Surf. Area, m^2/gm	Specific Rate, mole/Lmin m^2
$Fe_{80}B_{20}$	Abraded Single Roller Strip	Glass	5	41	53	7	0.76^2	0.6
$Fe_{80}B_{20}$	Single Roller Strip	Glass & Surface Crystallites	5	77	16	7	0.76^2	0.6
$Fe_{80}B_{20}$	Single Roller Strip	Crystal-lized Glass	10	70	24	6	0.76^2	0.6
Fe	Rolled Foil	α-Fe	5	69	24	7	0.01^3	300
Fe/ Graphite	Supported Fe on Graphite	α-Fe	1	77	17	5	2.76^2	0.2

[1] 1 atmosphere of 3:1 H_2/CO at a space velocity of 1800 hr^{-1} at 1.5% conversion.
[2] Measured by BET N_2 adsorption.
[3] Calculated geometric surface area.

308

constant 1.5% conversion. As can be seen, the abraded Fe-B strip exhibited a 53% selectivity for ethane, whereas the unabraded strip showed a selectivity similar to the results over crystallized $Fe_{80}B_{20}$ glass, Fe foil, and the supported Fe/graphite commercial catalyst. Such unique selectivity as exhibited by the metallic glass catalysts implies a corresponding uniqueness of the surface structure of such a glass. A difference which could account for this uniqueness is the lack of neighboring terrace sites on the glassy surface as compared to the ledge and kink sites on the crystalline surface even for small particles.

4. References

1. P.H. Gaskell, J. Non-Cryst. Solids 32, (1979) 207.
2. D.S. Boudreaux, T. Halicioglu, and G.M. Pound, J. Appl. Phys. 48, (1977) 5066.
3. M.G. Duffy et al., J. Non-Cryst. Solids 15, (1976) 435.
4. W. Kowbel and W.E. Brower, Jr., J. Non-Cryst. Solids 94, (1987) 70.
5. W.E. Brower, Jr., G.V. Smith, T.L. Pettit, and M.S. Matyjaszczyk, Nature 301, (1983) 497-499.
6. A. Yokoyama, H. Komiyama, H. Inoue, T. Masumoto, and H.M. Kimura, J. Catalysis 68, (1981) 355-361.
7. G.V. Smith, W.E. Brower, Jr., M.S. Matyjaszczyk, and T.L. Pettit, Proc. 7th Int. Cong. on Catalysis, (Ed. T. Seiyana, K. Tanabe), 355, (Elsevier New York, 1981).
8. W.E. Brower, Jr., E.E. Alp, W. Kowbel, and K.M. Simon, Proc. 8th Int. Cong. on Catalysis, (Verlag Chemie, Berlin, 1984).
9. G.V. Smith, O. Zahraa, M. Khan, B. Richter, and W.E. Brower, Jr., J. Catalysis 83, (1983) 238-241.
10. H. Yoshida, 4th Int. Cong. on Catalysis, (The Tanaguchi Foundation, Kobe, Japan, 1985).
11. B.B. Mandelbrot, Fractals: Form, Chance, Dimension, (Freeman, San Francisco, 1977).
12. T. Beecroft, A.W. Miller, and J.R.H. Ross, J. Catalysis 40, (1975) 281-285.

Part III

Adsorption on Metals

Energetics in Structure Transformation of Ir Clusters and Atomic Reconstruction of Ir Surfaces

Chong-lin Chen and Tien T. Tsong

Physics Department, The Pennsylvania State University,
University Park, PA 16802, USA

Abstract. On the Ir (111) and (001) surfaces, a three-atom Ir cluster can assume either a 1-D or a 2-D structure. From the temperature dependence of the relative probabilities of observing these structures, the differences in the binding energy and the entropy of the two structures are found to be $\Delta E_{12} = -0.098 \pm 0.004$ eV and $\Delta S_{12} = (7.0 \pm 0.3)$k on the (111). On the (001), $\Delta E_{12} = 0.33 \pm 0.02$ eV and ΔS_{12} is strongly dependent on the temperature. Although these data are taken below 420 K where surface reconstruction does not occur, the binding energy data can be used to explain why the (1×2) and (1×5) reconstructions of the Ir (110) and (001) surfaces are energetically favored, and no reconstruction of the Ir (111) can be expected.

I. Introduction

Fundamental to the atomistic understanding of many surface phenomena and crystal growth is how atoms and small atomic clusters behave on the surface and what are the energetics of various atomic processes. Field ion microscopy (FIM), with its atomic resolution and its ability to produce a well-characterized (1×1) surface by low temperature field evaporation, and the ease to control the surface temperature, is well suited for such studies.[1] We report a study of the energetics of the structure transformation of small Ir clusters on Ir surfaces.

II. Experimental Results

When diffusing adatoms encounter each other, they may combine into a cluster. As shown in Fig. 1, on the Ir (111) and (001) surfaces, a small Ir atomic cluster can assume either a 1-D chain structure or a 2-D island structure.[2] The probabilities of observing a cluster in the 1-D and 2-D structures, p_{1D} and p_{2D}, depend on the surface temperature. In general there is no well defined threshold temperature where a structure transition occurs. For an example, for the 6-atom clusters on the Ir (001), $p_{1D}/p_{2D} =$ 20/1 at 375 K, =21/9 at 490 K and =17/6 at 505 K. Large clusters often can assume many different structures, thus the energetics in the structure transformation is more complicated to study quantitatively. We choose to study the 3-atom Ir cluster which is the smallest cluster having a 1-D and a 2-D structure. We have made >6000 heating periods of observations.

Springer Series in Surface Sciences, Vol. 24   **The Structure of Surfaces III**
Editors: S.Y. Tong · M.A. Van Hove · K. Takayanagi · X.D. Xie
© Springer-Verlag Berlin, Heidelberg 1991

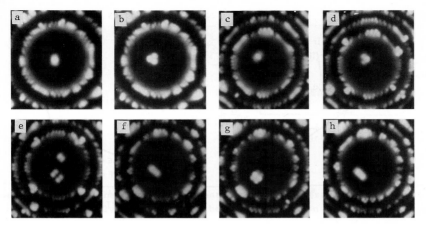

Fig. 1   (a) & (b): A 3-atom cluster on an Ir (111) surface with
the 1-D and 2-D structures.   (c) & (d):   A 3-atom cluster on an
Ir (001) surface.   (e) to (h): Formation of a 6-atom cluster on
an Ir (001) and 1-D to 2-D structure transformation at 470 K.

When a 3-atom Ir cluster changes its structure, which can occur below 420
K, the atomic structure of the substrate remains unchanged.   Therefore we
can expect p_{1D} and p_{2D} to follow the Boltzmann statistics according to

$$P_{1D}/P_{2D} = (W_{1D}/W_{2D}) \exp(\Delta E_{12}/kT),$$

where W_{1D} and W_{2D} are the statistical weights of the two structures which
are related to their entropies by $\Delta S_{12} = S_1 - S_2 = kT\ln(W_{1D}/W_{2D})$, and $\Delta E_{12} =$
$E_{1D} - E_{2D}$ is the difference in the total cluster binding energies
(magnitudes) of the cluster in the 1-D and 2-D structures.   Thus a
$\ln(p_{1D}/p_{2D})$ vs $1/T$ plot can be expected to be linear with a slope of $\Delta E_{12}/k$
and an intercept of $\ln(W_{1D}/W_{2D})$.   Indeed, for the 3-atom Ir cluster on the
Ir (111) surface, this plot exhibits a simple linear behavior as shown in
Fig. 2.   From the plot, we find $\Delta E_{12} = -0.098\pm0.004$ eV and $W_{1D}/W_{2D} =$
$1.1\times10^3\times(10)^{\pm1.1}$, or $\Delta S_{12} = (7.0\pm2.5)k$.   The linearity is 0.9892.   The 2-D
structure is more stable with a smaller entropy factor.

In contrast, the 3-atom Ir cluster shows a drastically different behavior
on the Ir (001) surface.   We have obtained two sets of data from two Ir
tips, one of which is shown in Fig. 3.   At high temperatures, the plot is
again linear.   The data derived are: $\Delta E_{12} = 0.348\pm0.008$ eV and 0.322 ± 0.021
eV, and $W_{1D}/W_{2D} = 1.6\times10^{-4}\times(10)^{\pm1.1}$ and $4.4\times10^{-4}\times(10)^{\pm1.3}$, or $\Delta S_{12} = -$
$(8.3\pm1.5)k$.   The linearities are 0.9860 and 0.9973.   Thus the 1-D structure
is more stable and has a smaller entropy factor.   Both sets of data also
show a significant deviation from the linear plot below ~386.5 K.   The
deviation part also fit well to a linear plot, but such a plot does not
have a clear physical significance.   It is better to describe the low
temperature part by a temperature dependent entropy factor as shown in Fig.

Let the transition rate from the 1-D to the 2-D structure to be k_{12} and that in the reversed direction to be k_{21}. In equilibrium, $P_{1D}k_{12} = P_{2D}k_{21}$. Or $P_{1D}/P_{2D} = k_{21}/k_{12}$. If the binding energy of the cluster at the transition state is represented by E_t, then $k_{12} = \nu_{12} \exp[-(E_{1D}-E_t)/kT]$ and $k_{21} = \nu_{21} \exp[-(E_{2D}-E_t)/kT]$. Therefore

$$P_{1D}/P_{2D} = (\nu_{21}/\nu_{12}) \exp[(E_{1D}-E_{2D})/kT] = (W_{1D}/W_{2D}) \exp(\Delta E_{12}/kT).$$

If the temperature is too low to have many structure changes within a heating period τ, then the probability that a 1-D cluster is found to have the 2-D structure after a heating, $p_{12}(\tau)$, and that to still have the 1-D structure, $p_{11}(\tau)$, are:

$$p_{12}(\tau) = [k_{12}/(k_{12}+k_{21})]\{1 - \exp[-(k_{12} + k_{21})\tau]\},$$

$$p_{11}(\tau) = 1 - p_{12}(\tau).$$

For starting with a 2-D structure, $p_{21}(\tau)$ and $p_{22}(\tau)$ have similar forms. It is clear that when $\tau \to \infty$ we have $p_{11}/p_{12} = k_{21}/k_{12} = P_{1D}/P_{2D}$. In general one would expect P_{1D} and P_{2D} to be dependent on τ since all the p_{ij} are dependent on τ. But they should still satisfy the difference equations:

$$P_{1D}(\tau) = P_{1D}(\tau)p_{11}(\tau) + P_{2D}(\tau)p_{21}(\tau),$$

$$P_{2D}(\tau) = P_{1D}(\tau)p_{12}(\tau) + P_{2D}(\tau)p_{22}(\tau),$$

where $P_{1D}(\tau) + P_{2D}(\tau) = 1$. From these equations one obtains

$$P_{1D}(\tau) = p_{21}(\tau)/[p_{12}(\tau)+p_{21}(\tau)], \text{ and } P_{2D}(\tau) = p_{12}(\tau)/[p_{12}(\tau)+p_{21}(\tau)] .$$

Therefore

$$P_{1D}(\tau)/P_{2D}(\tau) = p_{21}(\tau)/p_{12}(\tau) = k_{21}/k_{12} = P_{1D}/P_{2D}$$

is most surprisingly independent of τ because of the cancelation of the τ-dependent terms. In summary, as long as the data contains a statistically significant number of structure changes, they should represent well the equilibrium distribution. These data points should be well behaved without large, irregular fluctuations typical of the statistically unreliable data.

References

[1]. T. T. Tsong, Rpt. Prog. Phys. _51_, 759 (1988).
[2]. D. W. Bassett, Thin Solid Films _48_, 237 (1978).

Observations of Cluster Structures on fcc (111)

S.C. Wang and G. Ehrlich

Materials Research Laboratory and
Department of Materials Science and Engineering,
University of Illinois at Urbana-Champaign, Urbana, IL 61801, USA

Abstract. The ability of the field ion microscope (FIM) to locate single adatoms also makes it possible to map out the arrangement of binding sites, even on a complicated plane such as fcc(111), and therefore to examine the atomic arrangement of surface clusters. Using this capability, the behavior and properties of two-dimensional iridium clusters on Ir(111) have been examined. For clusters containing from 2 to 8 iridium atoms, the location of the atoms has been mapped out in relation to the substrate and it has been established that interatomic distances always correspond to roughly a nearest-neighbor spacing.

1. Introduction

The properties of clusters adsorbed on a crystal made of the same atoms are of interest in understanding crystal growth [1]; despite that, little detailed information is available about such clusters [2]. We have therefore undertaken a field ion microscopic [3] study on an atomically smooth surface, the close-packed (111) plane of iridium, a face-centered cubic metal, with the special aim of examining the structure, stability and diffusivity of clusters containing different numbers of iridium atoms.

2. Site Mapping

On an fcc(111) plane, there are actually two types of binding sites, shown in fig. 1. Addition of an adatom to a bulk site continues the fcc lattice; atoms placed at surface sites sit in an hcp arrangement and create a fault plane. Because of the close spacing of the (111) plane, and its smoothness on the atomic level, none of these features are revealed in the FIM. However, the location of a single adatom is clear, as in fig. 2(a). By repeated observation of the adatom after diffusion, it is therefore possible to plot the position of the sites at which the atom is adsorbed [4]. Shown in fig. 1(b) is a schematic of the fcc(111) plane, with the bulk sites connected by gridlines; a similar map of the surface sites is given in fig. 1(c). An atom that prefers bulk sites yet occasionally sits in surface sites will map out a

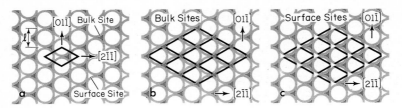

Fig. 1. Schematics of fcc(111) plane, showing different binding sites. (a). Map of outermost lattice plane; ℓ=nearest-neighbor spacing. (b). Grid expected for adatoms at bulk sites. (c). Grid for adatoms occupying surface sites.

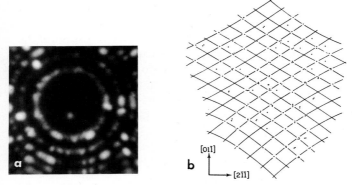

Fig. 2. (a). Field ion micrograph of Ir(111) surface with a single iridium adatom. (b). Map of locations of an Ir adatom on Ir(111), observed after diffusion for 5 sec at T=104 K. Points at the intersections of the gridlines mark surface sites; others, inside the unit cells, are bulk sites.

grid with an additional interior site, a surface site, toward the right of the unit mesh. In the opposite case, when surface sites are favored, but the atom occasionally occupies a bulk site, there will be an additional binding site at the left of the unit mesh. An actual map of sites on Ir(111), obtained for an iridium adatom after diffusion at 104 K, is shown in fig. 2(b). From a comparison with the grids in fig. 1, it is evident that the iridium atom prefers surface sites, but occasionally also sits at a bulk site.

This technique, although reliable and basic to site identification, requires extensive mapping and is quite time consuming. A more immediate indication of the type of site occupied is provided by the image of the adatom itself. In fig. 3, the image of an iridium adatom at surface as well as bulk sites clearly is triangular in shape [5]. At surface sites, the apex of the image triangle points toward [2$\overline{1}$$\overline{1}$], whereas at bulk sites, the apex points away from [2$\overline{1}$$\overline{1}$]. This unique relation between the orientation of the image spot and the type of binding site at which the adatom sits now provides a most convenient way of mapping the sites in a local region of the surface and makes it easy to explore the structure of iridium clusters.

318

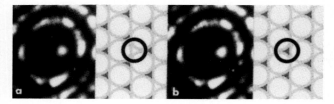

Fig. 3. (a). Triangular image of Ir adatom at a surface site, illustrated by schematic at right; apex of image triangle points to the right, along $[2\bar{1}\bar{1}]$. (b). Ion image and schematic of Ir adatom at bulk site; apex points to the left.

3. Atomic Clusters

On Ir(111), iridium atoms deposited from the vapor associate to form a variety of different clusters, which are clearly revealed in the FIM. Examples are given in fig. 4 , showing a single iridium adatom, a dimer, triangular trimer, linear trimer, tetramer, pentamer, hexamer, as well as clusters containing 7, 8, 12, and 13 atoms. The clusters were formed on an Ir(111) for which the location of the sites had previously been established. To determine the structure of the cluster it is then sufficient to map the location of the individual cluster atoms on the grid of binding sites.

Maps of the atom positions for clusters containing from 2 to 7 iridium atoms are shown in fig. 5. For Ir_2, Ir_3 and Ir_4, the images of the cluster atoms

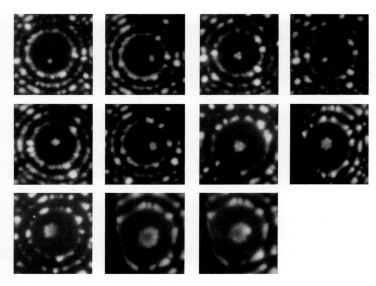

Fig. 4. Field ion images of iridium clusters on Ir(111). Shown in sequence are a single adatom, a dimer, triangular trimer, linear trimer, tetramer, pentamer, hexamer, heptamer, as well as Ir_8, Ir_{12}, Ir_{13}.

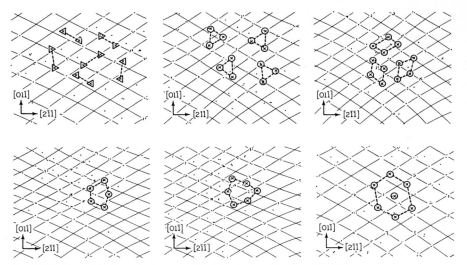

Fig. 5. Mapping of iridium clusters on grid of single atom binding sites. Dots give location of single Ir adatoms used to map sites, crosses mark the centers of cluster atoms. Cluster atoms are always at nearest-neighbor spacing from each other.

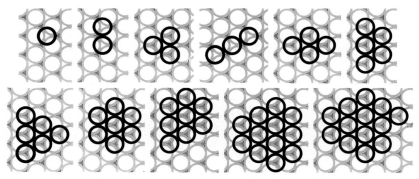

Fig. 6. Atomic arrangements of clusters imaged in Fig. 4, as deduced from detailed mapping experiments.

coincide with the previously mapped binding sites for iridium adatoms on this surface. For Ir_5, Ir_6, and Ir_7 the image points sometimes are displaced radially outward from the binding sites. However, this displacement is small compared to the interatomic distance on (111), and there can be no doubt that in all the clusters the atoms are separated from each other by roughly a nearest-neighbor distance.

From such detailed analyses we therefore arrive at the structures illustrated schematically in fig. 6 for the clusters previously imaged in fig. 4. At temperatures T < 460 K, all the clusters observed, except for linear

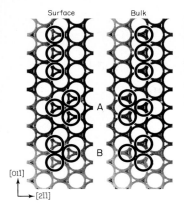

Fig. 7: Schematic of possible atomic arrangements of trimer atoms on surface and bulk sites. All of them are actually observed, but triangular trimers are slightly favored over linear ones, and type A occurs more frequently than B.

trimers, are in compact two-dimensional atomic arrangements, which maximize interactions with close neighbors. Because of the two types of sites on the (111) surface, a variety of different close-packed arrangements are possible. Iridium trimers, for example, exist in linear and triangular forms, and the cluster atoms can be either at surface or bulk sites, as illustrated in the schematics in fig. 7. Furthermore, regardless of which binding sites are occupied, we can distinguish between triangular trimers in which the center of the cluster is above a binding site (type A) or above an atom in the underlying surface layer (type B). All of these different arrangements have actually been observed. At temperatures $T < 250$ K, triangular trimers are found slightly more frequently than linear ones, and type A is more stable than type B.

After equilibration, triangular trimer atoms are found over surface and bulk site with roughly equal probability. For tetramers, only tetragonal clusters have been observed in three different orientations. Both surface and bulk sites can be occupied, but bulk sites are much favored by tetramer atoms. For Ir_5, Ir_6, and Ir_7, the cluster atoms are almost always found at bulk sites. From our observations it appears that as the size of the cluster increases, the cluster atoms prefer to occupy bulk sites [6] instead of the surface sites favored by adatoms [4] and also dimers, as summarized in table 1.

What should be clear even from this brief survey is that the ability of the FIM to reveal the location of individual adatoms at a surface makes it

Table 1 Effect of Cluster Size on Binding Sites: Ir on Ir(111)

Atoms in Cluster	1	2	3	4	7
Equilibration Temperature (K)	100	168	250	200	465
Percent at Bulk Sites	15	40	50	85	100

possible to gain considerable information about cluster structure, which is a prerequisite for a rational understanding of clusters properties.

This work was supported by the Department of Energy under contract DE-AC02-76ER01198.

References

1. R. Kern, G. LeLay, and J. J. Metois in: Current Topics in Materials Science, Vol. 3, Ed. E. Kaldis (North-Holland, Amsterdam, 1970) p. 295.
2. D. W. Bassett, in: Surface Mobilities on Solid Materials, Ed. V. Thien Binh (Plenum, New York, 1983) p. 83.
3. J. A. Panitz, J. Phys. E15 (1982) 1281; E. W. Müller and T. T. Tsong, Field Ion Microscopy, Principles and Applications (Elsevier, New York, 1969).
4. S. C. Wang and G. Ehrlich, Phys. Rev. Lett. 62 (1989) 2297.
5. S. C. Wang and G. Ehrlich, Surface Sci. 224 (1989) L997.
6. S. C. Wang and G. Ehrlich, Surface Sci. 217 (1989) L397.

Structure of Metal Overlayers by Low Energy Alkali Ion Scattering: Cu/Ru(0001) and Sn/Pt(111)

S.H. Overbury[1], *D.R. Mullins*[1], *M.T. Paffett*[2], *and B. Koel*[3]

[1]Oak Ridge National Laboratory, P.O. Box 2008,
  Oak Ridge, TN 37831, USA
[2]Los Alamos National Laboratory, Los Alamos, NM 87545, USA
[3]Department of Chemistry, University of Southern California,
  Los Angeles, CA 90089, USA

Structural studies of metal adsorption on metals by incident angle dependent, low energy alkali ion scattering are presented for the systems of Cu on Ru(0001) and Sn on Pt(111). Cu forms well ordered islands which are pseudomorphic for two layers. Sub-monolayers of Sn incorporate into the first Pt layer upon annealing, giving rise to a slightly buckled first layer.

1. Introduction

Continued studies of metal overlayer systems are driven in part by an interest in the surface chemistry of ultra-thin metal overlayers. The bimetallic systems of Cu-Ru and Sn-Pt have been shown to exhibit important catalytic properties. This has prompted studies of the surface chemistry on Cu/Ru(0001) [1] and Sn/Pt(111) [2] as models of the alloy catalysts. Complete understanding requires structural models of these systems which we have endeavored to clarify by studies of these surfaces by low energy alkali ion scattering.

2. Li$^+$ Scattering from Clean Pt(111) and Ru(0001)

Incident angle dependence of the single scattering from the hexagonal faces of clean Pt(111) and Ru(0001) is shown in Fig. 1. In one scan the plane of scattering is aligned along the [$10\bar{1}0$] azimuth of the Ru(0001) surface. Two scans are shown for the Pt(111) surface, obtained along two different azimuths located 60° apart. These azimuths correspond to the "long" direction in which atom chains have inter-atom spacing of 4.81 Å and 4.69 Å for Pt and Ru, respectively. The incident angle, ψ, is measured from the plane of the surface.

The scans exhibit an intensity edge at about 15°, corresponding to the onset of scattering from first layer atoms. Just above 20° the scattering is roughly constant, corresponding to the scattering from first layer atoms only. However, at higher angles, sharp peaks due to scattering from second layer and deeper atoms are observed. These peaks arise from a combination of structure dependent shadowing and blocking edges as indicated schematically in Fig. 1. For the Pt surface, a peak is observed near 54° for one long azimuth and a peak near 74° for the other, while for the Ru(0001), surface peaks at both angles are observed although with lower relative intensity. Near 28° there is a small peak for the [$\bar{1}2\bar{1}$] azimuth in Pt(111).

These features are consistent with the structures and expected differences between hcp(0001) and fcc(111) surfaces. An fcc(111) surface exhibits three-fold symmetry so that long azimuths separated by 60° appear different if both first layer and second layers are considered. For an hcp(0001) surface, adjacent domains separated by a step one layer high will have a registry between first layer and second layer atoms which is rotated by 60° for one

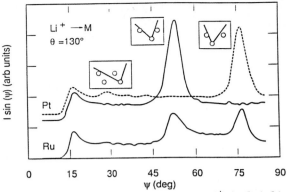

1. Intensity of single scattered Li⁺ (multiplied by sin ψ) as a function of incident angle are compared for Ru(0001) in [10$\bar{1}$0] azimuth and Pt(111) in [11$\bar{2}$] and [$\bar{1}$2$\bar{1}$] (dashed line) azimuths. Incident ion energy is 1000 eV.

domain compared to the other. Both types of domains contribute to the scattering in any long azimuth on the Ru(0001), giving rise to both peaks in the Ru scan. For the Pt(111) surface, the complete absence of a remnant of a peak near 54° for the [$\bar{1}$2$\bar{1}$] azimuth and similarly near 75° for the [11$\bar{2}$] proves that there are very few stacking faults between first and second layer.

3. Cu/Ru(0001)

Copper was deposited on Ru(0001) evaporatively with the crystal surface temperature near 300 K. A rough coverage calibration was established using comparison of our LEED and AES results with those of other reported LEED/AES studies of this system [3,4], especially the attenuation of the Ru AES peak, the occurrence of satellite LEED spots, and changes in the AES intensity caused by annealing for coverages above 2 ML. In addition, the Cu single scattering showed a sharp saturation at a coverage which by the LEED/AES calibration was about 1.2 ML. The discrepancy between this and the expected value of 1.0 ML is roughly within the uncertainty of the calibration.

Figure 2 shows scans of the Ru single scattering intensity along the long azimuth of Ru(0001), recorded as a function of Cu coverage. These scans were

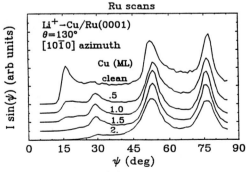

2. Intensity of Li⁺ single scattered from Ru atoms as a function of incident angle are shown for various coverages of deposited Cu.

324

recorded after Cu deposition without further annealing, although effectively identical results were obtained after annealing to 900 K. The effect of Cu is to cause attenuation of the Ru intensity at almost all angles where scattering occurs from only first layer atoms. However, the two peaks due to second layer scattering remain prominent for coverages up to above 2 ML. The fact that no new shadowing features appear is entirely consistent with pseudomorphic growth of Cu. In this case, it is expected that the Cu atoms located close to Ru lattice (overlayer) sites would shadow and block in the same way as Ru atoms, although with slightly smaller shadow cone diameters. A small region of angles near 28° is not as strongly attenuated, giving rise to a peak. This occurs because the Cu shadow cones are slightly smaller, allowing ions at this incident angle to pass between overlayer Cu atoms to scatter from first layer Ru atoms. The presence of both types of domains make it difficult to determine if first layer Cu deposits on hcp or fcc type of sites. A possibly significant shift in the positions of the two second layer peaks is observed between the clean and Cu covered surfaces. Such a shift could arise from a slightly different spacing between Cu and the Ru layers compared to the layer spacing in clean Ru, or from a different shadowing/blocking cone size of Cu compared to Ru. The analysis of this shift is complicated by the fact that each peak results from overlapping of both a shadowing and a blocking cone.

It is particularly interesting that the second layer features persist up to 2 ML. It has been previously suggested that the second Cu layer grows epitaxially as Cu(111) planes which are contracted with respect to the Ru(0001) surface and also presumably the first pseudomorphic Cu layer. This gives rise to a commensurate overlayer explaining satellite spots observed in LEED [3]. The lattice mismatch between the overlayer and the substrate would put Cu atoms in essentially random registry with respect to underlying Ru atoms and should occlude the holes leading into the Ru lattice. This should eliminate the sharp deeper layer features, contrary to the observations.

Additional information can be obtained from the incident angle dependence of the Cu single scattering such as shown in Fig. 3. The scans show a sharp shadowing edge at about 12°, corresponding to the onset of scattering from top layer Cu. The sharpness of the first layer onset is an indication that there are very few isolated Cu adatoms, since these would broaden this edge. The edge was sharp after deposition at room temperature and was unaffected by annealing to 900 K. This sharpness is indicative of 2-D island growth

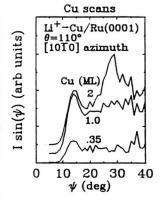

3. Intensity of Li$^+$ single scattered from Cu atoms as a function of incident angle are shown for various coverages of deposited Cu.

even at 300 K. Above 15°, the intensity is approximately constant for a 1 ML coverage. For 2 ML, however, the Cu scattering peaks at about 28°. This peak is due to scattering from the inner layer Cu and indicates that the top layer Cu atoms are in registry with next inner layer Cu atoms as expected, if both layers are pseudomorphic.

4. Sn/Pt(111)

In previous studies, it has been found that annealing of an Sn deposited surface results in a surface with characteristic Sn coverages and LEED patterns [5]. Depending upon the amount of deposited Sn and the length and temperature of annealing, two different structures exhibiting p(2×2) and p($\sqrt{3}$x$\sqrt{3}$)-R30° LEED patterns are obtained with coverages suggested to be 0.25 and 0.33 ML, respectively. Since Sn readily alloys with Pt, it has been suggested that both structures involve Sn incorporation into first layer Pt, but the possibility that the Sn is present as an ordered overlayer has been difficult to eliminate conclusively.

By monitoring the Pt single scattering for the two structures, it was found that Sn caused a small amount of attenuation but little or no change in the shape of the incident angle dependence for the long or the short azimuths. Distinguishing between an overlayer in pseudomorphic sites and incorporation into first layer lattice sites was best accomplished at intermediate azimuths using a lower scattering angle of 100° and an incident energy of 500 eV. Under these conditions, second layer Pt is not observed in the incident scans, which helps to accentuate any shadowing of first layer Pt atoms which might be caused by overlayer Sn atoms. Experimental and computed incident angle scans are shown in Fig. 4 for an azimuth 22° from the [$\bar{1}2\bar{1}$] direction. The experimental scan was obtained from a p($\sqrt{3}$x$\sqrt{3}$) structure and exhibits broadened shadowing and blocking edges near 20° and 80° accompanied by peaks, attributed to flux peaking, at 27 and 73°. Between the flux peaks the intensity is featureless. The shape of the dependence predicted for an incorporated and an overlayer type structure are also shown. These scans were calculated based upon predicted location and sizes of shadow and blocking cones and broadened by convoluting with a Gaussian profile to simulate the effects of thermal and instrumental broadening. The effects of flux peaking were not included in the calculation. It is seen that for the

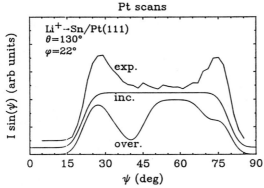

4. Experimental incident angle scan (exp.) is shown for the annealed p($\sqrt{3}$x$\sqrt{3}$) surface of Sn/Pt(111). It is compared with simulated scans computed for Sn incorporated into the Pt surface (inc.) and for Sn present as an overlayer (over). The experimental scan is scaled arbitrarily, and scans are offset for clarity.

overlayer model the simulation predicts a deep dip in intensity near 40° which is not observed experimentally. The incorporation model predicts a flat intensity dependence which agrees with experiment except the effects of flux peaking. This comparison therefore favors the incorporation model over the overlayer structure.

Analysis of the Sn scattering provides additional information. Upon deposition, the onset of first layer scattering is broad, indicating disorder or randomly adsorbed overlayer Sn adatoms. Upon annealing, the edge sharpens, indicating formation of a well ordered surface. From analysis of the angle associated with the edge, it is possible to distinguish a rippling in the surface. The Sn atoms appear to be buckled outward with respect to the Pt atoms by about 0.2 Å. Similar buckling has been observed in other strongly ordered alloys [6].

Acknowledgment. Research sponsored by Division of Chemical Sciences, Office of Basic Energy Sciences, U.S. Department of Energy, under contract DE-AC05-84OR21400 with Martin Marietta Energy Systems, Inc. BEK acknowledges support from Division of Chemical Sciences, Office of Basic Energy Sciences, U.S. Department of Energy under contract DE-FG03-90ER14117.

5. **Literature References**

1. D. W. Goodman and C. H. F. Peden, J. Chem. Soc. Fara. Trans. 1, 83, (1987), 1967.
2. M. T. Paffett, S. C. Gebhard, R. G. Windham, and B. E. Koel, Surf. Sci. 223, (1989), 449.
3. J. E. Houston, C. H. F. Peden, D. S. Blair, and D. W. Goodman, Surf. Sci. 167, (1986), 427.
4. C. Park, E. Bauer, and H. Poppa, Surf. Sci. 187, (1987), 86.
5. M. T. Paffett and R. G. Windham, Surf. Sci. 208, (1989), 34.
6. D. R. Mullins and S. H. Overbury, Surf. Sci. 199, (1988), 141.

Growth and Structure of Ordered Thin Films of Cu on Pd{001}

H. Li[1], S.C. Wu[1],*, D. Tian[1], J. Quinn[1], Y.S. Li[1], F. Jona[1], and P.M. Marcus[2]

[1]College of Engineering and Applied Science, State University of New York, Stony Brook, NY 11794, USA

[2]IBM Research Center, P.O. Box 218, Yorktown Heights, NY 10598, USA

1. Introduction

We have recently shown that thin films of Cu with thicknesses up to 10 layers can be grown epitaxially on Pd{001} at room temperature with a body-centered tetragonal (bct) structure (a=2.75 Å, c=3.24 Å) [1]. Since the lattice misfit between fcc Cu (a_0=3.61 Å, square cross section with side a=2.55 Å) and Pd (a_0=3.89 Å, a=2.75 Å) is 7.8%, a value usually considered too large for coherent epitaxy of 10-layer films, it is of interest to determine the equilibrium (i.e., the unstrained) phase of the epitaxial films. Linear strain analysis shows that the equilibrium phase cannot be bcc [1], while the hypothesis of a metastable α-bct phase [2] is negated by recent total-energy calculations that found no stable or metastable bct phase of Cu [3]. But the strain analysis is consistent with the assumption that the equilibrium phase is fcc Cu [4].

This success in establishing the structure of 10-layer films leads us to ask about thinner films. We report here on a LEED study of the early stages, and indirectly of the mode, of growth of Cu on Pd{001}, and on photoemission experiments on 5-layer films which confirm the fcc structure.

2. LEED study of the early stages of epitaxy

The sharp 1×1 LEED pattern of clean Pd{001} is only slightly worsened by the room-temperature deposition of up to 3 layer equivalents (L.E.) of Cu. LEED intensity data collected from coverages estimated to be between 0.5 and 1 L.E. could be fitted well with calculations based on a single flat layer of Cu{001} at 1.73±0.03 Å from the top Pd layer, and a very slightly expanded first interlayer spacing of Pd, 1.965±0.03 Å (the bulk Pd spacing is 1.945 Å, but the first interlayer spacing of a clean Pd{001} surface is 2.005±0.03 Å—about 3% expanded [5]). Figure 1 depicts experimental and theoretical spectra: the Van Hove-Tong R-factor

---

* On leave from the Department of Physics, Peking University, Beijing, The People's Republic of China.

Springer Series in Surface Sciences, Vol. 24   **The Structure of Surfaces III**
Editors: S.Y. Tong · M.A. Van Hove · K. Takayanagi · X.D. Xie
© Springer-Verlag Berlin, Heidelberg 1991

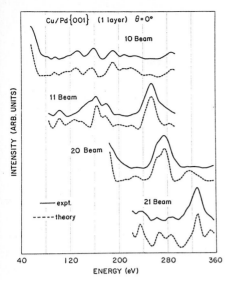

Fig. 1. Experimental and theoretical LEED I(V) spectra for 1 layer Cu on Pd{001}.

R_{VHT} [6] is 0.26, and the Zanazzi-Jona r-factor [7] is 0.05. An equally good fit is obtained with data collected for non-normal incidence of the primary electron beam, $\theta=10°$ and $\phi=0°$ (not shown).

Intensity data from thicker overlayers, nominally 2 to 3 L.E., could not be fitted well with calculations involving 2 or 3 full layers of Cu, although it could be established that I(V) curves calculated for Cu-Cu interlayer spacings between 1.6 and 1.7 Å were closer to experiment than all others. We conclude that it is very improbable that the available experimental data stem from either 2 or 3 *complete* layers of Cu. Averages of the calculations made for 2 and for 3 layers (25% 3-layers and 75% 2-layers, or 50%-50%, or 75%-25%) also failed to produce a satisfactory fit to experiment.

The conclusion is the following: on Pd{001}, Cu forms first a complete flat layer at a distance of 1.73 Å from the Pd top layer, and relieves most of the expansive relaxation that normally occurs on a clean Pd{001} surface. The formation of a full layer of Cu is consistent with the fact that the surface free energy of Cu (1.934 Jm^{-1}, ref.[8]) is slightly smaller than that of Pd (2.043 Jm^{-1}, ref.[8]). Further growth of the Cu occurs *probably* by island growth (Stranski-Krastanov mode), with Cu-Cu interlayers spacings close to 1.6 Å. After 5 or 6 L.E. the islands coalesce, the LEED I(V) spectra are stable, in the sense that they are no longer affected by the Pd substrate, and the film has the bct structure reported earlier [1].

3. Photoemission.

Angle-resolved electron distribution curves (AREDC's) were measured on a 5-layer film of Cu on Pd{001} with both s- and $(s+25\%p)$-polarized light in normal emission for photon energies ranging from 14 to 165 eV. Except for the presence of the Pd bands, the AREDC's are very similar to those obtained from a clean {001} surface of an fcc Cu crystal (compare refs. [9] and [10]). The dispersion of the Cu d-band, however, is different. In Fig. 2 the binding energy of the peak marked A has a minimum for $h\nu \cong 65$ eV and a maximum for $h\nu \cong 120$ eV [in fcc Cu{001} these extrema are 40 eV and 95 eV, respectively].

Peak A originates from the Δ_2-symmetry band in the band structure of fcc Cu (Fig. 3a). Transitions from the Δ_2 band are actually forbidden by the non-relativistic dipole selection rules, but are made possible by the mixing of additional orbitals in the relativistic-band basis set [10]. We show now that the positions of the extrema of peak A are consistent with the bulk interlayer spacing 1.62 Å as determined by LEED [1] (hence $\Gamma X = 2\pi/3.24 = 1.94$ Å^{-1}), but not with the bulk interlayer spacing 1.807 Å of fcc Cu. We assume as valid the free-electron (parabolic) relation

$$k_{f\perp} = 0.512(h\nu + E_F - E_b)^{1/2}, \qquad (1)$$

($k_{f\perp}$ in Å^{-1}, energies in eV) between final-state momentum $k_{f\perp}$ perpen-

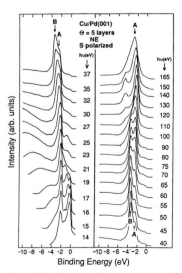

Fig. 2. Normal-emission EDC's from a 5-layer Cu film with s-polarized light and photon energies between 14 and 165 eV.

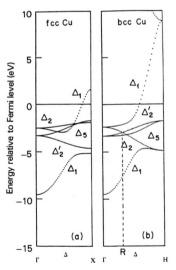

Fig. 3. Band structures of fcc Cu (a) and bcc Cu (b), after Moruzzi [11].

330

dicular to the surface inside the crystal, photon energy $h\nu$, Fermi energy E_F with respect to the bottom of the band, and (positive) binding energy E_b below E_F. From $E_F = |V_0| - \Phi$ (inner potential $V_0 = -8.6$ eV [9], work function $\Phi = 4.6$ eV] we find from Eq.(1) that with $E_b = 2.8$ eV, for $h\nu = 65$eV, $k_{f\perp} = 4.17$ Å $\cong 2\Gamma X$, while with $E_b = 2.5$ eV, for $h\nu = 120$ eV, $k_{f\perp} = 5.64$ Å $\cong 3\Gamma X$, as observed experimentally.

The peak marked B in Fig. 2, with larger binding energy than peak A, is due to transitions from the Δ_5 initial band. The calculated band structure of fcc Cu (Fig. 3a) shows that the Δ_2 and the Δ_5 bands are very close to each other when $k_{f\perp}$ is close to the X point, as observed in Fig. 2.

The calculated band structures of fcc and bcc Cu (Figs. 3a and 3b, ref. [11]) give further evidence for the fact that the films grown on Pd{001} have a distorted fcc structure and not a distorted bcc or bct structure. In bcc Cu the initial bands Δ_2 and Δ_5 cross each other for a value of $k_{f\perp}$ close to the point marked R in Fig. 3b. However, the experimental AREDC's in Fig. 2 exhibit no intersection of these bands when $k_{f\perp}$ is near the R point. Rather, peaks A and B approach one another for $h\nu \approx 100$ eV in accordance with the fact that in fcc Cu the Δ_2 and the Δ_5 bands are indeed close to one another when $k_{f\perp}$ is near the X point (see also ref. [12]). We also note in Fig. 3b that the dispersion of the Δ_2 band is very different in bcc Cu compared to fcc Cu. In fcc Cu, this band has the largest binding energy at the Γ point, and the Δ_5 band has the same dispersion at the Γ point. In bcc Cu, the Δ_2 band has the smallest binding energy at the Γ point, and its dispersion is different from that of the Δ_5 band at the Γ point. The experimental data depicted in Fig. 2 show that peaks A and B reach their largest binding-energy position for $h\nu \cong 65$ eV, which, as shown above, corresponds to $k_{f\perp} \cong 2\Gamma X$ and hence to the Γ point.

4. Conclusion

The LEED results indicate that the first Cu layer "wets" the Pd{001} surface and forms a complete layer, but successive layers *probably* grow by island formation. The photoemission results are clearly supportive of the LEED findings for thicker Cu films on Pd{001}, namely, that despite a large lattice misfit (7.8%), the Cu films do grow coherently to thicknesses of 8 to 10 layers with a structure that is tetragonally distorted from the equilibrium fcc phase.

Acknowledgements

We thank the National Science Foundation (Grant No. DMR-8709021) and the Department of Energy (Grant No. DE-FG02-86ER45239) for partial support of this work.

References

1. H. Li, S.C. Wu, D. Tian, J. Quinn, Y.S. Li, F. Jona and P.M. Marcus, Phys. Rev. B **40**, 5841 (1989-I).
2. I.A. Morrison, M.H. Kang and E.J. Mele, Phys. Rev. B **39**, 1575 (1989).
3. H. Jansen, personal communication (1990).
4. P.M. Marcus, H. Jansen and F. Jona, to be published.
5. J. Quinn *et al.*, to be published.
6. M.A. Van Hove, S.Y. Tong and M.H. Elconin, Surf. Sci. **64**, 85 (1977).
7. E. Zanazzi and F. Jona, Surf. Sci. **62**, 61 (1977).
8. L.Z. Mezey and J. Giber, Japan. J. Appl. Phys. **21**, 1569 (1982).
9. S.C. Wu, C.K.C. Lok, J. Sokolov, J. Quinn, Y.S. Li, D. Tian and F. Jona, Phys. Rev. B **39**, 13218 (1989-II).
10. S.C. Wu, J. Sokolov, C.K.C. Lok, J. Quinn, Y.S. Li, D. Tian and F. Jona, Phys. Rev. B **39**, 12891 (1989-I).
11. V.L. Moruzzi, private communication.
12. H. Eckardt, L. Fritsche and J. Noffke, J. Phys. F **14**, 97 (1984).

Electric and Magnetic Hyperfine Fields at Ni(111) Surfaces and Ni/In Interface Compound Formation

R. Platzer, X.L. Ding*, R. Fink, G. Krausch, B. Luckscheiter, J. Voigt, U. Wöhrmann, and G. Schatz

Fakultät für Physik, Universität Konstanz, Postfach 5560,
W-7750 Konstanz, Fed. Rep. of Germany
*Permanent address: Beijing Normal University, Beijing/P.R. of China

Electric and magnetic hyperfine fields have been measured at ^{111}In probes on Ni surfaces using the perturbed γγ-angular-correlation (PAC) method. The magnetic hyperfine field was found to be reduced over the entire measured temperature range compared to that of Ni-bulk. The probe lattice location was derived from simultaneous measurement of the electric-field-gradient tensor acting at the probe nuclei. This "fingerprint" for the location of the probe atoms was also used for studying interface reactions (i.e. interdiffusion and compound formation) in thin Ni/In films and In films on Ni surfaces in the monolayer range. The occurrence of Ni/In compounds has been observed.

1. Introduction

The PAC technique is like the Mössbauer spectroscopy based on the hyperfine interaction of probe nuclei with extranuclear electromagnetic fields. This feature can also be used for the investigation of surfaces. For these experiments a very low concentration of probe atoms is necessary which act as isolated observers. Thus it is possible to study any magnetic system. All experiments were done without applying an external magnetic field.

Interdiffusion at metal-metal or metal-semiconductor interfaces has been subject of numerous experimental studies, particularly because of its technical importance, for instance, in microelectronic devices. The formation of interface compounds, its stoichiometry and crystal structure may strongly influence the mechanical as well as electrical properties of the interface. Moreover, a microscopic study of the diffusion processes across the interface may lead to a deeper understanding of the growth kinetics and stability of such interface compounds. Here we report on an application of the perturbed γγ angular correlation (PAC) method to investigate compound formation in Ni/In thin film systems. Investigations of interdiffusion at interfaces by PAC have already been performed on the systems Cu-Ag-Au/In [1] and Sb/In [2].

For the present PAC work we used the isomeric nuclear state (spin $I=5/2$, half-life $T_{1/2}=84$ nsec) in [111]Cd, which is populated through electron capture from [111]In ($T_{1/2}=2.8$ days) and is intermediate between the 172 and 247 keV γ-transitions. Interaction of the nuclear magnetic dipole moment $[\mu=-0.7656(25)\mu_n]$ and the

Springer Series in Surface Sciences, Vol. 24 **The Structure of Surfaces III**
Editors: S.Y. Tong · M.A. Van Hove · K. Takayanagi · X.D. Xie
© Springer-Verlag Berlin, Heidelberg 1991

nuclear electric quadrupole moment [Q = 0.83(13)b] of the isomeric state with hyperfine fields leads to a splitting of this nuclear level, which causes a time dependency of the γγ-angular correlation. It is worth mentioning that the electric field gradient is measured at [111]Cd; however, the actual position of the probe atom is determined by the parent activity [111]In. This is due to the fact that the isomeric nuclear state in [111]Cd is populated within a few picoseconds after the electron capture. Thus, for the process studied in the present interface work the probe atom ^{111}In is chemically identical with the film constituent.

With a five γ-detector setup sixteen different coincidence spectra are recorded at the same time, from which eight counting-rate ratios R(t) are extracted. These R(t) spectra can be expressed as a superposition of cosine and sine modulations containing the transition frequencies ω_n. Fourier analyses of the R(t) spectra directly exhibit the transition frequencies; the set of hyperfine parameters, however, is extracted by a least-squares fit applied simultaneously to all eight time-dependent counting rate ratios. From these procedures the strength and direction of the magnetic hyperfine field as well as the strength, symmetry, and orientation of the electric-field-gradient tensor are obtained.

2. Ni(111) Surfaces

The Ni surface was produced by cutting a slice from a Ni(111) single crystal rod. After mechanical and electrochemical polishing the crystal was put into the UHV chamber. All investigations have been performed at a basic pressure better than 10^{-8} Pa. Further preparation was done by cyclic annealing up to 1250 K and Ar-sputtering. The following characterization of the Ni(111) surface was ensured by LEED and AES. After the preparation of the clean and well oriented (111)-surface ^{111}In probe atoms were evaporated with a concentration of about 10^{-4} monolayers at 77 K. This procedure also caused a small Cl contamination. Subsequent PAC measurements showed a disturbed signal due to not clearly defined probe positions. Only at temperatures above 350 K a well defined PAC signal was obtained, which then was stable over the whole measured temperature range.

In Fig. 1 (a + b) PAC spectra and corresponding Fourier transforms above and below the Curie temperature are shown. In the first case the magnetic field is switched off and the probe atoms perceive an electric field gradient. Due to the used probe-detector-geometry only two transition frequencies are visible. The observed frequency pattern shows clearly an axialsymmetric electric field gradient with the z-axis pointing perpendicular to the (111)-surface. In the second case the probe atoms are exposed to strong combined electric and magnetic hyperfine fields. It should be mentioned that if there is a magnetic field in addition to the electric field gradient at the probe atom position, up to 15 different transition frequencies can be observed instead of three for an intermediate state of I = 5/2. The amplitudes thereby depend on the probe-detector-geometry and the orientation of the magnetic hyperfine field with respect to the principal axis system of the field gradient. Fig. 1 (c) describes the situation at the ^{111}In atoms, when the surface is covered by 2.5 monolayers of Ni. The measured transition frequencies, the Larmor frequency and its first harmonic, show that the probes nearly experience the Ni bulk magnetic hyperfine field.

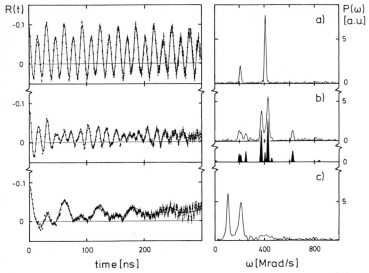

Fig. 1: *PAC spectra and corresponding Fourier transforms for ^{111}In probes on Ni(111) at different measurement temperatures (a) above the Curie temperature at T_M=632K and (b) below the Curie temperature at T_M=300K and (c) after a coverage of 2.5 monolayers Ni at T_M=100K. In (b) the Fourier transform of the fitted theory function is added for comparison.*

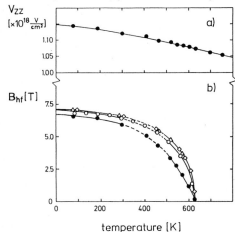

Fig. 2: *Temperature dependence of the electric field gradient and the magnetic hyperfine field at ^{111}In in the topmost layer of Ni(111) surfaces (solid circles), in the 3.5th monolayer (open circles), and in Ni bulk (open squares; open triangles).*

For the surface magnetic hyperfine field of the first monolayer on the Ni(111) surface we find an orientation in the surface-plane, most probably pointing along the six equivalent [110] directions, which are the second-easy axes of bulk magnetization. The easy axes in Ni bulk are along [111], which are not contained in the (111) surface plane. Strength, symmetry and orientation of the electric field gradient establish the assumption, that the ^{111}In probe atoms occupy substitutional terrace sites.

In Fig. 2 the results for the temperature dependence of the strength of both electric and magnetic quantities of ^{111}In probe atoms at the surface, in the 3.5th monolayer and in Ni bulk are represented.

The values for the magnetic field at the Ni surface is found to be slightly lower compared to the values for Ni bulk, whereas probes covered by 2.5 ML Ni reflect already bulk behaviour. A detailed discussion of the temperature dependence of the magnetic hyperfine field can be found in Voigt et al. [3].

3.   Ni/In Compound Formation

In a first experiment thin In film, homogeneously doped with ^{111}In, was evaporated on to a small sapphire plate at 100 K. Subsequently, at 77 K it was covered with a Ni film. The total thickness of the polycrystalline layers amounted to about 3000 Å with an atomic ratio Ni/In of 2:1. Finally, the sample was subjected to an isochronal annealing program (t_A = 10 min) within a temperature range from 230 K to 900 K and PAC spectra were taken at 300 K. Some typical PAC spectra and their corresponding Fourier transforms, clearly showing electric field gradients for Ni/In compounds, are shown in fig. 3. All Ni/In compounds are nonmagnetic.

Fig. 4 summarizes the results of the annealing cycle, showing the fractions of ^{111}In probe atoms which are exposed to well-defined electric field gradients. Up to annealing temperatures of about 350 K all probe atoms experience a single electric field gradient, characterized by a quadrupole coupling constant ω_Q = 20 Mrad/s. This electric field gradient is well-known and corresponds to ^{111}In on substitutional lattice sites in bulk indium.

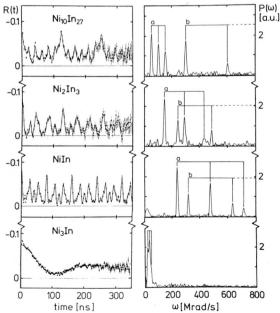

Fig. 3: PAC spectra and corresponding Fourier transforms clearly representing different Ni/In compounds.

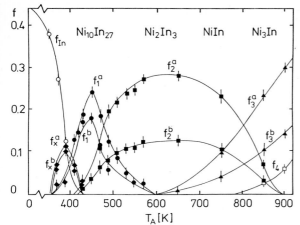

Fig. 4: *Fractions of ^{111}In probe atoms experiencing well-defined electric field gradients as a function of annealing temperature T_A for a Ni/In thin film couple.*

Beginning at a temperature of 250 K, f_{In} decreases on further annealing while two new fractions $f_x^{a,b}$ appear above 350 K. At temperatures exceeding 400 K an increase of two fractions $f_1^{a,b}$ is observed, which around 500 K convert into fractions $f_2^{a,b}$. These fractions remain stable over a wide temperature range. Finally, they are found to decrease at around 800 K, where two fractions $f_3^{a,b}$ occur. In addition, at 900 K one further fraction f_4 is found.

In order to clear up this rather complex picture, macroscopic PAC-samples were prepared with well-defined Ni/In stoichiometries. After PAC measurements the samples were pulverized and their structure was determined by Debye-Scherrer X-ray analysis. Consequently all electric field gradiends found in the film experiment, except $f_x^{a,b}$, could be identified as corresponding to certain Ni/In compounds. The fact that two different electric field gradients are found for $Ni_{10}In_{27}$, Ni_2In_3 and $NiIn$ is easily explained, in each of these compounds two non-equivalent In sites exist, whereas in Ni_3In only a single In site is possible.

In a second series of experiments the radioactive ^{111}In atoms were deposited again at very low concentration (10^{-4} ML) on a clean Ni(111) single crystal surface. After a clear Ni(111) surface signal had been obtained, the crystal was covered by In films of thicknesses of about 1 to 20 ML at 77 K. Then the samples were subjected to an isochronal annealing program in the same temperature range as before. In the beginning a fraction f_0 is visible, which is related to ^{111}In right at the interface, and decreasing above a temperature of 200 K. A small fraction f_{In} appears due to some probes propagating into the In film. Further annealing yields a thickness dependent sequence of formed intermetallic compounds. In contrast to the thin film couple case the fractions $f_1^{a,b}$ are not observed here.

The results of the interface experiment are summarized in Table 1. Comparing all these results with the phase diagram for Ni/In we find that the phase diagram has been run through from the indium-rich to the nickel-rich part with increasing annealing temperature. However, the fractions $f_x^{a,b}$ could only be observed in the thin film experiment. Additional investigations with Auger-electron spectroscopy suggest

Table 1: Formation temperatures for Ni/In compounds in dependence of film thickness for In films on Ni(111) and polycrystalline Ni/In films.

In thickness	$Ni_{10}In_{27}$	Ni_2In_3	NiIn	Ni_3In
1 ML	-	-	-	440
6 ML	-	-	470	600
20 ML	-	450	600	700
300 ML (poly-film)	370	430	650	800

the existence of an interface compound containing less than 27% Ni, which may not be stable as bulk alloy and does therefore not appear in the phase diagram.

In all experiments no ^{111}In probes were observed in Ni which would give rise to a strong magnetic hyperfine interaction. We therefore conclude that the formation of the interface compounds is caused by diffusion from Ni in In.

Conclusion

In our work on Ni(111) surfaces we have demonstrated that by means of the PAC method it is possible to measure the magnetic hyperfine field with high accuracy over the entire interesting temperature range. The position of our probes at the surface is well determined by the measured properties of the electric-field-gradient tensor. Application of PAC to the investigation of compound formation in thin films has again shown to be appropriate to detect interdiffusion and the appearance of intermetallic compounds. As a non-destructive method with high depth resolution, without limitation in film thickness, PAC has proven to be a powerful complementary tool to study thin film systems.

Acknowledgements

The authors would like to thank the Deutsche Forschungsgemeinschaft (SFB 306), Bonn, for the generous financial support of this work.

References

[1]   W. Keppner, R. Wesche, J. Voigt and G. Schatz, Thin Solid Films 143 (1986) 201
[2]   R. Wesche, W. Keppner, T. Klas, R. Platzer, J. Voigt and G. Schatz, Hyp. Int. 34 (1987) 573
[3]   J. Voigt, R. Fink, G. Krausch, B. Luckscheiter, R. Platzer, U. Wöhrmann, X.L. Ding and
       G. Schatz, Phys. Rev. Lett. 64 (1990) 2202

Structure and Surface Alloy Formation of the Fe(100)/Pt Overlayer System Studied with HEIS

G.W.R. Leibbrandt, R. van Wijk, and F.H.P.M. Habraken

Department of Atomic and Interface Physics, University of Utrecht,
P.O. Box 80.000, 3508 TA Utrecht, The Netherlands

Abstract. The structure of the Pt/Fe(100) overlayer system is studied employing High Energy Ion Scattering in combination with Shadowing and Blocking. It is shown that an elemental Pt layer is present after the evaporation of Pt on Fe(100) at room temperature, and after anneals at temperatures below 300°C. Annealing at higher temperatures results in the formation of an ordered $Pt_{0.5}Fe_{0.5}$ layer. The observed varying interlayer distances in the overlayers are related to the presence of thin overlayers of materials, that have in bulk situation a face centered cubic structure, on a body centered cubic crystal.

1 Introduction

A face-centered cubic (fcc) crystal can be transformed into a body-centered cubic (bcc) crystal by stretching the crystal in the [011] and [0$\bar{1}$1] directions and simultaneously contracting it in the [100] direction, while keeping the volume of the unit cell constant. The moment that the atomic distances in all these directions are equal the transformation is complete. These effects are expected to play a role in the Pt/Fe(100) overlayer system. Elemental Pt has a fcc crystal structure and Fe crystallizes in the bcc structure. The lateral Fe-Fe atom distance in a bcc Fe(100) surface is 3.5% larger than the lateral Pt-Pt atom distance in a fcc Pt(100) surface, whereas both surfaces have the same fourfold symmetry.

Heating a thin layer of Pt on a Fe(100) surface results in the formation of a $Pt_{0.5}Fe_{0.5}$ alloy layer (see section 3.2). This surface alloy layer may have a tendency to exhibit the fcc structure, since $Pt_{1-x}Fe_x$ bulk alloys have the fcc structure for $x < 0.82$ [1]. If so, the interatomic distance in the (100) plane of such an alloy will be smaller than the corresponding Pt-Pt interatomic distance, and consequently the lattice mismatch at the overlayer/Fe substrate interface will increase as a result of alloy formation. Assuming that the atoms at the overlayer/substrate interface in this overlayer system form a commensurate mesh, the lattice mismatch present causes a stretching of the overlayer in the [011] and [0$\bar{1}$1] directions and consequently, as indicated above, in a contraction in the [100] direction. The resulting structures can be conceived to be intermediate between fcc and bcc. In the present study we investigate the presence and the extent of this contraction employing high energy ion scattering in combination with shadowing and blocking (HEIS-SB) [2,3]. This investigation is part of a study of the structure and chemical reactivity of Pt and Pt based alloy layers on Ni [4,5] and Fe [6,7].

2 Experimental

The experiments are carried out in OCTOPUS which is a multichamber UHV system connected to one of the beamlines of the Utrecht University 3 MeV Van de Graaff accelerator. This system contains facilities for Rutherford Backscattering Spectrometry (RBS), HEIS-SB, Low Energy Electron Diffraction (LEED), Auger Electron Spectroscopy (AES), ellipsometry, Pt evaporation, low energy ion sputtering and annealing. The base pressure is in the low 10^{-10} mbar region. Further experimental details can be found in ref [4]. The Fe(100) surface is cleaned as described in ref. [6].

3 Results and discussion

3.1 Growth of Pt on Fe(100)

From the measurements of the Pt and Fe Auger peak heights as a function of the amount of room temperature deposited Pt, as determined using 2 MeV RBS analyses, it has been inferred that Pt grows in a layer by layer fashion on Fe(100) [7]. The absence of a LEED spot pattern for Pt coverages larger than 0.7 ML (monolayers) indicates that a disordered surface layer is the result of the Pt evaporation.

3.2 Alloy formation

Figure 1a shows the Fe_{703} and Pt_{233} Auger peak heights during 350°C annealing of a Fe(100) oriented crystal on which 3.5 ML Pt has been deposited. Initially, the Fe_{703} peak is very small, but its peak height increases with in-

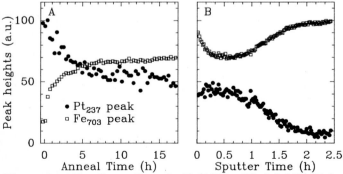

Figure 1. Alloy formation; **A**: Fe(100) with 3.5 ML Pt is annealed at 350 °C. **B**: Depth profile of the sample after subsequent anneal at 440°C.

creasing anneal time. At the same time the Pt_{233} peak height decreases. After 6 hours of annealing a steady state has been reached. In this steady state the Pt_{233} Auger peak has a height which is about half the value for the as deposited sample. This indicates a dilution of Pt in the surface region by approximately a factor 2 during the heat treatment.

Figure 1b shows a sputter depth profile measured at room temperature using 600 eV Ar ion sputtering. This profile shows that a mixed Pt/Fe layer of finite thickness has been formed during annealing of roughly 50%/50% composition.

It must be noted that during annealing at temperatures below or equal to 300°C no changes in the relevant AES peak heights have been observed.

3.3 Structure

For the investigations of the atomic structure of the as deposited Pt/Fe(100) overlayer system, the alloy system and possible intermediate states we will focus here on two series of measurements employing HEIS-SB. Firstly, we have deposited 0.8 ± 0.1 ML of Pt on Fe(100), and after studying the as deposited system, the sample has been annealed for 25 minutes at 440°C. This will be referred to as the 1 ML series. In the second series 3.0 ± 0.1 ML of Pt has been deposited. This system has been studied as deposited, after 30 minutes anneal at 150°C, 30 min. at 250°C and 30 min. at 350°C.

HEIS-SB measurements are performed by aligning a 750 keV He^+ beam along the $[\bar{1}11]$ direction of the Fe crystal and detecting the backscattered particles in the [111] direction (see inset in fig. 2). Under double alignment conditions the area of the Fe surface peak for the clean Fe(100) a in fig. 2 corresponds to 1.94 ± 0.10 visible Fe atoms per [111] row, or (2.37 ± 0.12) $\cdot 10^{15}$ Fe atoms/cm^2. The number of visible Pt atoms is determined from the area of the Pt peak (fig. 2b).

More information is obtained by varying the direction of the incoming beam around the $[\bar{1}11]$ directions and measuring energy spectra at different scattering angles. Blocking patterns depicting the number of visible Fe and Pt atoms at

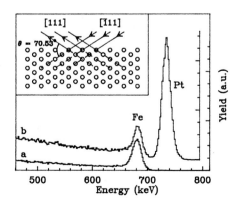

Figure 2. HEIS-SB energy spectra in double alignment of: a: clean Fe(100), b: Fe(100) covered with 1 ML of Pt. In the inset: the geometry for double alignment.

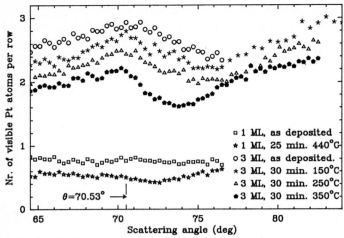

Figure 3. Blocking patterns of the Pt surface peaks. $\theta_{scatt} = 70.53°$ is the scattering angle where the beam is aligned with [$\bar{1}11$] direction (double alignment).

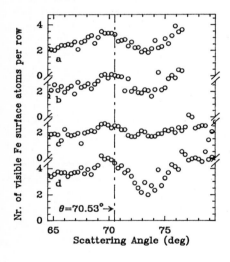

Figure 4. Blocking patterns of Fe surface peaks for Fe(100) with 3 ML Pt: **a**: as deposited, **b**: 30 min 150 °C, **c**: 30 min 250 °C, **d**: 30 min 350 °C. The scattering angle for double alignment is indicated.

each angle are shown in figure 3 and 4. There is no structure in the Pt blocking pattern of the as deposited 1 ML system and all Pt atoms are visible (fig. 3). After a 25 min. anneal at 440°C, the number of visible Pt atoms decreases and a minimum appears at a scattering angle $2° \pm 0.5°$ larger than the Fe bulk double alignment angle. The decrease of the number of visible atoms and the occurence of the minimum demonstrates that an alloy layer with some order has formed at the surface. The position of this minimum indicates that the interlayer distance in this alloy is larger by $9 \pm 2\%$ than the bulk Fe(100) interlayer distance [3]. On the 3 ML as deposited system we again see in double alignment ($\theta_{scatt} = 70.53°$) that all Pt atoms are visible. In this 3

342

Table 1. Increase of the interlayer distance in the surface region with respect to the Fe bulk value calculated from the Fe (Δ_{Fe}) and Pt (Δ_{Pt}) blocking patterns.

θ_{Pt}	Treatment	$\Delta_{Fe}(\%)$	$\Delta_{Pt}(\%)$
0	clean	4 ± 4	—
0.8	as deposited	4 ± 3	—
0.8	25 min. 440°C	10 ± 2	9 ± 2
3.0	as deposited	10 ± 2	—
3.0	30 min. 150°C	10 ± 2	21 ± 5
3.0	30 min. 250°C	10 ± 5	21 ± 4
3.0	30 min. 350°C	11 ± 1	12 ± 2

ML series the number of visible Pt atoms decreases with increasing anneal temperatures, indicating increasing order in the overlayer. After anneal at 350°C the minimum in the blocking pattern is broad and the depression of the yield in the minimum is small, indicating no complete order. From the shifts of the minima with respect to double alignment, the Pt interlayer distances in the overlayers are calculated. The Fe blocking patterns (figure 4) of the 3 ML as deposited system shows that with 3 ML of Pt the distance between the topmost layers of the Fe substrate is 10 % larger than the correeponding Fe bulk value. Furthermore disordered growth of Pt does not result in the disordering of the top Fe layer since the number of visible Fe atoms in the minimum is equal to the number of visible Fe atoms on clean Fe(100). This is in contrast with the observations for the Pt/Ni(100) system [5]. There is a distinct transition between the Fe blocking patterns of the sample annealed up to 250°C and the pattern of the sample annealed at 350°C.

In table 1 we have summarized the interlayer distances in the surface region related to the Fe bulk value, determined from shifts of the minima in the blocking patterns. The Pt interlayer distances are significantly different for the samples annealed at 150 and 250°C from the Fe interlayer distances, whereas for the samples annealed at 350 and 440°C these distances are about equal. From this observation we conclude the presence of an elemental Pt layer on top of the Fe(100) surface for the lower annneal temperatures and the formation of the Pt/Fe alloy overlayer at the higher anneal temperatures. If the interlayer distance in the Pt film should be equal to the bulk Pt interlayer distance in the same [100] direction, Δ_{Pt} should amount to 37%. The observed value is significantly smaller, which evidences the Fe substrate induced contraction of the Pt in the [100] direction as discussed in section 1. In agreement with the arguments given in section 1 the values for the interlayer distance for the alloy are smaller than for the elemental Pt film.

Summarizing, the HEIS-SB data indicate the presence of an elemental Pt overlayer with some order as a result of room temperature evaporation of Pt on Fe(100) and subsequent annealing at temperatures below 300°C. Annealing at higher temperatures results in the formation of an ordered $Pt_{0.5}Fe_{0.5}$ layer. HEIS-SB blocking patterns nicely reveal the structural aspects which are expected when a (100) oriented fcc Pt or $Pt_{0.5}Fe_{0.5}$ alloy film is brought into contact with a (100) oriented bcc Fe substrate.

343

References

1. M. Hansen: Constituents of Alloys, (McGraw-Hill, New York 1958)
2. P.F.A. Alkemade, S. Deckers, F.H.P.M. Habraken, W.F. van der Weg: Surf. Sci. **189/190**, 161 (1987)
3. J.F. van der Veen: Surf. Sci. Rep. **5**, 199 (1985)
4. S. Deckers: Ph.D. thesis, Utrecht University (1990)
5. S. Deckers, W.F. van der Weg, F.H.P.M. Habraken: This conference
6. G.W.R. Leibbrandt, P.C. Görts, S. Deckers, F.H.P.M. Habraken, W.F. van der Weg: Vacuum **41**, 13 (1990)
7. G.W.R. Leibbrandt, R. van Wijk, F.H.P.M. Habraken: To be published

Initial Growth and Structure of Pt on Ni(100)

S. Deckers[1], *W.F. van der Weg, and F.H.P.M. Habraken*[2]

Atomic and Interface Physics Department, University of Utrecht,
P.O. Box 80.000, 3508 TA Utrecht, The Netherlands

Abstract High Energy Ion Scattering in combination with Shadowing and
Blocking, ellipsometry, Auger Electron Spectroscopy and Low Energy Electron
Diffraction have been used to study the growth and annealing behaviour of
thin Pt layers on Ni(100). After room temperature deposition of Pt on the
Ni(100) surface a disordered layer-by-layer growth is observed. Above 250°C
crystallization effects occur and above 350°C a surface alloy starts to develop.
The results are compared with similar deposition studies on Ni(111).

1   Introduction

Both platinum and nickel are important metals for many reactions in heterogeneous catalysis. As part of a more general investigation of the deposition and chemical behaviour of thin Pt films on different metal surfaces [1,2], we report in this study on the growth of Pt on Ni(100) and the behaviour of the formed layer structure during annealing. The information about the structure of the overlayers is obtained from High Energy Ion Scattering in combination with Shadowing and Blocking (HEIS-SB), together with additional information from Low Energy Electron Diffraction (LEED) and Auger Electron Spectroscopy (AES).

2   Experimental

The Ni single crystal was prepared according to standard procedures. A detailed description of the crystal preparation and the Pt evaporator has been given elsewhere [2,3].

All coverages in this paper are expressed in monolayer equivalents (MLE), where one MLE of deposited Pt atoms is defined as the number of atoms in one bulk Ni(100) layer, i.e. $1.61 \cdot 10^{15}$ at/cm². The bulk (100) lattice constants for Pt and Ni are 0.3932 nm and 0.3524 nm respectively, so the lattice mismatch amounts to 11.6% [3].

In this study RBS will be used under double alignment conditions, see for example [4,5] applying a primary beam of 750 keV He ions. The backscattered particles are detected by a silicon surface barrier detector with an energy resolution of 12 keV. The scattering geometry used is shown in the insert of figure 1. Both the [0$\bar{1}$1] and the [011] directions are closely packed, resulting in an optimum shadowing and blocking configuration.

---

[1] Present adress: WA-14, Philips Research Labs, 5600 JA Eindhoven, the Netherlands
[2] To whom all correspondence must be adressed.

Springer Series in Surface Sciences, Vol. 24   **The Structure of Surfaces III**
Editors: S.Y. Tong · M.A. Van Hove · K. Takayanagi · X.D. Xie
© Springer-Verlag Berlin, Heidelberg 1991

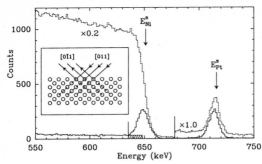

Figure 1: Energy spectra, random (dashed line) and in double alignment conditions (solid line). The energies of Pt and Ni surface peaks are shown in the figure. The hatched area indicates the background area that is subtracted from the Ni peak. The insert shows the scattering geometry used in this study.

3 Results

From Auger electron spectroscopy measurements as a function of Pt coverage, so called Auger uptake curves, a layer-by-layer growth of Pt on Ni(100) has been deduced [2,6]. The complete disappearance of the LEED pattern for Pt coverages below one MLE indicates beside a disordered Pt growth a distortion of the Ni lattice underneath. These results are in good agreement with results of similar reported investigations on Ni(111) [3,7]. By means of HEIS-SB measurements more information is obtained about the structure of the grown Pt layers. An example of a double alignment energy spectrum is shown in figure 1. The Pt and Ni peaks are well separated and the background below the Ni peak is less than 6% of the Ni peak height. A minimum yield for the clean crystal of 0.4% is observed. Energy spectra are recorded for different scattering angles ±6° around the angle where a blocking minimum occurs. An example of two so called angular scans is shown in figures 2a and b. In these figures, energy spectra of Ni(100) with 0.39 MLE of Pt, as deposited and after an annealing treatment for 2 hr at 420°C, are plotted for different scattering angles.

The contents of the Pt and Ni surface peaks in these spectra are converted using standard calibration techniques [4] to visible atoms per atom row and plotted versus

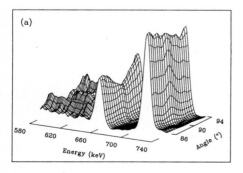
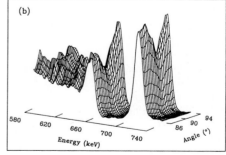

Figure 2: Perspective view of the energy spectra of two angular scans around the blocking minimum for Ni(100) with 0.39 MLE Pt before (a) and after (b) a 120 min annealing treatment at 420°C. The Ni bulk minimum is located at 90°, the central scattering angle.

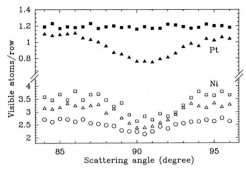

Figure 3: Blocking patterns of the Ni and Pt surface peaks for Ni(100), clean (Ni: ○), with 1.19 MLE Pt, after deposition (Pt: ■, Ni: □) and after a 120 min annealing treatment at 420°C (Pt: ▲, Ni: △).

scattering angle as so called blocking patterns. For a detailed description of background subtraction in HEIS-SB we refer to [4].

In figure 3 blocking patterns are shown for Ni(100), clean and with 1.19 MLE of Pt, both before and after an annealing treatment for 2 hr at 420°C. Before annealing, the content of the Pt peak remains constant over the whole range of angles, while the Ni peak shows a clear minimum around 91° scattering angle. After the annealing treatment the Pt peak reveals a strong blocking around 90.5°, and the Ni peak shows a similar behaviour. The observed shift of the minima in both the Pt and Ni surface peaks towards larger scattering angles compared to the bulk minima indicates that the lattice constant of the alloy is slightly larger than the bulk value of Ni. The occurrence of the Pt blocking minimum at the same scattering angle as the Ni blocking minimum is evidence for the formation of a surface alloy. The formation of Pt crystallites would have resulted in different positions of the minima. The HEIS-SB measurements have been performed at several Pt coverages, annealing periods and annealing temperatures. Table 1 gives a compilation of the HEIS-SB results. In this table, Δ_{Ni} and Δ_{Pt} denote the relaxations, i.e. the increase of the distance between the first and second crystalline layer relative to the Ni(100) bulk value.

4   Discussion

The data given in table 1 for the as deposited cases indicate that for all Pt coverages, the minimum yield has increased with respect to the clean surface. Furthermore, the Ni yield in the surface peak under the double alignment conditions is larger than for the clean surface. All Pt coverages measured under double alignment conditions are within the (statistical) errors equal to the Pt coverages determined from the 2 MeV random incidence RBS spectra. The number of visible Ni atoms/row compared to the clean surface increases with increasing Pt coverage for coverages lower than 1 MLE, and remains constant for larger coverages. The (1 × 1) LEED pattern disappears completely after deposition of more than one monolayer. Therefore it is concluded that the evaporation of Pt on Ni(100) results in the formation of a disordered homogeneous Pt layer. Furthermore, these results strongly indicate a disordering of the top Ni(100) layers upon Pt deposition. This is probably caused by a strong interaction between Pt and Ni atoms, as is evidenced by a negative mixing enthalpy upon alloying [8,9].

After annealing of the layer structure at 250, 350 and 420°C, crystallisation effects are observed. The occurrence of a blocking minimum in the blocking patterns of the Pt peak

Table 1: HEIS-SB measurements of Ni(100) covered with different amounts of Pt. Notes: (1) No visible minimum in the Pt blocking pattern. (2) Not measured.

θ	Treatment	χ_{min}(%)	Ni (at/r)	Pt (at/r)	Δ_{Ni} (%)	Δ_{Pt} (%)
0	clean	0.4 ± 0.2	2.2 ± 0.2	—	0 ± 0.6	—
0.39	as deposited	1.6 ± 0.2	2.5 ± 0.2	0.37 ± 0.1	0.2 ± 0.6	— [1]
0.39	2 h, 350°C	1.2 ± 0.3	2.6 ± 0.2	0.31 ± 0.1	1.1 ± 0.8	0.9 ± 0.5
0.70	as deposited	1.4 ± 0.2	2.7 ± 0.2	0.75 ± 0.1	— [2]	— [2]
0.97	as deposited	1.4 ± 0.2	3.0 ± 0.2	0.92 ± 0.1	2.2 ± 1.0	— [1]
0.97	0.25 h, 250°C	0.7 ± 0.3	2.8 ± 0.3	0.63 ± 0.2	0.2 ± 0.5	0.7 ± 0.7
1.19	as deposited	1.3 ± 0.2	2.8 ± 0.2	1.16 ± 0.1	1.7 ± 0.9	— [1]
1.19	2 h, 420°C	0.8 ± 0.2	2.4 ± 0.2	0.76 ± 0.1	0.6 ± 0.8	1.0 ± 0.6
4.70	as deposited	2.1 ± 0.3	3.4 ± 0.2	4.60 ± 0.1	1.0 ± 0.6	— [1]
4.70	0.25 h, 420°C	1.8 ± 0.3	3.1 ± 0.2	2.39 ± 0.1	1.7 ± 0.7	1.4 ± 0.4
4.70	2 h, 420°C	1.5 ± 0.3	2.9 ± 0.2	1.60 ± 0.1	1.9 ± 0.7	2.1 ± 0.4

evidences the presence of Pt in deeper layers of an ordered lattice, see also figure 3. The minimum yield in the energy spectra is decreased, as is the Ni peak content in the spectra taken under double alignment, both compared to the as deposited case. A (1×1) LEED pattern reappears after all annealing treatments, although the background intensity is larger than in the clean Ni(100) surface pattern. These observations confirm that the overlayer has become crystalline. Comparing the as deposited and annealed samples, especially at low coverages, the Pt peak reduction in the double alignment spectra and the changes in the AES spectra indicate the formation of a crystalline surface alloy in the top region of the crystal at all coverages. This is also observed in the case of Pt on Ni(111) [3]. In the latter case a Pt enrichment of the first atom layer is observed, in remarkable agreement with observations on the $Pt_{0.5}Ni_{0.5}$(111) bulk alloy surface [10].

The Ni blocking minima in the blocking patterns of the as deposited samples appear significantly shifted to higher scattering angles compared to the Ni bulk value, see figure 3. This indicates an outward relaxation of the crystalline surface layers, as indicated in the one but last column of table 1. No relaxation of the Pt layer could be measured, since the blocking pattern of these deposited Pt layers does not reveal any minimum. Both the Pt and Ni blocking minima in the blocking patterns after the different annealing treatments are also shifted to larger scattering angles. Due to the large uncertainties in the determination of the position of the blocking minimum, a comparison of the shifts between the different annealing procedures is not meaningful.

The observed relaxations can be compared to the relaxation expected on the basis of a lattice mismatch. For a 50%Ni–50%Pt alloy layer on top of Ni(100), Δ should amount to 5.7%. This is significantly larger than the observed relaxations. However, simple goniometry is used to obtain the relaxation values as shown in the last two columns of the table. Since in our double alignment measurements more than one atom layers is visible to the beam, the (relaxed) first layer atoms will only constitute a fraction of the surface peak, and therefore, the calculated relaxations will be too low. A more reliable conversion from shifts in the blocking pattern to realistic values of the relaxation can only be done by means of a best fit of Monte Carlo simulations to the data [11]. However, due to the large errors in the blocking patterns and due to the imperfections of the final lattice, these calculations do not result in more accurate results in the present case.

5 Conclusions

In this study we report on the room temperature growth of thin Pt films on Ni(100). The absolute Pt coverages were measured by means of Rutherford Backscattering Spectrometry. The structure of the grown layers is determined with HEIS-SB and LEED. The disappearance of the LEED pattern after less than 1 MLE of Pt deposition indicates that the grown Pt film is homogenous and disordered. This is confirmed by HEIS-SB results for Pt overlayer coverages up to 5 MLE. The HEIS-SB measurements show that the first Ni substrate layer is also disordered, which indicates a strong Pt-Ni interaction. Annealing above 250°C results in a ordered overlayer and above 350°C a surface alloy starts to develop. This surface alloy is found to have a lattice constant different from the Ni(100) value.

References

[1] G. W. R. Leibbrandt, P. C. Görts, S. Deckers, F. H. P. M. Habraken, and W. F. van der Weg, Vacuum **41**, 13 (1990).

[2] S. Deckers, Ph.D. thesis, Utrecht University, 1990.

[3] S. Deckers, S. H. Offerhaus, F. H. P. M. Habraken, and W. F. van der Weg, Surf. Sci. (1990), in press.

[4] P. F. A. Alkemade, S. Deckers, F. H. P. M. Habraken, and W. F. van der Weg, Surf. Sci. **189/190**, 161 (1987).

[5] S. Deckers, F. Bischop, D. de Jager, J. van Roijen, F. H. P. M. Habraken, and W. F. van der Weg, Surf. Sci. (1990).

[6] S. Deckers, F. H. P. M. Habraken, W. F. van der Weg, A. W. Denier van der Gon, B. Pluis, J. F. van der Veen, J. W. Geus, and R. Baudoing, Nucl. Instr. and Meth. **B45**, 416 (1990).

[7] J. A. Barnard, J. J. Ehrhardt, H. Azzouzi, and M. Alnot, Surf. Sci. **211/212**, 740 (1989).

[8] R. Hultgren, P. D. Desai, D. T. Hawkins, M. Gleiser, and K. K. Kelley, *Thermodynamic properties of binary alloys* (Am. Soc. for Metals, Metals Park OH, 1973).

[9] F. R. de Boer, R. Boom, W. C. M. Mattens, A. R. Miedema, and A. K. Niessen, *Cohesion in metals, transition metal alloys* (North-Holland, Amsterdam, 1989), Vol. 1.

[10] Y. Gauthier, Y. Joly, R. Baudoing, and J. Rundgren, Phys. Rev. B **31**, 6216 (1985).

[11] J. W. M. Frenken, R. G. Smeenk, and J. F. van der Veen, Surf. Sci. **135**, 147 (1983).

Structural Studies of Mercury Overlayers on Cu(001) by Atom Beam Scattering and LEED

Wei Li[1], J.-S. Lin[1], M. Karimi[2], C. Moses[3], P.A. Dowben[1], and G. Vidali[1]

[1]Syracuse University, Physics Department, Syracuse, NY 13244, USA
[2]Alabama A&M University, Physics Department, Normal, AL 35763, USA
[3]Utica College, Physics Department, Utica, NY 13502, USA

Abstract   We report investigations of ordered structures formed by a mercury layer adsorbed on Cu(001) using atom beam scattering (ABS) and LEED. Three of these structures (with symmetries of c(4x4), (3x3) and c(2x6)) are partially commensurate with the substrate; as revealed by ABS, these overlayer structures are not flat, but they have a supercell in which mercury atoms are inwardly or outwardly displaced with respect to the copper substrate. Simple models are proposed to obtain the relaxation height with respect to an ideally flat monolayer.

1. Introduction

There has been in the last few years an increased demand, for both technological and scientific reasons, for careful and detailed characterizations of the growth and structures of ordered phases of overlayers on crystalline substrates. Perhaps the best example of the successful applications of a variety of experimental and theoretical methods is the study of adsorption of rare-gas atoms and small molecules on the basal plane of graphite [1]. In fact, we could say that this class of systems is the best understood in surface physics.

On the other hand, despite numerous efforts, chemisorption systems, in which conspicuous electronic rearrangements take place between the adsorbate and the substrate atoms, are less understood [2].

The choice of our system, i.e. Hg on Cu(001), was dictated from the realization that a system in which the adsorbate binds more strongly than on graphite but still with little electronic rearrangement could lead to the discovery of interesting phenomena between the two classes of systems mentioned above.

In previous publications [3,4] we reported on different aspects of the growth, ordering, structural and electronic properties of Hg overlayers on Cu(001). Here we will concentrate on reporting experimental data on the structure of partially commensurate phases and on the influence that the substrate has on the arrangement of mercury atoms.

2. Experimental

Our apparatus consists of an atom beam scattering (ABS) set-up, a LEED system and an Auger electron spectrometer operating

Springer Series in Surface Sciences, Vol. 24   **The Structure of Surfaces III**
Editors: S.Y. Tong · M.A. Van Hove · K. Takayanagi · X.D. Xie
© Springer-Verlag Berlin, Heidelberg 1991

in the retarding field mode [3]. The helium beam is produced
by the expansion of 700 PSI of ultra-pure helium through a 5
micron nozzle. To improve velocity resolution ($\Delta v/v < 2\%$) and
minimize inelastic effects, the helium gas is cooled to about
82 K. The detector, a rotatable quadrupole mass spectrometer,
has an angular resolution of 0.7 degree. Triply distilled
mercury is admitted into the UHV scattering chamber (operating
pressure in the mid 10^{-10} torr range) via a leak valve. The
sample is mounted on a long-travel manipulator; a chromel-
alumel thermocouple measures the sample temperature. The
sample, an oriented Cu(001) disc, is cleaned prior to each run
by Ar ion sputtering followed by annealing. Mercury exposures
are calculated using the pressure read from an ion gauge
(uncorrected for gauge position and gas calibration). Absolute
coverages are calculated using the known ideal coverages
corresponding to structures observed by ABS or LEED and
assuming constant sticking coefficient (which is usually a
good approximation, as confirmed by Auger traces of the Hg
78eV peak) [3]. No evidence was found of alloying or
segregation effects [3].

3. Results

In Fig.1 we present the phase diagram for Hg on Cu(001)
obtained by adsorbing Hg at the indicated temperatures. The
3x3 and c(2x6) phases are metastable, i.e. they cannot be
obtained by adsorbing Hg at temperatures higher than 180 K and
then cooling the sample. Filled symbols indicate maxima of
intensities for ABS or LEED diffraction features. The central
hatched area indicates a co-existence of c(4x4) and (3x3)
phases. The cross-hatched vertical area is a disordered phase.

LEED pictures for the (3x3) and c(2x6) phases are shown in
Fig.2. Of all the possible structures compatible with the LEED
patterns, analysis of ABS data, such as shown in Fig. 2 and 3
for the c(4x4) and (3x3) phases respectively, enable us to
choose the structures shown in Fig. 2 and 3. We chose the
positions of the Hg atoms with respect to the Cu(001) lattice
according to our calculation discussed in the next Section.
Notice that the structures presented in Ref.(3c), based on
LEED evidence alone, should be considered incorrect.

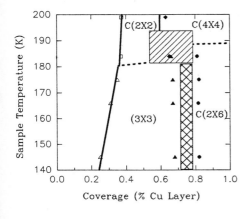

Figure 1. Phase
diagram for Hg on
Cu(001) from ABS and
LEED data.

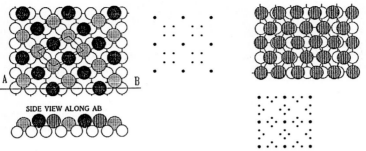

Figure 2. LEED pictures and proposed real space structures;
(3x3) left, c(2x6) right. Open circles are Cu atoms, others
are Hg atoms (see text).

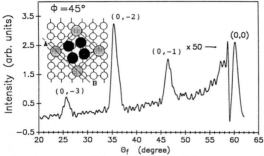

Figure 3. Helium diffraction scan for the c(4x4) phase. θ_i=60
deg.; scan taken along the line AB. Symbols as in Fig.2. [1]The
continuous lines indicate the supercell.

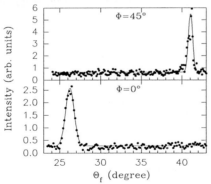

Figure 4. Helium
diffraction scan
for the (3x3) phase.
θ_i=60 deg.;
surface temperature
150 K.

   In Fig.3 we show a diffraction scan along the line AB
(<100>) in the inset for the c(4x4) phase; the azimuthal
angle ϕ is counted with respect to the <110> direction of bulk
copper. In reality we found that the c(2x2) and c(4x4) phases
co-exist; the peak labeled (0,-2) contains contributions from
both phases. As for the c(4x4) phase, the ABS diffraction peak
for ϕ=0 deg.(Fig.4) indicates that the (3x3) has a supercell
as shown in Fig.2 (solid lines).

352

4. Discussion

One of the striking results of this investigation is that the overlayer of mercury is not flat: from Figs.2 and 3 we see that Hg atoms occupy different positions with respect to the Cu(001) surface.

In order to interpret the ABS data, a reliable helium - overlayer+substrate potential has to be built; we also need to build a Hg-Cu(001) potential which allows us to predict where the Hg atoms will go with respect to the Cu lattice. Such a task is not easy even for the well studied rare-gas atoms adsorbed on graphite [1]; we are presently working on obtaining these potentials.

Nonetheless, we can obtain a semi-quantitative comparison with our experimental data in the following way. We built the Hg-Cu(001) and He-Hg+Cu(001) potentials, as explained in Ref.[3a], using He-Hg gas phase data and combination rules for Lennard-Jones pair potentials. The Hg-Cu(001) potential was found to have the correct binding energy for Hg on Cu(001), as given by isosteric heat data at zero coverage [3b].

If we have to place Hg atoms on the Cu(001) surface to form either a (3x3) or c(4x4) phase, they will occupy non equivalent sites, see Figs.2 and 3. For example, in the case of the c(4x4) phase, our calculation shows that four-fold hollow site Hg atoms (lightly shaded circles in Fig.3) are placed 0.15 Å lower than the other atoms. This structure influences the second layer adsorption, in which Hg atoms form a regular c(4x4) structure [4]. In the case of the (3x3), Hg atoms in alternate rows are, respectively, in four-fold hollow (lightly shadowed circles in Fig.2) and bridge sites (filled circles). The difference in outward displacements is about 0.18 Å.

It is known that the He atom is repelled by the electron charge density of the surface [5]. We used the standard approximation [5,6] of superimposing the atomic charge densities of Hg atoms to obtain the charge density contour that the He atom sees while approaching the surface [7]. Different profiles are obtained depending on the relative height position of the Hg atoms.

In this type of calculation, a difficulty arises in relating the interaction potential to the charge density contour [6]. Here the charge densities were not used to obtain the potential, but rather to give the overall shape function for a scattering calculation using the hard-wall model [8]. In order to decide which was the isodensity contour sampled by the helium atom, we matched the charge density contour with the hard-wall profile which was fitted to experimental data along the ϕ=18.4 deg. direction, where diffraction is not dominated by relaxation effects.

Our calculation shows that for the c(4x4), a relaxation below 0.1 Å or no relaxation at all are clearly incompatible with our results. For the (3x3) preliminary fits to the data indicate an outward displacement of about 0.2 Å.

In conclusion we have shown that Hg forms several ordered phases on Cu(001); using ABS we were able to obtain detailed information on these structures; for example we showed the departure of these structures from an ideal flat overlayer. This relaxation amounts to about 0.15 Å for the c(4x4) structure and 0.18 Å for the (3x3). This type of information is also important for the study of second layer adsorption [4] and correlations between structural and electronic properties of multilayer films [4,9].

Acknowledgments

Support for this research has been provided by NSF grant 8802512. We thank C.W.Hutchings for technical support.

References

1. N.D.Shrimpton, M.W.Cole, W.A.Steele and M.H.W.Chan in _Surface Properties of Layered Materials,_ Ed. by G.Benedek, (Kluwer Academic Publ., Dordrecht) in press.
2. E.Bauer, in _Structure and Dynamics of Surfaces II_, Ed. by W.Schommers and P. von Blanckenhagen, Topics in Current Physics (Springer Verlag) v.43, p.115 (1987); T.L.Einstein, CRC Critical Reviews in Solid State and Materials Sciences, $\underline{7}$, 261 (1978).
3. a) G.Vidali, C.W.Hutchings, P.A.Dowben, M.Karimi, C.Moses, and M.Foresti, J.Vac.Sci.Technol. A$\underline{8}$, 3043 (1990); b) C.W.Hutchings, W.Li, M.Karimi, C.Moses, P.A.Dowben and G.Vidali, Mat.Res.Soc.Symp.Proc. $\underline{153}$,133 (1990); c) P.A.Dowben, Y.J.Kime, C.Hutchings, W.Li, and G.Vidali, Surface Science $\underline{230}$, 113 (1990).
4. G.Vidali, W.Li, P.A.Dowben, M.Karimi, C.W.Hutchings, J.-S.Lin, C.Moses, D.Ila, and I.Dalins, Mat.Res.Soc.Symp. Proc. Symposium J, 1990 Spring Meeting, in press.
5. I.P.Batra, Surf.Sci. $\underline{148}$,1 (1984).
6. M.Karikorpi, M.Manninen, and C.Umrygar, Surf.Sci. $\underline{169}$, 299 (1986).
7. We used Hartree-Fock atomic charge densities as reported in: F.Herman and S.Skillman, _Atomic Structure Calculations_ (Prentice Hall, 1963).
8. T.Engel and K.H.Rieder, in: Springer Tracts in Modern Physics, v.91, _Structural Studies of Surfaces_, Ed. G.Hohler (Springer, Berlin, 1982).
9. W.Li, J.-S.Lin, P.A.Dowben, M.Karimi, C.Moses and G.Vidali, in preparation.

Yb on Al(001): A Two-Dimensional Disordered System

T. Greber, J. Osterwalder, and L. Schlapbach

Institut de Physique, Université de Fribourg, Pérolles,
CH-1700 Fribourg, Switzerland

Abstract: Between 0.3 and 1.5 monolayers of Yb have been investigated on an Al(001) substrate by means of photoelectron diffraction from Si Kα excited Al 1s and Yb 3d levels. At 300 K Yb does not diffuse into Al. At submonolayer coverage we find no well defined lateral position of Yb relative to the substrate. In this sense Yb forms a disordered two-dimensional system. In comparing the experimental results with a photoelectron single-scattering-cluster calculation one finds the Yb atoms to be 0.58 ± 0.04 aluminium lattice constants above the substrate. This compression of 14% relative to the corresponding Yb bulk interlayer distance indicates a 4f shell cracking of Yb on Al. The angle-resolved Yb $3d_{5/2}$ core-level photoemission current exhibits long-time-range fluctuations which disappear upon heating and above one monolayer Yb coverage. The anisotropies in the azimuthal scans 45° above the surface normal indicate a local ordering of the system above one monolayer.

1. Introduction

The competition between Coulomb repulsion and kinetic energy in lanthanide systems leads to peculiar ground states where the mean value of the valence can fluctuate between two integer values. Although most free lanthanide atoms are divalent, only europium and ytterbium remain divalent in their condensed bulk form. The tendency to increase the valence upon condensation can be studied at the surface where the coordination of an atom lies between the vacuum and the bulk value. In the case of samarium a divalent surface layer is observed [1,2], and in thulium systems low coordinated atoms are found to have divalent character [3].

Photoemission is among the few techniques which gives access to the question of valence in the first few layers of a lanthanide system. The strong Coulomb interaction between a core hole and the 4f shell resolves the core level photoemission final states for different 4f occupancies of a lanthanide atom. For submonolayer systems the different $3d^9 4f^n$ photoemission final states are particularly suitable to study the valence. Fäldt and Myers have investigated the interface of Sm on Al by means of core level photoelectron spectroscopy. They interpret the fact that a linear extrapolation for infinite Sm dilution on Al shows two different $3d^9 4f^n$ photoemission final states as an indication for a homogeneous mixed valence behaviour of Sm on Al [4,5].

Studying Yb on Al the findings are to some extent different and contradictory. In an investigation of Yb on Al (011) no significant Yb $4f^{12}$ multiplets in the valence band region have been reported [6]. A recent Yb 3d core level investigation however did exhibit a new type of core level shift behaviour for overlayers and a contribution in the Yb $3d_{5/2}$ photoemission spectra which can contain a $3d^9 4f^{13}$ final state contribution [7]. This was taken as an indication for a strong Yb compression and a cracking of its 4f shell on Al. In order to settle the question whether a submonolayer of Yb on Al is mixed valent or not we decided, in the spirit of early bulk experiments, to investigate the distance of an Yb ion above the substrate by means of photoelectron diffraction (PD).

Springer Series in Surface Sciences, Vol. 24   **The Structure of Surfaces III**
Editors: S.Y. Tong · M.A. Van Hove · K. Takayanagi · X.D. Xie
© Springer-Verlag Berlin, Heidelberg 1991

2. Experimental

The photoemission current was measured in a modified VG ESCALAB MarkII spectrometer under pressures better than $6 \cdot 10^{-9}$ Pa. The x-ray twin anode supplied Mg $K\alpha$ (1253.6 eV) and Si $K\alpha$ (1740 eV) radiation. The sample could be rotated about two perpendicular axes allowing polar and azimuthal scans of the emission direction above the surface. The electron acceptance cone was limited to less than 3° full opening. In a separate preparation chamber Yb was evaporated from a tantalum basket onto the clean substrate which was kept at room temperature. The pressure never rose above $5 \cdot 10^{-7}$ Pa. The deposition rates were between 0.3 and 1 monolayer per minute and were monitored by means of an INFICON quartz microbalance. Laterally scanning the sample with a spacial resolution of 1 mm the Yb $3d_{5/2}$ emission of a passivated surface indicated deviations from a homogeneous overlayer of less than 5%.

The Al (001) substrate was cleaned by argon ion bombardment and subsequent annealing to 600 K. Oxygen and carbon contamination was determined in comparing integrated intensity ratios normalized by atomic cross sections from Mg $K\alpha$ excited O 1s, C 1s, and Al 2s levels and was below 5% of a monolayer at the beginning of the experiments.

Since the Yb overlayer systems on Al are about 3 times more reactive to the residual gas in our ultra high vacuum chamber than the clean Al(001) face we chose a mode of data accumulation which optimizes acquisition time. The *total* intensities in the energy regions where the Al 1s (176-184 eV electron kinetic energy) and the Yb $3d_{5/2}$ (202-226 eV) were measured as a function of the polar angle θ ($\theta=0$ for normal emission). During the same scan the total intensity of the energy interval between 240 and 250 eV kinetic energy was recorded to provide a reference background intensity. In order to obtain a quantitative measure for the corresponding substrate and adsorbate emission, we subtracted an adequate portion of the background intensity such that the corrected intensities in normal emission were the same as those resulting from a linear background subtraction. A comparison with an Al 1s intensity scan from the clean substrate with a linear background subtracted immediately in the measuring process shows that the same features and similar anisotropies of the polar scan are observed in both data accumulation modes. The quantitative discrepancies are at the most 15% of the intensity at a given angle. However, the intensity in the [011] direction is the same for both procedures and the emission maxima do not shift in angle.

3. Results and Discussion

In Fig. 1a polar scans of Yb $3d_{5/2}$ emission in the (100) plane are given for four different Yb coverages. It shows an increase in Yb emission in going to grazing angles, as it is expected from a thin overlayer. The intensities follow those of a homogeneous overlayer with no atomic structure (dotted lines). The observed photoelectron diffraction patterns, i.e. deviations from the trend of the dotted lines, are weak. This suggests that at submonolayer coverage Yb stays on top of the substrate and does form a two dimensional system.

Polar scans on the clean substrate and the 0.6 ML Yb on Al system in the (110) and the (100) plane show that the Al 1s emission is strongly attenuated as a function of the Yb coverage and that the diffraction peaks become broader. It is an indication that not all Yb atoms occupy the same crystallographic site on the substrate: If this were the case we would expect an enhanced Al substrate emission along a new local bond direction because the heavy Yb atoms have an about two times more pronounced forward scattering amplitude. Fig. 1b shows polar scans of Al 1s emission in the (100) plane for different Yb coverages. The pronounced [011] emission decreases with increasing Yb coverage and becomes broader. The anisotropy A is defined as $(I_{max} - I_{min})/I_{max}$ where in this case I_{min} is determined between the surface normal and the [011] direction. It decreases

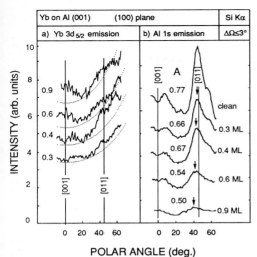

Yb on Al (001)	(100) plane		Si Kα
a) Yb 3d$_{5/2}$ emission	b) Al 1s emission		ΔΩ≤3°

POLAR ANGLE (deg.)

Figure 1 a). Yb 3d$_{5/2}$ photoemission intensity as a function of the polar angle θ which is measured with respect to the surface normal. No large diffraction effects are observed. This suggests a two dimensional overlayer of Yb on Al. The dotted lines give a behaviour which is expected from a homogeneous overlayer without atomic structure.
b) Polar scans of Al 1s photoelectrons in the (100) plane for different Yb coverages. The anisotropy A=(I$_{max}$ -I$_{min}$)/I$_{max}$ where I$_{min}$ is determined between the surface normal and the [011] direction decreases gradually. The [011] peak shifts toward the surface normal. This indicates a vertical distance of the Yb atoms larger than 0.5 Al lattice constants above the surface.

gradually with coverage. In addition to the broadening of the [011] peak we observe a significant shift toward the surface normal. This is an indication that the Yb atoms sit more than 0.5 substrate lattice constants above the Al top layer, which is not surprising since the lattice constant of Al (0.405 nm) and β–Yb (0.548 nm) differ considerably.

For the comparison of experiment with theory we chose the region within the polar scans with the most pronounced changes in intensity which is the [011] direction. In order to find a quantitative value for the Yb-Al bonding length which is expected to be the crucial parameter for the question whether the Yb atoms undergo a valence transition from a divalent to a mixed valent state, we compared the measured shifts to those obtained from single scattering cluster calculations. The theoretical shifts result from scattering simulations of Al 1s electrons coming from a 13 atom (001) surface cluster. The outward surface relaxation for the top and the second surface layer have been chosen +11% and +3% of a lattice constant, respectively [8]. The substrate has been kept unchanged for all calculations. If we take a mean Yb displacement parallel to the surface which is ±0.1 substrate lattice constants away from the fourfold hollow site we obtain a decrease in the anisotropy and a broadening of the [011] peak which are both in reasonable agreement with the experimental data set. The centroid position is rather sensitive to the position of the Yb atom above the surface. It has to be mentioned that in our theoretical simulations for the diffracted electrons no change in the inner potential has been incorporated. Changes in the order of 3 eV would yield refraction effects in the order of 0.5 degrees.

Comparing theory with experiment we get the best fit for a mean Yb-Al layer distance of 0.58±0.04 Al bulk lattice constants. This value can hardly be explained with divalent Yb atoms. It is smaller than 0.63 a$_o$ for β-Yb (fcc) at 4 GPa [9], where at 300 K the transition to γ-Yb (bcc) which has a cracked 4f shell [10], occurs. It is also larger than

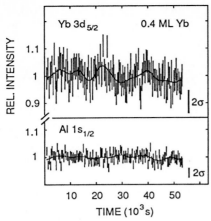

Figure 2. First observation of fluctuations in the Yb 3d and Al 1s photoemission currents as a function of time. The smooth line is a guide to the eye. It results in doing the Fourier back transformation for frequencies below 0.1 mHz.

0.52 a_0, the value for mixed valent $YbAl_3$. The layer distance in divalent β-Yb is 0.677 a_0. This reduction of almost 14% of the layer distance with respect to the Yb bulk value can be explained if the Yb atoms release a portion of their localized 4f electrons into the valence band. However, if we take the Yb contraction hypothesis of Ref. [7] i.e. a decreasing Yb-Al bonding length with increasing Yb coverage within the first Yb layer we get an even better agreement with the experimental data.

At this point we started to investigate the stability of a submonolayer Yb on Al. Figure 2 demonstrates that the angle resolved Yb $3d_{5/2}$ photoemission current normal to the surface fluctuates about its mean value on a time scale of 10^3 s. In all our data sets we find for Yb

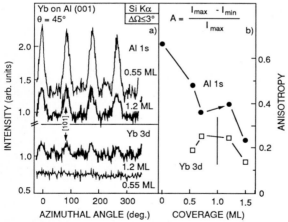

Figure 3. a) Two azimuthal scans 45° away from the surface normal for Al 1s and Yb $3d_{5/2}$ emission with 180 and 220 eV kinetic energy, respectively. For 1.2 ML Yb on Al a pronounced forward scattering feature in the direction of high substrate atom density is observed. Apparently the overlayer stabilizes toward the local order of the substrate.
b) Observed anisotropies A in the azimuthal scans. While the drop of the substrate anisotropy is not continuous around one monolayer, the Yb anisotropies seem to have a maximum at 1 monolayer coverage: This is taken as indication for a local ordering close to a full first Yb layer on Al

coverages below one monolayer at 300 K with sampling times of typically 100 s fluctuations which correspond to between 1 and 3% excess Gaussian noise for the Yb $3d^94f^{14}$ final state. The fluctuations stop when the sample is heated above 420 K, and they are not observed for systems above one monolayer coverage. For an attractive Yb-Yb interaction of Yb on Al we would expect the formation of two phases. The fluctuations suggest that at room temperature these phases are not stable. The disappearance of the fluctuations upon heating indicates that the two phases must be close to their critical point.

Since a full Yb layer should display a local order, we investigated the anisotropies of the Al 1s substrate and the Yb 3d adsorbate peaks. Fig. 3a shows that a submonolayer Yb does not have pronounced anisotropy directions in the Yb 3d emission, whereas 1.2 ML Yb on Al(001) exhibit clearly observable peaks in the high symmetry direction of the crystal. In Fig. 3b it can be seen that the anisotropy drop of the substrate peaks is not continuous in passing one monolayer coverage. This is taken as an indication that a small amount of the second layer stabilizes the interface and that locally induced order increases the anisotropies of the substrate signals.

4. Conclusion

We have performed photoelectron diffraction measurements of Yb on Al (001) systems. Submonolayers are found to be disordered two dimensional systems. A comparison of the experimental data with results from a single scattering cluster calculation yields a mean distance for Yb atoms of 0.58 Al lattice constants above the top substrate layer which supports the hypothesis that Yb ions are in a mixed valent state. At 300 K fluctuations in the adsorbate photoemission current are observed, which disappear in going to Yb coverages above one monolayer or to higher temperature. Anisotropies in azimuthal scans indicate that above one monolayer Yb on Al are in a stable configuration on Al(001).

Aknowledgement: This work has been supported by the Schweizerischen Nationalfond.

References

[1]   G. K. Wertheim and G. Crecelius, Phys. Rev. Lett. **40**, 813, (1978)
[2]   A. Stenborg, J. N. Andersen, O. Björneholm, A. Nilsson, N. Mårtensson, Phys. Rev. Lett. **63**, 187, (1989)
[3]   M. Domke, C. Laubschat, M. Prietsch, T. Mandel, G. Kaindl, and W. D. Schneider, Phys. Rev. Lett. **56**, 1287, (1986)
[4]   Å. Fäldt, H. P. Myers, Phys. Rev. B, **30**, 5481, (1984)
[5]   Å. Fäldt, H. P. Myers, Phys. Rev. B, **34**, 6675, (1986)
[6]   R. Nyholm, I. Chorkendorff, J. Schmidt-May, Surf. Sci. **143**, 177, (1984)
[7]   T. Greber, J. Osterwalder, L. Schlapbach, Phys. Rev. B, **40**, 9948, (1989)
[8]   J.J. Burton, G. Jura, J. Phys. Chem. **71**, 1937 (1967)
[9]   H. T. Hall and L. Merrill, Inorg. Chem. **2**, 618, (1963)
[10] G. Wortmann, K. Syassen, K. H. Frank, J. Feldhaus, G. Kaindl, in "Valence Instabilities", P. Wachter and H. Boppart ,Eds. North-Holland Publishing Company, (1982), pg. 159 ff

Order–Disorder Transition of 6 ML Fe Films on the Cu(100) Surface

C. Stuhlmann[1], *U. Beckers*[1], *J. Thomassen*[1], *M. Wuttig*[1], *H. Ibach*[1], and *G. Schmidt*[2]

[1]Institut für Grenzflächenforschung und Vakuumphysik,
Forschungszentrum Jülich, Postfach 1913,
W-5170 Jülich, Fed. Rep. of Germany
[2]Institut für Angewandte Physik, Lehrstuhl für Festkörperphysik,
Universität Erlangen-Nürnberg,
Erwin-Rommel-Str. 1, W-8520 Erlangen, Fed. Rep. of Germany

Abstract. Fe films, epitaxially grown on the Cu(100) surface, display a sequence of structures which are (5×1) and $(2 \times 1) - $p2mg at 3 ML and 6 ML coverage, respectively, according to a recent corrected thickness calibration. In order to understand the thermodynamics of the p2mg structure we recorded temperature dependent spot profiles of the (10), (11), (0 0) and $(1\frac{1}{2})$ spots. The peak intensity of the superstructure spots decreases exponentially with increasing temperature while the spot profile broadens.

1. Introduction

Ultrathin epitaxial Fe films on the Cu(100) surface have been extensively studied during the past few years with respect to their magnetic and structural properties [1-9]. Most of the authors found the films to grow in a $p(1 \times 1)$ structure up to a certain number of layers which varies from author to author, probably due to uncertainties in the thickness calibration. In addition to the $p(1 \times 1)$ structure at room temperature we observed a (2×1) reconstruction of the 6 ML films at 80 K which we already reported in an earlier paper [10]. In order to achieve a better understanding of the reconstruction we performed a systematic study of the transition between the $p(1 \times 1)$ structure at room temperature and the $(2 \times 1) - $p2mg structure at 80 K. In this paper we will give a short summary of our experimental observations.

2. Experiments

All experiments were performed in a UHV chamber equipped with a three grid LEED optics, a CMA Auger electron spectrometer and a high resolution electron energy loss spectrometer (HREELS). Sample and film preparation have been described in some detail in [10]. For thickness calibration we initially used Auger electron spectroscopy only. The dependence of the Fe 651-eV and Fe 703-eV Auger signals on deposition time was a sequence of linear segments with approximately equidistant break points. These break points were assigned to the completion of successive atomic layers[10]. A separate experiment, however, showed that this calibration was off by a factor of three. A quartz crystal monitor and MEED intensity oscillations were used to determine the Fe film thickness. The Auger intensity ratio of the Fe 703-eV and Cu 920-eV line and the superstructures, observed for 3 ML and 6 ML films in this experiment,

Springer Series in Surface Sciences, Vol. 24 **The Structure of Surfaces III**
Editors: S.Y. Tong · M.A. Van Hove · K. Takayanagi · X.D. Xie
© Springer-Verlag Berlin, Heidelberg 1991

were identical to those which we found at the first and second break point in the Auger calibration curves of our earlier experiments and which were therefore addressed as 1 and 2 ML, respectively, in our previous publications on this topic[10,11]. The 3 ML films have a (5×1) structure while the 6 ML films cooled down to 80 K reproducibly display the LEED pattern shown in fig. 1a. The symmetry of the LEED pattern is p2mg. The existence of a glide plane determines the underlying real space structure almost completely. The only variable parameters of the structure drawn in fig. 1b are the amplitude of the deviation of the surface atoms from their position in the unreconstructed surface and the layer(s) where the reconstruction takes place.

The temperature dependence of LEED reflexes was examined by two methods each of which has its advantages and its shortcomings.

- Spot profiles of the (10), (11) and $(1\frac{1}{2})$ spots in $[1\bar{1}0]$ direction were recorded by a Video-LEED system described elsewhere [12]. Profiles of the $(1\frac{1}{2})$ spots were also taken in the perpendicular direction.($[110]$ direction)

- The intensity of the $(1\frac{1}{2})$ spot and the (0 0) spot was also measured by HREELS.

HREELS combines a high energy resolution ($45cm^{-1} - 50cm^{-1}$ in our experiment) with a good resolution in $\mathbf{k} -$ space ($0.04 - 0.08\text{Å}^{-1}$ FWHM in the (0 0) beam). The scattered electrons are detected by a channeltron so that even very small intensities are determined rather accurately. Therefore HREELS is best suited for measuring the intensity of true elastic electrons in a Bragg peak. The disadvantage of the method is that Bragg peaks can be measured only along one fixed direction in $\mathbf{k} -$space because our sample holder does not provide polar rotations. Once we chose this direction along the line connecting the (0 0) reflex and the $(1\frac{1}{2})$ reflex (dashed line in fig. 1a)

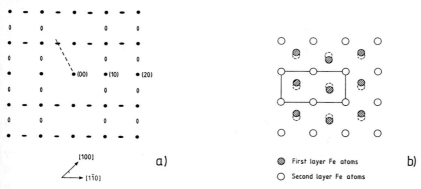

Fig.1   a) LEED pattern of 6 ML Fe/Cu(100) at 80 K. Filled symbols correspond to the domain orientation shown in b, open symbols to the domain rotated by 90°

b) Structural model for the reconstructed surface. Dashed circles mark the positions of the top-layer Fe atoms in the unreconstructed surface.

these were the only reflexes we could access by EELS. On the other hand Video-LEED permits arbitrary alignments of the "windows" on the screen image along which profiles are recorded, but energy resolution is poor, resulting in a high phonon background and a bad signal-to-noise ratio for weak Bragg peaks.

For the Video-LEED measurements the sample and the sample holder were cooled down by liquid N_2. When the final temperature of about 80 K was reached the sample was adjusted in front of the LEED screen. After that the sample was heated by electron bombardment without switching off the cooling so that the sample mounting remained at a low temperature. When the sample temperature had reached $280 - 300K$ the heating was interrupted and the Video-LEED was started. While the sample cooled down spot profiles along the specified directions and the sample temperature were stored every 8 s. The whole measurement took about 6 min. After that the sample was again heated and the whole measuring cycle was repeated several times. Good agreement between the data of different cycles verified the reversibility of the phase transition.

In the HREELS measurements the much lower rate of data acquisition (a single energy loss spectrum takes several minutes) demanded slower temperature changes so that a single temperature series took about $1\frac{1}{2}$ hours. Nevertheless, the measurements compare with the Video-LEED results very well. So, the observed transition is obviously not governed by dynamical processes such as adsorption and desorption. Temperature series were recorded by HREELS with increasing temperatures as well as with decreasing temperatures. Good agreement of the data in both cases again confirms that the phase transition is reversible.

3. Results

In fig. 2 the intensity of the (0 0) reflex and the $(1\frac{1}{2})$ reflex measured by HREELS is plotted versus the sample temperature. In both cases the intensity decays exponentially with increasing temperature (note the logarithmic intensity scale), but there is a residual intensity of the $(1\frac{1}{2})$ reflex even at 280 K. This intensity, however, is less than the intensity at 80 K by about a factor of ten and therefore the $(1\frac{1}{2})$ reflex is not visible on the LEED screen at room temperature.

So, as a first result, the room temperature structure is not really a $p(1 \times 1)$ structure, but it preserves -at least locally- the $(2 \times 1) - p2mg$ symmetry though with very faint superstructure spots. The second point is that the intensity of the (0 0) reflex does not fall off so extremely rapidly as that of the superstructure spot, but it goes down in a rather modest way (as do the other integral order spots). The exponential decrease of the (0 0) reflex can be described by a Debye-Waller factor e^{-2W} with

$$2W = 1.07 \times 10^{-3} \, T(K)$$

which is quite reasonable for metal surfaces. For the $(1\frac{1}{2})$ reflex, however, we obtain

$$2W = 1.07 \times 10^{-2} \, T(K)$$

Thus, in spite of the apparent similarity in the temperature behavior of the two re-

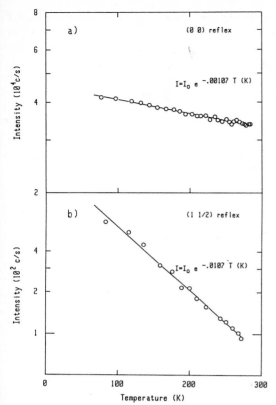

Fig.2   Peak intensity versus temperature
a) for the (0 0) reflex ($E_0 = 150$eV, $\vartheta_i = 65.4°$, $\vartheta_f = 65.4°$)
b) for the $(1\,\frac{1}{2})$ reflex ($E_0 = 150$eV, $\vartheta_i = 65.4°$, $\vartheta_f = 28.2°$)
Data were recorded by HREELS.

flexes - both display an exponential decrease typical of a Debye-Waller factor - there is an order of magnitude between the rates of decrease which wants explanation.

The same large Debye-Waller factor of the superstructure spots is found in the Video-LEED profile series. Concomitant with the reduction of peak intensity the spot profile broadens in k —space. Fig. 3 shows the full width of half maximum of the $(1\,\frac{1}{2})$ reflex in $[1\bar{1}0]$ direction - which is along the line connecting the (10) reflex and the (11) reflex - and in [110] direction - which is perpendicular to that line - as a function of temperature. In both directions there is a substantial broadening of the spot by a factor of 1.7 between 80 K and 200 K. At the same time the FWHM in [110] direction is always smaller than that in $[1\bar{1}0]$ direction. That means that the long range order along the zig-zag rows of displaced atoms in the reconstructed surface (cf. fig. 1b) is worse than that perpendicular to the zig-zag rows and that the long range order in both directions is reduced with rising temperature. An analogous measurement was also performed by HREELS. The intensity of elastic electrons was recorded while the angle of incidence ϑ_i was scanned, corresponding to a scan along the line marked in fig.1a. In this experiment, too, a broadening of the superstructure reflex was found though less pronounced than in the LEED spot profiles.

4. Discussion

From our data the root mean square displacement of surface atoms in z-direction at $T = 300$ K is calculated to be $\sqrt{< u_z^2 >_T} = 0.12$Å. This value is 6.4 % of the interlayer distance of the Cu(100) surface which is quite reasonable.

Apparently the $(1\frac{1}{2})$ reflex also displays a Debye-Waller like behavior, but applying the Debye-Waller picture one encounters some serious difficulties. First of all, the exponent $2W$ is about an order of magnitude larger than for integral order reflexes. The difference is hard to explain, since the dependence of the Debye-Waller factor on the wave vector transfer $\mathbf{K}$ is a rather structureless function. A possible solution of this problem would be the existence of a soft phonon in the layer where the reconstruction occurs. A more serious obstacle is the broadening of the reflex with increasing temperature which does not match the Debye-Waller picture. It rather indicates that the size of ordered regions in the surface decreases, the width of the reflex being a direct measure of this size. From fig. 3 the correlation length is estimated to be about 11 Å in [110] direction and about 3.7 Å in [1$\bar{1}$0] direction at 200 K, taking the width of the integral order reflexes as the instrument resolution. In the HREELS data the broadening is less pronounced so that a correlation length of about 18 Å is obtained in the direction of the angle scan (dashed line in fig.1), that is a direction which encloses an

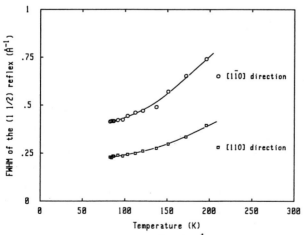

Fig.3   k −space width of the $(1\frac{1}{2})$ reflex versus temperature. Data were recorded by Video-LEED. For $T \gtrsim 200$ K the $(1\frac{1}{2})$ peak is not clearly resolved from the background and the peak width can not be determined.

angle of 26.6° with the [110] direction. Considering the better energy resolution of HREELS compared to LEED we conclude that the broadening in the Video-LEED profiles is at least partly due to inelastic scattering from phonons or other elementary excitations such as spin waves [13] with small momentum.

The small correlation lengths found in both experiments imply that the p2mg symmetry is conserved at 200 K only over a short range of 2 − 6 unit cells. At the same time the integral order reflexes do not broaden with increasing temperature and the low diffuse elastic intensity in phonon spectra confirms that long range order with the symmetry of the substrate (p4mm) is conserved even at room temperature.

In summary, we have shown that the local symmetry of 6 ML films on the Cu(100) surface is p2mg, at room temperature as well as at 80 K, but the size of ordered regions with this symmetry is reduced to few unit cells at room temperature.

The authors gratefully acknowledge enlightening discussions with D.L. Mills, T. Rahman, and L. Yang.

References

1. S.A. Chambers, T.J. Wagener, J.H. Weaver, Phys. Rev. B **36**, 8992 (1987)

2. S.D. Bader, E.R. Moog, J. Appl. Phys. **61**, 3729 (1987)

3. D. Pescia, M. Stampanoni, G.L. Bona, A. Vaterlaus, R.F. Willis and F. Meier, Phys. Rev. Lett. **58**, 2126 (1987)

4. M. Onellion, M.A. Thompson, J.L. Erskine, C.B. Duke and A. Paton, Surf. Sci. **179**, 219 (1987)

5. A. Clarke, P.J. Rous, M. Arnott, G. Jennings and R.F. Willis, Surf. Sci. **192**, L843 (1987)

6. D.A. Steigerwald, I. Jacob and W.F. Egelhoff, Jr., Surf. Sci. **202**, 472 (1988)

7. R. Germar, W. Dürr, J.W. Krewer, D. Pescia and W. Gudat, Appl. Phys. A **47**, 393 (1988)

8. S.H. Lu, J. Quinn, D. Tian, F. Jona and P.M. Marcus, Surf. Sci. **209**, 364 (1989)

9. Y. Darici, J. Marcano, H. Min, and P.A. Montano, Surf. Sci. **217**, 521 (1989)

10. W. Daum, C. Stuhlmann and H. Ibach, Phys. Rev. Lett. **60**, 2741 (1988).

11. C. Stuhlmann, U. Beckers, M. Wuttig, H. Ibach and G. Schmidt, in *Proceedings of IVC-11/ICSS-7*, Vacuum, in press

12. P.Heilmann, E. Lang, K. Heinz and K. Müller, in *Determination of Surface Structure by LEED*, edited by P.M. Marcus and F. Jona, (Plenum, New York, 1984), p.463

13. D.L. Mills, private communication

The Growth of Au on Ag(110): From Bilayer to Multilayer

P. Fenter and T. Gustafsson

Department of Physics and Astronomy and Laboratory for Surface Modification,
P.O. Box 849, Rutgers, The State University of New Jersey,
Piscataway, NJ 08855, USA

Abstract. We have studied the growth of Au on Ag(110) using medium energy ion scattering with channeling and blocking. At low coverages, Au grows in bilayers, while at intermediate coverages a (1x3) phase is observed. Finally, a (1x2) phase results, just as for the semi-infinite Au(110) surface. Not until the (1x2) phase is observed is the Ag surface completely covered.

1. Introduction

The Au on Ag system is an interesting model system in the study of epitaxial growth. The difference in lattice parameter between the two metals is unusually small (4.08 vs 4.09 Å), and the surface free energy of Au (γ_{Au} =1.6 J/m^2) is only slightly larger than that of Ag (γ_{Ag} =1.3 J/m^2) [1]. Therefore the Au/Ag system is expected to follow an ideal layer by layer (i.e., a Frank-van der Merwe (F-M)) growth model [2,3]. In this paper, we will describe results of an experimental study using medium energy ion scattering on the growth of Au on Ag(110). These data show that the growth of ultra-thin Au films on this surface exhibits a variety of interesting phenomena as a function of thickness. Initially, the growth is very unusual in that it occurs via island formation in *bilayers* [4]. At higher thicknesses, the symmetry of the Au films changes from (1x1) to (1x3) and finally to (1x2), with concurrent changes in growth properties. The (1x2) reflects the symmetry of the well known missing row reconstruction of the Au(110) surface [5]. We discuss the changes of symmetry in reference to the growth behavior.

2. Experimental

The strength of ion scattering as a structural probe is that it is quantitative; that is, the scattering cross section is well known. In addition, since Au is much heavier than Ag, ions that scatter off Au atoms will lose less energy in the collision, resulting in two distinct surface peaks (see fig. 1). This allows us to independently probe the structure of the substrate and the overlayer as a function of overlayer coverage (Θ). An ion beam incident in a channeling direction (in our case, normal to the surface) will be scattered almost exclusively by the top layer atoms due to *shadowing* [6], while a beam which is incident along a random direction will scatter off all

Springer Series in Surface Sciences, Vol. 24 **The Structure of Surfaces III**
Editors: S.Y. Tong · M.A. Van Hove · K. Takayanagi · X.D. Xie
© Springer-Verlag Berlin, Heidelberg 1991

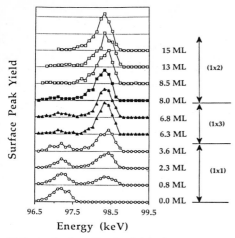

FIG. 1 Surface Peak yields versus Au coverage for a 100 keV proton beam incident along [11 0] and detected in the ($\bar{1}$11) scattering plane at $\vartheta_S = 120°$. The Au and Ag surface peaks are located at 98.4 and 97.2 keV respectively. The symbols indicate the observed symmetry, and are (1x1), (1x3) and (1x2) for open circles, triangles, and open squares respectively. The filled squares at 8.0 ML indicate a mixed phase.

atoms equally. Ions that are scattered by subsurface layers may exit the crystal freely, except in those directions where they are scattered again by other atoms in the surface region (*blocking*) resulting in lower yields. The position and depth of the 'blocking dips' provide direct structural and morphological information. Lastly, since the scattering cross section is known, a direct comparison may be made between the data and a computer simulation of the experiment (for a given structure), and Au coverages can be determined directly from the Au yield.

3.1 The Overall Growth Mode

We first demonstrate that the Au atoms occupy positions close to Ag lattice sites, i.e., *that the growth is epitaxial.* Figure 1 shows the dependence of the surface peak yields on Au coverage. The clean Ag surface exhibits a clearly defined surface peak at an energy of 97.2 keV. As Au is deposited, the size of this peak (i. e., the number of visible Ag atoms) decreases. This is due to Au atoms shadowing Ag, and shows that the growth is epitaxial. Because of the openness of the surface, 2 ML of Ag are completely visible to the ion beam for a clean Ag(110) surface. If the Au were growing in a layer by layer growth mode, the Ag yield should be nearly completely extinguished for Au coverages above 2 ML. Since a large fraction of the Ag yield is visible even at $\Theta_{Au} = 3.6$ ML, a layer by layer growth mode can be ruled out. Therefore the Au is growing epitaxially, but in a Volmer-Weber growth mode [4].

3.2 Low Coverage (Bilayer) Growth

In order to determine the structure of the Au layers below 1 ML we show both channeling and random incidence data (Fig. 2), taken at a Au coverage of 0.22 ML. Both sets of data are characterized by a deep blocking dip. This dip is due to ions which scatter off Au atoms and then are blocked by other Au atoms before exiting the crystal. The existence of this dip directly implies that the Au structure consists of at *least* two atomic layers. We can rule out Au atoms in third or higher layers by letting the ion beam be incident in a random direction (dashed line in fig. 2d). If there were Au atoms in higher layers, the yield should now increase, but there is no significant difference in the Au yield between the two sets of data. Therefore the Au structure consists of *only* two layers. The yield at the blocking dip minimum is very close to *one half* of the yield in the shoulders, which implies that one half of the Au atoms occupy *second* layer sites.

Aside from trivial scaling factors, the same results as in Fig. 2 are obtained from the very smallest coverages studied (0.05 ML) up to near 1 ML (where the surface is only half covered). We conclude that in this coverage range, Au grows epitaxially on Ag(110) in a *bilayer* form [4]. This is a very unusual growth structure and is not predicted by simple thermodynamic considerations.

Above 1 ML, the Au begins to grow in a three dimensional structure. To understand the morphology of the Au overlayer in detail at these coverages, we again show both channeling and random incidence data (fig. 3b-e). Although the total Au

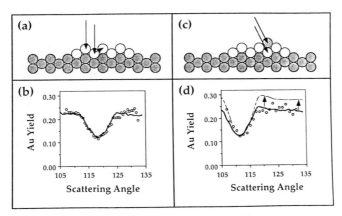

FIG. 2 (a) Side view of the crystal in the $(\bar{1}11)$ plane. Open (shaded) circles indicate Au (Ag) atoms. Arrows indicate the incident $[1\bar{1}0]$ ion direction. Note that one half of the ions are blocked upon exiting. (b) Au yield (ML) as a function of scattering angle for the geometry in (a). Circles are data points and the solid line is a Monte Carlo simulation for bilayer growth, assuming bulk like vibrational amplitudes for the Au atoms. (c) As in (a) but with the crystal rotated 7°. The hatched circle denotes atoms present in a possible 3-d growth. (d) As in (b) but for the configuration shown in (c). The dashed line indicates what is expected for 3-d growth.

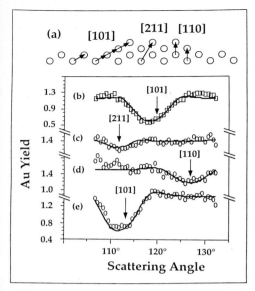

FIG. 3 (a) A schematic picture of the morphology of 1.4 ML Au (substrate not shown). The characteristic blocking directions ([211], [110], and [101]) are indicated. (b)-(e) Planar channeling data of Au yield (ML) taken in the ($\bar{1}$11) plane at incident ion beam directions of ϑ_i = (b) 0°, (c) 38°, (d) 53°, and (e) 7° (with respect to the surface normal). The blocking directions are indicated. The circles are data and the solid lines are simulations for perfect bilayer growth. The ion energy was 100 keV.

coverage is 1.41 ML, only 1.2 ML are visible to the ion beam in the channeling configuration (fig. 3b). Therefore the growth is three dimensional, and roughly 0.2 ML occupy third and higher layers.

The morphology can be directly measured by the random incidence data in figure 3c-e. For example, since all Au atoms are visible to the ion beam, the *depth* of the [211] blocking dip gives a direct measure of the fourth layer coverage (see fig. 3a) of ~0.15 ML. To quantify this interpretation, one performs a numerical analysis of the data using Monte Carlo simulations which take into account the details of the scattering events in the different blocking geometries. The solid lines (fig. 3b-e) are the result of such an analysis using a bilayer structure with 1st to 4th layer coverages of Θ_i of $\Theta_{1,2}$=0.56, and $\Theta_{3,4}$=0.145 and a total coverage of 1.41 ML. The agreement with the data is clearly satisfactory. The data then give a direct and consistent picture of the morphology of the Au layer that clearly leads to the conclusion of *bilayer growth*, not just in the first two layers, but *also subsequently* [4].

3.3 Multilayer Growth

In figure 1, we show the observed surface symmetry as a function of coverage. The Au overlayer initially grows with a (1x1) symmetry, and at high coverages exhibits

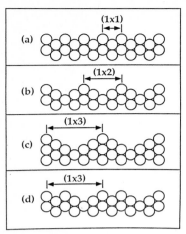

FIG. 4 A schematic picture of the structure of the Au film at different coverages. (a) The (1x1) structure. (b) The (1x2) missing row structure. (c,d) Two likely structures of the (1x3) phase. (c) is the random mixture of (1x1) and (1x2) phases, and (d) is a generalized missing row reconstruction.

the expected (1x2) symmetry of the reconstructed Au(110) surface [5] (see fig. 4). However, at moderate coverages the Au films show a (1x3) structure. It is well known that an *ordered* (1x1) → *ordered* (1x2) transition requires the introduction or removal of 0.5 ML [7]. Ours is a *growth* experiment and we are introducing atoms to the surface, but the (1x3) is still observed over a wide coverage range. This implies that the (1x3) phase is not due to to a simple kinetic limitation. Two likely structures for this phase are shown in fig 4. The first (fig. 4c) is a mixed (1x1) + (1x2) phase (presumably with intrinsic disorder), and the other is a higher order missing row reconstruction (fig 4d) which can be induced on the initially (1x2) reconstructed Au and Pt(110) surfaces [8]. Based upon the diffuseness of the (1x3) spots, and a preliminary analysis of the ion scattering data, we tentatively suggest that the (1x3) is due to the mixed phase (fig 4c).

A second aspect of the multilayer growth is that we do not observe the Au to wet the substrate until the (1x2) phase appears. Based upon energetic considerations, a film will wet the substrate when $\gamma_{Au} < \gamma_{Ag} - \gamma_{int}$ [2]. That the Au does not wet the surface until the (1x2) phase appears implies that this condition is met for the (1x2) phase, but not the (1x3) phase; i.e., γ_{Au} (1x2) < γ_{Au} (1x3). That is, the wetting of the surface has been achieved through the lowering of the surface energy by the introduction of the Au missing row surface reconstruction.

4. Conclusions

We have found that although Au grows epitaxially on Ag(110), the Au exhibits novel growth phenomena. These include bilayer growth at low coverages, and surface symmetry mediated wetting at high coverages. In addition, by studying the (1x1) → (1x2) transition as a function of Au coverage, we have found new insight into the mechanism of this phase transition.

Acknowledgement: This research was supported by NSF Grant DMR-8703897.

References.

1. L. Z. Mezey and J. Giber, Jap. J. Apl. Phys. **21**, 1569 (1982).
2. E. Bauer, Appl. Surf. Sci. **11**, 479 (1982).
3. R. J. Culbertson, L. C. Feldman, P. J. Silverman, and H. Boehm, Phys. Rev. Lett. **47**, 657 (1981).
4. P. Fenter and T. Gustafsson, Phys. Rev. Lett. **64**, 1142 (1990).
5. See e.g. M. Copel and T. Gustafsson, Phys. Rev. Lett. **57**, 723 (1986).
6. J. F. van der Veen, Surf. Sci. Rep. **5**, 199 (1985).
7. J. C. Campuzano, A. M. Lahee, and G. Jennings, Surf. Sci. **152/153**, 68 (1985).
8. P. Häberle, P. Fenter, and T. Gustafsson, Phys. Rev. B **39**, 5810 (1988).

Structure of Au and Pt Films on Pd(110): Deposition Temperature and Coverage Dependent (1×2) and (1×3) Reconstructions

P.J. Schmitz[1], H.C. Kang[1], W.-Y. Leung[2], and P.A. Thiel[1]

[1]Department of Chemistry and Ames Laboratory, Iowa State University,
Ames, IA 50011, USA
[2]Department of Physics, Florida International University,
Miami, FL 33199, USA

We present the results of a LEED study of the structures of Au and Pt films grown on Pd(110). We observe both (1x2) and (1x3) overlayer structures depending upon the coverage and deposition temperature. We explain the coverage and deposition temperature dependence of these reconstructions in terms of basic factors which control film growth processes.

1. Introduction

The growth and structure of ultra-thin metal films is an ongoing topic of research in our laboratory. We have been particularly interested in the growth of metal films whose bulk surfaces are known to reconstruct, and how the factors which control film growth can affect the bulk-like reconstructions of these films. The (100) and (110) low index faces of bulk Au, Pt and Ir are known to reconstruct [1]. Many studies of the growth of Au and Pt overlayers have appeared in the literature [e.g.,2-7]. However, all workers to date except Fenter and Gustafsson [8] have studied the growth of these metals on the more atomically-smooth fcc (111) and (100) crystal faces. We therefore have recently concentrated our efforts on the study of metal films on an fcc (110) substrate, Pd(110).

2. Experimental Results

We have measured the LEED intensity profiles for Au films as a function of coverage and annealing temperature. All data presented follow deposition at 130 K. For a coverage of 1 Au monolayer a (1x1) diffraction pattern is observed, and no fractional order spots or streaking are detected. For Au coverages greater than 1 monolayer, we observe the appearance of additional fractional order beams. These data are displayed in Fig. 1. At a coverage of 1.5 to 2.0 monolayers, Fig. 1(a) and 1(b) respectively, the Au films reconstruct irreversibly to a (1x2) at temperatures slightly above 300 K. For the 1.5 monolayer film of Fig. 1(a), the best (1x2) is observed at 530 K. At temperatures over 530 K, the half-order spots are lost and a (1x1) pattern is recovered. For the 2 monolayer film of Fig. 1(b) the half-order spots are broader for lower temperatures, but sharpen and give a good (1x2) pattern at 640 K. As the temperature is increased past 640 K the intensity of the half-order spots decrease and are finally lost after annealing to 770 K. When the Au coverage is increased further to 3 and 4 monolayers, Fig. 1(c) and 1(d) respectively, the half-order spots split continuously with increasing coverage and eventually a full (1x3) structure develops above 4 monolayers. For Au coverages

Springer Series in Surface Sciences, Vol. 24   **The Structure of Surfaces III**
Editors: S.Y. Tong · M.A. Van Hove · K. Takayanagi · X.D. Xie
© Springer-Verlag Berlin, Heidelberg 1991

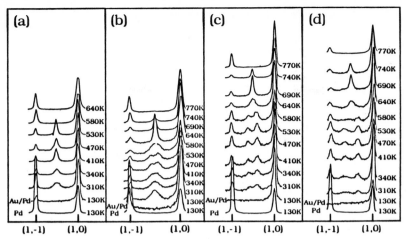

Figure 1. LEED intensity profiles for Au films on Pd(110) as a function of annealing temperature following deposition at 130 K. Profiles are taken along the [001] direction between the (1,0) and (1,-1) beams at a beam energy of 72 eV. Coverages shown are: a) 1.5 monolayers; b) 2 monolayers; c) 3 monolayers; and d) 4 monolayers.

in excess of 4 layers a (1x3) pattern is always observed after annealing, and follows the same evolution of superstructures with temperature as for the 3 and 4 monolayer cases. However, the quality of the diffraction pattern for these higher coverages degrades substantially. AES results indicate that the Au coverage begins to decrease at 530 K for all the films studied. This observation suggests dissolution of the film and thus explains the irreversibility of the (1x1) to (1x2) transition.

We have measured the LEED intensity profiles for a series of Pt films as a function of annealing temperature following deposition at 300 K. For a coverage of 1 Pt layer a (1x1) diffraction pattern is observed, and no fractional order spots or streaking are detected. As the Pt film thickness is increased to 2 monolayers heavy streaking develops. For higher Pt coverages we observe the appearance of additional fractional order beams that grow in intensity as the film is slowly annealed. These data are displayed in Fig. 2. At a coverage of 2 Pt layers, Fig. 2(a), some intensity centered around the half-order position is evident after annealing to 370-400 K, although it is very broad and is essentially featureless. Following deposition of 3 Pt layers, Fig. 2(b), the same streaking appears after annealing to 370-400 K as observed after annealing 2 monolayers of Pt to the same temperature. Increasing the temperature further results in a splitting of the half-order spot. The two components diverge as the temperature is raised and almost reach third order positions at 630 K. Annealing to temperatures higher than 630 K results in complete loss of the fractional-order components. The continuous splitting of the half-order spot as the film is annealed to higher temperatures is more apparent for thicker Pt films as evident in Figs. 2(c) and 2(d). For films of this coverage the splitting and the loss of the fractional-order spots occurs at slightly higher temperatures than those found for the 3 monolayer case. AES indicates that for all films, the Pt Auger signal begins to decrease at the same temperature where the half-order spots begin to split. This behavior suggests that dissolution and/or

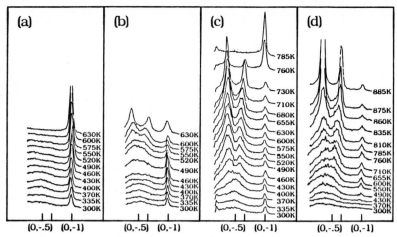

Figure 2. LEED intensity profiles for Pt films deposited at 300 K as a function of annealing temperature. Profiles are taken along the [001] direction between the (0,0) and (0,-1) beams at a beam energy of 35 eV. Coverages shown are: a) 2 monolayers; b) 3 monolayers; c) 5 monolayers; and d) 15 monolayers.

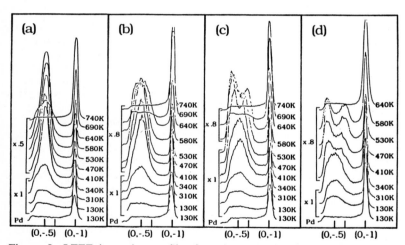

Figure 3. LEED intensity profiles for 3-layer Pt films deposited between 130 and 300 K as function of annealing temperature. Profiles are taken along the [001] direction between the (0,0) and (0,-1) beams at a beam energy of 35 eV. Deposition temperatures are: a) 130 K; b) 200 K; c) 225 K; and d) 300 K.

agglomeration of the Pt films is in some way associated with the spot splitting.

For the Pt films, we also observe a dependence of the evolution of the LEED superstructures with the deposition temperature as shown in Fig. 3. We find that for deposition temperatures below 200 K we can stabilize the (1x2) structure. A sharp and intense (1x2) develops upon annealing to ca. 410 K, only if the film is 3 monolayers deep and is deposited between 130 and 200 K, Figs. 3(a) and (b). If the same amount of Pt is deposited at higher substrate temperatures, 225 K, the half-order spot shows signs of splitting when annealed, Fig. 3(c). For deposition of 3 layers of Pt at 300 K, Fig. 3(d), annealing brings on the (1x3) structure. At all higher Pt coverages annealing brings on the (1x3) structure independent of the deposition conditions.

3. Discussion

We believe that the (1x2) and (1x3) structures for both the Au and Pt films on Pd(110) can be categorized as reconstructions of the films and are not due to formation of an ordered alloy. Supportive evidence for this is presented in more detail elsewhere [9,10]. The formation of the (1x3) by the continuous splitting of the half-order spot is analogous to the progression of LEED patterns observed for O/Ni(110) [11], which, by analogy with O/Cu(110), probably represents a reconstructive (added row) transformation [12]. Similar spot splitting for lattice gas systems has been explained by a statistically random distribution of (1x2) and (1x3) phases each having dimensions smaller than the coherence width of the LEED optics [13,14]. We adopt a similar explanation for our results.

Although the (1x2) structure has been reported most frequently for the bulk surfaces of Au and Pt (110), (1x3) structures have also been observed. However, STM results on a Au(110) surface show that the (1x3) occurs in regions of strong disorder [15], and for studies of bulk Pt(110) the (1x3) can only be stabilized by high temperature oxygen treatments [16,17]. These studies suggest that the (1x3) is not the most stable phase, and develops due to some type of disorder in the (1x2) structure. We therefore explain the development of the (1x3) as a break down in the two-dimensional order of the films as the coverage and deposition temperature changes, resulting from factors which affect film growth processes [9,10].

We have shown that Au on Pd(110) follows a Stranski-Krastanov growth mode, with a critical film coverage of 2 layers [9]. As the Au coverage exceeds 2 layers, the two-dimensional order of the films begin to break down due to the strain induced by the 4.8% lattice mismatch. This two-dimensional disorder induced by the defective nature of the film favors the formation of the (1x3) structure. For 3 monolayers of Au both the (1x2) and areas of (1x3) coexist resulting in the half-order spot splitting. By 4 monolayers of Au the (1x3) phase is the dominant species and results in the (1x3) pattern.

A few plausible explanations exist for the transition to the (1x3) in the Pt films [10]. Here we discuss a model similar to that used to explain the transition to the (1x3) in the Au films. For the Pt films, a disruption in the two-dimensional order may not only be induced by lattice strain but also by thermodynamic factors. The surface free energy for Pt is greater than that for Pd, 2.69 J\m² and 2.04 J/m² respectively [18]. These conditions may favor a break-down in the two-dimensionality of the film at higher temperatures [19]. However, dissolution of the film cannot be ruled out as a contributing factor. In any case, we propose that as the temperature is raised the two-dimensional quality of the film begins to deteriorate, due to agglomeration and/or dissolution, which favors the formation of the (1x3) structure. As the temperature increases the (1x3) grows at the expense of the (1x2) and at higher temperatures the (1x3) phase becomes the dominant species.

The deposition temperature dependence for the stabilization of the (1x2) may be explained by a model which consists of a combination of kinetic and thermodynamic effects which control film growth at the Pt-Pd interface [10]. As mentioned earlier, thermodynamic factors may force a break up in the two-dimensional order of the film as the temperature increases and equilibrium is approached [19]. Low temperature deposition may act to kinetically limit this transition due to a reduced adatom mobility. It is therefore possible that for deposition at 130 K the film is kinetically trapped into a smoother, more continuous film. Due to the better layer-by-layer quality following low temperature deposition the long range order of the (1x2) which forms is higher, and therefore the film produced under these conditions is more stable. Deposition at higher temperatures allows diffusion of the adatoms and results in some microscopic disruption in the two-dimensional quality of the film. This disruption results from roughness induced by the film's thermodynamic drive toward three-dimensional growth. The films deposited at higher temperatures are less stable due to the reduced long range order of the (1x2), and as temperature is increased disorder increases and the (1x3) is formed.

4. Conclusions

We observe both (1x2) and (1x3) reconstructions for the Au and Pt films on Pd(110). The (1x3) structure in both systems is found to evolve by the continuous splitting of the half-order spot. We propose for both of these systems that the transition of the (1x2) to the (1x3) results from a disruption of the two-dimensional order of the films which favors the formation of the (1x3) structure. We believe that the main differences in the (1x2) to (1x3) transitions for these two systems lies in the thermodynamic differences for each respective metal pair, and in the different strain energies involved at the interface resulting from the dissimilar lattice mismatches.

Acknowledgments

We thank R.J. Baird and G.W. Graham of Ford Motor Company for the use of their Pd(110) crystal. This work is supported by a grant from the Ford Motor Company, by a Presidential Young Investigator Award from the National Science Foundation, Grant No. CHE-8451317, and by a Camille and Henry Dreyfus Foundation Teacher-Scholarship (P.A.T.). In addition, some equipment and all facilities are provided by the Ames Laboratory, which is operated for the U.S. Department of Energy by Iowa State University under Contract No. W-7405-ENG-82.

References

1. P.J. Estrup, Chemistry and Physics of Solid Surfaces V, Eds. R. Vanselow and R. Howe (Springer, Berlin, 1984) p. 205.
2. S.L. Beauvais, R.J. Behm, S.-L. Chang, T.S. King, C.G. Olson, P.R. Rape and P.A. Thiel, Surf. Sci. 189/190 (1987) 1069.
3. J.W.A. Sachtler, M.A. Van Hove, J.P. Biberian and G.A. Somarjai, Phys. Lett. 45 (1980) 1601.
4. P.W. Davies, M.A. Quinlan and G.A. Somorjai, Surf. Sci. 121 (1982) 290.

5. D.L. Weissman-Wenocur, P.M. Stefan, B.B. Pate, M.L. Shek, I. Lindan and W.E. Spicer, Phys. Rev. B27 (1983) 3306.
6. Y. Kuk, L.C. Feldman and P.J. Silverman, Phys. Rev. Lett. 50 (1983) 511.
7. R.C. Jaklevic, Phys. Rev. B30 (1984) 5494.
8. P. Fenter and T. Gustafsson, Phys. Rev. Lett. 64 (1990) 1142.
9. P.J. Schmitz, H.C. Kang, W.-Y. Leung and P.A. Thiel, to be published.
10. P.J. Schmitz, H.C. Kang, W.-Y. Leung and P.A. Thiel, to be published.
11. L. H. Germer, J. W. May and R. J. Szostak, Surface Science 7 (1967) 430.
12. D. J. Coulman, J. Wintterlin, R. J. Behm and G. Ertl, Phys. Rev. Lett. 64 (1990) 1761; also J. Wintterlin, R. Schuster, D. J. Coulman, G. Ertl and R. J. Behm, J. Vac. Sci. Technol. (submitted 1990).
13. G. Ertl and J. Kuppers, Surf. Sci. 21 (1970) 61.
14. J.E. Houston and R.L. Park, Surf. Sci. 21 (1970) 209.
15. G. Binnig, H. Rohrer, Ch. Gerber and E. Weibel, Surf. Sci. 131 (1983) L379.
16. D. Fery, W. Moritz and D. Wolf, Phys. Rev. B38 (1988) 7275.
17. M. Salmeron and G.A. Somorjai, Surf. Sci. 91 (1980) 373.
18. L.Z. Mezey and J. Giber, Jpn. J. Appl. Phys. 21 (1982) 1569.
19. E. Bauer, Z. Kristallogr. 110 (1958) 372.

Faceting Induced by Ultrathin Metal Films: Pt and Au on W(111)

K.-J. Song, C.Z. Dong, and T.E. Madey

Department of Physics and Astronomy and
Laboratory for Surface Modification,
P.O. Box 849, Rutgers, The State University of New Jersey,
Piscataway, NJ 08855, USA

We have studied the effects of annealing temperature on the structure of two model metal-on-metal ultrathin film systems: Pt/W(111) and Au/W(111). Upon annealing to T>800K, the W(111) substrate undergoes a massive reconstruction to expose {112} type facets, with linear dimensions as large as 1000Å.

1. Introduction

The morphology of ultrathin metal films grown on atomically smooth, close packed metal substrates has long been of interest [1], and a few common growth modes have been identified [1,2]. In most of these cases the structural rearrangements occur mainly in the overlayer film, while the substrate undergoes little or no change. In contrast, the interaction of Pt and Au ultrathin films with a bcc W(111) surface, which is atomically rough and open, causes the W(111) substrate itself to reconstruct and form facets [3,4]. The experimental details and some of the results of Pt on W(111) that have been described elsewhere [3,4] will not be repeated here. Briefly, for *in situ* studies we use Auger electron spectroscopy (AES) to monitor surface chemical composition, low energy electron diffraction (LEED) to monitor structural transformations, and thermal desorption spectroscopy (TDS) together with AES to estimate film thickness. The morphologies of the faceted surfaces are measured using a scanning tunneling microscope (STM) which operates in air. In one case, high resolution scanning electron microscopy (SEM) has been used to compare with the STM results.

2. Results

For the Au/W(111) and Pt/W(111) cases, there are differences in detail but several common observations: (1) for coverages less than 1 monolayer, planar thin films are stable and remain mainly on top of W(111) at all temperatures below which thermal desorption is significant; (2) LEED spots due to {112} oriented pyramidal facets have been observed only for Pt or Au films above a threshold coverage of slightly more than 1 monolayer and only after annealing above 800K; (3) the faceted surface always reverts to its planar form when the Au or Pt is desorbed.

Springer Series in Surface Sciences, Vol. 24   **The Structure of Surfaces III**
Editors: S.Y. Tong · M.A. Van Hove · K. Takayanagi · X.D. Xie
© Springer-Verlag Berlin, Heidelberg 1991

2.1. Pt/W(111)

The LEED pattern of the clean W(111) surface at 101eV is shown in fig. 1a. In the experiments described below, about 2 monolayers of Pt are deposited onto clean W(111) at room temperature and the surface is subsequently annealed from 300K to 2100K in 100K intervals, for 3 minutes at each temperature. Deposition and annealing first cause reduction of the brightness of the (111) 1x1 LEED spots and an increase of the background. At 800K, the (111) 1x1 spots disappear completely. New diffuse spots are formed at T> 600K, and after annealing at 900K, can clearly be seen in Fig. 1b. As shown in Fig. 1c, these spots become as sharp as the original 1x1 spots of the clean surface after annealing at 1200K, and are identified as due to diffraction from {112} facets having a 1x1 structure [3]. The total disappearance of the (111) 1x1 spots below 1200K indicates that the whole surface is faceted while sharpening of the {112} spots indicates that the facets grow in size as the annealing temperature increases. As indicated by the markers in figures 1a and 1c, spots due to the planar (111) surface reappear after annealing at 1200K. As the annealing temperature increases further, these (111) spots gain intensity at the expense of spots due to {112} facets, although the latter become even sharper up to at least 1400K. This suggests that while some of the facets grow still larger, part of the faceted area converts to a planar form. Pt desorbs at T>1700K, and we obtain the clean surface LEED pattern after annealing at 2000K.

Images of samples annealed at different temperatures under ultrahigh vacuum conditions and subsequently characterized by STM in air are compatible with the LEED results [3,4]. At 880K, the surface consists completely of {112} oriented facets, with typical linear dimension of 15nm. At 1200K, the typical dimension increases to 30nm. At 1400K,

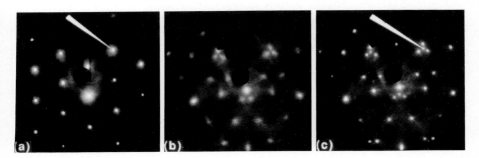

Fig. 1. LEED patterns at 101 eV for (a) the clean W(111) surface, (b), (c) W(111) surface covered with ~2 monolayers of Pt, after formation of {112} facets by annealing at (b) 900K and (c) 1200K. The marker in (c) indicates the reappearance of (111) 1x1 spots after annealing at 1200K.

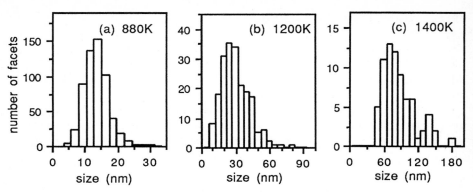

Fig. 2. Facet size distributions (number of pyramidal facets/unit area, expressed in arbitrary units, *vs.* linear size) for Pt/W(111) surfaces after being annealed at (a) 880K (b) 1200K and (c) 1400K.

very large pyramids with sizes > 100nm are seen on an otherwise planar surface. Images of the surface annealed to 1400K have also been obtained using a high resolution scanning electron microscope, which has the advantage of a large scanning area to provide better statistical information. The number densities of pyramidal facets as a function of linear size are plotted in Fig. 2 for different annealing temperatures.

2.2.   Au/W(111)

Fig. 3 shows two LEED patterns at 21 eV incident energy of a W(111) surface covered with about 1.5 monolayers of Au and annealed at different temperatures. The surface develops {112} facets that have 1×1 structures when annealed between 800K and 950K, as shown in Fig. 3a.   Annealing

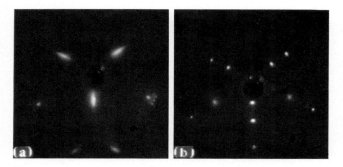

Fig. 3. Typical LEED patterns at 21 eV for W(111) surface covered with ~1.5 monolayers of Au, after being annealed (a) between 800K and 950K, with {112} facets formed and (b) between 950K and 1050K, with a 3×3 structure of the planar (111) surface.

above 950K causes the surface to assume a planar 3×3 structure, as shown in Fig. 3b. The phase transition at 950K between the faceted form and the 3×3 structure is completely reversible. Below 800K, this surface shows bistability and can assume either the 3×3 structure, or the faceted form, depending on the thermal history. Preliminary STM results indicate that the linear dimensions of the pyramidal facets formed are on the order of 5nm, compatible with the diffuse LEED spots observed. For multilayer Au films, thermal desorption of Au is detectable at 1050K. Using TDS, the threshold coverage of Au that induces faceting is found to be slightly larger than a monolayer [4].

3. Discussion

Faceting of surfaces is driven by the anisotropy in surface energy γ as a function of crystallographic orientation. In general, since the anisotropy in γ is small for clean metals, clean surfaces of pure metals are thermally stable and remain planar upon annealing, irrespective of their crystallographic orientations,(e.g., for clean W, the maximum anisotropy $\Delta\gamma/\gamma$ is ~ 3% [5]).

The presence of adsorbed layers on metals can lower the surface energy γ and can cause an increase in the anisotropy of γ. When facets are formed, the total energy $\int \gamma \, dA$ (where A is the surface area) is lowered by the formation of low energy facets even though the total surface area increases [6,7].

Whereas there have been many studies of faceting induced by adsorbed gases [7,8], faceting of metal surfaces induced by ultrathin metal films has not been widely reported [3,9]. Recently, Weinert et al. [10] performed total energy calculations for films of Pd and Ag on Nb(100) and Nb(110) that provide insights into the faceting process. They find that a single mono-layer of Pd or Ag lowers the surface energy γ of both surfaces, and also increases the anisotropy of γ to $\Delta\gamma/\gamma$ ~ 30%. We suggest that similar processes occur for Pt and Au on W(111), which induce the formation of stable {112} surfaces.

There are many outstanding questions that we are pursuing. A critical coverage of about one monolayer is necessary to induce faceting, but we do not yet know the fate of the excess Pt (Au): are surface alloys formed, or is the excess Pt (Au) incorporated into the pyramidal facets, or do clusters of Pt (Au) form elsewhere on the surface? What are the relative roles of kinetic and equilibrium processes in the formation of the faceted surfaces? What is the mechanism for continued growth of the pyramidal facets on an otherwise planar substrate with increasing temperature for Pt/W(111)? Detailed investigations are underway to identify the factors that influence the complicated morphological changes that occur during faceting.

4. Acknowledgements

The authors thank E. Garfunkel for valuable discussions and technical assistance with the STM. We also acknowledge valuable discussions with F. Cosandey, and thank W. MacGuire and J. Augustine from International Scientific Instruments, Inc. for valuable technical assistance with the high resolution SEM. This work is supported in part by the U.S. Department of Energy, Office of Basic Energy Sciences.

5. References

[1] E. Bauer, in "The Chemical Physics of Solid Surfaces and Heterogeneous Catalysis", V. 3, part B, Eds. D. A. King and D. P. Woodruff (Elsevier, Amsterdam, 1984) p. 1.

[2] J. P. Biberian and G. A. Somorjai, J. Vac. Sci. Tech. **16,** 2073 (1979).

[3] K.-J. Song, R. E. Demmin, C.-Z. Dong, E. Garfunkel and T. E. Madey, Surf. Sci. Lett. **227,** L79 (1990).

[4] T. E. Madey, K.-J. Song, C.-Z. Dong and R. E. Demmin, to be published in Surf. Sci. (1991).

[5] M. Drechsler and A. Mueller, J. Cryst. Growth **3/4,** 518 (1968).

[6] E. D. Williams and N. C. Bartelt, Ultramicroscopy **31,** 36 (1989).

[7] M. Flytzani-Stephanopoulos and L. D. Schmidt, Prog. in Surf. Sci. **9,** 83 (1979)

[8] J.C. Tracy and J. M. Blakely, Surf. Sci. **13,** 313 (1968).

[9] A. Cetronio and J. P. Jones, Surf.Sci. **40,** 227 (1973). S. Mroz and E. Bauer, Surf. Sci. **169,** 394 (1986).

[10] M. Weinert, R. E. Watson, J. W. Davenport, and G. W. Fernando, Phys. Rev. **B39,** 12585 (1989).

Core Level Photoemission for the Study of Metallic Interfaces and Two-Dimensional Compounds

J.N. Andersen[1], *A. Nilsson*[2], *O. Björneholm*[2], *and N. Mårtensson*[2]

[1]MAX-lab, University of Lund, Box 118, S-221 00 Lund, Sweden
[2]Uppsala University, Box 530, S-751 21 Uppsala, Sweden

Abstract. A model for the core level binding energy shifts in metallic systems is outlined with particular emphasis on the structural information contained in the shifts. A number of examples employing this model are presented. Finally a description of how the energetics of interface formation may be studied by core level photoemission is given.

1 Introduction

Consider an experiment where monochromatic light is incident on a metallic sample. The basic process which we consider is one in which a core electron in the sample absorbs a photon and is emitted after which it may be detected and energy analyzed. The term "core electron" means that the hole left behind is localized on one atomic site in the metal. The valence electrons of the metal will try to screen out the core hole which they experience as a positive charge. The core ionized atom can be viewed as an "impurity" atom with an increased nuclear charge. In a spectrum of the number of emitted electrons versus the electron energy a number of peaks appear which correspond to the core levels of the elements in the sample. The binding energy of a core level is defined simply as the difference in energy between the core level peak and the electrons emitted from states at the Fermi level. From this definition the binding energy is simply the difference in total energy of the sample before and after the core ionization i.e. it is the energy needed to transform one of the sample atoms into an "impurity". The core level binding energies of an element show small changes when the element is placed in different surroundings. Such shifts are often termed "chemical shifts". This name reflects that at the early stages of development of the method there was hope of deriving in a very direct way information about the chemical state of the atoms in the initial state material from these shifts. This hope, however, diminished early when it /1/ was realized that final state effects often play a significant role in determining the size and sometimes even the direction of the chemical shifts. This means that it is not a straightforward matter to obtain information about initial state properties, e.g. charge transfer in the initial state is not easily derived from core level binding energy shifts. However, it turns out that a wealth of information may be gained by realizing and fully utilizing the fact that the core level binding energy measures the change in total energy of the sample resulting from the core ionization of one of its atoms. Taking this total energy point of view

Springer Series in Surface Sciences, Vol. 24   **The Structure of Surfaces III**
Editors: S.Y. Tong · M.A. Van Hove · K. Takayanagi · X.D. Xie
© Springer-Verlag Berlin, Heidelberg 1991

the binding energy shifts can be related to fundamental thermodynamical quantities of the initial sample and of the sample with the core ionized "impurity" atom. Thus such thermodynamical quantities may be derived from the chemical shifts or conversely the core level binding energy shifts may be calculated on the basis of such termodynamical data and as discussed below used to obtain information on the surroundings of the ionized atom. Before continuing we will also stress that the detailed analysis of the binding energy shifts given below is not always necessary to exploit the powers of core level photoemission. The method is often useful simply as a fingerprinting technique.

2 A model for the chemical shift

Consider a metallic alloy made from the elements Z and M. The extension to alloys with more constituents is straightforward. The core level binding energy $E_B(Z)$ of one of the Z atoms in such an alloy is determined by the surroundings of that particular Z atom, i.e. by its coordination to other Z and M atoms. It will also depend on whether or not it is at the surface, that is, its total coordination. The reduction of coordination at the surface is formally treated as alloying with vacuum. It is expected that the binding energy is determined mainly by the first coordination shell of the Z atom and we may define effective concentration parameters C_Z, C_M and C_V which quantify the presence of the different kinds of neighbours of that atom. The binding energy $E_B(Z)$ may be written as

$$E_B(Z) = E_{B0} + C_Z \Delta E_Z + C_M \Delta E_M + C_V \Delta E_V ,$$

where E_{B0} is a reference energy and the last three terms describe the shifts caused by the coordination to other Z atoms, to M atoms and to vacuum, respectively. Since we are dealing with a metal the complete screening picture is valid /2,3/ and it may, by applying a Born-Haber cycle for the ionization process /3/, be demonstrated /4/ that the three partial shifts can be expressed as

$$\Delta E_Z = E_{SOL}(Z^* \text{ in } Z)$$
$$\Delta E_M = E_{SOL}(Z^* \text{ in } M) - E_{SOL}(Z \text{ in } M)$$
$$\Delta E_V = E_{COH}(Z^*) - E_{COH}(Z) ,$$

where E_{SOL} and E_{COH} represent the solution and cohesive energies, respectively, of the system. Z^* is the final state "impurity" atom created by the photoemission of a core electron. The properties of this Z^* impurity in a metallic system resemble to a good approximation that of a Z+1 impurity, a fact which is most helpful in estimation of core level shifts in metallic systems. From these formulae the partial shifts may be estimated from measured or from semiempirically calculated /5/ thermodynamical data. Another way is to obtain the partial shifts from binding energy shifts between situations of known geometry where it is possible to estimate the concentration parameters. Finally it should be noted that the complete screening assumption is expected to hold for any kind of adsorbate on metals, thus the above formalism is also applicable to such systems.

The exact relationship between the effective concentration parameters C_Z, C_M and C_V and the geometry might be quite complicated and we stress that we do not imply that the exact coordinations of the Z atom can be found from the core level binding energy by the formulae above. However, to a reasonable approximation the concentration parameters can be estimated from the structure in a straightforward way by applying, for example, a simple broken bond picture. In this way one may relate changes in the binding energy during e.g. interface formation to changes in the geometrical structure.

3 Applications of the shift model

As the first example we consider the Ni-Yb alloy system with emphasis on the core levels of the Yb. In this system there exists in addition to just shifts of the core level binding energies also the possibility of a valence change of the Yb. As a pure metal Yb is divalent ($[Xe]5p^6 4f^{14}(6s5d)^2$ configuration) but when forming bulk compounds with Ni it becomes trivalent ($[Xe]5p^6 4f^{13}(6s5d)^3$ configuration). For divalent Yb the partial shifts are /4/ $\Delta E_{Yb} = 0.7$ eV, $\Delta E_{Ni} = -1.5$ eV and $\Delta E_V = 2.8$ eV. The quite large negative value of the Ni related shift means that a large coordination to Ni results in a negative Yb 4f core level binding energy which as discussed in ref./4/ means that the Yb will become trivalent. It is seen that the vacuum term counteracts the influence of the Ni term, i.e. at the compound surface Yb may stay divalent.

The behaviour of Yb deposited on Ni(100) at 670 K /4,6/ illustrates these considerations. At the lowest Yb coverage all of the deposited Yb stays at the surface as divalent Yb with a quite low 4f binding energy of around 0.5 eV. The low 4f binding energy shows that the Yb is close to becoming trivalent, the vacuum term just barely overcomes the Ni term. At somewhat higher Yb coverage trivalent Yb starts to appear with a rather high 4f binding energy from which it is concluded that this trivalent Yb is also found at the surface (the surface shift is also positive for trivalent Yb /4/). At still higher Yb coverage alloying with the Ni sets in. This is indicated by the simultaneous appearance of trivalent Yb with a low 4f binding energy, and a new divalent component with a 4f binding energy of around 1 eV. This new divalent component is due to Yb adsorbed on top of an Yb-Ni compound layer giving a lower coordination to Ni than when adsorbed directly on the Ni surface. A most interesting result was the finding of two different binding energies for trivalent Yb, which implies two different sites, in a c(10x2) structure that is formed as the first ordered structure after alloying has begun. All of the above conclusions are consistent with those obtained by other methods /6/, however, some of them, e.g the existence of two trivalent sites in the c(10x2) structure would be hard to reach by other methods.

The understanding of the shifts in the Yb-Ni system was used /7/ to study 2-dimensional (2D) Yb-Ni layers formed by co-deposition of Yb and Ni on Mo(110).

In the submonolayer region the Yb stays divalent because of the vacuum term, whereas any 3D growth leads to trivalent Yb. Thus the 2D nature of the compound overlayers can easily be verified by core level photoemission. The Yb 4f spectra in the submonolayer region consist of 3 components. One of these is due to pure Yb, whereas the two others correspond to ordered 2D Yb-Ni compounds. From the deposited amounts the stoichiometry of the two compounds were determined to be Yb_2Ni and $YbNi_2$, respectively. The Yb 4f binding energies were consistent with these compositions and the proposed structural models of the compounds. The phase diagram of the co-adsorption system could be mapped by following the development of the three 4f components as the Ni and Yb amounts were varied. Another most interesting result was that the density of the pure disordered Yb layer could be studied by following the changes in the core level binding energy of this Yb.

Fig. 1 illustrates the changes in the Yb 4f spectra as multilayers of Yb are deposited on Mo(110). At a coverage of 1 ML only one spin-orbit doublet is present. As the coverage increases two new doublets appear, one with a lower and one with a higher binding energy than the 1 ML doublet and at 2 ML coverage only these two

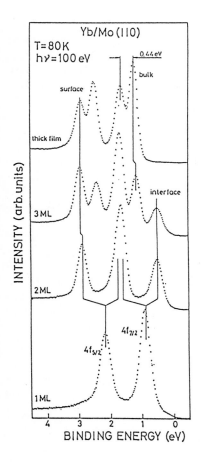

Fig. 1. Photoelectron spectra, showing the Yb 4f core levels for the indicated coverages of Yb on a Mo(110) surface. The photon energy was 100 eV.

doublets are found. The high binding energy component results from surface Yb which is now on top of Yb instead of Mo, and the low binding energy one comes from Yb in the interface towards Mo. Increasing the coverage further leads to the development of a new 4f doublet with a binding energy between that of the interface and of the surface. This component is due to "bulk" Yb, i.e. Yb which is completely surrounded by other Yb atoms. The small shift in the bulk peak from the 3 ML situation to the thick film illustrates the size of error expected from considering only nearest neighbours. Not only do such layer resolved measurements allow the growth to be followed, they also provide a unique way of measuring adhesion and segregation energies. It may be shown /8/ that the shift between the interface and the surface peak is equal to the difference in adhesion energy of the initial state Z and the final state Z^* materials on the substrate whereas the shift between the bulk and the interface (surface) peak is the interface (surface) segregation energy of the Z^* impurity in the Z material. Excellent agreement was found between experiment and values calculated from Miedema's semiempirical scheme for both Yb on Mo(110) /8/ as well as Al on Mo(110) /9/. Thus the energetics of interface formation may also be studied.

References

/1/   D. A. Shirley, J. Electron. Spectrosc. Relat. Phenom. **5** 135 (1974) and
      H. Basch, J. Electron. Spectrosc. Relat. Phenom. **5** 463 (1974).
/2/   J. F. Herbst, D. N. Lowy and R. E. Watson, Phys. Rev. **B6** 1913 (1972).
/3/   B. Johansson and N. Mårtensson, Phys. Rev. **B21** 4427 (1980).
/4/   A. Nilsson, B. Eriksson, N. Mårtensson, J. N. Andersen and J. Onsgaard,
      Phys. Rev. **B38** 10357 (1988).
/5/   A. R. Miedema, J. Less-Common Met. **32** 117 (1973).
/6/   J. N. Andersen, J. Onsgaard, A. Nilsson, B. Eriksson and N. Mårtensson,
      Surf. Sci. **202** 183 (1988).
/7/   J. N. Andersen, O. Björneholm, M. Christiansen, A. Nilsson, C. Wigren,
      J. Onsgaard, A. Stenborg and N. Mårtensson, Surf. Sci. **232** 63 (1990).
/8/   N. Mårtensson, A. Stenborg, O. Björneholm, A. Nilsson and J. N. Andersen,
      Phys. Rev. Lett. **60** 1731 (1988).
/9/   J. N. Andersen, O. Björneholm, A. Stenborg, A. Nilsson, C. Wigren and
      N. Mårtensson J. Phys.: Condens. Matter **1** 7309 (1989).

Temperature Dependence of the Work Function of Alkali-Metal Atoms Adsorbed on a Metal Surface

T. Kato[1] *and M. Nakayama*[2]

[1]Fukuoka Institute of Technology, Wajiro, Fukuoka 811-02, Japan
[2]College of General Education, Kyushu University,
 Ropponmatsu, Fukuoka 810, Japan

Abstract. The temperature dependence of the work function of alkali-metal adsorbed on metal surfaces is analyzed for typical cases by using a cell-CPA theory for a random adsorbate system based on the LCAO model with non-orthogonal orbitals. The monotonic change of the work function with the coverage at high temperature is explained by broadening of the overlayer band due to clustering of the adatoms.

1. Introduction

The coverage dependence of the work function in alkali-metals adsorbed on metals has been measured for various systems. So far, this phenomenon has been interpreted by the Gurney's picture [1-4]. Recently, Ishida and Terakura [5,6] proposed a different picture that the bonding character is covalent even in the low coverage, on the basis of their first-principles calculation. Which of the two pictures holds should depend on the material concerned. To describe the two regimes systematically, we proposed the LCAO model in which the non-orthogonality between the adatom and substrate orbitals is taken into account and analyzed the above problem by numerical calculations for several typical cases [7].

Recently, the work function change with the coverage was measured for the K/Al(100) and Na/Al(100) systems at different temperatures by Paul [8]. The experiment shows that at high temperature the work function decreases monotonically with the coverage in contrast to the usual behavior at low temperature. We consider that this qualitative difference comes from the difference of the adatom configurations, since the electronic properties do not seem to alter drastically. While the adatoms sit at a distance at low temperature, they occasionally adjoin

Springer Series in Surface Sciences, Vol. 24   **The Structure of Surfaces III**
Editors: S.Y. Tong · M.A. Van Hove · K. Takayanagi · X.D. Xie
© Springer-Verlag Berlin, Heidelberg 1991

each other at high temperature. The latter effect should change the broadening of the overlayer band, the charge transfer, and then the behaviour of the work function in the coverage.

In this paper, the temperature dependence of the work function change with the coverage is calculated for typical cases by a cell-CPA theory for a random adsorbate system based on the LCAO model with non-orthogonal orbitals.

2. Model and Formulation

We employ a LCAO model for the random adsorbate-surface system in which the non-orthogonality between the adatom orbital ϕ_a and the substrate orbital ϕ_b is taken into account. We construct orthogonalized bases $\tilde{\phi}_a = (\phi_a - S\phi_s)/\sqrt{1-S^2}$ and ϕ_s with $(\phi_s|\phi_a) = S$. The Hamiltonian $H = H_a + H_s + H_{as}$ is expressed, in terms of the orthogonalized basis, as

$$H_a = \sum_i (\tilde{\varepsilon}_a - v_p) a_i^+ a_i + \tilde{t} \sum_{i \neq j} a_i^+ a_j \; ,$$

$$H_s = \sum_k \varepsilon_k c_k^+ c_k \; , \quad H_{as} = \sum_{i,k} (\tilde{v}_{ik} a_i^+ c_k + h.c.) \quad .$$

Here i and k stand for $\tilde{\phi}_a(r-R_i)$ and $\phi_s(k) = (1/N) \sum_i \phi_s(r-R_i) e^{ikR_i}$, respectively, with R_i being the position coordinate of the i-th site and with N the number of the substrate atoms on the surface. In the numerical calculation, we assume the substrate band to be a one-dimensional surface band on the bulk metal, whose energy is given by $\varepsilon_k = -2|t_s|\cos(ka) \equiv -0.5 W_s \cos(ka)$ with a being the lattice constant, and this band is half-filled in the case of the clean surface. The v_p represents the depolarization potential due to other dipoles, which is self-consistently determined with the transferred electrons. It is assumed that the adatoms, whose coverage is c, randomly locate on the substrate atoms with some repulsive correlation. The quantities $\tilde{\varepsilon}_a$, $\tilde{t}$, and $\tilde{v}$ are related to the eigenenergy of ϕ_a, ε_a; the transfer energy of neighbouring adatoms, t_a; that of neighbouring substrate atoms, t_s; and the hybridization energy between the adatom and substrate atom, v; as follows.

$$\tilde{\varepsilon}_a = \varepsilon_a + \frac{1}{1-S^2}\{(\varepsilon_a - \varepsilon_s)S^2 - 2Sv\}, \quad \tilde{t} = \frac{1}{1-S^2}\{t_a + S^2 t_s\}, \quad \tilde{v}_{ik} = \frac{1}{\sqrt{N}} \tilde{v} e^{ikR_i}$$

with $\tilde{v} = v/\sqrt{1-S^2}$. The density of states is calculated in the
Green function formalism in the cell-CPA [9], the cell size of
which is set to three in a one-dimensional array. As the
details are given in our previous paper [4,7], only points
related to the temperature effect are stated. The Fermi distribution function is approximated by the step function at T=0.
The effect of the temperature enters only through the distribution function of the adatom configuration. The latter for the
m-th neighbour site g(m) is obtained in the lattice gas model
with an effective interaction V_a between the nearest neighbour
adatoms, which is expressed as

$g(m)=[g(1)+f+(-f)^m\{1-g(1)\}]/(1+f)$ with $g(1)=c-(1-c)f$ and
$f=1/\{\sqrt{0.25-(1-\exp[-V_a/k_BT])c(1-c)}+0.5\}-1$. In the calculated
result, T is expressed in units of k_B/V_a.

The change of the work function $\Delta\Phi$ is discussed later by a
reduced one $\Delta\phi$, which is defined as $\Delta\Phi \equiv -(4\pi e^2 d/S_0)\Delta\phi_0$, where S_0
is the area occupied by one substrate atom and d is the distance
between the adsorbate layer and the surface layer.

3. Numerical Results and Discussion

The position of the adatom valence level ε_a relative to the
Fermi level $E_F^{(0)}$, the hybridization energy $|v|$, and the degree
of non-orthogonality S are the important parameters for the
nature of the bonding of the overlayer. When $(\varepsilon_a-E_F^{(0)})$ is much
larger than $|v|$ and S is small, the ionic character should be
dominant for the bonding, while in the opposite case the covalent character should be dominant. Numerical calculations are
performed mainly for the typical cases: "ionic case (case I)"
and "covalent case (case C)". The parameters for the two cases
are listed in Table 1 and the value of $W_s/2|t_a|$ is set to 3.

Fig.1 shows the coverage dependence of the reduced work
function at various temperatures for the ionic case. It is seen
from this figure that the high temperature behaviour is almost
monotonic in contrast to the low temperature one. To examine

Table 1 Values of parameters in two typical cases

	$(\varepsilon_a- E_F^{(0)})/2|t_a|$	$v /2|t_a|$	S
Ionic case (case I)	1.0	− 0.3	0.1
Covalent case (case C)	0.3	− 0.9	0.3

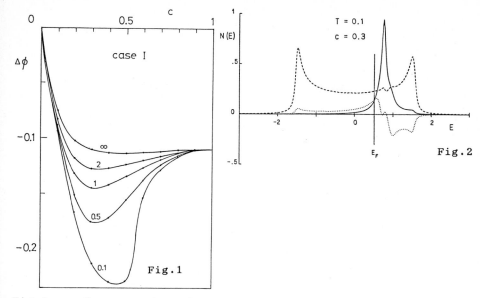

Fig.1　　　Coverage dependence of the reduced work function at various temperatures in the case I. The numerals denote the temperature in units of V_a/k_B.

Fig.2　　　Local density of states of the adsorbates ($N_a(E)$, solid line), the substrates ($N_s(E)$, broken line), and the bonding order ($N_b(E)$, dotted line) for c=0.3 at T=0.1 in the case I. Energy E and DOS N(E) are normalized by $2|t_a|$ and $1/2|t_a|$, respectively.

the origin of this difference, the local DOS curves are shown in Fig.2 for c=0.3 and in Fig.3 for c=0.6 in the low temperature case (T=0.1). At c=0.3, the DOS of the adsorbate $N_a(E)$ is composed of a sharp peak above the Fermi level and a small tail below the Fermi level. This means that the adatom is almost isolated from the other adatoms and most of the valence electrons flow into the substrate band. On the other hand, at c=0.6, $N_a(E)$ below the Fermi level becomes considerably large. This means that the cluster of the adatoms grows and the overlayer becomes metallic. In addition to the broadening of the adsorbate band, the large difference of the depolarization shift v_p between the two coverages ($\Delta v_p/2|t_a| \sim 0.2$) causes a sudden back flow of the transferred electrons and the sharp recovery of the work function. At high temperature (T=∞), the

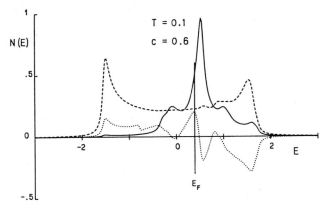

Fig.3   Local density of states for c=0.6 at T=0.1 in the
case I.   Notations are the same as Fig.2

$N_a(E)$ below the Fermi level is rather large even at the low
coverage, as shown in Fig.4 for c=0.3. Moreover, Δv_p is
negligible in the relevant two coverages in this case. These
facts mean that an electron backflow gradually occurs with
increasing coverage, and then $\Delta\phi$ tends to decrease monotonically at high temperature.

       Fig.5 shows the coverage dependence of the reduced work
function at various temperatures for the covalent case.   The
qualitative appearance is similar to the ionic case though
there are some differences in that even at the high temperature
a shallow minimum exists and the magnitude of $\Delta\phi$ in the inter-

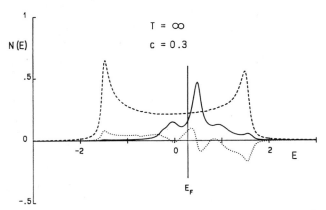

Fig.4   Local density of states for c=0.3 at T=∞ in the case
I. Notations are the same as Fig.2

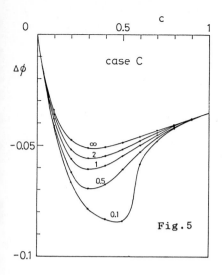

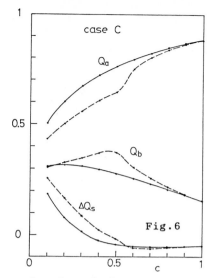

Fig.5   Coverage dependence of the reduced work function at various temperatures in the case I. The numerals denote the temperature in units of V_a/k_B.

Fig.6   Coverage dependence of the remaining charge on the adatom Q_a, the charge transferred onto the substrate ΔQ_s, and the bonding charge Q_b in case C, at T=0.1 (broken lines) and at T=∞ (solid lines).

mediate coverage region is much smaller. According to our calculated result of the DOS curves, $N_a(E)$ exhibits features of the metal even in the low coverage region both at the low and high temperature. In addition, the change of v_p with coverage is small. Fig.6 shows the coverage dependence of the three charges; the remaining charge on the adatom Q_a, the charge transferred onto the substrate ΔQ_s, the bonding charge Q_b (all are normalized per adatom), at high temperature (solid lines) and at low temperature (broken lines). At low temperature, Q_b becomes large even at low coverage. First Q_b increases slightly and rather steeply decreases at c>0.5. It is noticed that the adatom is nearly neutral for c>0.5, if a proper fraction of the bonding charge is ascribed to the adatom. It is concluded from these results that the work function change in this case is caused by the redistribution of the electrons in the bonding region, as observed by Ishida and Terakura [5]. It is seen that

393

at the high temperature Q_b changes monotonically. That is, the redistribution of the electrons occurs gradually and the behaviour of $\Delta\phi$ in the coverage becomes moderate.

From the above observation, it is found that the factors determining the change of the work function differ in the two typical cases though the qualitative behaviour is similar. For quantitative discussions, we only give an estimation of $\Delta\phi$. In the case of Na on Al(100), the maximum reduction of the work function is about 2 eV at 100K [8]. This value corresponds to 0.057 in $|\Delta\phi|$, if we use d=3 and S_0^2=5.4 in atomic units. It seems to suggest that the present system corresponds to the case C. However, it should be noted that the present calculation is not sufficient for qualitative discussions, because it assumes the one-dimensional surface state for the substrate band and neglects effects of the intra-atomic Coulomb interaction.

References

[1] R.W. Gurney, Phys. Rev. 47 (1935) 479.
[2] J.P. Muscat and D.M. Newns, J. Phys. C7 (1974) 2630.
[3] H. Ishida, N. Shima, and M. Tsukada, Surface Sci. 158 (1985) 438.
[4] T. Kato, K. Ohtomi, and M. Nakayama, Surface Sci. 209 (1989) 131.
[5] H. Ishida, K. Terakura, Phys. Rev. B36 (1987) 4510.
[6] H. Ishida, K. Terakura, Phys. Rev. B38 (1988) 5752.
[7] T. Kato, M. Nakayama, submitted to Surface Sci.
[8] J. Paul, J. Vac. Sci. Technol. A5 (1987) 664.
[9] M. Tsukada, J. Phys. Soc. Japan 26 (1969) 684.

A LEED Study of the Clean and Hydrogen-Covered W(310)-(1 × 1) Surfaces

D.L. Adams and S.P. Andersen

Institute of Physics, Aarhus University, DK-8000 Aarhus C, Denmark

Abstract. The structures of the clean and hydrogen-covered surfaces of W(310) have been studied by comparisons of experimental LEED intensity-energy spectra with spectra calculated by multiple-scattering theory. The structure of the clean surface exhibits a large contraction of the first interlayer spacing. Adsorption of hydrogen leads to multilayer relaxations.

1. Introduction

The structures of open metal surfaces are generally found to be relaxed from the ideal truncated-bulk geometry. The relaxations may involve several surface layers /1,2/. Quantitative agreement with the relaxations determined by LEED for a number of Al and Fe surfaces has been obtained in simple model calculations, in which the electrostatic energy of a system of point ions in a neutralizing background is minimised as a function of the layer positions and registries /3/. For these surfaces the general trends of the relaxations can be understood even more simply in terms of the electrostatic forces on the ions of the *unrelaxed* structure /2,4/.

For fcc metals the relaxations appear to be limited mainly to expansions and contractions Δd_i of the interlayer spacings d_i, whereas for open bcc metal surfaces the relaxations of the interlayer spacings may be accompanied by substantial relaxations Δr_i of interlayer registries r_i, that is, by relative shifts of the layer positions parallel to the surface /5,6/.

The purpose of the present work has been to investigate the relaxations of the clean W(310) surface and to study the effect of hydrogen adsorption on the surface structure.

2. Experimental measurements

The experimental LEED measurements were made using an Omicron reverse-view LEED optics in conjunction with a newly-developed video-LEED system. The latter is described in more detail elsewhere in these proceedings /7/.

Intensity-energy spectra were obtained by digitization of the LEED pattern at a set of discrete energies. The energy values were reproducible to ±0.1eV in repetitive energy sweeps, allowing averaging of sets of spectra to

Springer Series in Surface Sciences, Vol. 24   **The Structure of Surfaces III**
Editors: S.Y. Tong · M.A. Van Hove · K. Takayanagi · X.D. Xie
© Springer-Verlag Berlin, Heidelberg 1991

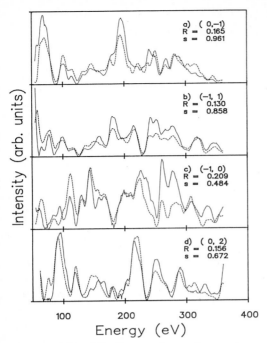

Fig. 1. (a) - (d) Comparison of experimental intensity-energy spectra (solid curves) with spectra (dotted curves) calculated for optimum values (Table 1) of the structural variables for clean W(310). The calculated spectra have been multiplied by a *single* scaling factor c (Eq. 1), derived from the analysis of 15 beams, in constructing the plots. The R values for the comparisons of the individual beams are noted on the plots, as are the factors s by which the intensities must be multiplied to bring all the beams on to the same intensity scale.

be carried out. The intensities of all the beams of interest were measured simultaneously (to within the 40ms digitization time) at each energy. The measured intensities were corrected for the background and for the variation in beam current with energy.

A number of sets of intensity-energy spectra were measured for the clean surface and for the hydrogen-covered surface for 16 diffracted beams (15 symmetry-inequivalent beams) at 326 energies in the energy range 50 - 375 eV at room temperature and at normal incidence. Each set of measurements required 110 secs. Preliminary measurements of intensity spectra made as a function of hydrogen exposure indicated that saturation coverage was obtained after about 2×10^{-5} torr sec. Plots of intensity-energy spectra for 4 of the 16 beams are shown in Figs. 1 and 2 for the clean and hydrogen-covered surfaces respectively. The former spectra are the result of averaging 10 sets of measurements. The latter are the result of averaging 5 sets of

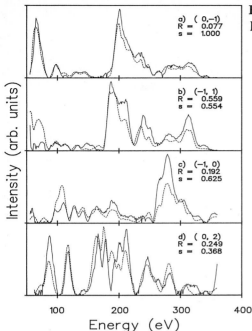

Fig. 2. (a) - (d) As Fig. 1 but for hydrogen-covered W(310)

a) (0,−1)
R = 0.077
s = 1.000

b) (−1, 1)
R = 0.559
s = 0.554

c) (−1, 0)
R = 0.192
s = 0.625

d) (0, 2)
R = 0.249
s = 0.368

Intensity (arb. units)

Energy (eV)

measurements. It can be seen that hydrogen adsorption leads to pronounced changes in peak positions and energies with respect to the spectra from the clean surface. Also shown in the figures are spectra calculated for the surface structures determined by the procedures described below.

3. Calculations of intensity-energy spectra

Sets of intensity-energy spectra were calculated via the layer-doubling method /8/ for different surface structures as specified by the first 5 interlayer spacings d_i and registries r_i. Scattering phase shifts for W were calculated from the muffin-tin band structure potential of Mattheis /9/. In the analysis carried out to date, the scattering due to adsorbed hydrogen has *not* been taken into account. Thus the structural conclusions for the hydrogen-covered surface are preliminary.

In order to obtain convergence in the intensity calculations, 196 partial waves (14 phase shifts) and up to 120 symmetry-adapted plane waves were used in the calculation of l-space and k-space scattering matrices respectively. As in previous calculations /2/, a large reduction in computational effort was realised by saving layer-scattering matrices, the reflection matrices for the bulk, and scattering matrices for slabs of surface layers, at all energies. A typical computational run, involving calculation of intensity-energy spectra for 11 values of a particular d_i or r_i required between 1 and 2 hours on a DEC 3100 work station.

4 Analysis of the experimental data

Determination of the structures of the clean and hydrogen-covered surfaces was carried out by an iterative minimization of a reduced χ^2 function for the comparison of the sets of experimental and calculated intensity-energy spectra for 15 symmetry-inequivalent diffracted beams. In each step of the process, the local optimum value of a particular variable d_i or r_i was determined for fixed values of the remaining variables. The process was iterated to convergence, which required 3 passes through the set of variables.

The function minimized was:

$$R = \sum_{hk,i} \left(\frac{I_{hk,i}^{exp} - c\, I_{hk,i}^{cal}}{\sigma_i} \right)^2 \Big/ \sum_{hk,i} \left(\frac{I_{hk,i}^{exp}}{\sigma_i} \right)^2. \tag{1}$$

It can be seen that R is a normalized χ^2 function, the denominator being constant. The single scaling parameter c is determined by $\partial R / \partial c = 0$, which leads to :

$$R = 1 - \left\{ \sum_{hk,i} \left(\frac{I_{hk,i}^{exp}}{\sigma_i} \right) \sum_{hk,i} \left(\frac{I_{hk,i}^{cal}}{\sigma_i} \right) \right\} \Big/ \left\{ \sum_{hk,i} \left(\frac{I_{hk,i}^{exp}}{\sigma_i} \right)^2 \sum_{hk,i} \left(\frac{I_{hk,i}^{cal}}{\sigma_i} \right)^2 \right\} \tag{2}$$

In these equations the σ_i are the *measured* standard deviations of the experimental intensities. It is emphasized that the scaling factor c applies to *all* beams. Individual-beam R values given in the plots are obtained from Eq. 1 using the value of c determined from the comparison of all 15 beams.

Plots of R versus d_i corresponding to the final variation of each of these structural variables are shown in Fig. 3 for both the clean and hydrogen-

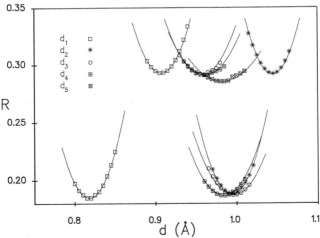

Fig. 3. Plots of R vs the interlayer spacings d_1 - d_5, constructed from the final variations of these variables. The upper curves are for the hydrogen-covered surface, and the lower curves are for the clean surface. The bulk value of the interlayer spacing is 1.001 Å

covered surfaces. Intensity-energy spectra calculated for the optimum values of d_i and r_i are shown together with the experimental spectra in Figs. 1 and 2. The level of agreement between experimental and calculated spectra can be seen to be worse for the hydrogen-covered surface than for the clean surface, as reflected in R values of 0.185 and 0.285 respectively.

5. Summary and discussion

The results of the structure determination are summarised in Table 1 together with previous results /5,6/ for W(211) and Fe(310). In view of these previous results, an unexpected finding of the present work is the small relaxation of the interlayer registries and the small relaxation of the deeper interlayer spacings for the clean W(310) surface. Adsorption of hydrogen on W(310), however, leads to larger multilayer relaxations of both interlayer spacings and registries.

Table 1. Relaxations Δd_i and Δr_i of interlayer spacings and registries for clean and hydrogen-covered W(310), together with results for clean Fe(310) and W(211) from Refs. 5 and 6 respectively.

	W(310)		W(211)	Fe(310)
	Clean	H-covered		
Δd_1	−18.3	−9.5	−12.4	−15.7
Δd_2	− 0.7	+4.5		+ 12.4
Δd_3	− 0.6	−4.2		− 3.6
Δd_4	− 1.7	−3.7		
Δd_5		−1.4		
Δr_1	− 1.8	+3.1	+ 9.8	+ 6.8
Δr_2	− 1.3	+3.0		+ 1.7
Δr_3	+ 2.3	−2.2		
Δr_4	− 0.8	−1.5		
Δr_5		+1.2		

Acknowledgements

This work has been supported by the Danish Natural Science Research Council and the Center for Surface Reactivity.

References

1. D. L. Adams, H. B. Nielsen, J. N. Andersen, I Stensgaard, R. Feidenhans'l and J. E. Sørensen, Phys. Rev. Lett. 49 669 (1982)
2. D. L. Adams, V. Jensen, X. F. Sun and J. H. Vollesen, Phys. Rev. B38 7913 (1988) , and references therein.
3. P. Jiang, P. M. Marcus and F. Jona, Solid State Commun. 59 275 (1986).
4. D. L. Adams, unpublished
5. J. Sokolov, H. D. Shih, U. Bardi, F. Jona and P. M. Marcus, Phys. Rev. B29 5402 (1984)
6. H. L. Davis and J. R. Noonan, Bull. Am. Phys. Soc. 29 221 (1984).
7. D. L. Adams, S. P. Andersen and J. Buchhardt, these proceedings
8. J. B. Pendry, *Low Energy Electron Diffraction* (Academic, London, 1974); M. A. Van Hove and S. Y. Tong, *Surface Crystallography by LEED* (Springer, Berlin, 1979); D. L. Adams, J. Phys. C16 6101 (1983)
9. L. F. Mattheis, Phys. Rev. A139 (1965) A1893

Coverage Dependent Hydrogen Induced Restructuring of Rh(110)

K. Heinz, W. Nichtl-Pecher, W. Oed, H. Landskron, M. Michl, and K. Müller

Lehrstuhl für Festkörperphysik, University of Erlangen-Nürnberg, Staudtstr. 7, W-8520 Erlangen, Fed. Rep. of Germany

Abstract. We show that today's accuracy of LEED structure determination is ready to determine adsorption sites of weak scatterers even in the case of missing superstructure spots or in the case of induced substrate reconstruction. As an example the adsorption system H/Rh(110) is presented. Hydrogen is found to make the Rh(110) buckle by local lifting of bonding rhodium atoms. It removes the considerable relaxation of the clean substrate linearly with coverage ending in an almost unrelaxed substrate at full hydrogen coverage.

1. Introduction

The structure determination of hydrogen adsorbed on a crystalline surface by means of low energy electron diffraction (LEED) suffers from the weak scattering strength of the adatoms. When no superstructure is observed the adsorption position is hidden in the substrate beam intensities dominated by the substrate geometry. In the case of superstructures the situation is only more favourable when there is no substrate reconstruction. However, for hydrogen this situation is rare and substrate atoms generally are induced to move /1/ causing contributions to the superstructure spots. This brings the situation back to the first case leaving the detection of hydrogen positions as a matter of precision for both theory and experiment of LEED. We demonstrate in this paper that today's standards of the LEED technique meet this problem.

As an example we present structure determinations for H/Rh(110). This system shows 5 commensurate adsorption phases with superstructures 1x3 and 1x2 with the number of H atoms in the surface unit cell varying with coverage/2,3/. Because of lack of space we concentrate here on the 1x3-H and 1x1-2H phases which represent prototypes for the two cases described above. Particularly we show which efforts have to be applied in order to retrieve the adsorbate structure. From an independent analysis we know that the clean Rh(110) surface is considerably relaxed by $\Delta d_{12}/d_0 = -6.9\%$ and $\Delta d_{23}/d_0 = +1.9\%$ /4/. Additional to the adsorption site special emphasis is on the detection of a change of this relaxation in the 1x1-2H phase and an adsorbate induced substrate reconstruction in the 1x3-H phase.

Springer Series in Surface Sciences, Vol. 24   **The Structure of Surfaces III**
Editors: S.Y. Tong · M.A. Van Hove · K. Takayanagi · X.D. Xie
© Springer-Verlag Berlin, Heidelberg 1991

2. Experimental and Computational

The Rh(110) surface was prepared as described earlier/4,5/. Hydrogen adsorbed at 90 K with the 1x3-H and 1x1-2H phases developing at exposures of 0.1 L and >10 L, respectively. The coverages 1/3 and 2 are in agreement with findings from TDS /6/. LEED intensities were taken using a TV amplifier camera able to detect the weak superstructure spots. The video signal is evaluated under computer control with automatic background correction. A spot is automatically followed with sweeping energy and so beam intensities are recorded with a rate of 20 ms/energy. Normal incidence of the primary beam was adjusted by comparing spectra of equivalent beams. They were finally averaged in order to reduce effects of a residual misalignment. Also, all spectra were multiply measured and averaged to increase the S/N-ratio. The total time of measurement for a spectrum was below 200 s to avoid residual gas adsorption at the working pressure of 10^{-10} mbar. All measures, such as background correction, multiple measurement and beam averaging as well as high speed measurement proved to be essential to produce reliable data.

The dynamical calculations applied standard computer programs /7/. A total of 8 phase shifts used for H and Rh were temperature corrected with Debye temperatures $\Theta(Rh)=480$ K and $\Theta(H)=3400$ K with the same vibration amplitudes for both atoms. Because of small distances between the H and top Rh layer they were treated as a composite layer. Bulk layers were handled by matrix inversion. Interlayer diffraction was calculated by layer doubling. For the 1x3-H phase up to 141 symmetrically inequivalent beams in the energy range 50-200 eV and for the 1x1-2H phase up to 21 symmetrically inequivalent beams in the energy range 50-206 eV were taken into account. Experimental and computed spectra are compared via the Pendry R-factor /8/ using 9 beams (1 0, 0 1, 1 1, 0 2, 0 2/3, 0·4/3, 1 1/3, 1 2/3, 0 5/3) for the 1x3-H phase and 7 beams (0 1, 1 0, 1 1, 0 2, 1 2, 2 0, 0 3) for the 1x1-2H phase respectively.

3. Structure Determination

3.1. The 1x3-H Phase: Adatom Position in the Case of Induced Substrate Reconstruction

Figure 1 a displays the structural model for the 1x3-H phase. Hydrogen rows ([110] direction) in a distance of 3 times the substrate periodicity produce third order fractional spots on their own. However, nearly independent of the position of the adatoms the computed average fractional order beam intensity is only 0.4% of the average substrate intensities in contradiction to the experimental value of 4.5%. This is a strong indication that the substrate was induced to reconstruct increasing the intensity level of the superstructure spots, which also explains the fact that extra spots are visible up to 1000 eV

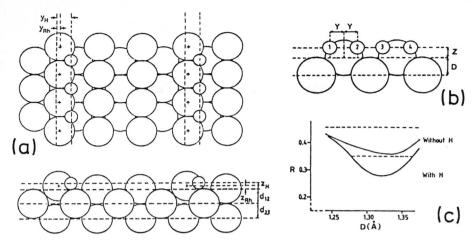

Fig.1: Structure models for (a) the 1x3-H phase and (b) the 1x1-2H phase. In (c) the R-factor varies as function of the first substrate layer distance for calculations with and without consideration of hydrogen in the 1x1-2H phase.

and that their R-factor for non-reconstruction models is always larger than 0.6. There is a best fit value for the threefold coordinated sites shown in fig. 1a.

With this key information from the intensity levels a number of different substrate reconstructions with additional adatom scattering were tried which we avoid describing in detail because of lack of space. A clear best fit reconstruction (R = 0.25) develops for the model displayed in fig. 1a with hydrogen in threefold coordinated sites. First layer rhodium atoms bonding to hydrogen are vertically lifted (z_{Rh} = 0.03 Å) and horizontally shifted (y_{Rh} = 0.04 Å) with hydrogen in positions z_H = 0.51 Å and y_H = 1.20 Å. This is equivalent to a Rh-H bond length of 1.86 Å or a hydrogen radius of 0.52 Å. The substrate layers relax to $\Delta d_{12}/d_0$ = -6.5% for non buckling rhodium atoms. We emphasize that besides the R-factor the ratio of average fractional and integer order spot intensities supplied independent information about having found the correct structure. This is important because most R-factor constructions normalise experimental and computed intensites to the same average level so that information from their true magnitude is lost. The reconstruction brings the ratio up from v = 0.4% for the unreconstructed substrate to v = 1.2% which is much nearer the experimental value (4.5%). Larger reconstructions improve the value of v further but simultaneously worsen the R-factor. This holds also for a second substrate layer reconstruction and variations of the surface Debye temperatures so that we must accept that rest of misfit. We want to stress, however, that the value R = 0.25 is rather good for this kind of structural complexity and that v is at least in the experimental order of magnitude. An example for the quality of the fit is given in fig. 2a for one substrate and one superstructure beam.

403

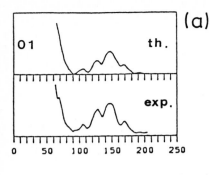

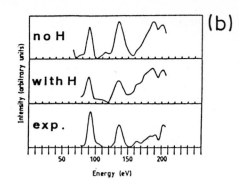

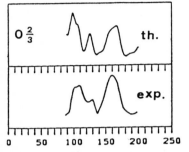

Fig.2: (a) Two best fit spectra for the 1x3-H structure. (b) Improvement of the 02 spectrum by consideration of hydrogen in the 1x1-2H phase.

3.2. The 1x1-2H Phase: Adatom Position in the Case of Missing Extra Spots

A key hint for the adatom site comes from the fact that the 02 beam spectrum is clearly more affected by hydrogen than other beams (peak shifts in all spectra indicate a change of the substrate relaxation). This is interpreted as the existence of a 1 x 1/2 structure of the adlayer, i.e. by equidistant hydrogen distances at coverage 2 /9/. Assuming equivalent adsorption sites for all adatoms as indicated by the absence of reconstruction, again only the three-fold sites are reasonable (fig. 1b). The adatom positions were varied around these sites. Though hydrogen scattering is weak, its inclusion improves the fit remarkably, as shown in fig. 1c. Figure 2b demonstrates the improvement for the 02 spectrum. The best fit R-factor is R = 0.28 at a substrate relaxation $\Delta d_{12}/d_0$ = -1.3%, $\Delta d_{23} \simeq 0$ and hydrogen position y = 0.98 Å and z = 0.78 Å corresponding to a nearly equidistant arrangement of hydrogen atoms. The H-Rh bond length is 1.84 Å equivalent to a hydrogen radius of 0.50 Å.

4. Discussion

It is evident from the above analyses that today's accuracy of LEED experiments and calculations is sufficient to retrieve weak scatterer positions even in the absence of superstructure spots or even when some substrate reconstruction contributes to superstructure spot intensities. We have shown that the average magnitude of fractional order intensities relative to substrate intensities is a key to information independent of and additional to that from the usual R-factor comparison. This is widely overlooked in comparable structure determinations.

Lack of space does not allow us to present the analyses of all phases of the H/Rh(110) system, but we want to mention the main features detected. As shown for the 1x3-H phase hydrogen adsorbs in (nearly) threefold coordinated sites and induces some substrate buckling of the order of 0.03 - 0.05 Å also for the other superstructures. This buckling is only local, i.e. concentrated to the substrate atoms to which the hydrogen atoms bond. Additionally there is also a small horizontal shift of these atoms (shift-buckling-reconstruction). When all rhodium atoms are occupied with two adatoms (fig. 2b) there is no reason for reconstruction. All first layer rhodium atoms are equally lifted entailing that the strong relaxation of the clean substrate is almost totally removed. As ascertained from studies of the intermediate coverages this removement proceeds linearly with coverage. Also, due to repulsive hydrogen-hydrogen interaction the vertical height of the hydrogen position increases with coverage while H-Rh bond lengths are almost unaffected.

Acknowledgements: We are indebted to Deutsche Forschungsgemeinschaft (DFG) and the Höchstleistungsrechenzentrum (HLRZ, Jülich).

References

/1/ K. Christmann, Surface Sci. Rept. 9 (1988) 1
/2/ W. Nichtl-Pecher, W. Oed, H. Landskron, K. Heinz and K. Müller, Vacuum, in press
/3/ W. Nichtl, L. Hammer, K. Müller, N. Bickel, K. Heinz, K. Christmann and M. Ehsasi, Springer Ser. Surf. Sci. 11 (1988) 201
/4/ W. Nichtl, N. Bickel, L. Hammer, K. Heinz and K. Müller, Surface Sci. 188 (1987) L 729
/5/ K. Lehnberger, W. Nichtl-Pecher, W. Oed, K. Heinz and K. Müller, Surface Sci. 217 (1989) 511
/6/ K. Christmann, M. Ehsasi, W. Hirschwald and J.H. Block, Chem. Phys. Lett. 131 (1986) 192
/7/ M.A. Van Hove and S.Y. Tong, Surface Crystallography by LEED, Springer Berlin (1979)
/8/ J.B. Pendry, J. Phys. C 13 (1980) 937
/9/ W. Oed, W. Puchta, N. Bickel, K. Heinz, W. Nichtl and K. Müller, J. Phys. C: Sol. State Phys. 21 (1988) 237

Atomic Beam Diffraction Studies of the Ordered Hydrogen Chemisorption Phases on Rh(110)

G. Parschau, E. Kirsten, and K.H. Rieder

Freie Universität Berlin, Fachbereich Physik,
Arnimallee 14, W-1000 Berlin 33, Fed. Rep. of Germany

Abstract. He- and Ne-diffraction investigations of the chemisorption of hydrogen on Rh(110) confirm the previous LEED-observations concerning the formation of five different ordered phases with periodicities (1×3), (1×2), (1×3), (1×2) and (1×1) and coverages of 1/3, 1/2, 2/3, 1.5 and 2 monolayers, respectively. He-diffraction intensity analyses were performed up to now for the clean surface as well as for three of the five hydrogen phases. In the best-fit corrugations the H-adatoms show up as pronounced hills with amplitudes of ∼ 0.3 Å and their arrangement in linear chains aside the close packed Rh-rows is directly visible. Charge density calculations performed to reproduce the observed corrugations yield for the low coverage (1×2) as well as for the (1×1)-saturation phase H-positions relative to the topmost close-packed metal-rows in good agreement with dynamical LEED.

1. Introduction

The chemisorption system H/Rh(110) is very rich in ordered phases as first observed with LEED [1] and in this respect resembles the system H/Ni(110) as first found and quantitatively analyzed with He-diffraction [2]. Due to the advent of Video-techniques [3] recently it became possible to study H-phases on heavy substrates quantitatively with LEED, whereby not only the substrate reconstructions but also the H-locations can be determined. For H/Rh(110) dynamical LEED-analyses were up to now performed for four of the five ordered phases by the Erlangen group [3-6]. It was therefore challenging to investigate this system with He-diffraction, which previously has proven to be able to give valuable structural information on H-adatom configurations without having to rely on model assumptions [2].

2. Diffraction Measurements and Intensity Analyses

Diffraction methods constitute a convenient means to study the development of ordered adsorption phases by tracing intensities of characteristic beams as a function of exposure. This procedure was often used up to now with LEED [1] and He-diffraction [2]. In this contribution we show for the first time that the formation of ordered chemisorption phases can also be followed with Ne-diffraction: Fig. 1 shows beams characteristic for the different periodicities observed in the H/Rh(110)-system as measured with Ne. Fig. 1 constitutes

Springer Series in Surface Sciences, Vol. 24 **The Structure of Surfaces III**
Editors: S.Y. Tong · M.A. Van Hove · K. Takayanagi · X.D. Xie
© Springer-Verlag Berlin, Heidelberg 1991

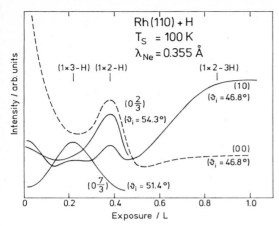

Fig. 1. Development of different ordered H-phases by following the intensities of characteristic beams in dependence of H_2 exposure detected with a Ne-beam

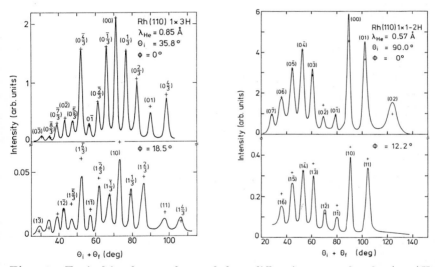

Fig. 2. Typical in-plane and out-of-plane diffraction scans for the (1×3)H-phase with a coverage of 1/3 ML (left side) and the (1×1)H-saturation phase (right side) on Rh(110). Solid lines: experimental scans, crosses: best-fit intensities corresponding to beam maxima.

convincing proof that with Ne-beams good diffraction spectra can not only be obtained for clean metal surfaces [7] but also for chemisorption systems; due to space limitations Ne-diffraction spectra will be published elsewhere. Here we turn to analyses of He-diffraction intensities for the low coverage (1×3) and the (1×1)-saturation phases, for which are shown in Fig. 2 typical He-diffraction scans (for a full discussion of the low coverage (1×2) see Ref. 8). The inten-

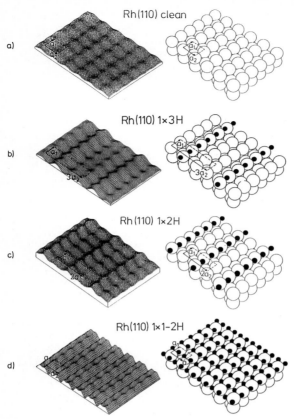

Fig. 3. Best-fit corrugations together with sphere models of (a) clean Rh(110), (b), (c) the right domains of the (1×3) and (1×2) H-phases with respective coverages 1/3 and 1/2 ML, and (d) the (1×1)H-saturation phase

sity analyses were performed on the basis of the hard corrugated wall model with the GR-method [9], a procedure valid in view of the small corrugation amplitudes of these phases. The analyses were performed both with general Fourier-representations for the corrugations to establish the basic structural elements in a model-free manner and by modelling the adatoms as Gaussian hills to arrive at more precise values on the locations of the adatoms relative to the substrate. The best-fit intensity maxima obtained in the latter way are shown as crosses in the figures. The best-fit corrugations of the clean surface as well as of all three H-phases analyzed quantitatively up to now are shown in Fig. 3; they yield direct pictures of the adatom configurations and confirm that the H-adatoms form linear chains slightly aside the close-packed metal rows (both right and left domains were taken into account in the best-fit searches of the asymmetric (1×3) and (1×2) H-phases with the domain distribution as an open parameter; the final results indicated equal distribution of both domains). In Fig. 3 we have also included corresponding sphere models of the respective surface structures.

3. Discussion

Apart from qualitatively confirming the H-adatom configurations of the various phases, present experience on H-adsorption systems based on He-diffraction on one hand and LEED and ion scattering on the other allows more quantitative exploitation of the best-fit corrugation functions. Using the known locations of the H-adatoms relative to the substrate atoms, the corrugation amplitudes of the adatom hills have to be reproduced in charge density calculations (performed conveniently by overlapping ionic charge densities), as the relevant charge density contours are directly related to the incoming He-energy [9]. For H on the (110)-surfaces of Ni, Pd and on Ni(100) an extra charge of -0.4e was found to be necessary to make the measured H-corrugation amplitudes [8-10] compatible with the measured bond lengths [11,12]. Applying the same charge transfer to the case of Rh we obtain for the (1×2)-phase a H-adsorption height of 0.82 ± 0.1 Å and a lateral distance to the close-packed Rh-rows of 0.97 ± 0.1 Å in very good agreement with the respective values of 0.84 and 1.08 Å deduced from dynamical LEED [4]. In the case of the (1×1)-phase the respective values are found to be very close to those of the (1×2)-phase again in good agreement with LEED [6]. Interestingly, in the rather open (1×3)-phase the H-adatoms seem to be closer to the metal rows than stated on the basis of LEED [3]. Further work to elucidate these discrepancies is in progress.

According to the LEED-results nearly three-fold coordinated sites (which can be regarded as shifted bridge sites) are occupied by the H-atoms as indicated in the sphere models of Fig. 3. Consequently, upon moving perpendicularly to the close-packed metal rows from one H-atom to another one should cross bridge positions above the unoccupied metal rows in the (1×2) and (1×3) phases. On the respective corrugations, however, one moves over the small hills characteristic of the substrate thus indicating shifted on-top sites as the H-locations. This appears highly unlikely as there is much evidence [11,12] for the occupation of threefold coordinated sites on unreconstructed fcc(110)-surfaces. We are thus faced with the problem of the "wrong substrate phase" relative to the adsorbate corrugation hills as observed previously also in the case of the low coverage c(2×6) H-phase on Ni(110) [2]. Anticorrugating effects giving rise to corrugation maxima at bridge sites along the close packed rows were theoretically anticipated by Annett and Haydock for Ni(110) [13] and could also play a role in the case of Rh. Since for Ne-diffraction anticorrugating effects are predicted to be less pronounced [14], we hope that with our Ne-data for the low coverage phases we may be able to resolve this problem.

References

1. K. Christmann, M. Ehsasi, W. Hirschwald, J.H. Block: Chem. Phys. Lett. **131**, 192 (1986)
2. K.H. Rieder, W. Stocker: Surf. Sci. **164**, 55 (1985)
3. K. Lehnberger, W. Nichtl-Pecher, W. Oed, N. Bickel, K. Heinz, K. Mueller: Surf. Sci. **217**, 511 (1987)
4. W. Puchta, W. Nichtl, W. Oed, N. Bickel, K. Heinz, K. Mueller: Phys. Rev. **B39**, 1020 (1989)
5. M. Michl, W. Nichtl-Pecher, W. Oed, H. Landskron, K. Heinz, K. Mueller: Surf. Sci. **220**, 59 (1989)
6. W. Oed, W. Puchta, N. Bickel, K. Heinz, W. Nichtl, K. Mueller: J. Phys. C (Solid State Physics) **21**, 237 (1988)
7. K.H. Rieder, W. Stocker: Phys. Rev. Lett. **52**, 352 (1984)
8. G. Parschau, E. Kirsten, K.H. Rieder: Surf. Sci. **225**, 367 (1990)
9. T. Engel, K.H. Rieder, in: Springer Tracts in Modern Physics, Vol. 91 (Springer, Berlin, 1982); K.H. Rieder: Surf. Sci. **117**, 13 (1982)
10. M. Baumberger, K.H. Rieder, W. Stocker: Appl. Phys. **A41**, 151 (1986);
11. W. Reimer, V. Penka, M. Skottke, R.J. Behm, G. Ertl, W. Moritz: Surf. Sci. **186**, 45 (1987); M. Skottke, R.J. Behm, G. Ertl, V. Penka, W. Moritz: J. Chem. Phys. **87**, 6191 (1987)
12. I. Stensgaard, F. Jacobsen, Phys. Rev. Lett. **54**, 711 (1985)
13. J.F. Annett, R. Haydock: Phys. Rev. Lett. **53**, 838 (1984)
14. J.F. Annett: Daresbury Lab. Inf. Q. Surf. Sci. **14**, 9 (1984)

Structural Studies of Highly Corrugated Systems with Atomic Beam Diffraction: Hydrogen-Chemisorption on Missing-Row Reconstructed Pt(110) and Ir(110)

E. Kirsten and K.H. Rieder

Freie Universität Berlin, Fachbereich Physik,
Arnimallee 14, D-1000 Berlin 33, Fed. Rep. of Germany

Abstract. He-diffraction results are presented for clean and H-covered (110)-surfaces of Pt and Ir. On both surfaces the basic reconstruction is of the missing row type. A dramatic increase of the rainbow angles upon H-chemisorption into the β_2-adstate indicates a strong increase of the corrugation amplitude on both metals. Structural implications of these findings are discussed on the basis of quantitative analyses of diffraction intensities for the Pt-case and leave H-occupation of either the bridge sites above the topmost metal rows or the first available octahedral subsurface sites open. The rainbow angle decreases upon filling the β_1-adstate, indicating population of sites deep in the troughs at higher coverages.

1. Introduction

Helium diffraction has been proven to be a valuable tool for surface crystallographic studies [1]. Due to the small particle energies of 20 – 200 meV, He-beams probe the topmost surface layer in an absolutely nondestructive manner and consequently allow investigation of very subtle adsorption phases. He-diffraction is especially valuable for investigations of light adsorbates on heavy substrates [2]. Intensity analyses yield the corrugation of the repulsive part of the potential, which — as a surface of constant total electron density — contains the information on the surface structure.

For surfaces with small corrugation amplitudes ($< 12\%$ of the lattice constant) details of the He-surface interaction potential do not play a role and analyses of diffraction intensities can be performed conveniently within the hard corrugated wall model with the computationally fast eikonal or GR-methods [1]. In this contribution we show on the basis of close coupled channels calculations that for large corrugation amplitudes ($\gtrsim 20\%$ of the surface periodicity) — like the ones encountered for the clean and H-covered missing-row reconstructed (110)-surfaces of Pt and Ir [3] — the distribution of diffraction intensities at various angles of incidence θ_i and wavelengths λ_i can only be described satisfactorily, if all parts of the interaction potential are properly accounted for: the corrugation and steepness of the repulsive part as well as the depth, width, shape and corrugation of the attractive part. The results deduced in this way from a large amount of experimental data for clean and H-covered Pt(110)1×2 yield a picture of the sequence of the population of available adsorption sites

scattering [9] with an extra charge of -0.4e on the H-adatoms. Applying the same charge transfer in the present case yields a reasonable H-Pt bond length of 1.8 Å for the bridge sites above the ridges; on-top sites yield too short and threefold coordinated sites too large values. (ii) β_2-H occupies sites below the topmost Pt-rows, whereby the initial $\approx 15\%$ inward relaxation changes into $\approx 20\%$ outward relaxation. In this case the H-Pt bond length would be 2.1 Å for H in the first octahedral subsurface sites in astonishigly good agreement with the value of 2.08 Å observed in Pt-hydride [10].

Based on the He-diffraction data alone we cannot decide between these two possibilities; recent ion scattering investigations [11] seem to indicate, however, that the outward relaxation of the Pt-atoms upon β_2-H chemisorption is only ~ 0.1 Å, making the H-location in bridge-sites above the Pt-ridges more likely.

At higher coverages a pronounced decrease of the rainbow angle (and consequently the corrugation amplitude) to a value close to that of the clean surface indicates population of deep trough sites with β_1-H-adatoms.

4. He-diffraction from Clean and H-covered Reconstructed Ir(110)

Typical He-diffraction results for clean and β_2-H-covered Ir(110) are shown in Fig. 3. In close analogy to Pt(110)1×2 strong rainbow scattering is observed for the clean surface at an angle of $\sim 70°$ from the specular indicating a corrugation amplitude of similar magnitude as for Pt(110)1×2. Upon β_2-H adsorption the rainbow angle shows again a dramatic increase of $\sim 20°$ evidencing a strong increase of corrugation amplitude. We may qualitatively argue from these observations that β_2-H-adatoms again occupy either sites near the topmost bridges or octahedral subsurface sites, both being able to account for the required increase in corrugation amplitude. The situation seems to be more complicated, however, since inspection of both the angular locations of the

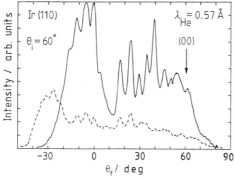

Fig. 3. Experimental He-diffraction scans for clean Ir(110) (full curve) and after saturation of the β_2-H adstate with 1 L H$_2$-exposure (broken line). The rainbow angle increases as in the case of Pt(110)1×2.

He-diffraction beams and the LEED-patterns indicate the presence of recon-structions on this surface with higher periodicities like (1×3) and (1×4) [12]. More data are required to elucidate this point further.

Acknowledgement. This work was supported in part by the Deutsche Forschungsgemeinschaft, SFB 6.

References

1. T. Engel, K.H. Rieder: *Structural Studies of Surfaces with Atomic and Molecular Beam Diffraction*, Springer Tracts in Modern Physics, Vol. 91 (Springer, Berlin, 1982) p. 55

2. K.H.Rieder, W. Stocker: Surf. Sci. **164**, 55 (1985)

3. W. Moritz, D. Wolf: Surf. Sci. **163**, L655 (1985); C.M. Chan, M.A. Van Hove: Surf. Sci. **171**, 226 (1986); M. Copel, P. Fenter, T. Gustafsson: Vac. Sci. Technol. **A5**, 742 (1987); P. Fery, W. Moritz, D. Wolf: Phys. Rev. **B38**, 7275 (1988)

4. J.R. Engstrom, W. Tsai, W.H. Weinberg: J. Chem. Phys. **87**, 3104 (1987); D.E. Ibbotson, T.S. Wittrig, W.H. Weinberg: J. Chem. Phys. **72**, 4885 (1980)

5. R. Ducros, J. Fusy: Surf. Sci. **207**, L943 (1988); K. Dueckers, K.C. Prince, H.P. Bonzel, V. Chab, K. Horn: Phys. Rev. **B36**, 6292 (1987)

6. E. Kirsten, K.H. Rieder: Surf. Sci. Lett. **222**, L837 (1989)

7. M. Baumberger, K.H. Rieder, W. Stocker: Appl. Phys. **A41**, 151 (1986); G. Parschau, E. Kirsten, K.H. Rieder: Surf. Sci. **225**, 367 (1990)

8. W. Reimer, V. Penka, M. Skottke, R.J. Behm, G. Ertl, W. Moritz: Surf. Sci. **186**, 45 (1987); M. Skottke, R.J. Behm, G. Ertl, V. Penka, W. Moritz: J. Chem. Phys. **87**, 6191 (1987)

9. I. Stensgaard, F. Jacobsen, Phys. Rev. Lett. **54**, 711 (1985)

10. G. Alefeld, J. Völkl, Eds., Hydrogen in Metals I and II (Springer, Berlin, 1982)

11. H. Niehus, R. Spitzel, G.Comsa: private communication

12. H. Bu, M. Shi, F. Masson, J.W. Rabalais: Surf. Sci. Lett. **230**, L140 (1990); W. Hetterich, W. Heiland: Surf. Sci. **210**, 129 (1989)

Temperature and Coverage Dependence of the Structure and Dynamics of Hydrogen on Pd(111)

C.-H. Hsu[1], B.E. Larson[1], M. El-Batanouny[1], C.R. Willis[1], and K.M. Martini[2]

[1]Department of Physics, Boston University, Boston, MA 02215, USA
[2]Department of Physics and Astronomy, University of Massachusetts, Amherst, MA 01003, USA

Abstract. We investigate the behavior of hydrogen on the Pd(111) surface by elastic scattering of a monoenergetic thermal He beam. Dramatic changes in the intensities of the diffraction peaks are observed as a function of sample temperature in range $140K$-$350K$, and are attributed to changes in H-coverage. A consistent explanation of the diffraction data is achieved when we consider the H system as a correlated quantum fluid where particle delocalization occurs at medium and low coverages.

1. Introduction

The hydrogen/transition metal adsorption system has been the subject of numerous studies [1]. Among the transition metals, Pd has always attracted a great deal of attention because of its high solubility for and diffusion of hydrogen. The structure of hydrogen overlayers on the Pd(111) surface has previously been studied by LEED [1,2]. Three ordered phases were observed : two H phases with $(\sqrt{3}\times\sqrt{3})R30°$ structure were found at 1/3 and 2/3 monolayer coverage when the sample temperature was below $110K$ [2], and above $110K$ a (1×1) structure is the only ordered phase. These structures as well as the disordered phases were considered in the framework of a conventional classical diffusive lattice gas localized in the confines of 3-fold hollow sites [1,2].

Although the high diffusive mobility of hydrogen on metal surfaces has been observed on several metallic surfaces [3], no delocalized quantum motion of hydrogen was recognized until Christmann *et al.* [4] proposed an 'atomic band' model in analogy with electronic band structure in order to interpret the state of chemisorbed hydrogen. In this model the motion of the hydrogen atoms are represented by wave functions localized normal to the surface but delocalized along the surface. Puska *et al.* [5] calculated the band energy levels and wave functions of the chemisorbed atomic hydrogen on Ni surfaces using the effective-medium theory. The idea was promulgated by Mate and Somorjai [6] in their interpretation of high-resolution electron-energy-loss spectroscopy measurements of the dynamics of H on a Rh(111) surface.

In this paper we present elastic He scattering data that support the existence of delocalized quantum motion of H on the Pd(111) surface. Furthermore, we have developed an extension of the sudden approximation [7,14] that allows the incorporation of the quantum mechanical wavefunction of the H system into the model calculation of the He diffraction intensities.

2. Experimental Set Up

Our experiments were conducted under ultra-high-vacuum conditions with a base pressure of 5×10^{-11} Torr. The details of the He-beam scattering facility that provides a

Springer Series in Surface Sciences, Vol. 24 **The Structure of Surfaces III**
Editors: S.Y. Tong · M.A. Van Hove · K. Takayanagi · X.D. Xie
© Springer-Verlag Berlin, Heidelberg 1991

momentum resolved He beam with $\Delta k < 0.05 \AA^{-1}$ have been described elsewhere [8]. The cleanliness and long-range order of the Pd(111) surface was acheived and monitored by conventional means. The intensity of the specular He-beam was used to monitor the final stages of sample cleanliness. Hydrogen was deposited by exposing the sample surface to an ambient partial H pressure of about 5×10^{-8}Torr for periods up to a few minutes (1-10L). The intensity of the specular beam was also used as a calibration of *surface* H coverage as demonstrated by Poelsema *et al.* [9].

3. Results and Discussion

In fig. 1.a we show a typical spectrum of elastically scattered He from a clean Pd(111) surface along the $< 11\bar{2} >$-direction. The intensity of the first-order diffraction peaks (FODP's) is extremely small ($< 10^{-3}$ of I_{00}). This reflects the smooth charge-density corrugations of the Pd(111) surface, and places an upper limit on the corrugation amplitude of about $0.02\AA$, typical of many close-packed metal surfaces. Fig. 1.b depicts spectra obtained after exposure to 5L of H at $T_S = 140K$. The growth of the Pd(111) FODP's together with the absence of any fractional order or incommensurate peaks present strong evidence that at saturation the H overlayer orders in a commensurate (1×1) state. The intensity ratio I_{10}/I_{00} reaches a value of $\simeq 0.06 - 0.05$ at saturation.

The behavior of the FODP's under sample rotation demonstrates the C_{3v} symmetry of the (1×1)H/Pd(111) system at saturation. An azimuthal rotation of $60°$ changes the intensity distribution of the six FODP's; while a further rotation of $60°$

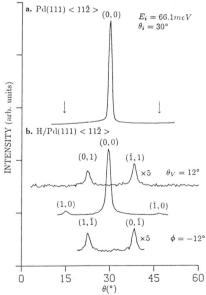

Fig. 1. Spectra of elastically scattered He beams for $E_i = 66.1$ meV along $< 11\bar{2} >$-direction at $\theta_i = 30°$. from : (a) clean Pd(111); (b) ordered (1×1) H/Pd(111) saturated at $140K$. The top (bottom) curve was taken at $\phi = \pm 12°$ above (below) the scattering plane.

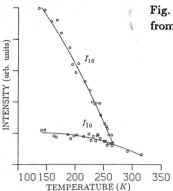

Fig. 2. The variation of I_{10} and $I_{\bar{1}0}$ with T_s starting from well-ordered (1 × 1) H/Pd(111) at $140K$.

restores the original distribution. He-scattering is sensitive to the atomic arrangement of *the topmost layer* only. Consequently, it can provide information about the distribution of H atoms <u>on</u> the Pd(111) surface. The observed C_{3v} symmetry implies global occupation of a preferred 3-fold hollow site : either the octahedral (fcc-stacked) or tetrahedral (hcp-stacked). Moreover, we determined the surface corrugation through a diffraction intensity calculation using a hard-corrugated-wall (HCW) model within the framework of the eikonal approximation (EA). The best intensity fit corresponded as well to the 3-fold hollow site configuration, with H-induced peak-to-peak corrugation amplitude = 0.13Å and an effective H corrugation radius of = 1.3Å. These values are almost identical to those obtained by Lee *et al.* [10] for H/Pt(111) employing the same method. It underlines the interesting similarities with H/Pd(111). Moreover, the extent of the H-induced corrugation reflects an adsorbate well-localized within the confines of the 3-fold hollow site.

Fig. 2 shows the dependence of I_{10} and $I_{\bar{1}0}$ on T_S starting from the saturated H(1 × 1) phase at $140K$. I_{10} is observed to decrease at a faster rate than $I_{\bar{1}0}$ up to $T_S \simeq 270K$. The ratio $I_{10}/I_{\bar{1}0}$ decreases from about 3 at $140K$ to 1 at $270K$. This indicates the recovery of inversion symmetry in the diffraction spectra. The appearance of this six-fold symmetry in the elastic spectrum could be explained by three possible models of the H configuration on the surface. In the first model, the H adatoms simultaneously occupy both 3-fold sites in the primitive cell (model I). However, the high repulsive H-H interaction energy resulting from the occupation of such close neighboring sites makes this model physically unacceptable. In the second model, equal numbers of randomly distributed domains coexist with either the *fcc* or the *hcp* site occupied (model II). In the third model, the H adatoms exhibit delocalized motion parallel to the Pd(111) surface, leading to an *apparent* simultaneous occupation of both 3-fold hollow sites in the primitive cell (model III).

In order to distinguish between models II and III we have to consider also the temperature dependence of I_{00} shown in fig. 3. Starting from the (1 × 1) phase at $140°K$, point A, a dramatic decrease in I_{00} with increase in T_S up to $270K$ is observed, path (a). I_{00} reaches a minimum at about $270K$, point B, almost two orders of magnitude smaller than the clean surface intensity. Upon cooling back to $140K$, path (b) no significant change in I_{00} was observed and the C_{6v} symmetry was maintained, point C. Moreover, path (b) is reversible. H deposition at point C leads to an increase in I_{00} to its initial value, A, and a change in symmetry to C_{3v}. Because

418

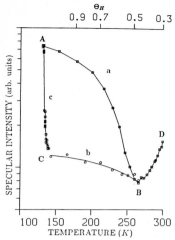

Fig. 3. Temperature dependence of the specular intensity I_{00} in the range of $140K$–$300K$ with $E_i = 66.1$ meV and $\theta_i = 30°$. Path (a), ($\mathbf{A} \to \mathbf{B} \to \mathbf{D}$), starts with a (1×1) H/Pd(111) at $140K$, $\Theta_H = 1.0$ ($\mathbf{A}$) and ends at $T_s = 300K$, $\Theta_H \simeq 0.3$ ($\mathbf{D}$). The H coverage Θ_H for the path (a) is also shown on the top. Path (b), ($\mathbf{B} \to \mathbf{C}$), is reversible. Path (c), ($\mathbf{C} \to \mathbf{A}$), is effected by further H deposition at C.

changes in H coverages can account for these observations, we attribute the temperature dependence solely to induced changes in H coverage. A similar behavior of I_{00} vs T_S has been reported by Poelsema et al. [9], who arrived at a similar conclusion for H/Pt(111) and established a unique calibration of H-coverage in terms of I_{00}. They attibuted the strong attenuation in I_{00} to a destructive interference between occupied and empty primitive cells distributed at random. The interference is strongest when the reflectivities of the two components are comparable. On Pt(111), where H absorption into the bulk is inhibited by a negative heat of solution, the attenuation of I_{00} is attributed to the thermal desorption of H. On Pd(111) the H depletion is due to both thermal desorption and bulk absorption through transfer to subsurface sites [2].

An acceptable model for the medium and low coverage regimes should consistently account for both the change in symmetry and the anomalous attenuation of the specular peak. We shall consider models II and III in the classical context of a HCW potential for He-surface scattering, and the EA. Although model II accounts for the symmetry change, it requires the introduction of a coverage-dependent increase in the corrugation amplitude of over $1Å$, in order to keep in step with the change in the I_{10}/I_{00} ratio. Such corrugation amplitudes lead to either an unphysical increase in the metal-H bond-length (MHBL) with temperature, or the possibilty of surface reconstruction which has been refuted by LEED measurements [2]. Another possibility within model II is to introduce H occupation of bridge sites; requiring corrugation amplitudes of up to $1.3Å$ and occupation numbers of about 20%. Both of these values are unreasonable since the bridge site is found to be unstable [11] and the corrugation corresponds to MHBL of about $2.6Å$. Based upon these results we rejected model II. At this point we attempt to reconcile delocalization (model III) with both changes in symmetry and the anomalous attenuation. Adopting the recipe of Poelsema et al. [9] we were able to fit our diffraction intensities by invoking a corrugation with amplitude of $0.3Å$ and an effective radius of 2.2Å, centered on the bridge site. In order to account for the large width of this corrugation, the scattered He atoms have to sample the occupation of neighboring hollow sites by the same H atom. The H hopping time must therefore be less than the duration of the He scattering event $< 10^{-13}$ sec. The corresponding H energy would then be $> 30\text{meV}$ and must be temperature independent as demonstrated by path (b) in fig. 3. Consequently, the hydrogen is out of thermal equilibrium with the metal surface over most if not all of the temperature range considered here and down to $80°K$ in ref. [9]. Thus a classical description of the H motion fails to account consistently for the experimental observations.

Since the EA is incapable of incorporating a quantum description of the H motion we adopted the framework of the sudden approximation [7] instead. In this approximation all coordinates except the relative ones involved in the scattering are slowly varying. All operators involving such coordinates are replaced by constants, and enter the problem only as *parameters*. The scattering phase shift is then computed for each value of the parameter-coordinates in the WKB approximation with the square of the many body wavefunction providing a probability density for different configuration of these coordinates. We considered the repulsive H-H interaction implicitly by neglecting overlap terms in the expansion of the many body density function. The He-H interaction potential is modelled by a simple hard-core repulsion. A detailed description of the method and the procedures used is given in ref. [14]. In computing the diffraction intensities we set the one particle reduced density to be an ellipsoidal function in the surface unit cell, with the major (along the $< 112 >$ direction) and minor axes denoted by r_{major} and r_{minor} respectively. Within the ellipsoid the reduced density $\rho_1(x)$ is a constant C_{in} times a normalized gaussian function along the z direction with a mean z_{in} and standard deviation σ. Outside the ellipsoid $\rho_1(x)$ is a constant C_{out} times another gaussian function with a different mean z_{out} and the same standard deviation σ. For simplicity, we take z_{out} to be more than 3Å below the hard step due to Pd, so $\rho_1(x)$ outside the ellipsoid does not contribute to the phase shift. We introduce C_{out}/C_{in} as the fifth independent parameter. With this model, we were able to reproduce the variations of specular and average first order intensities with coverage by the best fit values : $r_{major} = 0.8\text{Å}$, $r_{minor} = 0.5\text{Å}$, $\sigma = 0.2\text{Å}$, $z_{in} = -2.\text{Å}$, and $C_{out}/C_{in} = 0.065$. Notice that $2r_{major} = 1.6\text{Å}$ is reasonably close to 1.59Å, the separation between 3-fold hollow sites; while a z_{in} of -2Å corresponds to an average H height of 1Å above the Pd surface plane and a H-Pd bond length of 1.7Å. The convolution of the hard-core potential with the reduced single particle density yields an effective H contribution to the scattering potential which covers most of the unit cell. This delocalization of hydrogen results in the wide corrugation sampled by the scattered He atoms.

4. Summary

Elastic scattering intensities of thermal He beams from H/Pd(111) were measured as a function of surface temperature in the range $140K - 300K$. Two remarkable features were observed : the first feature is the presence of C_{3v} symmetry at (1×1) saturation coverage $(140K)$ and its transformation to C_{6v} symmetry at lower coverages $(270K)$. The second feature involves the anomalous attenuation of the specular He beam accompanying this transformation. Taken together, these features provide strong evidence of a fundamental change in the surface charge density corrugation. A classical interpretation of the motion of hydrogen either fails to reproduce the measured attenuation or leads to contradictory and unphysical conclusions regarding the metal-hydrogen bonding or the surface equlibrium. An alternative quantum mechanical model has been introduced, with diffraction intensities interpreted with the sudden approximation formalism. A consistent and satisfactory agreement with the variations in the measured diffraction intensities has been obtained. We have implicitly assumed strong repulsive H-H correlations at short distances, which can lead to low temperature ordered phases such as the observed $(\sqrt{3} \times \sqrt{3})$ phases, as well as to localization in the high coverage regime, in particular, the (1×1) phase.

Acknowledgements We would like to thank Prof. J.P. Toennies for lending us the Pd(111) crystal. This work has been supported by US DOE Contract DE-FG02-85ER45222. One of the authors (B.E.L.) is surpported by NSF Contract DMR-8914045.

References

1. K. Christmann, Surf. Sci. Rep. **9**, 1 (1988), and references therein.
2. T.E. Felter, S.M. Foiles, M.S. Daw, and R.H. Stulen, Surf. Sci. **171**, L379 (1986).
3. T.E. Felter, R.H. Stulen, M.L. Koszykowski, G.E. Gdowski, and B. Garrett, J. Vac. Sci. Technol. **A 7**, 104 (1989).
4. K. Christmann, R.J. Behm, G. Ertl, M.A. Van Hove, and W.H. Weinberg, J. Chem. Phys. **70**, 4168 (1979).
5. M. J. Puska and R. M. Nieminen, B. Chakraborty, S. Holloway, and J.K. Nørskov, Phys. Rev. Letts. **51**, 1081 (1983); M. J. Puska and R. M. Nieminen, Surf. Sci. **157**, 413 (1985).
6. C.M. Mate and G.A. Somorjai, Phys. Rev. **B34**, 7417 (1986).
7. R.B. Gerber, A.T. Yinnon, and J.N. Murrell, Chem. Phys. **31**, 1 (1978).
8. K.M. Martini, W. Franzen, and M. El-Batanouny, Rev. Sci. Instrum. **58** , 1027 (1987).
9. *Scattering of Thermal Energy Atoms from Disordered surfaces*, Springer Tracts in Modern Physics **115**, B. Poelsema and G. Comsa, Sec. 8.1 (Springer, 1989) and references therein.
10. J. Lee, J.P. Cowin, and L. Wharton, Surf. Sci. **130**, 1 (1983).
11. M.S. Daw and S.M. Foiles, Phys. Rev. **B35**, 2128 (1987).
12. C-H Hsu, B.E. Larson, M. El-Batanouny, C.R. Willis, and K.M. Martini, to be published.

Hydrogen Titration Studies of fcc Cobalt Layers on Cu(001)

M.T. Kief, G.J. Mankey, and R.F. Willis

Department of Physics, The Pennsylvania State University,
University Park, PA 16802, USA

Abstract: We have used hydrogen gas titration studies to establish the surface morphology of ultrathin layers of metastable fcc cobalt epitaxed to Cu(001) at different substrate temperatures. There are associated changes in thermal desorption behavior and the ferromagnetic properties of the films.

Recent work has shown that cobalt grows in a ferromagnetic fcc structure when epitaxed to copper(001) [1 - 4]. This is an ideal magnetic system in that the magnetic moment is stable unlike similar fcc iron films and the films show no complicating out of plane magnetic anisotropy or surface reconstructions. We have studied the effects of hydrogen titration. Analysis of the hydrogen desorption spectra reveal associated changes in the surface microstructure. Magnetic hysteresis behavior was monitored in situ by the Surface Magneto-Optic Kerr Effect (SMOKE).

The copper (001) substrate was prepared using standard techniques. Cobalt was evaporated from electron bombardment sources which could be isolated from the main chamber for extensive degassing. The evaporation sources were emission current stabilized to produce a constant flux at the sample equivalent to 1 monolayer of material in three minutes. The evaporation flux was monitored with a Quartz Crystal Microbalance (QCM) located at a quarter of the source/substrate separation. This configuration has the advantage that sixteen times more material is deposited on the QCM than on the substrate. Precise calibration of the flux was done by correlating the QCM thickness with RHEED oscillations. Film thickness and purity were also monitored with Auger electron spectroscopy.

We have determined that the cobalt films grown at substrate temperatures of 150 K, 300 K, and 450 K have very different microstructure. Films grown at 150 K show broad LEED spots with a high background and (01) spot width variation with energy, while RHEED measurements show a transmission diffraction pattern. These data suggest that the 150 K grown films are rough. Films grown at 300 K have good p(1x1) LEED patterns with a low background. Films prepared at 450 K have a very sharp p(1x1) LEED indistinguishable from the substrate. RHEED oscillations have been observed for both the 300 K and 450 K growth temperatures and this is generally thought to indicate layer by layer growth [5]. These data suggest that the optimal growth temperature to obtain uniform fcc cobalt films is 450 K. However, hydrogen titration experiments indicate complications due to surface segregation at the higher growth temperatures.

Springer Series in Surface Sciences, Vol. 24   **The Structure of Surfaces III**
Editors: S.Y. Tong · M.A. Van Hove · K. Takayanagi · X.D. Xie
© Springer-Verlag Berlin, Heidelberg 1991

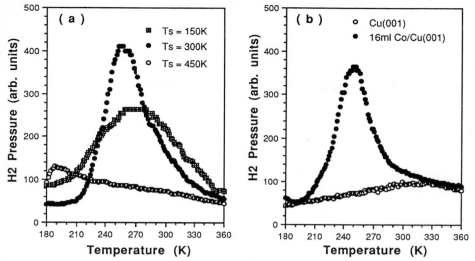

Figure 1: Programmed thermal desorption spectra for hydrogen saturation on (a) a 2 monolayer thick film of fcc Co/Cu(001) grown at different substrate temperatures, Ts = 150 K, 300 K, and 450 K; (b) the Cu(001) substrate compared with a 16 monolayer thick (bulk) cobalt film.

Figure 1a shows a comparison of hydrogen desorption spectra obtained from 2.0 monolayer thick films grown at substrate temperatures of 150 K, 300 K, and 450 K after dosing to saturation coverage at 150 K. The shape of the thermal desorption spectra, TDS, peak at 260 K shows the second order kinetics expected from a relatively smooth single-adsorbate-site surface for the film deposited at 300 K [6]. This peak is broadened and shifted to higher temperature for films deposited at the lower temperature of 150 K indicative of desorption from a rough multisite surface. In complete contrast, the very smooth films grown at 450 K show no distinct TDS peak. This is indicative of copper segregating atop the cobalt layer at these higher preparation temperatures, which we have confirmed by the comparison of hydrogen TDS spectra from bulk Cu(001) and a thick 16 monolayer film of cobalt shown in Figure 1b. The results clearly indicate that the films grown at substrate temperatures of 450 K are coated with a layer of copper. Furthermore, this copper overlayer is ordered, as evidenced by the evolution of the Cu(001) surface state at the M-bar point of the surface Brillouin zone,which we have observed in photoemission [7]. This conclusion is further supported by annealing the 150 K and 300 K films to 450 K and observing the copper surface segregation, both in the gas titration results(Fig 1a) and photoemission. These measurements strongly point to a substrate deposition temperature of the order of 300 K as the best compromise for growing thin smooth layers of fcc Co/Cu(001) but at the same time limiting surface segregation.

Magnetic hysteresis curves of magnetization vs. applied field were recorded in situ with a UHV SMOKE instrument [8]. We have measured the changes in surface magnetism simultaneously with hydrogen chemisorption. Figure 2 illustrates the effect of the hydrogen adsorption upon the magnetic hysteresis

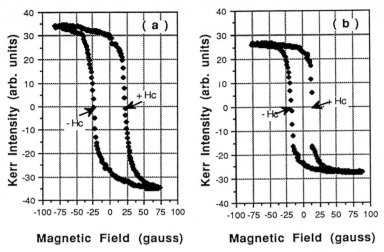

Figure 2: Magnetic hysteresis curves M vs. B showing the variations of the Kerr signal intensity with applied magnetic field for (a) a clean 2 monolayer cobalt film grown at 300 K and measured at 150 K, (b) the same film exposed to a saturation coverage of hydrogen.

loop of a 2.0 monolayer cobalt film grown at a substrate temperature of 300 K and measured at 150 K. We observe both a change in the coercivity, H_c , as well as the Kerr intensity which is proportional to the magnetization, M. Applications of higher external fields do not alter H_c or the maximum Kerr intensity indicating that the films are indeed saturated. It is known that hydrogen reduces the net magnetic moment of transition metal ferromagnets [9]. Band narrowing at the top of the ferromagnetic layer increases the surface magnetic moment. Theoretical calculations predict a surface magnetic moment of 1.78 μ_B compared to a bulk fcc cobalt moment of 1.64 μ_B [10]. Interaction of the hydrogen with the surface valence states is thought to fill the cobalt minority band and diminish the enhanced surface magnetic moment. For the 2 monolayer film illustrated in Figure 2, we find a reduction in film saturation magnetization of 19 percent with hydrogen saturation.

Changes in saturation magnetization, M_{sat}, and H_c with hydrogen exposure are shown in Figure 3 for a 2 monolayer films grown at 150 K, 300 K, and 450 K. Further results for 1.5 monolayer films and the effects of annealing are summarized in Table 1. We see that the changes in M_{sat} and H_c are most pronounced for the 150 K and 300 K films. The differing 150 K and 300 K film microstructure is seen to increase the H_c of the rougher 150 K films. The smoother 300 K films show increased sensitivity to hydrogen adsorption (Table 1). The films grown at 450 K remain insensitive to hydrogen adsorption due to the copper overlayer discussed above. Annealing of rough or island films is generally expected to reduce defects and thereby decrease the coercivity. The coercivity of the 150 K and 300 K films are reduced to a value near that of the 450 K films by both annealing to 450 K and hydrogen exposure.

424

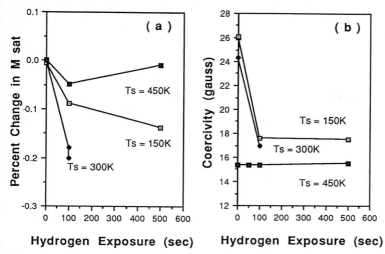

Figure 3: Variations of (a) the saturation magnetization, Msat, and (b) the coercivity of a 2 monolayer film of fcc Co/(Cu(001) as a function of hydrogen exposure, for three films deposited at substrate temperatures of Ts = 150 K,300 K, and 450 K then measured at 150 K.

Table 1: The effects of hydrogen saturation exposure and annealing to 450 K on the coercivity, Hc, and saturation magnetization, Msat , of 1.5 and 2.0 monolayer cobalt films grown at substrate temperatures Ts = 150 K, 300 K, and 450 K. All measurements were made at 150 K except '*' which were measured at 210 K. Coercivities are in gauss and changes in Msat, Δ Msat are relative to the clean film.

Growth Temperature (K)	Film Thickness (ml)	ΔMsat	ΔMsat annealed	Hc	Hc with H2	Hc annealed	Hc with H2 annealed
150	1.5	-0.14	-0.11	18	13	14	13
	2.0	-0.12	-0.09	26	17	20	16
300	1.5	-0.54		15	14		
	2.0	-0.19		24	17		
450	1.5	+0.03		14	14		
	2.0	-0.03	+0.03	15	15	14*	14*

In summary, our results indicate that the morphology of ultrathin cobalt films on Cu(001) is very sensitive to substrate deposition temperature. The hydrogen titration studies show clearly that smoother more uniform films are produced at substrate temperatures of 300 K; lower temperatures produce rougher surfaces; higher substrate temperatures promote surface segregation which lowers the surface free energy [11]. A similar surface segregation growth mode has recently been reported for the Rh / Ag(100) system [12], which is also an immiscible system but with a larger difference in surface free energies. These changes in film microstructure are reflected in the M vs. H hysteresis behavior, particularly in response to hydrogen chemisorption. Exposure to hydrogen reduces the saturation magnetization and also decreases the coercivity for the 150 K and 300 K films. This indicates that hydrogen chemisorption diminishes the cobalt magnetic moment but also exerts a subtle effect on surface two-dimensional magnetic structure which is yet to be understood.

Acknowledgement: This work was supported by grant NSF-DMR 881884-01.

References

1 A. Clarke, G. Jennings, R.F. Willis, P.J. Rous, J.B. Pendry, Surf. Sci. **187**, 327 (1987).
2 Hong Li, B.P. Tonner, Phys. Rev. **B40**, 10241 (1989).
3 J.J. de Miguel, A. Cebollada, J.M. Gallego, S. Ferrer, R. Miranda, C.M. Schneider, P. Bressler, J. Garbe, K. Bethke, J. Kirschner, Surf. Sci. **211**, 732 (1989).
4 C.M. Schneider, P. Bressler, P. Schuster, J. Kirschner, J.J. de Miguel, R.Miranda, Phys. Rev. Lett. **64**, 1059 (1990).
5 W.F. Egelhoff, Jr., I. Jacob, Phys. Rev. Lett. **62**, 921 (1989).
6 P.A. Redhead, Vacuum **12**, 203 (1962).
7 G.J. Mankey, M.T. Kief, R.F. Willis, to be published.
8 M.T. Kief, R.F. Willis, to be published.
9 M. Weinert, J.W. Davenport, Phys. Rev. Lett. **54**, 1547 (1985); M. Streszewski, C. Jedrzejek, Phys. Rev. B **34**, 3750 (1986).
10 Chun Li, A.J. Freeman, C.L. Fu, J. Magn. Magn. Mat. **83**, 51 (1990).
11 A.R. Miedema, Z. Metallkde. **69**, 455 (1978); P.M. Ossi, Surf. Sci. **201**, L519 (1988).
12 P.J. Schmitz, W.-Y. Leung, G.W. Graham, P.A. Thiel, Phys. Rev. **B40**, 11477 (1989).

Disordered Adsorption of O and S on Ni(100) Studied by Diffuse LEED

U. Starke[1],*, W. Oed[1], P. Bayer[1], F. Bothe[1], G. Fürst[1], P.L. de Andres[2], K. Heinz[1], and J.B. Pendry[3]

[1]Lehrstuhl für Festkörperphysik, University of Erlangen-Nürnberg,
 Staudtstr. 7, W-8520 Erlangen, Fed. Rep. of Germany
[2]Instituto de Materiales, C.S.I.C., c/Serrano 144, E-28006 Madrid, Spain
[3]The Blackett Laboratory, Imperial College, London SW7 2BZ, UK
*Present address: Materials and Chemical Sciences Division,
 Lawrence Berkeley Laboratory, 1 Cyclotron Road, Berkeley, CA 94720, USA

Disordered adsorption for both oxygen and sulfur on Ni(100) at low coverages and temperatures were studied by the technique of diffuse LEED (DLEED). The results support the idea of two coexisting adsorption clusters: isolated atoms in four-fold hollow sites and atoms in asymmetric sites due to incomplete dissociation.

1. Introduction

The chemisorption of oxygen and sulfur on clean Ni(100) in each case shows two ordered overlayer structures after annealing at elevated temperatures. However, adsorption at low temperatures of 80 K without annealing leads to a disordered arrangement of adatoms on the surface, which shows diffuse intensity distributions in low energy electron diffraction (DLEED). Recently it has been shown that DLEED intensities can be reliably measured with modern data acquisition techniques [1]. Models of local geometry can be obtained by comparison of the experiment with multiple scattering calculations inside a finite adsorbate-substrate cluster [2]. Using improvements of the experimental and theoretical tools, this paper reinvestigates the local structure of disordered O/Ni(100) using a larger data base than was available earlier [3] and for the first time looks into the geometry of disordered S/Ni(100).

2. Experimental details

The Ni(100) single crystal was the same as used in an analysis of the ordered overlayers and so the same cleaning procedure was applied as described recently [4], including heating and sputtering cycles. To prepare the adsorbate layers the crystal was exposed to molecular oxygen and hydrogen sulfide, respectively, at temperatures of 80 K and low coverages $\Theta \leq 0.25$ ML, where both molecules are known to dissociate and the atoms remain disordered on the surface.

The diffuse LEED intensities arising from both structures were measured with a conventional LEED optics. The two-dimensional diffraction patterns were taken from the fluorescent screen using a video camera operated under computer control. A detailed description of the data acquisition system and instrumental parameters in the experiment has been published elsewhere [1]. In the experiment the mutual arrangement of adsorbate atoms is unknown. So undetermined interference effects influence the diffuse intensity distributions. To extract local structure information Y-function maps have to be generated

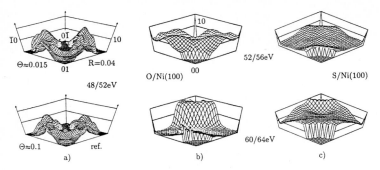

Figure 1: Experimental Y-functions. The displayed reciprocal space is indicated by beam positions.
- a) Different oxygen coverages ($\Theta \approx 0.015$ and $\Theta \approx 0.1$). R-factor comparison of the maps shown at 48/52 eV yields 0.04.
- b) Energy sets 52/56 eV and 60/64 eV for oxygen ($\Theta \approx 0.1$).
- c) Same as b) for sulfur.

from intensity maps [5]. In fact these maps are found to be independent of coverage below 0.25 ML as illustrated in fig. 1a, indicating that the local structure is fairly stable in this coverage range. To exclude any contributions to the intensities from multiple scattering processes between different adsorbates a coverage of 0.1 ML was selected for the final experiments. A total of 8 intensity maps was taken between 48 eV and 76 eV, each separated by 4 eV. Neighbouring pairs were used to generate 7 Y-function maps from 50 eV to 74 eV . At two energies these experimental Y-functions are displayed for oxygen and sulfur, respectively, in fig. 1b and c.

3. Model calculations

The theoretical calculations were performed using the three step model established by Pendry [2] that determines electron scattering at the adsorbate and surrounding substrate atoms with an XANES like cluster approach and considers preceding and successive interactions in the periodical substrate by means of conventional LEED algorithms. Additionally local reconstructive movements of substrate atoms were taken into account using the Tensor LEED technique [6]. The phase shifts were adopted from the ordered phase analyses [4,7]. Because of the low energies only $l_{max} = 5$ was necessary. The electron attenuation was modeled by $V_{0i} = $ -4eV. To find the most likely model experimental and theoretical data were compared in a trial and error procedure by means of the Pendry R-factor [8].

The enlarged experimental data base enabled us to vary a large number of parameters in the geometry models. As in our earlier analysis of disordered oxygen both the adsorbate height d_{01} and a lateral displacement s out of the hollows towards the bridge (pseudo-bridge model) were varied. The experience from the analyses of the ordered phases of both O and S on Ni(100) [4,7] guided the selection of reconstructive parameters, which we included: In addition to an overall relaxation of the interlayer spacing D_{12} of the substrate we have considered perpendicular movements dA' of the four first layer Ni atoms A' forming the hollows (see fig. 2). Perpendicular deviations dA, dB and dC of second layer atoms of type A, B and C were allowed in the model calculations. The parameters D_{12}, dB and dC had only little influence on the theoretical intensities and were varied independently in order to save computer time.

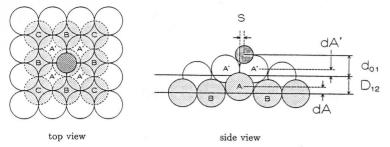

top view   side view

Figure 2: Model and geometry parameters considered in the calculation. Substrate atoms are labeled in the top view. The side view corresponds to a cut in [110] direction identical to the direction of the displacement s. The adsorbate is drawn in hatched pattern.

4. Structure analysis

As best fit result for oxygen we found a pseudo-bridge site yielding an R-factor of 0.198. Adsorbed at $d_{01} = 0.86$ Å above the surface the oxygen is displaced towards the bridge by $s = 0.45$ Å. While the first layer of the substrate is unaffected by the adsorbate ($dA' = 0$) a reconstructive movement was detected in the second layer where atom A is lifted by $dA = 0.07$ Å. Additional lateral movement of the two Ni atoms A' that are forming the pseudo-bridge, off the oxygen by up to 0.15 Å did not impair the fit. Even more it is very likely as it brings the oxygen nickel bond lengths into an acceptable range. Now the distances of oxygen to the five nearest nickel atoms are $d_{OA'near} = 1.85$ Å, $d_{OA'far} = 2.23$ Å and $d_{OA} = 2.56$ Å. The behaviour of the R-factor is shown in two-dimensional maps in fig. 3a and b as function of s versus d_{01} and dA', respectively. However, in addition to the pseudo-bridge at a slightly higher R-factor of 0.241, we found a local minimum for a symmetrical hollow position including a vertical lift of all the five atoms A and A' out of the surface by 0.035 Å. For both models $D_{12} = 1.74$ Å and $dB \approx dC \approx 0$ was found by a separate variation.

For sulfur adsorption we found a similar behaviour. The data also favour a pseudo-bridge site with $d_{01} = 1.2$ Å and $s = 0.6$ Å without evidence for reconstruction in the substrate. The R-factor of this model is 0.225. A second minimum of R = 0.308 appears for the hollow site at the same adsorption height of 1.2 Å with a small upwards movement of atom A by $dA = 0.035$ Å. No first layer reconstruction ($dA' = 0$) is found in either

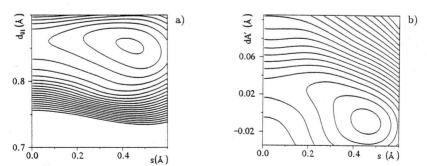

Figure 3: R-factors for O/Ni(100) pseudo-bridge model as function of adsorption height (a) and first layer reconstruction (b) versus oxygen displacements.

429

model, while further parameters were fixed at $D_{12} = 1.74$ Å and $dB = dC = 0$. As in the ordered adsorption phases, sulfur shows a smaller influence on the substrate geometry.

5. Discussion

Pseudo-bridge adsorption of oxygen on Ni(100) was first proposed for a c(2x2) structure [9]. However, a refined analysis did not confirm this result and favoured the symmetric hollow site [4]. The pseudo-bridge site appeared only as a local minimum of the fit at a higher R-factor. So caution seems to be advisable as the present analysis again produces this model. Nevertheless, it is quite conceivable that at low temperatures both oxygen and sulfur occupy low-symmetry adsorption sites. Certainly both molecules used for preparation, O_2 and H_2S, dissociate at 80 K [10,11], but it is questionable if the thermal energy is sufficient that after the dissociation the atoms separate completely by diffusion. In the case of oxygen adsorbed in next neighbour hollows the two atoms presumably would repel each other. Logically a displacement out of the perfect four fold coordination would be the consequence, implying the pseudo-bridge model. Indeed recently a high resolution electron energy loss spectroscopic (HREELS) investigation [12] found vibrational modes that were explained by a mixture of two different adsorption clusters. One was proposed to be the incomplete dissociated pair of atoms, the other a single oxygen in a hollow site. We assume that we have determined the geometry of these two adsorption sites coexisting at 80 K. Consequently we tried to fit the experimental data with a mixture of calculated data of both models. Unfortunately, whether by adding amplitudes or intensities, we could not determine a clear minimum for a specific mixing ratio; but the R-factors did not worsen. However, it has to be kept in mind that a basic assumption in the present DLEED method is that only identical adsorption sites are occupied; this allows the extraction of local structure information by means of Y-functions [1]. So mixing of different adsorption sites possibly should be modeled by a more accurate procedure than averaging amplitudes or intensities of single adsorbate clusters. We also have to admit that even for the pure pseudo-bridge model there are 4 orthogonal off-center displacements which are equally probable but symmetrically inequivalent. So, interference effects may not cancel in the Y-function construction and possibly some systematic error is introduced.

To support the idea of asymmetric adsorption in the case of sulfur the same arguments could be given as for oxygen. If the hydrogen atoms don't migrate off the sulfur they certainly would influence the binding of the sulfur atom. However, experimental and theoretical data do not agree as well as in the oxygen case. Additionally no information from other methods is presently available for the low coverage phase. To finally resolve the structure of disordered sulfur on Ni(100), further attempts appear to be necessary.

Acknowledgements

The authors are indebted to M.A. van Hove for helpful discussions. Support by the Deutsche Forschungsgemeinschaft and the Höchstleistungsrechenzentrum für Wissenschaft und Forschung is gratefully acknowledged.

Literature

[1] U.Starke, P.Bayer, H.Hloch and K.Heinz, Surf. Sci. <u>126</u>, 325 (1989)
[2] J.B.Pendry and D.K.Saldin, Surf. Sci. <u>145</u>, 33 (1984)
[3] U.Starke, P.L.de Andres, D.K.Saldin, K.Heinz and J.B.Pendry, Phys. Rev. B <u>38</u>, 12277 (1988)

[4] W.Oed, H.Lindner, U.Starke, K.Heinz and K.Müller, Surf. Sci. 224, 179 (1989)
[5] K.Heinz, D.K.Saldin and J.B.Pendry, Phys. Rev. Lett. 55, 2312 (1985)
[6] P.J.Rous, J.B.Pendry, D.K.Saldin, K.Heinz, K.Müller and N.Bickel, Phys. Rev. Lett.
     57, 2951 (1986); An advanced version has been implemented into the program.
[7] W.Oed, H.Lindner, U.Starke, K.Heinz, K.Müller, D.K.Saldin, P.L.de Andres and
     J.B.Pendry Surf. Sci. 225, 242 (1990); W.Oed, U.Starke, F.Bothe and K.Heinz,
     Surf. Sci., in press; U.Starke, F.Bothe, W.Oed and K.Heinz, Surf. Sci., submitted
[8] J.B.Pendry, J. Phys. C 13, 937 (1980)
[9] J.E.Demuth, M.J.DiNardo and G.S.Cargill, Phys. Rev. Lett. 50, 1373 (1983)
[10] eg. M.Rocca, S.Lehwald and H.Ibach, Surf. Sci. 163, L738 (1985)
[11] A.G.Baca, M.A.Schulz and D.A.Shirley, J. Chem. Phys. 81, 6304 (1984); R.McGrath,
     A.A.MacDowell, T.Hashizume, F.Sette and P.H.Citrin, Phys. Rev. B 40, 9457 (1989)
[12] R.Franchy, M.Wuttig and H.Ibach, Surf. Sci. 215, 65 (1989)

A Directed LEED Search for Many Structural Parameters: Substrate Relaxations and Buckling in Mo(100)-c(2 × 2)-C and -S

P.J. Rous*, D. Jentz, D.G. Kelly**, R.Q. Hwang***, M.A. Van Hove, and G.A. Somorjai

Department of Chemistry, University of California, Berkeley, and
Center for Advanced Materials, Materials and Chemical Sciences Division,
Lawrence Berkeley Laboratory, 1 Cyclotron Rd., Berkeley, CA 94720, USA
*Present address: The Blackett Laboratory, Imperial College,
  London SW7 2BZ, U.K.
**Present address: Rohm and Haas Co., Research Laboratories,
  727 Norristown Road, Spring House, PA 19477, USA.
***Present address: Institut für Kristallographie und Mineralogie,
  Universität München, Theresienstrasse 41,
  W-8000 München 2, Fed. Rep. of Germany

Abstract A newly developed automated structural search method is applied to the determination of as many as 15 unknown structural parameters for two different adsorption structures on Mo(100). The method is based on tensor LEED coupled with a steepest descent search that locates a minimum in the R-factor in a high-dimensional parameter space. The results are very stable despite the large number of fitted parameters and give very reasonable structures that exhibit adsorbate-induced first and second layer relaxations in the substrate. A second-layer buckling is detected due to S adsorption. It is shown that an experimental angle of incidence accidentally set off-normal by about 1° can induce artificial lateral relaxations in the surface atoms of about 0.1Å.

1 Introduction

Low-energy electron diffraction (LEED) has traditionally employed a trial-and-error search controlled at each step by a human intervention [1]; an approach which becomes very cumbersome and unreliable when it is applied to complex surfaces with a large number of unknown structural parameters. Recently, automated optimization procedures for LEED have been presented [2,3], which allow such complex structures to be solved far more efficiently and reliably.

The method which we have recently proposed [3] combines the complementary techniques of tensor LEED ·and a numerical search algorithm. The significant advantages of this approach have been described in detail elsewhere [3], and have been illustrated there by application to the solution of simple low-dimensional problems with 3 to 4 unknown structural parameters.

Here we apply the same method to relatively simple, but nevertheless high-dimensional, problems containing 15 unknown structural parameters. This allows us to explore the stability, reliability and performance of the method which are essential to the proper solution of complex structures, especially when even the qualitative nature of their structures are not easily guessed.

Springer Series in Surface Sciences, Vol. 24   **The Structure of Surfaces III**
Editors: S.Y. Tong · M.A. Van Hove · K. Takayanagi · X.D. Xie
© Springer-Verlag Berlin, Heidelberg 1991

We have chosen the systems Mo(100)-c(2x2)-C and Mo(100)-c(2x2)-S. Based on many previous similar adsorbate systems, these are expected to display hollow-site adsorption with bond lengths close to covalent values. An earlier, trial-and-error, LEED analysis of Mo(100)-c(2x2)-S [4] determined 4 structural parameters. That LEED analysis, as well as an earlier ion scattering study [5] and total-energy calculations [6], implicate hollow-site adsorption. This provided us with an excellent starting point for our multi-dimensional analysis.

A surprising result of our more recent LEED analysis [4] was lateral displacements of the sulfur atoms in which the adatoms were found to reside about 0.18Å away from the hollow sites towards the nearest bridge sites. Other studies [5,6] found no evidence for such an asymmetrical site. This encouraged us to analyze possible lateral relaxations in the substrate, requiring a multiparameter fit (clean Mo(100) reconstructs with such lateral relaxations [4]). We were also aware that in a recent study of Ni(100)-c(2x2)-O [7], it emerged that lateral relaxations could be artificially induced by neglecting perpendicular relaxations in the second substrate layer.

We have also explored another possible cause of the surprising asymmetries found earlier. We investigated the effect of an unintentional deviation from normal incidence. If experimental data are measured at off-normal incidence and compared with calculations that assume normal incidence, one may pay for that inconsistency in the form of artificial structural results. This becomes especially likely when one allows many structural parameters to relax freely.

For this reason a new set of LEED IV spectra for Mo(100)-c(2x2)-S was measured, with closer attention to setting normal incidence accurately. As we shall report below, this resulted in elimination of the asymmetry found in the structural parameters.

2   Experimental

Experimental details have been described elsewhere [4,8]. Briefly, the Mo(100) sample was cleaned by heating in oxygen and flashing to 2000K to remove the oxide. Sulfur was deposited on the Mo surface from an electrochemical cell (Pt/AgI/Ag$_2$S) and the surface annealed to obtain a sharp LEED pattern. Carbon was deposited by dissociation of hydrocarbons.

For the analysis of Mo(100)-c(2x2)-C structure we used 10 symmetrically inequivalent beams over an energy range of 70-310eV for comparison with theory. For Mo(100)-c(2x2)-S we used 8 beams over an energy range of 60-250eV. Only data measured at (or very near) normal incidence were used.

3   Theoretical

Our approach [3], combines tensor LEED to explore distortions from a guessed surface structure with an automated search to optimize the unknown parameters. Tensor LEED [9] expands the LEED amplitudes linearly in terms of deformations of a surface structure from a reference structure for which a full dynamical LEED calculation is performed . With tensor LEED, the intensities for any distortion of this reference structure can be explored with little computational effort provided that no atom is moved more than 0.4Å from its position in the reference surface (beyond 0.4Å the tensor LEED approximation becomes unreliable).

In our approach tensor LEED is coupled to an automated R-factor optimization scheme which minimizes the misfit between the calculated and measured IV spectra (in this study we used the Pendry R-factor R_P). Automated searching greatly enhances the efficiency of the search, particularly in high-dimensional problems, since a relatively very small part of the large structural phase space is actually sampled. No such search strategy can distinguish between local minima and the global R-factor minimum. However, for the structures examined in this paper, we only found multiple local minima that are mutually related by rotation and mirror symmetry operations of the surface: they are thus actually only one and the same minimum, repeated symmetrically.

The fact that an unknown error in the angle of incidence might be expected to lead to artificial asymmetric structures is easily appreciated in the kinematic limit where an error in $\delta k_\parallel$ can be compensated for by a false lateral displacement of a surface atom $\delta \mathbf{r}_\parallel$.

Recently we have demonstrated [10] that such artifacts can occur in the full dynamical limit by performing the following controlled computer experiment for Ni(100)-c(2x2)-O. First we calculated IV spectra of a surface in which the O atom occupied the symmetric hollow site for angles of incidence between 0° and 3° towards the bridge position. These "experimental" spectra were then compared to those calculated assuming normal incidence but in which the adsorption position of the O atom was free to vary both parallel and perpendicular to the surface. Despite averaging over the four symmetrically equivalent domains, an angle of incidence error of 1° produced an artificial minimum in the R-factor for a lateral displacement of the adatom by 0.1Å towards the bridge site. For this artificial structure the Pendry R-factor R_P is 5% below that for the (correct) hollow site while the Zannazzi-Jona R-factor is reduced by 10% for a lateral displacement of 0.2Å. Clearly, the potential for structural artifacts exists if the angle of incidence is not controlled to much better than 1°. In this paper we present experimental confirmation of this effect in the case of Mo(100)-c(2x2)-S.

4 Results and Discussion

The structure search for both Mo(100)-c(2x2)-C and Mo(100)-c(2x2)-S allowed the 3 topmost atomic layers to relax fully and also optimized the muffin-tin zero. In each c(2x2) unit cell, one adatom and two Mo atoms in each of the first and second substrate layers could adjust their 3 coordinates freely: a total of 1 non-structural and 15 structural parameters. A typical search sequence examined 1000 trial structures, after which the coordinates of each atom were stationary to within 0.001Å. Many searches were initiated from different reference structures and base points, resulting in a total of close to 50,000 trial structures being tested. The search seemed remarkably stable, in each case finding the same R-factor minimum (within 0.005Å) provided that the reference structure was located within the 0.4Å radius of convergence of

Table 1. Optimized structural parameters for Mo(100)-c(2x2)-C and Mo(100)-c(2x2)-S labelled according to figure 1. All lengths are in Å.

Surface	d_{AMo}	d_{12}	d_{23}	b_2	R_P
Mo(100)-c(2x2)-S	1.00	1.54	1.57	0.16	0.234
Mo(100)-c(2x2)-C	0.43	1.56	1.57	0.00	0.464

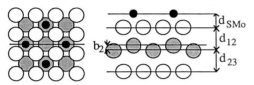

Figure 1. Schematic diagram of the Mo(100) c(2x2) adsorption structures showing a top (left panel) and side view (right panel) through the (010) plane.

the tensor LEED. Thus our structural results for both Mo(100)-c(2x2)-C and Mo(100)-c(2x2)-S appear to exhibit "noise" on a scale of much less than 0.01Å despite the finite data set and experimental as well as theoretical uncertainties.

The optimum structural parameters are summarized in Table 1. Mo(100)-c(2x2)-C appears to be a fairly conventional adsorption structure with no evidence for either asymmetric adsorption or substrate buckling. Lateral displacements of the C atom and Mo atoms within the first two substrate layers were less than 0.005Å from the symmetric sites. In the case of Mo(100)-c(2x2)-S we find no evidence for asymmetric adsorption; the lateral displacements of both the S atom and Mo substrate atoms were less than 0.005Å. There is, however, evidence for a 0.16Å buckling of the second Mo layer in which the atom below the sulfur adatom is pulled towards the surface. However, the R-factor is not very sensitive to this second-layer buckling which lowers R_P from 0.246 to 0.234. These structural results for both Mo(100)-c(2x2)-C and Mo(100)-c(2x2)-S are very satisfactory when compared with other known surface structures and bond lengths [1].

In the light of the absence of lateral relaxations in Mo(100)-c(2x2)-S it seems likely that the incorrect alignment of the angle of incidence in our previous study [4] was responsible for the asymmetric adsorption site seen there. This is confirmed by the significantly lower R-factor value of 0.234 (compared to 0.461) obtained in this analysis.

5 Conclusions

We have exhibited the stability and reliability of a 15-parameter structural fit of LEED IV curves to solve two relatively simple surface structures. This achievement generates confidence in the application of this approach to solve more complex structures where the structure is less easily guessed a priori.

An asymmetry in the Mo(100)-c(2x2)-S structure (which we reported earlier [4] and could reproduce with the present method) was probably induced by an inaccurate experimental alignment of normal incidence. This illustrates the need for accurate data when fitting more free parameters since additional free parameters can artificially compensate for any experimental errors.

Acknowledgements. PJR thanks the U.K. SERC for support. This work was supported by the Director, Office of Energy Research, Office of Basic Energy Sciences, Materials Science Division of the US Department of Energy, Contract No. DE-AC03-76SF00098. Supercomputer resources were provided in part by the Office of Energy Research of the Department of Energy.

References

1. J.B. Pendry, **Low Energy Electron Diffraction**, (Academic Press, London, 1974); M.A. Van Hove, W.H. Weinberg and C.-M. Chan, **Low-Energy Electron Diffraction: Experiment, Theory and Surface Structure Determination**, Springer-Verlag (Heidelberg), 1986.

2. P.G. Cowell, M. Prutton and S.P. Tear, Surf. Sci. **177** L915 (1986); G. Kleinle, W. Moritz, D.L. Adams and G. Ertl, Surf. Sci. **219** L637 (1988).

3. P.J. Rous, M.A. Van Hove and G.A. Somorjai, Surf. Sci. **226** 15 (1990).

4. D.G. Kelly, R.F. Lin, M.A. Van Hove and G.A. Somorjai, Surf. Sci. **224** 97 (1989).

5. B.M. Dekoven, S.H. Overbury and P.C. Stair, J. Vac. Sci. Technol. **A3** 1640 (1985).

6. X.W. Wang and S.G. Louie, Surface. Sci. **226** 257 (1990).

7. W. Oed, H. Lindner, U. Starke, K. Heinz, K. Müller and J.B. Pendry, Surf. Sci. **224** 179 (1989).

8. D.G. Kelly, Ph.D. Dissertation, University of California, Berkeley (1987).

9. P.J. Rous, J.B. Pendry, D.K. Saldin, K. Heinz, K. Müller and N. Bickel, Phys. Rev. Lett. **57** 2951 (1986); P.J. Rous and J.B. Pendry, Surf. Sci. **219** 355 (1989); *ibid* **219** 373 (1989);

10. P.J. Rous, M.A. Van Hove and G.A. Somorjai to be published.

Lattice Gas Models for Order–Disorder Behaviour of Oxygen on Ruthenium(001)

K. De'Bell[1,*], *H. Pfnür*[2], *and P. Piercy*[3]

[1]Department of Physics, Memorial University,
St. John's, Newfoundland, Canada
[2]Fakultät für Physik E20, Technische Universität München, W-8046 Garching,
and Institut für Festkörperphysik, Universität Hannover,
W-3000 Hannover 1, Fed. Rep. of Germany
[3]Department of Physics, University of Ottawa, Ottawa, Ontario, Canada
*Permanent address: Physics Department, Trent University, Peterborough,
Ontario, Canada K9J 7B8

Abstract. The chemisorption of oxygen on the Ru(001) surface shows ordered p(2x2) and p(1x2) phases at coverages of 1/4 and 1/2 monolayer, respectively, that undergo apparently continuous transitions to a disordered state upon heating. We report here on lattice gas model calculations aiming to explain the observed phase diagram, using Monte Carlo simulations based on a phenomenological Hamiltonian describing pairwise interactions between adatoms. Restricting adsorption of oxygen atoms to the triangular lattice of the hcp-type hollow sites as seen experimentally, repulsive first- and second-nearest neighbour interactions are sufficient to drive the observed ordering behaviour at coverages $\theta \leq 0.4$. This model is then generalized to allow oxygen adsorption on both hcp- and fcc- hollow sites, with a difference in binding energy between the two types of sites. Repulsive interactions within the adsorbate lead to a small spill-over of atoms onto the sites of higher energy, markedly reducing the transition temperature at 1/2 monolayer coverage, in agreement with the experimental results on O/Ru(001).

1. Introduction

Order-disorder phenomena in chemisorbed, submonolayer systems provide unique examples of two-dimensional phase transitions, and studies of their phase diagrams have provided phenomenological understanding of lateral interactions within the adsorbate layer.[1] The adsorption of oxygen on the ruthenium(001) surface displays well-characterized ordering behaviour that may be interpreted in a lattice gas picture, permitting a simplified, empirical modeling of the energetics of the adsorbate interactions. The critical exponents of order-disorder transitions in this system have also been studied in detail recently.[2]

At coverages of less than 1/2 monolayer, oxygen is found to chemisorb on the hcp-type hollow sites of the Ru(001) surface,[3,4] with a binding energy of $\approx 6eV$.[5] Ordering in a p(2x2) pattern at a coverage $\theta = 1/4 ML$ gives way to a p(1x2) structure as the coverage is increased to $1/2 ML$, where a buckling type of reconstruction reduces the symmetry of the hollow sites, as found by recent LEED I-V studies.[3] The degree of order in the adsorbate may be monitored in a LEED experiment by measuring the the intensity of the 1/2-order diffraction beams due to the oxygen superstructure, as a function of coverage and temperature. At low temperatures, the LEED intensity of the 1/2-order spots is peaked at coverages of 1/4 and 1/2 ML.[2,5] (The oxygen coverage was determined principally by Auger and LEED measurements.[2])

The experimentally determined phase diagram for the O/Ru(001) system is shown in Fig. 1.(a). The phase boundary was determined by observing the decrease in inte-

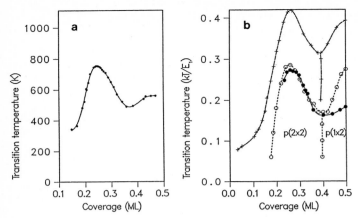

1. Order-disorder phase boundary. (a) O/Ru(001) experiment, as determined by LEED. (b) Open circles: Triangular lattice gas model with first- and second-neighbour interactions $E_2 = 0.3E_1$; Crosses: As above but including third-neighbour interaction $E_3 = -0.1E_1$; Filled circles: Honeycomb lattice gas model as described in text.

grated intensity of an oxygen-induced LEED beam as the temperature was increased from the ordered to disordered phase, and shows peak transition temperatures at coverages of 1/4 and 1/2ML.[6] (Quantitative measurements of the LEED intensities were made using a Faraday cup controlled by stepping motors.[7]) Similar measurements of diffraction beam angular profiles show no broadening at temperatures below the order-disorder transition line, indicating single-phase ordered regions with no island-type structures.[7] However, experiments were not possible at temperatures below about 300 K, due to slow ordering kinetics in the adsorbate. Note that no $(\sqrt{3} \times \sqrt{3})$ R 30° phase was observed, in contrast to the phase diagram for the O/Ni(111) system.[8]

In this paper we attempt to understand the O/Ru(001) phase diagram in terms of lattice gas models in which oxygen atoms order due to effective pair-wise interactions between neighbours. The use of a lattice gas model to study adsorption assumes that the adsorbed species are bound to localized sites, in both ordered and disordered phases, and that the transition states crossed by an adatom in hopping from one site to another are insignificantly occupied. Experimental evidence of lattice gas behaviour is found in recent VLEED studies of the O/Ru system.[4] VLEED (very low energy electron diffraction) can be sensitive to binding site and to local reconstruction, and shows oxygen adsorption restricted predominantly to hcp-hollow sites, in both ordered and disordered phases.[4]

2. Lattice Gas Models

Two lattice gas models will be discussed here.[8-10] In the first model considered, oxygen atoms are restricted to a triangular lattice of hcp-type hollow sites, since LEED structural studies do not detect a significant population of another site-type. Since the many-body energy surface describing the motion of oxygen atoms at a Ru(001) surface is not known, we will approximate it in terms of effective, pair-wise interactions within the adsorbate. The Hamiltonian[9]

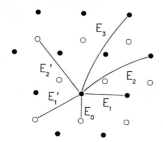

2. Lattice gas models. The interaction energies between the first five nearest neighbours are shown. Numerical values for the energies are given in the text. Filled and open circles represent the two types of hollow sites.

$$H = E_1 \sum_{(ij)_1} n_i n_j + E_2 \sum_{(ij)_2} n_i n_j + E_3 \sum_{(ij)_3} n_i n_j \tag{1}$$

includes pair-wise interaction energies E_m between the mth-neighbours $(ij)_m$, where m = 1,2,3 for first, second and third nearest neighbours, as shown in Fig. 2, and n_i is the occupation variable of the ith site.

In the second theoretical model considered, we allow oxygen adsorption at both hcp- and fcc-type hollow sites[8], by explicitly including a small difference in binding energy F between these two sites.[11] The phenomenological Hamiltonian

$$H = E_0 \sum_{(ij)_1} n_i n_j + E_1 \sum_{(ij)_2} n_i n_j + E_1' \sum_{(ij)_3} n_i n_j + E_2' \sum_{(ij)_4} n_i n_j + E_2 \sum_{(ij)_5} n_i n_j + F \sum_i{}' n_i \tag{2}$$

includes pair-wise interactions up to fifth-nearest neighbours on the honeycomb lattice of hollow sites, as depicted in Fig. 2. Simultaneous occupation of nearest neighbour hollow sites on the honeycomb lattice, which are separated by only 1.6 Å, is suppressed, effectively setting $E_0 = \infty$. In the final term in Eq. (2), the primed sum runs over all sites on the triangular sublattice of the less favoured adsorption site. This form of Hamiltonian was applied previously to the O/Ni(111) system by Roelofs et al[8].

The thermodynamics of these lattice gas systems is computed using Monte Carlo (Metropolis) algorithms.[12] Simulations based on Eq. (1) used Kawasaki-type kinetics, while those for Eq. (2) used Glauber-type kinetics.[10] We use an order parameter whose magnitude, ψ, is proportional to the square root of the intensity of the 1/2-order diffraction spot, taken in the Born approximation and averaged over the three symmetry directions, as employed previously by Glosli and Plischke[9]. It is noted that the order parameter does not decrease precisely to zero as the order-to-disorder transition is crossed in a system of finite size, so we define the transition temperature by the inflection point of the graph of ψ versus temperature.[13] The simulations were carried out on lattices of 1600 to 5929 sites, computing 5,000 to 200,000 MCS at each temperature and coverage, using either cyclic or free boundary conditions.[10]

3. Results

With oxygen adsorption restricted to the triangular lattice of hcp-hollow sites, ordering is governed by the Hamiltonian in Eq.(1). Repulsive interactions between first and second nearest neighbours (i.e. E_1, $E_2 > 0$) have been shown previously to lead to p(2x2) and p(1x2) phases near coverages of 1/4 and 1/2 ML, respectively.[9] Setting $E_2 = 0.3E_1$ and $E_3 = 0$ in Eq.(1) and computing the order parameter as a function of temperature and coverage leads to the theoretical phase diagram shown in Fig. 1.(b). The ratio E_1/E_2 was chosen so that the ratio of the order-disorder transition temperatures at coverages $\theta = 0.25$ and $\theta = 0.4ML$ agrees roughly with that of the experiment. Choosing $E_1 = 0.23eV$ then fits the absolute transition temperatures to that of the experiment. While the actual energies of interaction are not known for the O/Ru system, energies of the magnitude proposed above are indeed plausible. While this model is sufficient to describe the experimental phase diagram in the coverage region $\theta \leq 0.4$, the predicted transition temperature at $\theta = 0.5ML$ is much larger than was found experimentally, suggesting that a more complex model might be justified.

The effect of an attractive interaction between third neighbours on the phase diagram is shown in Fig. 1 (b), for the case $E_3 = -0.1E_1$. The 2x2 and 1x2 ordered phases extend to higher temperatures, and a low temperature phase of 2x2-islands grows in at low coverages $\theta < 1/4$. The existence of such a mixed phase has not been seen using LEED on O/Ru(001), but cannot be ruled out at temperatures below 300 K which were inaccessible experimentally. In fact, small islands have been seen at very low coverages in recent scanning tunneling microscope experiments.[14]

If the binding energy difference F of an oxygen atom between hcp- and fcc-hollow sites is not much greater than the nearest-neighbour repulsive interaction E_1, occupation of both site-types becomes important, and the honeycomb lattice model is applicable. In fact, if F is small enough, occupation of the second site-type may be sufficiently large to favour the (2x2)-honeycomb structure over p(1x2) ordering at $\theta = 1/2ML$. Choosing for example the interaction energies $E_2 = E_1' = 0.3E_1$, $E_2' = 0$ and $F = 2.2E_1$ in Eq. (2) yields a phase diagram in good agreement with that of the experiment, as shown in Fig. 1 (b). The order-disorder transition temperature at $\theta = 1/2ML$ is lowered, in comparison to the triangular lattice case, and is accompanied by a spill-over of adatoms onto the less-favoured site-type. The relative occupation of this higher energy site is $\approx 10\%$, at the transition temperature. As F is decreased, at $\theta = 1/2ML$, graphs of order parameter versus temperature show a less abrupt transition that becomes almost obscured at $F = 1.7E_1$ due to competition between 1x2 and honeycomb ordered structures. At the coverage $\theta = 1/4ML$ on the other hand, $\approx 0.1\%$ of the adatoms reside on the higher energy site-type at the transition temperature, which is only slightly lowered. (Note that the slight shift in the position of the peak transition temperature from $\theta = 0.25$ to 0.26 in the honeycomb lattice simulations is due mainly to the use of free boundary conditions, and is a finite-size artifact that is of no concern here.)

Finally, we note that apparently continuous order-disorder transitions are seen at all coverages in our simulations, in agreement with the O/Ru(001) experiment. The critical behaviour of these transitions will be studied in a forthcoming paper.[10]

4. Conclusions

In summary, we find that ordering of atomic oxygen on the Ru(001) surface may be understood in terms of a lattice gas model dominated by repulsive 1st and 2nd nearest neighbour interactions, with energies $E_1 = 0.23eV$ and $E_2 = 0.3E_1$, respectively. At coverages $\theta \leq 0.4ML$, this triangular lattice model for adsorption on hcp-hollow sites describes the observed phase diagram (and critical exponents[10]); a 3rd neighbour attractive interaction $E_3 = -0.1E_1$ is also consistent with the experiment.

It has been shown that the markedly lower transition temperature at $\theta = 1/2$, in comparison with that at $\theta = 1/4ML$, can be caused by a small spill-over of a few percent of the adatoms onto the fcc-type hollow site. The O/Ru phase diagram is quite well described by this model, for binding energies differing by $\approx 0.5eV$ between fcc- and hcp-hollow sites.

5. Acknowledgements

This work was supported by the Deutsche Forschungsgemeinschaft (Sonderforschungsbereich 128), and by the Natural Sciences and Engineering Research Council of Canada.

References

1. E. Bauer in *Structure and Dynamics of Surfaces II*, W. Schommers and P. von Blanckenhagen ed. (Springer-Verlag, Berlin, 1987),p.115.
2. H. Pfnür and P. Piercy, Phys. Rev. **B40**, 2515 (1989); Phys. Rev. **B41**, 582 (1990).
3. M. Lindroos, H. Pfnür, G. Held, and D. Menzel, Surf. Sci. **22**, 451 (1989); H. Pfnür, G. Held, M. Lindroos, and D. Menzel, Surf. Sci. **220**, 43 (1989).
4. M. Lindroos and H. Pfnür, in preparation.
5. T. E. Madey, H. A. Engelhardt, and D. Menzel, Surf. Sci. **48**, 304 (1975).
6. P. Piercy, M. Maier, and H. Pfnür, in *The Structure of Surfaces II*, J. F. van der Veen and M. A. Van Hove ed. (Springer-Verlag, Berlin, 1988), p.480.
7. M. Maier, Diplomarbeit, Technische Universität München, 1986 (unpublished).
8. L.D. Roelofs, A. R. Kortan, T. L. Einstein, and R. L. Park, J. Vac. Sci. Technol. **18**, 492 (1981); L.D. Roelofs in *Chemistry and Physics of Solid Surfaces IV*, R. Vanselow and R. Howe ed. (Springer-Verlag, Berlin, 1982), p.219.
9. J. Glosli and M. Plischke, Can. J. Phys. **61**, 1515 (1983).
10. K. De'Bell, H. Pfnür, and P. Piercy, in preparation.
11. A. B. Anderson and M. K. Awad, Surf. Sci. **183**, 289 (1987).
12. K. Binder ed., *Applications of the Monte Carlo Method in Statistical Physics*, 2^{nd}ed. (Springer-Verlag, Berlin, 1987).
13. N. C. Bartelt, T. L. Einstein and L. D. Reolofs, Phys. Rev. **B32**, 2993 (1985).
14. R.J. Behm, private communication.

Phase Separation of the O/W(110) $p(2 \times 1) + p(2 \times 2)$ System

M.C. Tringides

Department of Physics, Iowa State University, Ames, IA 50011, USA

The growth of order is measured in the O/W(110) p(2×1) + p(2×2) system as is quenched from an initial "infinite" temperature to temperatures within the coexistence region. The average domain size grows according to a power law $L = A(T)t^x$ with $x = 0.31 \pm 0.03$ in agreement with the Lifshitz-Slyozov theory of phase separation. The activation energy of diffusion $E = 0.58$ eV is deduced from the growth rate $A(T)$.

The study of adsorbate systems on surfaces is strongly related to the collective thermodynamic behavior of two-dimensional systems. At equilibrium, this behavior leads to the formation of ordered phases below a critical temperature, and to the universality of critical phenomena which is well documented. Recently, there is strong interest [1] to extend the understanding of such collective behavior for systems under non-equilibrium conditions. A system is quenched from a high to a low temperature and, evolves in time to attain equilibrium at the new temperature. If this temperature is within the ordered region, then ordered domains are formed and grow to reach a size determined by the final state. The evolution is expected to be a power law

$$L = A(T)t^x \tag{1}$$

where L is the average domain size, T the temperature, and t the time. A(T) is a growth rate related to the diffusion barrier and x a characteristic growth exponent, expected to be universal i.e. it does not depend on the details of the system but on some general properties.

These growth processes are non-linear because they involve adsorbate-adsorbate interactions. Despite the microscopic complexities that inhibit analytic approach to the problem, one can ask two questions: 1) What is the growth exponent x and does it verify the expectations of universality? 2) What is the activation energy of the growth rate A(T) and how is it related to the diffusion activation energy?

On the theoretical side, different models have been extensively studied with Monte Carlo simulations to determine the universal values of the growth exponent x. Although this work requires a substantial amount of computation to ensure good statistics, it has been concluded that x = 1/2

Springer Series in Surface Sciences, Vol. 24   **The Structure of Surfaces III**
Editors: S.Y. Tong · M.A. Van Hove · K. Takayanagi · X.D. Xie
© Springer-Verlag Berlin, Heidelberg 1991

for models [2] with non-conserved and x = 1/3 for models [3] with conserved order parameter [3].

The O/W(110) p(2×1) + p(2×2) system at high coverage θ > 0.5 provides an opportunity to test the universality of the phase separation process in a two-dimensional system with two novel features. Both phases have finite densities, the (2×1) phase has density θ = 0.5 and the (2×2) phase has θ = 0.75. An additional complexity is the four-fold degeneracy of each phase. Very recent simulations [4] on models with these phases found x = 1/3. This is in agreement with more general arguments [5] that for models whose order parameter is not conserved but is coupled to a conserved field, the coverage, the exponent is 1/3. The growth exponent is determined by the requirement of long range mass transport from smaller to larger domains.

We have performed experiments on the O/W(110) p(2×1) + p(2×2) system by dosing with oxygen at a coverage θ = 0.68 as measured with Auger spectroscopy. The crystal temperature is approximately 130K, sufficiently low to exclude adsorbate diffusion, which is evidenced from the absence of super-structure spots. When the crystal is heated in the range 350-390K, the superstructure spots appear gradually. Diffraction patterns are collected at different times with a SIT video camera, digitized and stored on an IBM-AT computer for further analysis to extract the average domain size.

The results are shown in Fig. 1. The inverse square root of the second moment of the (1/2 1/2) spot profile defined by

$$q(t) = [\frac{\Sigma q^2 S(q,t)}{\Sigma S(q,t)}]^{-1/2}$$

$$(2)$$

which is proportional to the average domain size is plotted vs. t^x to test for power law dependence. We prefer this plot that magnifies deviations from a power law instead of a log-log plot which compresses the scale and minimizes the deviations. We use q(t) as a measure of growth because instrumental effects are less pronounced as compared to instrumental effects on the peak intensity. The exponent determined from these plots is x = 0.31 ± 0.03 except at the highest temperature when growth is extremely fast and the initial time interval is uncertain.

This measured value of the growth exponent x = 0.31 is consistent with the Lifshitz-Slyozov growth mechanism [3], which was proposed originally to explain the phase separation of a dense phase (θ = 1) at low concentration out of a matrix of the dilute phase (θ = 0). Monomers evaporate from islands of the dense phase with a rate inversely proportional to their radius and they diffuse away until they encounter other islands of the dense phase and condense. In the O/W p(2×1) + p(2×2) system, one of the two phases acts like a "dense" phase and growth is accomplished by long range diffusion of monomers through the "sea" of the other phase. The condition of long range transport of particles is sufficient to guarantee the universality of x = 1/3.

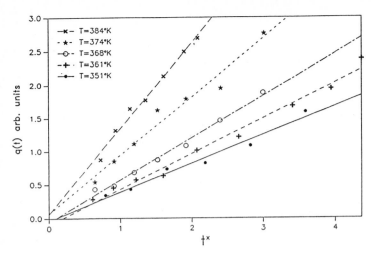

1. Plot of the inverse square root of the second moment of the spot profile, q(t) vs. t^x for several temperatures shown, in the O/W(110) p(2×1) + p(2×2) system. The slopes of the lines give the growth rate A(T).

Figure 1 provides, in addition, the growth rates A(T). How is the activation energy of A(T) related to the diffusion activation energy? As pointed out previously [6], dimensional analysis of Eq. (1) suggests that since the left-hand side of the equation has only units of length and no units of time, the right-hand side cannot have units of time either. If we assume that the only time dependence in A(T) is through the diffusion coefficient D, it follows that $D \alpha A^{1/x}$. Under this assumption, we obtain for the growth rate activation $E_A = 0.19eV$ which implies that the diffusion activation energy is $E_D = 0.58eV$ with 12% experimental uncertainty.

Explicit expressions for the relation between D and A exist for two models: the first [2] has non-conserved order parameter (Lifshitz-Allen-Cahn law) and the second [3] has conserved order parameter (the Lifshitz-Slyozov growth law mentioned above). Also, simulations [7] on a model with non-conserved order parameter, two-fold degeneracy of the ground state and Kawasaki dynamics confirm that $D \alpha A^{1/x}$. The ordered phase is c(2×2) and the system is quenched from an initial random configuration to a temperature well within the ordered region (J is the nearest neighbor repulsive interaction, the range of quench temperature is -3 > J/kT > -4 with the critical temperature at $|J|/kT_c = 1.76$). Growth is measured from the peak intensity increase. The peak intensity is proportional to the square of the average domain size and since in Fig. 2 the growth of the peak intensity is linear in time, it suggests that x = 1/2 as expected. By using the relation $D \alpha A^{1/x}$, we obtain the diffusion activation energy

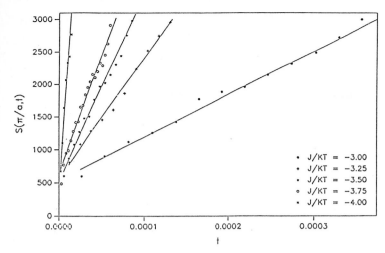

2. Plot of the peak intensity vs. time for a model with nearest neighbor repulsion J in the ordered region for several temperatures shown. The slopes of the lines provide the diffusion activation rates, because the growth exponent is x = 1/2.

2.90J. Simulations on the same model to obtain the diffusion coefficient directly, give the value of 2.56J for the diffusion activation energy in good agreement with the coarsening experiment, within the 10% accuracy of the estimates.

In these growth experiments, as pointed out before, the system is in a strongly non-equilibrium state. Diffusion experiments can also be performed after the system is fully annealed at high temperature to establish equilibrium. Equilibrium experiments were performed [8] on the O/W(110) system with the use of the current fluctuation method.

In the equilibrium experiments, the activation energy measured is strongly coverage dependent. It is 0.6eV for $\theta < 0.3$ and it rises rapidly to 1eV for $\theta = 0.56$. Previous non-equilibrium measurements [6] at coverages $\theta = 0.25, 0.5$ gave very weak coverage dependent activation energies of 0.57eV. In the present experiment with $\theta = 0.68$, the activation energy is 0.58eV, almost identical to the other non-equilibrium values. It is clear that the activation energy of the equilibrium experiments agrees with the non-equilibrium ones at low coverage, but it is 0.42eV higher at high coverages. This difference can be understood in terms of the microscopic adsorbate configuration: in the non-equilibrium experiments, one has predominantly random configurations but in the equilibrium ones, the (2×1) and (2×2) superstructures are developed. Adsorbate-adsorbate interactions contribute differently in the two experiments since the microscopic configurations are different. Such differences have also been seen in simulations [7] on the model with nearest neighbor repulsive interaction J that has c(2×2) as the ordered

445

phase. The equilibrium experiment gave activation energy 0.73J which in this case is less than the non-equilibrium one 2.56J because the formation of the c(2×2) in the equilibrium experiment, with the absence of nearest neighbors, neutralizes the effect of interactions.

In the O/W p(2×1) + p(2×2) system, the final domain size is 30Å approximately as deduced from the final full-width-at-half maximum of the (1/2 1/2) spot. No quantitative fit of the profiles was attempted to extract exact domain sizes but it is clear that the final size is less than the maximum possible size. Such limitations on the final domain size attained, have been observed in almost all other experimental studies [9] of growth kinetics in adsorbed systems. The usual explanation (suggested) is the presence of impurities that pin down the growth of domains.

Experimentally we find that the average domain size slows down at earlier times as the quench temperature is increased. Also, it reaches a higher value at the higher temperature. Is this onset of slowing consistent with the presence of impurities? Simulations [10] on models with controlled number of impurities have demonstrated that if the impurities are immobile, then the onset of slowing occurs at a later time (in units of Monte Carlo Steps per spin, MCS) as the quench temperature increases. On the other hand, if the impurities are diffusing, then slowing occurs at an earlier time at the higher temperature. However, in the simulations only the contribution of the adsorbate-adsorbate interactions is included in the jumping probability and the unit of time (MCS) is temperature dependent given by $(\nu_0 e^{-E_0/kT})^{-1}$ where ν_0 is the attempt frequency and E_0 the adsorbate-substrate potential. So actually even in the case of non-diffusing impurities, the onset of slowing down occurs earlier at the higher temperature if the same time unit is used at all temperatures. So in both cases for diffusing and non-diffusing impurities, growth slows down earlier at higher temperatures and it would be difficult to distinguish either case experimentally.

In the Auger analysis of the clean surface, no impurities are detected. When the surface is heated at a higher temperature by at least 100K, then growth resumes and the domain size becomes at least twice as large. If we assume that growth is limited at the lower temperature because the domain size becomes comparable to the average distance between impurities, it follows that heating at a higher temperature increases the average spacing between impurities or that the impurities are diffusing less. The spacing between the immobile impurities is greater than 70Å (based on the domain size achievable at this higher temperature). This limits their number to less than 1 impurity per 400 substrate atoms, a low fraction. Since the growth limitation and subsequent resumption of growth at a higher temperature is observed in all growth experiments [9] it would be difficult to believe that in so many disimilar systems, if there were impurities, they become always less mobile than the adatoms as the temperature is raised. Since this inability of the system to grow to a larger size is so general, it might be related to other effects. In any case, in the present experiment there is enough growth before the onset of slowing, over two decades in time, to determine an accurate value of the growth exponent.

In summary, we have performed growth kinetics experiments in the O/W(110) p(2×1) + p(2×2) system at high coverage θ = 0.68. that verifies the validity of the Lifshitz-Slyozov mechanism on a system with finite density and multidegeneracy of the separating phases. This strengthens the claim of universality to be valid in non-equilibrium processes. At the same time, we measure the non-equilibrium diffusion activation energy E = 0.58eV which is less than the corresponding equilibrium value.

REFERENCES

1. J. D. Gunton, M. San Miguel and P. S. Sahni in <u>Phase Transitions and Critical Phenomena</u>, ed. by C. Domb and J. L. Lebowitz (Academic, NY, 1983).
2. L. M. Lifshitz, Zh. Eskp. Teor. Fit. 42, 1354 (1962), Sov. Phys. JETP 15, 939 (1962), S. M. Allen and J. W. Cahn, *Acta. Metall.* **21**, 1085 (1979).
3. L. M. Lifshitz and V. V. Slyozov, J. *Chem. Phys. Solids* **15**, 35 (1961).
4. Ole G. Mouritsen, P. J. Shah and J. V. Andersen, preprint (1990).
5. L. Jörgenson, R. Harris and M. Grant, *Phys. Rev. Letters* 63, 1693 (1989).
6. P. K. Wu, M. C. Tringides and M. G. Lagally, *Phys. Rev. B* **39**, 7595 (1989).
7. M. C. Tringides, *J. Chem. Phys.* **92(3)**, 2077 (1990).
8. J. R. Chen and R. Gomer, Surf. Sci. 79, 413 (1979); M. Tringides and R. Gomer, *Surf. Sci.* **155**, 254 (1985).
9. M. C. Tringides, P. K. Wu, M. G. Lagally, Phys. Rev. Letters 59, 315 (1987); W. Witt and E. Bauer, Ber. Bunsengs. Phys. Chem. 90, 248 (1986); R. J. Behm, G. Ertl and J. Wintterlin, Ber. Bunsenges. Phys. Chem. 90, 294 (1986); H. M. Clearfield, Ph.D. dissertation, University of Wisconsin, Madison, unpublished (1984).
10. O. G. Mouritsen and P. J. Shah, *Phys. Rev. B* **40**, 11445 (1990); D. J. Srolovitz and G. N. Hassold, *Phys. Rev. B* **35**, 6902 (1987).

STM Studies of the Initial Stage
of Oxygen Interaction with Ni(100)

G. Wilhelmi, A. Brodde, D. Badt, H. Wengelnik, and H. Neddermeyer

Institut für Experimentalphysik der Ruhr-Universität Bochum,
Postfach 10 21 48, W-4630 Bochum, Fed. Rep. of Germany

Abstract. We have studied Ni(100) in its clean state and for oxygen exposures up to 5000 Langmuir at room and elevated temperatures (up to 270° C) by using STM, LEED and AES. The STM images from clean Ni(100) show a complicated terrace structure due to (in general) monoatomic steps, which are related to the actual misorientation of the surface against (100) and to the presence of bulk imperfections. After oxygen exposure we could obtain atomically resolved STM images for the p(2x2) and c(2x2) symmetry in the low coverage range (up to exposures of a few 10 Langmuir) and less resolved images from surfaces, which have developed an "epitaxial" thin NiO layer after an exposure of 5000 Langmuir. The latter surface showed the presence of NiO islands with typical lateral dimensions of 50 Å, which are preferentially oriented along [110]-like directions. In the low-coverage range we have observed ordered structures in the p(2x2) symmetry, which produced a c(2x2) LEED pattern.

1. Introduction

The recent interest in the system oxygen on Ni(100) concentrates on the low-coverage regime, where a fully ordered adsorbate/substrate structure has not yet been formed [1-3] and on surfaces, where a thin layer of Ni oxide has been grown [4-7] by using oxygen exposures in the order of 1000 Langmuir (L) (1 L is approximately equivalent to an exposure at 10^{-6} mbar for 1 s).

The low-coverage studies of Ref. 1 and 2 have been performed at low temperatures (80 K). In both cases the diffraction pattern did not indicate the presence of long-range order. From fitting their experimental diffuse low energy electron diffraction (DLEED) data Starke et al. [1] located the adsorption sites for O slightly shifted (by 0.4 Å) against the ideal 4-fold hollow site of Ni(100). Franchy et al. [2] interpreted high-resolution electron energy loss spectra in case of low temperature adsorption by the existence of pairs of O atoms in neighboring 4-fold hollow sites, which were also believed to act as initial nuclei for further growth of Ni oxide.

Springer Series in Surface Sciences, Vol. 24   **The Structure of Surfaces III**
Editors: S.Y. Tong · M.A. Van Hove · K. Takayanagi · X.D. Xie
© Springer-Verlag Berlin, Heidelberg 1991

In Ref. 3 an unexpected adsorption geometry was found by Monte Carlo simulation, which should give rise to a c(2x2) diffraction pattern although the local geometry was of the p(2x2)-type.

With an exposure of approximately 1000 L of oxygen the Ni(100) surface should develop a kind of epitaxial Ni oxide with only a few layer's thickness in the (100) orientation. Such a surface is used as model system for adsorption studies on an oxide surface without having the difficulty of charging if electron spectroscopic techniques are applied.

The use of scanning tunneling microscopy (STM) for this system in the disordered low-coverage range and in the beginning stage of oxidation is certainly tempting. Previous STM measurements on O/Ni(100) have been reported by Binnig et al. [8] and Behm [9]. Atomically resolved STM images of an ordered O/Ni(100) structure have not been described in Ref. 8, however. Moreover, the role of the electronic states of the adsorbed O for the STM images has been computed only but recently [10] and should be explored experimentally. We have therefore measured O on Ni(100) in the range from 1 L up to a few 1000 L the substrate being at room temperature or slightly heated (270° C) during the exposure. We have obtained STM images for "clean" Ni(100), p(2x2)O-Ni(100), c(2x2)O-Ni(100) and for "epitaxial" NiO(100) the latter surface exhibiting a rather broadened LEED pattern with 4-fold symmetry. The main results will be described here and discussed in the light of recent literature.

2. Experiment

Details of the instrumental equipment used for the present measurements have been described elsewhere [11]. After mechanical and chemical pol-ishing the Ni(100) substrates were cleaned in situ by repeated cycles of argon bombardment and annealing up to 800° C until in the Auger elec-tron spectra contamination by C, S and O could no longer be detected within the noise of the data. The sensitivity was estimated to be in the order of 1 % of a monolayer. A clear LEED pattern with 4-fold sym-metry according to clean Ni(100) was finally observed. The adsorption experiments were performed in the STM chamber.

In the STM measurements constant current topographies (CCT's) and local I/U characteristics have been recorded for sample bias voltages between -1 and 1 V and a tunneling current in the range of 1...20 nA. For the clean surface and in the low-coverage range (up to an oxygen exposure of 50 L) atomically resolved images could be obtained, although a large number of experiments have been found necessary to obtain satisfactory results. In many cases, however, the presence of oxygen on the sample surface and on the tip led to instabilities in the tunneling current or to insufficient sharpness of the tip and to images without atomic resolution. For clean Ni(100) and under optimum conditions an atomic corrugation due to the Ni atoms of less than 0.1 Å was observed. The rows of atoms appeared as straight lines and the entire surface showed the presence of defect-like structures in form of depressions with a concentration of about 1 % of a monolayer. The position of these defects appeared in 4-fold hollow or in bridge sites and can therefore not be caused by missing Ni atoms. Instead, we assign them to locally reacted Ni atoms due to contamination by C, O or S.

The macroscopic topography of the surfaces, i.e., the terrace and step structure could always be resolved. Since the surface was oriented within an accuracy of 0.5° for monoatomic steps a mean terrace width of 200 Å due to the misorientation was expected and actually observed. On some parts of the surface additional steps were found due to lattice imperfections in the bulk. These steps cross the misorientation-related steps mostly without any noticeable interaction and appear normally completely straight in a low-index symmetry direction of the crystal. In some cases a screw dislocation was found to be the origin of these steps. Since steps provide preferential sites for oxide growth (see below) the possible existence of a rather complicated step structure on Ni(100), which not only results from the misorientation of the crystal, has to be considered for interpretation of the results.

3. Results

In Fig. 1 the gross effects upon O exposure of Ni(100) are demonstrated. In Fig. 1 (a) the step structure due to the misorientation of the surface is shown for the clean surface. The noise in the image is in the order of 0.3 Å and an atomic corrugation not visible. The results shown in Fig. 1 (b) are representative for effects after an exposure up to approximately 10 L. The step structure consists of steps due to the misorientation of the sample (the average direction is rotated slightly compared to that in Fig. 1 (a) due to measuring on a different part of the sample surface). Two additional steps (marked with a "v") originate in screw dislocations at s. The influence of O adsorption leads to the appearance of smaller protrusions on the entire surface. It is evident from the image reproduced in Fig 1 (b) (and other measurements not shown here) that some of the additional protruding structures have already grown to islands with lateral dimensions in the order of 50 Å (e.g., near o). These larger features are ascribed to NiO islands. It will be shown below that similar structures cover the surface entirely, when the initial oxide layer has been completed by an exposure of approximately 1000 L.

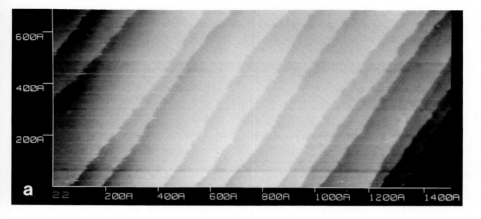

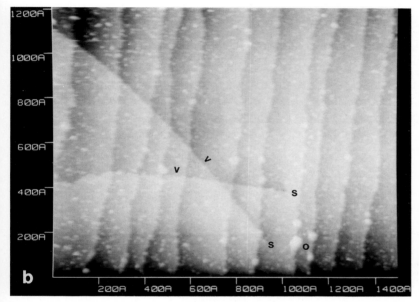

Fig. 1. STM images from clean Ni(100) (a) and Ni(100), which has been exposed to 10 L of oxygen (b). A sample tip bias of $U = 1$ V (sample positive) has been applied and a tunneling current $I = 1$ nA (a) and $I = 0.5$ nA (b).

The LEED pattern of this surface shows a clear c(2x2) symmetry, (1,0) spots of p(2x2) symmetry are hardly visible in this case. Noise and resolution in this image do not allow the observation of the initial building blocks of a periodic surface structure. If the measurements are restricted to a small area of the surface (e.g., a larger terrace without steps), and the

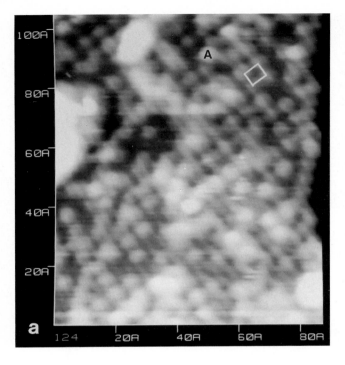

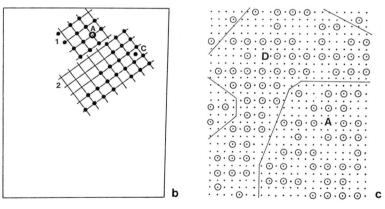

Fig. 2. (a) CCT from 10 L oxygen on Ni(100), $U = -0.5$ V and $I = 2$ nA. (b) Schematic representation of the depressions in the right upper half of (b). (c) A section of the low-coverage p(2x2) phase giving rise to a c(2x2) LEED pattern as computed by Monte-Carlo simulation by Bartelt et al. [3]. The lines separate the two favorable domains A and D obtained in Ref. 3.

tip has been conditioned for atomic resolution, images like that shown in Fig. 2 (a) could be recorded. In Fig. 2 (b) a schematic representation of a part of the surface is displayed. Although the arrangement of species still does not exhibit long-range order, regular building blocks and their periodic arrangement are clearly visible. The local structure (as indicated by the little square) is generally of the p(2x2)-type. The protruding features occupy a square lattice with twice the lattice parameter (approximately 5 Å) of clean Ni(100). In addition to the regular structures of Fig. 2 (a) more protruding features like the O-induced structures seen in Fig. 1 (b) are also identified. In some cases they exhibit the tendency of sitting atop of O atoms (see next paragraph, one of these locations is marked with A in Figs. 2 (a) and (b)) and may therefore be indeed a nucleus for NiO.

An important question is the localization of the adsorbed O atoms. According to our results of the "clean" surface, where the positions of residual contaminant atoms are observed as depressions, the assignment of the little depressions on Fig. 2 (a) to the adsorption sites of the O atoms is very suggestive. This assignment is further supported by the fact that the number of these 2x2 ordered depressions increases with the O exposure. It is also known from other surfaces exposed to O that the initial sites for O appear as depressions and not as protruding features [9]. In Ref. 10 a reversal of the corrugation resulting from an adsorbed O atom has been predicted as a function of the sample-tip distance. In particular, for small distances O should appear as a depression while for larger distances it should be observable as a protrusion. We could not confirm this effect. Using a large variety of operating conditions for tunneling a reversal of the contrast was never observed.

An interesting aspect of the results shown in Fig. 2 is that the LEED pattern of this sample was essentially of c(2x2) symmetry. We explain this observation by the prediction of Bartelt et al. [3] that in the low-coverage range a c(2x2) diffraction pattern may result from two domains of local p(2x2) structures, which are shifted against each other by the Ni lattice parameter. In Fig. 2(c) a section of the computed surface structure of Ref. 3 has been redrawn. The general agreement of the arrangement of adsorbed species (i.e., a lack of long-range order on a large scale) is evident in both the theoretical and experimental results. Moreover, the shift of the local p(2x2) cells in both directions by one lattice parameter in neighboring domains (labeled 1 and 2 in Fig. 2 (b)) is also visible. Note, however, that differences exist in the width of the domain walls obtained in the work of Bartelt et al. and in our measurement.

In addition to the local p(2x2) structures we also see in some cases the formation of local c(2x2) structures (marked with c in Fig. 2 (b)), which differ from the p(2x2) square by the missing protrusion in the centered position. We have to mention in this context that we have also obtained images, where the entire surface was covered with a lattice of c(2x2) structures. These structures are easily distinguishable from the p(2x2) lattice, since they are rotated by 45° against each other and differ in their lattice parameter by $\sqrt{2}$. The main difference of the experimental conditions to see either the p(2x2) or the (ordered) c(2x2) structure in the STM results was not the O dose but the status of the Ni surface and eventually the substrate temperature during adsorption. With increasing cleanliness and order of the Ni(100) sample during the course of the experiments the probability to observe the c(2x2) structure also became larger. We have to note that we could not obtain atomically resolved images in the very low exposure range (e.g., for 1 L) possibly due to diffusion processes of the adsorbate.

We finally describe one result from a surface which has been exposed to 5000 L of O at a temperature of 270° C. According to the literature a thin oxide layer with (100) orientation should then be grown on the

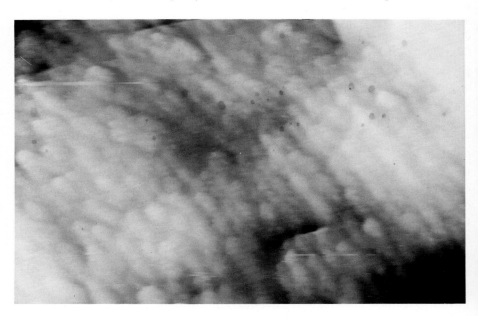

Fig. 3 CCT from Ni(100) which has been oxidized by an oxygen exposure of 5000 L at a temperature of 270° C. The image has been obtained at $U = -0.1$ V and $I = 0.5$ nA and covers an area of $\approx 600 \times 1000$ Å^2. The gray tone scale from dark to bright corresponds to about 50 Å.

surface. In the LEED pattern of the sample the integral order spots from Ni(100) were still visible and those from a c(2x2) lattice. The presence of "epitaxial" NiO was indicated by the expected LEED beams (the larger lattice parameter of NiO compared to Ni(100) consistently led to smaller distances of the corresponding LEED spots, which, however were much broader than those from Ni(100) and c(2x2)O-Ni(100). In this particular case the pattern of the NiO-related beams was slightly tilted with respect to the Ni(100) surface normal indicating nonparallelity of the NiO lattice compared to Ni(100). The part of the surface seen in Fig. 3 is characterized by a nearly regular arrangement of small islands with typical lateral dimensions of 50 Å, which show a preferential orientation in the [110]-like directions, i.e., parallel to the edges of the p(2x2) structure (see. Fig. 2). The overall corrugation of the entire surface is in the order of 100 Å and therefore much larger than for clean Ni(100). This means that the grown NiO does not form a thin homogeneous film but rather a fairly rough surface structure of oriented NiO crystallites. It has to be mentioned that we did not yet succeed in resolving periodic atomic structures on the NiO crystallites. This may be a general difficulty of STM experiments on rough and not well conducting surfaces, where a sharp and stable tip structure cannot be preserved during the measurement of one image.

Acknowledgment

This work has been supported by the Deutsche Forschungsgemeinschaft.

References

[1] U. Starke, P.L. de Andres, D.K. Saldin, K. Heinz and J.B. Pendry, Phys. Rev. B **38**, 12227 (1988).

[2] R. Franchy, M. Wuttig and H. Ibach, Surface Sci. **215**, 65 (1989).

[3] N.C. Bartelt, L.D. Roelofs and T.L. Einstein, Surface Sci. **221**, L750 (1989).

[4] R.S. Saiki, A.P. Kaduwela, J. Osterwalder and C.S. Fadley, Phys. Rev. B **40**, 1586 (1989).

[5] K. Hono, T. Iwata, H.W. Pickering and T. Sakurai, Surface Sci. **209**, L109 (1989).

[6] W.-D. Wang, N.J. Wu and P.A. Thiel, J. Chem. Phys. **92**, 2025 (1990).

[7] H.-J. Freund (private communication).

[8] G. Binnig, H. Fuchs and E. Stoll, Surface Sci. **169**, L295 (1986).

[9] R.J. Behm (private communication).

[10] G. Doyen, D. Dravoka, E. Kopatzki and R.J. Behm, J. Vac. Sci. Technol. **A6**, 327 (1988).

[11] Th. Berghaus, A. Brodde, H. Neddermeyer and St. Tosch, Surface Sci. **184**, 273 (1987).

O$_2$ Induced (1×3) → (1×1) Phase Change of an Ir{110} Surface from TOF-SARS

H. Bu, M. Shi, and J.W. Rabalais

Department of Chemistry, University of Houston, Houston, TX 77204, USA

1. Introduction

Low energy ion scattering has been developing into a sensitive surface structural analysis technique [1-3]. It has recently been shown that the Ir{110} surface, when annealed to 1400°C and cooled to room temperature, is reconstructed into major (1x3) and minor (1x1) structural units [4]. Theoretical calculations [5] have predicted the existence of higher-order reconstructions of the type (1xn) where n>2. The major faceted (1x3) structures have two missing first-layer rows and one missing second-layer row in the center of the trough. The minor (1x1) structures have adjacent first-layer rows. It has been shown from low energy electron diffraction (LEED) [6-7] and medium energy ion scattering [8] studies that the reconstructed Ir{110} surface undergoes a phase transition upon O$_2$ chemisorption to form the oxygen stabilized (1x1) bulk structure. In this paper we report the investigation of the (i) (1x3)→(1x1) phase change on Ir{110} upon O$_2$ chemisorption and (ii) surface periodicity of clean and O$_2$ stabilized surfaces. The purpose of this paper is to demonstrate the ability of low energy ion scattering as a probe for surface short range structures and periodicity. Fig. 1 shows such (1x3) and (1x1) structures of an Ir{110} surface.

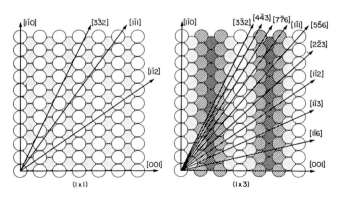

Fig. 1. Models of the (1x1) and (1x3) phases of fcc{110} surfaces.

Springer Series in Surface Sciences, Vol. 24   **The Structure of Surfaces III**
Editors: S.Y. Tong · M.A. Van Hove · K. Takayanagi · X.D. Xie
© Springer-Verlag Berlin, Heidelberg 1991

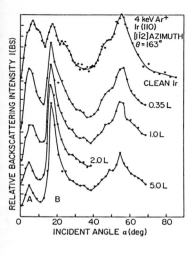

Fig. 2. Scattering intensity I(S) versus incident angle α scans along the [1Ī2] azimuth with θ = 163° for different O_2 exposures.

2. Experimental

The time-of-flight ion scattering and recoiling spectrometry (TOF-SARS) technique and its applications to structural analysis have been described previously [9]. The experimental parameters used herein are: pulsed 4 keV Ar^+ primary ion beam; pulse width ≈30 nsec; pulse rate ≈30 kHz; average current density <0.1 nA/mm^2. TOF spectra were acquired with a dose of ≈10^{-4} ion/target atom. The Ir{110} sample was cleaned by O_2 treatment followed by repeated annealing to 1700 K. Cleanliness was verified by the absence of H, C, and O recoils in the forward scattering TOF-SARS spectrum. For preparation of the Ir{110}-(1x1) surface, O_2 was adsorbed and the sample was then annealed at 650 K for several minutes, after which a sharp c(2x2) LEED pattern was observed.

3. Experimental Results

3.1. Scattering Intensity I(S) Versus Incident Angle α Scans as a Function of O_2 Coverage

I(S) vs. α scans are shown in Fig. 2 for the clean surface and the surfaces exposed to different O_2 doses. The interpretation of the clean surface α scans has been presented previously [4-5]. Only the points relevant to this work are provided. Along the [1Ī2] azimuth the α=6° peak (denoted as A) results from the focusing of Ar trajectories by first-layer Ir atoms onto their first-layer neighbors across the troughs in a (1x3) structure. The α=18° peak (denoted as B) results from both first-second layer interactions across the (1x3) trough and also first-layer neighboring interactions in the (1x1) structure. These appear at similar α angles and are not resolved. Upon exposure to O_2, the peak height of B (H_B) increases and that of A (H_A) decreases until A is finally reduced to a small peak at 5L exposure. Since H_A is representative of the amount of (1x3) structure and H_B is

457

representative of the amount of (1x3) plus (1x1) structures, the ratio H_B/H_A is related to the amounts of coexisting (1x1) and (1x3) structures on the O_2 dosed surface. The H_B/H_A ratio is constant for doses <0.2L, increases in the range 0.3-4L, and remains almost constant above 5L. This shows that the ratio (1x1)/(1x3) of the abundances of the two different structures increases with oxygen coverage, i.e. oxygen stimulates removal of the reconstruction. However, both structures are observed to coexist on both the clean and O_2 saturated surfaces, i.e. there is never 100% conversion into a single structure. The maximum H_B/H_A ratio was obtained for a surface prepared by four cycles of 5L O_2 adsorption followed by annealing to ≈670 K. This surface exhibited a sharp c(2x2) LEED pattern and peak A in the I(BS) vs. α scan was very small, indicating that (1x1) structures dominated the surface.

3.2. I(S) Versus δ Scans for the Ir(110)-c(2x2)-O Surface

I(S) versus δ scans are shown in Fig. 3 for the Ir{110}-c(2x2)-O surface. Scan (3a) is taken under specular scattering conditions. Five prominent peaks in I(S) are observed at δ = 0°, ≈35°, ≈55°, ≈65°, and 90°. These angles correspond to alignment of the scattering plane along the [001], [1$\bar{1}$2], [1$\bar{1}$1], [3$\bar{3}$2], and [1$\bar{1}$0] azimuths of the (1x1) surface as shown in Fig. 3. Under these specular scattering conditions, clear intensity maxima are observed along the low index azimuths. These intensity maxima are due to focusing and semichanneling. Scan (3b) is taken under off-specular scattering conditions for the same surface condition. In contrast to (3a), minima rather than maxima in I(S) are observed along the low index azimuthal directions. These minima are due to defocusing of trajectories at large β along the principal azimuths.

3.3. I(S) Versus δ Scans for the Ir(110)-(1x3) Surface

I(S) versus δ scans are shown in Fig. 4 for the clean Ir{110}-(1x3) surface. Scan (4a), taken under the same specular scattering conditions as scan (3a), exhibits only two maxima which are located along the [001] and [1$\bar{1}$0] azimuths; the

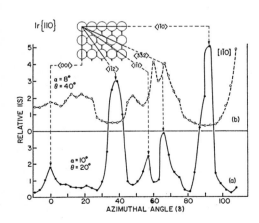

Fig. 3. I(S) versus azimuthal angle δ for the Ir{110}-(1x1) surface which is stabilized by O_2 chemisorption.

458

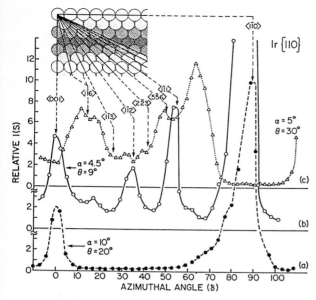

Fig. 4. I(S) versus δ for the clean Ir{110}-(1x3) surface.

structure along the intermediate azimuths is not observable. These structures are observed for different specular scattering conditions in scan (4b) with the exception of the peak corresponding to the [3$\bar{3}$2] azimuth at ≈65°. The necessity for using different specular scattering conditions to observe the peaks along intermediate azimuths and the absence of the [3$\bar{3}$2] peak reflect the difference in the (1x1) and (1x3) dominated structures. With the off-specular scattering conditions of scan (4c), minima are observed along the principal low index azimuths. In addition, extra shallow minima are observed corresponding to the [1$\bar{1}$6], [1$\bar{1}$3], and [2$\bar{2}$3] directions of a (1x3) structure as shown in Fig. 1.

4. Discussion

When the incident angle α is small enough, i.e. smaller than Linhard critical angle $α_L$, the scattering potential can be approximated as a multi-atom continuous potential due to the chains of atoms along the close-packed rows. For such a case, focusing is enhanced at the specular scattering condition, i.e. α = β [10]. The concept of continuous chains of atoms is no longer valid for α > $α_L$ and focusing is lost due to scattering from individual atomic centers. $α_L$ is directly dependent on the interatomic spacing d. For the (1x1) structure, $α_L$'s [11] are larger than the experimental α (α = 10°) used in scan (3a). As a result, five peaks are observed corresponding to specular focusing along the principal low index azimuths. For the (1x3) structure, the $α_L$'s for the [1$\bar{1}$2], [1$\bar{1}$1], and [3$\bar{3}$2] azimuths are close to or less than 10° [11]. Consequently, peaks along these azimuths are not observed in scan (3a) where the same α and β angles are used as for the (1x1) structure. Lowering α so that it is

459

below α_L allows these peaks along intermediate azimuths to be observed for the (1x3) structure, with the exception of the [3$\bar{3}$2] peak. The most likely reason is that the large d (≈35 Å) and proximity of off-planar atoms to the trajectory along this azimuth for a (1x3) structure do not provide good focusing conditions.

Using a glancing α and large β as in scans (3b) and (4c), minima rather than maxima are observed. Under these conditions, along the close-packed azimuths the first-layer atoms lie within the shadow cones of their neighbors so that no small impact parameter (large scattering angle) collisions are possible and the scattered flux is focused along the forward scattering specular direction. At intermediate azimuths where first-layer atoms are not aligned, interatomic spacings are large, small impact parameter collisions are possible, and there is no preferential focusing effect; this produces the maxima between the low index azimuths observed in scans (3b) and (4c). For δ values between the [001] and [1$\bar{1}$1] azimuths, minima corresponding to all of the low index azimuths of a (1x3) structure are observed except the [5$\bar{5}$6] direction. This may be due to disorder in the (1x3) structure and minor domains of (1x1) structures. These interpretations are supported by trajectory simulations [11]. Also, surface semichanneling may occur at glancing angles [9,10], which greatly enhances the scattering intensity along "semichannels". As a result, the intensity maxima in scans (3a), (4a), (4b) are partially attributed to this effect [11].

5. Summary

The clean Ir{110} surface is dominated by (1x3) structures and the O_2 saturated surface is dominated by (1x1) structures; both structural units coexist under all conditions investigated. The relative amount of (1x1) to (1x3) structures can be obtained by monitoring specific peaks in the α scans. The δ patterns can be interpreted in terms of specular and off-specular scattering and semichanneling effects, providing a measure of surface periodicity.

6. Acknowledgment

This material is based on work supported by the National Science Foundation under Grant No. CHE-8814337.

7. References

1. M. Aono and R. Souda, **Jpn. J. Appl. Phys. 19**, 1249 (1985).
2. S. H. Overbury, **Nucl. Instrum. Methods B27**, 65 (1987).
3. H. Derks, W. Hetterich, E. Van de Riet, H. Niehus, and W. Heiland, **Nucl. Instrum. Methods B48**, 315 (1990).
4. H. Bu, M. Shi, F. Masson, and J. W. Rabalais, **Surface Sci. 230**, L140 (1990).
5. K.-M. Ho and K. P. Bohnen, **Phys. Rev. Lett. 59**, 1833 (1987); M. S. Daw, **Surface Sci. 166**, L161 (1986); V. Heine and L.D. Marks, **Surface Sci. 165**, 65 (1986).
6. J. L. Taylor, D. E. Ibbotson, and W. H. Weinberg, **Surface Sci. 79**, 349 (1979).

7. V. P. Ivanov, V. I. Savchenko, and V. L. Tataurov, **Sov. Phys. Tech. Phys. 26**, 237 (1981).

8. M. Copel, P. Fenter, and T. Gustafsson, **J. Vac. Sci. Technol. A5**, 742 (1987).

9. O. Grizzi, M. Shi, H. Bu, and J. W. Rabalais, **Rev. Sci. Instrum. 61**, 740 (1990).

10. E. S. Mashkova and V. A. Molchanov, "Medium Energy Ion Reflection From Solids", (North-Holland, Amsterdam, 1985).

11. H. Bu, M. Shi, and J. W. Rabalais, **Surface Sci.**, submitted.

Oxygen Induced Reconstruction of Cu(110) and Cu(100) Studied by STM

F. Jensen, F. Besenbacher, E. Lægsgaard, and I. Stensgaard

Institute of Physics, University of Aarhus, DK-8000 Aarhus C, Denmark

Abstract. The nucleation and growth of oxygen induced reconstructions on the Cu(110) and the Cu(100) surfaces has been studied by means of scanning tunneling microscopy. The Cu(100)-$(2\sqrt{2}x\sqrt{2})$R45°O phase is a missing row structure formed by "squeezing out" every fourth [001] surface row of Cu atoms, resulting in the formation of Cu islands which agglomerate at elevated temperature. This growth mode is opposed to the (2x1)O reconstruction on Cu(110) which shows up as "added" Cu-O-Cu rows, formed by removal of Cu atoms from step edges. A general picture evolves in which both reconstructions, as well as the Cu(110)-c(6x2)O phase, are stabilized by Cu-O-Cu chains directed along the [001] surface direction.

1. Introduction

For adsorbates which interact strongly with the substrate, the chemisorption process is often accompanied by the breaking of several nearest-neighbor bonds within the substrate lattice, resulting in a reconstructed surface phase with a substantially altered atomic density in the topmost layer. Oxygen chemisorption on Cu surfaces is a prototype of this category of reconstructions which has recently provoked a considerable experimental and theoretical interest [1-10]. By means of scanning tunneling microscopy (STM) it is possible to study the microscopic mechanism for nucleation and growth of such adsorbate induced reconstructions, a subject which is open for vigorous dispute. It appears that such information is of utmost importance also in the understanding of the static surface structure [2,3,10,12].

2. Experimental

The present experiments were performed with a fully automated STM [11] implemented in UHV. The images shown below, recorded in the constant current mode, were obtained in ≈5 sec. The Cu surfaces were prepared using standard techniques [3,10].

3. Experimental results and discussion

3.1 Cu(110)-(2x1)O

It is known that molecular oxygen chemisorbs dissociatively on Cu(110), and that the LEED pattern shows a (2x1) structure at an O coverage $\Theta = 0.5$ ML. However, after more than 20 years of research, the detailed atomic structure of this Cu(110)-(2x1)O surface is still open for debate. The two most common structural models proposed in the past are the "missing row" model, where every second [001] row in the surface is absent, and the "buckled row" model, where every second [001] row is shifted outward [1-6].

Springer Series in Surface Sciences, Vol. 24   **The Structure of Surfaces III**
Editors: S.Y. Tong · M.A. Van Hove · K. Takayanagi · X.D. Xie
© Springer-Verlag Berlin, Heidelberg 1991

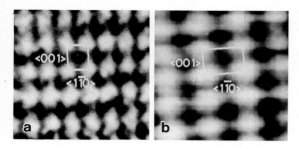

Fig.1. (a) Atomically resolved STM topograph of a 15x15 Å^2 region of a bare (1x1) Cu(110) surface. (b) STM topograph of a 15x15 Å^2 region of a fully developed (2x1) structure. The height scale from black to white corresponds to 0.3 Å and 0.45 Å for (a) and (b) respectively; surface protrusions are white, while depressions are black.

The STM topographs of Figs. 1(a) and 1(b) show how the clean, atomically resolved (1x1) surface reconstructs to a (2x1) structure with a periodicity which is doubled in the [1$\bar{1}$0] direction after an oxygen exposure of ≈10 L (Θ=0.5 ML).

From the STM images it appears that the (2x1) reconstruction is initiated when the surface is exposed to oxygen at ≈100 °C at exposures ranging from 0.1 to 1 L, and that the reconstruction shows up as "added" rows of atoms (interpreted as Cu-O-Cu chains) along the [001] direction (Fig. 2(a)). At higher exposures (≈ 1-2 L), resulting in Θ ≈ 0.1-0.2 ML, these "added" rows are found in islands with typical dimensions of 100-200 Å in the [001] direction and 15-20 Å in the [1$\bar{1}$0] direction, corresponding to a preferential growth in the [001] direction. This is consistent with the observation of a streaky (2x1) LEED pattern which indicates a lack of order in the [1$\bar{1}$0] surface direction. For exposures to ≈10 L, most of the surface is covered with the reconstructed phase (Fig. 2(b)).

The dynamical growth of the reconstructed phase was studied by imaging a region of two terraces separated by a monatomic step [3]. Images taken during room temperature (RT) exposure at a pressure of 1x10^{-8} mbar showed that Cu atoms were removed exclusively from the step edge of the upper terrace between the reconstructed (2x1) islands, and that the rate of removal differed at different points of the edge. A simultaneous growth of added rows, which later clustered to reconstructed islands on the terraces, was observed. From this and the observed great mobility of both single and groups of Cu-O-Cu chains, one can conclude that the reconstructed "added row" phase grows on top of the terraces by nucleation of Cu atoms (supplied from step edges) and O atoms diffusing on the surface [2-4]. The added row model is identical to the previously adopted missing row model at the saturation coverage

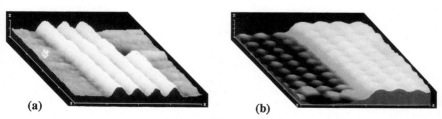

Fig 2. STM grey scale images, (40x40 Å^2), showing the formation of (a) "added rows" after an exposure of ≈1 L, and (b) the (2x1)O reconstructed phase at 10 L (Θ≈0.5).

463

$\Theta = 0.5$ ML, but the two models differ significantly in terms of mass transport. For the added row model, the Cu atoms are supplied from step edges whereas the missing row model would lead to a mass transport from terraces to step edges.

In a very recent STM study [12] of the Cu(110)-c(6x2)O phase formed at high exposures we have shown that the dynamic recording of images during adsorption can be crucial for a determination of the number of Cu atoms in the reconstructed unit cell. Cu-O-Cu chains were also in this case found to be a basic part of the structure.

3.2 Cu(100)-(2√2x√2)R45° O

A detailed LEED study has shown that the only well ordered and thermodynamically stable structure formed by oxygen adsorption on Cu(100) is the (2√2x√2)R45° structure, with an oxygen saturation coverage of 0.5 ML [7].

In Figure 3(a) is shown an STM image of the Cu(100) surface prior to oxygen exposure. As seen, we are indeed able to resolve the single Cu atoms on the (1x1) surface. The corrugation along a <011> surface row is quite small (≈ 0.1 Å).

To create a fully developed (2√2x√2)R45°O structure, with a sharp LEED pattern, the Cu(100) surface was exposed to 1000 L oxygen at 300 °C and at a pressure of 2.5×10^{-6} mbar, followed by an anneal at 300 °C for 5 min [7,8]. Figure 3(b) shows a topograph of this surface over an area of 15x15 Å^2. The [001] and [010] directions are equivalent on the Cu(100) surface, and thus the reconstruction is twinned into two different domain orientations which appear to be randomly distributed. In the following, we shall for simplicity use the orientational notation shown in Fig. 3(b).

The chains of bright spots along the [001] direction have a corrugation and periodicity of ≈ 0.2 Å and 3.6 Å, respectively, the latter being the interatomic distance along the [001] direction. The chains appear to be grouped in pairs of two (Fig. 3(b)), and the distances from A to B and from B to C are 2.9 Å and 4.3 Å, respectively, implying that the structure in the [010] direction repeats itself for every 7.2 Å. This allows us to superimpose a unit cell on Fig. 3b. We interpret the chains of bright spots along the [001] direction as Cu-O-Cu chains/bonds, equivalent to the observations for the Cu(110)-(2x1)O structure [3]. The [010] periodicity of 7.2 Å can be explained by removal of every fourth row of Cu atoms, as will be discussed below.

The nucleation and growth of the reconstructed phase was studied by imaging continuously (every ≈ 10 sec) during RT oxygen exposure at 1.5×10^{-5} mbar, two terraces

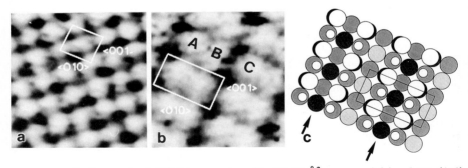

Fig.3. Atomically resolved STM topographs of a 15x15 Å^2 region of (a) a bare (1x1) Cu(100) surface, and (b) a fully developed (2√2x√2)R45° structure. The height scale from black to white corresponds to 0.3 Å (a) and 0.45 Å (b). (c) Atomistic model of the Cu(100)-(2√2x√2)R45° O reconstructed phase. The small and large white circles represent the O atoms and the remaining Cu surface atoms, respectively, whereas the grey and dark circles represent Cu atoms in the 2nd and 3rd layer. The arrows indicate the missing row of Cu atoms. In Figs. (b) and (c) a unit cell is shown.

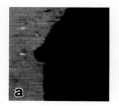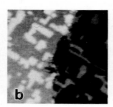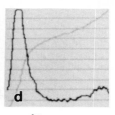

Fig.4. STM grey scale topographs recorded over an area of 500x500 Å² prior to (a) and after (b) oxygen exposure. The height scale is 4.5 Å from black to white. Fig. (c) shows a 1000x1000 Å² area without a step, and (d) depicts the differential and integral height distributions of (c).

separated by a monatomic step (Fig.4) [10]. When the oxygen is adsorbed at RT a higher exposure is required before saturation occurs [6]. Fig. 4(a) was recorded prior to, and Fig. 4(b) after an oxygen exposure of 6.2x10⁴ L. We immediately conclude that the dramatic change depicted in the topographs reflects the nucleation and growth of small islands on both terraces, whereas the step is essentially intact. The islands grow preferentially along the [010] and [001] directions, at saturation (Fig. 4(c)) they cover 25% of the surface area as determined from a plot of the height distribution (Fig. 4(d)), and their height is 1.8 Å (the layer distance for Cu(100)). These observations give the first direct proof that the oxygen induced reconstruction of the Cu(100) surface is of the missing row type, with one quarter of the Cu [001] rows missing, and that these extra atoms nucleate and grow epitaxially in small islands on top of the Cu surface. The $(2\sqrt{2}x\sqrt{2})R45°$ LEED pattern after RT oxygen saturation was diffuse, but after an anneal to ≈300 °C for ≈5 min., the fairly small islands in Figs. 4(b) and 4(c) have agglomerated, and the LEED pattern has become a bright $(2\sqrt{2}x\sqrt{2})R45°$.

The driving force for the oxygen induced reconstruction on both surfaces might in fact be the formation of the Cu-O-Cu chains. This is supported by recent theoretical calculations based on the effective medium theory [1] which show, that if the coordination number of the Cu atoms is decreased in the reconstruction process, this will enhance the oxygen binding energy to the reconstructed surface by more than it costs to reconstruct the clean surface, thereby stabilizing the oxygen-induced reconstructions.

Finally, we return to the highly resolved STM topograph of the $(2\sqrt{2}x\sqrt{2})R45°$ structure (Fig. 3(b)). Along the [010] direction, the observed periodicity of 7.2 Å can be explained by removing every fourth row of Cu atoms, consistent with the conclusion drawn from Fig. 4, with theoretical calculations [1] and with both LEED [8] and X-ray diffraction [9] experiments. The apparent height in STM images like e.g., Fig. 3(b) is a convolution of surface topography and electronic structure represented as the density of states near the Fermi level. In fact, we obtained only highly resolved images like Fig. 3(b) for very low gap voltage, and we tentatively interpret the bright spots along the [001] direction as the anti-bonding states for the Cu-O-Cu bonds. These anti-bonding states are fairly delocalized along the [001] direction [1], and thus we cannot associate the white protrusions with either O or Cu atoms but rather they must be associated with the Cu-O-Cu bonds. The pairing of the Cu-O-Cu states in the [010] direction can be explained by displacing the Cu and/or O atoms next to the missing row ≈0.35 Å towards the missing row, consistent with the recent experimental [9,8] and theoretical [1] findings.

4. Conclusion

The direct imaging obtained with the STM enabled us to study the dynamics of the nucleation and growth of the oxygen induced reconstruction of Cu surfaces by imaging the surface with atomic resolution during oxygen adsorption. The dynamic information gained in this way was found to be very crucial, also for the subsequent understanding the static surface structure. For the Cu(110)-(2x1) and Cu(100)-(2$\sqrt{2}$x$\sqrt{2}$)R45° phase, as well as for Cu(110)-c(6x2)O, the formation of Cu-O-Cu chains along [001] surface direction was found to be the driving force.

Acknowledgments: Many discussions with J. K. Nørskov are gratefully acknowledged. This research has been supported by the Danish Research Councils through "Center for Surface Reactivity", and by the Knud Højgaard Foundation.

References:

[1]   K. W. Jacobsen and J. K. Nørskov, to be published.
[2]   D. J. Coulman, J. Wintterlin, R. J. Behm, and G. Ertl, Phys. Rev. Lett. **64**, 1761 (1990)
[3]   F. Jensen, F. Besenbacher, E. Lægsgaard, and I. Stensgaard, Phys. Rev. **B41**, 10233 (1990).
[4]   F. M. Chua, Y. Kuk, and P. J. Silverman, Phys. Rev. Lett. **63**, 386 (1989), and Y. Kuk, F. M. Chua, P. J. Silverman and J.A. Meyer, Phys. Rev. **B41**, 12393 (1990).
[5]   R. Feidenhans'l, F. Grey, R. L. Johnson, S. G. J. Mochrie, J. Bohr, and M. Nielsen, Phys. Rev. **B41**, 5420 (1990).
[6]   S. R. Parkin, H. C. Zeng, M. Y. Zhou and K. A. R. Mitchell, Phys. Rev. **B41**, 5432 (1990).
[7]   M. Wuttig, R. Franchy, and H. Ibach, Surf. Sci. **224**, L979 (1989), Surf. Sci. **213**, 103 (1989).
[8]   H. C. Zeng, R. A. McFarlane, and K. A. R. Mitchell, Surf. Sci. **208**, L7 (1989).
[9]   I. K. Robinson, E. Vlieg, and S. Ferrer, to be published.
[10] F. Jensen, F. Besenbacher, E. Lægsgaard and I. Stensgaard, to be published.
[11] E. Lægsgaard, F. Besenbacher, K. Mortensen, and I. Stensgaard, Jour. of Microscopy, Vol. **152**, 663 (1989).
[12] R. Feidenhans'l, F. Grey, M. Nielsen, F. Besenbacher, F. Jensen, E. Lægsgaard, I. Stensgaard, K. W. Jacobsen, J. K. Nørskov and R. L. Johnson, to be published.

Geometrical Structure of Molecular Adsorbates from Core-Level Binding Energies

A. Nilsson, H. Antonsson, A. Sandell, and N. Mårtensson

Department of Physics, Uppsala University, Box 530,
S-751 21 Uppsala, Sweden

Abstract. The adsorption of CO and NO on clean and hydrogen covered Ni(100) has been studied using high resolution XPS. The chemical shifts of the adsorbate core levels depend sensitively on the local structure of the adsorbate. For CO the C1s and O1s binding energies decrease with increasing coordination to Ni. In the case of NO the N1s binding energy is sensitive to the molecular angle relative to the surface whereas the O1s binding energy is more sensitive to the adsorption site.

XPS (X-ray Photoelectron Spectroscopy) is one of the major analytical techniques for investigating adsorbed atoms and molecules. Core level photoelectron spectroscopy is first of all used to identify what elements are present on the surface. The chemical shifts of the core levels can furthermore be used to determine the chemical state of the adsorbate species. This has been used as fingerprints of different types of adsorbates, such as for instance to monitor the dissociation of a molecular adsorbate. One can also obtain direct information on the local geometry of an adsorbate from the binding energy positions. We will in the present contribution show how different adsorption sites and different molecular orientations relative to the surface may lead to different core level binding energies. The quantitative aspect of the technique also makes it possible to determine the number of adsorbed molecules in different geometrical arrangements (provided that the satellite intensities are properly included). Furthermore, the temperature dependent dynamical distribution of adsorption sites can also be probed since the time scale for the core ionization process is much shorter than the time scale for the vibrational motion of adsorbates [1-3].

The binding energy of a core electron corresponds to the change in total energy of the system as the core electron is removed. When considering binding energy shifts between different situations it is therefore important to consider the changes in total energy of <u>both</u> the initial and final states. For an adsorbate on a metal surface screening effects lead to final states where charge (often close to one electron) has been transferred to the core ionized site. A core ionization leads to similar changes of the valence electron distribution as would occur with an increase of the nuclear charge by one (Z+1 approximation). For an adsorbed molecule the binding energy will therefore depend on the difference in adsorption energy between the original Z-atom molecule and the Z+1 atom molecule.

Adsorption of CO on Ni(100) leads at $\theta=0.5$ (substrate units) to the formation of a c(2x2) structure in which all CO molecules occupy equivalent on top sites with the

---

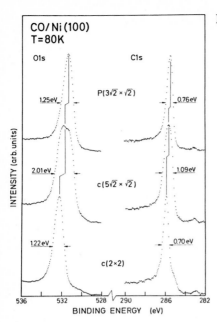

CO/ Ni (100)
T=80K

O1s   C1s

P(3√2 × √2)

1.25eV   0.76eV

c(5√2 × √2)

2.01eV   1.09eV

1.22eV   0.70 eV

c(2×2)

536   532   528 290   286   282
BINDING ENERGY (eV)

Fig.1 C1s and O1s spectra for CO on Ni(100) in three different structures.

molecular axis perpendicular to the surface and the carbon end downwards. For higher CO coverages two new structures appear [4]. At saturation (θ=0.67), a p(3√2x√2)R45 structure is obtained for which vibrational spectroscopy indicates the population of only bridge sites[4]. At θ=0.6, a c(5√2x√2)R45 structure develops for which there is a mixture of on top and bridge sites [4].

Fig.1 shows C1s and O1s spectra for the three different CO phases. The overall resolution was 0.4 eV [5]. In the c(2x2) structure the C1s and O1s peaks appear at 285.9 eV and 532.2 eV binding energy, respectively. For the c(5√2x√2)R45 structure there is a shift towards lower binding energies and both lines become much broader. In the p(3√2x√2)R45 structure the line widths are about the same as in the c(2x2) structure but the C1s and O1s binding energies are lower by 0.5 eV and 0.9 eV, respectively, showing that the bridge sites have lower binding energies. The increased line width for the intermediate structure is due to overlapping peaks corresponding to bridge and top sites.

Further information on the site dependent shifts in adsorbed CO can be obtained by studying the coadsorption system CO+H on Ni(100) since a wider range of adsorption sites becomes accessible, see Fig. 2. For this system CO may be obtained in on top, bridge and hollow adsorption sites as proposed by vibrational spectroscopy [6]. The coadsorbed phases were prepared by first producing a p(1x1)H structure at 80 K and then adsorbing CO at the same temperature with subsequent annealings to different temperatures.

At the bottom of Fig. 2 the spectrum from the c(2x2) overlayer of CO on Ni(100) is shown for comparison. Coadsorption of CO and H at 80K also leads to a c(2x2) structure with CO in on top positions. The C1s and O1s peaks shift to higher

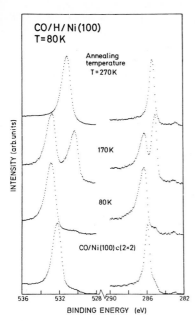

Fig.2 C1s and O1s spectra for CO in three different phases on a hydrogen precovered surface. The annealing temperatures to obtain the pure coadsorbed phases are shown. Spectra from a c(2x2) overlayer of CO on clean Ni(100) is shown for comparison.

binding energies by 0.4 eV and 0.7 eV, respectively, due to the coadsorbed hydrogen. The binding energy positions are similar to the c(2x2)CO phase on Cu(100) which has also on top adsorption sites. Elevating the substrate temperature to 170 K yields a c(2√2x√2)R45 structure. Vibrational spectroscopy shows that hollow adsorption sites are also populated in this phase and that some of the atomic hydrogen has recombined to yield molecular adsorbed hydrogen [6]. In the photoelectron spectra double peaks are seen with splittings of 1.2 eV and 2.6 eV for the C1s and O1s spectra, respectively. The relative intensities are 1:1 which shows that 50% of the molecules in this phase populate hollow sites. Increasing the temperature to 270 K results in some desorption of CO and H_2 (Σ-desorption). This leaves the remaining CO in bridge sites as seen by vibrational spectroscopy. The photoelectron spectra show peaks with energies between those corresponding to the on top and hollow positions. The binding energies are almost the same as for the p(3√2x√2)R45 structure of the pure CO overlayer which also contains only bridge CO.

The results clearly show that there are strong site dependent binding energy shifts for adsorbed CO. The shifts are about twice as large for O1s than for C1s. The binding energies decrease with increasing coordination to the substrate Ni atoms, i.e. in the order on top - bridge - hollow. This is an empirical relationship which could be used in the study of CO adsorption on Ni and other surfaces. In vibrational spectroscopy a similar empirical rule has been used for many years based on the CO stretch frequency. The CO stretch energy is lowered when the coordination to the Ni substrate atoms is increased.

The shifts can be understood from some total energy considerations. In the initial state (before ionization) there are only small differences in adsorption energy (a few 100 meV) between different sites. Therefore the major contribution to the shift has to come from changes in the energy of the core ionized state. Within the Z+1 approximation, the C1s ionization produces an NO-like final state and the O1s ionization leads to a CF-like final state. The carbon part of the CF molecule could rehybridize to form "three unpaired" electrons available for bonding (compare CH$_3$F), while the "fluorine" part of the molecule is completely saturated due to the large difference in electronegativity between the two atoms. For a free carbon atom with four unpaired electrons the adsorption site is such that the coordination to Ni is optimized. On Ni(100) the interaction is so strong that a reconstruction of the Ni lattice occurs with the carbon atoms in the hollow position. CF can be anticipated to maximize the coordination in a similar way. It is therefore most reasonable that O1s ionized CO has its equilibrium adsorption site in the hollow position. In order to move the CF molecule to bridge and on top sites the spectra show that energies of 1 and 2 eV are required. NO (C1s ionized CO) is expected to adsorb in a manner more similar to CO with slightly larger adsorption energies due to an extra unpaired valence electron. Therefore the variation in the final state energy is smaller for C1s than for O1s. These chemical arguments have recently been justified using cluster calculations [7] simulating the ionization at on top and hollow adsorption sites of CO.

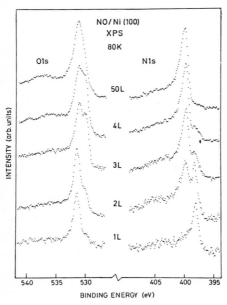

Fig.3 N1s and O1s spectra of NO adsorbed on Ni(100) at 80 K with increasing exposures.

The adsorption of NO on Ni(100) is in general much more complex than the adsorption of CO. There are no ordered molecular overlayer structures formed in this system. Fig. 3. shows how the N1s and O1s spectra develop for different doses of NO at 80 K. At 1L only one peak can be seen in each spectrum indicating a single adsorption site with N1s and O1s binding energies of 398.1 eV and 531.4 eV, respectively. For a dose of 2L a second peak is seen in the N1s spectrum at 1.8 eV higher binding energy. With increasing coverage the second N1s peak becomes more pronounced and the first one vanishes completely for the saturated surface. In the O1s spectra a second peak is observed at intermediate coverages with 1 eV lower binding energy. At saturation this O1s peak decreases in intensity and results only in an asymmetry of the line towards lower energies. There is no clear connection between the shifts in the N1s and O1s spectra. The shifts of the two core levels must therefore have different origins.

The first adsorption state at low coverage has been found to be a lying down or highly tilted NO species based on X-ray photoelectron diffraction and UPS studies[8]. The second N1s peak is related to a perpendicular NO species [8]. The large N1s binding energy shift of 1.8 eV can be understood from the Z+1 approximation within which the final state molecule can be replaced by an adsorbed O_2 molecule. For molecularly adsorbed oxygen only lying down geometries have been found on different surfaces [9]. Therefore the lying down final state must have the lowest total energy in agreement with our results. From the spectra we can estimate that an energy of nearly 2 eV is required in order to raise molecular oxygen into a perpendicular geometry.

In the O1s spectra the second peak at lower binding energies grows at intermediate coverages when the perpendicular NO adsorption geometry becomes populated but nearly vanishes at saturation coverage. From a comparison with vibrational spectra of this system we propose that the second peak is related to a hollow site adsorption and the first one to a bridge site [10]. For the perpendicular adsorbate, the hollow site should then be populated first and with increasing coverage the molecules are transferred to bridge sites. This is also consistent with the conclusions from the spectra for adsorbed CO. From a Z+1 analysis of the C1s shifts it can be concluded that NO has the lowest total energy in the hollow site. Applying the Z+1 approximation to the O1s final state for NO results in an NF final state molecule. The observed O1s shifts indicate that the site dependence of the adsorption energy for this molecule is similar to CF.

In summary we have shown that the core level binding energies of adsorbates depend on the local structure of the adsorbate-substrate complex. For CO adsorbed on Ni(100) the binding energy decreases with increasing Ni coordination. For adsorbed NO ionization of the different atoms provides complementary information. The N1s energy is mainly sensitive to the molecular angle relative to the surface whereas the O1s energy is more sensitive to the adsorption site.

References

1. A. Nilsson and N. Mårtensson, Solid State Commun. **70**, 923 (1989).
2. A. Nilsson and N. Mårtensson, Phys. Rev. Lett. **63**, 1483 (1989)
3. N. Mårtensson and A. Nilsson, J. Electron Spectr. **52**, 1 (1990).
4. P. Uvdal, P. A. Karlsson, C. Nyberg, S. Andersson and N. V. Richardson, Surf. Sci. **202**, 167 (1988).
5. A. Nilsson and N. Mårtensson, Phys. Rev. **B40**, 10249 (1989).
6. L. Westerlund, L. Jönsson and S. Andersson, Surf. Sci. **199**, 109 (1988).
7. I. Panas, P. Siegbahn, H. Antonsson, A. Nilsson and N. Mårtensson, to be published.
8. A. Sandell, A. Nilsson and N. Mårtensson, to be published
9. J. L. Gland, B. A. Sexton and G. B. Fisher, Surf. Sci. **95**, 587 (1980).
10. G. Odörfer, R. Jaeger, G. Illing, H. Kuhlenbeck and H. J. Freund, Surf. Sci. , in press.

Lateral Surface Stark Effect in Chemisorbed Molecules: CO on Metal Surfaces

B. Gumhalter[1], K. Hermann[2] and K. Wandelt[3]

[1]Institute of Physics of the University of Zagreb, Y-41001 Zagreb, Yugoslavia
[2]Fritz-Haber-Institut der Max-Planck-Gesellschaft,
  Faradayweg 4–6, W-1000 Berlin 33, Fed. Rep. of Germany
[3]Institute of Physical and Theoretical Chemistry, University of Bonn,
  W-5300 Bonn 1, Fed. Rep. of Germany

Abstract. The influence of localized surface fields arising from local surface potential differences near structural and chemical surface irregularities on chemisorbed molecules is simulated by ab initio SCF-LCAO cluster calculations with a linear AlCO cluster in the presence of an external inhomogeneous field. Our exploratory study substantiates the notion of a local surface Stark effect induced shifting and splitting of the CO $2\pi^*$-type orbitals which suggests an increased charge back-donation effect and, hence, an increased CO dissociation probability near surface defects, in accord with corresponding experimental observations.

1. Introduction

Dissociation of molecules upon adsorption on solid surfaces has attracted considerable attention because the understanding of this process may provide a clue for the microscopic description of various surface reactions and phenomena among which heterogeneous catalysis is one of utmost importance [1]. Experimental studies of dissociation processes involved many different surfaces and adsorbates [2,3].

Chemisorption of CO on clean metallic substrates is often studied as a prototype of molecular adsorption of both dissociative and nondissociative kind, the latter depending on the underlying substrate. Thus, it is known that CO chemisorbs nondissociatively on smooth (low index) planes of e.g. Cu, Pt, Ni, Pd whereas on W, Fe, Co it chemisorbs dissociatively [3]. Furthermore, it has been observed that the presence of steps, kinks, heteroatoms and similar irregularities and inhomogeneities may give rise to dissociative CO chemisorption on some of the above surfaces although the corresponding flat surfaces do not exhibit the same property [4a].

The occurrence of CO dissociation at metallic surfaces is usually interpreted as being initiated by partial filling of the $2\pi^*$ derived molecular electronic states of chemisorbed CO by electron charge transfer from the metal valence band (back-donation mechanism). Since the $2\pi^*$ molecular orbitals of free CO are antibonding relative to the C-O molecular bond a large filling of these states upon chemisorption weakens the C-O bond considerably and may even lead to its dissociation [4b].

Within the $2\pi^*$ resonance model of CO chemisorption [5a] the large back-donation induced charge transfer [5b] will take place if the center $\bar{\epsilon}_{2\pi}$ of the adsorbed CO $2\pi^*$ derived resonance is located close enough to the substrate Fermi level ϵ_F (c.f. Fig. 1). This can occur either by shifting the substrate Fermi energy upwards towards the vacuum level ϵ_V, which may be achieved e.g. by coadsorption of other species which reduce the work function of the substrate, or by shifting the $2\pi^*$ derived levels

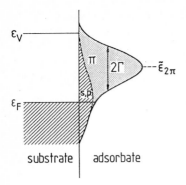

Fig. 1: Sketch of a CO $2\pi^*$ derived valence resonance centered at energy $\bar{\epsilon}_{2\pi}$ and of full width at half maximum 2 Γ. The projected densities of states of metal s, p character (hatched) and of adsorbate character (dotted) are indicated schematically.

down towards ϵ_F by some local perturbation such as a field at the surface. Localized electric fields are expected to originate from surface irregularities like steps, kinks and adatoms which lead to local surface potential differences. As shown by PAX experiments, these potential differences near such surface sites may be as large as 1-1.5 eV [6], falling off rapidly over a few Ångstroms distance. The corresponding localized electric fields at these surface irregularities may have a large lateral component oriented parallel to the surface.

A heteronuclear diatomic molecule like CO which adsorbs at a stepped surface of a transition metal may sit either on a terrace where the field direction is dominantly perpendicular to the surface (and parallel to the molecular axis) or close to a step edge where it may be subject to an electric field with major components perpendicular to the molecular axis. Such local fields will perturb the valence levels of the adsorbate leading to their splitting and shifting on the energy scale. This phenomenon is known as the Stark effect for gas phase atoms and molecules in external electric fields [7]. In the case of chemisorbed CO these perturbations may eventually also affect the occupancy of the CO $2\pi^*$ valence states close to ϵ_F and, hence, the stability of the molecule.

2. Theory

In the resonance picture of nondissociative CO chemisorption on Cu and transition metals [5a] the $2\pi^*$ derived level of the adsorbate is broadened into a resonance by the bonding chemical interaction with the substrate. According to abundant experimental evidence, this electronic resonance is centered above and is overlapping with the substrate Fermi level (c.f. Fig.1) with full width at half maximum (FWHM) 2Γ being of the order of 1 eV.

The action of the lateral electric field on the $2\pi^*$ derived resonance of chemisorbed CO may be most easily visualized within low order perturbation theory. Due to hybridization of the CO $2\pi^*$ orbital with the substrate valence band (in the present example we shall consider only substrate s, p bands), the states determining the localized $2\pi^*$ derived resonance will be predominantly of s, p character near the resonance bottom and of predominantly $2\pi^*$ character near the upper resonance edge with the admixture of both types of symmetries near the resonance center (see Fig.1). Hence, the wavefunctions of electrons occupying the $2\pi^*$ derived resonance will appear as linear combinations of wavefunctions with metal s, p and CO π symmetry

$$| \Psi^{res} \rangle = \sum_{i=s,p,\pi} a_i | i \rangle \tag{1}$$

Let us consider a field of strength E oriented along the y direction in a coordinate system in which the substrate occupies the halfspace $z < 0$ and in which the molecular axis of chemisorbed CO is perpendicular to the surface plane $z = 0$ pointing along the z-axis. Upon switching on the lateral field $\mathbf{E}$ all the electronic molecular levels ϵ_j undergo a shift $\delta\epsilon_j$ depending on the strength of the applied field. Due to the mixed symmetry character of the wavefunction (1) diagonal matrix elements of the form

$$\langle \Psi^{res} | \mathbf{E} \cdot \mathbf{r} | \Psi^{res} \rangle = E \langle \Psi^{res} | y | \Psi^{res} \rangle \tag{2}$$

(giving the average dipole moment of the x and y derived contribution to the $2\pi^*$ resonance induced by $\mathbf{E}$) will not vanish. Therefore, by applying first order perturbation theory we find that the field induced shifts $\delta\epsilon_{2\pi_x}$ and $\delta\epsilon_{2\pi_y}$ of the resonance level increase linearly with $| E |$ in analogy to the linear Stark effect. In addition, as a consequence of the symmetry reduction by the lateral electric field $\delta\epsilon_{2\pi_x}$ and $\delta\epsilon_{2\pi_y}$ will be different and the differential shift $\Delta\epsilon = \delta\epsilon_{2\pi_x} - \delta\epsilon_{2\pi_y}$ contributes to a broadening of the resonance. Thus, the existence of a negative potential close to a surface step and of a lateral field across a chemisorbed CO molecule may give rise to a considerable downward shift of its valence $2\pi^*$ derived resonance and its simultaneous broadening. Both effects lead to an *increase of the resonance occupation* through the back-donation mechanism. In turn, since the $2\pi^*$ derived levels retain their predominant antibonding character with respect to the CO molecular bond, this should result in a Stark effect induced weakening of the adsorbed CO molecular bond.

In order to substantiate these qualitative arguments we simulate the CO-metal interaction by a linear AlCO cluster with the metal atom modelling the influence of the substrate s, p electrons on the adsorbed CO. The interatomic distances of AlCO are taken from optimizations previously performed on larger $Al_n CO$ clusters [8a] where the Al-C distance (2.09 Å) and the C-O distance (1.15 Å) were found to give a reasonable CO-metal binding picture. In our exploratory study of the electronic response of adsorbed CO to inhomogeneous electric fields at surface steps we consider the interaction of AlCO with an electric field originating from an external positive point charge. The charge of magnitude q (measured in fractions of the electron charge e) is placed at a distance d away from the center of the C atom along the y direction (the molecular axis of AlCO points along the z-axis). In the calculations we have varied both q ($0 \le q \le 1$) and d/D ($1 \le d/D < 2$) with D = 2.861 Å being the nearest neighbor distance on Al(100). Hence, the magnitude of the field at the center of the C atom $E(C)=q/(d/D)^2$ may be considered as a measure of the strength of an external perturbation produced by surface inhomogeneities and acting on the adsorbate electronic structure.

The electronic states of the AlCO cluster with and without the point charge are obtained from ab initio SCF-LCAO calculations using flexible all-electron basis sets of contracted Gaussian-type orbitals [9]. Previous studies on AlCO clusters [8] yielded an electronic ground state for AlCO without electric field which is described in $C_{\infty v}$ symmetry by $^2\Pi(9\sigma^2 3\pi^1)$. Here all AlCO cluster orbitals can be characterized as arising from either the Al or the CO subunit with the singly occupied 3π cluster orbital representing a mixture of Al 3p with CO $2\pi^*$ contributions. Thus, the 3π orbital models the metal to CO back-donation in the cluster. The partial occupation of the CO 2π orbital in the chemisorbed state affects the C-O bond and can lead to a quite different behavior of chemisorbed CO, as opposed to free CO in the presence of external electric fields [10]. In linear AlCO without external field the π_x- and π_y- type orbitals are energetically degenerate due to the rotational symmetry of the system. However, in the presence of the inhomogeneous field this symmetry is no longer conserved and the degeneracy of the π_x and π_y levels is removed which is most pronounced in the 3π cluster orbital.

3. Results and Discussion

The valence electron spectrum of CO in the presence of the Al metal and the external field due to a point charge q is characterized by respective Hartree-Fock one-electron energies $\epsilon(q)$ in the present model calculations. Their shifts $\delta\epsilon(q)$ due to the point charge q are used to model the modification of the adsorbate level spectrum in the presence of an external electric field. This approximation is not expected to yield quantitative results to be compared with the real adsorbate system but should describe the mechanism underlying the interaction of the adsorbate with an external local field on the surface in a physically meaningful way.

Table 1 shows the field dependence of the energy levels $\epsilon(q)$ and level shifts $\delta\epsilon(q)$ of the CO-type valence orbitals 4σ, 5σ, $1\pi_x$ and $1\pi_y$ in AlCO and of the AlCO $3\pi_x$.

Table 1: Field-dependent Hartree-Fock one-electron energies $\epsilon(q)$ (in parenthesis) and level shifts $\delta\epsilon(q) = \epsilon(q) - \epsilon(0)$ of the CO 4σ, 5σ, 1π derived and of the 3π back-donation orbitals in linear AlCO. The last column contains results for the differential splitting $\Delta\epsilon_\pi = \delta\epsilon(q, 3\pi_x) - \delta\epsilon(q, 3\pi_y)$. The data refer to a positive point charge of magnitude q placed at a distance D away from the C center (see text). All energies are given in eV.

q	$[\epsilon(q)]$		$\delta\epsilon(q)$				
	4σ	5σ	$1\pi_x$	$1\pi_y$	$3\pi_x$	$3\pi_y$	$\Delta\epsilon_\pi$
0.0	[-21.27]	[-17.61]	[-16.66]	[-16.66]	[-6.61]	[-6.61]	0.00
0.3	-1.18	-1.21	-1.18	-1.20	-1.33	-1.47	0.15
0.5	-2.00	-2.04	-1.99	-2.02	-2.22	-2.46	0.25
0.7	-2.83	-2.89	-2.82	-2.85	-3.10	-3.46	0.35
1.0	-4.10	-4.19	-4.09	-4.12	-4.44	-4.96	0.52

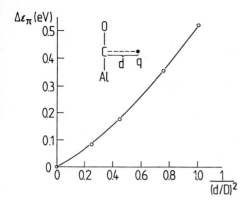

Fig. 2: Field induced splitting $\Delta\epsilon_\pi$ of the 3π cluster level of linear AlCO as a function of the lateral field q/d^2 caused by a point charge ($q=1$) placed at a distance d away from the C center (see inset). For a definition of D see text.

and $3\pi_y$ cluster orbitals. Here the field strength is represented by the magnitude of the point charge q at a fixed distance D which should give a reasonable separation of the chemisorbed CO molecule from a presumed step at the surface.

From Table 1 one immediately observes the strong downward shifts of all the levels due to the positive charge q. These shifts scale linearly with q, i.e. with the strength of the applied field. The $3\pi_y$ level shift is larger than that of $3\pi_x$ because the charge q lies within the yz plane. The differential 3π shift defined by $\Delta\epsilon_\pi = \delta\epsilon_{3\pi_x} - \delta\epsilon_{3\pi_y}$ is also linear in q and amounts to about 10% of the basic shift of all the orbitals at that distance. Such a linear dependence on q signifies the predominant role of the linear Stark effect in the splitting of the cluster 3π levels. In the resonance picture this splitting would be equivalent to an additional field induced broadening $\Delta\Gamma(E) = \beta|\,E\,|$ of the resonance by the applied external field E. Hence, we may deduce from Table 1 that both the field induced downward shift and the broadening of the actual $2\pi^*$ derived resonance will be predominantly affected by the *linear Stark effect*. This conclusion becomes even more obvious upon inspecting Fig.2. Here, we show the dependence of the differential shift $\Delta\epsilon_\pi$ of the 3π cluster level on the strength of the field q/d^2 at the center of the C atom at variable distances d/D. This dependence is quite close to linear. Higher order corrections of the dependence are due to the fact that the electric field of the point charge in our model is inhomogneous in the vicinity of the CO molecule. Furthermore, our calculations are not restricted by the first order perturbation theory.

In conclusion, our exploratory investigation of the effect of lateral electric fields on the degenerate $2\pi^*$ derived valence levels of chemisorbed CO molecules indicates that there exists a strong dependence of the level shifts and level broadening on the applied field, dominantly linear in our model. This particular feature which we call the *lateral surface Stark effect* is intimately connected with the existence of atomic-scale structural irregularities of a concrete surface.

We have also investigated within our model the charge transfer from the metal to CO as a function of the electric field (obtained from population analyses). We found that the transfer is directly combined with the CO 2π contribution to the cluster 3π orbitals and increases with the increase of the lateral electric field produced by the charge q. This means that the surface Stark effect tends to destabilize or even dissociate the CO molecules chemisorbed in the vicinity of steps or kinks. Such properties of stepped surfaces have, indeed, been observed experimentally [4a]. Thus the mechanism of the lateral surface Stark effect may also provide a clue to the understanding of some elementary processes of heterogeneous catalysis.

Acknowledgements: This work has been supported in part by the 'Internationales Büro der KFA Jülich' (Project "Surface Physics", No. 32.2 A.F.), YUZAMS Office, Zagreb and the National Science Foundation Grant JF-798. Support from the Sonderforschungsbereich 126 Clausthal/Göttingen and the Sonderforschungsbereich 6 Berlin is also acknowledged.

References

[1] See e.g., **Catalysis, Science and Technology**, J.R. Anderson, M. Boudart (Eds.) Springer Verlag, Berlin.

[2] See e.g., **The Chemical Physics of Solid Surfaces and Heterogeneous Catalysis**, D.A. King, D.P. Woodruff (Eds.), Elsevier, Amsterdam.

[3] T.N. Rhodin, J.W. Gadzuk, in: **The Nature of the Surface Chemical Bond**, North-Holland, Amsterdam, 1981.

[4] a) G.A. Somorjai, **Chemistry in Two Dimensions: Surfaces**, Cornell University Press, Ithaca, 1981.

b) N.D. Lang, S. Holloway, and J.K. Nørskov, Surf. Sci. **150**, 24 (1985).

[5] a) B. Gumhalter, K. Wandelt, and Ph. Avouris, Phys. Rev. B **37**, 8048 (1988).

b) K. Hermann, P.S. Bagus, and C.J. Nelin, Phys. Rev. B **35**, 9467 (1987).

[6] K. Wandelt, in **Thin Metal Films and Gas Chemisorption**, Ed. P. Wißmann, Elsevier, Amsterdam, 1987; p. 280 and references therein.

[7] e.g. H. Levebre-Brion, R.W. Field, **Perturbations in the Spectra of Diatomic Molecules**, Academic Press, New York, 1986.

[8] a) P.S. Bagus, C.J. Nelin, and C.W. Bauschlicher, Phys. Rev. B **28**, 5423 (1983).

b) K. Hermann, H.J. Hass, and P.S.Bagus, Z. Physik D **3**, 159 (1986).

[9] For these calculations the Hartree-Fock cluster program CLUSTER developed at the Technical University Clausthal was used. Gaussian basis sets were identical to those used in P.S. Bagus, K. Hermann, and C.W. Bauschlicher, J. Chem. Phys. **80**, 4378 (1984).

[10] D.T. Clark, B.J. Cromarty, and A. Sgamellotti, Chem. Phys. **26**, 179 (1977).

Part IV

Clean Semiconductors

Deposition and Annealing of Silicon on Cleaved Silicon Surfaces Studied by Scanning Tunneling Microscopy

R.M. Feenstra and M.A. Lutz

IBM Research Division, T.J. Watson Research Center,
Yorktown Heights, NY 10598, USA

Abstract. The formation of ordered surface structures, as a function of annealing, is studied on cleaved Si(111)2×1 surfaces covered with 0.1 monolayer of deposited silicon. The evolution of the observed surface structures is compared with previous results, obtained using no deposited silicon. At intermediate temperatures ($\simeq$ 400°C) we find significant amounts of 9×9 reconstruction on the deposited surface, whereas the non-deposited surface consists mainly of 5×5 structure. At higher temperatures both surfaces convert entirely to the 7×7 reconstruction. The results demonstrate that the availability of atoms on the surface plays a dominant role in determining the intermediate structures formed during the transition from the 2×1 to the 7×7 structure.

1. Introduction

When a Si(111) surface is formed by cleavage at room temperature, the resulting surface structure has a 2×1 unit cell[1]. This structure is metastable, and it transforms irreversibly to a 7×7 structure upon annealing at temperatures near 600°C. An intermediate phase which is sometimes observed in this transition is a 5×5 reconstruction[1,2]. Using the scanning tunneling microscope (STM), we have recently studied the transformation from 2×1 to 7×7, and the formation of the intermediate 5×5 structure[3]. We proposed a model in which the transition is separated into nucleation and growth stages, and during the growth stage the 5×5 structure is strongly favored over 7×7 because of conservation of the atom density on the surface. In this work, we report the results of an experiment which provides a test for this model of the transformation. Specifically, we deposit 0.1 monolayer (ML = 1.57×10^{15} atom/cm²) of silicon on a cleaved silicon surface, and then subject it to a succession of annealing cycles. We find that the presence of this additional silicon on the surface suppresses the formation of the 5×5 structure, thus supporting our model. Moreover, we observe significant amounts of 9×9 reconstruction on the surface, the atomic density of which is greater than that of 5×5 or 7×7. Thus we find that the availability of atoms on the sur-

Springer Series in Surface Sciences, Vol. 24   **The Structure of Surfaces III**
Editors: S.Y. Tong · M.A. Van Hove · K. Takayanagi · X.D. Xie
© Springer-Verlag Berlin, Heidelberg 1991

face plays an important role in determining the intermediate structures which form during the transition from 2×1 to 7×7 structures.

2. Experimental

The STM, and procedures used for image acquisition, are identical with those previously described[3]. Deposition of silicon was accomplished by thermal evaporation of a Si bar. A crystal thickness monitor was used for monitoring the deposition. Significant noise spikes were observed in this monitor (possibly arising from thermal fluctuations of the Si source) which limited the accuracy of the determined thickness to only a factor of about 3. Within this range, the precise thickness was determined directly from the STM images. For the experiment reported here, the deposited layer had a thickness of 0.09 ± 0.01 ML, which is denoted simply by 0.1 ML in this paper. This sample was subjected to a series of 60 s anneals, at successively increasing temperatures. For the results from a non-deposited sample discussed below, 10 s anneals were used.

3. Results & Discussion

Before presenting the results of the annealing experiments on the deposited surface, we first summarize our previous model for the transformation from 2×1 to 7×7 on non-deposited surfaces. This model is pictured in Fig. 1. The surface transformation nucleates at steps, or at domain boundaries of the 2×1 reconstruction. Initially, excess atoms migrate away from the nucleation boundary, forming ordered arrays of adatoms such as 2×2, c4×2, and c2×8 arrangements, which we collectively denote by "a/1×1". These simple adatom structures order, forming 5×5 and 7×7 structures, with atomic arrangements described by the dimer-adatom-stacking fault

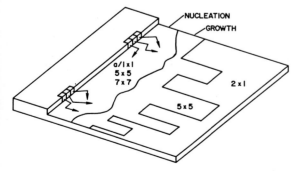

FIG. 1. Schematic view of the transformation of the Si(111)2×1 surface, at intermediate temperatures near 400°C.

481

(DAS) model[4]. Since the 7×7 structure requires 4% more atoms than the original 2×1 surface, the supply of excess atoms on the surface is eventually depleted. Then, the growth of the transformed surfaces proceeds by forming exclusively a 5×5 structure, which contains the same number of atoms as the 2×1. These 5×5 regions form "finger-like" growth patterns, growing parallel to the chains of the 2×1 structure. On a surface containing few nucleation boundaries, the transformed surface can consist almost entirely of 5×5 structure.

Typical results for the evolution of the surface structures, on a sample with no silicon deposition, are presented in Fig. 2(a). These results represent surface averages, obtained from sampling numerous areas over the surface. The cleaved surface consists completely of 2×1 structure. Near 300°C, the surface starts to transform to other structures. In the range 350−400°C, a large amount of 5×5 structure is seen, along with some 7×7 and simple adatom a/1×1 arrangements. At higher temperatures, the 5×5 structure gradually disappears, and the 7×7 dominates. At all temperatures we observe less than 1% of 9×9 structure on the surface. The "OTHER" structures indicated in Fig. 2 include disordered adatom arrangements and partially ordered DAS structures.

We now turn to the results obtained on deposited Si surfaces, as shown in Fig. 2(b) and Fig. 3. The 0.1 ML deposited silicon forms small clusters on the surface, as seen in Fig. 3(a). As the surface is annealed, these clusters grow together and form ordered adatom arrangements.

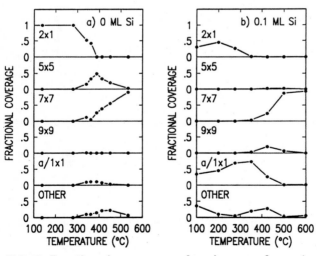

FIG. 2. Fractional coverage of various surface structures, as a function of the annealing temperature, for (a) no Si deposition, and (b) 0.1 ML Si deposition.

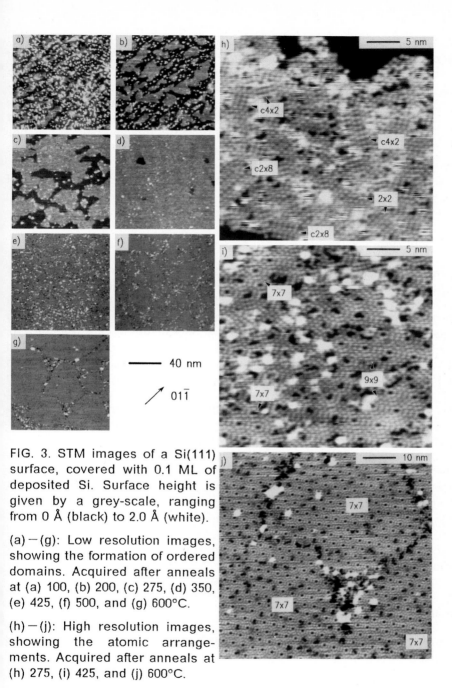

FIG. 3. STM images of a Si(111) surface, covered with 0.1 ML of deposited Si. Surface height is given by a grey-scale, ranging from 0 Å (black) to 2.0 Å (white).

(a) – (g): Low resolution images, showing the formation of ordered domains. Acquired after anneals at (a) 100, (b) 200, (c) 275, (d) 350, (e) 425, (f) 500, and (g) 600°C.

(h) – (j): High resolution images, showing the atomic arrangements. Acquired after anneals at (h) 275, (i) 425, and (j) 600°C.

Fig. 3(h) shows a high resolution STM image of these simple adatom arrangements, obtained after annealing at 275°C; we see small domains of 2×2 and c4×2 structure, along with some c2×8 (which is composed of an alternating arrangement of 2×2 and c4×2 unit cells[5]). At higher temperatures, these simple adatom structures order into DAS arrangements. Near 400°C we find a significant amount of 9×9 structure, covering about 20% of the surface. Fig. 3(i) shows some of the 9×9 regions. At higher temperatures, the 9×9 regions disappear in favor of 7×7 domains. These 7×7 domains grow in size, as can be seen by comparing Figs. 3(f) and (g). At all temperatures, we find that less than 2% of the surface consists of 5×5 structure.

4. Conclusions

From a comparison between the results for non-deposited and deposited surfaces, Figs. 2(a) and (b), it is clear that the deposition of silicon has dramatically affected the formation of structures on the surface. We find that the Si deposition suppresses the formation of the 5×5 structure, and produces significant amounts of 9×9 reconstruction. This tendency is consistent with what is expected by consideration of the atom density of these structures: relative to the top monolayer of the π-bonded chain model of the 2×1 surface[6], the DAS 5×5, 7×7, and 9×9 structures contain 0%, 4.1%, and 6.2% more atoms, respectively. Thus, we conclude that number of atoms locally available on the surface determines the structures formed during the surface transformation. We also conclude that, kinetically, the formation of 9×9 or 5×5 structure is preferred over the formation of 7×7 plus additional 2nd layer or missing 1st layer atoms, respectively. Both of these conclusions imply that the energy barriers for formation of the various DAS structures are almost equal, which is not surprising considering the similarity in the underlying DAS structures. These results support our model for the surface transformation on non-deposited samples, in which we proposed that the growth of the 5×5 structure is favored because it has the same atom density as the 2×1 surface.

References

1. J. J. Lander, G. W. Gobeli, and J. Morrison, J. Appl. Phys. *34*, 2298 (1963).
2. R. I. G. Uhrberg, E. Landemark, and L. S. O. Johansson, Phys. Rev. B*39*, 13525 (1989).
3. R. M. Feenstra and M. A. Lutz, submitted to Phys. Rev. B; and submitted to Surf. Sci.

4. K. Takayanagi, Y. Tanishiro, S. Takahashi, and M. Takahashi, Surf. Sci. *164*, 367 (1985); Phys. Rev. B*34*, 1034 (1986).
5. R. S. Becker, J. A. Golovchenko, and B. S. Swartzentruber, Phys. Rev. Lett. *54*, 2678 (1985).
6. K. C. Pandey, Phys. Rev. Lett. *47*, 1913 (1981).

Simulation and STM Studies of Equilibrium Properties of Vicinal Surfaces[*]

T.L. Einstein[1], N.C. Bartelt[1], J.L. Goldberg[1], X.-S. Wang[1], E.D. Williams[1], and B. Joós[2]

[1]Department of Physics, University of Maryland, College Park, MD 20742, USA
[2]Department of Physics, University of Ottawa,
 Ottawa, Ontario, Canada K1N 6N5

Abstract. In conjunction with LEED and STM experiments at Maryland, we have carried out a variety of theoretical studies on simple models of vicinal surfaces using Monte Carlo and free-fermion techniques. Our goal is to determine how experimentally accessible quantities can be interpreted in terms of fundamental energies. A general theme is the inadequacy of one-dimensional models in describing equilibrium properties. Among the problems we have studied are 1) what the splitting of diffraction beams at the out-of-phase condition reveals about the degree of real-space order, 2) how the scaling of the diffraction beam profiles perpendicular to the direction of splitting can be used to gauge the energy of a kink, and 3) how to characterize numerically and analytically the terrace-width distributions one might observe with STM.

1. Introduction

With experimental progress, it is now becoming possible to obtain quantitative information about the structure of vicinal surfaces.[1] If these stepped surfaces are in <u>equilibrium</u>, then we can make many predictions about the morphology of these surfaces by studying relatively simple models. Such models lend focus to the experiments by directing attention to characteristic lengths, temperatures, and energies; they also provide a means to correlate the information about long-range order coming from diffraction probes with the microscopic details of short-range behavior from STM. Specifically, we have considered 1) the terrace-step-kink (TSK) model, in which the only excitations from a "perfect staircase" are kinks in the step edge, each with energy ε, and 2) the solid-on-solid (SOS) model, which allows additional terrace excitations.[2] (In the simple-cubic version, the elementary terrace adatom or vacancy has energy 4ε. At temperatures T well below $T_R(\hat{z})$ [the roughening temperature of the terraces], the two models become essentially equivalent.) These models do not include any explicit energetic interactions between steps, so that the interstep repulsions arise solely from entropic effects.[3] In actual surfaces, we expect that energetic contributions due e.g. to elastic interactions[4] will play a significant role; we have made some progress in taking such effects into account.

As part of an ongoing effort to study these models numerically and to bring these results, and the preceding work of many others,[2,3,5,6,7] to bear on current experiments, we have studied several aspects of these problems. In this proceedings paper, we collect together and summarize the results[8-12] for equilibrated stepped surfaces set down in several preprints.

[*]Supported in part by NSF grant DMR 88-02986. Computations performed at Pittsburgh Supercomputer Center.

Springer Series in Surface Sciences, Vol. 24   **The Structure of Surfaces III**
Editors: S.Y. Tong · M.A. Van Hove · K. Takayanagi · X.D. Xie
© Springer-Verlag Berlin, Heidelberg 1991

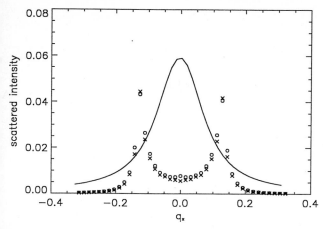

Fig. 1 From ref. 8, diffraction profiles at an out-of-phase condition compu-
ted by Monte Carlo for several models, with an average terrace width <L>=8.
The crosses and circles are for the SOS and TSK model, respectively, at k_BT =
0.7ε. To within the accuracy of the calculation, these curves are the same.
The solid line is for straight, non-interacting steps, as in a 1-d model (or
a 2-d model at T=0). The units of the abscissa are π/(lattice constant).

2. What the Splittings in LEED Profiles Reveal about Local Order[8]

Applying Monte Carlo techniques to the models, we simulated in the kinematic
approximation the structure factor of a vicinal surface containing wandering
steps. Consistent with expectations based on the analogy of stepped surfaces
with incommensurate phases,[13] we find sharp split diffraction beams at
out-of-phase conditions, characteristic of a well-defined step-step
periodicity, until $T \lesssim T_R(\hat{z})$. In Fig. 1 we show results for $T = 0.7\varepsilon/k_B \cong$
$0.6T_R(\hat{z})$. These curves contrast with the results inherent in common
one-dimensional models of step positional disorder[14]: straight steps with
short-ranged interactions give rise to broad unsplit diffraction features at
out-of-phase conditions. Thus the meandering of the steps induces "step
order", i.e. uniform spacings between the steps. Even near $T_R(\hat{z})$, when Monte
Carlo "snapshots" show dramatic local disorder, the structure factor still
shows sharp split features, albeit at substantially reduced intensity.

The shape of the split beams is much sharper than a lorentzian; often
difficult to distinguish from delta-functions, these beams actually diverge
like k to a negative power. We have explored the behavior of the associated
exponent of the power law. We have related it to the amplitude of the
logarithmic divergence of the height-height correlation function and studied
dependencies on beam energy, misorientation angle, and temperature. We have
assessed the validity of a popular analytic approximation using our numerical
results.[15]

3. Using Scaling Properties of LEED Profiles to Estimate Kink Energies[9]

In the TSK model, the intensity of the split beams discussed above depends
only weakly on T, even above $T_R(\hat{z})$ of the corresponding SOS model. However,
in the direction $\hat{y}$ perpendicular to the splitting, i.e. perpendicular to the
misorientation direction and along the average direction of the step edges,

487

the beam shape has considerable thermal dependence, due to the Boltzmann
factors associated with step wandering. Since the important characteristic
length d in this direction is the mean distance between collisions of
adjacent steps, we are led to a scaling relationship which says that the onl
way T enters the structure factor $S(k_y, T)$ is through d; hence S should depene
only on the single variable $k_y d$. In Fig. 2 we illustrate this idea by first
plotting $S(k_y, T)$, normalized by $S(0, T)$, for several temperatures. Then by
taking these curves to be functions of $\lambda(T)k_y$ and adjusting λ in each case,
we find that the curves have a universal form. From an Arrhenius plot of λ,
we can extract an estimate of the kink energy. Since the thermal range of
this procedure ranges from the SOS $T_R(\hat{z})$ to twice that temperature, the
viability of this procedure requires that terrace excitations be relatively
more costly energetically than in the SOS model.

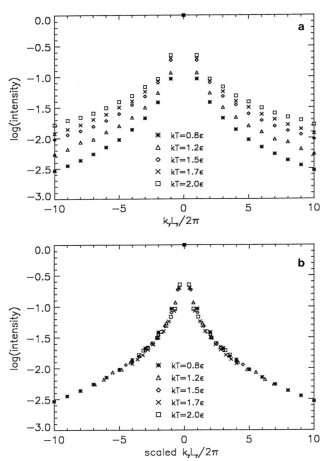

Fig. 2 a) $\mathrm{Log}(S(k_y, T)/S(0, T))$ for several values of T, computed in the TSK
model. On the abscissa k_y runs perpendicular to the direction of beam
splitting, and $k_y=0$ is the center of one of the two peaks of the split beam.
b) The same data as in Fig. 2a, plotted as $\log(S(\lambda(T)k_y)/S(0, T))$, with λ
adjusted for each T so that the curves coincide. Evidently the scaling
relation is quite accurate, even at these high temperatures. From ref. 9.

4. Analysis of the Terrace-Width Distribution Function Using Free-Fermion Techniques, and Comparison with STM Data[10-12]

Using STM for vicinal Si (111) misoriented in the [2$\overline{1}\overline{1}$] direction, we have obtained extensive information about the distribution P(L) of spacings L between neighboring steps.[10] Since the equilibration temperature of 500-700°C is well below the T_R(111), we expect a TSK model to be a good approximation.

The simplest analysis follows the Gruber-Mullins[16] approximation of allowing a single step to wander between two rigid straight steps separated by 2<L> in the $\hat{x}$ direction. This statistical problem can be recast into the elementary quantum mechanics problem of a particle (the step) trapped in an 1-d infinite-barrier well. The wandering of the step in the $\hat{y}$ direction becomes the time evolution of the position of the particle. P(L) is obtained by averaging over time, and so we lose any information about the amount of wandering. Accordingly, we quickly retrieve the Gruber-Mullins result P(L) = <L>$^{-1}$cos^2(πL/2<L>), *independent of T*. By including energetic interactions between steps to lowest order, we next encounter a particle in a 1-d harmonic oscillator potential. Now the elementary result is that P(L) has the form of a Gaussian with a width that depends explicitly on T as well as on the amount of wandering. Thus, we expect that if P(L) exhibits sizable T-dependence, energetic interactions between steps must be playing a substantial role.

Turning to the full problem in which all steps wander, we see that we now have a problem in *many-body* quantum mechanics. To account for the physical condition that steps do not cross (preventing overhangs),the particles are taken to be [spinless] fermions. If we neglect [energetic] interactions, these particles are *free* fermions, which can be analyzed using a 1-d tight-binding model.[17] To find P(L), the distribution of spacings between <u>nearest</u> steps (fermions), we need a many-particle correlation function, since there must be no particles between the fermions separated by L. The

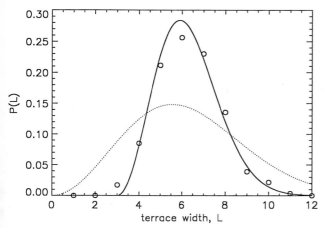

Fig. 3 Probability distribution P(L) for finding terraces of width L, for well-annealed Si(111) misoriented by 1.3° toward the [2$\overline{1}\overline{1}$] direction. The unit of L is 23 Å, the length of the perpendicular bisector of one of the two triangles comprising the (7x7) unit cell, so that <L>=6. In comparison with the data (circles), P(L) of the TSK model with the same value of <L> (dotted curve) significantly overestimates the occurrence of terraces with small L (and so, also large L). The solid curve, which describes the data rather well, is a TSK model modified to exclude L≤3; it is P(L-3) for <L>=3.

calculation[11,12] involves the determinant of [Toeplitz] matrices as large
as $(L+1) \times (L+1)$, but $P(L)$ need only be computed out to about $3<L>$, beyond
which it is negligible. Moreover, we find that the scaled form $<L>P(L)$ vs.
$L/<L>$ is nearly independent of $<L>$, even at the smallest value $<L>=2$,
indicative once again of insensitivity to the amount of wandering, and so to
ϵ and T. (Correspondingly, from Monte Carlo simulations we find that this
free-fermion result is a good approximation until remarkably high T.[11])
While this "universal" scaled form is not well described by either a
Lorentzian or a Gaussian, we can write an excellent analytic approximation
in terms of algebraic and trigonometric functions by using the scaling idea
and considering up to 4-particle correlations.[12]

Fig. 3 shows how the distribution compares with that measured by STM for
vicinal Si(111) misoriented by $1.3°$ in the $[2\bar{1}\bar{1}]$ direction.[10] STM shows
that the (7x7) reconstruction grows to the step edge and that the kink has
the form of one edge of the reconstructed unit cell, so that the unit of L is
23Å. There are significantly fewer terraces with small widths in the
experiment than occur in the free-fermion model. By assuming short-range
exclusions over two unit spacings and repulsions at three units separation,
one can account for the deviation.[10] We are currently investigating,
analytically[18] and numerically, the role of [long-range] elastic repulsions
in modifying this distribution: we can describe their effects quite well
using the Gaussian distribution to which we alluded above.

References

1. R.J. Phaneuf and E.D. Williams, Phys. Rev. Lett. <u>58</u>, 2563 (1987); Phys.
    Rev. B<u>41</u>, 2991 (1990); R.J. Phaneuf, E.D. Williams, and N.C. Bartelt,
    Phys. Rev. B<u>38</u>, 1984 (1988); B.S. Swartzentruber, Y.-W. Mo, M.B. Webb,
    and M.G. Lagally, J. Vac. Sci. Technol. A<u>7</u>, 2901 (1989); G.P. Kochanski,
    CRC Critical Reviews in Materials Science (1990).
2. A recent review discussing these models is H. van Beijeren and I. Nolden,
    in <u>Structure</u> <u>and</u> <u>Dynamics</u> <u>of</u> <u>Surfaces</u> <u>II</u>, ed. by W. Schommers and P. von
    Blanckenhagen (Springer, Berlin, 1987), p. 259.
3. M.E. Fisher and D.S. Fisher, Phys. Rev. B<u>25</u>, 3192 (1982); M.E. Fisher,
    J. Chem. Soc. Faraday Trans. 2, <u>82</u>, 1569 (1986).
4. O.L. Alerhand, D. Vanderbilt, R.D. Meade, and J.D. Joannopoulos, Phys.
    Rev. Lett. <u>61</u>, 1973 (1988), and references therein.
5. C. Jayaprakash, C. Rottman, and W.F. Saam, Phys. Rev. B<u>30</u>, 6549 (1984);
    W.F. Saam, Phys. Rev. Lett. <u>62</u>, 1632 (1989).
6. W. Selke and A.M. Szpilka, Z. Phys. B<u>62</u>, 381 (1986).
7. S.T. Chui and J.D. Weeks, Phys. Rev. B<u>23</u>, 2438 (1981).
8. N.C. Bartelt, T.L. Einstein, and Ellen D. Williams, "Diffraction from
    Stepped Surfaces in Thermal Equilibrium," accepted for publication in
    Surface Sci.
9. N.C. Bartelt, T.L. Einstein, and Ellen D. Williams, "Extracting Kink
    Energies of Steps from Thermal Scaling of Diffraction Profiles,"
    preprint.
10. X.-S. Wang, J.L. Goldberg, N.C. Bartelt, T.L. Einstein, and Ellen D.
    Williams, "Terrace Width Distributions on Vicinal Si(111)," preprint
    submitted for publication in Phys. Rev. Lett., May 1990.
11. N.C. Bartelt, T.L. Einstein, and Ellen D. Williams, "The Influence of
    Step-Step Interactions on Step Wandering," accepted for publication in
    Surface Sci. Lett.
12. B. Joós, T.L. Einstein, and N.C. Bartelt, "Distribution of Terrace
    Widths on a Vicinal Surface in the 1-d Free-Fermion Model," submitted for
    publication to Phys. Rev. B(1), Aug. 1990.

13. D. A. Huse and M. E. Fisher, Phys. Rev. B $\underline{29}$, 239 (1984); C Jayaprakash, W. F. Saam, and S. Teitel, Phys. Rev. Lett. $\underline{50}$, 2017 (1983).

14. P. R. Pukite, C. S. Lent, and P. I. Cohen, Surface Sci. $\underline{161}$, 39 (1985), and references therein; L. H. Zhao, E. Z. Luo, and M. Henzler, Appl. Phys. A $\underline{50}$, 595 (1990).

15. J. Villain, D. R. Grempel, and J. Lapujoulade, J. Phys. F $\underline{15}$, 809 (1985).

16. E. E. Gruber and W. W. Mullins, J. Phys. Chem. Solids $\underline{28}$, 875 (1967).

17. M. den Nijs, in Phase Transitions and Critical Phenomena, vol. 12 (Academic, London, 1988), p. 210.

18. F. J. Dyson, Comm. Math. Phys. $\underline{19}$, 235 (1970).

The Precipitation of Kinks on Stepped Si(111) Surfaces

Jian Wei, N.C. Bartelt, and E.D. Williams

Department of Physics, University of Maryland, College Park, MD 20742, USA

Abstract. We have used high resolution low-energy electron diffraction to study the temperature dependence of a vicinal (stepped) Si(111) surface with a polar angle of 6° from (111) along an azimuth rotated 10° away from the high symmetry $[\bar{1}\bar{1}2]$ direction. The resulting kink density on the step edge causes novel behavior: we find that the (1×1) to (7×7) reconstruction causes kinks to precipitate into a phase with high step density and increased rotation from the $[\bar{1}\bar{1}2]$ direction, leaving behind a $[\bar{1}\bar{1}2]$ oriented phase with triple-layer height steps and (7×7) reconstruction. The inclination and azimuth of the orientation of the kinked phase changes continuously with temperature. This phase separation (faceting) is reversible and can be understood thermodynamically by analogy with phase separation in a two component fluid.

1. Introduction

The work reported here is part of a continuing study of the orientational stability of vicinal Si(111) surfaces. In the past, we have studied vicinal Si(111) surfaces misoriented towards the $[1\bar{1}0]$, $[\bar{2}11]$, and $[\bar{1}\bar{1}2]$ directions. For surfaces misoriented in the $[1\bar{1}0]$ and $[\bar{2}11]$ directions [1], the (1×1) to (7×7) reconstruction induces reversible phase separation of the surfaces into reconstructed (111) planes and unreconstructed stepped surfaces. The associated orientation-temperature phase diagrams have been mapped out in detail [1, 2, 3]. In contrast, surfaces misoriented towards the $[\bar{1}\bar{1}2]$ direction undergo a step-height tripling with no obvious phase separation at low temperature [4]. In this paper, we show that this apparent orientational stability is strongly dependent on the precision of the azimuth of the surface misorientation. Surfaces misoriented by 10° from $[\bar{1}\bar{1}2]$ are thermodynamically unstable. The mechanism for this instability seems to be an unfavorable interaction between <u>kinks</u> in the step edges and the (7×7) reconstruction: the insolubility of kinks in the (7×7) reconstructed surface causes them to precipitate out into an unreconstructed phase. Another interesting feature of the observed phase separation is that both the inclination angle and azimuthal angle of the kinked surface phase change with temperature: the phase separation is analogous to phase separation in a <u>two</u> component fluid.

2. Experiment

The sample used in this work was inclined by approximately 6° from the (111) plane towards a direction 10° from $[\bar{1}\bar{1}2]$. We cleaned the sample by heating to 1250°C for 90 seconds in ultra-high vacuum. During this time the background pressure did not increase above 4×10^{-10} torr. This procedure yields well-ordered surfaces with undetectable levels of O and C contamination [2, 5]. Our high resolution LEED instrument is described in

Springer Series in Surface Sciences, Vol. 24 **The Structure of Surfaces III**
Editors: S.Y. Tong · M.A. Van Hove · K. Takayanagi · X.D. Xie
© Springer-Verlag Berlin, Heidelberg 1991

Ref. [6]; it has an instrumental resolution of 500Å. Temperature was measured with a thermocouple attached to the face of the sample. The thermocouple was calibrated by comparison with an optical pyrometer.

3. Results

Figure 1 shows the sequence of diffraction patterns we observe as a function of temperature near the terrace specular beam. The incident energy corresponds to an out-of-phase condition for single-layer height steps. At high temperature only a single pair of split beams are observed. These are due to a uniform array of single height steps, similar to what is found at high temperatures on all the vicinal Si(111) surfaces examined previously [1, 4, 7]. As indicated in the figure, the direction of the splitting is rotated by approximately 10° from the $[\bar{1}\bar{1}2]$ azimuth. Analysis of the splitting and energy dependence of the beams show the polar angle with respect to the [111] direction is 6°. At approximately 840°C, more closely spaced beams appear, oriented along the $[\bar{1}\bar{1}2]$ azimuth. Analysis of these beams reveals they are due to regions of the surface which contain triple-height steps. The polar angle of the surface giving rise to these beams is again close to 6° and the average orientation of these regions does not change with temperature. Coincident with the appearance of these beams, diffraction features associated with the (7×7) reconstruction appear. The correlation between the appearance of the (7×7) reconstruction and the closely spaced beams is shown in Fig. 2. The one-seventh order beams are also split in the $[\bar{1}\bar{1}2]$ direction, with the same splitting as the beams due to the step tripling. As the sample is cooled further, the splitting, k_s, of the original beams associated with the high temperature phase increases, indicative of a decrease in average step separation. The trajectory of the coordinates of these diffraction beams is shown in Fig. 3. As the temperature decreases the beams rotate to larger azimuthal angles away from the $[\bar{1}\bar{1}2]$ direction. The energy dependence of the beams indicates no change in step height. All changes are reversible on heating.

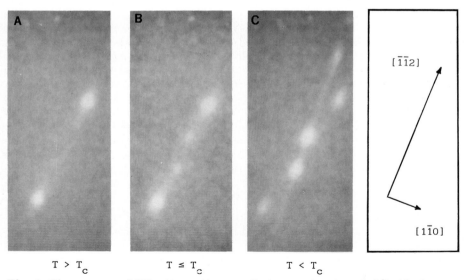

$$T > T_c \qquad T \lesssim T_c \qquad T < T_c$$

Fig. 1 The sequence of diffraction patterns as the temperature is varied [incident energy is 72eV, incident angle is 34°]

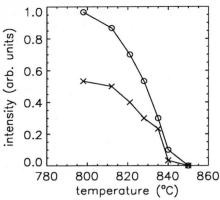

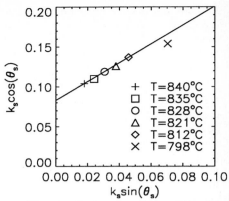

Fig. 2 The temperature dependence of the triple step beam intensity (circles) compared with the intensity of a seventh-order feature (crosses).

Fig. 3 Trajectory of the diffraction beam corresponding to the kinked, unreconstructed phase as the temperature is lowered. The vertical axis points in the $[\bar{1}\bar{1}2]$ direction.

4. Discussion

The physical picture of the structural transition reported above involves primarily changes in kink density at step edges. Above the transition, the single height steps separate unreconstructed terraces and have a kink density of ~ 0.18 to accommodate the 10° azimuthal misorientation. An unfavorable interaction between the kinks and the (7×7) reconstruction causes a phase separation when the (7×7) reconstruction and the related formation of triple-height steps in the $[\bar{1}\bar{1}2]$ direction occurs. (From STM studies [8], it is known that the step-tripling occurs because of a direct repulsive interaction between single-height steps when the (7×7) reconstruction is present.) This work shows that this step-tripling is also accompanied by a strong preferential alignment along the $[\bar{1}\bar{1}2]$ azimuth, which can only be accomplished by reducing the kink density at the step edges. Since the macroscopic surface orientation must be preserved, the kinks are expelled to a different region of the surface with <u>increased</u> azimuthal misorientation. In this case, the kinks segregate into the original unreconstructed phase causing the observed beam rotation. The continuous increase in the beam rotation (i.e. azimuthal angle) and related change in splitting (polar angle) result from an increasing <u>area</u> of the step-tripled phase (which has fixed orientation) with decreasing T.

For a surface of macroscopic surface normal $\hat{n}_0$ and area A, separating into regions with surface normals $\hat{n}_{s,t}$, conservation of surface orientation requires

$$A\hat{n}_0 = A_s\hat{n}_s + A_t\hat{n}_t \quad , \tag{1}$$

Using one of the vector components of this equation we can define fractional compositions $f_{s,t}$ of the phases containing single-height (unreconstructed) and triple-height (reconstructed) steps:

$$f_s = \frac{A_s \cos \phi_s}{A \cos \phi_0}, \tag{2a}$$

$$f_t = 1 - f_s = \frac{A_t \cos \phi_t}{A \cos \phi_0}, \tag{2b}$$

494

where $\phi_{0,s,t}$ are the polar angles with respect to the [111] direction. The remaining vector components of Eq. (1) can then be expressed in lever rule form:

$$\tan \phi_0 \cos \theta_0 = f_s \tan \phi_s \cos \theta_s + (1 - f_s) \tan \phi_t \cos \theta_t \qquad (3a)$$

$$\tan \phi_0 \sin \theta_0 = f_s \tan \phi_s \sin \theta_s + (1 - f_s) \tan \phi_t \sin \theta_t \qquad (3b)$$

where $\theta_{0,s,t}$ are the azimuthal angles with respect to the $[\bar{1}\bar{1}2]$ direction. This equation is exactly analogous to the equation for phase separation in a two-component fluid. Our experimental results show that $\theta_t = 0$, ϕ_t=const. Using this result, eliminating f_s between Eqns. 3a and 3b and using $k_s \propto \tan \phi_s$, we predict a linear trajectory for the beams due to the single height phase:

$$k_s \cos \theta_s = k_s \sin \theta_s \left(\frac{\cos \theta_0 - \tan \phi_t / \tan \phi_0}{\sin \theta_0} \right) + C \qquad (4)$$

The experimental data show such a linear trajectory as in Fig. 3 with the slope giving a value of $\tan \phi_0 / \tan \phi_t = 1.2$, in fair agreement with the measured value of ≈ 1.08.

The phase separation we observe is evidently driven by the (7×7) to (1×1) reconstructive transition. As emphasized by Cahn [9], orientational phase separation requires sharp edges in the equilibrium crystal shape; i.e. discontinuities in the free energy associated with the angular variables. It is necessary to postulate that sharp edges evolve between the (1×1) and (7×7) phases with decreasing temperature to explain the dependence of the surface orientation of the coexisting phases we observe.

That kinks are rare on reconstructed surfaces, suggests a "knife edge" singularity in $[\bar{1}\bar{1}2]$ direction in the orientational dependence of the free energy. This would give rise to a cylindrical section in the equilibrium crystal shape. Although this possibility was raised long ago by Herring [10], we know of no theoretical or experimental study of it.

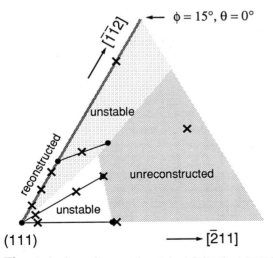

Fig. 4 A phase diagram for vicinal Si(111) at 820°C which is consistent with the results of this work and several previous studies [1, 2, 4]. Following the analogy with a two component fluid given by Eq. (3), this diagram is a polar plot of $\tan \phi$. The crosses mark the studied surface orientations. The filled circles mark the observed boundaries of the coexistence regions, and the solid lines are the associated tie lines.

Figure 4 collects together the information we have obtained [1, 2, 4] for vicinal Si(111) at one particular temperature (820°). The crosses mark the net surface orientations we have studied; the circles mark the borders of the observed coexistence regions. The solid lines which join the circles are tie lines. As the temperature is lowered, the regions labeled "unstable" (i.e., where faceting occurs) get larger. The results of this study, in particular Fig. 3, show how the unstable region near the $[1\bar{1}2]$ azimuth evolves with temperature.

A more extensive discussion of this phase separation, including direct SEM and STM images of the coexisting phases, is in preparation [11].

References

[1] R.J. Phaneuf and E.D. Williams, Phys. Rev. Lett **58**, 2563 (1987); R.J. Phaneuf, E.D. Williams, and N.C. Bartelt, Phys. Rev. B **38**, 1984 (1988).

[2] R.Q. Hwang E.D. Williams, and R.L. Park, Phys. Rev. B **40**, 11716 (1989).

[3] N.C. Bartelt, E.D. Williams, R.J. Phaneuf, Y.Yang, and S. Das Sarma, J. Vac. Sci. Technol. A **7**, 1898 (1989).

[4] R.J. Phaneuf and E.D. Williams, Phys. Rev. B **41** 2991, (1990).

[5] B.S. Swartzentruber, Y.-W. Mo, M.B. Webb, and M.G. Lagally, J. Vac. Sci. Technol. **7**, 2901 (1989).

[6] R.Q. Hwang, E.D. Williams and R.L. Park, Rev. of Sci. Instrum. **60**, 2945 (1989).

[7] B.Z. Olshanetsky and A.A. Shklyaev, Surf. Sci. **82**, 445 (1979); B.Z. Olshanetsky, V.I. Mashanov, and A.I. Nikiforov, Surf. Sci. **111**, 429 (1981).

[8] X.-S. Wang, J.L. Goldberg, N.C. Bartelt, T.L. Einstein and E.D. Williams, preprint (1990).

[9] J. Cahn, J. Phys. (Paris) Colloq. **12** C6-199 (1982).

[10] C. Herring, Phys. Rev **82**, 87 (1951).

[11] J. Wei, X.-S. Wang, N.C. Bartelt, R.T. Tung, and E.D. Williams, in preparation.

Disordering of the (3 × 1) Reconstruction of Si(113): Realization of the Chiral Three-State Potts Model

Y.-N. Yang, N.C. Bartelt, T.L. Einstein, R.L. Park, and E.D. Williams

Department of Physics, University of Maryland, College Park, MD 20742, USA

Abstract. Using high-resolution LEED we have studied the disordering of the (3x1) reconstruction of Si(113). This disordering occurs via a continuous transition at about 850 K. Above this transition temperature T_c, the positions of the superlattice diffraction beams shift away (continuously) from their commensurate [but non-high-symmetry] positions by an amount q. The direction and magnitude of $\bar{q}$ is approximately independent of the third-order beam index for any particular incident electron energy. The shift is observed to be proportional to the broadening of the beams, i.e. to the inverse correlation length ξ^{-1}. The observed behavior is consistent with that expected for a system in the universality class of the chiral three-state Potts model. The limiting behavior of $\bar{q} \cdot \xi$ is consistent with theoretical predictions. Critical exponents α, β, γ, and ν were also measured. Their values are approximately equal to those of the pure three-state Potts model, also consistent with previous numerical studies.

According to the Lifshitz condition (the first of three rules in the mean-field-like Landau theory of continuous phase transitions),[1] in order for an overlayer or reconstruction pattern to order <u>continuously</u> to a <u>commensurate</u> structure (as opposed to an incommensurate, floating array), the position of the "extra" spot in reciprocal space must be at a high-symmetry position of the surface Brillouin zone, i.e. at a corner or at the center of an edge. In the general framework of Landau theory, there are just four universality classes in two dimensions;[1] the Lifshitz condition is naturally satisfied by the several examples in surface science of these classes. However, a remarkable feature of 2-d phase transitions, with their strong fluctuations, is that two of these classes actually violate the second Landau rule. One might then wonder whether there are any exceptions to the Lifshitz condition. Huse and Fisher[2] suggested several years ago that in addition to these four universality classes, there is another associated with the chiral Potts model.[3] The surface-science realizations of this class are (3x1) overlayers or reconstructions on rectangular or centered rectangular substrates. The attendant extra spot lies in the interior of the surface Brillouin zone, at a non-high-symmetry position. As we reported earlier[4], we have found what we believe to be the first physical realization of this universality class, the (3x1) reconstruction of Si(113) and studied its details using high-resolution LEED. In this paper we summarize the results[4] and amplify on some aspects.

First we discuss why the adjective "chiral" is appropriate to this model. Suppose one labels the columns of the underlying rectangular or centered rectangular by $A, B, C, A, B, C, A, B, C, \ldots$ In (3x1) ordering, columns with one of these letters are preferentially occupied. Inspection shows that the $A|B$ walls (between domains with A columns preferentially occupied and domains with B columns occupied to the right) are different from $B\|A$

*Supported by NSF grant DMR 88-02986.

walls. However, $A|B$, $B|C$, and $C|A$ walls are the same, as are $B\|A$, $C\|B$, and $A\|C$ walls. Based on several theories, it has been shown that this chirality or handedness in one spatial direction ("uniaxial") is a relevant perturbation in renormalization-group language, and so should change the universality class of the transition from that of a model in which all six walls are identical, namely a 3-state Potts model. {Note that in the $(\sqrt{3}x\sqrt{3})R30°$ overlayer on a triangular lattice, there is a similar handedness in the wall structure, but in three directions ("triaxial chirality"); consequently, this chirality is believed to be irrelevant, leaving this transition in the 3-state Potts class.[2]}

Disordering of the (3x1) structure involves formation of domain walls. Thus, as the temperature T rises above the critical temperature T_c, the peak of each extra beam moves away from the position of long-range order (1/3 of the way along the line between a substrate-beam position and the next substrate-beam position in the "uniaxial" direction) by an amount $\bar{q}(T)$. Since we are above T_c, the beam is due to diffuse scattering. The magnitude of the shift is predicted to be proportional to the broadening of the beams along the direction of the line (called here the ⊥ direction), i.e. to the inverse correlation length ξ^{-1}. This behavior is quite remarkable: $\bar{q}\cdot\xi$ approaches 0 near a transition involving commensurate disordering; for an incommensurate [floating] phase, this product is ∞. Below the transition, the delta-function-like scattering due to long-range order is locked onto the 1/3 position, but the peak of the critical scattering again shifts away by an amount $\bar{q}$ which is proportional to the broadening of the critical-scattering contribution. The direction of this shift is the *same* as the direction of the shift above T_c, so that $d\bar{q}(T)/dT$ is discontinuous at T_c. While this behavior can be readily observed in Monte Carlo simulations[5], it is likely to be masked in most experiments, as it is in the one reported here.

The direction of the shift can be understood in terms of the domain wall structure. From the wetting behavior[6] anticipated for these walls,[2] we realize that ordered domains of two sorts are separated by only one of the two possible kinds of wall, | or ‖: any unfavorable wall will split into two walls of the favorable kind (with an intervening phase of the third type of domain). To illustrate how the shift depends on the walls, we present a very crude model: We consider a chain of sites spaced by distance a, with periodic boundary conditions (i.e. a ring) on which there are three domains, one of each kind, each with five atoms. For a ring with configuration *AAAAABBBBBCCCCC*, the walls will have an extra [empty] site, giving 48 sites in all. In Fig. 1, we show a) the pair correlation function

$$G(r) = \sum_{r'}n(r')n(r+r') \qquad (1)$$

and b) its Fourier transform S(k). In G(r) we see that in addition to the contributions at 3a and 6a from the ordered domains, there are smaller contributions at 4a and 7a. (Because of the cyclicality of the ring, the values at r > 8 are not independent.) This effective shift of each peak to larger r translates in reciprocal space to a shift of the peak to a value of k less than 1/3, as in Fig. 1b. Conversely, if we create a chain with the opposite type of domain walls, for instance *AAAAACCCCCBBBBB*, by removing an empty site between each domain (i.e. with 42 sites), we find in G(r) the same contributions at 3a and 6a as before, but with the smaller contributions now at 2a and 5a, respectively, indicative of a shift of the peaks of G(r) to smaller values, as depicted in Fig. 1a. Correspondingly, as shown in Fig. 1b, the peak of the structure factor shifts to a k greater than 1/3. Thus we can very qualitatively understand the chiral shifts to larger k as due to a decreased mean unit cell length when thicker domain walls form, and to smaller k when thin domain walls form. (If we contruct this configuration using a ring with 51 sites and *two* extra empty sites between domains, G(r) is less straightforward than in Fig. 1a, but S(k)

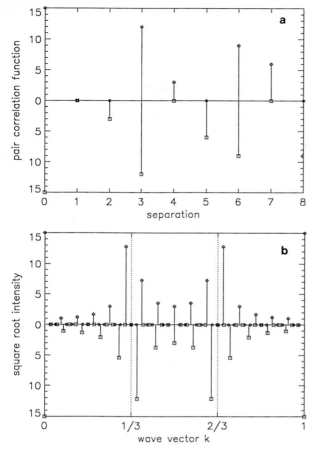

1. Illustrations for the crude model described in the text.

   a) Pair correlation function G(r) vs. separation in units of the lattice
      spacing a. The upper portion [diamonds] is for a 48-site ring, with
      $A|B$ walls (extra empty site: 3 empty sites between 5-atom domains).
      The lower portion [squares], plotted downward, is for a 42-site ring,
      with $B\|A$ walls (only 1 empty site between 5-atom domains).

   b) S(k) for Fig. 1a. The units of the abscissa are $2\pi/a$. For a single
      ordered domain, there would be intensity only at 0, 1/3, 2/3, and 1.

is very similar to Fig. 1b.) We also note that the magnitude of the shift
varies roughly inversely with the size of the ordered domains.

    In the actual experiment, contour plots of HRLEED intensity from
Si(113) were measured as the temperature was raised through $T_c \sim 844K$, as
shown in Fig. 2 of Ref. 4. The center of the beam was indeed found to shift
continuously away from the 1/3 position, in the direction corresponding to
formation of $A\|C$ walls. To ascertain that the shift is not a simple
consequence of steps or multiple scattering (cf. ftnt. 16 of Ref. 4), we
checked that the same behavior occurred for many different beams. In Table
1, we list the shift in peak position between 730K and 932K for the 8 beams

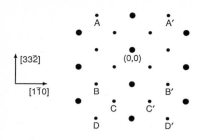

2. Positions in the surface Brillouin zone of beams listed in Table 1

[33$\bar{2}$]

[1$\bar{1}$0]

Table 1: Shifts of all the third-order beams (depicted in Fig. 2) appearing on the LEED screen when the electron energy is 40eV and the incident angle is normal to the surface. The peak positions are the location of the maximum in the image data files. All shifts are in ±[1$\bar{1}$0]. Due to the resolution of the video camera, the uncertainty in the peak position is about 1 pixel.

	Beam Position (pixel)		Shift
Beams	T = 730 K	T = 932 K	(pixel)
A	274, 143	274, 146	0,3
A'	274, 316	274, 313	0,-3
B	152, 143	152, 145	0,2
B'	152, 317	152, 315	0,-2
C	122, 186	122, 182	0,-4
C'	122, 272	122, 275	0,3
D	90, 140	90, 142	0,2
D'	88, 318	88, 314	0,-4

indicated in Fig. 2. Evidently our results are not due to such vagaries.

Careful analysis[4] shows that, as predicted, $\bar{q}$ is proportional to ξ^{-1}. The limiting value of $\bar{q} \cdot \xi_{\perp}$ as T approaches T_c, slightly above 1/2, is consistent with Monte Carlo computations[5] for a generic model in this class and for transfer-matrix analysis of a one-dimensional model,[2b] which found a value of $\cot(\pi/3) \cong 0.577$. By fitting the profiles to a Lorentzian form, we could extract the magnitude and the width above T_c and thereby gauge the susceptibility and the inverse $\perp$ correlation length, respectively. To obtain critical exponents, we then used a log-log plot, as illustrated in Fig. 3 for the latter. We find $T_c = 844K$, $\gamma = 1.03\pm0.19$, and $\nu = 0.99\pm0.18$. From measurements below T_c we also estimate $\beta = 0.11\pm0.04$. These numbers are consistent with those of the 3-state Potts model ($\gamma = 13/9$, $\nu = 5/6$, and $\beta = 1/9$). Similar consistency with pure 3-state Potts exponents was found in various numerical studies.[5,7]

Using a low-resolution LEED system, we also measured the *integrated* intensity.[4] Such data can be used to extract the specific heat exponent α;[8,9] we find $\alpha = 0.32\pm0.06$, consistent with the 3-state Potts value of

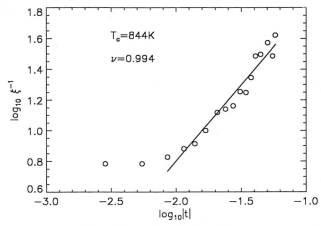

3. Log-log plot of beam broadening (inverse ⊥ correlation length) vs. reduced temperature, used to determine the critical exponent ν.

1/3. This number was obtained by choosing the set of parameters found in numerical simulations[9] of a generic $(\sqrt{3}\times\sqrt{3})R30°$ overlayer to give the best agreement with 3-state Potts numbers. The value of T_c produced by this analysis, 900K, gives an idea of the difficulties of obtaining accurate *absolute* temperature measurements in this experiment. In this analysis of the data, we also estimated the critical amplitude ratio of the free energy above and below T_c to be slightly below one, unexpectedly consistent with the pure 3-state Potts model, in contrast to simulations[5] for the (3x1) in which this ratio was 3±1. These issues should not distract the reader from the crucial, remarkable observation that the shift of the peak is proportional to the broadening, which is the novel feature of this transition.

References

1. E.g. M. Schick, Prog. Surface Sci. 11, 245 (1981) or T.L. Einstein, in Chemistry and Physics of Solid Surfaces IV, ed. by R. Vanselow and R. Howe (Springer, Berlin, 1982), p. 251, and references in both.
2. D.A. Huse and M.E. Fisher, a) Phys. Rev. Lett. 49, 793 (1982); b) Phys. Rev. B 29, 239 (1984).
3. S. Ostlund, Phys. Rev. B 23, 2235 (1981); 24, 398 (1981).
4. Y.-N. Yang, E.D. Williams, R.L. Park, N.C. Bartelt, and T.L. Einstein, Phys. Rev. Lett. 64, 2410 (1990); for more details, see Y.-N. Yang, Ph.D. thesis, University of Maryland, College Park (1990), unpublished.
5. N.C. Bartelt, T.L. Einstein, and L.D. Roelofs, Phys. Rev. B35, 4812 (1987).
6. S. Dietrich, in Phase Transitions and Critical Phenomena, vol. 12, ed. by C. Domb and J.L. Lebowitz (Academic, New York, 1988), p. 1; M.E. Fisher, J. Chem. Soc., Faraday Trans. 2 (Faraday Symposium 20), 82, 1569 (1986); A.L. Stella, X.-C. Xie, T.L. Einstein, and N.C. Bartelt, Z. Phys. B67, 357 (1987).
7. W. Selke and J. Yeomans, Z. Phys. B 46, 311 (1982).
8. N.C. Bartelt, T.L. Einstein, and L.D. Roelofs, in The Structure of Surfaces, ed. by M.A. Van Hove and S.Y. Tong (Springer, Berlin, 1985), p. 357.
9. N.C. Bartelt, T.L. Einstein, and L.D. Roelofs, Phys. Rev. B32, 2993 (1985).

Current Effects on Clean Si(111) and (001) Surfaces Studied by Reflection Electron Microscopy

A. Yamanaka, N. Ohse, H. Kahata, and K. Yagi

Physics Department, Tokyo Institute of Technology,
Oh-okayama, Meguro, Tokyo 152, Japan

Abstract. Step bunching and its disappearance on Si(111) and (001) surfaces depending on the direction of a current fed through the specimen were studied in detail by reflection electron microscopy. On (001)2x1 surfaces, the changes of the surface structure were observed together with conversion of the major domain between the 2x1 and 1x2 domains. On (111) surfaces the observations in a wide range of temperatures across the phase transition temperature between the 7x7 and 1x1 structures showed that the changes are temperature dependent.

1. Introduction

In addition to surface electromigration of metal atoms deposited on Si(111) surfaces [1-4], a current fed through Si crystal has been found to induce structure changes of clean surfaces. The phenomenon was first noted by LEED [5] on Si(001)2x1 surfaces and was studied independently by three groups by reflection electron microscopy (REM) [6,7] and scanning REM (SREM) [8]. Current effect on (111)1x1 surfaces was first studied by REM [9]. The phenomena on the (001) surfaces has been discussed in terms of electro-migration of adatoms including anisotropy of diffusion on Si(001) surfaces [10] and current-direction-dependent instability of the steps [8]. Stoyanov [11] studied the phenomena theoretically including anisotropic diffusion and relative stability of the steps but not current-direction-dependent instability of the steps.

At present the phenomena have not been physically explained. In the present paper our later observations on (001) surfaces with a current not only along the [110] direction but also along the [100] direction, and observations on (111) surfaces, which had started independently, are described. In the latter case the current was along <112> directions as in the previous report by Latyshev et al. [9]. However, the [$\bar{1}\bar{1}2$] and [11$\bar{2}$] directions differs geometrically on (111) surface. So that we differentiated two directions by RHEED patterns and checked this anisotropy effect, which had not been carried out in the previous paper [9].

2. Experimental

A UHV electron microscope used in the previous report [2,3,7,10] was used. Specimens, 11x1x0.3m^3 in size, with (111) and (001) surfaces were cut from wafers. The cutting was done in such a way that the current directions were <110> and <100> on (001) surfaces and <11$\bar{2}$> on (111) surfaces. Temperatures of the specimens are 700-900°C in the case of (001) surfaces and 800-1250°C in the case of (111) surfaces.

Springer Series in Surface Sciences, Vol. 24   **The Structure of Surfaces III**
Editors: S.Y. Tong · M.A. Van Hove · K. Takayanagi · X.D. Xie
© Springer-Verlag Berlin, Heidelberg 1991

3. Results

3.1 (001) surfaces

Figure 1 reproduces REM images of the same area of a Si(001) surface, which show changes of the surface structure depending on the current direction (1.0 A, about 830 °C). The imaging beam was along the [110] and the current was parallel to the [1$\bar{1}$0], parallel and perpendicular to the dimer direction on the terraces. The surface steps up to the left and the current is step-down in (a) and step-up in (b). Two different current effects were noted. In areas marked by A, wide bright terraces in (a) are 1x2 domains, where dimers are perpendicular to the step-down current. By a reversal of the current from (a) to (b) the wide 1x2 terraces changed to dark 2x1 terraces by motions of single steps. This behavior agrees with what was reported previously [7].

In areas marked by B, however, wide bright terraces of 1x2 domains in (a) disappear in (b) to form uniformly spaced terraces. By a further reversal of the current bright wide terraces reappeared in the regions B like those shown in (a). This means that in the region B step bunching was produced by a step-down current (a) and wide terraces produced as a result of bunching are 1x2 domains, which agree with the major domain produced by a step-down current [7]. It should also be noted that the bright 1x2 domains are still wider in regions B in (b).

Figure 2 reproduces REM images of a Si(001) surface whose behavior towards the current was similar to, in one sense, but different from, in the other sense, that of the area B in Fig. 1. The similarity is step bunching and its disappearance depending on the current direction. However, the major domains are always 2x1 domains and the step bunching is

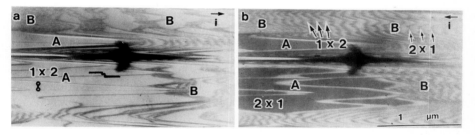

Figure 1. REM images of a Si(001)2x1 surface showing changes of surface structure by a reversal of the current direction. A step-down current causes step bunching in regions marked B. The current is in the [1$\bar{1}$0] direction.

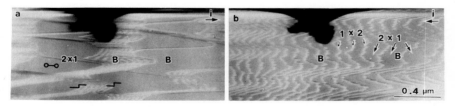

Figure 2. REM images of a Si(001)2x1 surface showing step bunching by a step-up current in (a).

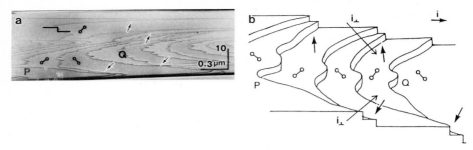

Figure 3. A REM image of a Si(001) surface and its schema, showing structure caused by a step-down current parallel to the [100] direction.

produced by a step-up current ((a)). The step bunching produces wide terraces and they are 2x1 domains, which agree with the major domain produced by a step-up current [7].

Figure 3 shows a REM image of Si(001) surface imaged along the [100] direction. When REM images are taken along this direction domain contrast is lost and single steps are clearly seen but the directions of dimers on the terrace can be known from the fine structure of the steps [12]. The current was along the [010] direction and successive steps perpendicular to the current (vertical steps in the image) are equivalent. It was observed that the steps in the [010] direction are uniformly spaced irrespective of the current direction. However, pairing of the steps as indicated by arrows in (a) is noted where the steps deviate from the vertical direction and are close parallel to the [110] and [110] directions. This is schematically shown in (b). Components of a step-down current perpendicular to the steps are different in upper and lower regions of the micrograph but on wider terraces dimer directions are perpendicular to the step-down current in either region. A reversal of the current produced different pairing of the steps in these regions as was expected.

3.2 (111) surfaces

Figure 4 reproduces REM images of the same area of a Si(111) surface which shows changes of surface structure depending on the current direction. They were taken at $840^{\circ}C$, $10^{\circ}C$ above the phase transition temperature T_c between the 7x7 and 1x1 structures. A step-up current in (a) produced uniformly spaced steps, while a step-down current in (b) produced step bunching.

A mark in the lower right in (b) denotes an orientation of the 7x7 unit cell, when the surface is cooled below T_c; the dark triangle means the

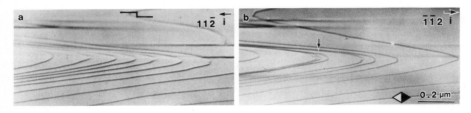

Figure 4. REM images of a Si(111) surface, showing step bunching by a step-down current.

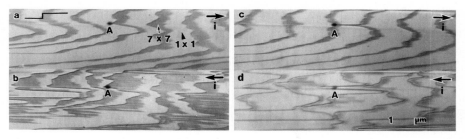

Figure 5. REM images of a Si(111) surface at T_c showing step bunching by a step-down current. Dark regions are the 7x7 structure regions formed along outer edges of terraces.

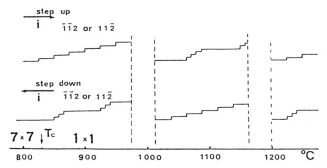

Figure 6. A schematic illustration of temperature dependence of current effects on (111) surfaces.

stacking fault part of the DAS structure. We searched an area of the same specimen where the the surface stepped up to the left and we noticed step bunching by a step-down current (see Fig. 1 (a)). This means that difference of the structures of [$\bar{1}\bar{1}2$] and [11$\bar{2}$] steps does not affect the phenomenon.

Figure 5 shows similar observation at T_c, where the direction of the current was changed three times. The step-down current produces step bunching in (b) and (d), indicating that the difference in the surface structures of terraces and steps do not affect the phenomenon.

Our observations of step bunching induced by a step-down current at around T_c differ from those at 1000°C reported by Latyshev et al. [9]. We studied the phenomenon at elevated temperature differentiating the [$\bar{1}\bar{1}2$] and [11$\bar{2}$] directions and results are schematically shown in Fig. 6. The relation between the current direction relative to steps and step-bunching reverses twice. Latyshev et al. reported another reversal at higher temperature, around 1350°C [9].

4. Summary and Discussion

A new phenomenon was found of an effect of the current on the structure of Si(001)2x1 surfaces with relatively narrow terraces in addition to that reported previously on wide terraces [5-8] (Figs. 1 and 2). In one such region, a step-down current produces step bunching, while in the other regions a step-up current does so. A notable fact is that the major domain

505

depends on the type of regions irrespective of the current direction (note relative sizes of the 2x1 and 1x2 domains in Figs. 1 (b) and 2 (b)). Strain dependent domain structures recently observed by STM [13] may not be responsible because (1) steps are widely spaced in the present case and (b) reversals of the sense of strain at short distance, say 100 μm in the specimen of 300 μm thick, is hard to imagine.

The observation for a current along the [100] direction Fig. 3 is consistent with the previous one [7] when we consider an effect of the component of the current perpendicular to the steps along the [110] and [$\bar{1}$10] directions. The fact that the current reversal did not affect steps parallel to the [010] is due to the fact that the structures of successive single steps are the same and the relative direction of the current with dimer directions is equivalent on the successive terraces.

On (111) surfaces, we found another transition of the current induced structure at around 1000°C and that overall phenomena do not depend on the structures of steps ([$\bar{1}\bar{1}$2] and [11$\bar{2}$] steps) and of terraces (7x7 and 1x1).

All of the observations are considered to be related to surface electromigration of adatoms emitted from the steps including anisotropy of diffusion and relative stability of steps against current direction (emission of atoms from the steps including the Shwoebel effect). It is necessary to solve the problem analytically or numerically (Monte Carlo simulation) using various parameters. At the same time further detailed observations are needed including the current direction dependent growth of two dimensional islands on terraces, which is in progress using the present microscope.

This work was supported by a Grant in Aid from the Ministry of Education, Science and Culture (No. 63609507 and 01609501).

References

[1] See for the review paper by H. Yasunaga and references therein. Proc. 26th Yamada Conference on "Surface as a New Material"(Osaka, 1990. Surface Sci. **242**(1991) 171).
[2] A. Yamanaka, K. Yagi and H. Yasunaga: Ultramicroscopy **29**(1989) 161.
[3] A. Yamanaka and K. Yagi: Proc. 26th Yamada Conference on "Surface as a New Material" (Osaka, 1990. Surface Sci. **242**(1991) 181).
[4] A. Yamanaka, Y. Tanishiro and K. Yagi: Proc. 3rd NEC Symposium "Ordering at Surfaces and Interfaces" (Oct. 1990)(Springer Verlag (1991 in press).
[5] T. Enta, S. Suzuki, S. Kono and T. Sakamoto: Phys. Rev. **B39** (1989) 5524.
[6] A. V. Latyshev, A. B. Krasilnikov and A.V. Aseev: J. Electron Microscope Technique (to be published) and references therein.
[7] H. Kahata and K. Yagi: Jpn. J. Appl. Phys. **28** (1989) L585.
[8] M. Ichikawa and T. Doi: Proc. 7th ICSS (Korn, Sept. 1989).
[9] A. V. Latyshev, A. L. Aseev, A. B. Krasilnikov and S. I. Stenin: Surface Sci. **213** (1989) 157. 10. H. Kahata and K. Yagi; Jpn. J. Appl. Phys. **28** (1989) L1042.
[11] S. Stoyanov: Jpn. J. Appl. Phys. **29** (1990) L659.
[12] H. Kahata and K. Yagi: Surface Sci. **220** (1989) 131.
[13] B. S. Swartzentruber, Y. W. Mo, M. B. Webb and M. G. Lagally: J. Vac. Sci. Tech. **A8** (1990) 210.

Morphology and Electron States Along Clean Si(100) Vicinal Surfaces

M. Khial, J.-P. Lacharme, and C.A. Sébenne

Laboratoire de Physique des Solides, URA 154 au CNRS,
Université Pierre et Marie Curie, F-75252 Paris Cedex 05, France

Abstract : Photoemission yield spectroscopy has been used to compare the detailed shape of the filled surface state band of various LEED-characterized Si(100) and vicinal surfaces. It is shown that vicinal surfaces cut along an angle in the [011] direction display a regular array of double steps with a single domain 2 x 1 reconstruction and have a surface state band almost identical to that of the smoothest Si(100) surfaces. On the contrary vicinal surfaces cut in the [010] direction show a high density of kinks along a two domain surface and the surface state band is deformed, bringing more states into the gap and displacing the Fermi level away from the valence band edge at the surface.

Introduction

The crystallographic and electronic properties of the clean Si(100) surface have been the object of numerous studies in the last three decades and new precisions have been associated with new techniques. If the 2 x 1 reconstruction has been recognized very early from LEED observations, the actual atomic arrangements based on so-called symmetric or tilted dimers has been established with scanning tunneling microscopy (1). Recent works on slightly misoriented surfaces have shown that single-plane high steps have a preferential orientation with respect to the dimers : the dimerizing bond between two Si surface atoms fits well when parallel to the step edge (2).

The corresponding electron surface states are well known both theoretically (3) and experimentally (4) : the filled band closest to Fermi level is a dispersive band giving two peaks in the density of states roughly 0.3 and 0.7 eV below the valence band edge E_{vs} (which is about 0.4 eV below the Fermi level position at the surface).

The influence of surface roughness has been studied : the detailed shape of the filled surface state band strongly depends on the preparation of the clean surface both outside and inside the vacuum vessel (5). In particular we have shown that whatever the degree of roughness of a clean surface under vacuum, it can be improved by exposure to a low oxygen dose followed by annealing at most at 920°C (6). Besides, a slightly oxidizing chemical treatment prior to vacuum loading is essential in getting smooth clean surfaces (7) and the smoothest surfaces are obtained through a combination of both. This has been shown using the high sensitivity of photoemission yield spectroscopy (PYS) which allows to detect changes in the gap surface state density as low as 10^{11} cm^{-2} eV^{-1} (7).

In the present paper we deal with vicinal surfaces of Si(100) prepared by the procedure leading to the smoothest surfaces, and compare them to the smoothest Si(100) surfaces cut at nominal angle. Two vicinal surfaces are considered. In one case, the cut is made along a plane making an angle of 6° with (100) along the [011] direction : this is the case where the surface dimers are either parallel or perpendicular to the nominal step edge direction. In the other case the cut is made along a plane making an angle of 5° with (100) along the [010] direction : surface dimers are at 45° from the nominal step edge direction.

Some LEED and ELS studies (8) as well as angle resolved UV photoemission spectra (9) have been published : we compare them to our own results which associate LEED and PYS measurements, taking advantage of the high sensitivity and energy resolution of the latter.

Experimental

The multitechnique ultrahigh vacuum system has been described earlier (10). Besides standard LEED and AES equipment, the photoemission yield spectrometer is made for use in the 3.8-6.7 eV photon energy range. Proper calibration allows to measure the absolute yield value Y(hv), number of emitted electrons per incident photon, as a function of photon energy. Then, since a photoyield spectrum is the addition of both surface and bulk filled state contributions, a procedure is needed to separate the effective surface density of states. In the case of a clean (100) surface of low doped silicon (10^{15} cm^{-3}), the procedure has already been explained in detail (5). We call effective density of surface states $N_s^*(E)$ a quantity proportional to :

$$d\,[Y(E)]\,/\,dE - c\,(E-E_i)^{5/2}$$

where E is the energy, E_i the ionization energy of the material (5.3 eV in clean Si) and c a constant adjusted in the upper photon energy range.

The samples were industrial grade silicon wafers cut as rectangles 20 x 5 mm^2, 0.5 mm thick and either chemico-mechanically polished for the 0° and the 6° [011] or mechanically polished for the 5° [010]. Before loading into the UHV chamber, they underwent the same oxidizing chemical treatment (11). Once under a pressure of 1 x 10^{-10} Torr, the sample was Joule heated at a temperature of 900°C, as measured with a calibrated optical pyrometer, for 30 s. The resulting surface was clean, as checked by AES, and the LEED pattern was usually good. However the surface state band of a reference 0° sample had not the shape expected for the smoothest surfaces and one or two smoothing cycles were used : each cycle consisted of a 2 langmuirs oxygen exposure followed by Joule heating at 800°C (7). This procedure was also used for the vicinal surfaces.

Results and discussion

Typical LEED diagrams of the two vicinal surfaces under study are compared to a reference 0° (100) surface in Fig. 1. In the case of the 6° [011], the spots are split in two along a direction perpendicular to the steps : it proves a very regular array of steps, the height of which corresponds to twice the distance between adjacent (100) crystallographic planes (8). Moreover the surface is a single domain 2 x 1 in the stable configuration where the dimer bond is parallel to the step edges.

In the case of the 5° [010] surface, the LEED diagram is much less sharp. However the splitting is clearly observable, looking like a crescent roughly around the perpendicular to the nominal step edge direction : the step height is mostly a single distance between adjacent (100) planes and the surface is a two domain one. Moreover the step edges must have some kind of regularity since the terrace widths show a statistical constancy.

The densities of filled surface states as deduced from the photoemission yield spectra for the three cases under consideration are displayed in Fig. 2 after being normalized to constant area. Both linear and semilogarithmic scales are used in order to show the general shape of the band and the tail of states in the gap respectively. The first striking result is the near superposition of the bands for the reference 0° surface and the 6° [011] one which has the very regular array of double steps : since the edge atoms represent 16 percent of the surface atoms, they do not induce a

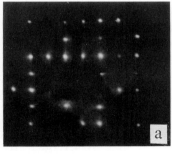

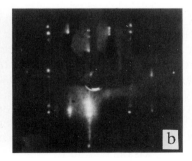

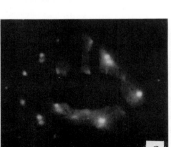

Fig. 1 : LEED diagrams of Si(100) and vicinal surfaces. a : smoothed Si(100) 0° reference surface at $E_p = 100$ V ; b : vicinal surface 6° off the (100) plane along [011] at 74 V ; c : vicinal surface 5° off the (100) plane along [010] at 57 V.

509

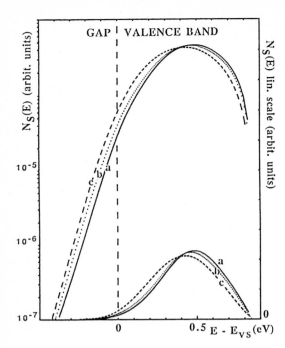

Fig. 2 : Densities of filled surface states deduced from photo-emission yield spectra (see text) as a function of energy referred to the valence band edge E_{VS} in semilogarithmic (left scale) and linear (right scale) representations ; a : reference 0° Si(100) smoothed surface ; b : vicinal surface 6° off the (100) plane along [011] ; c : vicinal surface 5° off the (100) plane along [010].

specific structure in the filled surface state band. Reciprocally the remaining steps in the reference 0° surface, the presence of which is proved by the two domain LEED diagram, do not influence strongly the density of surface states. On the contrary the band of the 5° [010] surface shows an increase of the density of surface states on both sides of the valence band edge and a decrease in the former maximum region, giving an apparent translation of 0.1 eV of the maximum. The overall change concerns about 15 percent of the total area, which is compatible with a high density of kinks along single steps. It is therefore very tempting to attribute to such defects the increase of the density of surface states around the valence band edge which is observed upon roughening of a clean Si(100) surface and which is able to push the Fermi level away from the valence band edge E_{VS} by as much as 0.3 eV with respect to its position (0.35 eV above E_{VS}) on smooth 0° surfaces. Our present results are slightly different from angular resolved UPS measurements (9) which showed a displacement of the surface state band by as much as 0.15 eV towards the gap upon increasing the step density : this overestimation of the step effect is probably due to the surface preparation. The spherical samples used in (9) did not undergo the most effective smoothing procedure which is essential to make quantitative evaluation. Some theoretical support would be now needed to optimize the atomic arrangement and determine the actual position of the corresponding surface states.

References

1. R.J. HAMERS, R.M. TROMP and J.E. DEMUTH, Phys. Rev. B **34**, 5343 (1986).
2. A.J. HOEVEN, J.M. LENSSINCK, D. DIJKKAMP, E.J. VAN LOENEN and J. DIELEMAN, Phys. Rev. Lett. **63**, 1830 (1989).
3. J. POLLMANN, A. MAZUR and M. SCHMEITS, Physica B **117/118**, 771 (1983).
4. I.G. URHBERG, G.V. HANSSON, J.M. NICHOLLS and S.A. FLODSTROEM, Phys. Rev. B **24**, 4684 (1981).
5. I. ANDRIAMANANTENASOA, J.P. LACHARME, C.A. SEBENNE and F. PROIX, Semicond. Sci. Technol. **2**, 145 (1987).
6. I. ANDRIAMANANTENASOA, J.P. LACHARME, and C.A. SEBENNE, J. Vac. Sci. Technol. A **5**, 902 (1987).
7. C.A. SEBENNE, J-P. LACHARME, I. ANDRIAMANANTENASOA and M. KHIAL, Appl. Surface Sci. **41/42**, 352 (1989).
8. R. KAPLAN, Surface Sci. **93**, 145 (1980).
9. CHEN Xiao-Hua, XU Ya-Bo, W. RANWKE, LI Hai-Yang and JI Zheng-guo, Phys. Rev. B **35**, 678 (1987).
10. D. BOLMONT, CHEN Ping, C.A. SEBENNE and F. PROIX, Phys. Rev. B **24**, 4552 (1981).
11. T. TABATA, T. ARUGA and Y. MURATA, Surface Sci. **179**, L63 (1987).

2 × 1 Reconstructions of Si(001) Interfaces: Dimer and Non-dimer Models Derived from Grazing X-Ray Diffraction Data

N. Jedrecy[1], *N. Greiser*[1], *M. Sauvage-Simkin*[2], *R. Pinchaux*[3], *J. Massies*[4], and *V.H. Etgens*[1]

[1]LURE, CNRS-MENJS-CEA, Bat. 209d, Centre Universitaire,
F-91405 Orsay, France
[2]Laboratoire de Minéralogie-Cristallographie, 4 Place Jussieu,
F-75252 Paris Cedex 05, France
[3]Université Pierre et Marie Curie, 4 Place Jussieu,
F-75252 Parix Cedex 05, France
[4]LPSES-CNRS, rue B. Grégory, F-06560 Valbonne, France

Abstract. The grazing incidence X-ray diffraction (GIXD) method is used to perform structural analyses of several Si(001) 2x1 interfaces. We find an asymmetric dimer buckled by 7.4° on the clean Si(001) 2x1 surface, and confirm the symmetric conformation of the As dimer after adsorption of one monolayer (ML). The Ga/Si(001) 2x1 interface data are also compatible with a dimer model, whereas the Pb/Si(001) 2x1 structure is interpreted by means of 3 Pb atoms per 2x1 unit cell. A three dimensional model is envisaged to understand the growth mode of this interface.

Dimerization of silicon atoms on the clean (001) surface is known to significantly reduce the surface energy, thereby leading to long-range ordered 2x1 reconstructed areas and to regions of local c(4x2) or p(2x2) symmetry [1]. The 2x1 structure may be interpreted either by symmetric or by asymmetric dimers, whereas the As/Si(001) 2x1 surface is accepted to consist of symmetric dimers [2]. GIXD is a unique tool to assess the atomic positions in the surface cell as well as to provide the registry of the top-most layer with respect to the bulk [3]. A small amount of buckling (7.4°) for the Si dimers on the clean surface was thus deduced from the diffraction data, while adsorbed As dimers were confirmed to be symmetric [4]. Our data on the Ga/Si(001) 2x1 surface (1 ML) are in good agreement with a dimer model, but we cannot conclude about their conformation. We investigated the Pb/Si(001) 2x1 interface, and precluded the formation of Pb dimers. The projected structure implies three lead atoms per 2x1 unit cell, corresponding to a 1.5 ML surface coverage. We propose a three-dimensional (3D) overlayer model, which explains the Stranski-Krastanov growth mode, the second layer being initiated on top of a strained (110) fcc type facet.

The GIXD measurements were performed at the LURE synchrotron radiation facility (Orsay, France), in an ultra-high vacuum (UHV) 4-circle diffractometer, coupled to a molecular beam epitaxy (MBE) growth chamber [5]. All the nominally flat Si(001) samples were Shiraki etched, and presented a sharp two-domain 2x1 LEED pattern after flash-annealings up to 1000° C.

The diffracted intensities $|F_{hk}|^2$ were obtained from the integration of in-plane angular transverse scans performed at the critical incidence angle for total reflection

Springer Series in Surface Sciences, Vol. 24   **The Structure of Surfaces III**
Editors: S.Y. Tong · M.A. Van Hove · K. Takayanagi · X.D. Xie
© Springer-Verlag Berlin, Heidelberg 1991

($\alpha_c = 3.8 \ 10^{-3}$ rad for $\lambda = 1.488$ Å). The surface deterioration was monitored through a reference surface peak intensity. Only in-plane reflection data were used in the least-square refinement, based on the χ^2 residual defined by

$$\chi^2 = (N\text{-}p)^{-1} \sum_{h,k} (\ F_{hk}(\text{obs}) - |F_{hk}(\text{calc})|\ )^2 / \sigma_{hk}^2\ ,$$

where h, k refer to indices of the in-plane momentum transfer q, with basis vectors corresponding to the LEED conventional reciprocal lattice for the Si(001) surface, N is the number of reflections, p is the number of optimized parameters, and σ_{hk} is the experimental error on F_{hk}.

1. The Si(001) 2x1 surface structure

The substrate was kept at room temperature with a base pressure of 10^{-9} mbar in the diffractometer chamber. We measured 38 in-plane integrated intensities, which reduced to 11 independent values, for reasons of symmetry and equivalence between the two domains found in the same proportion. The surface coherent domain sizes ranged from 1000 to 2000 Å. The structure factor values were first fitted considering a symmetric dimer and lateral inward displacements from bulk-like positions in the second layer. This led to a minimal χ^2 value of 5.1. Shifting the dimer midpoint from the center of the bulk unit cell, which corresponds in our projected scheme to a dimer buckled with respect to the surface plane, made χ^2 fall to 1.8 (see fig. 1). Allowing asymmetric displacements in the second layer, and incorporating a Debye-Waller (DW) factor for the dimer atoms (keeping for the other Si atoms the bulk value 0.45 $Å^2$) further improved the χ^2 value to 1.3. The DW factor of 1 $Å^2$ for the dimer atoms is in agreement with the value proposed on the basis of perturbation theory [6]. If we assume a back-bond length value of 2.35 Å (bulk value), our projected structure corresponds to a dimer tilt of 7.4° with a true bondlength of 2.32 Å. The consequent 0.3 Å height difference between dimer atoms is quite compatible with the STM image protrusions [1]. Moreover, a recent calibrated core-level spectroscopy experiment [7] has concluded in favour of a small charge transfer between dimer atoms ; this parameter influences the STM corrugation profiles strongly, and calculations which had led to a symmetric dimer conclusion had been performed in a large transfer assumption. Alerhand and Mele [8] had already investigated the

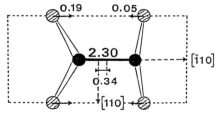

Fig. 1. Projection of the Si(001) 2x1 unit cell. Distances are indicated in Å.

correlation existing between the charge transfer and buckling, introducing in their total energy calculations an on-site e^--e^- repulsion term, which was found to reduce much more the charge transfer than the tilt. Nevertheless, the reasons for the gap opening in the electronic surface states, together with the driving force towards higher order reconstructions, are still open questions.

2. The As/Si(001) 2x1 analysis

The 2x1 reconstructed surface was obtained by exposing a Si(001) 2x1 sample to the As flux, after the temperature was brought down from 750°C to 350° C. The FWHM reduction observed on identical fractional reflections before and after As deposition (by transferring the sample from the MBE chamber to the diffractometer without breaking UHV), clearly demonstrates the better ordering of the 2x1 surface. The average increase in the intensity of reflections confirmed the chemisorption of a single As layer. We collected 17 independent intensities. The symmetric dimer model led to a χ^2 of 3.94 with a dimer bond length of 2.55 Å in full agreement with previous theoretical and experimental work [2]. The fit with an asymmetric conformation was always worse. For instance, a dimer midpoint shift of 0.35 Å from the symmetric position (the same as in the clean Si reconstruction model) makes the χ^2 increase to 10. The final structural parameters ($\chi^2 = 0.77$) are a bond length of 2.55 Å for the As dimer, inward displacements of 0.15 Å for the Si atoms in the first substrate layer, and a DW factor of 2.2 Å^2 for the dimer As atoms.

3. The Ga/Si(001) 2x1 reconstruction

After calibration of the Ga evaporation rate by Auger Electron Spectroscopy (AES), the following sequence has been performed : deposition of 1 ML at 330° C which led to a faint 8x1-1x8 RHEED pattern, further deposition of 2 ML at 120° C which converted the 2D pattern to a 3D one, and thermal desorption at 480° C until a sharp 2x1-1x2 pattern was observed. The dimer model [9] is compatible with our 6 reflections set, but we obtain a dimer bond length of 2.18 Å which appears quite small in view of the 2.48 Å shortest distance found in the bulk gallium structure. We cannot investigate the buckled dimer model with our presently reduced set of data.

4. The Pb/Si(001) 2x1 interface

Although metal-semiconductor interfaces have attracted much interest, particularly concerning the Si(111) surface and Schottky-barrier formation [10], few studies deal with the adsorption of metals on the particular Si(001) surface. Among these, the Pb/Si(001) 2x1 reconstructed interface has already been studied by ultra-violet photoemission [11]. The preparation conditions have been reproduced in our MBE chamber. The Pb deposition growth rate was calibrated by AES, which clearly denoted the transition from a 2D overlayer to 3D Pb crystallites. The sample held at

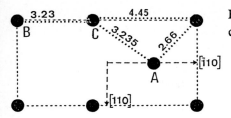

Fig. 2. Projection of the Pb/Si(001) 2x1 unit cell. Pb-Pb distances are indicated in Å.

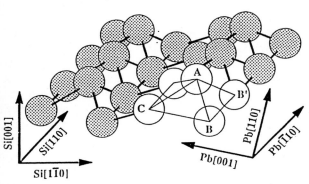

Fig. 3. 3D model of the Pb overlayer on the Si(001) surface. The depicted pyramid helps to recognize the fcc (110) facet.

290° C was first exposed for 60 s to the Pb flux, and then annealed for 15 min at 350°C to obtain a sharp 2x1-1x2 pattern. The X-ray measurements were carried out with the sample held at 270° C to reduce the surface contamination. 11 independent intensity values have been collected, from two sets of data showing a good reproducibility for common reflections. The coherent domain sizes ranged from 500 to 800 Å. We first considered the 9 half-order reflections, incorporating in our refinement only Pb atoms owing to their strong scattering power (Z=82) compared to Si atoms (Z=14). A Pb dimer model definitely does not fit the data, and 3 Pb atoms labelled A, B, C in figure 2, leading to a honeycomb-like projected structure have to be introduced to decrease significantly the χ^2 value. $\chi^2 = 5.56$ is obtained with the projected interatomic distances respectively equal to 3.23, 3.235, and 2.66 Å. A DW factor of 5 Å^2 has been found, which is quite compatible with the value obtained at room temperature from GIXD experiments on the Pb/Ge(111) system [12]. This value was considered as a fixed parameter in further refinements.

Thus, the onset of island growth takes place after 1.5 ML coverage. A 3D atomic arrangement may be proposed to explain this nucleation (see fig. 3). Indeed, the surface is known from spectroscopic data to become metallic beyond the monolayer coverage regime, and the A atom could be considered as the first atom of the second layer deposited on top of a distorted fcc (110) facet, thus establishing a continuous Pb-Pb bond mesh. The direction of non doubled periodicity on the Si (001) surface would correspond here to the Pb [110] direction, while B and C atoms would be aligned in a [001] like direction. If we assume the true bond lengths between Pb

515

atoms to be all nearly equal, we find a common length of 3.4 Å (the nearest neighbour distance in the Pb bulk structure is 3.5 Å) ; the A atom is located at 1.0 Å above the C atom, situated itself at 1.1 Å above the B atom. To the dilation of 10 % in the [110] direction corresponds a contraction of 8 % in the [001] direction. The facet tilt (13.5°) implies a somewhat loose sticking of the Pb overlayer on the Si surface. We considered the registry of the Pb 2x1 unit cell with respect to the underlying Si atoms, first set at bulk-like positions, by including 2 integer order reflections in our refinement. A χ^2 of 5.0 is obtained with quite the same Pb-Pb distances, and B-Si and C-Si projections at near 90° from the B-C line. In fact, this implies that the Pb-Si bonding occurs primarily with B-type atoms, the C atom being displaced upward. A minimal χ^2 value of 1.6 may be reached by allowing displacements in the Si substrate first layer, but the large values obtained (0.05 and 0.8 Å) cannot be understood without considering distortions in the deeper layers. The present limited data set does not allow us to introduce that many parameters. The preservation of the Si dimer reconstruction below the Pb overlayer was also tested. We know by core level photoemission that the dangling bonds on the dimerized Si atoms are saturated with adsorbed Pb atoms after the ML coverage [13]. In our model, this would occur through Si and type-A atoms, and we have little confidence in such an asymmetric scheme with two quite different Si-A lengths. Owing to our quite limited set of data, we are not able to go deeper into the Pb-Si bonding scheme.

References

[1] R.J. Hamers, R.M. Tromp, and J.E. Demuth, Phys. Rev. B **34**, 5343 (1986)

[2] R.I.G. Uhrberg, R.D. Bringans, R.Z. Bachrach, and J.E. Northrup, Phys. Rev. Lett. **56**, 520 (1986) ; R.D. Bringans, M.A. Olmstead, R.I.G. Uhrberg, and R.Z. Bachrach, Phys. Rev. B **36**, 9569 (1987)

[3] R. Feidenhans'l, Surf. Sci. Reports **10**, 105 (1989), *and references therein*

[4] N. Jedrecy, M. Sauvage-Simkin, R. Pinchaux, J. Massies, N. Greiser, and V.H. Etgens, Surf. Sci. **230**, 197 (1990)

[5] P. Claverie, J. Massies, R. Pinchaux, M. Sauvage-Simkin, J. Frouin, J. Bonnet, and N. Jedrecy, Rev. Sci. Instrum. **7**, 2369 (1989)

[6] O.L. Alerhand, J.D. Joannopoulos, and E. J. Mele, Phys. Rev. B **39**, 12622 (1989)

[7] D.H. Rich, T. Miller, and T.C. Chiang, Phys. Rev. B **37**, 3124 (1988)

[8] O.L. Alerhand and E.J. Mele, Phys. Rev. B **35**, 5533 (1987)

[9] B. Bourguignon, K.L. Carleton, and S.R. Leone, Surf. Sci. **204**, 455 (1988)

[10] R. Heslinga, H.H. Weitering, D.P. van der Werf, T.M. Klapwijk, and T. Hibna, Phys. Rev. Lett. **64**, 1589 (1990)

[11] G. Le Lay, K. Hricovini, and J.E. Bonnet, Phys. Rev. B **39**, 3927 (1989)

[12] J.S. Pedersen, R. Feidenhans'l, M. Nielsen, F. Grey, and R.L. Johnson, Surf. Sci. **189/190**, 1047 (1987)

[13] K. Hricovini and G. Le Lay, private communication

Kinematic Low-Energy Electron Diffraction Study of the Atomic Structure of the Si(001)2 × 1 Surface

R.G. Zhao, Jinfeng Jia, Yanfang Li, and W.S. Yang

Department of Physics, Peking University, Beijing 100871, P.R. of China

With the constant-momentum-transfer-averaging (CMTA) procedure, which has been tested with the Cu(100) surface to produce the kinematic low-energy electron diffraction (KLEED) intensity curves, we have collected 10 such curves from the Si(001)2X1 surface. Using the KLEED program, which has been reported in our earlier works, comparing with the experimental curves, and adopting the Van Hove-Tong R factor as the optimization criterion, we have reached the optimized model with $R_{VHT}=0.22$. From the present mode the Si(001)2X1 surface consists mainly of weakly-buckled dimers in 2X1 symmetry and highly-buckled dimers in c(4X2) symmetry, being very similar to recent STM results. Besides, there also exists an oscillatory multi-layer-spacing relaxation in the surface.

1. INTRODUCTION

Because of the basic scientific interest and technological importance of the Si(001) 2X1 surface [1], a great deal of effort has been devoted to the understanding of it in the last decade [2-11]. As a result, it is generally accepted that dimerization at the first layer and multi-layer relaxation at the deeper layers are two major structural features of the surface [2-11]. Besides, thanks to recent scanning tunneling microscopy (STM) studies on the surface [7], it seems to be quite clear that the energy differences between buckled and non-buckled dimers as well as between different symmetries are not significant. This point has also received support from the theoretical side [9,10].

Despite this progress, both science and technology want to know more concretely about the dimers and the multi-layer relaxation. However, the STM has not been capable enough to tell if the 2X1 reconstruction of the surface consists of weakly-buckled or non-buckled dimers, while a theoretical work favors the former [10]. On the contrary, LEED is known for being sensitive to structural details, especially those in the normal direction of the surface, i.e. buckling and multi-layer-spacing relaxation. Being trapped in a local minimum is a common problem in optimizing a model with many parameters, which may also happen in total-energy minimization calculations guided with Hellmann-Feynman forces [9]. Carrying out an as complete as possible structural search might be the only way to avoid this happening. However, that could not be done with dynamic LEED calculations, at least at present. Consequently, we use KLEED [12] to search for the optimized structural model of the surface. In order to make sure that our CMTA

Springer Series in Surface Sciences, Vol. 24   **The Structure of Surfaces III**
Editors: S.Y. Tong · M.A. Van Hove · K. Takayanagi · X.D. Xie
© Springer-Verlag Berlin, Heidelberg 1991

procedure [13] can indeed give kinematic intensity curves and our KLEED program is correct, we have collected four CMTA curves, i.e. (10), (11), (20), and (21), from the Cu (100) surface [14] and carried out a structural search with our KLEED program. It turned out that the optimized model of our search is exactly the same model of Davis and Noonan [15], a result of full dynamic LEED calculations. The R_{VHT} of the model is 0.14, indicating an excellent agreement. The CMTA curves of Si(001) 2X1 used in this work were collected in the same way.

2. CALCULATION

The KLEED program used in this work and the approximations made in it have been described earlier [12,17]. The Van Hove-Tong R factor [16] is used in the optimization process, which is expected to give less possibility of being trapped in local minima.

Three different models have been optimized. They were the 2X1 symmetric dimer model (Model 1), the 2X1 asymmetric dimer model (Model 2), and the (2X1)-plus-c(4X2) model (Model 3), which was defined as:
Intensity (Model 3)=0.75 • Intensity (asym.2X1)+0.25 • Intensity (asym.c(4X2)).
This is in accordance with the STM result [7] which shows that on the sample surface the total c(4X2) area was slightly smaller than that of 2X1. All of the models consisted of six reconstructed surface layers and enough 1X1 layers.

To keep ourselves from being trapped in a local minimum, the following measures were taken: (i) Several different geometries were used as the starting point of each model. For Model 1, they were the Batra model [9], AB model [18], and bulklike 2X1 model. For Model 2, the same symmetric dimer models as well as some asymmetric dimer models [3,4] were used. For Model 3, the optimized Model 2 was used as the starting point of its 2X1 portion, and the c(4X2) portion was optimized firstly. (ii) The atoms in the upper layers were optimized prior to those in the deeper layers, and the vertical coordinates prior to the lateral. The optimization process was continued for many rounds until an as low as possible R_{VHT} was reached for a model. (iii) As LEED is not very sensitive to lateral shifts and there is still controversy over the y-shifts [9,10], none of our models included any y-shifts. (iv) During the process, the bond lengths were monitored to avoid resulting in obviously unreasonable structures. All bonds in our 2X1 models have a strain less than 3% and in c(4X2) less than 5%.

The energy range of the data is between 30 and 250 eV and a constant inner potential V_0 and mean free path λ are assumed within this range. The value of V_0 and λ, however, are varied to optimize the agreement. The optimized V_0=12.5 eV and λ=4.0 Å.

3. RESULTS AND DISCUSSION

The optimized parameters of the present models are listed in Table 1 and explained with Fig.1. In Table 2, the topmost 8 averaged layer spacings, the first-layer buckling, and the R_{VHT} of the models tested in this work are listed. Fig.2 is arranged to give a feeling on the agreements between the experimental curves and calculated curves of these models. Finally, Fig.3 shows the excellent agreement of the Model 3, which we think approximately reflects the real situation of the Si(001) 2X1 surface.

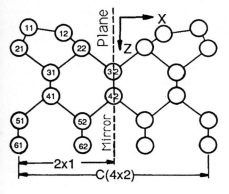

Fig.1: Schematic side view of the Si(001)c(4X2) surface. For the top view as well as the symmetry elements see Ref.12.

Table 1. The relative displacements required for the present optimized models. No y-displacements are present for these models. For description of the models see text. For numbering of the atoms see Fig.1.

Atom	Model 1		Model 2		Model 3			
					2x1		c(4x2)	
	ΔZ	ΔX	ΔZ	ΔX	ΔZ	ΔX	ΔZ	ΔX
11	0.00	0.80	0.16	0.95	0.14	0.82	0.42	1.12
12	0.00	-0.80	-0.04	-0.60	0.00	-0.67	-0.14	-0.48
21	-0.08	0.18	-0.03	0.17	0.02	0.19	-0.12	0.19
22	-0.08	-0.18	-0.07	-0.15	-0.03	-0.14	-0.12	-0.19
31	0.16	0.00	0.22	0.00	0.20	0.00	0.27	0.00
32	-0.25	0.00	-0.17	0.00	-0.18	0.00	-0.19	0.00
41	0.22	0.00	0.26	0.00	0.25	0.00	0.28	0.00
42	-0.24	0.00	-0.13	0.00	-0.15	0.00	-0.08	0.00
51	-0.05	-0.12	-0.05	-0.12	-0.01	-0.12	-0.07	-0.11
52	-0.05	0.12	-0.05	0.12	-0.01	0.12	-0.07	0.11
61	-0.03	-0.03	-0.02	-0.06	-0.02	-0.07	-0.06	-0.05
62	-0.03	0.03	-0.02	0.06	-0.02	0.07	-0.06	0.05
B7	-0.03	0.00	-0.04	0.00	-0.04	0.00	-0.04	0.00
B8	0.00	0.00	0.00	0.00	0.00	0.00	0.00	0.00

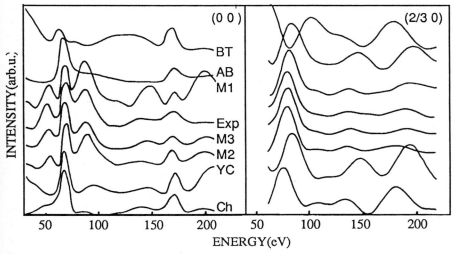

Fig.2. comparison of the kinematically calculated LEED I-V curves of different models with the CMTA experimental curves. Exp--experiment, BT--Batra [9], AB--Abraham and Batra [18], Ch--Chadi [3], YC--Yin and Cohen [4], M1,M2, and M3-- Model 1,2, and 3 (see text), respectively.

Table 2. Topmost eight averaged layer spacings, buckling of the first layer, and total R_{VHT} of the models tested in this work. B-Batra [9], AB--Abraham and Batra [18], Ch--Chadi [3], YC--Yin and Cohen [4], Model 1,2, and 3 --see text.

MODEL		B [9]	AB [18]	Ch [3]	YC [4]	Present Models		3	
						1	2	2 x 1	c(4 x 2)
LAYER SPACING (Å)	1	1.00	1.24	1.15	1.06	1.28	1.25	1.28	1.10
	2	1.35	1.38	1.38	1.37	1.40	1.43	1.38	1.52
	3	1.31	1.36	1.37	1.35	1.39	1.40	1.40	1.42
	4	1.33	1.36	1.36	1.34	1.32	1.25	1.30	1.19
	5	1.36	1.36	1.36	1.36	1.38	1.39	1.35	1.37
	6	1.36	1.36	1.36	1.36	1.36	1.34	1.34	1.38
Buckling(Å)		0.00	0.00	0.56	0.31	0.00	0.20	0.14	0.56
R_{VHT}		0.66	0.46	0.46	0.39	0.30	0.26	0.22	

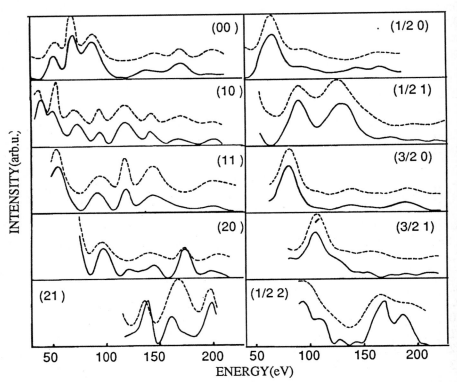

Fig.3. Comparison of the kinematically calculated LEED I-V curves of the best model of this work (Model 3, see text) with the CMTA experimental curves.

Based on comparison of the R_{VHT} of the Model 1 and Model 2, it is concluded that the surface is mainly consisted of buckled dimers in 2X1 symmetry. There is another important structural feature in the surface, i.e. an oscillatory multi-layer-spacing relaxation shown in Table 2. From the point of view of LEED, this feature is probably more important than the buckling. This conclusion is drawn from the fact that model tested here can be grouped into two, one with and the other without the oscillatory multi-layer-spacing relaxation. Those with the relaxation, the present optimized models, have significantly lower R_{VHT} than those without, the remaining models in Table 2. Although in both groups there are both buckled and non-buckled dimer models. It is interesting to note that such relaxations have also been reported for Si(111) 7X7 [17] and for Si(001) c(4X2) [12].

Another significant fact is that the R_{VHT} of the Batra model [9] is much higher than that of the AB model [18]. This indicates that the first layer spacing in the former model is too far from the real situation, even if it was important to the low total energy of the model [9].

The best model reached in this work, Model 3, is compatible with recent STM results [7], since:(i) The 2X1 portion consists of weakly-buckled dimers with a buckling of only 0.14 Å, while the c(4X2) portion consists of highly-buckled dimers with a buckling of 0.56Å. (ii) The constants of 0.75 and 0.25 in the Model 3 are in accordance with the roughly equal amounts of non-buckled and buckled dimers together with the fact that only part of the buckled dimers were in c(4X2) symmetry [9].

1. R.E. Schlier and H.E. Farnsworth, J.Chem.Phys. 30, 917 (1959).
2. J.A.Appelbaum and D.R.Hamann, Surf.Sci. 74,21 (1978).
3. D.J.Chadi, Phys.Rev.Lett. 43, 43(1979); Phys.Rev.B19, 2074(1979).
4. M.T.Yin and M.L.Cohen, Phys.Rev.B24, 2303(1981).
5. W.S.Yang, F.Jona, and P.M.Marcus, Slid State Commun 43,847(1982); Phys.Rev.B 28, 2049 (1983).
6. K.C.Pandey, in Proceedings of the 17th International Conference on the Physics of Semiconductors, edited D.J.Chadi and W.A.Harrison (Springer, Berlin, 1984), p.55.
7. R.M.Tromp,R.J.Hamers,J.E.Demuth,Phys.Rev.Lett. 55, 1303(1985); R.J.Hamers, R.M.Tromp, and J.E.Demuth, Phys. Rev. B24, 5343(1986).
8. E.Artacho and F.Yndudr in, Phys.Rev.Lett. 62, 2491 (1989).
9. I.Batra, Phys.Rev.B41, 5048 (1990).
10. F.Bechstedt and D.Reichardt, Surf.Sci. 202, 83(1988).
11. Zizhong Zhu, Nobuyuki Shima, and Masaru Tsukada, Phys, Rev.B40, 11868(1989).
12. R.G.Zhao and W.S.Yang, Phys.Rev.B33, 6780(1986).
13. M.G. Lagally, T.C. Ngoc, and M.B. Webb, Phys. Rev. Lett. 26, 1557(1971); J.B.Pendry, Low Energy Electron Diffraction, (Academic Press, London, 1974).
14. to be published.
15. H.L.Davis and J.R.Noonan, J.Vac.Sci.Technol. 20, 842(1982).
16. M.L.Xu and S.Y.Tong, Phys.Rev.B31, 6332(1985).
17. W.S.Yang and R.G.Zhao, Phys.Rev.B30, 6016 (1984).
18. F.F.Abraham and I.P.Batra, Surf.Sci.163, L752(1985).

X-Ray Photoelectron Diffraction Study of a High-Temperature Surface Phase Transition on Ge(111)

T.T. Tran, D.J. Friedman, Y.J. Kim, G.A. Rizzi*, and C.S. Fadley

Department of Chemistry, University of Hawaii, Honolulu, HI 96822, USA
*Present address: Department of Chemistry, University of Padua,
  Padua, Italy

Abstract. X-Ray photoelectron diffraction (XPD) data reveal a reversible disordering transition at the (111) surface of germanium that is completed at about 1050 K or 160 K below the bulk melting point. Azimuthal XPD data at $\theta=19°$, including nearest-neighbor forward-scattering directions and yielding high surface sensitivity, show an abrupt drop in intensity of 40% over the interval of 850-1050 K. Data taken at $\theta=55°$ for which second-nearest-neighbor scattering directions and more bulk sensitivity are involved also exhibit distinct changes through this temperature range. In conjunction with model single-scattering cluster calculations, these data suggest that the disordering process involves more than the surface double layer of Ge atoms.

1. Introduction

Evidence for a reversible surface phase transition on Ge(111) has been found previously by McRae and coworkers [1,2] in temperature-dependent LEED experiments. This transition occurs near 1050 K or about 160 K below the bulk melting point. It has been proposed [1] that this is not a surface melting or surface roughening transition but rather a domain disordering transition in which the domains are laterally strained to a depth of one double layer. However, beyond the suggestion based upon molecular mechanics modeling that there is loss of registry between strained domains and deeper layers [2], no precise structural conclusions have been possible concerning the type of disorder involved.

In an attempt to more quantitatively understand this phase transition, we have studied this system with x-ray photoelectron diffraction (XPD), a surface structure probe that is primarily sensitive to short-range order in the first 3-5 shells of neighbors around each emitter [3]. This technique thus provides complementary information to LEED, a long-range-order probe over distances of the order of 100 Å. We have examined the Ge 3d photoelectron intensity from Ge(111) as a function of azimuthal angle of emission and temperature.

2. Experiment

The azimuthal dependence of Ge 3d core-level intensities emitted from a Ge(111) surface at a kinetic energy of 1456 eV was studied as a function of temperature from

Springer Series in Surface Sciences, Vol. 24   **The Structure of Surfaces III**
Editors: S.Y. Tong · M.A. Van Hove · K. Takayanagi · X.D. Xie
© Springer-Verlag Berlin, Heidelberg 1991

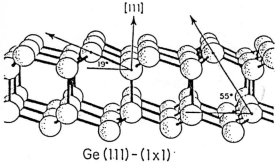

Ge (1.11) - (1×1)

Fig. 1   An unreconstructed Ge(111) surface showing near-neighbor scattering events at takeoff angles of θ=19° and θ=55°.

298 K to 1120 K using monochromatized Al Kα radiation as an excitation source. In order to vary surface sensitivity, azimuthal scans were performed at two polar angles of emission relative to the surface of θ=19° and 55° as shown in Fig. 1. For an estimated mean free path of 22.0 Å, the mean depths from which elastically scattered electrons can be emitted are approximately 2 and 6 Ge(111) double layers, respectively.

The experimental system used was a Hewlett-Packard 5950A spectrometer combining XPS, XPD, and LEED capabilities that is described elsewhere [3]. Heating was done by a noninductively wound resistive button heater and temperatures were measured with a thermocouple-calibrated infrared pyrometer. The sample was a mirror-polished Ge wafer (n-type, Sb doped, 5-30 Ω-cm) oriented to within ±1.0° of (111). Surface cleaning involved sputtering (10^{-5} torr Ar$^+$, 800 eV, 850 K, 45° off normal incidence, 20 min.) and annealing (970 K, 30 min.). This leads to a highly ordered surface exhibiting a very sharp c(2x8) LEED pattern against a low background that permitted easy observation of the quarter-order beams. Surface cleanliness was monitored by XPS core-level peaks, and no detectable contaminant peaks were found after a continuous series of azimuthal scans at different temperatures.

3. Results and Discussion

Fig. 2(a) shows five azimuthal scans taken at temperatures from ambient to about 20 K above the transition and at a surface sensitive polar angle of 19° (cf. Fig. 1). As temperature is increased, the azimuthal curves gradually lose and/or change some of their fine structure. More specifically, going above the transition point, the two peaks in the azimuths [11$\bar{2}$] (φ=0°) and [$\bar{1}$21] (φ=60°) are found to be much reduced in both absolute and relative intensities. The fine structure in the middle angles from about φ=14° to 52° also is reduced in intensity and changes its form. In Fig. 2(b), the intensity of the [11$\bar{2}$] peak corresponding to nearest-neighbor scattering is plotted against temperature. Upright-arrow points correspond to increasing temperatures, whereas inverted-arrow points are decreasing temperatures. It is clear that an abrupt and reversible drop of about 40% in intensity occurs over the interval 850-1050 K.

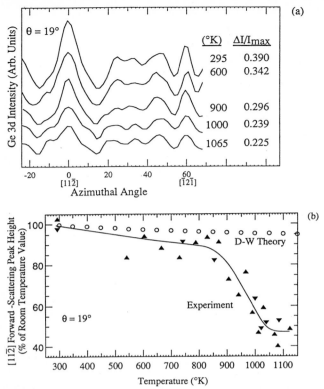

Fig. 2   (a)Temperature-dependent azimuthal XPD data for Ge 3d emission at 1456 eV from Ge(111) at a low takeoff angle of θ=19° (b)The detailed temperature dependence of the height of the nearest-neighbor forward-scattering peak along φ=0°. Upright triangles represent increasing temperature and inverted triangles decreasing temperature.

The overall anisotropy $(I_{max} - I_{min})/I_{max} = \Delta I/I_{max}$ also drops significantly from 0.30 to 0.23, or by 23%, over this range.

This drop cannot be explained by thermal vibrations. Simple Debye-Waller calculations within a single-scattering model with correlated vibrations [4] also are shown in Fig. 2(b). These calculations yield, as expected, a smooth, in fact linear, temperature dependence in the intensity of the forward-scattering peak. Although multiple-scattering effects might be expected to increase the slope of this curve, it is still not reasonable to expect such calculations to predict the step-like transition observed. Thus, Debye-Waller effects may explain the experimental behavior below 850 K, but the transition itself must be due to a larger-scale motion of the surface atoms.

McRae et al. [1] have measured the intensities of several LEED beams for the same system as a function of temperature, and their data are similar to Fig. 2(b) in dropping sharply toward 1060 K and leveling off thereafter. Some of the LEED intensities drop more rapidly than the curve in Fig. 2(b) near 1060K; some have a broader form,

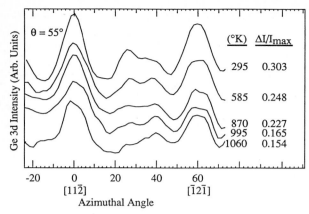

Fig. 3 Temperature-dependent azimuthal XPD data for Ge 3d emission at 1456 eV from Ge(111) at a high takeoff angle of θ=55°.

approaching the shape of the XPD curve. Thus, it can be concluded that the same transition is observed in both sets of data, even though the LEED measurement is expected to be sensitive to longer-range order on a scale of approximately 100 Å, whereas XPD should primarily probe distances of the order of 10-20 Å.

Fig. 3 is analogous to Fig. 2(a) in showing five azimuthal scans at increasing temperatures, but this time at a more bulk sensitive polar angle of 55°. The polar angle chosen here also causes the emission direction to sweep through second-nearest-neighbor forward-scattering directions (cf. Fig. 1). Again, with increasing temperatures, the mid-angle fine structure in the azimuthal curves gradually gets washed out and there is a corresponding decrease in the intensity of the forward-scattering peaks along the [11$\bar{2}$] and [$\bar{1}$2$\bar{1}$] azimuths. The overall anisotropy here decreases from 0.23 to 0.15, or by 35%, over the transition. These data thus suggest that the same disordering process is observed when a greater number of layers (averaging about 6 double layers) of the Ge(111) surface are probed.

In comparing the θ=19° and the θ=55° sets of azimuthal curves, the decrease in intensity of the main peak of the 55° series along the [11$\bar{2}$] azimuth is noticeably less than that of the 19° series over the temperature range studied. Changes in the mid-angle fine structure, are however, more pronounced in the 55° series. As indicated in Fig. 1, the more surface sensitive 19° case is expected to probe <u>intra-double-layer</u> disordering. That is, both atoms forming the 19° bond with respect to the surface are within the same double layer. The two atoms forming the 55° scattering direction, are, on the other hand, in different double layers, so that <u>inter-double-layer</u> disordering should be probed in this data. This comparison thus qualitatively suggests that, although the disordering process involves more than one double layer, the effect of intra-layer disordering is more dominant.

Further evidence from model single-scattering cluster (SSC) calculations also supports this multilayer disordering mechanism. We performed SSC calculations with spherical-wave scattering for three clusters of Ge atoms with a 19° takeoff angle.

525

Results from a first cluster in which the top double layer is compressed by 2% on the vertical axis so as to reflect the difference between the densities of the solid (5.22 g/cm^3) and the liquid (5.53 g/cm^3) at the melting point, and those from a second cluster in which the top double layer is completely flattened, thus neglecting all 19° scattering events in this double layer, are not found to differ significantly from results obtained with a third cluster of ideally terminated, non-reconstructed Ge(111).

A more detailed study of the temperature dependence of the forward-scattering peaks and diffraction fine structure for various takeoff angles is currently in progress. These data will be combined with further model calculations at both single and multiple scattering levels to better understand the disordering process and its depth of penetration. However, our experimental and theoretical results thus far suggest a multilayer disordering mechanism different than that proposed by McRae and coworkers in which domains are laterally strained to a depth of only one double layer of crystalline Ge(111).

1. E.G. McRae and R.A. Malic, Phys. Rev. Lett. 58, 1437 (1987); E.G. McRae and R.A. Malic, Phys. Rev. B38, 13183 (1988).
2. E.G. McRae, J.M. Landwehr, J.E. McRae, G.H. Gilmer, and M.H. Grabow, Phys. Rev. B38, 13178 (1988).
3. C.S. Fadley, Physica Scripta T17, 39 (1987); C.S. Fadley in Synchrotron Radiation Research: Advances in Surface and Interface Science, ed. by R.Z. Bachrach, (Plenum Press, New York, 1990) to appear.
4. M. Sagurton, E.L. Bullock, and C.S. Fadley, Surface Science 182, 287 (1987).

Surface Mosaic of Multi-Component Systems

V.L. Avgustimov[1], D.I. Bidnyk[1], L.A. Bystraya[2], and S.P. Kostenko[2]

[1]MICROPROCESSOR Science & Manufacturing Unified 1,
  Severo-Syretskaya St., 252136 Kiev, USSR
[2]SENSOR Dept. Ref. Instrum. Diagn., 63, Kirov St., 270045, Odessa, USSR

Abstract. On different multi-component surfaces similar component interactions have been revealed, often caused by the presence of oxygen atoms in the surface layer. Any sort of intervention, e.g. alteration of the amount of surface oxygen, changes the conditions of surface structure self-formation and thus the properties of the surface.

1. Introduction

The construction of new computers from the fundamentals of molecular electronics and the realization of novel devices, such as biosensors and multi-sensors with specified reactions, require surface fragments (even of Si or Ge) to be considered as multi-component systems. Environmental modifications of surface phases formed in different ways may produce very similar substructures because the interactions of the components of a complex system, or of its parts, obey the same laws of physics. If a solid volume is reduced to tens or hundreds of atoms, new material phases are formed, named clusters [1].

Among the variety of surface attributes, the above features most often result from the quantity of atomic oxygen in the surface layers [2]. This study focused on the atomic structure of the surface microvolumes and on oxygen in its intimate interplay with its immediate environment in multi-component systems.

2. Experimental

The specimen investigations were performed in an 09|/|0C-005 Auger spectrometer. Repeated irradiation of the same surface site and simultaneous recording of the secondary Auger electrons are the main features of the method used. Thus, the time of irradiation by the electron beam ($E = 5\,\text{keV}$) can be made large enough while the experimental conditions remain constant.

Springer Series in Surface Sciences, Vol. 24   **The Structure of Surfaces III**
Editors: S.Y. Tong · M.A. Van Hove · K. Takayanagi · X.D. Xie
© Springer-Verlag Berlin, Heidelberg 1991

3. Results and Discussion

3.1 Indium Phosphide: InP(Fe), InP(S), InP(S):Yb⁺

A characteristic of the initial InP surface reconstruction is diffusion of In atoms towards the surface to form indium oxides. These samples are most remarkable for the intermediate valency of iron in combinations with oxygen.

A nonstoichiometric InP(S) layer results from the superposition of cluster formations which preserve order in the first coordination sphere but hardly in the

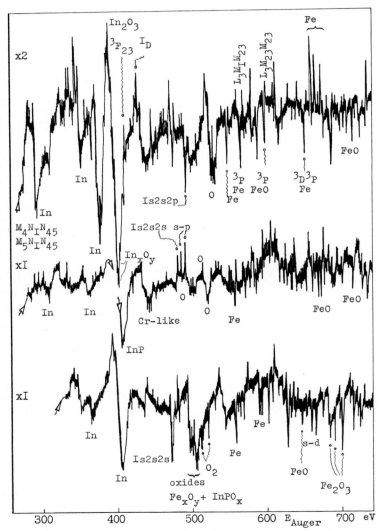

Fig. 1. Auger spectra from an InP⟨Fe⟩ surface as a function of the duration of electron irradiation (in situ)

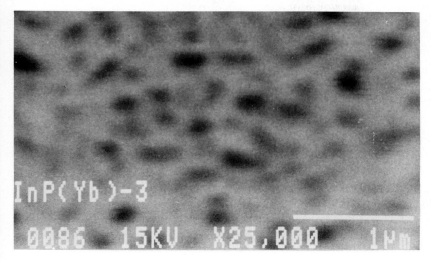

Fig. 2. Microstructure of an InP(S) surface implanted with Yb^+. Dose: $10^{13}\,cm^{-2}$

second. Indium dominate at the surface in different phases: In, In_xO_y, InS, In_2S_3, InP, In_2O, $InPO_x$. The phosphorus deficit is compensated by atoms of sulphur and oxygen. The latter compete in exhibiting their properties. The surface structure evolves as follows. Electron bombardment activates the "excess" oxygen to oxidize the impurity particles of $3d$ metals, which tend towards the electron flux. Sulphur segregates at the surface as a substitute for a fraction of the oxygen atoms and generates stable cluster compounds with In. The relatively uniform mixture of cluster particles that is formed initially starts to demix into monolayers. If the irradiation time is prolonged to say 30–40 min, desorption becomes predominant, primarily of oxygen. The number of metal indium microclusters increases (Fig. 1). The InP(S):Yb^+ surface layers have more homogeneous structures.

The physical pattern of the surface evolution may be presented as follows: the Yb_xO_y oxide phase controls the processes on the surface, allowing the predominant formation of metal-oxide phases as microclusters (Fig. 2).

Thus we may conclude:

1) Metal atoms in the surface layers occur as metal-oxide cluster particles.

2) Perturbative treatments initiate relative motion not of Fe atoms but of microclusters of FeO, $FeO+Fe_2O_3$, and Fe_2O_3. This may be extended to other metals with $3d$ and $4d$ shells.

3) The Yb^+ ions which have not found places in the cationic sites of the lattice after the post-implantation annealing generate cluster associations in the surface layers, which are coordinated by Yb^+ and the crystal field.

4) A range of characteristics is very often due to oxygen atoms, which may be controlled by a rare-earth element.

5) Introducing f elements into InP, Yb^+ in particular, causes the reduction of $3d$ metals, for example, iron is reduced from the oxide phases and then forms phosphide phases.

3.2 $CdGa_2Se_4(I)$, $CdGa_2Se_2(Cl)$

The atomic structure of $CdGa_2Se_4$ contains microclusters of several types with cadmium, selenium and iron intercoordinated by atomic chlorine, iodine, oxygen and carbon (Fig. 3).

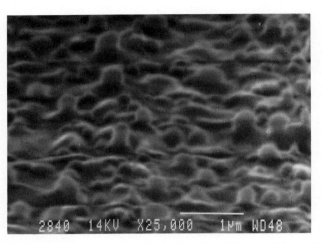

Fig. 3. The surface microclusters in $CdGa_2Se_4\langle Cl\rangle$, comprising mainly metallic gallium

3.3 Al-Mg Alloy

The main feature of Al*, a special aluminium, was the presence of small-grained inclusions about 0.2–0.3 μm in size in the homogeneous layer of the Al-Mg substrate for the special aluminium (Fig. 4). An X-ray study of the inclusions' microstructures revealed Mn, Fe and Ni microclusters with additional Cu, Lu and Os atoms. Macrovariations involve more Mg atoms in the Al* surface. This feature results in an increased quantity of oxygen and changes in the mechanical properties. Electron irradiation increases the surface concentration of atomic oxygen (coming from the bulk perhaps) with increasing magnesium concentration.

Hence, magnesium "controls" the rate of alloy surface oxidation and is a "key" for self-formation of the mosaic surface of the alloy. It also provides the adsorptivity of the surface towards carbon, chlorine and sulphur.

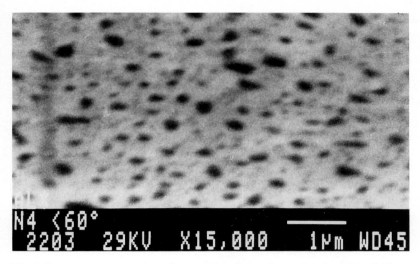

Fig. 4. The structure of certain surfaces and interfaces of a special aluminium (Al*)

References

1   M.A. Duncan, D.H. Rouvray: Microclusters. Sci. Am. 261 (Dec. 1989)
2   V.N. Andreyev et al.: JETP Lett. **15** (8), 65 (1989)

Electron–Hole Counting Rule at III–V Surfaces: Applications to Surface Structure and Passivation

D.J. Chadi

Xerox Palo Alto Research Center, 3333 Coyote Hill Road,
Palo Alto, CA 94304, USA

Abstract. The reconstructions of most III-V semiconductor surfaces are shown to be consistent with constraints imposed by a simple "electron-hole" counting rule proposed by Pashley. The rule ensures that the predicted surfaces are as close as possible to being nonmetallic, nonpolar, and metastable since the compensation of "donorlike" surface electrons leaves no occupied states in the upper part of the band gap which can easily induce other reconstructions. Applications of the method to reconstructed (111), ($\bar{1}\bar{1}\bar{1}$), and (100) surfaces of GaAs and to the problem of (100) surface passivation are examined.

1. Introduction

Our understanding of the surface atomic structure of compound semiconductors such as GaAs has advanced greatly during the past decade. As a result, the nature of the relaxed (110)-1×1, and reconstructed (111)-2×2, and (100)-2×4 surface atomic geometries of the principal low index surfaces of GaAs are now known with a high degree of confidence. The determination of each of these surface structures occurred as a result of major experimental and theoretical efforts. Since the three surfaces do not bear any outward resemblance to each other, the solution of their surface atomic structures does not seem, at first sight, to pave the way towards a solution of other longstanding surface problems. The atomic arrangements of the surface atoms give rise to: a zig-zag chain of threefold coordinated Ga and As atoms on the (110) surface [1-4], an ordered array of $\frac{1}{4}$ monolayer of Ga vacancies on the Ga-(111)-2×2 surface [5-8], and, As dimers and missing dimers (Fig. 1) on the As-stabilized (100)-2×4 surface [9-11]. Despite the disparities in their atomic structure, the three surfaces share important features which, as shown in this paper, can be exploited to guide us in the solution of other surface structural problems.

The most important similarity between the three surfaces is in their electronic structure. Whether in an ideal or relaxed 1×1 configuration, the (110) surface, has *nonmetallic* surface electronic bands with the filled surface states being primarily As-derived and lying near the bulk valence-band-maximum and the empty states

Springer Series in Surface Sciences, Vol. 24 **The Structure of Surfaces III**
Editors: S.Y. Tong · M.A. Van Hove · K. Takayanagi · X.D. Xie
© Springer-Verlag Berlin, Heidelberg 1991

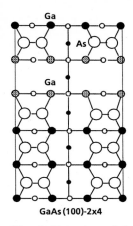

GaAs (100)-2x4

Fig. 1. Top view of the As-terminated (100)-2×4 surface of GaAs. The *threefold coordinated* Ga atoms on the *second* layer are shown as shaded circles. Even though this surface has six threefold coordinated As atoms versus only four similar Ga atoms, the numbers of donor and acceptor states match exactly.

being Ga-derived and lying near the bulk conduction-band-minimum. Unlike the (100) surface, the polar (100) and (111) surfaces would be metallic in a 1×1 configuration. However, the observed reconstructions on both surfaces lead to nonmetallic surface bands. On the Ga-terminated (111)-2×2 surface, the ordering of $\frac{1}{4}$ monolayer of Ga vacancies leads to the formation of three dangling-bonds on the second layer As atoms. The surface and second layer dangling bonds are arranged in six-membered rings in each 2x2 cell. Microscopically, the bonding is nearly identical to that on the (110) surface [5-6]. For the (100) surface, the conditions that the surface electronic structure should be nonmetallic and that there should be no high energy surface states near the conduction-band-minimum were used to predict the most likely bonding topology for the 2×4 reconstructed surface [9], and this prediction was later confirmed by scanning-tunneling-microscopy (STM) studies [10,11]. Farrell *et al*. [12] independently used a similar argument to propose a missing row model for the Ga-terminated (100)-4×2 surface. A similar model for the 4×2 surface of InP was proposed earlier by Hou *et al*. [13] on the basis of P-H and In-H induced states seen in their photoemission spectra.

In a recent paper, Pashley [14] has shown that the nonmetallicity condition can be formalized into a very simple electron-hole counting rule (or "electron counting model"). According to this rule the stable surfaces of a semiconductor such as GaAs correspond to those structures in which the number of high energy electrons in "donorlike" states (e.g., Ga-dangling bonds) is as close as possible to the number of available (i.e., empty) low energy "acceptorlike" states (e.g., As-dangling bonds).

He points out that the structures of nearly all III-V and II-VI surfaces which have been determined up to now are consistent with this rule and that the 2×4 reconstruction of (100)-As surfaces with missing dimer rows occur as straightforward consequences of this condition. As importantly, the rule simultaneously ensures that the predicted surfaces are nonmetallic, nonpolar, and at least, metastable since the compensation of the donor electrons leaves no occupied states in the upper part of the band gap which can easily induce other reconstructions. The nonpolarity condition for the stability of reconstructed polar surfaces was initially proposed by Harrison [15]. The new electron-hole counting rule represents a generalization and a simplification of this rule in that atomic configurations not envisioned previously can occur.

The purpose of this paper is to show that the electron-hole counting approach is a powerful tool for surface structural analysis. Although it generally does not lead to a unique structural model, it provides an extremely simple procedure for testing the reasonableness of any proposed surface geometry. A description of the method is given in Sec. 2. Application to $(\bar{1}\bar{1}\bar{1})$ surface reconstructions, and, in particular, to the $\sqrt{19} \times \sqrt{19}$ surface reconstruction of the As-terminated $(\bar{1}\bar{1}\bar{1})$ surface of GaAs is made in Sec. 3. It is shown that the notions of donor and acceptorlike states need to be expanded in order to deal with the structure of this surface revealed by the recent STM experiments of Biegelsen et al. [16]. For the (100) surface, the electron-hole counting rule is shown in Sec. 4 to be very useful in analyzing the passivation of (100) surfaces by Group VI atoms. The conclusions of this work are summarized in Sec. 5.

2.   Electron-Hole Counting Scheme

2.1   Fractional Charges

In order to implement the electron-hole counting scheme it is essential to have a simple method for calculating the number of electrons and holes in dangling-bond orbitals. Such a method was first introduced by Appelbaum *et al.* in surface electronic structure studies of GaAs [17]. They showed that the correct surface band fillings could be obtained easily and without any need for detailed calculations by assigning *fractional* electronic occupations to the dangling-bond orbitals of surface atoms. The fractional charges at dangling bond orbitals for GaAs can be obtained by noting that in the bulk, each Ga (As) atom makes four bonds with its As (Ga) nearest neighbors. Since Ga has three valence electrons, it contributes, on average, $\frac{3}{4}$ electron to each bond. Therefore, at an ideal Ga-terminated (111) surface, where the surface Ga atoms are threefold coordinated, each Ga dangling bond orbital should contain, on average, $\frac{3}{4}$ electron. Similarly, since As has five valence electrons, there would be, on average, 5/4 electrons in each As dangling bond orbital at an ideal As-terminated (111) surface. The same fractional occupancies of the dangling-bond orbitals occur for the (110)-1×1 surface in which each surface atom is threefold coordinated and there are no like-atom (*i.e.*, Ga-Ga or As-As) bonds. All fourfold coordinated atoms which do not involve any like-atom bonds are to be left out of the counting. However, as discussed below in Secs. 2,3 and 2.4, in the presence of such "wrong" bonds, the electron counting needs to be properly modified.

On (110)-1×1 and Ga (111)-2×2 surfaces which contain nearest-neighbor Ga and As dangling-bonds, there is, generally, some charge flow from the more electropositive atoms to the more electronegative ones. For the case of GaAs, a number of calculations have shown that As dangling-bonds generally have energies near the bulk valence-band-maximum whereas Ga dangling-bonds are near the bulk conduction-band-minimum. On the (110) surface, therefore, Ga atoms behave as "donors" with each dangling bond donating $\frac{3}{4}$ electron whereas the partially filled As orbitals behave as "acceptors". Each As dangling-bond orbital can accommodate two electrons and since it is only 5/4 full it is a $\frac{3}{4}$ e acceptor. There is, therefore, a perfect balance between the donor and acceptor "centers" at the (110) surface and after charge transfer the low energy As orbitals are in a closed shell environment and the high energy Ga orbitals are empty. For the Ga (111) terminated surface of GaAs such a balance is established as a result of vacancy formation at the surface [5,6]. As a result, the 2×2 unit cell contains three Ga atoms each with one dangling bond and an equal number of threefold coordinated As atoms. As in the case of the (110) surface, perfect balance between the donor and acceptor states is established. Vacancy creation has the effect of transforming the surface from a polar (metallic) into a nonpolar (nonmetallic) one. This can be easily seen from the close microscopic similarity between the (110) and (111) surfaces [6,18] or from a more formal treatment based on evaluating dipoles [15]. It should be noted that for both surfaces the actual character of the surface state wave functions corresponds to a mixture of Ga and As dangling bond orbitals because of the relatively significant nearest-neighbor hopping matrix element between these orbitals.

2.2   Surface Dimers

More complex bonding situations which involve like-atom bonds occur on the (100) surface where Ga-Ga or As-As dimerizations can take place. The evaluation of the orbital occupancy of the dangling-bond orbitals is simple in this case too. For the case of the dimerized As (100)-2×1 surface, for example, each As atom of the dimer makes two bonds to substrate Ga atoms and one bond to another As atom. Just as in the bulk, each bond with a Ga atom requires a contribution of 5/4 electron from the As atom. The As-As bond contains two electrons and because of symmetry each As contributes one electron to the bond. Each surface As atom, therefore, must have $(5-5/4-5/4-1)$ or 1.5 electrons in its dangling-bond orbital and can accept another $\frac{1}{2}$ electron from Ga dangling-bond orbitals (if they could be found). The fact that a 2×1 periodicity leaves the As orbital only partially full (leading to a metallic surface structure) suggests that this dimer arrangement should be unstable and explains why it is not seen on the (100) surfaces of III-V semiconductors. A similar approach for a 2×1 dimerized Ga (100) surface gives a Ga dangling-bond occupation of $(3-\frac{3}{4}-\frac{3}{4}-1)=0.5$ electrons. Each Ga atom of the dimer would be, on average therefore, a 0.5 electron donor. It can be seen that for similar geometries, Ga and As atoms are counterparts of each other as far as donor or acceptor activity is concerned, i.e., in the above examples the As atoms of the dimer can each accept 0.5 electron whereas the Ga atoms of the dimer can each donate 0.5 electron.

2.3 Other Types of Like-Atom Bonds

From the above discussion it may appear that surface As atoms will always act as acceptors and Ga atoms as donors. While this turns out to be true for most surfaces there are exceptions, which need to be examined. These occur in the presence of "antisite-like" centers at or near the surface. Antisites occur when As atoms occupy Ga atomic sites (or vice versa). A simple example of antisites near a surface is shown in Fig. 2. Here the top *two* layers of the (100) surface consist of As atoms and they are arranged in a c-4×4 arrangement [11,16,19]. The *fourfold*

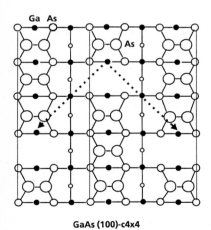

GaAs (100)-c4x4

Fig. 2. Top view of the c4×4 reconstructed As-rich surface of GaAs. The dashed lines denote the primitive translation vectors of the surface. The density of missing dimers is the same as that for the 2x4 structure in Fig. 1.

coordinated As atoms in the second layer make bonds to both Ga and As atoms. Similar to the situation of As antisites in the bulk these As atoms act as donors and not as acceptors. Their donor activity can be easily calculated. Starting with their five valence electrons, $2 \times 5/4$ electrons are used in making the two bonds with the Ga substrate atoms. Each bond with the As atoms on top requires one electron from the second layer As atom and one electron from the top layer As atom. Each fourfold coordinated second layer As atom, therefore, needs $(5/4 + 5/4 + 1 + 1)$ or 4.5 electrons to satisfy all its bonding requirements. Since it has five valence electrons the remaining $\frac{1}{2}$ e must go into a high energy antibonding state [20]. These As atoms will act, therefore, as $\frac{1}{2}$ *donors*. If the surface coverage is not complete, some of the second layer As atoms will be threefold coordinated. These atoms donate $2 \times 5/4$ electrons to the bonds with the third layer Ga atoms and use one electron to bond with a top layer As atom. The dangling bond has $(5 - 2.5 - 1) = 1.5$ electrons in it and has room for an additional $\frac{1}{2}$ electron. These As atoms will act, therefore, as $\frac{1}{2}$ e acceptors. The threefold coordinated top layer As atoms are in a closed shell environment and are neither acceptors nor donors [21].

3. As-rich (1̄1̄1̄) Surfaces of GaAs

3.1 2x2 Surface Reconstruction

The As-rich (1̄1̄1̄) surface of GaAs exhibits 2×2, 3×3, and $\sqrt{19} \times \sqrt{19}$ surface reconstructions [16,22-27]. The 2×2 reconstruction is observed after MBE growth, [16,22-24] the 3×3 is seen sometimes after ion bombardment and annealing, [25] and the $\sqrt{19} \times \sqrt{19}$ surface occurs results after annealing an MBE grown sample in UHV [16,22-24]. The 2×2 to $\sqrt{19} \times \sqrt{19}$ transition is reversible during MBE growth when the surface can exchange atoms with the incoming Ga and As beams. The 2×2 and $\sqrt{19} \times \sqrt{19}$ surfaces have been recently imaged by STM [16]. The 2x2 structure apparently arises from an As trimer (i.e., three As atoms forming a triangle above the As covered surface) leaving the fourth As atom (the "rest atom" of the surface) uncovered. The three As atoms in the trimer act as donors and passivate the As rest atom which acts as an acceptor. Theoretical studies [16] based on *ab initio* pseudopotential calculations indicate that this bonding configuration is much more stable than either a Ga trimer or simple adatom or vacancy-induced [28,29] reconstructions.

3.2 $\sqrt{19} \times \sqrt{19}$ Surface Reconstruction

3.2.1 Structure from STM

The $\sqrt{19} \times \sqrt{19}$ surface of GaAs has been recently imaged via STM by Biegelsen et al. [16]. The observed surface has an unusual reconstruction shown in Fig. 3. For

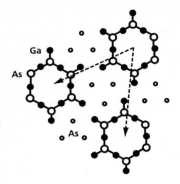

Fig. 3. Top view of the As-terminated ($\bar{1}\bar{1}\bar{1}$)-$\sqrt{19}\times\sqrt{19}$ reconstructed surface of GaAs seen via STM [16]. The dashed lines indicate primitive translation vectors of the lattice.

the top bilayer *all the atoms* are either threefold or twofold coordinated; there are no fourfold coordinated atoms. The structure suggested by STM has 13 As-dangling-bonds (six at the surface and seven at the third layer) and 18 Ga-dangling-bonds per unit cell [30]. The total dangling bond density at this surface is about 9% *higher* than that for a simple 2x2 vacancy model [6] for the Ga-terminated (111)-2x2 surface. Clearly, lattice relaxation and orbital rehybridization must play an important role in stabilizing this surface.

The large ringlike features of the $\sqrt{19}\times\sqrt{19}$ surface are similar to those which had been previously predicted [28,29] for the 2x2 surface. The main advantage of such a structure is that it automatically ensures that the threefold coordinated Ga atoms in the second layer are in a planar configuration. After some upward relaxation of the surface As atoms, these Ga atoms come close to a perfect sp^2 bonding. The upward relaxation of the As atoms is also energetically favorable since it leads to a reduction of the tetrahedral angles and to an s^2p^3 type of rehybridization.

We have examined the atomic and electronic structure of the $\sqrt{19}\times\sqrt{19}$ surface via tight-binding calculations. The optimal surface atomic coordinated were obtained via an iterative total-energy-minimization approach. The calculations reveal several interesting features concerning the atomic and electronic properties of this surface. First, the relaxation of the six surface As and the six twofold coordinated Ga atoms has a threefold symmetry axis. The relaxations of the inequivalent Ga and As surface atoms are substantially different. Three of the As and three of the adjacent Ga atoms have large lattice relaxations whereas the other three As and three Ga atoms have much smaller displacements. The bond angles around the first group of As atoms is 89° , 89°, and 100° while the corresponding values for the second group is 104°, 104°, and 101°. Similarly, the angles for the two groups of Ga atoms are 150° and 113° respectively. The third layer threefold coordinated As atoms also have substantial relaxations leading to average bond angles of approximately 103° at these sites. An eight layer slab periodic in two dimensions was used in the calculations.

A study of the surface states associated with the $\sqrt{19} \times \sqrt{19}$ surface relaxation shows that the acceptorlike states of the third layer threefold coordinated As atoms are compensated by the three second layer twofold coordinated Ga atoms with large lattice relaxation (i.e., the ones with bond angles of 150°). The large atomic relaxations at these sites are consistent with the loss of charge and the resultant tendency of these atoms to seek an sp bonded configuration. The electrons on the other three Ga atoms which do not undergo a large lattice relaxation also lower their energy through electronic rehybridization. The linear combinations $(\phi_1 \pm \phi_2)/\sqrt{2}$ of the two sp^3 hybrids on these atoms lead to an orbital with pure p symmetry and to an sp orbital. The dangling-bond electrons can be accommodated entirely in the low energy sp orbitals. The present tight-binding calculations suggest that as a result of this rehybridization, the surface states associated with these Ga atoms are about 0.6 eV above the valence-band-maximum instead of being near the conduction-band-minimum.

3.2.2 Other Structural Possibilities

As discussed above, of the six top layer As atoms and six twofold coordinated Ga atoms within the $\sqrt{19} \times \sqrt{19}$ cell, only half undergo a large lattice relaxation. The reason for this is that there are more Ga than As dangling bonds on the surface, therefore, not all the Ga surface atoms can transfer charge to the As dangling bonds, and this prevents large atomic relaxations at these sites. If the stabilization of the $\sqrt{19} \times \sqrt{19}$ surface is primarily a result of the large orbital rehybridizations occurring on the other atoms, then it is not clear why the surface would choose a $\sqrt{19} \times \sqrt{19}$ periodicity. For example, assuming the same ringlike topology, if the periodicity were 5x5 instead of $\sqrt{19} \times \sqrt{19}$, then there would be six extra threefold coordinated third layer As atoms, thereby balancing more closely the numbers of Ga and As dangling bonds at the surface (at values of 18 and 19 respectively). This would allow the remaining three Ga and three As surface atoms within the rings to undergo a large lattice relaxation to significantly lower their energies. We have tested the possibility of such a 5x5 periodicity via tight-binding total-energy and structural optimization calculations and do indeed find that the larger unit cell allows all, and not just half, the atoms in the ring to assume their optimal atomic configurations.

The experimental fact that a 5x5 periodicity is not observed on the ($1\bar{1}\bar{1}$) surface may be an indication, therefore, that the $\sqrt{19} \times \sqrt{19}$ surface may have a slightly different structure than the one shown in Fig. 3. We have considered a few alternatives. One possibility is to allow the twofold coordinated Ga atoms to dimerize resulting in three Ga-Ga dimers per unit cell. The initial distance between the Ga atoms is too large for any effective bonding between them. Starting from a "guess" geometry in which these atoms are displaced towards each other but all other atoms are in their ideal positions, the total-energy is found to decrease very rapidly as lattice relaxation over the entire cell is allowed to occur. The relaxation energy is of order 15 eV/cell. For an eight layer thick slab, the fully optimized structure shows a larger than normal Ga-Ga bond length of 2.77 Å, which is about 10% greater than the sum of covalent radii of the two atoms.

Another possibility is the introduction of extra Ga atoms between the second layer twofold coordinated Ga atoms and the third layer threefold coordinated As atoms. Structural optimization via total-energy-minimization was carried out for a geometry in which 6 extra Ga atoms were added to each unit cell. The positions of the six top layer As atoms are consistent with STM results, but the shallower depth of the valleys between the rings seems inconsistent with the data. The previously proposed structure in Fig. 3 [16] appears to be, therefore, the best model for the $\sqrt{19} \times \sqrt{19}$ surface at the present time.

4   GaAs (100) Surfaces

4.1   Surface Reconstructions

The (100) surfaces of GaAs and other III-V semiconductors exhibit a very large number of reconstructions as a function of growth temperature, deposition conditions, and surface stoichiometry [10,11,32,33]. The structure of the 2×4 {and the very closely related c-2×8} reconstruction of the As-rich surface, has been unambiguously determined [10,11] by STM and is well understood theoretically [9,12,31]. In addition, X-ray diffraction analysis [19] and, more recently, STM studies [11] indicate that the other As-rich surface with a c-4×4 reconstruction arises from a simple rearrangement of As-dimers and missing dimers. The major difference is that the As-dimers on the c-4×4 surface appear to grow on top of an As instead of a Ga layer. In the intermediate As-coverage limit STM reveals [11] a 2×6 reconstruction with an As surface coverage of about $\frac{1}{3}$. The surface unit cell contains two As-dimers and several rows of missing dimers on the outermost layer. At a lower As coverage the surface is Ga rich and 4×2 (and a closely related c-8×2) and 4×6 reconstructions are observed in low-energy-electron-diffraction studies [32,33]. The application of the electron-hole counting method to the (100)-2×4 and (100)-c4×4 surfaces was discussed above in Secs. 2.2 and 2.3.

4.2   Impurity Stabilization of the (100) Surface

Impurities often play an important role in the stabilization of surfaces. For example, the (111) and (100) surfaces of Si have been successfully stabilized and made chemically inert through the addition of a monolayer of As to the surface layer. The surface passivation in both cases occurs when *all* the surface atoms are in a closed shell configuration, *i.e.*, all dangling bond orbitals are full and there are no acceptor or donor like states at the surface. Moreover, the filled dangling bond states lie close to the valence band maximum and not high in the gap which would make the system susceptible to instabilities. In this section we address the problem of the passivation of GaAs (100) surfaces. As shown below, it should prove easier to passivate the As-rich than the Ga-rich surface.

We consider first the case of the As-rich surface. From a purely dangling bond electron counting scheme, column VI atoms such S, Se, and Te are expected to be the most effective in passivating the surface. A possible structure with 2×1 periodicity is shown in Fig. 4. In this structure $\frac{1}{2}$ of the surface As atoms have been replaced by S atoms. A sulfur atom satisfies its back bonding in the same way as the As atom it replaces: it contributes 5/4 electron to each of its two back bonds with Ga. Each S atom also contributes one electron to each dimer bond. This leaves a S atom with $(6-5/4-5/4-1)=2.5$ valence electrons which have to be accommodated in its dangling bond orbital. Since the orbital can contain at most two electrons, each S atom acts as a donor. A dimerized As atom, on the other hand has $(5-5/4-5/4-1)=1.5$ electrons in its dangling bond orbital and can accept the extra $\frac{1}{2}$ electron from the S atom. It can be seen that for this structure both the S and As atoms are now in a closed shell configuration and the surface should be nonmetallic. At the same time, both As and S (or Se) give rise to low energy dangling bond states near or below the valence band maximum of GaAs. This suggests that the S or Se passivated 2×1 surface in Fig. 4 should be stable and chemically inert. The use of sulfides to passivate III-V surfaces has been amply

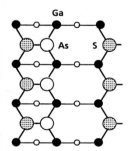

Fig. 4.   Passivation of the GaAs (100) surface. In the 2x1 structure (top view) shown above, all dangling bond orbitals contain two electrons and are in a closed shell configuration thereby making the surface unreactive.

demonstrated during the last two years [34,35] and the 2x1 model has been examined in more detail by Ren and Chang [36].

For the Ga-terminated (100) surface a S passivated structure, similar to that in Fig. 4 but with the As and Ga atoms exchanged, leads to a situation which is electronically similar to that of the S passivated As surface described above. For this case, three electrons (i.e., $4 \times \frac{3}{4}$) are needed for bonding to the second layer and two electrons are needed for the Ga-S bond. This leaves a net of four remaining electrons which have to go to the dangling bonds of the surface S and Ga atoms. Since the dangling bond state of Ga has a high energy close to the conduction-band-minimum, this configuration, although closed shell and inactive, should be highly unstable. The electron counting scheme suggests that the best choice of an impurity to passivate the Ga terminated surface should be a Column IV atom. This choice leads to a filling of the low energy S dangling bond state and leaves the Ga dangling bond state empty. The presence of the empty Ga orbital and the fact that the atomic potentials of such Group IV atoms as Si and Ge are not as strong as those of S and Se suggests that the passivation of the Ga-rich surface by impurities will passivate the surface but it will not lead to a chemically unreactive surface.

5.   Conclusions

We have shown that the *ansatz* [14] that the stable and metastable atomic configurations at III-V semiconductor surfaces correspond to those geometries which, as far as possible, give rise to equal numbers of donor and acceptor states provides a very useful tool in surface structural determination. For the (100) surface the 2×4 and c-4×4 reconstructions and the passivation of the surface were examined from a donor-acceptor passivation viewpoint. The atomic structure of the $\sqrt{19} \times \sqrt{19}$ reconstructed $(\bar{1}\bar{1}\bar{1})$ surface of GaAs was also examined via tight-binding based total-energy calculations. We expect the electron-hole counting scheme will prove useful in future studies on vicinal surfaces.

6.   Acknowledgments

I would like to thank David Biegelsen, Ross Bringans, and John Northrup for helpful discussions and for making available their STM results prior to publication. This work is supported in part by the Office of Naval Research through Contract No. N00014-82-C-0244.

References

1. S.Y. Tong, A.R. Lubinsky, B. J. Mrstik, and M.A. Van Hove, Phys. Rev. B **17**, 3303 (1978).

2. R.J. Meyer, C.B. Duke, A. Paton, A. Kahn, E. So, J.L. Yeh, and P. Mark, Phys Rev. B **19**, 5194 (1979).

3. D.J. Chadi, Phys. Rev. B **19**, 2074 (1979); Phys. Rev. Lett. **41**, 1062 (1978).

4. R.M. Feenstra, J.A. Stroscio, J. Tersoff, and A.P. Fein, Phys. Rev. Lett. **58**, 1192 (1987); R.M. Feenstra and J.A. Stroscio, J. Vac. Sci. Technol. B **5**, 923 (1987).

5. S.Y. Tong, G. Xu, and W.N. Mei, Phys. Rev. Lett. **52**, 1693 (1984); S.Y. Tong, H. Huang, and C.M. Wei, in *Chemistry & Physics of Solid Surfaces*, Vol. 8, R. Vanselow and R. Howe, eds., Springer Verlag, Berlin (1990), pp. 395-417.

6. D.J. Chadi, Phys. Rev. Lett. **52**, 1911 (1984); J. Vac. Sci. Technol. A **4**, 944 (1986).

7. J. Bohr, R. Feidenhans'l, M. Nielsen, M. Toney, R.L. Johnson, and I.K. Robinson, Phys. Rev. Lett. **54**, 1275 (1985).

8. E. Kaxiras, Y. Bar-Yam, J. D. Joannopoulos, and K. C. Pandey, Phys. Rev. B **35**, 9625 (1987).

9. D.J. Chadi, J. Vac. Sci. Technol. A **5**, 834 (1987).

10. M.D. Pashley, K.W. Haberern, W. Friday, J.M. Woodall, and P.D. Kirchner, Phys. Rev. Lett. **60**, 2176 (1988); M.D. Pashley, K.W. Haberern, W. Friday, and J.M. Woodall, J. Vac. Sci. Technol. B **6**, 1468 (1988).

11. D.K. Biegelsen, R.D. Bringans, J.E. Northrup, and L.-E. Swartz, Phys. Rev. B **41**, 5701 (1990).

12. H.H. Farrell, J.P. Harbison, and L.D. Peterson, J. Vac. Sci. Technol. B **5**, 1482 (1987); D.J. Frankel, C. Yu, J.P. Harbison, and H.H. Farrell, J. Vac. Sci. Technol. B **5**, 1113 (1987).

13. X. Hou, G. Dong, X. Ding, and X. Wang, J. Phys. C **20**, L121 (1987).

14. M.D. Pashley, Phys. Rev. B **40**, 10481 (1989).

15. W.A. Harrison, J. Vac. Sci. Technol. **16**, 1492 (1979).

16. D.K. Biegelsen, R.D. Bringans, J.E. Northrup, and L.-E. Swartz, (to be published).

17. J. A. Appelbaum, G. A. Baraff, and D. R. Hamann, Phys. Rev. B **14**, 1623 (1976).

18. The zig-zag chains of threefold coordinated Ga and As atoms on the (110) surface are turned into sixfold rings on the (111)-2x2 surface. After atomic relaxation is taken into account, the nearest neighbor environments of the atoms on the two surfaces become even more alike.

19. M. Sauvage-Simkin, R. Pinchaux, J. Massies, P. Calverie, N. Jedrecy, J. Bonnet, and I.K. Robinson, Phys. Rev. Lett. **62**, 563 (1989).

20. The antibonding states are much higher in energy than the dangling bond states.

21. Each of these As atoms uses three electrons to make three bonds with its nearest-neighbor As atoms. The remaining two valence electrons completely fill up the dangling bond orbitals.

22. J.R. Arthur, Surf. Sci. **43**, 449 (1974).

23. A.Y. Cho, J. Appl. Phys. **41**, 2780 (1970).

24. K. Jacobi, C. v. Muschwitz, and W. Ranke, Surf. Sci. **82**, 270 (1979).

25. M. Alonso, F. Soria, and J.L. Sacedon, J. Vac. Sci. Technol. A **3**, 1598 (1985); A.U. Mac Rae, Surf. Sci. **4**, 247 (1966).

26. A metastable $\sqrt{3} \times \sqrt{3}$ surface is sometimes seen also. See Ref. 24.

27. R.D. Bringans and R.Z. Bachrach, Phys. Rev. Lett. **53**, 1954 (1984).

28. D.J. Chadi, Phys. Rev. Lett. **57**, 102 (1986).

29. E. Kaxiras, Y. Bar-Yam, J.D. Joannopoulos, and K.C. Pandey, Phys. Rev. Lett. **57**, 106 (1986).

30. Each of these twofold coordinated Ga atoms is counted as having two dangling bonds.

31. G.-X Qian, R.M. Martin, and D.J. Chadi, Phys. Rev. B **38**, 7649 (1988); Phys. Rev. Lett. **60**, 1962 (1988).

32. P. Drathen, W. Ranke, and K. Jacobi, Surf. Sci. **77**, L162 (1978).

33. R.Z. Bachrach, R.S. Bauer, P. Chiaradia, and G.V. Hansson, J. Vac. Sci. Technol. **19**, 335 (1981).

34. C.J. Sandroff, R.N. Nottenburg, J.-C. Bischoff, and R. Bhat, Appl. Phys. Lett. **51**, 33 (1987); C.J. Sandroff, M.S. Hedge, L.A. Farrow, C.C. Chang, and J.P. Harbison, Appl. Phys. Lett. **54**, 362 (1989).

35. E. Yablonovitch, B.J. Skromme, R. Bhat, J.P. Harbison, and T.J. Gmitter, Appl. Phys. Lett. **54**, 555 (1989).

36. S.-F. Ren and Y.C. Chang, Phys. Rev. B **41**, 7705 (1990).

The Dimerization of the Reconstructed Clean and Chemisorbed (100) Surfaces of Si and SiC

B.I. Craig and P.V. Smith

Department of Physics, University of Newcastle,
Shortland, NSW 2308, Australia

In this paper we discuss the dimerization of the (100)2x1 reconstructed surfaces of both elemental silicon and β-SiC. In particular, the differences in the behaviour of the silicon dimer within the silicon terminated surface of partially ionic β-SiC, as compared with the surface of elemental silicon, is examined. Changes which result from the adsorption of gases are also discussed.

1. Introduction

In this paper we describe the application of the periodic MINDO (modified intermediate neglect of differential overlap) method [1] to the study of the reconstruction and chemisorption of semiconductor surfaces [2,3]. The main advantage of this method is that it allows one to optimize the surface structure with respect to a very large number of atomic co-ordinates within the surface unit cell. Moreover, it can readily incorporate any charge transfer within a surface via its self-consistent iterative technique. As a result, the chemisorbed bond lengths and angles, the reconstructed surface structure, and the displacements of the subsurface atoms from their ideal bulk positions, can all be determined to the same accuracy with which the MINDO method can describe molecular structures [4].

Application of this periodic MINDO scheme to the covalent solids silicon and diamond, and to partially ionic binary semiconductors such as SiC, has given a good description of their bulk electronic and lattice dynamical properties [1,5,6]. To study the surface structure of these materials we employ a slab geometry characterised by the two dimensional periodicity of the surface unit cell. In the following work a film of 14 layers of semiconducting material is employed. We find

Springer Series in Surface Sciences, Vol. 24   **The Structure of Surfaces III**
Editors: S.Y. Tong · M.A. Van Hove · K. Takayanagi · X.D. Xie
© Springer-Verlag Berlin, Heidelberg 1991

.this is sufficient to produce bulk-like character near the middle of the film and effectively eliminate any interaction between the two surfaces.

2. Theoretical Method

In the periodic MINDO method [1] the total energy E is obtained from the expression

$$E = \frac{1}{2N} \sum_{k} \sum_{pq} P_{pq}(\underset{\sim}{k}) \left[F_{pq}(\underset{\sim}{k}) + H_{pq}(\underset{\sim}{k}) \right] + \sum_{A>B} C(\underset{\sim}{R}_A, \underset{\sim}{R}_B)$$

where H, F and P are the core Hamiltonian, Fock and electron density matrices respectively. The Brillouin Zone summations over $\underset{\sim}{k}$ are evaluated to convergence using a N=32 special wavevector set. The second term describes the core-core repulsions between all sites A and B. In the self-consistent field approach the electron density matrix P is solved iteratively, via the eigenvectors of F, until convergence for E is obtained. The surface topology is derived from the minimisation of E with respect to all of the atomic co-ordinates within the first four layers of the semiconducting material. Any chemisorbed atoms provide additional co-ordinates for this minimisation process.

3. Results

In Table 1 we present our derived bond lengths for a range of clean and chemisorbed (100)2x1 structures. The ideal unreconstructured surface contains two dangling bonds per surface atom. Sharing these bonds with adjacent atoms leads to the formation of the dimer topology. The nature of the dimer bond determined by our MINDO method varies, however, with a double bonded symmetric dimer being obtained for the elemental Si(100) surface whilst the Si-terminated surface of β-SiC yields a single bonded dimer buckled at 5.0°. The C-terminated surface of β-SiC has a double bonded symmetric dimer structure. The formation of such double bonds arises from both strong σ bonding between orbitals aligned along the dimer bond and a π bonding of the dangling bond orbitals perpendicular to the surface. This π bonding is weak for dimer bond lengths close to the characteristic single bond length but becomes significant at the shorter double bond lengths.

Table 1. Structural bond lengths as shown in Fig. 1.

Substrate	Surface X	Adsorbate A	Adsorbate B	Bond lengths dimer (X-X)	(Angstrom) X-A	X-B
Si	Si	clean		2.130		
Si	Si	H	H	2.373	1.530	1.530
Si	Si	H	Cℓ	2.405	1.532	2.045
Si	Si	H	F	2.406	1.529	1.706
Si	Si	Cℓ	Cℓ	2.451	2.050	2.050
Si	Si	F	F	2.859	molecular	
SiC	Si	clean		2.329		
SiC	Si	H	H	2.476	1.547	1.547
SiC	C	clean		1.359		
SiC	C	H	H	1.556	1.136	1.136

In the case of the elemental Si surface the dimer bond is 90.4% of the silicon bulk nearest neighbour distance of 2.352 Å. Similarly, the dimer bond length of the C-terminated surface of β-SiC is only 88.0% of the diamond nearest neighbour distance of 1.544 Å.

The buckling of the dimer leads to a weakening of the π bond with electronic charge being transferred from the lower atom, which becomes positively charged, to the higher atom, which becomes negatively charged. This behaviour is important in understanding why the dimer structure for the Si-terminated surface of β-SiC is buckled. As a result of the more electronegative character of C, a partially ionic behaviour for bulk β-SiC is obtained, with the valence electron populations of C and Si being 4.44 and 3.56, respectively [6]. The negatively charged C layer lying below the silicon terminated surface therefore energetically favours dimer buckling with the lower Si atom electrostatically attracted to the C layer and the raised atom repelled. The valence electron populations determined by the MINDO method for these raised and lowered Si dimer atoms are 4.59 and 3.61, respectively.

The polar nature of β-SiC also explains some of the unexpected features of our results. For example, because of the electronegativity of the underlying C layer, each Si dimer atom for the hydrogenated Si-terminated β-SiC surface is predicted to have a nett charge of 0.20 e compared to just 0.04 e for the Si(100)2x1:H system. As a

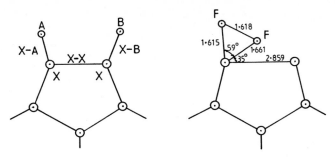

Figure 1. Side view of the dissociative chemisorption of
molecule AB and molecular adsorption of F_2 onto Si(100).

result, these atoms repel each other to yield a Si-Si
dimer bondlength 0.103 Å greater than that for the
hydrogenated elemental surface, and contrary to simple
arguments based on lattice constants. The fact that the
C-C double bond for the C-terminated β-SiC surface
(relative to its bulk nearest neighbour distance) is
shorter than the Si-Si double bond for the elemental Si
surface, and hence under relatively less tensile stress,
is also a direct consequence of the polar vs non-polar
nature of the underlying layers.

The chemisorption of H, HCℓ, HF and Cℓ on the Si(100)
surface reduces the dimer double bond to a single bond
value. Similar behaviour is found for the chemisorption
of H onto the C-terminated surface of β-SiC. On the
Si-terminated surface of β-SiC, hydrogen chemisorption
removes the buckling of the surface dimer to produce
a weaker but symmetric dimer bond. These types of
structures are illustrated in Figure 1. Of most interest
is the quite different structure obtained for F as
compared to the other adsorbate systems, including Cℓ.
This molecular-type topology indicates substantial
weakening of the dimer reconstruction by fluorine and is
consistent with the experimental observations that
fluorine and chlorine exhibit quite different reactivities
on silicon [7].

The structures presented in this paper illustrate the
application of the periodic MINDO method in determining
the reconstruction of both clean and covered
semiconducting surfaces. The work reported here provides
a description of the basic structure of the 2x1 topologies
of the various chemisorbed surfaces of silicon that have
been experimentally observed [8,9]. The applicability of a

single configurational approach such as our SLAB MINDO method to the controversial clean Si(100) surface [10,11], however, remains to be determined. It may well be that correlation plays an important role, in terms of the unattached dangling bond orbitals, in determining both the precise surface dimer topology, and the electron surface state dispersions, of this particular surface.

References

[1] B. I. Craig and P. V. Smith, Phys. Stat. Sol. (b)146, 149(1988)
[2] B. I. Craig and P. V. Smith, Surface Sci. 218, 569(1989)
[3] B. I. Craig and P. V. Smith, Surf. Sci. Lett. 226, L55(1990)
[4] R. C. Bingham, M. J. S. Dewar and D. H. Lo, J. Am. Chem. Soc. 97, 1285(1975)
[5] B. I. Craig and P. V. Smith, Phys. Stat. Sol(b)147, K31(1988)
[6] B. I. Craig and P. V. Smith, Phys. Stat. Sol(b)154, K127(1989)
[7] M. Seel and P. S. Bagus, Phys. Rev. B. 28, 2023(1983)
[8] L. S. O. Johansson, R. I. G. Uhrberg and G. V. Hansson, Phys. Rev. B 38, 13490(1988)
[9] A. L. Johnson, M. M. Walczak and T. E. Madey, Langmuir 4, 277 (1988)
[10] R. Kaplan, Surface Sci. 215, 111(1989)
[11] D. H. Rich, T. Miller and T. -C. Chiang, Phys. Rev. B. 37, 3124 (1988)

Towards an Understanding of the Composition of the Reconstructed α SiC(000$\bar{1}$) Surface

P. Badziag*, M.A. Van Hove, and G.A. Somorjai

Materials and Chemical Sciences Division, Lawrence Berkeley Laboratory,
1 Cyclotron Road and
Department of Chemistry, University of California, Berkeley, CA 94720, USA
*Permanent address: Department of Physics, University of South Africa,
   P.O. Box 392, Pretoria 0001, RSA

Abstract. Selfconsistent total energy quantum chemical (AM1) calculations were performed on clusters modelling unreconstructed as well as $(\sqrt{3} \times \sqrt{3})R30°$ reconstructed α SiC(000$\bar{1}$) surfaces. Calculated surface energies favour the Si adatom terminated $(\sqrt{3} \times \sqrt{3})R30°$ reconstructed surface with the Si atoms directly under adatoms (second layer in the unreconstructed surface) substituted by carbon. The mechanism which favours this particular composition is also discussed.

1. Introduction

After annealing, the SiC(000$\bar{1}$) (carbon terminated) surface is known to assume the $(\sqrt{3} \times \sqrt{3})R30°$ reconstruction pattern hereafter referred to as R3 [1,2]. The process is accompanied by the depletion of silicon, which suggests possible excess of carbon in the reconstructed surface. Nevertheless the final composition of the reconstructed surface has not been resolved experimentally. This situation then leaves room for theoretical investigation.

In this contribution we report results of the total energy selfconsistent quantum chemical (AM1) [3] cluster calculations modeling the unreconstructed relaxed (1 × 1) surface and the different compositions of the R3 reconstructed surface. The calculations assume the adatom induced reconstruction. In the absence of convincing experimental evidence regarding the atomic arrangement in the R3 reconstructed surface, the idea of the adatom induced reconstruction is appealing because of the analogy with the (adsorbate induced) R3 reconstructions of the (111) surfaces of silicon and germanium [4-6]. Moreover, the results presented later in this paper further support the claim that the simple adatom induced structure makes a feasible model for the R3 reconstruction of the SiC(000$\bar{1}$) surface, at least as far as the energetics of the process is concerned.

2. Comparison of surface energies

Using hydrogen and the adjusted hydrogen-like pseudoatom described elsewhere [7] as saturators terminating the clusters, we looked at the relative stabilities of four different SiC(000$\bar{1}$) surfaces: bulk-like terminated (1 × 1), R3 with Si adatoms (re1), R3 with C adatoms (re2) and R3 with Si adatoms and with the second layer Si atoms under adatoms replaced by C atoms (re3). The investigated clusters contained 16 to 17 C and 16 to 17 Si atoms plus the necessary saturators and represented four layers of SiC (plus a possible adatom).

Springer Series in Surface Sciences, Vol. 24   **The Structure of Surfaces III**
Editors: S.Y. Tong · M.A. Van Hove · K. Takayanagi · X.D. Xie
© Springer-Verlag Berlin, Heidelberg 1991

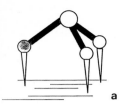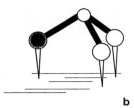

Figure 1. Carbon rich (a) and silicon rich (b) islands. Atoms indicated by full circles are in the text examined as possible sources of additional carbon/silicon atoms necessary to form the investigated reconstructed surfaces. The dangling bonds on all island atoms but on those indicated by full circles were in the calculations saturated by hydrogen. Larger circles denote Si atoms, smaller C atoms. Triangles denote bonds between the islands and the surface. The individual surface layer atoms are not shown.

The comparison of surface energies of surfaces with different compositions is a delicate issue. Basically, the problem is in determining chemical potentials of the auxiliary atoms at their source [8]. In particular, to compare surface energies of the reconstructed surface with adatoms and of the unreconstructed surface one has to calculate:

$$\Delta E = F_{re} - F_{1\times1} = E_{slabr} - [E_{slab1} + \mu_{ad} \cdot n_{ad}]. \tag{1}$$

Here F_{re} and $F_{1\times1}$ are surface energies of the reconstructed (with adatoms) and the 1×1 (unreconstructed) surfaces respectively, E_{slabr} and E_{slab1} denote energies per surface unit cell (SUC) of the infinite slabs terminated by the reconstructed and 1×1 surfaces respectively, μ_{ad} denotes the chemical potential of the adatoms at their source, while n_{ad} is the number of adatoms per SUC in the reconstructed surface.

The experimentally observed depletion of silicon which accompanies the surface reconstruction [2,9] indicates that the reconstructing surface has, at least locally, an excess of carbon atoms. Therefore one can expect that the additional carbon atoms necessary to produce re2 and re3 reconstructed surfaces come from carbon rich islands, rather than from an ideal stoichiometric bulk. On the other hand, evaporation of Si from the surface suggests absence of analogous Si rich islands in the R3 reconstructing surface. Therefore, while comparing surface energies of different surfaces, we assumed that the necessary additional C atoms came from the carbon rich islands while the Si atoms come from the stoichiometric bulk. To keep the numerical calculations within manageable limits, only small islands indicated in fig. 1 were investigated in the reported calculations.

The relevant calculated surface energies, together with some geometrical parameters optimised in the total energy calculations, are listed in table 1. The results show that re3 is clearly the most stable of the investigated compositions.

For comparison we also calculated the surface energies of the different compositions of the R3 reconstructed surface following the assumption that the additional Si atoms come from the Si rich islands (fig. 1b) while the additional C atoms come from the stoichiometric bulk. This changed the energies of re1, re2 and re3 to 0.06eV, 1.65eV and 0.18eV per 1×1 SUC respectively. The result for re1 shows that even if one assumes unrealistically favourable conditions for the formation of re1, this composition of the R3 reconstructed surface is still of slightly higher energy than the unreconstructed relaxed surface. On the other hand, the numbers for re2 and re3 indicate that without the favourable conditions produced by the depletion of Si from the reconstructing surface (formation of carbon rich

Table 1. Surface energies in eV per 1×1 surface unit cell relative to the relaxed 1×1 C-terminated surface and some geometrical parameters in Å for different compositions of the R3 reconstructed α SiC(000$\bar{1}$) surface. Notation as in fig.2.

Surface composition	re1	re2	re3
surface energy	0.82	1.24	-.23
h(2B)-h(2A)	0.47	0.68	0.37
Δx	0.14	0.32	0.33
h(3B)-h(3A)	0.29	0.43	0.01

islands), all compositions of the R3 reconstructed surface are probably of higher energy than the unreconstructed surface.

3. Mechanism which favours the re3 composition

One can regard the re3 composition as an ideally cleaved surface with a 1/3 monolayer of additional carbon atoms substituting every third Si atom in layer 2. The substituted Si atoms in the reconstructed surface become adatoms, directly above the additional (second layer) carbon atoms. In this respect the structure is analogous to the Si(111)R3:B surface, where there is a strong evidence for boron to occupy the B_5 location (layer 2, below adatoms) [5]. The mechanism which favours the particular atomic arrangement in the SiC(000$\bar{1}$)R3 surface is also likely to be similar to the one which is responsible for the particular B adsorption site on the Si(111) surface.

We believe that the re3 composition is stabilized by the particular relation between the bulk bond length and the geometry of bonds in the vicinity of adatoms. To increase the bonding strength between the adatom and the surrounding layer 1 atoms (later referred to as L1), the L1 atoms have to rotate their (basically sp^3) hybrids, so that the hybrid which participates in the bond with the adatom points to the adatom rather than perpendicularly away from the surface. Three geometrical effects can contribute to this rotation: compression of 2A-3A bonds, stretching of 2B-3B bonds and compression of L1-2A bonds (notation as in fig.2). In re1 and to even larger extent in re2 the combination of these effects produces a considerable stress in the structure, clearly indicated by the different elevations of atoms 3A and 3B (cf. Table 1). On the other hand the fact that the C-C bond is some 0.4Å shorter than the Si-C bond makes the combination of the first two geometrical effects natural in re3. Moreover, the fact that the $2s$ and $2p$ orbitals of carbon have lower energies than the $3s$ and $3p$ orbitals of silicon supports some charge transfer from the 2B-3B bond to the C-C bonds which in turn weakens the 2B-3B bond and allows stretching it by about 0.03Å without significant cost in energy.

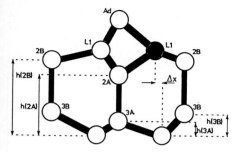

Figure 2. Geometry in the vicinity of an adatom, projection on the plane spanned by 3A–2A–marked L1.

4. Concluding remarks

To summarize, we examined relative stabilities of different compositions of the $(\sqrt{3} \times \sqrt{3})R30^0$ reconstruction of SiC(000$\bar{1}$) surface. The pattern of energies shows that the experimentally observed depletion of Si accompanying the R3 reconstruction is energetically important for the formation of the reconstructed surface. Without it the adatom terminated R3 reconstructed SiC(000$\bar{1}$) appears to be of higher energy than the relaxed unreconstructed surface.

Among the adatom terminated R3 reconstructed surfaces the surface with Si adatoms and C atoms substituting for Si in the H_5 locations is the most stable one. The result can be associated with the carbon covalent radius being considerably smaller than the silicon covalent radius. This particular relation between the two radii favours the re3 surface composition by relieving much of the stress present in chemical bonds in the vicinity of adatoms. Large differences between energies of different compositions of the α SiC(000$\bar{1}$) R3 reconstructed surface indicate that the qualitative picture of surface energetics which applies to this surface should also apply to the similar R3 reconstructed β SiC(111) C-rich surface.

Acknowledgements. We would like to thank R. Kaplan, W.S. Verwoerd and J. Powers for valuable discussions. One of us (P.B.) acknowledges a travel grant from the South African Foundation for Research Development. This work was supported in part by the Director, Office of Energy Research, Office of Basic Energy Sciences, Material Sciences Division, U.S. Department of Energy under contract No. DE-AC03-76SF00098. Supercomputer time was provided by the University of California - Berkeley.

References

[1] A.J. van Bommel, J.E. Crombeen and A. van Tooren, Surface Sci. 48 (1975) 463.

[2] R. Kaplan and T.M. Parill, Surface Sci. 165 (1986) 38.

[3] M.J.S. Dewar, E.G. Zoebisch, E.F. Healy and J.J.P. Stewart, J. Am. Chem. Soc. 107 (1985) 3902.

[4] J.E. Northrup, Phys. Rev. Lett. 53 (1984) 683; J.M. Nicholls, P. Martensson, G.V. Hansson and J.E. Northrup, Phys. Rev. B32 (1985) 1333.

[5] P. Bedrossian, R.D. Meade, K. Mortensen, D.M. Chen, J.A. Golovchenko and D. Vanderbilt, Phys. Rev. Lett. 63 (1989) 1257; I. -W. Lyo, E. Kaxiras and Ph. Avouris, Phys. Rev. Lett. 63 (1989) 1261.

[6] J.S. Pedersen, R. Feindenhans'l, M. Nielsen, K. Kjær, F. Grey and R.L. Johnson, Surf. Sci. 189/190 (1987) 1047.
[7] P. Badziag, Surface Sci., to be published.
[8] G.-X. Qian, R.M. Martin and D.J. Chadi, Phys. Rev. Lett. 60 (1988) 1962 and Phys. Rev. B38 (1988) 7649; J.E. Northrup, Phys. Rev. Lett. 62 (1989) 248.
[9] R. Kaplan, Surface. Sci. 215 (1989) 111.

Structure of the GaAs(-1-1-1) 2×2 and GaAs (-1-1-1) $\sqrt{19} \times \sqrt{19}$ Surfaces

R.D. Bringans, D.K. Biegelsen, L.-E. Swartz, and J.E. Northrup

Xerox Palo Alto Research Center, 3333 Coyote Hill Road,
Palo Alto, CA 94304, USA

Abstract. The 2x2 and $\sqrt{19}$x$\sqrt{19}$ reconstructions of the GaAs (-1-1-1) surface have been investigated with photoemission spectroscopies, low energy electron diffraction (LEED), scanning tunneling microscopy (STM) and total energy calculations. Analysis of the STM images for the 2x2 surface shows that the unit cell contains an adatom or cluster of adatoms, with total energy calculations favoring an added triplet of As atoms. Annealing this surface to around 500 °C leads to a sharp transition to the $\sqrt{19}$x$\sqrt{19}$ reconstruction which is seen in STM images to have a two-layer hexagonal ring as its main building block.

1. Introduction

During and after growth with molecular beam epitaxy (MBE), GaAs (-1-1-1) surfaces are found to exhibit a 2x2 reconstruction and, following vacuum annealing, a transition to a $\sqrt{19}$x$\sqrt{19}$ reconstruction. Although these surfaces have been investigated for some time, there has been no definitive determination of their atomic structure. This situation has been further complicated by the suggestion that there are several stable reconstructions with a 2x2 symmetry[1].

GaAs surfaces reconstruct in order to reduce the dangling bond derived density of states within the bulk band gap. It has been shown that the surface energy can be minimized by filling all of the As dangling bonds and leaving Ga dangling bonds empty [2]. While it is possible in principle to remove all of the electrons from the region of the band gap in the 2x2 case, it can only be part of the story for the $\sqrt{19}$x$\sqrt{19}$ reconstruction which has an odd number of electrons in its surface unit cell.

2. Results

Results for the GaAs(-1-1-1) 2x2, the $\sqrt{19}$x$\sqrt{19}$ surface and the transition between them will be discussed separately.

Springer Series in Surface Sciences, Vol. 24   **The Structure of Surfaces III**

Editors: S.Y. Tong · M.A. Van Hove · K. Takayanagi · X.D. Xie

© Springer-Verlag Berlin, Heidelberg 1991

2.1 GaAs(-1-1-1) 2x2

Total energy calculations have been carried out for a number of models for the 2x2 reconstruction, including Ga or As adatoms, vacancies and trimers, among others [3,4]. The relative energies of these vary depending on the local chemical potentials of As and Ga. Our recent results [4] show that an As trimer model has the lowest energy for As-rich surfaces and a single Ga adatom is stable at higher Ga coverages. We have used STM and angle resolved photoemission (ARPES) to determine which of these structures is present after MBE growth and x-ray photoemission (XPS) to detect whether more than one 2x2 structure is present.

After oxide removal, a 300 nm layer of GaAs was grown onto a polished GaAs(-1-1-1) sample at 640 °C. After the growth ended, the sample was held in the As_4 beam while its temperature was reduced. LEED patterns from as-grown surfaces and those subsequently annealed to 300 °C had sharp 2x2 spots. The corresponding STM image [4] in Fig. 1 has unit cells consisting of a high topographic region completely surrounded by a low region and is inconsistent with vacancy models. Although we cannot rule out tip effects, the highest region appears to be triangular and consistent with the As-trimer model shown in Fig. 1(b).

The electronic structure of the GaAs(-1-1-1)2x2 surface has been examined previously with ARPES [5,6]. The dispersions of surface related spectral features were found to show a 2x2 periodicity. Figure 2 was obtained by taking the results from [6] and replotting them into a single Brillouin zone. Comparison of this experimental band structure with those calculated for models of the surface is complicated because the surface states are resonances, overlapping with projected bulk bands. Nevertheless, for the As-trimer model we find states with close to the

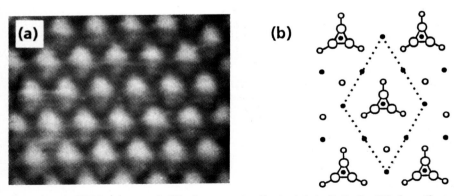

Fig. 1. (a) STM image of filled states in GaAs(-1-1-1) 2x2. Higher (lower) features are shown as light (dark) parts of the image. (b) As trimer model of the surface with Ga atoms shown as black and As atoms as open circles.

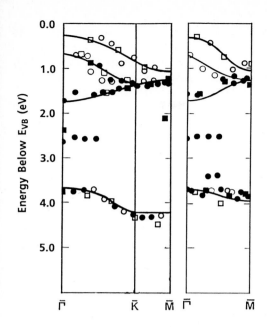

Fig. 2. Surface bands for GaAs(-1-1-1) 2x2 determined by angle resolved photoemission

experimental splittings. Qualitatively, the bands with energies at Γ of 0.3, 1.7 and 3.7 eV correspond to the rest atom lone pair, the trimer lone pair and As-As back bonds respectively.

2.2 Transition from 2x2 to $\sqrt{19}$x$\sqrt{19}$

The transition from 2x2 to $\sqrt{19}$x$\sqrt{19}$ in the diffraction pattern took place over a narrow range of annealing temperatures around 500 °C. The ratio of Ga to As atoms at the surface can also be seen in Fig. 3 to vary rapidly as the transition takes place. The constant surface composition before the transition takes place, provides strong evidence that the 2x2 reconstruction obtained after growth is stable until the $\sqrt{19}$x$\sqrt{19}$ occurs and that there is no other intermediate 2x2 structure with a different surface composition. We cannot rule out the possibility that sputter annealing produces another type of 2x2 structure, but LEED i-v data from the samples in the present study appear to be very similar to those seen for sputter-annealed surfaces [7].

2.3 GaAs(-1-1-1) $\sqrt{19}$x$\sqrt{19}$

STM images of this surface showed large triangular terraces with bilayer step heights, consistent with the notion that the 2x2 to $\sqrt{19}$x$\sqrt{19}$ transition requires motion of As and Ga atoms on the surface in order to conserve the (non-volatile) Ga atoms as the surface composition changes.

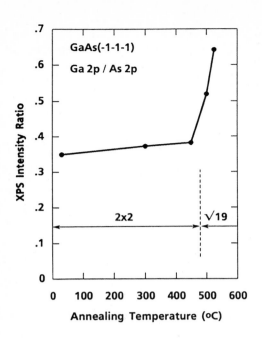

Fig. 3. The ratio of photoemitted intensities in the Ga 2p and As 2p core levels as a function of annealing temperature. The measurement was made with 1486 eV photons.

Figure 4 shows a high resolution image in which we can see that each unit cell has one hexagonal ring as its main building block.

Because of the rotational orientation of these hexagonal units with respect to one another and to the bulk, the number of possible models is severely limited. Fig. 4 (b) is an example of a model consistent with the image. Any model of a unit cell with an odd number of atoms cannot fulfill the requirement that all As (Ga) dangling bonds are occupied (unoccupied). For the model in Fig 4(b) there are 3.75 electrons in Ga orbitals. These electrons could either be localized in a narrow band at E_F

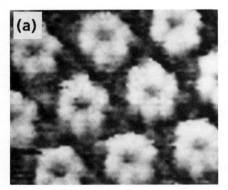
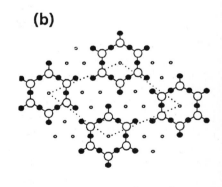

Fig. 4. (a) STM image of filled states in GaAs(-1-1-1) $\sqrt{19}$x$\sqrt{19}$. (b) Possible model of the surface with As atoms shown as open circles.

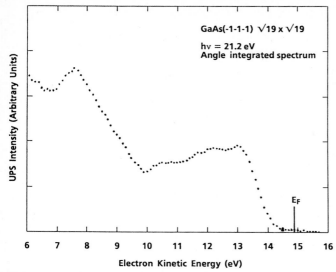

Fig. 5. Angle integrated UPS spectrum of GaAs (-1-1-1) √19x√19.

or be localized in a much broader band. With angle integrated UPS we have made an estimate of the density of states at E_F. The spectrum in Fig. 5 shows no obvious peak at E_F and by comparing the intensity at E_F with that for the valence band and assuming a 3 nm escape depth for electrons coming from the bulk valence band, we estimate that there is less than about one electron per √19x√19 unit cell within 0.5 eV of E_F.

References

[1]   M. Alonso, F. Soria and J. L. Sacedon, J. Vac. Sci. Technol., **A3** 1598, (1985).

[2]   See, for example, the discussion by D. J. Chadi in this volume.

[3]   E. Kaxiras, Y. Bar-Yam, J. D. Joannopoulos and K. C. Pandey, Phys. Rev. Lett., **57**, 106 (1986).

[4]   D. K. Biegelsen, R. D. Bringans, J. E. Northrup and L. -E. Swartz, to be published.

[5]   K. Jacobi, C. v. Muschwitz and W. Ranke, Surf. Sci., **82**, 270 (1979).

[6]   R. D. Bringans and R. Z. Bachrach, Phys. Rev. Lett., **53**, 1954 (1984)

[7]   S. Y. Tong, private communication.

A New Reconstruction of the GaP($\bar{1}\,\bar{1}\,\bar{1}$) Surface

X.Y. Hou, X.K. Lu, P.H. Hao, X.M. Ding, P. Chen, and X. Wang

Surface Physics Laboratory, Fudan University, Shanghai, P.R. of China

On GaP($\bar{1}\bar{1}\bar{1}$) clean surface, we observed a complicated LEED pattern with the periodicity of ($\sqrt{247}\times\sqrt{247}$)R22.7°. A computer simulation was used to find the possible atomic structural model.

1. Introduction

On the polar surfaces of III-V compound semiconductors, many reconstructions related to the surface stoichiometry have been found. For the GaP($\bar{1}\bar{1}\bar{1}$) surface, besides the (1X1) and (2X2) reconstructions, a very complicated low energy electron diffraction (LEED) pattern can be observed after ion sputtering and subsequent annealing. In previous literature, this structure has been assigned as a (17X17) reconstruction[1,2]. By carefully comparing the unit mesh of the (17X17) reconstruction with the observed LEED pattern, it has been found that the diffraction spots are not at the exact positions predicted by a (17X17) unit mesh. In this work, we did a re-examination of this surface reconstruction and found that a ($\sqrt{247}\times\sqrt{247}$)R22.7° might be a correct designation of this surface structure. To derive the atomic structure in real space, a computer simulation based on the kinetic scattering theory was carried out and the suggested structural model was tested by the optical simulation which could reproduce similar diffraction to that of the LEED experimental observation.

2. Experimental

The sample used was mechano-chemically polished n-type GaP($\bar{1}\bar{1}\bar{1}$) single crystal wafer with the doping concentration of 5×10^{17} cm^{-3}. The surface treatment was carried out in the ultra-high vacuum chamber of an electron spectrometer by 1keV Ar$^+$ ion bombardment for 40 min. and subsequent thermal annealing(IBA). After several cycles of IBA, the surface becomes a clean and ordered one without C, O contaminations as illustrated by the X-ray photoelectron spectroscopy measurements. The surface structures were observed by a four grid LEED facility.

After IBA treatment, the sample surface usually reveals a complicated LEED pattern with very sharp contrast as shown in Fig. 1. The pattern is quite stable and reproducible if the annealing temperature is chosen within the suitable range of 480-520°C. Above this temperature range, the surface becomes facetted after annealing, while below this temperature range, only a (1x1) LEED pattern with a relatively strong background can be obtained.

Springer Series in Surface Sciences, Vol. 24 **The Structure of Surfaces III**
Editors: S.Y. Tong · M.A. Van Hove · K. Takayanagi · X.D. Xie
© Springer-Verlag Berlin, Heidelberg 1991

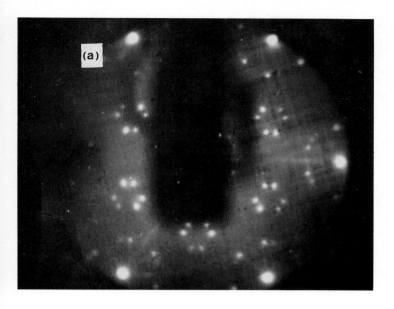

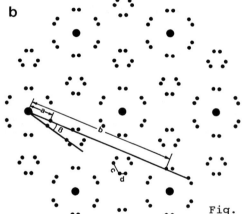

Fig. 1: The LEED pattern (a) and its schematic diagram (b).

In the LEED pattern, the fractional-order spots are all twins, which are believed to be caused by the contribution of two different reconstruction domains.

3. LEED pattern designation

In previous work[1,2], people did observe a similar LEED pattern of the GaP($\overline{1}\overline{1}\overline{1}$) surface, but the primary energy of electron beam was not adjusted suitably to show all the characteristics of the LEED pattern. Moreover, they assigned inadequately this LEED pattern as a (17x17) surface reconstruction. However, if one compares carefully the positions of all observed LEED spots with the reciprocal unit mesh of a (17x17) structure, the overall coincidence is poor. It is difficult to find out the reconstruction unit mesh directly from the observed LEED pattern because only a small portion of fractional-order spots could be seen on the picture. To compare the LEED pattern quantitatively with known reciprocal lattices, we used three parameters to characterize the unit mesh: ß, n=b/a and m=c/d, where ß, a, b, c and d are denoted in Fig. 1(b). Among a hundred types of reconstructions from the simplest ($\sqrt{3}X\sqrt{3}$)R30° to a largest one (17X17), the best fit between the parameter values obtained by the observed LEED pattern and the prediction could be reached by the ($\sqrt{247}X\sqrt{247}$)R22.7° as shown in Table 1.

Table 1: Experimental and prediction parameters

	Experimental values	($\sqrt{247}$)R22.7°	(17X17)	($\sqrt{112}$)R19°	($\sqrt{43}$)R7.6°	($\sqrt{199}$)R7°
ß	14.5°±0.4°	14.6°	13.2°	11.2°	15.2°	14.1°
n	6.22±0.05	6.25	*	5.5	6	6.5
m	1.97±0.09	2.0	2.0	2.0	4.5	1.3

*: a and b do not coincide with each other

4. Computer simulation of the atomic structure

In principle, a ($\sqrt{247}X\sqrt{247}$)R22.7° reconstruction contains many more diffraction spots than that in the observed LEED pattern. This means that there must exist 246 fractional-order spots in a reciprocal unit mesh for each reconstruction domain. But only 12 fractional-order spots actually appeared in a $\sqrt{247}$ unit mesh on the LEED pattern. All missing spots were independent of electron energy and might be caused by the fact that the intensities of these spots are very weak. The absence of about 95% diffraction spots is believed to be due to the structure factor of the atomic geometry in the $\sqrt{247}$ unit mesh. The GaP($\overline{1}\overline{1}\overline{1}$) face prepared by IBA might not be a perfect P-terminated one. P vacancies might be created. The atomic structure of $\sqrt{247}$ reconstruction is assumed to be a P vacancy model. In a ($\sqrt{247}X\sqrt{247}$)R22.7° unit mesh, the possible combination of P vacancies accounts for $2^{246}=1.13X10^{74}$. Taking the six-fold symmetry of the first P atomic layer into consideration, the arbitrary combination of (247-1)/6=41 atoms rather than (247-1) atoms should be taken into account. So, the search for correct P vacancy distribution in a $\sqrt{247}$ unit mesh has to be carried out within $2^{41}=2.2X10^{12}$ possible combinations.

To do such a tedious job, only the kinetic scattering theory could be used as a first order approximation. In order to reduce the computation time, we first transform the ($\sqrt{247}X\sqrt{247}$)R22.7° unit mesh to a quasi-one dimensional ("1"X247) unit mesh, in which one of the unit vectors coincides with the direction of the original (1X1) unit vector, another vector is chosen as shown in Fig. 2. The unit vectors of ("1"x247) can be written as **OA**=247**a**, **OB**=68**a**+**b**, where **a** and **b** are the unit vectors of (1x1)

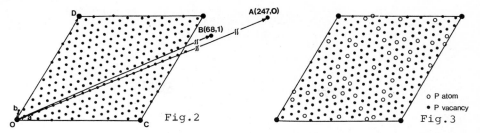

Fig. 2: The diagram of both (√247x√247)R22.7° and ("1"x247) unit
vectors

Fig. 3: 48 atoms arrangement in the (√247x√247)R22.7° unit mesh

unit mesh. **OC** and **OD** are the unit vectors of (√247x√247)R22.7°
reconstruction. After the transformation, the 247 atoms are now
equidistantly arranged on the long "247" unit vector in the
("1"x247) cell and the corresponding reciprocal unit mesh is
also a quasi-one dimensional unit. The 12 observed two
dimensional distribution of fractional-order spots contributed
by a single domain in one unit mesh can now be transferred into
a one dimensional array with the intensity peaks appearing at
following lattice points: 28, 44, 56, 72, 88, 103, 144, 159,
175, 191, 203 and 219, which are named as X series.

The computer simulation was carried out by using an IBM
compatible EC-386/33MHz microcomputer with 80387 co-processor.
In order to increase the calculation speed, the major programs
were made by assembly language. The procedure is as follows:
First, we choose a specific atomic distribution in the ("1"x247)
unit mesh. Then we calculate the structure factor of such an
atomic arrangement to see whether it could give rise to the
diffraction intensity distribution as the X series above. If
not, choose another atomic distribution and test again. Up to
now, 2.47×10^8 different models have been tried. Among them, the
best fit could be reached by a model containing 48 atoms
distributed in the outermost layer as shown in Fig. 3. The
calculated structure factor shows that the diffraction
intensities of those spots in the X series are at least five
times stronger than those outside the X series.

5. Optical simulation of the diffraction pattern

The suggested surface atomic structure of (√247x√247)R22.7°
reconstruction derived from the computer simulation has been
checked by the optical diffraction technique[3]. A diffraction
grating containing several unit cells of two different domains,
each of them has the geometry of Fig. 3, was prepared by
photographic film. By using this grating and a He-Ne laser as
incident beam, an optical diffraction pattern has been shown in
Fig. 4. In spite of some comparatively weak diffraction spots,
the main feature of Fig. 4 coincides well with the observed LEED
pattern in Fig.1.

Fig. 4: Optical simulated (√247x√247)R22.7° diffraction pattern spots

6. Discussion

The P vacancy model derived above is in principle consistent with other experimental facts. The XPS measurements of the P to Ga ratio of a clean GaP($\overline{1}\overline{1}\overline{1}$) surface show that the coverage of surface P atoms is less than unity. In high resolution electron energy loss spectroscopy, both the Ga-H and P-H vibrational modes could be observed upon hydrogen adsorption. The appearance of Ga-H mode is an indication of the presence of P vacancies. These show that a P vacancy model might be a reasonable one for GaP($\overline{1}\overline{1}\overline{1}$) √247 surface.

However, the vacancy model shown in Fig. 3 is not the unique one because only the scattering by the first P atomic layer has been considered. A reliable structural model can only be derived from the dynamical theory taking into account the multiple scattering by several outermost atomic layers. But the (√247x√247)R22.7° unit mesh is too large to be handled by the LEED dynamical calculation. The above computer simulation technique could provide a fairly satisfactory explanation of the absence of a large number of diffraction spots in the observed pattern.

Yang and Jona[4] have discussed in detail the origin of the disappearance of diffraction spots in LEED patterns. For those small reconstruction unit meshes, the missing spots distribute regularly to form a "missing-spot net" with their positions predicted exactly by the structural model, from which one could deduce some information about the reconstruction structure. But for the (√247x√247)R22.7° reconstruction, it is not easy to determine the positions of all the missing spots by using Yang's method, since about 95% of the fractional-order spots in a (1X1) reciprocal unit mesh is missing and no "missing-spot net" can be found. From the above computer and optical simulations, it is concluded that the structure factor of a large reconstruction unit mesh could cause the cancellation of the intensities of many diffraction beams and leave a few fractional spots to form a complicated diffraction pattern.

7. Conclusion

After IBA treatment the GaP($\overline{1}\overline{1}\overline{1}$) surfaces show a ($\sqrt{247}x\sqrt{247}$) R22.7° reconstruction rather than (17x17) reconstruction. Computer and optical simulations have shown that the absence of many fractional spots is due to the structure factor for such a large ($\sqrt{247}$x$\sqrt{247}$)R22.7° reconstruction.

This work was supported partly by the National Natural Science Foundation and partly by the Natural Science Foundation of Shanghai.

References

[1] Yoshihiro Kumazaki, Yasuo Nakai and Noriaki Itoh, Surface Sci. **184**, L445(1987).
[2] A. J. Van Bommel and J. E. Grombeen, Surface Sci. **93**, 383(1980).
[3] W. P. Ellis: in Optical Transforms, ed. by H. Lipson (Academic Press, London 1972).
[4] W. S. Yang and F. Jona, Phys. Rev. **B29**, 899(1984).

Atomic Geometry of the CdS(11$\bar{2}$0) Surface

A. Kahn[1] and C.B. Duke[2]

[1]Department of Electrical Engineering, Princeton University,
  Princeton, NJ 08544, USA
[2]Xerox Webster Research Center, 800 Phillips Road 114,
  Webster, NY 14580, USA

The atomic geometry of the CdS(11$\bar{2}$0) surface is determined by dynamical analysis of low energy electron diffraction intensities measured from surfaces cleaved in ultra-high vacuum and cooled below 50K. The agreement between measured and calculated intensities is optimized for a structure characterized by a bond-length conserving relaxation in the top atomic plane and a 0.1Å contraction of that plane toward the bulk. This structure is in quantitative agreement with predictions from total energy minimization calculations for this surface.

1. Introduction

Tight binding total energy minimization calculations have been recently applied to the determination of the atomic geometry and electronic structure of the (10$\bar{1}$0) and (11$\bar{2}$0) cleavage surfaces of wurtzite structure II-VI compound semiconductors [1]. These calculations predict bond-length conserving surface relaxations driven by the lowering of the surface electronic energy, a result similar to that obtained for the (110) cleavage surfaces of zincblende structure III-V and II-VI compound semiconductors [2]. The analyses of low energy electron diffraction (LEED) [3] and low energy positron diffraction (LEPD) [4] intensities from the (10$\bar{1}$0) and (11$\bar{2}$0) CdSe surfaces have produced atomic geometries which are in qualitative agreement with the predicted structures. To expand the scope of our experimental results, we present here a determination of the atomic geometry of CdS (11$\bar{2}$0) by dynamical analysis of LEED intensities. We find that the surface geometry is relaxed with respect to the bulk truncated structure and is in quantitative agreement with the predicted geometry.

2. LEED Experiment and Calculations

The CdS (11$\bar{2}$0) surfaces were cleaved in ultra-high vacuum (10^{-10} Torr) and cooled below 50K to reduce surface atomic vibrations. The LEED analysis was performed on the average of two sets of intensity vs. energy profiles (I-V) of fifteen diffracted beams measured for energies between 35eV and 250eV. The LEED pattern exhibited the (hk)=(h$\bar{k}$) symmetry and the missing (h0) beams for h odd imposed by the glide plane symmetry of the (11$\bar{2}$0) surface. A diagram of the surface unit cell geometry is given in Fig. 1.

Springer Series in Surface Sciences, Vol. 24   **The Structure of Surfaces III**
Editors: S.Y. Tong · M.A. Van Hove · K. Takayanagi · X.D. Xie
© Springer-Verlag Berlin, Heidelberg 1991

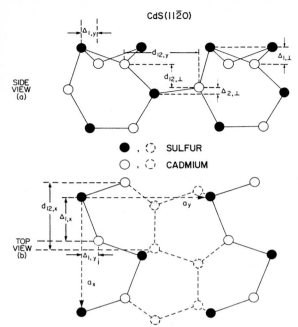

CdS(11$\bar{2}$0)

SIDE VIEW (a)

●, ◌ SULFUR
○, ◌ CADMIUM

TOP VIEW (b)

Figure 1 -- Schematic representation of the wurtzite lattice showing the side view and top view of the (11$\bar{2}$0) unit cell with relaxed geometry.

The structure determination was done by multiple-scattering analysis of the experimental I-V profiles. Details concerning the calculation of the potential and phase shifts which describe the scattering are given elsewhere [5]. The structure search was conducted with six phase shifts, a slab of six atomic layers, full dynamical computation for the top three layers, and the inelastic collision mean free path λ_{ee} equal to 10Å . To insure convergence of the results, the computation was later expanded to a slab of seven layers and full dynamical computation for the top four layers. The procedure utilized for determining the surface structure was that of minimizing the X-ray R-factor (R_X [6]). The integrated intensity R-factor (R_I [7]) was used to measure how the calculation reproduces the relative intensities of various beams.

3. Results and Discussion

The initial structure search was conducted for a bond-length conserving relaxation characterized by a rotation of the top layer anions and cations away and toward the bulk, respectively. The unit cell includes two anions and two cations and the relaxation induces a strong puckering of the surface. It is described in terms of ω, the angle between the normal to the surface and the normal to the planes defined by the Cd-S-Cd or S-Cd-S triplets of the surface zig-zag chains. The specification of ω defines the structural parameters shown in Fig. 1. The

567

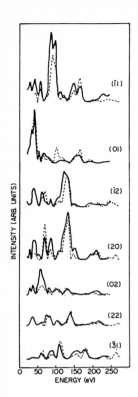

Figure 2 -- Comparison between calculated (solid line) and measured (dashed line) intensities of various beams. The calculated intensities correspond to the "best fit" structure specified in Table 1.

structure search was conducted by varying ω in order to minimize R_x. A minimum $R_x=0.25$ was found for $\omega=32°$. The corresponding structure was used as a starting point for a bond-length distorting contraction of the first layer toward the bulk. A minimum value of R_X equal to 0.21 was found for $\omega=30°$ and a 0.1Å contraction. No significant improvement was obtained by distorting the second layer. The "best fit" structure is described in Table 1. The truncated bulk structure and the structure predicted by tight binding energy minimization are also given for comparison. The I-V profile of selected beams computed with the best fit structure are compared with experimental I-V profiles in Fig. 2.

The "best fit" structure described in Table 1 is in quantitative agreement with the structure predicted from total energy minimization. It is also qualitatively similar to the structure of CdSe($11\bar{2}0$) determined by LEED and LEPD [4]. Furthermore, the local bonding conformation of the surface atoms resulting from the relaxation is similar to that found on the wurtzite ($10\bar{1}0$) [1,3] and the zincblende (110) [2] surfaces. On all these surfaces, the anion and cation rotate into a p^3-like and an sp^2-like configuration, respectively, regardless of the chemical nature of the constituent species. A common characteristic of these surfaces is the possibility of relaxing the surface geometry through low energy-cost bond-length conserving atomic motions which result in a rehybridization of the surface bonding and the elimination of the dangling-bond surface states of the

Table 1: Structural parameters specifying the atomic geometry of CdS($11\bar{2}0$). All are defined in Fig. 1. Except for ω, all quantities are in $\overset{\circ}{A}$.

Structure	a_x	a_y	$\Delta_{1\perp}$	Δ_{1y}	$d_{12\perp}$	d_{12y}	$\Delta_{2\perp}$	ω	R_x	$R_\perp$
Unrelaxed	6.749	7.162	0	0	2.067	3.58	0	0	0.27	0.29
"Best fit"	6.749	7.162	0.646	0.509	1.50	3.907	0	30°	0.21	0.083
Theory	6.749	7.162	0.691	0.53	1.468	3.705	0.106			

truncated bulk geometry. The present study reinforces therefore the concept that a common relaxation mechanism operates on the cleavage surfaces of zinc-blende and wurtzite tetrahedrally coordinated compounds. It is characteristic of the local structure and atom connectivity at these surfaces, and is dominated by the lowering of the surface electronic energy.

Acknowledgement

This work is supported by a grant of the National Science Foundation (DMR-8709531)

References

1. Y.R. Wang and C.B. Duke, Phys. Rev. **B37** 6417 (1988).

2. C.B. Duke, in *Surface Properties of Electronic Materials*, edited by D.A. King and D.P. Woodruff (Elsevier, Amsterdam, 1988), pp.69-118.

3. C.B. Duke, A. Paton, Y.R. Wang, K. Stiles and A. Kahn, Surf. Sci. **197** 11 (1988).

4. T.N. Horsky, G.R. Brandes, K.F. Canter, C.B. Duke, S.F. Horng, A. Kahn, D.L. Lessor, A.P. Mills,Jr., A. Paton, K. Stevens and K. Stiles, Phys. Rev. Lett. **62** 1876 (1989).

5. A. Kahn and C.B. Duke, (to be published)

6. E. Zanazzi and F. Jona, Surf. Sci. **62** 61 (1977).

7. C.B. Duke, S.L. Richardson and A. Paton, Surf. Sci. **127** L135 (1983).

Strain Relaxation in Metastable Cubic CdS Epilayers

D.W. Niles and H. Höchst

Synchrotron Radiation Center, University of Wisconsin–Madison,
3731 Schneider Drive, Stoughton, WI 53589, USA

Abstract. Thin films of CdS were grown at 250 °C by molecular beam epitaxy (MBE) on highly ordered GaAs(100) substrates. Reflection high energy electron diffraction and photoemission spectroscopy was used to investigate the structural and electronic properties of the CdS epilayer. RHEED shows that CdS grows pseudomorphically in the metastable zinc-blende structure with biaxial compressive strain parallel to the (100) plane. The CdS overlayer is initially strained by ~3.2 % but misfit dislocation starts at ~10 Å causing a gradual relaxation of the in-plane lattice strain. At a film thickness of ~200Å the strain relaxes to ~0.8 %. We estimate a strain energy density at the critical thickness of $d_{crit.}$~10 Å of 3.7 meV/Å^2. According to the variation in the width of the RHEED structures with increasing film thickness we estimate the misfit dislocation formation to be confined within ~50 Å around the interface region.

1. Introduction

The capability of growing high quality epitaxial films which are under considerable tensile or compressive strain generated a whole new generation of semiconductor materials with novel technological applications.[1-3] In addition to the film thickness strain generated by mismatched lattice constants, strain can be used as an additional parameter to fine tune the band gap and lift the degeneracy of heavy and light hole bands at the Γ-point of the semiconducting heterostructure.[4]

However, despite the advantages in using strain as an additional tool to tailor device properties, it may also create problems which are of a disadvantage. In structures where the lattice mismatch is confined to an electronically inactive buffer layer, there is the potential problem that dislocations may spread and adversely alter the device performance. The dynamics of misfit dislocation formation is even more serious in cases where device performance depends on strain as an active ingredient, as is the case in any strained layer semiconductor superlattice system.

Due to the special interplay of the density of misfit dislocation length (which is independent of the the film thickness but depends linearly on the the lattice mismatch) with the strain areal density (which depends linearly with the film thickness but varies quadratically with the lattice mismatch) it is possible to accommodate coherent strain only up to a critical film thickness. Stability curves based on the energy minimization developed by Matthews and Blakeslee which assumed isotropic elastic constants and ignored entropy effects were very successful in describing the phase diagram of strained $Si_{1-x}Ge_x$ epilayers.[5,6]

Springer Series in Surface Sciences, Vol. 24 **The Structure of Surfaces III**
Editors: S.Y. Tong · M.A. Van Hove · K. Takayanagi · X.D. Xie
© Springer-Verlag Berlin, Heidelberg 1991

Depending on the thermodynamics of the misfit dislocation formation, there seems to be a possibility of considerably exceeding a critical thickness calculated on simple energy arguments. Combining special growth conditions such as temperature and growth rate with the presence of a misfit dislocation barrier can result in a thick coherently strained metastable structure. It is predicted that as long as the interface strain is smaller than a critical value it should be possible to grow metastable structures of arbitrary film thickness.[7]

The present paper reports on the growth and strain relaxation of metastable cubic CdS films on GaAs(100) substrates. We used *in situ* reflection high energy electron diffraction (RHEED) and surface sensitive core and valence band photoemission spectroscopy to quantitatively study the strain relaxation and its influence on the evolution of the interface and overlayer crystallographic and electronic structural properties.

2. Experimental

A clean stoichiometric GaAs(100) surface was prepared by annealing the substrate in the MBE growth chamber to desorb an As cap until a well developed RHEED typical for an As terminated GaAs(100) surface appeared. Surface sensitive *in situ* core level spectroscopy was also used to confirm the cleanliness of the substrate after desorbing the protective As overlayer .

CdS films were grown from a single nitrogen shrouded MBE effusion cell on the GaAs substrate at ~250 °C with typical growth rates of ~1Å/min. Additional details of the CdS growth conditions were reported earlier. [8]

Improvement in the substrate's structural quality during the initial cleaning procedure and the development of the CdS overlayer were monitored by RHEED. The RHEED images were recorded with a computerized video framegrabber which allowed for a quantitative analysis of the digitized diffraction pattern at the end of a growth sequence. Figure 1 shows an intensity profile from a digitized GaAs(100) RHEED image. The full image consists of 480 lines with 640 pixels per line. The intensity distribution shown in Fig. 1 was measured with an electron beam

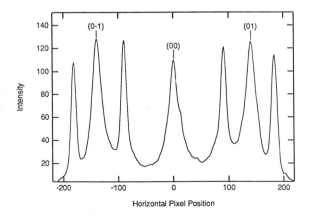

Fig. 1: RHEED intensity profile of GaAs(100) along the [011] azimuth.

of 25 keV along the [011] azimuth. The line profile shows strong reflection peaks from the specular reflected (00) beam as well as from the (0-1) and (01) reciprocal lattice rods. As one can see from Fig. 1, fractional order peaks are also very pronounced indicating the high degree of structural perfection of the substrate surface. The reason for having strong 1/3 order diffraction peaks only near the {01} lattice points but not to the left and right of the (00) reflection is caused by our technique of measuring a line profile across a certain part of the RHEED pattern, rather than following the intensity distribution of the diffraction pattern along a circular section, mapping intensity variations along the same Laue-ring.

The electronic structure of the CdS/GaAs interface and the overlayer growth was also monitored with high resolution photoemission experiments utilizing the synchrotron radiation from the 1 GeV electron storage ring ALADDIN. A discussion of how strain affects the splitting and dispersion of valence states and a comparison with our experimental data will be given elsewhere.[9]

3. Results and Discussion

The GaAs substrate has a lattice constant which is 3.17% smaller than that of a bulk cubic zinc-blende CdS crystal. Under epitaxial growth conditions the in-plane lattice constant $a_{||}$ of the CdS epilayer will be the same as that of the substrate during the initial coherent overlayer growth but will eventually relax as misfit dislocation starts to relieve some of the accumulated overlayer strain energy. The separation of the integer RHEED reflections g which are inversely proportional to the in-plane lattice constant by $a_{||}=2\pi/g$ can be used to precisely measure the lattice spacing of the overlayer with increasing film thickness. Normalizing the RHEED data to that of the GaAs substrate provides a procedure for determining changes in lattice spacing with a very high relative accuracy independent of the exact knowledge of the geometrical set up of the RHEED experiment.

Figure 2 shows the compressive in-plane strain relaxation as a function of CdS thickness. The film thickness is given in units of the unstrained CdS lattice constant which is $a_0=5.832$ Å Along the [100] growth direction the zinc-blende structure has 4 layers per unit cell which results in a thickness for the single

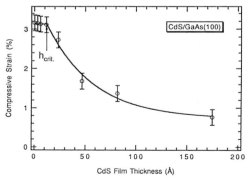

Fig.2: Strain relaxation as a function of CdS film thickness.

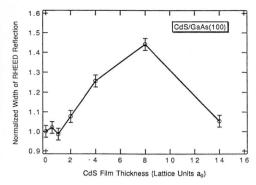

Fig.3: Normalized width of the (01) RHEED reflection versus CdS thickness.

layer of 1.458Å. Our data show that coherently strained CdS films can be grown up to a critical thickness of $d_{crit.} \sim 12$ Å. For thicker CdS films the in-plane lattice constant relaxes towards the equilibrium lattice constant of bulk CdS. The strain relaxation which does follow an exponential behavior as a function of film thickness is not complete and seems to level off at ~0.8% compressive strain for a films >150 Å.

The onset of strain relaxation and the concomitant generation of misfit dislocations can also be seen in the width of the RHEED reflex. Fig. 3 shows the full width at half maximum (FWHM) of (01) reflection of the CdS overlayer RHEED diffraction pattern which is normalized against the width of the unstrained GaAs substrate. Increased scattering and inhomogeniety in the overlayer causes a broadening of the (01) reflection for films exceeding the critical thickness. Misfit dislocation formation broadens the reflex by 45% for a coverage of ~8 lattice units. For thicker films the structural quality of the overlayer shows significant improvements and the FWHM of the (01) reflex is only slightly larger than that of the GaAs substrate. Even though we are not pursuing a quantitative analysis linking epilayer strain distribution with diffraction properties, it seems quite obvious from Fig. 3 that such a relationship exists and that RHEED is sensitive enough to provide information which relates to the strain induced structural modifications of epilayer surfaces.

The variation of the thickness dependent broadening of the RHEED reflex suggest that misfit dislocations do not migrate across the entire thickness of the film but are confined within a region of ~50 Å away from the interface. The dynamics of misfit dislocation formation in the compressively strained CdS/GaAs heterostructure behaves completely differently to what we observed for the formation of the highly tensile strained CdS/CdTe heterostructure.[9]

4. Conclusions

The usefulness of RHEED for quantifying the relaxation of strain at a lattice mis-matched heterostructure which was recently demonstrated by Whaley and Cohen [10] could be verified with our experiments of the heteroepitaxial growth of cubic CdS on GaAs(100). The first 12 Å of CdS grown on GaAs do not give rise to a change in the spectrum of the RHEED streaks, indicating that CdS has grown pseudomorhically on the substrate. Further CdS deposition yields an

573

increase in the width of the RHEED streaks, leading us to to conclude that misfits have formed within a region of ~50 Å from the interface and that the CdS overlayer has begun to relax towards its natural lattice constant.

Acknowledgements

We would like to thank the staff of the Synchrotron Radiation Center for technical support and M. A. Engelhardt for his assistance with the experimental set-up. DWN would also like to acknowledge partial support from the Alexander von Humboldt Stiftung while the actual experiments were in progress.

References.

1. G. C. Osbourn, J. Vac. Sci. Technol. **B4**, 1423 (1986).
2. R. People, IEEE Journal of Quantum Electrodynamics Vol. **QE-22**, 1696 (1986).
3. C. Mailhiot, and D. L. Smith, Critical Review in Solid State and Material Sciences, **CRC 16**, 131 (1990).
4. C. Mailhiot, and D. L. Smith, Phys. Rev. **B 37**, 10415 (1988).
5. J. Matthews, and A. E. Blakeslee, J. Cryst. Growth **27**, 118 (1974).
6. J. C. Bean, L.C. Feldman, A. T. Fiory, S.Nakahara, and I. K. Robinson, J. Vac. Sci. and Technol. **A 2**, 436 (1984).
7. J. Y. Tsao,B. W. Dodson, S. T. Picraux, and D. M. Cornelison, Phys. Rev. Lett. **59**, 2455 (1987).
8. D. W. Niles, and H. Höchst, Phys. Rev. B **41**, 12710 (1990).
9. D. W. Niles, and H. Höchst, unpublished.
10. G. J. Whaley and P. I. Cohen, Appl. Phys. Lett.**57**, 144 (1990)

Part V

Adsorbates on Semiconductors

Overview of Metal/Semiconductor Interfaces

Xie Xide

Department of Physics, Fudan University, 200433 Shanghai, P.R. of China

Abstract. Recent developments in both experimental studies and theoretical studies of metal/Si (111), metal/Si (100) surfaces and metal/GaAs (110) surfaces are reviewed. Emphasis is put on adsorptions of the group III and group V elements, noble metals and Pd on Si (111) surfaces. Results on geometrical configurations, electronic structures and Fermi-level pinning are discussed. Recent results on the adsorption of alkali metals on Si (100) 2x1 and the co-adsorption of oxygen and alkali metals on Si (100) surfaces as well as submonolayer adsorption of metals on GaAs (100) will be also mentioned.

1. Introduction

The metal semiconductor interface plays an important role in microelectronic technology, since in most devices the quality and stability of metallization affect the reliability of devices. It is of both practical significance and theoretical interest to understand the atomic positions, the nature of bonding and the electronic structures of the interfaces by using various experimental techniques and theoretical methods. Recently with the advent of scanning tunneling microscopy (STM), a deeper understanding of the metal/semiconductor interfaces has been achieved. Since it is almost impossible to cover all the existing results in this review, only some of the controversial issues and recent developments on submonolayer coverages of certain systems will be discussed.

2. Adsorption of Metal/Si (111) Surfaces

The reconstruction structures of the Si (111) surface have been studied for more than two decades. It has been well established that there is an annealed Si (111) 7x7 reconstructed structure and a room temperature cleaved 2x1 structure. At present the 7x7 reconstruction can be explained by the dimer-adatom-stacking fault (DAS) model [1]. Recently Fan et al. [2] reported a new vacancy model in which a first layer Si atom in each $\sqrt{3}\times\sqrt{3}$ unit cell is missing for the Si (111) $\sqrt{3}\times\sqrt{3}R30°$ reconstruction

Springer Series in Surface Sciences, Vol. 24 **The Structure of Surfaces III**
Editors: S.Y. Tong · M.A. Van Hove · K. Takayanagi · X.D. Xie
© Springer-Verlag Berlin, Heidelberg 1991

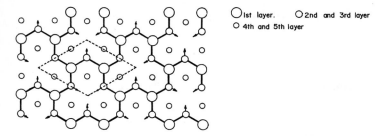

Fig. 1.   Schematic top view diagram for the vacancy model taken
from Ref.2. Broken lines denote the √3X√3 R30° unit cell.

(thereafer referred to as Si (111) - √3). It can be seen from Fig.1 that
there exist 1/3 monolayer (ML) vacancies. The vertical distance between
the first and the second layer atoms is highly compressed and the atoms in
the third layer are laterally displaced along the arrows shown in Fig.1.

2.1   Adsorption of Group III and Group V Elements

It has been discovered that for most metal / Si (111) interfaces, the Si
(111) √3 structures are most frequently found before the formation of
other images. For group III elements, it was found both experimentally and
theoretically that the adsorbed atom can sit on the well known H_3 site
which is usually referred to as the threefold hollow site or on the T_4
threefold filled site in which the adatom is placed above the second layer of
Si atoms with four neighbors, three in the surface layer, one in the second
layer directly below. For Al/Si (111), theoretical results of Northrup [3]
preferred the T_4 model and the result was also consistent with the
photoemission studies by Hansson et al.[4] and the LEED studies by Fan et
al. [2].
For In/Si(111) results from the recent impact-collision ion scattering
spectroscopy(ICISS) studies [5] also favored the T_4 sites. As for boron
which is the smallest of the group III elements, instead of occupying the T_4
site, it has been found by combining the STM and spectroscopy [6],first
principle calculation [7] and synchrotron X ray diffraction studies [8] that
boron occupied a new B_5 site in which the boron atom reverses its position
with Si in the T_4 configuration. As for gallium which is larger than B and
Al, it was found by Chen et al. [9] that at the 1/3 ML coverage, the
diffraction pattern was 3-Ga, new spots which could be associated with
the 6.3x6.3 incommensurate superlattice structure appeared by increasing
the coverage to 1 ML and subsequent annealing. The real space images of
such structure were obtained by STM.
For group V elements, the most well understood adsorption is the As/Si
(111) system. It was found that arsenic adsorbed upon the complicated Si

(111) 7x7 with the surface temperature at 600 °C converts the 7x7
surface to a new ideal 1x1 surface [10].

2.2   Adsorption of Noble Metals and Pd on Si (111)

Among   various interfaces the Si (111) √3X√3 Ag(simply √3-Ag) surface
has been the most extensively investigated.[11-19].   However , there still
exist conflicting results on both coverages and the surface atomic
geometry.   Using the STM Wilson and Chiang [13,14] first reported a
honeycomb H structure based on an array of Ag atoms embedded in
threefold   hollow sites of Si surfaces (Fig.2 H) with a 2/3 ML coverage
of Ag at room temperature.   There are 2 Ag atoms and 3 Si atoms per unit
cell.   By counting the valence electrons per unit cell the surface should be
metallic if uncharged.   Van Loenen et al.[15] using the STM and
current-imaging tunnel spectroscopy(CITS) determined the local electronic
structure and the stoichiometry of the √3 -Ag surface.   From the nature of
the local electronic structure and the spatial character of the wave
functions, a structure consisting of Ag trimers(Fig.2T) was suggested.
There are two Si atoms and 3 Ag atoms per unit cell and the surface is
semiconducting which agrees with most experimental results.   Kono et al.
[16] suggested another Ag honeycomb structure with the Ag atoms
embedded in a missing top layer structure (Fig.2 MTL).   A negatively
charged Ag surface was favored by analyzing the photoemission data and
it was also concluded that the surface should be semiconducting.   From the
results of ICISS scans Porter et al. [17] also showed that the top two
layers have the arrangement of Si honeycomb on top of Ag trimers. Results
of Oura et al. [18] from the high energy ion channeling suggested that for
√3-Ag vertical displacements of Si are larger than for a 7x7 surface and
the lateral displacement is small.   Results of Copel and Tromp [19] from

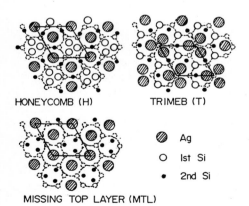

HONEYCOMB (H)   TRIMEB (T)

MISSING TOP LAYER (MTL)

⦸ Ag
○ Ist Si
• 2nd Si

Fig. 2.   Structure models for the Si (111) -√3 - Ag surface taken from
          Ref. 18. Dashed circles denote the oroginal positions of Si atoms.

the medium-energy ion scattering suggested that the model should include 1 ML of Si laterally displaced and 2 ML of Si vertically displaced from the bulk sites. With all the above mentioned controversies, it seems that much more research work remains to be done in order to get a deeper understanding of this mysterious surface.

For Au/Si (111) and Cu/Si (111), the situation is more complicated since there is a stronger tendency of intermixing with Si at room temperature. Recently it was reported by St. Tosch et al. [20] that a 5x5 noncommensurate structure was found by STM studies for Cu/Si (111). As for the Au/Si(111) interface, recent work by Chester et al. [21] showed a $\sqrt{3}$ LEED pattern at coverages from 0.85 to 1.0 ML which is quite sharp and can be explained in terms of a twisted trimer model suggested by Akiyama et al. for Pd on Si(111)[22]. As an example of another metal-semiconductor system, the Pd-Si interface will be given here since it has attracted much attention due to the fact that Pd is reactive and the silicide Pd_2Si formed is used as Ohmic contacts in semiconductor devices. Like many other metals, LEED patterns allow the formation of $\sqrt{3}$-Pd structure at monolayer coverages. Akiyama et al. [22] using the ultra high vacuum transmission electronic spectroscopy (UHVTEM) and transmission electron diffraction (TED) have proposed a most plausible structure $\sqrt{3}$-Pd which is a twisted trimer similar to the Si-rich lattice plane of Pd_2Si except for a rotation of the Pd trimers by about 6°.

3.   Adsorption of Alkali Metals on Si (100)

In recent years the alkali/semiconductor systems have attracted great interest partly due to the fact that those systems possess certain unique characteristics totally different from that of other systems mentioned previously, and partly due to the fact that alkali metals on Si (100) or Si (111) serve as electronic promoters for both oxidation and nitridation which are commonly used in semiconducting processing techonology. Among all alkali metals potassium has been studied extensively [23-25]. However, there still exists certain controversy regarding whether the adsorbates form a conducting chain on the dimerized Si (100) 2x1 surface or a non-conducting chain in which case the Si surface gains charge from potassium atoms and changes into a metallic state. Recently Ye et al.[26] investigated a variety of possible adsorption sites using the total energy local density formulism and a cluster model with the number of atoms up to 89. Fig.3 and Fig.4 give respectively the various possible adsorption sites on Si(100) and the clusters used to determine the most plausible site by the minimization of the total energy. Previous theoretical studies all used the pedestal site (see Fig.3) proposed by Levine [23] for Cs/Si(001)2x1. However, it was found by Ye et al. that the cave site shown in Fig.3 is the most stable one. This seems to be a plausible feature since towards the cave site dangling bonds of the two nearest Si atoms are directed.

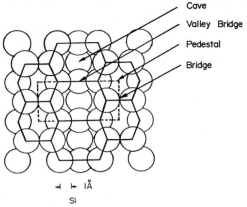

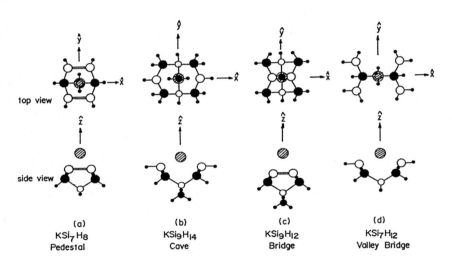

Fig. 3.   Model of the reconstructed Si (001) surface with the possible adsorbed sites indicated by arrows taken from Ref. 26.

Fig. 4.   Cluster models: (a) a single pedestal site cluster, (b) a single cave site cluster, (c) a single bridge site cluster and (d) a single valley site cluster taken from Ref. 26.

For the effect of co-adsorption of oxygen molecule and potassium, Ye et al. [27] carried out a first extensive study by using the self consistent local density approximation and a molecular cluster approach, the cluster used is shown in Fig. 5 which consists of 96 atoms (O_2 K_6 Si_{44} H_{46}) with four layers of Si, saturated below and on the lateral sides with hydrogen atoms, the six K atoms all sit on the cave sites, whereas the oxygen molecule is allowed to adsorb on the middle bridge site.   Studies were carried out for both individually adsorbed system and co-adsorption case. It was found for

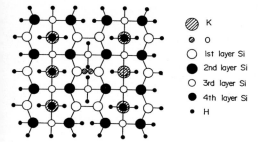

Fig. 5.   Cluster model for the study of the co-adsorption of oxygen
           molecule and potassium taken from Ref. 27.

the co-adsorption system, the oxygen molecule moves somewhat closer to
the surface and become less strongly bound. Those results are tentative
indications for the possibility of the alkali-promoted oxidation of the
Si(100) surface and for the desorption of K atoms.

4.   Adsorption of Metals on GaAs(110)

Although   there have been a wide range of studies of metal/GaAs(110)
surfaces during the past two decades, however, there is a renewed
interest of the adsorption of metals with very low coverages in order to
get a better understanding of the mechanism of Fermi-level pinning and the
formation of Schottky barriers. Only with the STM First et al. [28] and
Feenstra [29,30] were able to observe the atomic configurations of Cs and
Au on the GaAs(110) at coverage as low as 0.1 ML. It was found that the
Ga site is preferred to the As site for isolated adatoms. The experimental
results were supported by the theoretical studies carried out by Klepeis
and Harrison[31] and Klepeis et al.[32].

Acknowledgement

The author is very grateful to Professor Kaiming Zhang and Ms Jihuang Hu
for their help during the preparation of the manuscript and is much obliged
to Professor Ling Ye et al. for letting me use some of their unpublished
results.

Reference

1.   K.Takayanagi, T.Tanishiro, M.Takahashi and S.Takahashi, J.
          Vac. Sci. Technol. **A3**,1502 (1985).
2.   W.C. Fan. A Ignatiev, H.Huang and S.Y.Tong, Phys. Rev. Lett.
          62,1516 (1989).

3. J. E. Northrup, Phys. Rev. Lett. **53**, 683(1984).

4. G. V.Hansson, J. N. Nicholis, P. Maritensson and R.I.G. Uhrberg, Surf. Sci. **168**,105 (1986).

5. D. M. Cornelison, C. S. Chang and I. S. T.Tsong, Bull. Am. Phys. Soc. **33**, 459 (1990).

6. L. W. Lyo, E. Kaxiras and Ph. Avouris, Phys. Rev. Lett. **63**, 1261 (1989).

7. P. Bedrossian, R. D. Meade, K. Mortensen, D. M. Chen, J.A. Golovchenko and D. Vanderbilt, Phys. Rev. Lett. **63**, 1257 (1989).

8. R.L. Headrick, J.K. Robinson, E. Vlieg and L.C. Feldman, Phys. Rev. Lett. **61**, 1253(1989).

9. D. M. Chen, J.A.Golovchenko, P. Bedrossian and K. Mortensen, Phys. Rev.Lett. **61**, 2867 (1989)

10. M. A. Olmstead, R. D. Bringans, R. I. G. Uhrberg and R. Z. Bachrach, Phys. Rev. **B34**, 6401(1986).

11. G. Le Lay, Surf. Sci. **132**, 169(1983).

12. J. Stohr, J Jaeger, G.Rossi, T. Kendekewicz and I. Lindau, Surf. Sci. **134**, 813 (1983).

13. R.J. Wilson and S. Chiang, Phys. Rev. Lett. **58**,369 (1987).

14. R.J. Wilson and S. Chiang, Phys. Rev. Lett. **59**,2329 (1987).

15. E. J. Van Loenen, J. E. Demuth, R. M. Tromp and R. J. Hamers, Phys. Rev. Lett. **58**,373(1987).

16. S. Kono, K. Higashiyama, T. Kinoshita, T. Miyahara, H. Kato, H. Ohsawa, Y. Enta, F. Maeda and Y. Yaegashi, Phys. Rev. Lett. **58**,1555 (1987).

17. T. L. Porter, C. S. Chang and S. T. Tsong, Phys. Rev. Lett. **60**, 1739 (1988).

18. K. Oura, M. Watamori, F. Shoji and T. Hawana, Phys. Rev. **B38**, 10146 (1988)

19. M. Copel and R. M. Tromp, Phys. Rev. **B39**, 12688(1989).

20. St. Tosch and H. Neddermeyer, Surf. Sci. **205**,177 (1988).

21. M. Chester and T. Gustafsson, Bull. Am. Phys. Soc. **35**,450 (1990).

22. K. Akiyama. K. Takayanagi and Y. Tanishiro, Surf. Sci. **205**, 177 (1988).

23. J. D. Levine, Surf. Sci. **34**,90 (1973).

24. S. Ciraci and I. P. Batra, Phys. Rev. Lett. **56**,877 (1986).

25. P. Soukiassian, T.M. Gentle, M.H. Bakashi, A.S. Bommannavar and Z. Hurych, Physica Scripta, **35**,757(1987).

26. L. Ye , A. J. Freeman and B. Delley, Phys. Rev. **B39**, 10144 (1989)

27. L. Ye, A. J. Freeman and B. Delley (private communication).

28. P. N. First, J. A. Stroscio, R.A. Dragoset, D.T. Pierce and R.J. Celotta, Phys. Rev. Lett. **63**,1416 (1989).

29.   R.M. Feenstra, J. Vac. Sci. Technol. **B7**, 925 (1989).

30.   R.M. Feenstra, Phys. Rev. Lett. **63**,1412 (1989).

31.   J. E. Klepeis and W.A. Harrison, Phys. Rev. **B40**, 5810 (1989).

32.   J. E. Klepeis amd M. Schilfgaarde, Bull. Am. Phys. Soc. **35**, 269 (1990).

Scanning Tunneling Microscopy Investigation of the K/Si(100)2 × 1 Stepped (4°) Surface

P. Soukiassian[1,] and J.A. Kubby[2]*

[1]Commissariat à l'Energie Atomique, Service de Physique des Atomes et des Surfaces, Centre d'Etudes Nucléaires de Saclay,
F-91191 Gif-sur-Yvette Cedex, France and
Département de Physique, Université de Paris-Sud,
F-91405 Orsay Cedex, France
[2]Xerox Webster Research Center, 800 Phillips Road 114,
Webster, NY 14580, USA
*Also at Department of Physics, Northern Illinois University,
De Kalb, IL 60115, USA

We present the first investigation of potassium adsorption in the monolayer (Ml) range, on a Si(100)2x1 surface by scanning tunneling microscopy (STM). Standing waves measurements indicate lowering of the local surface work function as a result of K deposition. Topographic results with atomic resolution show that at 1 Ml, the K atoms are adsorbed on one dimensional chains distant by 7.68 Å and parallel to the Si dimer rows. Below 0.5 Ml, the growth mode is found to be perpendicular to the Si dimer rows with K atoms sitting in various coexisting sites, thereby indicating the existence of a phase transition in which the adsorbate-adsorbate interaction appears to be the leading driving force. We discuss the results within the existing structural models of alkali metal ordering on Si(100)2x1.

The alkali metal adsorption on semiconductor surfaces is presently intensively studied in surface science. In fact, from a fundamental point of view, it represent a model in the formation of metal-semiconductor interfaces as well as in surface catalysis.[1] While the electronic properties of alkali metal/semiconductor interfaces were studied with the use of the most recent experimental and theoretical techniques, the structural properties were mainly investigated by more classical tools such as electron diffraction.[2-4] It is only very recently that new structural investigations using more recent techniques like photoelectron diffraction (PED),[5] surface and/or photoemission extended x-ray adsorption fine structure (SEXAFS)[6] and scanning tunneling microscopy (STM)[7] were performed on these systems.

More than 15 years ago, Levine proposed from low electron energy diffraction - LEED measurements on the Cs/Si(100) that the Cs atoms are adsorbed at pedestal sites (see figure 1 for the various sites of adsorption) as one dimensional chains.[2] This model was found to explain rather well the corresponding properties of this system for negative electron affinity (NEA).[2] Later, this picture was also proposed by Tochihara for the corresponding K/Si(100).[2] However, this model was recently questionned by Abukawa and Kono who, from PED measurements, proposed that Cs and K atoms are adsorbed on two different sites, the pedestal and the valley bridge, with the existence of two alkali layers.[5] Using ab-initio total energy DMOL calculations, Ye Ling, Freeman and Delley investigated the stability and the distance of each possible site of adsorption.[8] While they found that the bridge, the valley bridge and the

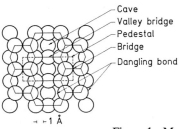

Cave
Valley bridge
Pedestal
Bridge
Dangling bond

⊢ ⊣ 1 Å
Si

Figure 1: Model of the reconstructed Si(100) surface (after reference 8).

Springer Series in Surface Sciences, Vol. 24   **The Structure of Surfaces III**
Editors: S.Y. Tong · M.A. Van Hove · K. Takayanagi · X.D. Xie
© Springer-Verlag Berlin, Heidelberg 1991

pedestal sites were not stable, they shown that the K-Si bond length was the same for the pedestal and the cave sites at 3.22 Å,[8] in excellent agreement with the measured value of 3.14 Å (which is the exact sum of K and Si covalent radii) given by SEXAFS experiments.[6] These experimental and theoretical values were also in very good agreement with an ab-initio total energy calculation using a pseudofunction method for the same system giving a K-Si bond length of 3.30 Å.[10] Very recently, a STM experiment on the Li/Si(100) system performed at very low lithium coverages (below 0.1 Li monolayer) claims that the dangling bonds are the site of adsorption just like hydrogen atoms. Furthermore, the mode of growth was found to be perpendicular to the Si rows, instead of the expected parallel one reported before for Cs and K.[2] However, one should note that Li is the smallest alkali metal and that, due to rather close size and electronic structure, it is possible to find a similar situation as for hydrogen atoms.

Investigations on the electronic properties of alkali metal on silicon surfaces [1,11] have shown that the nature of the bond between the alkali adsorbate and the silicon substrate is covalent as a result of the hybridization between alkali metal s and silicon 3p (dangling bond) valence electrons,[11,12] in contrast with the classical ionic model of Langmuir.[13] Recent theoretical ab-initio calculations were found to be in very good agreement with this model.[8-10]

Since the mode of growth (parallel versus perpendicular to the Si dimer rows) is one of the major issues, we have decided to use a 4° stepped Si(100) surface which is now well known at the atomic scale.[14] This allows us to have a surface with step edges facilitating the determination of chain orientation. Furthermore, the terraces are sufficiently large to represent the Si(100)2x1 surface and are separated one from each other by 2 layers with the Si dimer rows oriented in the same direction (perpendicular to the steps) for each terrace.[14,15]

The experiments were performed in an ultra-high vacuum chamber equipped with a third generation small-size scanning tunneling microscope using a cylindrical piezoelectric ceramics for scanning.[14,15] The alkali metal covered surfaces are well known to be highly sensitive to contamination, even at a very low level.[1] In fact, the site of adsorption of potassium is the result of a very delicate balance between adsorbate-substrate versus adsorbate-adsorbate interactions and traces of contamination by oxygen, atomic hydrogen, water or carbon-monoxide could result in the increase of this latter. Therefore, it is very important to stress this point here in order to be in the position to solve some of the numerous controversies concerning the structural properties of these systems. During the experiments, the pressure was at $1.2 \ 10^{-10}$ torr including mainly a partial pressure of molecular hydrogen ($8 \ 10^{-11}$ torr), but also CO ($1.6 \ 10^{-11}$ torr), H_2O ($5.7 \ 10^{-12}$ torr) and CH_4 ($5.5 \ 10^{-12}$ torr). Another important point is to outgass very carefully the SAES Getters K source used which otherwise would release atomic H and molecular H_2, CO and oxygen.[1] The pressure increase during K deposition always remains below $1 \ 10^{-11}$ torr. Since molecular H_2 is basically inert with alkali metal covered surfaces, we feel confident that our experimental conditions would leave a surface lifetime long enough to perform safely STM measurements. The silicon surface was prepared using the method described in reference 14. The measured corrugation was better than 1.5 Å and 0.7 Å when tunneling into the filled and empty states respectively. In order to make the things simpler, we will label in all this work the saturation coverage as 1 monolayer - Ml ($\approx$2 atoms of Si for 1 atom of K) which corresponds in some other works to 0.5 Ml.[6] All the other experimental details could be found in references 1, 6, 11, 14 and 15.

Figure 2 exhibits the standing waves (conductivity versus distance tip-surface) measurements performed for the clean (a) and 1 Ml K covered (b) silicon surface as previously described elsewhere.[16] The first maximum is shifted by 1.8 V as a result of the decrease of the substrate local work function upon potassium deposition. Figure 3 and 4 displays the STM pictures for the 1 Ml K/Si(100)2x1 stepped surface by tunneling into the filled and empty states (bias ±1.2 Volt) respectively. As seen from figure 3, the K adsorbate appears as an oval adsorbed in one dimensional chains perpendicular to the step, i.e. parallel to the Si rows. The distance between two chains is 7.68 Å, i.e. the distance between two Si rows. Also of interest, is the presence in the lower part of figure 3 of an excess of potassium at the step edge which could be consistent with the existence of a surface defect allowing additional K adsorption. This effect might also result from the higher sticking probability of the step edges as claimed for Cs/Si(100).[17] However, in this latter case, a more pronounced effect should be expected.

The question is now to know if there is a second layer between these two K chains that could not be detected due to possible limitation in the resolved corrugation. We have calculated the height of the potassium atom above the Si surface for various possible sites of adsorption. We use as a basis the experimental value of the K-Si bond length of 3.14 Å measured by SEXAFS.[6] This give heights of 2.19 Å for the pedestal site, 1.65 Å for the cave site and 3.05

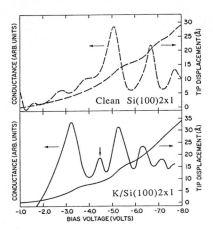

Figure 2: Standing wave measurements for the clean (a) and 1 Ml K covered (b) Si(100)2x1 surface.

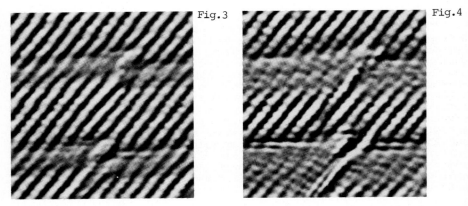

Figure 3: STM image of 1 ML of potassium on the Si(100)2x1 - 4° - stepped surface ($V_t = +1.2$ V). The distance between 2 potassium chains is 7.68 Å.

Figure 4: STM image of 1 ML of potassium on the Si(100)2x1 - 4° - stepped surface ($V_t = -1.2$ V).

Å for the dangling bond (DB) site. In this latter, we used a K position tilted by 14° from the vertical as shown to be energetically favorable for the dangling bond DB site.[18] In contrast to the very low coverage regime, the 2 dangling bonds are likely to be bonded to K atoms at the monolayer coverage.[7] Therefore, in the case of DB sites, there will be two layer chains in the same plane distant respectively by 3.87 Å and 3.81 Å which should be seen in the STM picture. Since we only see one chain here, this rules out the possibilty that the dangling bond could be the site of adsorption at the monolayer coverage. We still have to know whether a second chain could be "buried" in the cave between two pedestal chains as suggested recently by a theoretical calculation for Na/Si(100).[19] In this case, the extreme situation would be the coexistence of pedestal and cave sites which would have the largest height difference as compared e.g. to pedestal + valley bridge sites. If we have two layers, the height difference between pedestal and cave sites would be 2.19 - 1.65 = 0.54 Å. Since, as mentioned above, the measured corrugation would go up to 1.5 Å and 0.7 Å for filled and empty states respectively, a difference of only 0.54 Å between the two sites would be, in principle, clearly visible in this experiment on both figures 3 and 4 which would display two layers (distance between 2 chains

= 3.84 Å) instead of one (distance between 2 chains 7.68 Å). The height difference between the pedestal (P) and the valley bridge (VB) site will be even smaller than 0.54 Å. So, our results would in principle favor a single site of adsorption. In this case, the only two possible sites of adsorption for K atoms at the monolayer are either cave (C) [8] or pedestal (P) [2] along one dimensional chains parallel to the Si dimer rows. However, despite the very good contrast, one cannot totally exclude possible enhanced corrugation (rather unlikely) clouding a second layer.

It is interesting to remark that the K atoms seem to generally appear as ovals (fig.3). Tunneling into the filled state occurs within the alkali "s" orbitals which have been shown to hybridize to Si 3p dangling bonds leading to the formation of a weak and polarized covalent bonding.[1,11] There is a significant amount of the charge density located between the K and the Si atoms at the interface. Therefore, tunneling will also occur into the hybridized s-p electronic states which will give rather an oval picture of the K atom and its bonding with Si in the case of a K-Si bonding close from the horizontal plane. Interestingly, this would precisely be the case for the cave (C) site, but not for the pedestal (P) ones. So the observation of an oval would in principle favor the cave (C) site model of adsorption for the K atoms.[8] However, one cannot totally exclude that the "soft" alkali s orbital would be somehow deformed by the field induced by the scanning tip, despite the fact that such a deformation does not occur in the case of Si atoms of the clean silicon surface. Anyway, additional work is needed to make a more definitive conclusion between cave (C) and pedestal (P) sites at the monolayer coverage.

So far, we have limited the discussion to the one monolayer K coverage regime. Figure 5 displays the picture for a submonolayer K coverage (< 0.4 Ml) by tunneling into the filled states (bias +2.75 Volt). In this picture, only a part of the Si surface is covered which could allows in principle, the identification of adsorption sites. First, it looks from figure 5 that, at coverage below half a monolayer, the mode of growth for the K adsorbate appears to be in chains perpendicular to the Si dimer rows, in agreement with the previous STM work on the Li/Si(100)2x1 at very low coverage (< 0.1 Li monolayer).[7] However, since the streaks run along the scan direction, we cannot totally rule out a tip-induced motion of the adsorbate. At low coverages, the K atoms have an attractive interaction with the Si substrate leading to the formation of linear chains perpendicular to the Si dimer rows. Furthermore, it is surprising to see individual K atoms adsorbed on various sites (pedestal, cave, dangling bond). This behavior obviously indicates that the sites of adsorption for K are metastable and that this coverage is likely to be an intermediate step in the middle of a phase transition. This interpretation is consistent with the prediction of structural changes suggested on the basis of the electronic properties of the Cs/Si(100) and Na/Si(100) systems.[11] It also stresses that, on the clean surface, potassium atoms just adsorb on various sites without any long range order. When the coverage is increased, thereby reducing the K-K distance, the adsorbate-adsorbate interactions become more important than the weak [1,11] substrate-adsorbate ones. This provokes

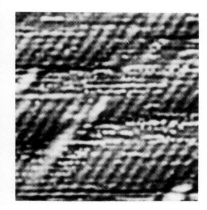

Figure 5:   STM image of potassium (coverage < 0.4 Ml) on the Si(100)2x1 - 4° - stepped surface (V_t = +2.75 V).

a general rearrangement of the K atoms, leading to an ordered structure in one dimensional chains as observed above (figures 3 and 4). This indicates that the K-K interaction is the leading driving force of this ordering transition. The behavior below half a monolayer (with coexistence at the same time of various adsorption sites) as well as at one monolayer, is consistent with the constant observation at all coverages of the same 2x1 LEED pattern.[2]

In conclusion, we have presented the first STM study of a K covered Si(100) surface in the monolayer range. At 1 Ml, K atoms adsorbed along one dimensional chains parallel to the Si dimer rows (110) direction and distant by 7.68 Å. Below half a monolayer, K atoms adsorbed in chains perpendicular to the Si rows with various adsorption sites and no long range order. This suggests the existence of a phase transition leading to the formation of ordered chains parallel to the Si dimer rows at the monolayer coverage. The adsorbate-adsorbate interaction appears to be the principal driving force in this ordering transition.

We want to acknowledge useful discussions with Shirley Chiang and Randall Feenstra. This work was supported in part by the U.S. National Science Foundation under contract DMR No. 88-07754 through the Northern Illinois University and by the Commissariat à l'Energie Atomique, CEA - DPh/G. One of us (P.S.) is grateful to the Xerox Webster Research Center for its hospitality and friendly atmosphere.

References

1- P. Soukiassian and H.I. Starnberg in Physics and Chemistry of Alkali Metal Adsorption, Elsevier Science Publishers B.V., Amsterdam, H.P. Bonzel, A.M. Bradshaw and G. Ertl editors, Monographs in Materials Science **57**, 449 (1989) and references therein.
2- J. Levine, Surf. Sci. **34**, 90 (1973); H. Tochihara, Surf. Sci. **126**, 523 (1983).
3- G.S. Glander and M.B. Webb, Surf. Sci. **222**, 64 (1989); ibid, **224**, 60 (1990).
4- S. Kohmoto, S. Mizuno and A. Ichimiya, Appl. Surf. Sci. **41/42**, 107 (1989).
5- T. Abukawa and S. Kono, Phys. Rev. B **37**, 9097 (1988); Surf. Sci. **214**, 141 (1989).
6- T. Kendelewicz, P. Soukiassian, R.S. List, J.C. Woicik, P. Pianetta, I. Lindau and W.E. Spicer, Phys. Rev. B **37**, 7115 (1988).
7- T. Hashizume, Y. Hasegawa, I. Kamiya, T. Ide, I. Sumita, S. Hyodo, T. Sakurai, H. Tochihara, M. Kubota and Y. Murata, J. Vac. Sci. Technol. A **8**, 233 (1990).
8- Ye Ling, A.J. Freeman and B. Delley, Phys. Rev. B **39**, 10144 (1989).
9- H. Ishida and K. Terakura, Phys. Rev. B **40**, 11519 (1989).
10-R.V. Kasowski and M.H. Tsai, Phys. Rev. Lett. **53**, 372 (1988).
11-P. Soukiassian, M.H. Bakshi, Z. Hurych and T.M. Gentle, Surf. Sci. Lett. **221**, L 759 (1989).
12-P. Soukiassian, Surf. Sci. Lett. **172**, L 507 (1986).
13-K.H. Taylor and I. Langmuir, Phys. Rev. **21**, 380 (1923); J. Topping, Proc. Roy. Soc. A **114**, 69 (1927).
14-P.E. Wierenga, J. Kubby and J.E. Griffith, Phys. Rev. Lett. **59**, 2169 (1987).
15-J.E. Griffith, J. Kubby, P.E. Wierenga, R.S. Baker and J. Vickers, J. Vac. Sci. Technol. A **6**, 493 (1988).
16-R.S. Baker, J.A. Golovchenko and B.S. Swartzentruber, Phys. Rev. Lett. **55**, 987 (1985).
17-S. Kennou, M. Karamatos, S. Ladas and C. Papageorgopoulos, Surf. Sci. **216**, 462 (1989).
18-S.P. Tang and A.J. Freeman, private communication.
19-I.P. Batra, Phys. Rev. B **39**, 3919 (1989).

Lead Adsorption on Ge(001) and Si(001) Studied by Core-Level Spectroscopy

K. Hricovini[1], *G. Le Lay*[2], *A. Kahn*[3], *A. Taleb-Ibrahimi*[1], *and J.E. Bonnet*[1]

[1]LURE, Bât. 209d, F-91405 Orsay Cedex, France
[2]CRMC2-CNRS, Campus de Luminy, Case 913,
  F-13288 Marseille Cedex 09, France
[3]Department of Electrical Engineering, Princeton University,
  Princeton, NJ 08544, USA

Abstract. We have studied the evolution of the surface structure of Pb/Ge(001) and Pb/Si(001) interfaces during the early stages of formation using synchrotron radiation. These experiments have been performed at liquid nitrogen and room temperatures. We have observed a complete removal of the Si(001) surface peak in the Si 2p spectra for lead coverages above 1 ML. In the case of Pb/Ge(001) a new surface peak is induced by lead overlayer. Our photoemission results point out that with increasing lead coverage the dimers on the germanium (001) surface are progressively disrupted.

1. Introduction

Lead overlayers on germanium or silicon have been extensively studied since they form perfectly abrupt interfaces without intermixing [1]. The interest in studying these systems increased recently as evidenced by reports on group IV semiconductors concerning the structure dependent Schottky-barrier height [2] and the photovoltage effects observed on clean Si surfaces [3] and metal/Si interfaces [4].

Only few studies of metal/Ge(001) interfaces are available in the literature. In the case of the Ag/Ge(001) interface, Miller et al. [5] bring the evidence that the Ag-Ge interaction is very weak and that the nucleation of Ag atoms begins at 0.33 monolayer (ML) coverage. The Ge 3d core-level spectra are not altered at all by the presence of silver atoms; in particular the dimer peak contribution persists even beyond 3 ML of Ag.

To our knowledge no study of the adsorption of lead on the Ge(001) surface has been performed until now. The aim of this paper is to present a core-level study of the formation of the Pb/Ge(001) interface and to compare this interface with the Pb/Si(001) system. Previous low energy electron diffraction (LEED), Auger electron spectroscopy (AES), scanning electron microscopy (SEM) and photoemission studies of the Pb/Si(100) interface have concentrated on room temperature (RT) measurements [6]. They showed that the growth at RT proceeds according to a layer-plus-islands mode and that the surface becomes metallic beyond the monolayer coverage range. Here, we have completed these studies by measurements at liquid nitrogen temperature (LNT).

2. Experimental

Part of the LEED and AES observations were carried out in Marseille and a series of photoemission measurements were

---

Springer Series in Surface Sciences, Vol. 24   **The Structure of Surfaces III**   589
Editors: S.Y. Tong · M.A. Van Hove · K. Takayanagi · X.D. Xie
© Springer-Verlag Berlin, Heidelberg 1991

performed at the SUPERACO storage ring of the Orsay Synchrotron
Radiation Facility (LURE). Germanium single crystals (p-type)
oriented along the (001) direction were heated to 500°C during
argon ion bombardment. After the cleaning procedure, the
crystals were annealed at 700°C in order to recover the surface
reconstruction. Silicon crystals were just heated to 950°C and
the cleanliness of the surface checked directly by photoemission
. The base pressure in the experimental chamber was better than
$2*10^{-10}$ mbar. During lead evaporations, done with a Knudsen
cell, the pressure remained in the low 10^{-9} mbar range because
of the high pressure of lead vapors. The Pb deposition was
monitored with a quartz-crystal oscillator. The ML scale is
referred to the density of Ge (Si) atoms on the ideal Ge (Si)
surface. In the photoemission experiments, the incidence angle
of the photon beam was 45°. Photoelectrons were collected at
normal emission with a hemispherical analyzer (acceptance angle
< 2°). The total resolution (monochromator and analyzer) was 250
meV.

Details of the deconvolution procedure have been published
elsewere [7]. We just mention the parameters needed for the
deconvolutions : 0.63 (branching ratio), 0.6 eV (spin-orbit
splitting), 0.18 eV (Lorentzian width), 0.4 eV (Gaussian width).

3. Results and Discussion

3.1 Pb 5d Core-levels

We have observed the persistence of a 2x1 LEED pattern upon RT
deposition of about 1 ML Pb on both clean Ge(001) and Si(001)
surfaces. Thicker overlayers gave poor (2x1) patterns with a
high background. At LNT, a very clean pattern could be
observed in the case of Pb on Ge(001)c(2x4). At around a
monolayer a clear 2x1 pattern was still visible. For depositions
of 10 ML, we obtain a good epitaxial Pb (111) film with two
orthogonal orientations, as expected for a two-domain c(4x2)
initial reconstruction, with Pb(111) // Ge(001) and common <110>
directions. Pb/Si interface looks amorphous at LNT.

The lead atoms were always deposited on "cold" (RT or LNT)
surfaces. Unlike for the Ag/Ge(001) system [5], we found no
evidence of lead clustering from the variation of the
intensities of the Pb 5d peaks in the sub-monolayer range.
During the photoemission experiments, we systematically recorded
the Pb 5d, Ge 3d and Si 2p spectra just after deposition of Pb
and also after annealing the semiconductor surfaces. This
allowed us to follow the ordering process through the variation
of the core-level width and position. One direct evidence, as
already mentioned in previous studies on Pb/Si(001), Pb/Si(111)
and Pb/Ge(111) [1], is the absence of any pronounced chemical
shift. We observe only a broadening of the full width at half
maximum (FWHM) of the Pb 5d peaks, indicating that there are
several inequivalent sites for Pb atoms with slightly different
chemical environments and that the charge tranfer is weak. As
the electronegativities of germanium and lead are close, one
expects the bonding to be nearly covalent.

There is no unique solution for the deconvolution of such
spectra without complementary information . However, we can draw
several conclusions when comparing the spectra of the same lead
deposition recorded before and after annealing the samples. A
typical situation is shown in fig. 1a where we display the Pb 5d

590

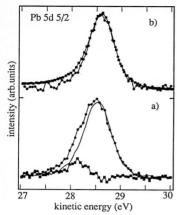

Fig. 1. Pb 5d 5/2 spectra of lead deposited on Ge(001) surface.
a) 0.5 ML deposited at RT before (circles) and after (full
line) the annealing. The squares show the difference curve. b)
Pb 5d 5/2 for 2.3 ML (squares) and 0.02 ML (circles). The
spectra are aligned to the same kinetic energy.

spectra for the Pb(0.5 ML)/Ge(001) interface. The annealing
reduces the FWHM and induces a 0.08 shift to higher kinetic
energies. The reduction of the FWHM brings evidence of an
ordering process in the way that only one peak contributes to
the Pb 5d spectrum. It seems that one site is preferred with a
slightly different binding, as deduced from the evolution of the
shift.

The shifts are presented on table I. The sign of the shift
with increasing Pb coverage and, for the same deposition, after
annealing, is always positive (in terms of photoelectron kinetic
energy). Both shift and narrowing of the FWHM decrease
progressively with increasing Pb overlayer and disappear
completely beyond 1ML.

It is known [8] that in the special case of lead, the core-
level shift is not always proportional to the positive charge of
the lead atoms. It is therefore difficult to deduce the sign of

Table I.

The positions of Pb 5d core-level peaks on the Pb/Ge(001)
interface refered to the peak position for the lowest coverage.

coverage (ML)	relative kin.energy (eV)	deposition conditions
0.02	0	LNT
0.1	0.25	RT
0.1	0.35	RT, annealed
0.5	0.3	RT
0.5	0.38	RT, annealed
0.7	0.3	LNT
1.6	0.38	LNT
2.3	0.38	RT

the charge transfer from the positions of lead 5d peaks alone. This will be discussed below in connection with Ge 3d core-level measurements.

The LNT and RT depositions are comparable from the standpoint of Pb 5d core-level spectroscopy. This is probably due to a high mobility of Pb atoms and it seems necessary to reach lower temperature in order to observe any differences in the Ge-Pb interaction.

One clear evidence of the development of interface metallicity is the appearance of an asymmetry in the Pb 5d peaks. They reach the Doniach-Sunjic form for coverages ~1.5 ML. This is shown in fig. 1b where we compare the 5d 5/2 peaks of lead for 2.3 ML(squares) and 0.02 ML (circles). It is interesting to note that lead core-level peaks have the lowest FWHM (0.6 eV), an entirely symmetrical shape, and the lowest kinetic energy for the lowest coverages (0.02 ML). In this range of coverage, the Pb atoms are probably so dispersed that the core-level peak approaches the atomic-like shape. On table I., the evolution of the Pb 5/2 peak position in energy can be understood in the same way : it is known that the binding energy diminishes by about 2 eV when passing from atomic to condensed matter [9].

In the sub-monolayer range, the FWHM of Pb 5d is always about 20% wider in the case of deposition on the Si(100) surface and reaches its bulk value (0.6 eV) only for thick coverages (beyond 4 ML). The differences in FWHM reflect a higher energy separation between Pb-Pb and Pb-substrate bonds, thus a larger interaction of Pb with the Si(001) surface than with the Ge(001) one.

3.2 Ge 3d and Si 2p Core-levels

The atomic structures of Ge(001) and Si(001) surfaces are now well established. The building block of the (2x1) reconstruction observed at RT is a surface dimer formed by dangling bond pairing of adjacent atoms in the top-most layer. There is a consensus that the majority of dimers on both Ge(001) and Si(001) surfaces are asymmetric.

In the core-level photoemission, the dimer atom contribution can be seen as a shoulder on the higher kinetic energy side of the spectra. In fig. 2a we show the Ge 3d spectrum for the clean Ge(001) surface taken with photon energy of 52 eV. In this energy range the escape depth of 5.4 Å for the photoelectrons has been determined [10]. The deconvolution results into bulk (B) and surface (S) components separated by 0.46 eV. This is consistent with previous results concerning Ge(001) [5].

After lead deposition at RT the dimer surface peak decreases progressively and is replaced by a new component (S') which is closer to the bulk one (0.36 eV). S' is the only surface peak persisting above 1.5 ML coverage as shown in fig. 2c. The deposition at LNT is, from the point of view of core-level spectroscopy, the same as at room temperature. At both temperatures, the new surface component appears at very low coverages (~0.02 ML). The position in energy of the new peak remains the same for all coverages, whereas its intensity increases roughly by a factor of two when compared with the original dimer peak. This implies an increase of the number of electrons involved in the new bonds.

The presence of the new surface peak S' in the Ge(001) spectra points out that the charge transfer estimated to be 0.2

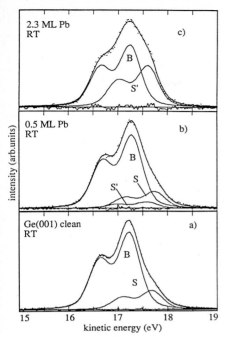

Fig. 2. a) Ge 3d spectra of clean Ge(001) (2x1) surface, b) after the deposition of 0.5 ML and c) 2.3 ML of lead. The deconvolution shows the bulk (B) and surface (S,S') contributions. Photon energy of 52 eV has been used.

electrons/Ge atom occurs from lead atoms to the germanium surface . For each deposition, the annealing enhances the new surface peak S' to the detriment of the peak S.

The charge transfer, the ordering observed through the Pb 5d core-levels and the increase of peak S', allow us to conclude that during the deposition of lead atoms the dimer bonds on the Ge(001) surface are progressively replaced by Ge-Pb bonds. The dimers disruption is completed for lead coverages of about 1.5 ML.

In the case of the clean Si(001) surface, we also observe two components in the spectrum. The surface peak is shifted by 0.51 eV to higher kinetic energies relatively to the bulk peak. The spectra given by the silicon surface after a thick lead layer (beyond 1.5 ML) deposition show no surface contribution. Therefore, the charge transfer must be stronger than for Ge(001) in order to saturate completely the dangling bonds, so that silicon atoms feel a similar chemical environment as in the bulk material. Again, the most straightforward description of the interface consists in progressively breaking the dimers and bonding the surface atoms to lead.

4. References

[1] G. Le Lay, K. Hricovini, and J.E. Bonnet: Appl.Surf.Sci. 41/42, 25 (1989).

[2] D.R.Heslinga, H.H.Weitering, D.P. van der Werf, T.M.Klapwijk, and T.Hibma: Phys.Rev.Lett., 64, 1589 (1990).

[3] J.E.Demuth, W.J.Thompson, N.J.DiNardo, and R.Imbihl: Phys.Rev.Lett. 56, 1408 (1986).

[4] G. Le Lay, M. Abraham, A.Kahn, K. Hricovini, and J.E. Bonnet: to appear in Physica Scripta.

[5] T.Miller, E.Rosenwinkler, and T.-C.Chiang: Phys.Rev. B 30, 570 (1984).

[6] G. Le Lay, K. Hricovini, and J.E. Bonnet: Phys.Rev. B 39, 3927 (1989).

[7] K. Hricovini, G. Le Lay, M. Abraham, and J.E. Bonnet: Phys.Rev. B 41, 1258 (1990).

[8] K.S.Kim, T.J.O'Leary, and N.Winograd: Anal.Chem. 45, 2213 (1973)

[9] H. Siegbahn and L.Karlson, Photoelectron Spectroscopy, in : Encyclopedia of Physics, Ed. S. Flüge, vol. XXXI, Springer-Verlag 1982.

[10] R.D.Schnell, F.J.Himpsel, A.Bogen, D.Rieger, and W.Steinmann, Phys.Rev. B 32, 8052 (1985).

594

Location of Ag in Si(111)-($\sqrt{3} \times \sqrt{3}$)R30°-Ag from X-Ray Standing Waves

E. Vlieg, E. Fontes, and J.R. Patel

AT&T Bell Laboratories, Murray Hill, NJ 07974, USA

The X-ray standing-wave technique has been used to determine the positions of the Ag atoms in the Si(111)-($\sqrt{3}\times\sqrt{3}$)R30°-Ag reconstruction. Using the (111) reflection, the Ag atoms are found to be located in one plane at a height of 3.44 ± 0.02 Å above the center of the last Si(111) bilayer. From the (11$\bar{1}$) reflection the most likely arrangement of the Ag atoms is found to be in the form of triangles, implying that there are three Ag atoms per unit cell.

1. Introduction

The ($\sqrt{3}\times\sqrt{3}$)R30° reconstruction induced by Ag on the Si(111) surface has become one of the more challenging structures in surface science. No consensus has been reached on this structure, despite studies involving almost all structural techniques used in surface science [1]. One of the reasons for this lack of agreement may be the fact that many techniques provide an overall picture of the structure, and the results are hard to interpret for someone unfamiliar with the technique or without access to the appropriate analysis programs.

Simple structural parameters are therefore very important. One such parameter is the position of the Ag atoms with respect to the Si substrate. The X-ray standing wave technique is able to determine distances from specific reflection planes with high accuracy [2]. We have therefore performed X-ray standing wave measurements on the Ag on Si(111) system, using the (111) and (11$\bar{1}$) Bragg reflections.

2. X-ray standing-wave experiment

In the X-ray standing wave experiment the fluorescence yield from the Ag atoms is measured in conjunction with the rocking curve of a Bragg reflection. In a normalized form, the general shape of a fluorescence yield curve is given by [2]:

$$Y(\theta) = 1 + R(\theta) + 2\sqrt{R(\theta)}\, F \cos[v(\theta) - 2\pi P], \tag{1}$$

where the reflectivity R and the phase factor v can be computed as a function of the reflection angle θ. The coherent fraction F and coherent position P are the two parameters determined from an X-ray standing-wave measurement. These two

Springer Series in Surface Sciences, Vol. 24   **The Structure of Surfaces III**
Editors: S.Y. Tong · M.A. Van Hove · K. Takayanagi · X.D. Xie
© Springer-Verlag Berlin, Heidelberg 1991

Table I   Average coherent position P and coherent fraction F for the (111) and (11$\bar{1}$) reflections.

reflection	P	F
(111)	1.096 ± 0.005	0.98 ± 0.02
(11$\bar{1}$)	0.871 ± 0.01	0.6 ± 0.1

parameters are related to the structure of the unit cell by:

$$F\, e^{-2\pi i P} = \frac{1}{N} \sum_j e^{-2\pi i H \cdot r_j}, \tag{2}$$

where H is the reciprocal lattice vector of the reflection of interest, r_j is the position of the j-th atom in the unit cell, and N is the number of atoms. Eq. (2) is used to calculate the coherent position and fraction for a structure model, which then can be compared with the experimentally determined values.

The experiment was performed at the AT&T beam line X15A at the National Synchrotron Light Source in Brookhaven. Ag was deposited from a Knudsen cell while the substrates were held at 500°C. The total Ag coverage was estimated to lie between 0.6 and 0.9 monolayer for the four different samples prepared. In the experiment the Ag L_α fluorescence (2.98 keV) was measured, while cycling the substrate many times through a Bragg reflection. The (111) reflection was measured on all four samples, the (11$\bar{1}$) reflection on two. By fitting the rocking curves and fluorescence yields, the coherent position and coherent fraction are obtained. The average values for both reflections are listed in Table I.

3. Discussion

The X-ray standing-wave data for the (111) reflection is a direct measure of the vertical distance d of the Ag atoms above the center of the last bilayer. The measured coherent fraction for this reflection is very close to 1, indicating that the Ag atoms are located in one plane. The small deviation from 1 is due to an enhanced thermal vibration amplitude of the Ag atoms. The average $P_{(111)}$ value of 1.096 ± 0.003 corresponds to a height of the Ag atoms above the center of the last bilayer of 3.44 ± 0.02 Å, modulus a (111)-lattice spacing (see fig. 2). The data from the (11$\bar{1}$) reflection is sensitive to both height and lateral position of the Ag atoms, and therefore we have to use eq. (2) to compare the experimental results with a model.

A large number of structure models has been proposed for the √3 structure. Since the X-ray standing wave results are only sensitive to the Ag atoms, we can ignore the reconstruction of the Si substrate, and focus on the Ag atoms. Then only a few models remain, the two most important ones are shown in fig. 1. The first model has the Ag atoms arranged in a honeycomb, and contains two Ag atoms per unit cell. This Ag-honeycomb model was proposed a long time ago [3], and has recently been inferred from STM [4] and, in a two-domain variant, from photo-electron

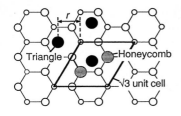

Fig. 1 Top view of the Si(111) surface, showing two proposed arrangements of the Ag atoms in the √3 unit cell. r is the distance from the origin in the Triangle model, expressed as a fraction of the √3 lattice parameter (6.65 Å).

diffraction [1]. The height of the Ag atoms above the center of the last bilayer has been estimated to be 2.5 Å [1], giving $P_{(111)} \approx 0.8$. This is in disagreement with the experimental value of 1.096. The coherent fraction of the Ag-honeycomb model is 1, which agrees with our data. Forcing the honeycomb to be at the correct height ($d = 0.096$), we calculate for the $(11\bar{1})$ reflection $P_{(11\bar{1})} = 0.86$ and $F_{(11\bar{1})} = 1$. For this reflection the coherent position agrees reasonable with the data, but the coherent fraction does not. We conclude that the Ag honeycomb model does not agree with our data.

The second model has three Ag atoms arranged in a triangle which is characterized by the vertical distance d and by the radial distance r of the Ag atoms from the origin (Triangle in fig. 1). By allowing r to be both positive and negative, two orientations are obtained, which differ by a rotation over 60°. This type of model was initially proposed with $|r| \approx 0.24$, corresponding to a Ag trimer at the origin [5]. X-ray diffraction experiments [6] found $|r| \approx 0.44$; this type of model has been called the honeycomb-chained trimer (HCT) model [6a]. The Ag atoms in the Triangle model are located in one plane, in agreement with the coherent fraction from the (111) reflection.

For the Ag-trimer model ($|r| \approx 0.24$) no height of the Ag has been proposed in the literature. For the $(11\bar{1})$ reflection we calculate for this model $P_{(11\bar{1})} \approx 0.86$ and $F_{(11\bar{1})} \approx 0.6$, which is in agreement with our data. For the HCT model, values for the height were given in studies using X-ray diffraction [6a] and RHEED [7], giving a $P_{(111)}$ of 1.05 and 1.07 respectively. This is in good agreement with our data. The HCT model gives for r positive $P_{(11\bar{1})} \approx 0.89$ and $F_{(11\bar{1})} \approx 0.6$, which is also in reasonable agreement with the X-ray standing-wave data.

Thus two variants of the Triangle model give a fair agreement with the $(11\bar{1})$ data, implying that we cannot distinguish clearly between these models on the basis of our data alone. However, the Triangle arrangement with $|r| \approx 0.24$ was ruled out by X-ray diffraction [6b] and ion scattering [8], and therefore the only remaining candidate is the Triangle arrangement with parameters given in refs. [6,7]. We should note that in the literature other Ag arrangements have also been proposed [8,9], but none of these agrees with our data.

The Triangle model as favoured here is shown in fig. 2. The registry of the Ag atoms is in agreement with one X-ray diffraction study [6a], but not with the other [6b,10]. The X-ray standing-wave technique cannot determine the positions of the Si atoms, but X-ray diffraction [6b] found that in addition to the Ag triangle there is a Si trimer, which is also shown in fig 2. Various techniques have claimed that a Si

597

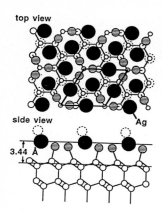

Fig. 2 Schematic showing the position of the Ag atoms (black circles) with respect to the Si(111) substrate in both a top and side view. The hatched atoms are reconstructed Si atoms forming a trimer as suggested by other techniques, see text. The dashed circles indicate the positions of a possible Si layer that would form a honeycomb structure on top of the surface.

honeycomb forms the top-most layer of the surface [6b,8,11], but there is also a large amount of data that indicates that this is not true [1,12]. We have therefore indicated this possible Si-honeycomb structure by dashed circles in fig. 2.

4. Conclusions

By measuring the Ag-fluorescence yield while rocking through the (111) and (11$\bar{1}$) reflections, the position of the Ag atoms in the $\sqrt{3}$ reconstruction on Si(111) has been determined using the X-ray standing-wave technique. The Ag atoms are located in one plane, at a height of 3.44 ± 0.02 Å above the center of the last Si bilayer. There are three Ag atoms per unit cell that form a triangle. The error bars do not allow an exact determination of the size of the triangle, but the data is consistent with the sizes determined from X-ray diffraction [6] and RHEED [7]. The model that emerges for the $\sqrt{3}$ reconstruction is supported by X-ray standing waves, X-ray diffraction [6], ion scattering [8,13], RHEED [7] and has a registry that is in agreement with an STM determination [4]. The main technique that disagrees with the current model is photo-electron diffraction [1,9c].

NSLS is supported by the United States Department of Energy under contract DE-AC02-76CH00016.

References

[1]   For a recent list of references, see: E.L. Bullock, G.S. Herman, M. Yamada, D.J. Friedman and C.S. Fadley, Phys. Rev. **B41**, 1703 (1990).
[2]   N. Hertel, G. Materlik and J. Zegenhagen, Z. Phys. **B58**, 199 (1985).
[3]   G. Le Lay, M. Manneville and R. Kern, Surf. Sci. **72**, 406 (1978).
[4]   R.J. Wilson and S. Chiang, Phys. Rev. Lett. **58**, 369 (1987); id. **59**, 1555 (1987).
[5]   F. Wehking, H. Beckermann and R. Niedermayer, Surf. Sci. **71**, 364 (1978).

[6] (a) T. Takahashi, S. Nakatani, N. Okamoto, T. Ishikawa and S. Kikuta, Jpn. J. Appl. Phys. **27**, L753 (1988); (b) E. Vlieg, A.W. Denier van der Gon, J.F. van der Veen, J.E. Macdonald and C. Norris, Surf. Sci. **209**, 100 (1989).

[7] A. Ichimiya, S. Kohmoto, T. Fujii and Y. Horio, Appl. Surf. Sci. **41/42**, 82 (1989).

[8] M. Copel and R.M. Tromp, Phys. Rev. **B39**, 12688 (1989).

[9] (a) T.L. Porter, C.S. Chang and I.S.T. Tsong, Phys. Rev. Lett. **60**, 1739 (1988); (b) W.C. Fan, A. Ignatiev, H. Huang and S.Y. Tong, Phys. Rev. Lett. **62**, 403 (1989); (c) S. Kono, T. Abukawa, N. Nakamura and K. Anno, Jpn. J. Appl. Phys. **28**, L1278 (1989).

[10] 'Model II' from ref. [6b] gave also reasonable fits to the X-ray diffraction data, and has the registry as favoured here.

[11] E.J. van Loenen, J.E. Demuth, R.M. Tromp and R.J. Hamers, Phys. Rev. Lett. **58**, 373 (1987).

[12] (a) M. Aono, R. Souda, C. Oshima and Y. Ishizawa, Surf. Sci. **168**, 713 (1986); (b) R.S. Williams, R.S. Daley, J.H. Huang and R.M. Charatan, Appl. Surf. Sci. **41/42**, 70 (1989); (c) K. Sumitomo, K. Tanaka, Y. Izawa, I. Katayama, F. Shoji, K. Oura and T. Hanawa, Appl Surf. Sci. **41/42**, 112 (1989).

[13] K. Oura, M. Watamori, F. Shoji and T. Hanawa, Phys. Rev. **B38**, 10146 (1988); id Surf. Sci. **226** (1990) 77.

Study of ($\sqrt{3} \times \sqrt{3}$)R30° Ag on Si(111) by Photoelectron Diffraction

G.S. Herman[1], A.P. Kaduwela[1], D.J. Friedman[1], M. Yamada[1],
E.L. Bullock[1], C.S. Fadley[1], Th. Lindner[2], D. Ricken[2], A.W. Robinson[2],
and A.M. Bradshaw[2]

[1]Department of Chemistry, University of Hawaii, Honolulu, HI 96822, USA
[2]Fritz-Haber-Institut der Max-Planck-Gesellschaft,
Faradayweg 4-6, W-1000 Berlin 33, Fed. Rep. of Germany

Abstract. Photoelectron diffraction (PD) data for Ag 3d emission from ($\sqrt{3}\times\sqrt{3}$)R30° Ag on Si(111) have been obtained in polar-, azimuthal-, and scanned energy- modes. Single and multiple scattering cluster calculations have been used to analyse these data. It is found that the Ag atoms in this structure cannot be located at greater than 0.5Å below the surface, nor can they be present in a diffuse interface. An R-factor analysis has been used to compare experiment and theory for a large number of proposed or possible structures. A two-domain missing-top-layer model with Ag at ~0.2Å below the surface is found to give the best description for these results.

1. Introduction

The ($\sqrt{3}\times\sqrt{3}$)R30° structure of Ag on Si(111) has been studied by practically every technique in surface science. However, controversy still remains concerning the atomic structure, and there is not even agreement on the Ag coverage required to produce this surface [1,2]. In several cases, separate studies using the same experimental technique have given conflicting results for this surface [3,4]. A recent impact collision ion scattering spectroscopy study by Williams et al. [5] has concluded that these results can be described by only two structural models. One is the two-domain missing-top-layer (MTL) Ag honeycomb structure proposed earlier by Bullock et al. [6], while the other is a Ag honeycomb-chained-trimer model (HCT) [7(a)]. In analyzing the photoelectron diffraction data to be discussed here, we have studied every proposed structure in the literature as well as many other possible models using single scattering cluster calculations and R factor analysis to compare theory to azimuthal x-ray photoelectron diffraction data. In addition, further analysis was performed using multiple scattering calculations on the two-domain MTL and HCT models as compared to both azimuthal and recently obtained scanned energy data.

2. Experiment

The azimuthal and polar dependence of Ag $3d_{5/2}$ core-level intensities emitted from ($\sqrt{3}\times\sqrt{3}$)R30° Ag on Si(111) at a kinetic energy of 1126 eV was studied using monochromatized Al Kα radiation as an excitation source. The experimental system

Springer Series in Surface Sciences, Vol. 24   **The Structure of Surfaces III**
Editors: S.Y. Tong · M.A. Van Hove · K. Takayanagi · X.D. Xie
© Springer-Verlag Berlin, Heidelberg 1991

used was a Hewlett-Packard 5950A spectrometer combining XPS, XPD, and LEED capabilities that is described elsewhere [8]. The sample was a mirror-polished Si wafer (B doped, 0.25 Ω-cm) oriented to within ±0.4° of (111). This sample was chemically cleaned in a multistep process. Such surfaces exhibited excellent (7x7) LEED patterns after ion bombardment and annealing in vacuum to about 1150°C. The (√3x√3)R30° Ag structure was formed by depositing about 1.3 ML of Ag on a surface heated to 550°C. Very sharp (√3x√3) LEED patterns were observed.

The energy dependence of analogous Ag 3d core-level intensities from Ag/Si(111) surface was studied at the BESSY synchrotron radiation facility in West Berlin. The radiation was obtained from the Fritz-Haber-Institute's HE-TGM, and permitted scanning the kinetic energy over 95-595 eV. The photoemission spectra were recorded in a Vacuum Generators ADES 400 spectrometer equiped with LEED for characterisation of the sample surface. The samples were prepared in the same method as mentioned above.

3. Results and Discussion

Polar-scan data of the Ag $3d_{5/2}$ intensity from the (√3x√3)R30° Ag surface are shown in Fig. 1 for several high-symmetry azimuths, with $\phi = 0°$ being defined as the [11$\bar{2}$] direction. The polar angle is defined with respect to the surface, and instrumental effects cause a decrease of the intensity as the polar angle goes to zero. The smooth featureless data seen here indicate that there are no diffraction effects due

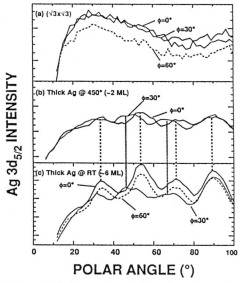

FIG 1. Polar scans of Ag $3d_{5/2}$ intensity from (a) the (√3x√3)R30° Ag structure formed after an anneal to 550° C, (b) a Ag overlayer of approximately 2-ML average thickness at 450° C, and (c) a thick Ag overlayer of approximately 6-ML thickness at ambient temperature.

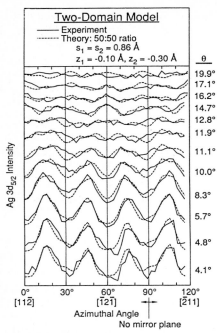

Two-Domain Model
— Experiment
······ Theory: 50:50 ratio
$s_1 = s_2 = 0.86$ Å
$z_1 = -0.10$ Å, $z_2 = -0.30$ Å

θ

19.9°
17.1°
16.2°
14.7°
12.8°
11.9°
11.1°
10.0°
8.3°
5.7°
4.8°
4.1°

Ag $3d_{5/2}$ Intensity

0°   30°   60°   90°   120°
[11$\bar{2}$]   [$\bar{1}$2$\bar{1}$]   [$\bar{2}$11]

Azimuthal Angle
No mirror plane

FIG 2. Experimental and calculated azimuthal scans of Ag $3d_{5/2}$ intensity at various polar angles with respect to the surface from 4° to 20°. The calculated curves include correct d-to-p+f emission and are for the fully optimized two-domain model shown in Fig. 3(b) with 50% domain 1, 50% domain 2, $z_1=-0.1$ Å, $z_2=-0.3$ Å, and $s_1=s_2=0.86$ Å. These curves yielded an R factor of 0.138.

to forward scattering from atoms overlying the Ag. Previous experimental and theoretical studies of overlayer growth by x-ray photoelectron diffraction, together with the present data, indicate that the Ag atoms for this system are not buried beneath other Ag atoms in microclusters of ≥ 2 ML thickness nor present to any degree as species deeply buried under the Si surface [9].

The azimuthal scan data in Fig. 2 permit a more quantitative indicator of the maximum depth at which Ag can be below the surface-layer atoms. Forward scattering peaks are present in this data only for low take off angles from 4° up to about 10°, where they appear as the quartet of peaks in the bottom experimental curves of Fig. 2. These forward scattering peaks become more complex as the polar angle above the surface gets larger, and the overall diffraction anisotropy $\Delta I/I_{max}$ falls to only a few % by θ = 20°. Both of these facts argue that the relevant Ag-nearest neighbor angle is not larger than 10° relative to the surface. For typical expected Ag-Si and Ag-Ag bond distances, this yields a maximum Ag depth of z=-0.5Å .

The analysis of our azimuthal data consisted of calculating diffraction curves using both single scattering and multiple scattering cluster formalisms for different

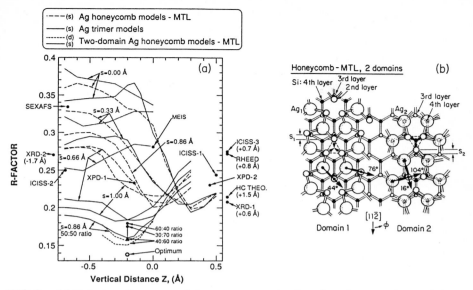

FIG 3. (a) R factors for some of the structures tested against the experimental data of Fig. 2. Included are z scans for various choices of the contraction parameter s in the MTL Ag honeycomb (Ref. 11), the two-domain MTL Ag honeycomb (Ref. 6), and the embedded Ag trimer model suggested by STM [Ref. 3(a)], as well as single points for structures based upon several different techniques as indicated: ICISS-1 [Ref. 4(a)], ICISS-2 [Ref. 4(b)], ICISS-3 [Ref. 4(c)], MEIS (Ref. 12), RHEED (Ref. 13), SEXAFS (Ref. 14), XPD-1 (Ref. 11), XPD-2 (Ref. 2), XRD-1 [Ref. 7(a)], XRD-2 [Ref. 7(b)], and a theoretically determined structure for a full-double-layer Ag honeycomb (HC-theo.) (Ref. 15). (b) The two-domain missing-top-layer Ag honeycomb model proposed for the (√3x√3)R30° Ag structure. In the bottom half of the figure are shown the two sets of nearest-neighbor-Si forward scattering peaks that produce the four-peak structure seen at low θ values in Fig. 2.

structures at different θ values and then comparing the experimental and theoretical diffraction curves by means of R factors. Figure 3(a) summarizes our search over many structures, with each curve representing a variation of Ag height relative to the first Si layer. Prior work using only single scattering cluster calculations gave a lowest R factor for a two-domain MTL Ag honeycomb model with a z distance of about -0.20Å (i.e 0.20Å below the Si surface atoms) and a compression factor s for the Si trimers of 0.86Å [6]. The two types of domains are illustrated in Fig. 3(b). In domain 1, the Ag atoms of the honeycomb do not have a 4th-layer Si atom beneath them, but in domain 2 they do. Single domains cannot correctly predict the strong four peak structure at low θ values, since only two strong features at φ≈44° and 76° are seen due to scattering from the Si nearest neighbors in each compressed trimer for domain 1, and only two strong features are observed at φ≈16° and 104° for domain 2. These are shown in Fig. 3(b) with the arrows indicating the azimuthal

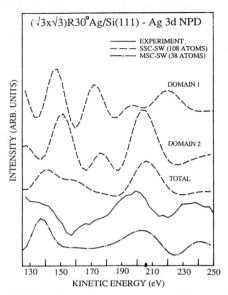

FIG 4. Experimental and calculated scans of Ag $3d_{5/2}$ scanned-energy photoelectron diffraction data for normal emission. Calculations were performed on the optimized two-domain MTL model.

directions through the nearest-neighbor scatterer. The final optimized curves for this model are given in Fig. 2, where generally excellent agreement is seen. Analogous comparisons of experiment and theory for any of the other structures previously proposed for this surface are significantly worse visually, and they also have much higher R factors as illustrated in Fig. 3(a). Furthermore, <u>multiple scattering</u> calculations for the two-domain MTL model are found to best describe the experimental data for z, s and mixing values of the two-domains which are very similar to those determined from the single-scattering analysis [10].

Finally, we present in Fig. 4 very recently obtained scanned-energy photoelectron diffraction data. The analysis of this data is complicated by Auger interferences at several points in the spectra, but there is a region from 125 to 250 eV which is free of such complications, and it is this region which we will analyze. These data are for normal emission from the surface. In the upper portion of Fig. 4 are single scattering cluster calculations performed on our optimized two-domain MTL model. The general shape of the experimental and theoretical curves for a 50:50 mix of the two domains match well with one another. The addition of two domains instead of using just one or the other greatly improves agreement. Further improvement in agreement is seen in the multiple scattering calculations for this model.

Thus, among all of the models proposed to date both scanned-angle and scanned-energy photoelectron diffraction data are most consistent with the two-domain MTL model, although we will continue to test these data against other models as they are proposed.

Acknowledgments

This work has been supported by the National Science Foundation under Grant CHE83-20200, the office of Naval Research under contract N00014-87-K-0512 and Grant N00014-90-J-1457, and the Fritz-Haber Institut. Multiple-scattering calculations were performed at the San Diego Supercomputer Center.

References

1. T. Doust, F.L. Metcalfe, and J.A. Venables, Ultramicroscopy **31**, 116 (1989).
2. S. Kono, T. Abukawa, N. Nakamura, and K. Anno, Jpn. J. Appl. Phys. **28**, 302 (1989)
3. (a). E.J. van Loenen, J.E. Demuth, R.M. Thomp, and R.J. Hamers, Phys. Rev. Lett. **58**, 373 (1987); (b) R.J. Wilson and S. Chiang, J. Vac. Sci. Technol. A**6**, 800 (1988).
4. (a) M. Aono, R. Sonda, C. Oshima, and Y. Ishizawa, Surf. Sci. **168**, 713 (1986); (b) T.L. Porter, C.S. Chang, and I.S.T. Tsong, Phys. Rev. Lett. **60**, 1739 (1988); (c) R.S. Williams, R.S. Daley, J.H. Huang, and R.M. Charatan, Appl. Surf. Sci. **41/42**, 70 (1989).
5. R.S. Williams, R.S. Daley, J.H. Huang, and R.M. Charatan, to appear in Phys. Rev. B.
6. E.L. Bullock, G.S. Herman, M.Yamada, D.J. Friedman, and C.S. Fadley, Phys. Rev. B**41**, 1703 (1990).
7. (a) T. Takahashi, S. Wakatani, N. Okamoto, T. Ishikawa, and S. Kikuta, Jpn. Appl. Phys. **27**, L753 (1988); (b) E. Vlieg, A.W. Denier van der Gon, J.F. van der Veen, J.E. Macdonald and C. Norris, Surf. Sci. **209**, 100 (1989).
8. C. S. Fadley, Prog. Surf. Sci. **16**, 275 (1984).
9. (a) R.A. Armstrong and W.F. Egelhoff, Surf. Sci. **154**, L225 (1985); [b] C.S. Fadley, Phys. Scr. T**17**, 39 (1987).
10. A.P. Kaduwela, D.J. Friedman, Y.J. Kim, T.T. Tran, G.S. Herman, C.S.Fadley, J.J. Rehr, J. Osterwalder, H. Aebischer, and A. Stuck, to appear in these proceedings.
11. S. Kono, K. Higashiyama, and T. Sagawa, Surf. Sci. **165**, 21 (1986).
12. M. Copel and R. Tromp, Phys. Rev. B**39**, 12688 (1989).
13. A. Ichimiya, S. Kohmoto, T. Fujii, and Y. Horio, Appl. Surf. Sci. **41/42**, 82 (1989).
14. J. Stohr, R. Jaeger, G. Rossi, T. Kendelewicz, and I. Lindau, Surf. Sci. **134**, 831 (1983).
15. C.T. Chan and K.M. Ho, Surf. Sci. **217**, 403 (1989).

Electronic and Atomic Structure of the Cu/Si(111)-Quasi-5 × 5 Overlayer

T.N. Rhodin and D.D. Chambliss*

School of Applied and Engineering Physics, Cornell University,
Ithaca, NY 14853, USA
*Present address: IBM Almaden Research Center, San Jose, CA, USA

Abstract. We have performed detailed angle-resolved UV photoemission (ARUPS) measurements and *ab initio* band-structure calculations to investigate the atomic structure of the quasi-5x5 layer and the unique bonding behavior it embodies. In electronic structure calculations using the pseudofunction method of Kasowski *et al.*, the CuSi model agrees much better than the $CuSi_2$ model with ARUPS. The formation of Si(p)-Cu(d) bonding hybrid orbitals appears to be important in making the CuSi structure stable, but the Cu(4s) orbitals also play a significant role in hybridizing with Si(3p) states.

1. Introduction and Results of Angle-Resolved Photoemission

The quasi Cu/Si(111)(5x5) structure has been well studied using LEED [1,2], STM [3,4], helium diffraction [5], reflection electron microscopoy [6-9]. Two atomic models seem most likely for the dominant (1x1) regions of the quasi-5x5 overlayer as indicated by Auger electron diffraction, and electron loss spectroscopy [10], a CuSi substitutional structure or a $CuSi_2$ interstitial-like structure [11]. With band-structure calculations it is possible to determine which overlayer structure best explains the photoemission measurements.

Angle-resolved photoemission spectra were measured for three samples: a clean Si(111)-(7x7) surface; a "half-layer" sample with 0.6 ML Cu (nom.), annealed to 600°C, which displayed both (7x7) and quasi-5x5 LEED spots; and a "full-layer" sample with 1.1 ML Cu (nom.), annealed, whose LEED pattern was pure quasi-5x5.

The Cu(3d) manifold can be well described as the superposition of three peaks. These peaks, much broader than the instrumental resolution of 0.2 eV, are fit well by Gaussian peaks with full width at half maximum of about 0.7 eV. The peaks at normal emission are determined using photon polarization to be two states of Λ_3 symmetry at –3.1 eV and about –3.8 eV and one of Λ_1 symmetry at –2.5 eV. The absence of true periodicity in the layer, demonstrated by scanning tunneling microscopy and inhomogeneous broadening in photoemission, means that the photoemission peaks are due to a number of related states. The dispersion of peak maxima with emission angle were measured as a function of $k_{||}$. Most of the features are identified as bulk states in the underlying Si [12,13]. In addition, a dispersive feature with minimum energy –1.2 eV is seen at emission angles around 40°, which corresponds to the boundary of the surface Brillouin zone. This state is a surface state 0.4 eV above the projected bulk-band maximum at this $k_{||}$, which originates in a bulk state of L_3 symmetry.[14] The surface state appears to be split off from the bulk band by the different bonding environment for Si. The photoemission results indicate that the strongest interactions of the Cu atoms are with in-plane

Springer Series in Surface Sciences, Vol. 24 **The Structure of Surfaces III**
Editors: S.Y. Tong · M.A. Van Hove · K. Takayanagi · X.D. Xie
© Springer-Verlag Berlin, Heidelberg 1991

neighbors, for this gives rise to the observed ordering of the Cu(3d) state energies ($d_{x^2-y^2}$ lowest). More specific structural conclusions utilize a detailed comparison of peak energies with calculated band structure for different models.

2. Calculations: Methods and Results

The electronic structure for several model atomic structures was computed within the local density approximation (LDA) using the pseudofunction method developed by Kasowski et al.[15,16]. The atomic models used for the calculations had (1x1) periodicity and were used to reproduce the essential features of local bonding. For comparisons which yield energy values relative to E_F the *double-sided slabs* shown in Fig. 1 are used; these consist of two Cu-Si mixed layers with one or two Si bilayers between, to represent bonding to the substrate. Our models representing the $CuSi_2$ model for the quasi-5x5 layer bonded to the substrate did not use a planar geometry. Instead the models used the structure of a Si bilayer with Cu added at different levels. Comparison of experiment with calculation focuses on the relative energies of Cu(3d) levels. Good agreement between the photoemission data and the calculated results is shown in Fig. 2 for dispersion of the parallel wave vector of the Cu(3d) states at 22 eV corresponding to the contracted CuSi layer.

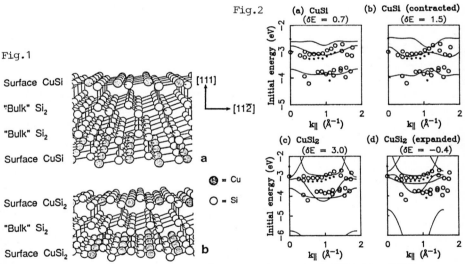

Fig.1

Surface CuSi

"Bulk" Si_2

"Bulk" Si_2

Surface CuSi

Surface $CuSi_2$

"Bulk" Si_2

Surface $CuSi_2$

Fig.1. Double-sided models for electronic structure calculations of Cu/Si(111)-quasi-5x5. (a) Cu_2Si_6 unit cell model for CuSi overlayers. (b) Cu_2Si_6 unit cell model for $CuSi_2$ overlayers. Models can be regarded as composed of stacks of bi- and trilayers, as labeled. About 30 unit cells are viewed approximately along the $[1\bar{1}0]$ direction parallel to the surface. The actual model extends to $\pm\infty$ in x and y.

Fig. 2. Comparison of ARUPS results (circles and dots) for even-parity states with calculated bands (lines) for isolated-layer models, along the line $\bar{\Gamma}$-$\bar{M}$-$\bar{\Gamma}$ in the surface Brillouin zone. Open circles are strong peaks and dots are shoulders (measured). (a) CuSi commensurate layer. (b) CuSi layer, contracted 20%. (c) $CuSi_2$ layer. (d) $CuSi_2$ layer, expanded 20%. Computed bands (solid lines) are shifted in energy by δE values indicated.

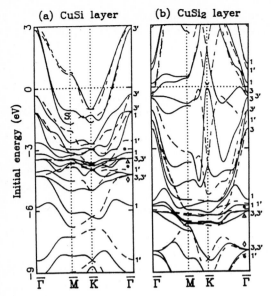

(a) CuSi layer   **(b) CuSi₂ layer**

Fig. 3. Energy-band dispersion for models of the quasi-5x5 structure with a Si substrate. The surface structures are (a) CuSi and (b) CuSi₂. Cu(3d) states are marked as follows: ◊: $d_{x^2-y^2}/d_{xy}$ character; Δ: d_{xz}/d_{yz} character; •: states at $\overline{\Gamma}$ with d_{z^2} character, split by substrate interaction. "S" marks surface states at $\overline{M}$ of CuSi layer. Labels at right of each graph denote symmetry of states at $\overline{\Gamma}$. The notation is that used for L in the bulk Brillouin zone. (Point group D_{3d}: $1=L_1=A_g$; $1'=L_1'=A_u$; $3=L_3=E_g$; $3'=L_3'=E_u$.) The Λ_1 representation of the (111) axis symmetry contains both L_1 and L_1'; Λ_3 state, both L_3 and L_3'. Solid and broken lines denote even and odd parity, respectively. Along $\overline{\Gamma}-\overline{M}$, parity is with respect to $(1\overline{1}0)$ mirror plane of crystal. Along $\overline{M}-\overline{K}$ and $\overline{K}-\overline{\Gamma}$, parity is with respect to two different C_2 axes of the model which are not characteristic of the true surface.

Calculated energy bands for isolated CuSi and CuSi₂ layers (not shown) indicate the effect of bonding geometry on Cu(3d) energy levels. For the CuSi layer, the separation between the two d-state energies is about 0.8 eV, close to the experimental value of 0.7 eV. For a CuSi₂ layer, the d-state behavior is qualitatively similar, but the d-state separation is much greater. Peak energies for different isolated-layer models are also compared with experiment. An isolated CuSi layer produces relatively good agreement with ARUPS results, while a CuSi₂ layer does not.

Calculations for double-sided slabs confirm the CuSi planar structure and clearly show that discrepancies in the isolated-layer results can be attributed to the effect of overlayer-substrate bonding. The effect of the Si substrate on the Cu/Si overlayer is calculated using two-sided models with Cu₂Si₆ unit cells and with lattice constant equal to that of Si(111). The main conclusion drawn from the isolated-layer calculations is confirmed with the two-sided models: the Λ_3-state separation at gamma-bar is correct for the CuSi model but too large by a factor of 2 or 3 for a CuSi₂ model. The dispersion curves for the two-sided models (Fig. 3) do, however, reveal changes from the isolated-layer calculations. The Cu(3d) bands for the two-sided models have more dispersion than those for the isolated-layers. This difference between calculation and experiment is attributed to the lack of periodicity in the real system.

3. Discussion and Conclusions

The experimental evidence suggests that the quasi-5x5 structure consists largely of a planar overlayer that is locally well ordered. This is indicated by the dominance of two features with distinct angular momentum in the Cu(3d) photoemission results, as well as by the amplitude of Auger electron anisotropies and the (1x1) corrugation of STM images with which quasiperiodic features align. These photoemission results indicate that the dominant part of the overlayer has CuSi composition.

The discrepancies between the $CuSi_2$ model and the photoemission results are well outside our error bounds. Thus this model does not represent the majority of the Cu/Si(111)-quasi-5x5 layer. For a CuSi overlayer agreement of theory with photoemission is much more convincing. The discrepancies are accounted for by variability in the overlayer-substrate bonding configuration in the complex long-range structure.

The CuSi-layer model also explains the observed nonreactivity of the quasi-5x5 layer. Intuition suggests that the Si atoms of the $CuSi_2$ layer that are not bonded to the substrate would have an outward dangling bond, and this is confirmed by the calculations. In the CuSi model, on the other hand, the states near E_F are predominantly of p_x and p_y character. Thus the $CuSi_2$ layer should be fairly reactive, while the CuSi layer will be mostly inert as is observed.

(Measurements were made at Tantalus; research supported by Cornell Materials Science Center and assistance from R. Kasowski, W. Hoehler and J. Demuth.)

References

1. E.Daugy, P. Mathiez, F. Salvan, J.M. Layet, Surf. Sci. **152/153**, 1239 (1985).
2. E. Daugy, P. Mathiez, F. Salvan and J.M. Layet, Surf. Sci. **154**, 267(1985).
3. R.J. Wilson, S. Chiang and F. Salvan, Phys. Rev. B **38**,12696 (1988).
4. J.E. Demuth, U. Koehler, R.J. Hamers, P. Kaplan, PRL. **62**, 641 (1989).
5. R.B. Doak and D. B. Nguyen, Phys. Rev. B **40**,1495 (1989).
6. T. Ishitsuka, K. Takayanagi, Y. Tanishiro and K. Yagi, Proc. Int. Cong. Elect. Microscopy XI, Kyoto, p. 1347 (1986).
7. See, e.g., T. Kinoshita, S. Kono and T. Sagawa, Phys. Rev. B **34**, 3011 (1986).
8. K.Takayanagi, Y.Tanishiro, M. Takahashi, S. Takahashi, Surf. Sci. **164**,367 (1985).
9. D.D. Chambliss, Ph.D. thesis, Cornell University, 1989.
10. S.A. Chambers, S.B. Anderson and J. H. Weaver, Phys. Rev. B **32**, 581 (1985).
11. S.A. Chambers, M.del Giudice, M.W. Ruckman, S.B. Anderson, J.H. Weaver, and G.J. Lapeyre, J. Vac. Sci. Technol. A **4**, 1595 (1986).
12. S.A. Chambers and J.H. Weaver, J. Vac. Sci. Technol. A **3**, 1929 (1985).
13. For reference on ARUPS from Si(111)-(7x7) see R. I. G. Uhrberg, G. V. Hansson, U.O. Karlsson, J.M. Nicholls, P.E.S. Persson, S. A. Flodström, R. Engelhardt and E.-E. Koch, Phys. Rev. B **31**:3795 (1985).
14. The experimental value of $E(L_3)$ is $E_{VBM} - 1.6\pm0.1$ eV [Ref. 15]. The work function increase of 0.6 eV puts $E_{VBM} \approx E_F$.
15. R.V.Kasowski, T.N.Rhodin, M.-H.Tsai, D.D.Chambliss, PR B**34**, 2656 (1986).
16. M.-H. Tsai, R.V. Kasowski and T.N. Rhodin, Surface Science **179**, 143 (1987); M.-H. Tsai, J.D. Dow and R.V. Kasowski, Phys. Rev. B **38**, 2176 (1988).

Structures Formed by Co-deposition of Two Metals on Si(111)7×7 Surfaces Studied by Reflection Electron Microscopy

I. Homma, Y. Tanishiro, and K. Yagi

Physics Department, Tokyo Institute of Technology,
Oh-okayama, Meguro, Tokyo 152, Japan

Abstract. Co-deposition of Au and Cu on Si(111) produced a new adsorbate structure of $\sqrt{3}\times\sqrt{3}$ composed of two metals, while that of Ag and Cu produced two spatially separated adsorbate structures which are formed by deposition of individual metals. A $\sqrt{21}\times\sqrt{21}$ structure was sometimes formed in the latter case.

1. Introduction

A wide variety of adsorbate structures formed by deposition of metals on Si(111) surfaces has been studied [1]. However, few works have been reported on the structures formed by co-deposition of metals. In such cases, two extreme cases are conceivable. One is to form an ordered two dimensional alloy adsorbate and the other is to form two spatially separated adsorbates which are formed by deposition of individual metals.

In the present study structures formed by co-deposition of Cu-Au and Cu-Ag were studied by ultra-high vacuum reflection electron microscopy (UHV-REM), which can characterize structures with spatial resolution [2]. Figure 1 reproduces superlattice structures formed by deposition of Au [1], Cu [3] and Ag [1] on Si(111)7x7 surfaces. These metals form different adsorbate structures at their initial stages of deposition: Au; 5x2, Cu; incommensurate phase (IC), and Ag; $\sqrt{3}\times\sqrt{3}$ structure. Au and Cu mix well in the bulk, while Cu and Ag do not. These are the reasons for choosing the above systems.

In the Au-Cu system a new alloyed adsorbate was noted, while in the Cu-Ag system two adsorbates formed by deposition of individual metals were observed. Details of the former case were reported separately [4]. Here, the results of the Ag-Cu case are described together with some results of the Cu-Au case for comparison.

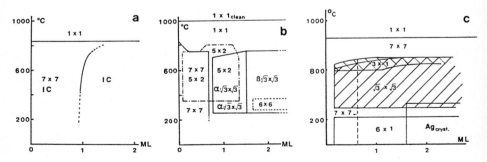

Figure 1. Superlattices observed on Si(111) surfaces induced by (a) Au, (b) Cu and (c) Ag depositions.

Springer Series in Surface Sciences, Vol. 24 **The Structure of Surfaces III**
Editors: S.Y. Tong · M.A. Van Hove · K. Takayanagi · X.D. Xie
© Springer-Verlag Berlin, Heidelberg 1991

2. Experimental

Details of the experimental procedures are in ref. 4. Here, they are described shortly. Two metals were deposited in-situ in a UHV-electron microscope on cleaned Si(111)7x7 surfaces simultaneously or successively. The amount of the deposit of each metal was monitored by a quartz oscillator. In order to enhance surface diffusion to mix two metal atoms, the substrates were kept at high temperature, say 700°C. However, this may enhance desorption of metals from the surfaces. Therefore, REM observations were carried out at lower temperatures.

3. Results

3.1. Cu-Au system

Figure 2 reproduces REM images of the same area of a Si(111) surface taken (a) before and (b) after a simultaneous deposition of Cu and Au (2/3 monolayer (ML) each) at 700°C and a reflection high energy electron diffraction (RHEED) pattern (c) taken after the deposition. The imaging beam was along the <112> direction. (b) and (c) were taken at 550°C. The pattern (c) indicates that the surface structure transformed from the clean 7x7 structure to a mixture of the 7x7 structure (small arrow) and a √3x√3 structure (large arrow) (hereafter denoted by √3 structure). It should be noted that the √3 structure is not formed at initial stages of Au or Cu

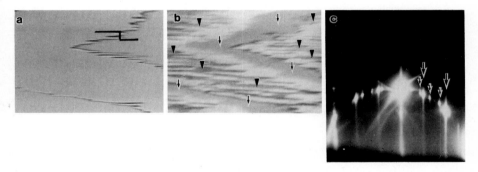

Figure 2. REM images taken before and after simultaneous deposition of Au and Cu (2/3 ML each) on a Si(111)7x7 surface and a RHEED pattern from the surface shown in (b).

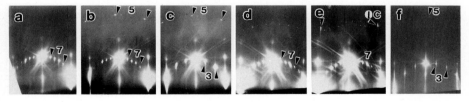

Figure 3. RHEED patterns taken during successive deposition of Au and Cu ((a)-(c)) and Cu and Au ((d)-(f)). Marks 7, 5, 3 and IC indicate reflections from the Si(111)7x7, Si(111)5x2-Au, Si(111)IC-Cu and Si(111) √3x√3-CuAu, respectively.

deposition and the √3 structure formed by the Au deposition alone does not coexist with the 7x7 structure (see Fig. 1). A dark field image with a spot from the √3 structure indicated that the dark regions uniformly formed along outer edges of the steps (arrows) and dark islands formed on terraces close to inner edges of the steps (arrow heads) are √3 structure regions. Bright regions are the 7x7 structure regions. These facts strongly suggest a formation of an alloy adsorbate formed by some sort of interactions between the Au and Cu atoms.

Figure 3 shows changes of RHEED patterns from Si(111) surfaces during successive depositions of Au and Cu ((a) to (c)) and of Cu and Au ((d) to (f)). It is noted that the 7x7 structure of (a) and (d) partially transformed to the 5x2 (indicated by 5 in (b)) and IC (indicated by IC in (e)) by Au and Cu depositions, respectively. However, these adsorbate structures and the 7x7 structure transformed to the √3 structure (indicated by 3 in (c) and (f)) by the second depositions of Cu and Au. Thus, it can be said that the √3 structure is a stable adsorbate structure formed only when Au and Cu atoms coexist on the surface.

3.2 Cu-Ag system

Figure 4 reproduces REM images ((a) and (b)) and RHEED patterns ((e) and (f)) taken before and after a simultaneous deposition of Cu (1/3ML) and Ag (2/3ML) at 560°C. The RHEED pattern (f) indicates that the surface is composed of regions of the 7x7, √3x√3 (due to Ag adsorption) and the IC phases (due to Cu adsorption). Dark field images due to reflections from the √3x√3 and IC structures are shown in (c) and (d), respectively. They indicate that dark regions in (b) formed along outer edges of terraces are composed of regions of the two adsorbate structures (the IC and √3x√3 structures spatially separated on the surface. Thus, Cu and Ag do not form an alloy adsorbate structure.

A series of REM images in Fig. 5 shows a case of successive deposition of Cu (b) and Ag (c)(2/3ML each). The IC phase seen in (b) as dark regions along outer edges of terraces retains in (c) and deposited Ag forms darker regions along the step and along the boundaries between the IC phase and the 7x7 structure regions as schematically illustrated in (e). The REM image (d) was taken after annealing the surface shown in (c) at 600°C. Due to desorption of Ag from the surface the image (d) is quite similar to that

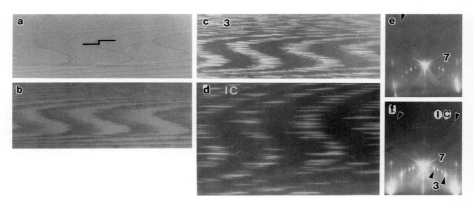

Figure 4. REM images (a) and (b) and RHEED patterns (e) and (f) of a Si(111) surface taken before and after simultaneous deposition of Cu(1/3ML) and Ag(2/3ML). Dark field images (c) and (d) are due to spots 3 and IC in (f).

612

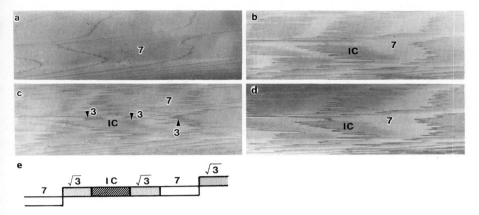

Figure 5. REM images of a Si(111) surface taken before and after successive deposition of Cu and Ag ((a) to (c)). The image (d) was taken after desorption of Ag from the surface shown in (c). A schema (e) illustrates the structure of the surface shown in (c).

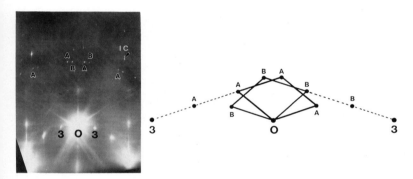

Figure 6. A RHEED pattern and its illustration showing formation of a $\sqrt{21}\times\sqrt{21}$ structure of two domains A and B by annealing a surface on which Cu (1.2ML) and Ag (2/3ML) was deposited successively.

shown in (b) taken before the Ag deposition, which indicates little interaction between Ag and Cu atoms.

In many cases the co-deposition gave the above mentioned two adsorbate structures. However, in some cases of excess Cu deposition in comparison with Ag and subsequent annealing, extra spots appeared as shown in Fig. 6. The surface was formed by Cu deposition of 1.2ML, Ag deposition of 2/3ML and subsequent annealing for 18 min at 480°C. The extra spots in the RHEED pattern were found to be due to a $\sqrt{21}\times\sqrt{21}$ structure as illustrated in Fig. 6. The $\sqrt{21}\times\sqrt{21}$ structure should show a domain structure and in fact dark field images (not shown here) due to these spots A and B showed orientational domains, which indicates p3 symmetry. They also showed that those domains grew mainly in regions where step density was high. It should be noted that the $\sqrt{21}\times\sqrt{21}$ structure is a superlattice ($\sqrt{7}\times\sqrt{7}$) of the $\sqrt{3}\times\sqrt{3}$ structure due to Ag adsorbate.

4. Discussion

The present REM study clearly showed that studies of co-deposition of metals on Si surfaces open an interesting field of surface science such as two dimensional alloy phases adsorbed on surfaces.   Structures are considered to be determined by an interaction of individual metal atoms with substrate and that between different metal atoms (a direct one and an indirect one through the substrate are conceivable).   The direct interaction between different metal atoms on the surface may be similar or dissimilar to that in the bulk.   The present two systems Au-Cu and Cu-Ag are cases of similar interactions, because the alloy adsorbate phase was easily formed in the Au-Cu system as in the bulk, while phase separation was generally noted in the Cu-Ag system.   In the latter system the formation of the $\sqrt{21}$x$\sqrt{21}$ structure was noted but for this structure further studies are needed.

Finally, it should be noted that for this kind of work in-situ real space observations of surfaces are indispensable because various phases may exist on the surface spatially separated as is seen in Figs. 4 and 5. The REM-RHEED technique used in the present study as well as TEM-TED (transmission electron microscope and diffraction) [4] are quite useful.

This work was supported by a Grant in Aid from the Ministry of Education, Science and Culture (No. 63609507 and 01609501).

References

[1] S. Ino: in Reflection High-Energy Electron Diffraction and Reflection Imaging of Surfaces. Eds. P. K. Larsen and P. J. Dobson (Plenum Press, New York, 1988) p.3.
[2] K. Yagi: J. Appl. Cryst. **20** (1987) 147.
[3] T. Ishizuka, K. Takayanagi, Y. Tanishiro and K. Yagi: Proc. 12th Int. Cong. EM (Kyoto, 1986) Vol.2,p.1347.   T. Ishizuka: Master Thesis (Tokyo Institute of Technology, 1987).
[4] I. Homma, Y. Tanishiro and K. Yagi: Proceedings of 26th Yamada Conference on "Surface as a New Material" (Osaka, 1990, July) Surface Sci. **242**(1991) 81.

Catalytic Effect of Metals (Sn, Ag, and Pb) on Homoepitaxial Growth of Ge and Si

K. Fukutani*, H. Daimon, and S. Ino

Department of Physics, Faculty of Science, University of Tokyo,
Hongo, Bunkyo-ku, Tokyo 113, Japan
*Present address: Institute for Solid State Physics,
 University of Tokyo, Roppongi, Minato-ku, Tokyo 106, Japan

Homoepitaxial growth of Si and Ge on metal covered surfaces, such as Si(111)-Ag, Ge(111)-Ag and Ge(111)-Sn surfaces, are studied by observing RHEED oscillation and by monitoring the surface elements by TRAXS. The observed RHEED oscillations indicate the layer-by-layer growth via 2D-nucleation. At the initial stage, the Si growth on the surface of Si(111)-(3×1+6×1)-Ag was promoted compared with that on the clean surface, whereas the growth on the Si(111)$\sqrt{3} \times \sqrt{3}$-Ag was disturbed. For the Ge growth, the epitaxial temperature on the silver covered surfaces was lowered compared with that on the clean surface. RHEED patterns induced by metals were always seen during the growth of Si or Ge at higher temperatures, therefore the segregation of metals throughout the growth was strongly suggested. Temperature dependence of Sn segregation to the surface is shown by TRAXS.

1. Introduction

Molecular beam epitaxy (MBE) combined with reflection high energy electron diffraction (RHEED) is a widely used technique for synthesizing good quality materials with more complex structure and composition. RHEED gives us real-time information about the structure of a sample surface. Besides intensity oscillations which occur in RHEED spots during MBE growth, have been of great interest and have stimulated both experimental and theoretical investigations for the past decade [1-6]. These oscillations are believed to take place when the growth proceeds in a layer-by-layer fashion via two dimensional (2D) nucleation. Thus this technique has been applied to monitoring and controlling in-situ the growth of epitaxial films on an atomic scale. It is well-known that 7×7 and c(2×8) reconstructed surfaces are formed on the clean Si(111) and Ge(111), respectively. Growth of Si on Si(111) was studied by LEED [7] and RHEED [6,8-10], which pointed out the presence of the anomalous period at the initial stage of the growth; the growth of Si on Si(111) does not start smoothly from the beginning, which seems attributed to the 7×7 structure. On the other hand, such an anomalous period was not observed in the growth of Ge on Ge(111) [9,11,12], which is considered to have a simple surface structure compared to the Si(111)7×7.

In the present paper, we have studied the homoepitaxial growth of Si and Ge on metal covered surfaces that are modified in atomic structure and electronic property. We found that the epitaxial growth is promoted or disturbed by the pre-deposited metal atoms depending on the sorts of metals and substrate.

Editors: S.Y. Tong · M.A. Van Hove · K. Takayanagi · X.D. Xie

Another point of this study is whether the metal layer deposited before growth stays at the interface between the substrate and grown layer. If the metal layer stays at the interface, Si/metal/Si or Ge/metal/Ge sandwich structures are formed, and if it doesn't metal atoms diffuse in the bulk or segregate at the surface. Total reflection angle X-ray spectroscopy (TRAXS) combined with RHEED [13,14] was used for the composition analysis of the surface.

2. Experimental

The experiments were performed in a UHV chamber equipped with a RHEED and TRAXS system. Both Si and Ge were evaporated from 2kW electron-guns, and tungsten filaments were used for the evaporation of metals (Ag, Sn and Pb). The typical evaporation rate of Si and Ge was about two monolayers/min in the experiments of intensity oscillations, which was monitored by a quartz oscillator. The intensity of a diffraction spot on the fluorescent screen was obtained by a photodiode through a lens and an optical fiber. X-rays emitted from the sample during RHEED observation were detected by using a SSD (Solid State Detector) with a Si(Li) that is placed almost on the same plane of the sample surface. When an X-ray is detected at a small take-off angle that coincides with the total reflection angle of the X-ray, the sensitivity to the surface is considerably improved owing to the refraction and absorption of X-rays [13,14].

The Ge(111) sample (Sb-dope, 0.35mm thick) used in this study was cleaned by Ar^+ ion bombardment (3keV, $1\mu A$) and subsequent annealing at about 700°C until a sharp c(2×8) RHEED pattern was visible. The Si(111) sample (N-type, 0.4mm thick) was cleaned by flashing at 1200°C for several times to reveal a sharp 7×7 RHEED pattern. The heating of both samples was carried out resistively by passing current directly through the sample.

3. Results and Discussion

3.1 Si Growth on Si(111)-Ag Surface

When silver atoms were deposited on a clean Si(111)7×7 surface at around 500°C, a $\sqrt{3} \times \sqrt{3}$ structure was formed at the coverage of about one monolayer(ML). By annealing this sample at about 700°C, the $\sqrt{3} \times \sqrt{3}$ structure changed to a mixed structure of 3×1 and 6×1 owing to the desorption of Ag [15-17]. When Si was grown on this Si(111)-(3×1+6×1)-Ag surface from around 300°C up, a $\sqrt{3} \times \sqrt{3}$ was observed along with the 3×1 and the 6×1 for about 50 monolayer growth. At room temperature, however, this (3×1+6×1) structure gradually disappeared changing into a 1×1, which also disappeared with further growth (about until fifty monolayers). When Si was grown on the Si(111)$\sqrt{3} \times \sqrt{3}$-Ag surface, the initial structure was maintained above 300°C for about 50 monolayers and changed into an amorphous-like structure at room temperature via 1×1 in the same way as the growth on the Si(111)-(3×1+6×1)-Ag. Since the $\sqrt{3} \times \sqrt{3}$, 3×1 and 6×1 structures are believed to be induced by silver, it is considered that the deposited Si atoms are successively replaced by

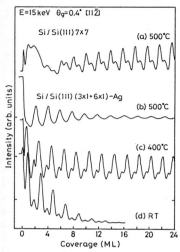

Fig. 1 Intensity oscillations of the specular spot during growth of Si in the [11$\bar{2}$] incidence. The electron energy is 15keV and the glancing angle is 0.4°. (a) Si growth on clean Si(111)7×7 at the substrate temperature of 500°C. (b)-(d) Si growth on Si(111)-(3×1+6×1)-Ag. The substrate temperatures are (b)500°C, (c)400°C and (d)room temperature.

Ag atoms and that Ag atoms are always present on the surface of the grown layer throughout the growth.

Fig. 1 shows some results of intensity oscillations which occurred in the RHEED specular spots during the Si growth (a) on the clean Si(111) surface and (b)-(d) on the Si(111)-(3×1+6×1)-Ag surface. The electron beam energy was 15keV, the glancing angle was 0.4° and the incident azimuth was [11$\bar{2}$]. The substrate temperatures were (a) 500°C, (b) 500°C, (c) 400°C, and (d) room temperature. These results indicate the growth takes place in a layer-by-layer fashion via 2D-nucleation. The basic periods of all four curves are equivalent to two monolayer growth of Si. While only bilayer(2ML) mode oscillation is seen in the curve (b), single layer mode oscillations are observed in (a), (c) and (d). The result of the growth on the clean surface (a) is roughly consistent with the earlier studies [6,8-10]. As seen in Fig. 1(a), anomalous behavior of the intensity was observed, that is, the growth of Si on the clean Si(111) does not start smoothly, which was indicated by a LEED study [7]. However this anomalous period at the initial stage of the growth is invisible in (b)-(d), thus the smooth growth is brought about by Ag atoms. In other words Ag atoms assist the Si growth like a catalyst.

Fig. 2 shows some results of intensity oscillations of the specular spots during the growth of Si on the Si(111)$\sqrt{3}$ × $\sqrt{3}$-Ag surface at the same diffraction condition as Fig. 1. The substrate temperatures were (a) 500°C, (b) 450°C, (c) 400°C and (d) room temperature. The layer-by-layer growth via 2D-nucleation is also indicated in this temperature region. Each oscillation period corresponds to a two monolayer growth; single layer mode oscillation

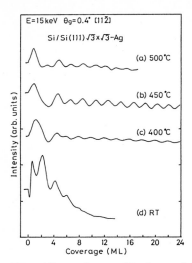

Fig. 2 Intensity oscillations of the specular spot during growth of Si on Si(111) $\sqrt{3} \times \sqrt{3}$-Ag in the [11$\bar{2}$] incidence. The electron energy is 15keV, the glancing angle is 0.4° and the substrate temperatures are (a)500°C, (b)450°C, (c)400°C and (d)room temperature.

in Fig. 1 is not observed. Comparing with the results of Fig. 1, the oscillation amplitudes are considerably reduced, however in Fig.2, strong and long oscillations are exhibited only in the beginning, of which period corresponds to about four monolayers ((a)-(c)). If we assume that these changes of RHEED spot intensities directly reflect the growth mechanism, bilayer mode growth starts after the first four monolayer growth is completed. Furthermore small oscillation amplitudes seem to indicate the disturbance of the growth in terms of the layer-by-layer growth via 2D-nucleation. The major difference between Figs. 1 and 2 is the surface superstructure, that is, the mixed structure for Fig. 1 and $\sqrt{3} \times \sqrt{3}$ for Fig. 2. The growth is accompanied with a considerable movement of Ag atoms, hence the $\sqrt{3} \times \sqrt{3}$ may be a complicated structure compared with the 3×1 and 6×1.

3.2 Ge Growth on Ge(111)-Ag Surface

When silver atoms were deposited on the clean Ge(111) surface at room temperature and the sample was annealed at about 400°C, the c(2×8) disappeared and instead 4×2 or $\sqrt{3} \times \sqrt{3}$ superstructures were formed depending on the coverage, which is roughly consistent with earlier studies by LEED [18,19]. When Ge was grown on the Ge(111)-(c(2×8)+4×2)-Ag or Ge(111)$\sqrt{3} \times \sqrt{3}$-Ag surfaces, these initial structures were maintained at about 120°C for about 50 monolayer growth. However these structures gradually disappeared at room temperature changing into an amorphous-like surface via 1×1. These results suggest the presence of Ag atoms on the growing surface at 120°C in the same way as the Si growth on the silver covered Si(111) surface.

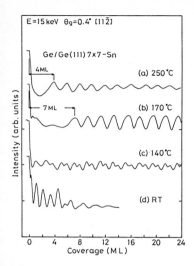

Fig. 3 Intensity oscillations of the specular spot during growth of Ge on clean Ge(111) c(2×8) ((a) and (b)), and on Ge(111)-(c(2×8)+4×2)-Ag ((c) and (d)). The electron beam is incident at a glancing angle of 0.4° along the [11$\bar{2}$] incidence with the primary energy of 15keV. The substrate temperatures are (a)180°C, (b)140°C, (c)140°C and (d)room temperature.

Intensity oscillations were also observed during the growth of Ge on these surfaces, hence the growth mode is considered to be layer-by-layer via 2D nucleation. Fig. 3 shows some results of intensity oscillations of the specular spot during the growth of Ge on the clean Ge(111)c(2×8) surface ((a) and (b)), and on the Ge(111)-(c(2×8)+4×2)-Ag surface ((c) and (d)) at the same diffraction condition as Fig. 1. The substrate temperatures were (a) 180°C, (b) 140°C, (c) 140°C, and (d) room temperature. Basic periods of oscillations are all two monolayers and the monolayer mode oscillations are observed in the curves (b)-(d). As shown in Figs. 3(a) and (b), the oscillation damps quicker at 140°C than at 180°C. In fact, it was confirmed that the oscillation at 180°C lasts for about fifty monolayers with little damping. As for the growth on Ge(111)-(c(2×8)+4×2)-Ag surface, the optimal temperature was about 140°C in terms of the duration of the intensity oscillation. Further the optimal temperature was found to be still lower on the Ge(111)$\sqrt{3}$ × $\sqrt{3}$-Ag surface. We can say that silver atoms lower the temperature for the epitaxial growth of Ge, that is, Ag atoms act as a catalyst.

3.3 Ge Growth on Ge(111)-Sn Surface

When about a half monolayer of tin atoms were deposited on the clean Ge(111) surface at about 400°C, a 7×7 superstructure appeared instead of the c(2×8), which is nearly consistent with the previous study [20]. When Ge was grown on this 7×7 surface, several kinds of surface structures were observed during

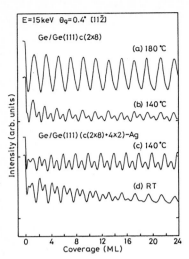

Fig. 4 Intensity oscillations of the specular spot during growth of Ge on Ge(111)7×7-Sn in the [11$\bar{2}$] incidence. The electron energy is 15keV, the glancing angle is 0.4° and the substrate temperatures are (a)250°C, (b)170°C, (c)140°C and (d)room temperature.

the growth. At room temperature the 7×7 has changed into a 1×1 after about two monolayer growth and the 1×1 was gradually weakened afterwards. At about 140°C and 170°C, a diffuse half order streak pattern appeared that is similar to the pattern observed during the Ge growth on the clean Ge(111) [9,11,12]. On the other hand, a $\sqrt{3} \times \sqrt{3}$ was observed at about 250°C and the mixed structure of the 7×7 and 5×5 was seen at about 310°C throughout growth. In the structure formation diagram of the Ge(111)-Sn system [20], it is reported that a metastable $\sqrt{3} \times \sqrt{3}$ and a stable 5×5 appear at a coverage of a half monolayer and more than a half monolayer, respectively. Therefore it is considered that tin atoms are present on the growing surface.

Fig. 4 shows some results of RHEED intensity oscillations of the specular spot during the growth of Ge on the Ge(111)7×7-Sn surface at the same diffraction condition as Fig. 1. The substrate temperatures were (a) 250°C, (b) 170°C, (c) 140°C and (d) room temperature. Similar to the above results of sections 3.1 and 3.2, the growth of Ge on this surface in a layer-by-layer fashion via 2D nucleation is strongly suggested. All the oscillations basically exhibit a period corresponding to bilayer and the monolayer mode is seen at low temperatures ((c) and (d)). Comparing with the results of Ge growth on the clean Ge(111) in Fig. 3, the oscillation amplitudes are rather small, and anomalous behavior of the intensity at the initial stage is observed at higher temperatures ((a) and (b)); the normal oscillation starts after deposition of several monolayers as shown in Fig. 4. Thus, we can say that the growth of Ge is disturbed by tin atoms.

Performing the TRAXS measurement for composition analysis during the growth of Ge, the intensity of the GeLα characteristic X-ray was found to be

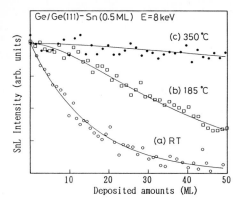

Fig. 5 Changes of SnLαβ intensity during growth of Ge on Ge(111)7×7-Sn(0.5ML). The electron energy is 8keV, the take-off angle of X-rays is 0.75° and the substrate temperatures are (a)room temperature (open circle), (b) 185°C (square) and (c)350°C (filled circle).

maintained almost at a constant value, whereas SnLαβ has undergone an notable change as the deposited amounts of Ge increased. Fig. 5 shows some results of the change of the SnLαβ intensity during the growth of Ge at various temperatures. The substrate temperatures are (a) room temperature, (b) 185°C, and (c) 350°C. The evaporation rate of Ge was about 0.7ML/min. As the temperature gets higher, the decrease rate of the Snαβ intensity becomes smaller and at above 350°C the intensity of SnLαβ remains almost constant during the growth. Taking into consideration the attenuation coefficient of the SnLα X-ray for Ge, the tin is thought to remain at the interface between the substrate and the grown layer at 350°C. On the other hand, the tin atoms are present in the surface layer at higher temperatures, which is consistent with the RHEED observations mentioned above.

4. Conclusion

We have performed measurements of RHEED intensity oscillations during the homoepitaxial growth on metal covered surfaces of Si(111) and Ge(111). It was found that modifying the surfaces with different metals is very effective for the epitaxial growth. Silver atoms, in particular, promote the growth of both Si and Ge, though the Si growth is disturbed at a high coverage (one monolayer). Tin atoms were found to disturb the growth of Ge. TRAXS measurements allowed a direct analysis of surface composition, and the results showed that tin atoms are more segregated at the surface during the Ge growth as the substrate temperature is raised.

5. References

1. J. J. Harris, B. A. Joyce and P. J. Dobson: Surf. Sci. Lett. **103**, L90(1981).
2. C. E. C. Wood: Surf. Sci. Lett. **108**, L441(1981).
3. J. H. Neave, B. A. Joyce, P. J. Dobson, and N. Norton: Appl. Phys. **A31**, 1(1983).
4. J.M. Van Hove, C.S. Lent, P. R. Pukite, and P. I. Cohen: J. Vac. Sci. Technol. **B1**, 741(1983).
5. T. Sakamoto, H. Funabashi, K. Ohta, T. Nakagawa, N. J. Kawai and T. Kojima: Japan. J. Appl. Phys. **23**, L657(1984).
6. T. Sakamoto, N. J. Kawai, T. Nakagawa, K. Ohta, and T. Kojima: Appl. Phys. Lett. **47**, 617(1985).
7. R. Altsinger, H. Busch, M. Horn and M. Henzler: Surf. Sci. **200**, 235(1988).
8. M. Ichikawa and T. Doi: Appl. Phys. Lett. **50**, 1141(1987).
9. J. Aarts and P. K. Larsen: Surf. Sci. **188**, 391(1987).
10. H. Nakahara and A. Ichimiya: J. Cryst. Growth **95**, 472(1989).
11. L. Daweritz, O. P. Pchelyakov, V. I. Mashanov, L. V. Sokolov, S. I. Stenin and H. Berger: Surf. Sci. Lett. **230**, L162(1990).
12. K. Fukutani, H. Daimon and S. Ino: to be published.
13. S. Hasegawa, S. Ino, Y. Yamamoto, and H. Daimon: Japan. J. Appl. Phys. **24**, L387(1985).
14. S. Ino, S. Hasegawa, H. Matsumoto, and H. Daimon: Proceeding of ICSOS-2, The Structure of Surfaces II (Springer, Berlin, Heidelberg, New York, London, Paris, and Tokyo, 1987)p. 334.
15. Y. Gotoh and S. Ino: Japan. J. Appl. Phys. **17**, 2097(1978).
16. G. Le Lay: Surf. Sci. **132**, 169(1983).
17. S. Hasegawa, H. Daimon and S. Ino: Surf. Sci. **186**, 138(1987).
18. G. Le Lay, G. Quentel, J. P. Faurie and A. Masson: Thin Solid Films **35**, 289(1976).
19. E. Suliga and M. Henzler: J. Vac. Sci. Technol. **A1**, 1507(1983).
20. T. Ichikawa and S. Ino: Surf. Sci. **105**, 395(1981).

REM and RHEED Studies of Lead Adsorption on Silicon (111) Surfaces

Y. Tanishiro, M. Fukuyama, and K. Yagi

Department of Physics, Tokyo Institute of Technology,
Oh-okayama, Meguro-ku, Tokyo 152, Japan

Abstract. Adsorption processes of lead on Si(111)7x7 surfaces and the phase transitions between the adsorbate structures are studied by REM and RHEED. Formation of a two-dimensional(2D-) contracted Pb(111) layer and two kinds of $\sqrt{3}\times\sqrt{3}$ structures, $\sqrt{3}$-H and $\sqrt{3}$-L, and the phase transitions between them are described. As the 2D-Pb(111) layer forms, the 7x7 structure of the clean surface transforms to the so-called δ-7x7 structure.

1. Introduction

Structure changes of silicon surfaces by metal adsorption and by heat treatment are of great interest in surface physics and semiconductor-device technology. Lead adsorption on Si(111)7x7 has been studied by low-energy electron diffraction(LEED)[1-3], ion scattering spectroscopy[2] and ellipsometry[4], and $\sqrt{3}\times\sqrt{3}$ (abbreviated $\sqrt{3}$ hereafter) and "1x1" structures and a two-dimensional(2D-) Pb(111) layer have been reported. The structures and the phase transitions between them, however, have not been clarified yet.

We observed the structural changes by the deposition of lead on Si(111)7x7 and heat treatment by reflection electron microscopy combined with diffraction(REM-RHEED) and found a 2D-contracted Pb(111) layer, two kinds of $\sqrt{3}$ structures named $\sqrt{3}$-H and $\sqrt{3}$-L, and incommensurate monolayers named α and α'. The formation of the 2D-contracted (111) layer and the $\sqrt{3}$-H and $\sqrt{3}$-L structures and the phase transitions between them are described here, and the two-dimensional "phase diagram" of the adsorbate structures including the incommensurate ones will be published separately[5].

2. Experimental

Experiments were performed using a UHV electron microscope (modified JEM100B)[6]. Si(111) substrate crystals were heated and cleaned by passing a DC current through them. Lead was evaporated from a tungsten filament and the amount deposited was monitored by a quartz oscillator. The deposition rate was 1-3 ML/min. The azimuthal direction of the incident electron beams for the REM and RHEED observations was the $\langle 11\bar{2} \rangle$ direction.

3. Results

Adsorption processes were observed at substrate temperatures between 160 and 400°C. Different structures were formed depending on the substrate temperature; low(T<240°C), high(T∼400°C) or intermediate(240<T<360°C) temperature.

Springer Series in Surface Sciences, Vol. 24   **The Structure of Surfaces III**
Editors: S.Y. Tong · M.A. Van Hove · K. Takayanagi · X.D. Xie
© Springer-Verlag Berlin, Heidelberg 1991

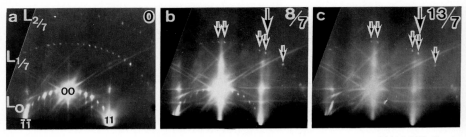

Fig.1   A series of RHEED patterns during Pb adsorption on a Si(111)7x7 at 225°C.   Formation of 2D-Pb(111) layer and the structure change from the 7x7 to δ-7x7 structure are noticed.

Figure 1 shows a series of RHEED patterns during the adsorption of lead on a Si(111)7x7 surface at 225°C. The amount deposited is indicated in monolayer(ML) units in the upper-right corner of each figure. In fig.1a the superlattice reflections of the 7x7 structure of the clean surface are seen on several Laue zones, concentric circular arcs marked L_0, $L_{1/7}$ and $L_{2/7}$, in addition to the fundamental reflections marked (0 0) and (±1 ±1). The reflections on the zeroth Laue zone(L_0) are indexed by (r,r) and are indicated by the r values hereafter. After the deposition of lead, extra streaks appeared outside the fundamental reflections from the silicon surface and became strong as indicated by big arrows in figs.1b and 1c. This streak indicated by the arrow is the (1 1) reflection from the Pb(111) surface. Thus, islands of 2D-Pb(111) layer grew on the surface epitaxially with Pb[1$\bar{1}$0]//Si[1$\bar{1}$0]. A notable fact is that the intensity distribution of the 7x7 superlattice reflections changed from fig.1a to 1c: Only the superlattice reflections indicated by small arrows on the 1/7-th Laue zone($L_{1/7}$) in figs.1b and 1c are seen to remain strong, the indices of which are ($\bar{1}$/7 0), (0 1/7), (6/7 1), (1 8/7) and (10/7 11/7). All these reflections are on the lines connecting the neighboring fundamental reflections. This intensity distribution is similar to that for the δ-7x7 structure which has been reported to be formed by the adsorption of alkali metal or hydrogen atoms on Si(111)7x7[7,8]. The r value of the streak due to the 2D-Pb(111) layer was r=1.14(=8/7), being shifted from the value of r=1.096 for the bulk Pb(111) surface. Hence, the 2D-Pb(111) layer is contracted by 4% from the bulk (111) plane. This is caused by lattice matching of the 2D-Pb(111) layer with the 7x7 superlattice of the substrate surface, as mentioned in ref.2.

By further deposition on the 2D-Pb(111) layer at low temperature, three-dimensional(3D-) Pb(111) islands of the bulk lattice parameter grew on it(the S-K growth mode).

Figures 2a and 2b show RHEED patterns after the deposition of 0.5 and 1 ML of lead, respectively, on a Si(111)7x7 surface at 400°C. The extra streaks due to a $\sqrt{3}$ structure, indicated by arrows in fig.2a, appeared and those intensities became stronger as the amount deposited increased. The $\sqrt{3}$ structure is labeled $\sqrt{3}$-H hereafter. This structure was stable upon cooling to room temperature. By heating above 400°C lead atoms desorbed from the surface and the surface returned to the clean 7x7 structure.

Figure 3 shows a series of REM images and corresponding RHEED patterns during the adsorption of lead on a Si(111)7x7 surface at 350°C. The wavy lines marked P, Q, and R in fig.3a are the steps. The images are foreshortened by about 1/50 in the vertical direction. First, small domains were formed as seen in fig.3b. Correspondingly, the 7x7 superlattice spots on the zeroth Laue zone in fig.3a' changed to streaks in fig.3b', while no extra spots were formed due to the domains. By further deposition the 7x7 superlattice reflections weakened and streaks due to a

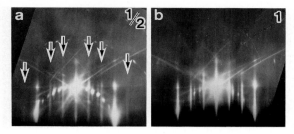

Fig.2 Formation of the $\sqrt{3}$-H structure by Pb deposition on a Si(111)7x7 at 400°C.

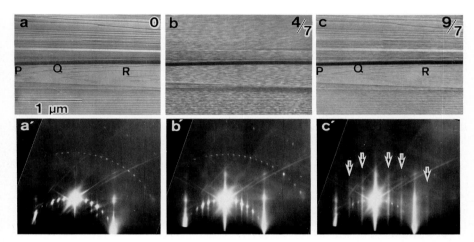

Fig.3 A series of RHEED patterns and corresponding REM images during Pb adsorption on a Si(111)7x7 at 350°C. The $\sqrt{3}$-L structure was formed in (c).

$\sqrt{3}$ structure labeled $\sqrt{3}$-L in distinction from the $\sqrt{3}$-H structure appeared as indicated by arrows in fig.3c'. No changes of step positions by the adsorption have been recognized in the REM image of fig.3c as compared to fig.3a.

The $\sqrt{3}$-L structure was also formed by annealing of the 2D-Pb(111) layer, as shown in figs.4a and 4b which were taken at 240°C and 330°C, respectively. The RHEED patterns of the $\sqrt{3}$-L after cooling again are shown in figs.4c and 4d which were taken at 225°C and 195°C, respectively. The intensity of the streaks due to the $\sqrt{3}$-L decreased below 200°C, unlike the $\sqrt{3}$-H structure. This is characteristic of the $\sqrt{3}$-L structure.

By heating up to 360°C the $\sqrt{3}$-L transformed irreversibly to the $\sqrt{3}$-H structure, as shown in fig.5. The intensities of the $\sqrt{3}$ streaks once weakened(fig.5b) on the way of the phase transition from the $\sqrt{3}$-L(fig.5a) to the $\sqrt{3}$-H(fig.5c) structure.

By further deposition on the $\sqrt{3}$-L or $\sqrt{3}$-H structure at low temperature, the surface changed to the incommensurate layer, α, and then 3D-Pb(111) islands of the bulk lattice parameter grew on it (the S-K growth mode)[5].

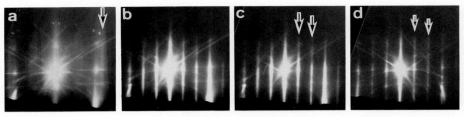

Fig.4   Phase transition from (a)2D-layer with the δ-7x7(240°C) to (b)the √3̄-L structure (330°C) by heating. (c) and (d) were taken at 225°C and 195°C, respectively.   Note decrease of streak intensities due to the √3̄ structure in (d).

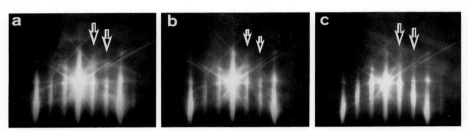

Fig.5   Phase transition from (a)the √3̄-L(240°C) to (c)the √3̄-H structure (360°C) by heating.   Note decrease of streak intensities due to the √3̄ structure on the way of the transition around 300°C in (b).

4. Summaries and Discussion

Adsorption processes of lead on Si(111)7x7 and the phase transitions between the surface structures were observed.   Surface structures formed by the adsorption depend on the substrate temperature.   Changes of the positions and the shapes of the steps due to the adsorption have not been recognized.

The 2D-Pb(111) layer always coexisted with the δ-7x7.   It is, thus, considered that the deposited lead atoms partially destroy the DAS structure[9] of the clean 7x7 surface and form the 2D-Pb(111) layer on the δ-7x7, leaving the stacking-fault layer of the DAS structure at the interface.   The 2D-Pb(111) layer is contracted by 4% from the bulk Pb(111) plane so as to make the lattice matching with the δ-7x7 structure.   Since the r values of the streak from the 2D-Pb(111) layer was 8/7, 8x8 lead atoms in the 2D-Pb(111) layer are considered to sit on the 7x7 unit mesh. This is caused by the influence of the undestroyed 7x7 reconstructed structure of the substrate surface on the adsorption or growth of lead. Upon heating this structure transformed irreversibly to the more stable √3̄-L, the δ-7x7 structure being destroyed.

Two kinds of √3̄ structures, the √3̄-H and the √3̄-L structures are found. In the previous reports, several kinds of √3̄ structures have been reported[5]: √3̄-I(1/3ML) and √3̄-II(4/3ML) in ref.1; √3̄-β(1/3ML) and √3̄-α (4/3ML) in ref.2; √3̄$_1$(1/3ML), √3̄$_2$(2/3ML) and √3̄$_3$(1ML) in ref.3.   The coverage of lead for each structure proposed is indicated in parentheses. All the extra streaks due to the √3̄-H and the √3̄-L structures had similar intensities.   These intensity distributions for the √3̄-H and the √3̄-L structures are similar to that for the √3̄-I, √3̄-β or √3̄$_2$ structure.   The intensity decrease of the streaks due to the √3̄-L structure upon cooling below 200°C is a new finding.   Although the reason for it has not been clarified yet, the intensities of the streaks decreased due to the rearrangement of the atoms during the transition from the $\sqrt{3}$-L to the $\sqrt{3}$-H structure, where the coverage of lead atoms may decrease due to sublimation.

Small domains which did not give extra spots in the RHEED pattern shown in fig.3b, were formed as shown in fig.3b'. Although the unit mesh of their structure is unknown, the pattern for this structure is similar to a LEED pattern for the $\sqrt{3}_1$ structure reported, in which any extra spots are missing. The structure model proposed in ref.3, however, is incompatible with the RHEED pattern. The structure analysis is a future problem.

The two-dimensional "phase diagram" of the adsorbate structures including incommensurate ones will be given in a separate paper[5].

References

[1] P.J.Estrup and J.Morrison, Surf.Sci.**2**(1964)465.
[2] M.Saitoh, K.Oura, K.Asano, F.Shoji and T.Hanawa, Surf.Sci.**154**(1985)394.
[3] G.Le Lay, J.Peretti, M.Hanbücken and W.S.Yang, Surf.Sci.**204**(1988)57.
[4] G.Quentel, M.Gauch and A.Degiovanni, Surf.Sci.**193**(1988)212.
[5] Y.Tanishiro, M.Fukuyama and K.Yagi, in preparation.
    The structure reported as the $\sqrt{3}$ structure labeled $\sqrt{3}$-II[1], $\sqrt{3}$-α[2] or $\sqrt{3}_3$[3] is shown not to be the $\sqrt{3}$ structure but to be an incommensurate structure.
[6] K.Takayanagi, K.Kobayashi, K.Yagi and G.Honjo, J.Phys.E**11**(1978)441.
[7] H.Daimon and S.Ino, Surf.Sci.**164**(1985)320.
[8] A.Ichimiya and S.Mizuno, Surf.Sci.**191**(1987)L765.
[9] K.Takayanagi, Y.Tanishiro, S.Takahashi and M.Takahashi, Surf.Sci.**164**(1985)367.

Electron Accumulation Near the α-Sn/CdTe(111) Interface Region?

I. Hernández-Calderón[1] *and H. Höchst*[2]

[1]Physics Department, Centro de Investigación y Estudios Avanzados del IPN, Apdo. Postal 14-740, 07000 Mexico, D.F., Mexico
[2]Synchrotron Radiation Center, University of Wisconsin–Madison, Stoughton, WI 53589, USA

Abstract. *In situ* angle resolved photoemission with synchrotron radiation and reflection high energy electron diffraction were employed to investigate the electronic and structural properties of the heterostructure α-Sn/CdTe(111). Initial deposition of α-Sn moved the valence band edge to the Fermi level with a density of states considerably higher than that of the metallic β-Sn phase. Further deposition of Sn produces valence band edges characteristic of thick α-Sn films. This effect is attributed to the formation of a tetrahedrally coordinated $Sn_xCd_{1-x}Te$ interface alloy layer with electron excess. A possible contribution from a surface fotovoltage effect is analyzed. We did not find any evidence of quantum size effects on the thin α-Sn films.

1. Introduction

It has been shown that diamond structured metastable α-Sn films can be grown above the bulk $\alpha \rightarrow \beta$ transition temperature of 13.2 ^{o}C on closely matched substrates [1,2]. CdTe is a zinc-blende type semiconductor with a typical band gap E_g=1.5 eV. Because α-Sn is a zero-gap semiconductor, the α-Sn/CdTe system presents large valence band offsets. Previous studies of α-Sn/CdTe interfaces indicated valence band offsets (ΔE_v) around 1.1 eV for (110) and (100) substrates [3,4]. From a theoretical background, electron and hole confinement in α-Sn films would result in quantum levels promoting the opening of the bulk band gap for very thin films [5]. This effect would allow the design of quantum well structures with high degree of carrier confinement for their application in novel semiconductor devices. We did a careful examination of the photoemission spectra in order to identify the presence of quantum size effects. However, we did not find any conclusive evidence of quantization. Our results are in contrast with the work of Takatani and Chung [6], who claimed the observation of band gaps in the range from 420 to 230 meV for α-Sn films between 50 and 80 Å. They inferred those values, indirectly, from the loss onsets in experiments of high resolution electron energy loss spectroscopy. Having in mind these results, Craig and Garrison employed a modified linear combination of atomic orbitals to calculate the value of the band gap as a function of layer thickness [7]; they found a maximum gap of 430 meV for a film 40 Å thick.

Springer Series in Surface Sciences, Vol. 24   **The Structure of Surfaces III**
Editors: S.Y. Tong · M.A. Van Hove · K. Takayanagi · X.D. Xie
© Springer-Verlag Berlin, Heidelberg 1991

We present the results obtained from experiments of angular-resolved synchrotron radiation photoemission spectroscopy (ARSRPES) performed in molecular beam epitaxy (MBE) α-Sn thin films grown onto CdTe(111) crystaline substrates. The electronic properties of the α-Sn/CdTe interface were investigated through the analysis of Te, Cd and Sn 4d core levels as a function of Sn coverage.

2. Experimental Details

Mechanically polished CdTe substrates with the [111] axis perpendicular to the Cd surface were prepared in ultra-high vacuum until a well developed Cd(111)(2 × 2) reflection high energy electron diffraction (RHEED) pattern was observed. MBE α-Sn films were deposited at growth rates around 1.4 Å/s. Deposition rates were measured with a quartz monitor which could be positioned ∼ 6 cm closer than the sample, so actual film thickness is slightly smaller than the nominal value given hereafter. The photoemission experiments were performed at the University of Wisconsin's 1 GeV storage ring Aladdin with photons from a 6 meter toroidal grating monochromator (TGM). For more details see reference [8]. Te, Cd and Sn 4d levels were measured at 50 eV kinetic energy to obtain higher surface sensitivity.

3. Results and Discussion

Figure 1a presents the normal valence band photoemission spectrum of the clean CdTe(111) surface. Due to the high energy of the incident photons the spectrum strongly reflects the density of initial states, the two main peaks corresponding to the Σ_1^{min} and L^v_6 critical points of the bulk electronic band structure [9]. The valence band edge (VBE) is located 0.7 eV below E_F. After deposition of 10 Å tin, the VBE shifts ∼ 1.1 eV to lower binding energies, showing intense emission around E_F, stronger than that of the metallic β-Sn, as illustrated in Fig. 1b. The spectrum from β-Sn was taken during the same experiment from the tin deposited onto the stainless steel sample holder. Further deposition of α-Sn on the CdTe substrate brings back the VBE to the position and intensity characteristic of thick α-Sn films [2-4]; after growth of ∼ 40 Å tin no important changes were observed. At first sight, three origins of the sudden increase in emission around E_F after 10 Å Sn deposition could be argued: i) the formation of β-Sn; ii) upward band bending effects, and iii) a surface photovoltage effect [10]. We will briefly show that none of these effects causes the most important contribution to the VBE shift. Since β-Sn is the stable room temperature phase of tin, it would not be possible to grow α-Sn tin on top of the β phase, as indicated by the lack of emission around E_F for the 100 Å thick film. Besides this fact, the α-Sn 4d core levels are broader than the β-Sn 4d levels [2-4], and that was always the case for Sn onto CdTe, Fig. 2. Upward band bending or surface photovoltage

629

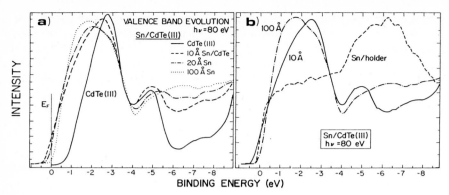

Figure 1. a) Normal valence band photoemission spectra of the α-Sn/CdTe heterostructure. After deposition of 40 Å Sn, no important changes were observed in the spectra. b) Comparison of valence band spectra of α-Sn deposited on CdTe and β-Sn deposited on the stainless steel sample holder.

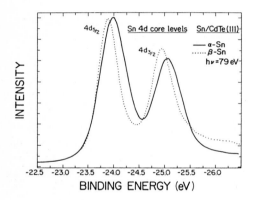

Figure 2. Typical 4d core levels of the α and β phases of tin.

effects would produce similar results: roughly the same shift for the VBE, Cd 4d and Te 4d core levels. That was not the case, as illustrated in Fig. 3, where the VBE is referred to E_F, Cd and Te core levels are referred to the values of the clean surface and Sn core levels relative to 10 Å coverage.

The initial behavior of the α-Sn/CdTe(111) system could be explained in terms of the formation of a tetrahedrally coordinated α-$Sn_{1-x}Cd_xTe$ interface alloy layer [11]. Sn has been deposited on the cadmium CdTe(111) (2×2) surface and it is known that Cd vacancies are native defects of CdTe [12]. Assuming that Cd vacancies in the surface region are filled by Sn atoms and taking into account that each Sn atom contributes 2 extra electrons per filled vacancy, an excess of electrons would be located at the interface region. The fact that the Te 4d core levels do not present a significant chemical shift after the initial Sn coverage could be an indication that the excess electrons are associated to delocalized states, producing a metallic-type confined layer

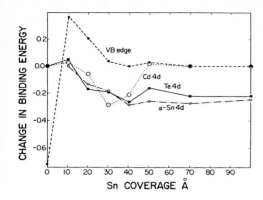

Figure 3. Changes in the VBE and 4d core levels with Sn coverage. The VBE is referred to E_F; Cd and Te to the values of the clean surface; for Sn relative to the initial 10 Å coverage.

which is the cause of the enhancement of the emission at E_F. The changes between 10 and 40 Å are then the result of the evolution of the alloy towards α-Sn. Around 40 Å coverage the VBE and Te and Sn core levels seem to reach their final state, in contrast with the Cd cores.

Yuen et al. recently reported an observation of a two dimensional electron gas (2DEG) in an α-Sn/InSb(100) system [13]; the authors speculated about the origin of this phenomenon. We cannot make any statement about similarities with our findings. However, the existence of an inversion layer at the α-Sn/CdTe(111) interface, producing accumulation of electrons, can be easily discarded since the VBE shift would be in the opposite direction.

Since at the begining of the growth process there was no significant change in binding energies for Te and Cd, another possible explanation for the similar VBE and core level shifts between 10 and 40 Å coverage, Fig. 3, is the removal of a surface photovoltage effect (~ 0.35 V) which was originally present in the CdTe clean surface. In this case, the correction of the binding energies will result in an almost flat behavior with coverage, characteristic of an abrupt α-Sn/CdTe(111) interface. If this were the case, it is not clear why the α-Sn cores are affected in the same way as the CdTe levels.

The results of our work do not show evidence of quantum size effects as those claimed in refs. 5 and 6. Firstly, changes of Sn 4d cores between 10 and 40 Å tin deposition are similar to those of Te and Cd, moreover, an opening of the α-Sn fundamental band gap would produce a shift towards lower binding energies of the Sn 4d core levels at the higher coverages, but after 40 Å they reach almost a final state.

4. Conclusions

The electronic structure of the α-Sn/CdTe(111) interface was investigated by ARSRPES. Enhanced emission at E_F for a coverage of 10 Å Sn was explained in terms of the formation of a α-Sn$_x$Cd$_{1-x}$Te interface layer with an electron excess due to the filling of Cd vacancies by Sn atoms. No evidence of quantum size effects were found. The possible contribution from a surface photovoltage effect to the shift of core levels was also considered.

Acknowledgments

IHC wants to thank Consejo Nacional de Ciencia y Tecnología (CONACyT) of Mexico for partial support within a CONACyT-NSF(USA) program. We thank M. A. Engelhardt for technical support during the experiments.

References

1. R. F. Farrow, D. S. Robertson, G. M. Williams, A. G. Cullis, G. R. Jones, I. M. Young, and P. N. J. Dennis, J. Cryst. Growth **54**, 507(1981).
2. H. Höchst, and I. Hernández-Calderón, Surf. Sci. **126**, 25(1983).
3. H. Höchst, D. W. Niles, and I. Hernández-Calderón, J. Vac. Sci. Tech. **B6**, 1219(1988)
4. M. Tang, D. W. Niles, I. Hernández-Calderón, and H. Höchst, Phys. Rev. **B36**, 3336(1987).
5. B. A. Tavger and V. Ya. Demikhovskii, Soviet Phys. USPEKHI (Engl. Transl.) **11**, 644(1968).
6. S. Takatani and Y. W. Chung, Phys. Rev. **B31**, 2290(1985).
7. B. I. Craig and B. J. Garrison, Phys. Rev. **33**, 8130(1986).
8. H. Höchst, M.A. Engelhardt, and I. Hernández-Calderón, Phys. Rev. **40**, 9703(1989).
9. H. Höchst, D.W. Niles and I. Hernández-Calderón, Phys. Rev. **B40**, 8370 (1989).
10. H. C. Gattos and J. Lagowski, J. Vac. Sci. Tech. **10**, 130(1973).
11. H. Höchst, D.W. Niles, M. A. Engelhardt, and I. Hernández-Calderón, J. Vac. Sci. Tech. **A7**, 775(1989).
12. K. Zanio, in *Semiconductors and Semimetals,* Vol. 13, edited by R. K. Willardson and A. C. Beer (Academic Press, New York, 1978).
13. W. T. Yuen, W.K. Liu, S.N. Holmes, and R. A. Stradling, Semicond. Sci. Technol. **4**, 819(1989).

Successive Oxidation Stages of Adatoms on the Si(111)7 × 7 Surface Observed with Scanning Tunneling Microscopy and Spectroscopy

J.P. Pelz and R.H. Koch

IBM Watson Research Center, P.O. Box 218,
Yorktown Heights, NY 10598, USA

Abstract. Scanning tunneling microscopy and spectroscopy reveal at least a two-stage reaction process which occurs at adatom sites on Si (111) 7x7 surfaces exposed to oxygen at 300 K. . Reacted adatoms in the first stage are characterized by small ($\sim 1/2$ eV) electronic energy shifts, and show a preference for the faulted half and the corners of the 7x7 unit cell. Second stage adatoms appear similar to "missing" or "lowered" adatoms in topographic images. Annealing of the partially-oxidized surface to ~ 625 K causes large changes in the number and site preference of the two stages.

For many years, a variety of experimental techniques [1-3] and theoretical approaches [4, 5] have been applied to understand the initial oxidation of Si surfaces. Despite this effort, fundamental questions remain concerning the actual atomic structure and electronic properties of the initial oxidation stages. Recently, this system was studied with atomic resolution [6, 7] using scanning tunneling microscopy (STM). These studies reported a particular type of reacted site which appeared to nucleate near defects on the surface [6, 7].

In this paper, we report new measurements of the room-temperature oxidation of the Si(111) 7x7 surface using STM and the related technique scanning tunneling spectroscopy (STS) [8] which indicates the existence of at least a *two-stage* reaction process of individual Si surface atoms. The first stage appears to nucleate preferentially at certain symmetry sites within the 7x7 unit cell. We also find that subsequent annealing at relatively low temperatures (~ 625K) has a large effect on the number and spatial distribution of the two stages. Part of this work has been discussed elsewhere [9, 10].

Experiments were performed in an ion-pumped UHV chamber with base pressure of roughly 1×10^{-10} Torr. Oxygen dosing was performed with all hot filaments off, and O_2 pressure (typically $\sim 3 \times 10^{-9}$ Torr during a dose) was determined by monitoring the ion-pump current. Clean 7x7 surfaces were prepared by quickly heating n -type Si (111) wafers for a few seconds to ~ 1500 K. Our custom-built ultra-high vacuum STM has been described elsewhere [9]. Constant-current topographies (CCT's) were measured with the tunnel current $I_t \cong 1$ nA, and sample voltage V_s between ± 3 V. We performed STS by measuring an array of fixed-gap current-voltage (I-V) curves [8, 11, 12] simultaneously with some of the CCT's.

Springer Series in Surface Sciences, Vol. 24   **The Structure of Surfaces III**
Editors: S.Y. Tong · M.A. Van Hove · K. Takayanagi · X.D. Xie
© Springer-Verlag Berlin, Heidelberg 1991

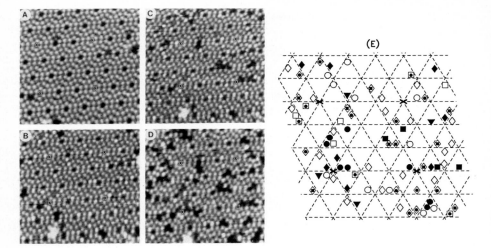

FIG. 1. (A) - (D): Grey-scale CCT's of particular sample area for 0-, 0.3-, 0.4, and 0.6 L total O_2 dose, measured at $V_s \cong 2$ V. Small x's mark specific points for position reference. Images cover $\sim$20x20 (nm)^2. (E): Schematic diagram of same area, where open (filled) symbols represent S1 (S2) sites first observed in (A) (triangle), (B) (diamond), (C) (square), or (D) (circle).

Figure 1 shows that we can sequence O_2-induced changes on the surface at room temperature with atomic resolution. Fig. 1(A) shows a CCT of a $\sim$ 20 × 20 nm^2 region of the clean 7x7 surface, showing several native defects. The Si atoms most obvious in this image are called "adatoms", and the dark holes are know as corner holes [13]. Figs. 1(B) -1(D) show CCT's of roughly the same region after the surface was exposed to a total O_2 dose $D_T \cong 0.3$ L, 0.4 L, and 0.6 L respectively, where 1 L = 10^{-6} Torr-s. During each dose the tip was pulled back $\sim$ 200 nm from the surface and V_s was set to 0. We note that O_2 exposure has caused certain adatoms to react: some appear to be "raised" or brighter than the rest, while others appear as "lowered" or "missing" adatoms. We denote these two species as "Stage 1" (S1) and "Stage 2" (S2) adatoms, respectively, and also denote unreacted adatoms as "Stage 0" (S0). The S2-type of reacted adatom has been observed in prior studies [6, 7], but the S1-type adatoms have not been previously reported. Measurements similar to those discussed here have also been made by another group [14]. Our STS measurements [9, 10] indicate that the DOS at S1 adatoms is characterized by a $\sim$ 1/2 eV gap at the Fermi level, with strong filled and empty states at $\pm$ 0.7 eV. We note here that the appearance of S1 adatoms in CCT's varies greatly with different tips and/or sample bias V_s. With some tips, the S1 adatoms appear indistinguishable from S0 in CCT's at certain V_s, but are often detectable at different sample bias or from I - V measurements [9]. S2 adatoms generally appear "lower" than both S0 and S1 for $-3V < V_s < 3V$, indicating a generally reduced DOS close to E_F as compared with S0 and S1.

634

Figure 1 also shows that S1 adatoms have a high probability of further reacting into S2 adatoms with increasing O_2 dose. Fig. 1(E) tabulates the sequence of creation of S1 adatoms and conversion to S2 seen in Figs. 1(A) - 1(D). We find that S1 adatoms are more than 5 times as reactive as unreacted adatoms, and that more than half the S2 adatoms seen in Fig. 1(D) were observed to have an S1 intermediate state. Hence, there appear to be at least two *successive* oxidation stages of Si adatoms which can be observed, which coexist on the surface even at low O_2 exposures. In the absence of additional O_2, S1 and S2 adatoms generally appear quite stable at 300 K in successive images for periods up to several hours, with an exception to be discussed later.

S1 adatoms also tend to nucleate preferentially at specific adatom sites within the 7x7 unit cell. By measuring the spatial distribution of S1 adatoms within the unit cell as observed on samples with $D_T \leq 0.2$ L, we find that S1 adatoms show a preference for the *faulted* vs. the *unfaulted* half of the unit cell [13] by a ratio which ranges from 2:1 to 4:1, and a preference for *corner* adatom sites (adjacent to a corner hole) vs. *center* sites (not adjacent to a corner hole) by a ratio of 1.5:1 to 2:1. We note here that a previous STS study of the *clean* 7x7 surface [12] has identified a filled dangling bond (DB) state on the adatoms at ~ 0.3eV below E_F, which appears to have a higher occupation on the faulted (vs. unfaulted) half of the unit cell, and at corner (vs. center) sites. This ordering is the same as we observe for the adatom reactivity, suggesting that the reaction site preferences are directly related to differences in this adatom DB between the various sites.

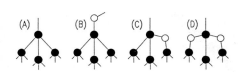

FIG. 2. Possible bonding structures. Si (O) atoms shown as filled (empty) circles. (A) unreacted Si adatom, (B) O atom on-top, (C) O in back-bond, (D) two O's in back-bonds.

We now consider the probable atomic structure of these reacted sites. Figure 2 shows several possible bonding configurations for atomic oxygen at a Si adatom. In an earlier report, we argued for a model in which the S1 adatom corresponds to Fig. 2(C), i.e. a single oxygen atom inserted into a *back-bond* configuration, while an S2 adatom has two (or possibly three) oxygen atoms in back-bonds as in Fig. 2(D). This simple assignment is consistent with our observation of a two-step reaction process at 300 K, where each step requires exposure to oxygen. It is also consistent with prior XPS [2] work which indicates that atomic oxygen is predominantly bonded (by a 4:1 ratio) to two Si atoms rather than one, and recent Transmission Electron Diffraction (TED) work [15] which indicates a prevalence of back-bonded O's at corner sites.

However, we have recently made measurements of the annealing behavior of the partially oxidized surface which casts some doubt on this model. Figure 3(A) shows a CCT of a sample dosed at 300 K with 0.15 L of O_2. Fig. 3(B)

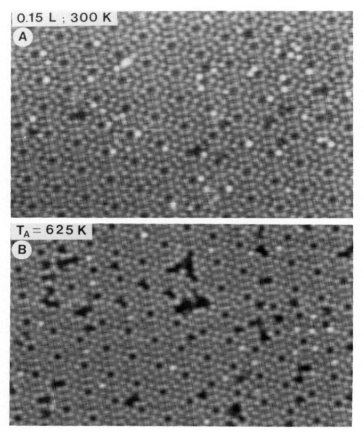

FIG. 3. Grey-scale CCT′s measured at $V_s \cong 2$ V of sample annealing behavior. (A): surface of sample A after 0.15 L O_2 dose at 300 K. (B): Different area of same sample after 60 s anneal at $T_A \cong 625$ K.

shows a different area of the same sample after it was subsequently annealed at $T_A \cong 625$ K. We see that annealing causes a decrease in the density of S1 adatoms, an increase in S2, and a net shift in their preferred location from corner sites to center sites within the unit cell. We have counted the number and position of reacted sites from large ($\sim 75 \times 75$ nm²) areas of several samples before and after annealing. Some of these results are summarized in Table I. One conclusion here is that for $T_A \leq 680$K, the ratio $|\Delta\overline{S2}/\Delta\overline{S1}|$ is within statistical uncertainties equal to unity. This suggests that mild heating may cause a 1-to-1 conversion of S1 to S2. We note here that even at 300 K, we have observed in very rare occasions a direct conversion of S1 to S2 in subsequent images without exposing the surface to O_2.

TABLE I. Annealing behavior of samples dosed with O_2 at 300 K to $D_T \sim$ 0.15 L. $\overline{S1}$ and $\overline{S2}$ are average number per 100 unit cells of S1 and S2, respectively. $\Delta\overline{S1}$ and $\Delta\overline{S2}$ are decrease (increase) in $\overline{S1}$ ($\overline{S2}$) due to annealing at T_A. Uncertainties determined from counting statistics.

Sample	T_A	$\overline{S1}$	$\overline{S2}$	$\Delta\overline{S1}$	$\Delta\overline{S2}$	$\lvert\Delta\overline{S2}/\Delta\overline{S1}\rvert$
A	300 K	52(3)	22(2)	--	--	--
	625 K	29(2)	46(2)	-23(3)	24(3)	1.04(18)
	680 K	14(1)	61(3)	-38(3)	39(3)	1.03(11)
B	300 K	58(3)	37(2)	--	--	--
	625 K	39(2)	59(3)	-20(4)	22(3)	1.14(27)
C	300 K	54(3)	30(2)	--	--	--
	730 K	4(1)	68(3)	-50(3)	38(3)	0.77(8)

Since the back-bonded structure shown in Fig. 2(C) appears to be more stable than the "on-top" structure shown in Fig. 2(B) [2], then according to the above model the only simple ways that mild heating could eliminate S1 adatoms are by combination of two (or more) S1's to form an S2 (i.e., migration of two separated back-bonded O atoms to a single adatom site), or by a back-bonded O-atom (S1) combining with an existing S2 or going sub-surface. All these processes require that *at least* two S1's be eliminated for the creation of each new S2. This would predict that $\lvert\Delta\overline{S2}/\Delta\overline{S1}\rvert \leq 0.5$, in disagreement with our findings. We are presently considering other models and experiments to try to resolve this issue.

In summary, we have observed successive oxidation of adatoms on the Si(111) 7x7 surface at 300 K, with the first stage nucleating preferentially on the faulted half and at corner sites in the 7x7 unit cell. A model which assigns these two stages to Si adatoms back-bonded to one and two oxygen atoms, respectively, is most consistent with our measurements and previous work [2, 15], but is less consistent with the apparent 1-to-1 conversion of S1 to S2 caused by mild heating.

We thank R. Hamers for the use of image processing software and helpful advice, and D. Abraham, J. Boland, J. Clabes, J. Demuth, R. Laibowitz, I.-W. Lyo, K. Markert, and A. Samsavar for useful discussions.

References

1. A. J. Schell-Sorokin and J. E. Demuth, Surf. Sci. **157**, 273 (1985).
2. G. Hollinger, J. F. Morar, F. J. Himpsel, G. Hughes, and J. L. Jordan, Surf. Sci. **168**, 609 (1986).
3. P. Gupta, C. H. Mak, P. A. Coon, and S. M. George, Phys. Rev. B **40**, 7739 (1989).
4. S. Ciraci, S. Ellialtioglu, and S. Erkoc, Phys. Rev. B **26**, 5716 (1982).

5. I. P. Batra, P. S. Bagus, and K. Hermann, Phys. Rev. Lett. **52**, 384 (1984).
6. F. M. Leibsle, A. Samsavar, and T. Chiang, Phys. Rev. B **38**, 5780 (1988).
7. H. Tokumoto, K. Miki, H. Murakami, H. Bando, M. Ono, and K. Kajimura, J. Vac. Sci. Technol. A **8**. 255 (1990).
8. R. J. Hamers, Annu. Rev. Phys. Chem. **40**, 531 (1989).
9. J. P. Pelz and R. H. Koch, Phys. Rev. B **42**, 3761 (1990).
10. J. P. Pelz, to be published.
11. R. M. Feenstra, W. A. Thompson, and A. P. Fein, Phys. Rev. Lett. **56**, 608 (1986).
12. R. J. Hamers, R. M. Tromp, and J. E. Demuth, Phys. Rev. Lett. **56**, 1972 (1986).
13. K. Takayanagi, Y. Tanishiro, M. Takahashi, and S. Takahashi, J. Vac. Sci. Technol. **A3**, 1502 (1985).
14. Ph. Avouris and I.-W Lyo, to be published.
15. J. M. Gibson, *Atomic Scale Properties of Interfaces, MRS Symposia Proceedings, Vol. 159.* (Materials Research Society, Pittsburgh, 1990), p. 179.

Full Ab Initio Determination of the Structural Parameters of H on GaAs(110)

C.M. Bertoni[1], F. Finocchi[1], M. Buongiorno Nardelli[2], F. Bernardini[1], and E. Molinari[3]

[1]Dipartimento di Fisica, Università di Roma "Tor Vergata",
 Via E. Carnevale, I-00173 Roma, Italy
[2]SISSA-ISAS, Strada Costiera 11, I-34014 Trieste, Italy
[3]CNR, Istituto "O.M. Corbino", Via Cassia 1216, I-00189 Roma, Italy

Abstract. We present a theoretical description of a fully ordered chemisorbed layer of hydrogen on GaAs(110). This is a realistic model of the structure observed at the completion of the first-stage non-erosive chemisorption. By accurate calculations of total energy and of Hellmann-Feynman forces we find a stable configuration. H bonding removes the clean surface relaxation, creating a further slight counterelaxation of the first GaAs plane. Bond lengths, angles and stretching frequencies are also calculated.

1. Introduction

The H-covered cleavage surface of GaAs has been studied intensively both as a prototype system of chemisorption on semiconductors and as a reference system for the comparison of electronic and structural properties with respect to the clean relaxed surface. Experimental investigations on the chemisorption of *atomic* hydrogen on GaAs(110) indicate that, when the H coverage reaches the value of about one monolayer (two H atoms per unit cell), a (1×1) phase is present [1,2]. The surface periodicity and flatness is lost with further exposition to H. A direct knowledge of the geometrical parameters is still lacking, due to the difficulty in detecting hydrogen by the usual structural probes. Some indications exist that H adsorbs on both surface sites [1] and for the possible removal of the substrate relaxation already at the very early stage of adsorption [2]. The interpretation of photoemission [3] and energy loss data [4] in terms of the electronic structure calculated for selected geometries [5], was supporting these indications. A first selection of the geometry on the basis of a full *ab-initio* theoretical calculation, based on the density functional scheme in the local density approximation [6,7] has been performed only recently [8]. This method is predictive only when applied in connection with first-principle *norm-conserving* pseudopotentials [9]. It requires, for surface systems, a very large number of basis functions and a good numerical accuracy in the self-consistency procedure. The calculation of the

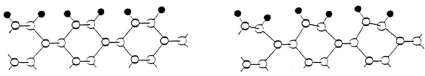

Fig.1. Sketch of two of the trial chemisorption geometries used at the initial stage of the calculation.

forces acting on the atoms for each configuration allows to approach step by step a local-minimum geometry if it exists.

2. Calculation of the equilibrium geometry

The surface system is described by a repeated-slab consisting of 7 GaAs(110) planes plus the equivalent of 6 planes containing the H atoms and the vacuum region. Plane wave (PW) basis sets with a kinetic energy cutoff up to 10 Ryd (corresponding to 2000 PW's in the diagonalization of the hamiltonian matrix and 16000 PW's in the expansion of the potential and charge density) were used. Iterative techniques were adopted to obtain the lowest occupied states in the diagonalization and the computer codes were conceived to take full advantage from the use of vector processing. The exchange and correlation energy and potential are described by the Ceperley-Alder [10] values for the homogeneous electron gas. One single k-point at the centre of the irreducible part of the two dimensional Brillouin zone (BZ) is used in the iterative procedure.

Due to the large number of choices for the values of the possible configuration parameters, we compared at the beginning some extreme geometries of chemisorption. In Fig.1 we show two of them: chemisorption on the ideal surface, and on the relaxed surface assuming the substrate geometry described in ref. 11. The H-Ga and H-As bond lengths were kept fixed at 1.59$\mathring{A}$ and 1.52$\mathring{A}$ respectively, and their directions were chosen in order to give equal angles with the other neighbouring bonds. The computed total energy difference is in favour of the former unrelaxed model with a gain of 0.73 eV per unit surface cell. Then we started from this configuration to relax the positions of the H atoms and of the GaAs outermost layer with the aid of the calculation of Hellmann-Feynman forces. The first steps were performed using a 5 Ryd cutoff in the PW's kinetic energy. An equilibrium configuration is reached when the computed force moduli are less than 5×10^{-3} $Ryd/a.u.$. Fig.2 shows

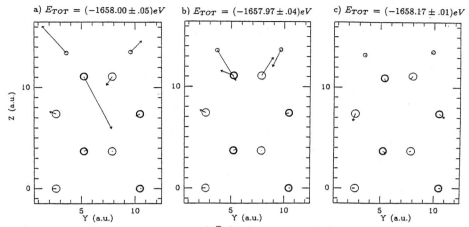

a) $E_{TOT} = (-1658.00 \pm .05)eV$ b) $E_{TOT} = (-1657.97 \pm .04)eV$ c) $E_{TOT} = (-1658.17 \pm .01)eV$

Fig.2 Atomic positions on a $(1\bar{1}0)$ plane perpendicular to the surface. The $Z = 0$ plane is the central (110) plane of the supercell. The thick and thin large circles represent the As and the Ga atoms respectively, while the H atoms are drawn as small circles. Atomic forces are drawn, the scale is fixed being $|\vec{F}|_{max} = 7.3 \times 10^{-2} Ryd/a.u..$. The corresponding total energies are also indicated. Figures 2a and 2b represent the case of the substrate atoms in their ideal configurations with two different trial positions for H atoms, while in fig.2c the case of our optimized counterrelaxed geometry is shown.

the results for the first two configurations used and the last optimized one. It is interesting to note that the values of the total energy (per unit cell containing 7 Ga, 7 As and 4 H atoms) of the three configurations differ only slightly: so the use of the forces is much more helpful in searching the stable geometry. A further refinement was then performed through some more accurate calculations with a 10 Ryd cutoff for PW's. Using configurations which differ with respect to each other within the elastic limit, it is possible to obtain also the frequencies of the localized modes associated to H motion. The calculated values for the stretching frequencies can be compared with the ones obtained by energy loss measurements [1]. In table 1 we report the values of bond lengths, the angles θ_{H-As} and θ_{H-Ga} between the bonds and the surface normal, the tilt angle θ_{Ga-As} of the first sublayer, the stretching and bending frequencies ω_1 and ω_2 evaluated in the two equilibrium geometries with 5 Ryd and 10 Ryd cutoffs respectively. The comparison with the experimental values of ω_1 is also given.

Table 1. Structural parameters and frequencies for an ordered H monolayer on GaAs(110).

		5 Ryd	10 Ryd	exp. [1]
H-Ga	bond length ($\overset{\circ}{A}$)	1.624	1.549	
	θ_{H-Ga}	38.1°	38.0°	
	ω_1 stretching (cm^{-1})	1932	1757	1890
	ω_2 bending (cm^{-1})	512	517	
H-As	bond length ($\overset{\circ}{A}$)	1.571	1.534	
	θ_{H-As}	−39.3°	−39.0°	
	ω_1 stretching (cm^{-1})	2360	1950	2150
	ω_2 bending (cm^{-1})	519	536	
	θ_{Ga-As}	−4.4°	−5.0°	

3. Conclusions

We have shown that a stable structure with optimized geometrical parameters can be found assuming that the (1×1) surface periodicity is maintained. The surface relaxation of the clean surface is removed, and a small counterrelaxation occurs with a −5° tilt angle. We observe that this seems to be a general feature of the ordered monolayer adsorption on the non-polar surfaces of III-V semiconductors. Additionally, the energy gain in counterrelaxing the surface in the adsorbed phase is much larger than the one calculated passing from the ideal to the relaxed clean surface. The values of the stretching frequencies are in good agreement with the experimental data obtained close to the monolayer coverage [1]. Also the electronic structure calculated in the equilibrium configuration favourably compares with the spectroscopical measurements [3-5,8].

   This work was supported in part by INFM and CNR through Progetto Finalizzato Sistemi Informatici e Calcolo Parallelo (grant n. 89.00011.69).

References

1. H. Luth and R. Matz, *Phys. Rev. Lett.* **46**, 1652 (1981)
2. O. M'Hamedi, F. Proix and C. Sebenne, *Semicond. Sci. Technol.* **2**, 418 (1987)

3. F. Antonangeli, C. Calandra, E. Colavita, S. Nannarone, C. Rinaldi and L. Sorba, *Phys. Rev.* **B 29**, 8 (1984)

4. S.Nannarone, C.Astaldi, L.Sorba, E.Colavita, and C.Calandra, *J. Vac. Sci. Technol. A* **5**, 619 (1987)

5. F. Manghi, C. M. Bertoni, C. Calandra and E. Molinari, *J. Vac. Sci. Technol.* **21**, 358 (1982)

6. P.Hohenberg and W.Kohn, *Phys. Rev.* **136**, B864 (1964)

7. W.Kohn and L.J.Sham, *Phys. Rev.* **140**, A1133 (1965)

8. C.M.Bertoni, M.Buongiorno Nardelli, and E.Molinari, *Vacuum*, **41**, 663 (1990)

9. G.B. Bachelet, D. R. Hamann and M. Schlüter, *Phys. Rev.* **B 26**, 4199 (1982)

10. D.M.Ceperley and B.J.Alder, *Phys. Rev. Lett.* **45**, 566 (1980)

11. S.B.Zhang and M.L.Cohen, *Surface Sci.* **172**, 754 (1986)

Part VI

Oxides

Observation of a Corrugated Structure
for the TiO$_2$(100) 1×3 Surface Reconstruction

P. Zschack

Oak Ridge Associated Universities, National Synchrotron Light Source,
x14, Brookhaven National Laboratory, Upton, NY 11973, USA

Abstract. Grazing incidence x-ray diffraction and LEED were used together to demonstrate that the TiO$_2$ (100) 1x3 surface reconstruction incorporates micro-facets with (110) character to form a corrugated surface.

1. Introduction

The various low index surfaces of rutile TiO$_2$ have been of interest since the early 1970's when the photo-active properties of rutile were first reported [1]. Additionally, dispersed noble metal catalysts supported on TiO$_2$ substrates were found to exhibit the strong-metal support-interaction (SMSI) [2]. Since then, there have been many experimental and theoretical efforts to characterize and understand the physical and electronic properties of these surfaces [3-14]. Until now, no information concerning the detailed surface crystallography of any TiO$_2$ surface has been available. So, theoretical methods applied to calculate the electronic properties of these surfaces have depended on assumptions concerning the specific atomic geometry.

Traditional LEED techniques can be used to solve surface structures in the usual way by comparison of experimentally measured I-V curves with I-V curves generated from a model for the surface structure. This approach generally has good sensitivity to perpendicular positions the surface atoms, but rather poor sensitivity to their lateral registries.

As grazing incidence x-ray diffraction techniques for surface structural determinations are now becoming well established, excellent sensitivity to the lateral positions of surface atoms can be attained. X-ray scans along the surface diffraction rods to obtain information about the perpendicular atom positions are experimentally difficult, and can be unrealistic in light atom systems such as this, where compton scattering and TDS from the bulk produce substantial background intensities.

It is then quite natural to employ both grazing incidence x-ray scattering and normal incidence LEED as complementary techniques and exploit the better resolution features of each. In this study, grazing incidence x-ray diffraction was first used to determine the lateral positions of the surface atoms in the 1x3 reconstruction of the rutile (100) surface. By projecting the electron density along the surface normal, a model for the surface was constructed. This model was refined using

Springer Series in Surface Sciences, Vol. 24   **The Structure of Surfaces III**
Editors: S.Y. Tong · M.A. Van Hove · K. Takayanagi · X.D. Xie
© Springer-Verlag Berlin, Heidelberg 1991

LEED crystallography, including I-V curve analysis. In this way, a three dimensional representation for the reconstructed surface was obtained.

The 1x3 reconstruction was found to incorporate micro-facets with a (110) character to form a corrugated surface structure. This structure is appealing from an energy consideration since the (110) is the most stable TiO_2 surface, and helps to explain the similarities in the electronic properties observed for both the (100) 1x3 and the (110) 1x1 surfaces [3-14].

2. Experimental

To obtain both electron and x-ray diffraction information from the same surface, a portable UHV compatible surface-science chamber was designed and constructed [15]. A (100) TiO_2 single crystal was mechanically ground and lapped with successively finer diamond paste compound, and was finely polished with 0.05 μm Al_2O_3. To remove gross surface contaminants, a 5 minute etch with 6 N NaOH was used prior to mounting the sample on the annealing stage of the surface-science chamber. After bakeout, the sample was heated to 875 K, sputtered with 2 KeV Ar^+ for 30 minutes, annealed at 875 K for 30 minutes, and slowly cooled to ambient. This procedure was repeated once to obtain a good, sharp 1x3 LEED pattern.

X-ray integrated intensities from 19 surface reflections were obtained using 1.5 mrad of focused bending magnet synchrotron radiation at the x18a beamline facility at the NSLS. The sample and chamber were mounted to the 4-circle diffractometer, and to minimize the background intensity, the glancing angle was set to approximately the critical angle for total reflection from rutile (4 mrad). A monitor of the specular beam was used to assure a constant glancing angle, and the integrated intensities were obtained from rocking curve scans through the fractional order reflections. Any small change in the glancing angle produced an approximately linear background (primarily due to compton scattering from the bulk) that was subtracted from the integrated intensities. For this study, the surface diffraction rods were measured at the $0kl$ position in reciprocal space.

LEED beam I-V curves were obtained using the same sample and chamber, and an electron pulse-counting data acquisition system [15,16]. The LEED data were collected by storing digital LEED images for incident electron beam energies between approximately 20 and 100 eV at 3 eV increments. Each observed LEED beam was then integrated to obtain the experimental I-V curves for 7 symmetry independent diffraction rods.

3. Results

A Fourier synthesis using only the $0kl$ fractional order structure factors produces a projection of surface electron density along the [100] surface

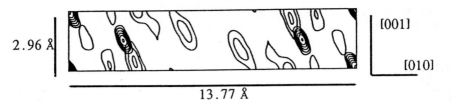

2.96 Å

[001]

[010]

13.77 Å

Figure 1. Electron density projected along the surface normal.

normal [17]. Because integer $0kl$ reflections contain crystal truncation rod effects, and because certain integer $0kl$ reflections are allowed by the bulk, these reflections were omitted. To obtain an electron density representation of the diffraction information, it was necessary to assign a phase angle to each structure amplitude. In a centrosymmetric structure, this phase assignment is reduced to a choice of 0 or π for each measured F_{0kl}.

The structure factor amplitudes were first used to construct a Patterson function from which the approximate lateral Ti positions were evident. The position of the relatively heavy Ti atoms were then used to determine the phase angles. It was a straightforward exercise to generate an electron density map from the set of phased structure factors. Since only $0kl$ structure factors were used for the Fourier inversion, the resulting electron density map represents a projection of electron density along the [100] surface normal, and since the integer order reflections were omitted, only atoms with translational symmetry different from that of the bulk are revealed. This electron density map is shown in figure 1.

From the electron density projection, one can construct a model for the 1x3 reconstruction. This model, shown in figure 2, is formed by terminating the (100) surface with alternating (110) and (-110) micro-facets. These micro-facets produce a surface with a corrugation approximately one unit cell deep, with troughs along the [001].

Actually, the lateral positions of the surface atoms are slightly displaced from the ideally terminated corrugated surface. To determine the lateral positions more precisely, an atomic parameter least squares refinement was performed. No systematic x-ray structure factor extinctions were observed, and since 2-fold rotational symmetry is imposed on the diffraction pattern by the presence of two surface domains (caused by surface steps and verified by comparison of the (10) and (-10) LEED beams), only the $p2$ and pmm crystallographic plane groups are possible representations of the unit cell symmetry. Since pmm is the symmetry of the bulk projection to (100), it was selected for the refinement. To account for the average thermal displacements, an effective isotropic Debye-Waller factor was included [18,19]. The results of the least squares refinement (where R = 0.079) appear in table 1, where x is the fractional coordinate along the [010] axis of the surface unit cell, and each atom resides on a mirror plane perpendicular to [001].

Now, with the lateral positions of the surface atoms determined, and a model for the surface constructed, a LEED experiment including a full

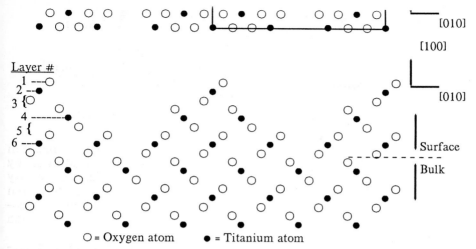

Layer #

O = Oxygen atom   • = Titanium atom

Figure 2. View of the (100) 1x3 reconstruction along the [100] (top) and along the [001] (bottom). Atoms are in projection, and only the surface atoms are projected to the (100). Also shown is the layer stacking sequence used in the LEED analysis.

dynamical calculation of the diffraction rod profiles was used to refine the perpendicular positions. For convenience and flexibility, the Tong and Van Hove LEED computational package [20] was chosen to perform the theoretical I-V curve computations. Alternating oxygen and titanium composite layers were considered (see figure 2), and matrix inversion was used to calculate the scattering from each layer. To consider the complete stack of layers that form the surface structure, the renormalized forward scattering perturbation was applied. The first 5 interlayer spacings were allowed to vary in typically 0.05 Å increments, and over 320 variations of the model were tested. The perpendicular positions of atoms in the composite layers were held at their bulk values, and only rigid perpendicular displacements of the layers were allowed.

To determine the best structural model, an average R-factor was used as the selection criterion [21]. The surface model with the lowest R-factor can be summarized as follows: $\Delta_{12} = 0.05$, $\Delta_{23} = 0.05$, $\Delta_{34} = 0.10$, $\Delta_{45} = -0.20$, and $\Delta_{56} = 0.00$, where Δ_{ij} represents the change in Å in the spacing between layers i and j. The bulk spacing is 0.918 Å between all layers. This structure has an average R-factor = 0.24 (Zanazzi-Jona R-factor = 0.34), and taken together with the lateral positions from Table 1, allows a detailed description of the complete 1x3 reconstruction.

It should be noted that another model (a simple missing row type structure) was consistent with the Patterson function, but was rejected by consideration of both the x-ray and LEED R-factors.

Table 1. Refinement of lateral positions of surface atoms

	Wyckoff Notation	Point Symmetry	Fractional Coordinates	Ideal Termination	Refined Values
Ti Atoms	1 a	mm	0,0	x = 0	x = 0.0 (not refined)
	1 a	mm	0,0	x = 0	x = 0.0 (not refined)
	2 e	m	$^+/-$ x,0	x = 0.333	x = 0.331$^+/-$ 0.002
	2 f	m	$^+/-$ x,1/2	x = 0.167	x = 0.184$^+/-$ 0.002
O Atoms	2 e	m	$^+/-$ x,0	x = 0.100	x = 0.144$^+/-$ 0.010
	2 e	m	$^+/-$ x,0	x = 0.233	x = 0.239$^+/-$ 0.007
	2 f	m	$^+/-$ x,1/2	x = 0.067	x = 0.021$^+/-$ 0.012
	2 f	m	$^+/-$ x,1/2	x = 0.067	x = 0.067 (not refined)
	2 f	m	$^+/-$ x,1/2	x = 0.267	x = 0.275$^+/-$ 0.008
	2 f	m	$^+/-$ x,1/2	x = 0.400	x = 0.410$^+/-$ 0.010

4. Summary

Inspection of the refined interlayer spacings suggests that the topmost layers are expanded away from the bulk. Because there are lateral displacements as well, a more important parameter is the average Ti - O bond length between each layer. These are summarized as follows (in Å): b_{12} = 1.79, b_{23} = 2.01, b_{34} = 2.07, b_{45} = 1.82, b_{56} = 2.02, and b_{67} = 1.959 (= bulk). So, the Ti - O bond length is actually smaller at the first layer, and oscillates toward the bulk value with increasing depth.

This study has demonstrated that the (100) 1x3 TiO_2 surface is a corrugated structure, with (110) micro-facets and equal numbers of 5-fold and 6-fold coordinated Ti atoms. The equilibrium surface involves atomic displacements from the ideally terminated corrugated structure. It was also demonstrated that both grazing incidence x-ray diffraction and normal incidence LEED can be utilized effectively together to solve complicated surface structures.

Acknowledgements. Support for this work was provided by the US DOE through contract DE-FG02-85-ER45183. Additional financial assistance was provided by IBM and by the Northwestern University Murphy Endowment. The help and advice of Professors J.B. Cohen and Y.W. Chung is gratefully acknowledged.

References

[1] A. Fujishima and K. Honda, Nature 238, 37 (1972).
[2] S.J. Tauster and S.C. Fung, J. Catal. 55, 29 (1978).
[3] Y.W. Chung, W.J. Lo, and G.A. Somorjai, Surf. Sci. 64, 588 (1977).
[4] R.V. Kasowski and R.H. Tait, Phys. Rev. B 20, 5168 (1979).
[5] S. Munnix and M. Schmeits, Phys. Rev. B, 2202 (1984).
[6] S. Munnix and M. Schmeits, Surf. Sci. 126, 20 (1983).
[7] C.R. Wang and Y.S. Xu, Surf. Sci. 219, L542 (1989).
[8] V.E. Henrich and R.L. Kurtz, Phys. Rev. B 23, 6280 (1981).

[9] V.E. Henrich, Prog. Surf. Sci. **9**, 143 (1979).
[10] V.E. Henrich, Prog. Surf. Sci. **14**, 175 (1983).
[11] R.H. Tait and R.V. Kasowski, Phys. Rev. B **20**, 5178 (1979).
[12] W. Gopel, G. Rocker and R. Freierabend, Phys. Rev. B **28**, 3427 (1983).
[13] W. Gopel, J.A. Anderson, D. Frankel, M. Jaehnig, K. Phillips, J.A. Schafer and G. Rocker, Surf. Sci. **139**, 333 (1984).
[14] P. Zschack, J.B. Cohen and Y.W. Chung, to be published.
[15] P. Zschack, J.B. Cohen and Y.W. Chung, J.Appl.Cryst.**21**, 466 (1988).
[16] P.C. Stair, Rev. Sci. Instrum. **51**, 132 (1980).
[17] R. Feidenhans'l, Surf. Sci. Reports **10**, 105 (1989).
[18] P. Zschack, Ph.D. Thesis, Northwestern University 1989.
[19] J.K. Burdett, T. Hughbanks, G.J. Miller, J.W. Richardson Jr. and J.V. Smith, J. Am. Chem. Soc. **109**, 3639 (1987).
[20] M.A. Van Hove and S.Y. Tong, *Surface Crystallography by LEED. Springer Series in Chemical Physics 2.*, Springer-Verlag. (1979, 1984 revisions)
[21] M.A. Van Hove, W.H. Weinberg and C.M. Chan, *Low Energy Electron Diffraction. Springer Series in Surface Science 6.* Springer-Verlag. (1986)

Simultaneous Epitaxial Growth of Two Differently Oriented Ni Island Structures upon TiO$_2$(110)

Ming-Cheng Wu and P.J. Møller

Department of Physical Chemistry, H.C. Ørsted Institute,
University of Copenhagen, Denmark

The growth of ultrathin (0–80 Å) layers of Ni upon a (1×1) TiO$_2$(110) surface
at room temperature is studied by low-energy electron diffraction (LEED) and
Auger electron spectroscopy (AES). A comparison of experimentally observed
LEED patterns to those from a computer simulation based upon Ewald-sphere
analysis indicates that two types of Ni islands are formed on TiO$_2$(110) simultaneously: hexagonal structures oriented parallel and inclined to the substrate
plane. Following the attenuation of the substrate Auger signals the growth is
determined to be of the Stranski-Krastanov mode and the density of Ni islands
on the surface is estimated on the basis of an isotropic-growth model.

The growth of thin metal films on a number of metal oxides has long been
a subject of fundamental interest. However, the geometrical structures
of metals on well-characterized single-crystal metal oxides have been
studied only in recent years [1-5], and it is not clear yet how many
factors are crucial in determining the interfacial geometrical structures
and what is the epitaxial relationship.

This article presents a study of the epitaxial growth of Ni overlayers
upon (1×1) TiO$_2$(110) surfaces by means of low-energy electron diffraction (LEED) and Auger electron spectroscopy (AES). The studies were
carried out in a multidetector system with a base pressure of 6×10^{-9}
Pa. It performs a number of surface electron spectroscopies and LEED,
as described elsewhere in detail [6]. A quartz crystal microbalance moni-
tors the film thickness of deposited metal from a shutter-controlled beam
of metal vapor, produced by electron beam evaporation. The Ni deposition was performed at room temperature at rates of 1-2 Å/minute using
99.995 % pure nickel. The surface treatment and characterization of
TiO$_2$(110) specimens have been described previously [7].

The growth of Ni upon (1×1) TiO$_2$(110) at room temperature has
been studied for coverage ranging from 0 to 80 Å. At lower Ni coverages
(below 10 Å) the substrate LEED pattern was attenuated gradually

Springer Series in Surface Sciences, Vol. 24   **The Structure of Surfaces III**
Editors: S.Y. Tong · M.A. Van Hove · K. Takayanagi · X.D. Xie
© Springer-Verlag Berlin, Heidelberg 1991

with increasing Ni deposition thickness, d_{Ni}. No visible extra LEED spots were observed during the initial Ni deposition. Three weak lines along the substrate [$1\bar{1}0$] direction began to appear at $d_{Ni}=10$ Å and the intensity of the lines increased with increasing Ni deposition. The substrate pattern completely disappeared when $d_{Ni}>30$ Å.

Upon slightly annealing the surface the three-line LEED patterns from Ni overlayers changed into many extra spots lying in the lines. Fig. 1a shows the reappearance of the substrate pattern in coexistence with the extra spots from the Ni overlayer after the surface that was covered with 80 Å of Ni had been annealed at 520 K for 7.5 minutes. Fig. 1b is the corresponding simulated LEED pattern which will be discussed later. Three sets of diffraction spots from the Ni overlayers were recognized after careful analysis to a series of the LEED patterns, among which the two sets of spots are actually symmetrical with respect to the substrate [001] line in reciprocal space.

It is of interest to notice that the two sets of extra spots move in a complex way with increasing LEED electron-beam energy. The corre-

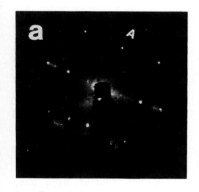

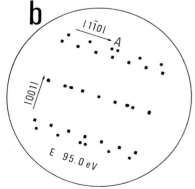

Figure 1　The LEED pattern from Ni/TiO$_2$(110) at the beam energy of 95.0 eV after the surface was annealed at 520 K. (a) The observed pattern; (b) the simulated one, the black squares represent the substrate diffraction spots, and the black circles the diffraction spots from the Ni overlayers.

653

sponding symmetrical diffraction spots from the two sets, for example, move away from each other with increasing the beam energy. These extra spots exhibit an anomalous behaviour in which the change of the LEED pattern from the overlayers, as a function of the beam energy, does not behave normally as does the substrate pattern.

This usually indicates that the overlayer plane is tilted away from the substrate plane. A computer simulation based upon Ewald-sphere analysis has been carried out in order to obtain structural information from the LEED patterns. We notice that the observed complex movement of the diffraction spots occurs only in the lines along the $[1\bar{1}0]$ direction of the substrate. The relation between the overlayer- and substrate-diffraction spots in the [001] direction is somehow similar to the case of Cu on $TiO_2(110)$, in which a hexagonal close-packed Cu superlattice was observed [5]. It thus suggests to us to consider using a hexagonal-structure lattice in simulation of the LEED patterns from the overlayer.

It is assumed that Ni atoms are hexagonal close-packed, i.e. that nickel grows in a (111)-oriented overlayer, and that its $[10\bar{1}]$ direction is

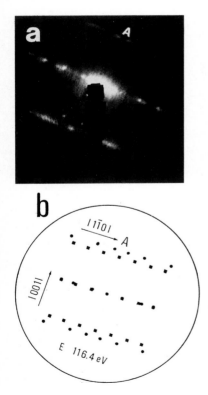

Figure 2 The LEED pattern from $Ni/TiO_2(110)$ at the beam energy of 116.4 eV after the surface was annealed at 520 K. (a) The observed pattern; (b) the simulated pattern

parallel to the [001] direction of the substrate. Let the surface normal of the Ni(111)-oriented overlayers be tilted an angle α with respect to that of the substrate plane. The [001] direction of the substrate lies along the axis of the tilt. A detailed scheme and formulas for the simulation is given in the appendix.

The patterns are simulated by superposition of patterns from the equivalent plane (tilted by an angle $-\alpha$) and with patterns from the plane parallel to the substrate plane. We have compared the observed LEED patterns to those simulated for the beam energies ranging from 50 eV to 180 eV in steps of approximately 10 eV. It was found that the best fit between the observation and the simulation was obtained for an angle (α) of 27°.

Figures 1-3 show some typical LEED patterns at several beam energies, where the simulated patterns are also inserted. Let us here give one example to describe how the spot (marked A in figures 1-2 just above the (1,0) spot of the substrate) develops with the electron-beam energy, E_p.

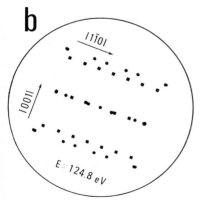

Figure 3 The LEED pattern from Ni/TiO$_2$(110) at the beam energy of 124.8 eV after the surface was annealed at 520 K. (a) The observed pattern; (b) the simulated pattern

The flat broad spot A at E_p=95.0 eV splits into two spots at E_p=116.4 eV, while on the simulated pattern the already split spots at E_p=95.0 eV separate more in distance from each other at E_p=116.4 eV.

The above results indicate that two types of Ni islands are formed simultaneously upon the initial monolayer on the TiO_2(110) surfaces. The first is a hexagonal structure oriented parallel to the substrate, and the second a hexagonal structure whose direction of growth is inclined with reference to the substrate plane. In the former case the Ni grows in registry with the $[1\bar{1}0]$ direction but incommensurately in the other two-dimensional directions of the substrate, similarly to the case of Cu/TiO_2(110) [5], leading to the growth of normal f.c.c. bulk structured nickel. The Ni lattice constant at the interface is derived to be 2.50 Å, slightly larger than the bulk 2.49 Å value of nickel. In the latter case the (111) plane of Ni is inclined to the substrate plane by an angle 27°. It is interesting to see whether this tilted angle has physical meaning because one of Ni crystal planes has to touch the substrate under such strict geometrical conditions. We found that the (131) plane of the Ni crystal satisfies the requirement: the angle between the (111) and (131) planes is 29.5°, in good agreement with the value of 27°. The next two crystal planes adjacent to the (131) plane of Ni have an angle difference with reference to 27° larger than 10°. The two types of Ni islands with the f.c.c. structure upon (1×1) TiO_2(110) can be characterized as follows:

$$(111)_{Ni} \parallel (110)_{TiO_2}, \; [10\bar{1}]_{Ni} \parallel [001]_{TiO_2}$$

$$(131)_{Ni} \parallel (110)_{TiO_2}, \; [10\bar{1}]_{Ni} \parallel [001]_{TiO_2}$$

The growth of Ni upon the room-temperature (1×1) TiO_2(110) surfaces in the coverage range from 0 to 16 Å has also been followed by AES measurements. The substrate O(KLL) and Ti($L_3M_{23}M_{23}$) Auger peak-to-peak heights as a function of d_{Ni} ($AS - d$ plots) are plotted in fig. 4, respectively. The solid curves in the figure are simulated by a theoretical form and will be discussed later.

The first-monolayer break appears around 2.5 Å as read by the quartz-crystal microbalance monitor. In the first-monolayer coverage range attenuation of the substrate signals shows approximately linear dependence on d_{Ni}. The straight lines in fig. 4 represent the numerical fits based upon the least-square analysis. The completion of the first monolayer thus suggests that the growth mode of Ni films upon TiO_2(110) obeys at least the Stranski-Krastanov (SK) mode (layer-plus-islands). The layer-by-layer (FM) mode is excluded because the substrate LEED

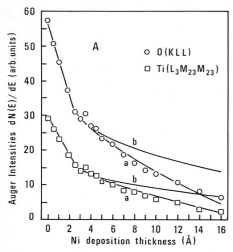

Figure 4   O(KLL) and Ti($L_3M_{23}M_{23}$) Auger intensities dN(E)/dE versus deposition thickness d_{Ni} of Ni deposited at 293 K upon a TiO$_2$(110) surface. The first break point at d_{Ni}=2.5 Å corresponds to Ni monolayer coverage. The solid curves over the 2.5–16 Å coverage range are simulated on the basis of the theoretical forms, respectively: curve (a) the quasi-isotropic growth model, and curve (b) the anisotropic growth model.

pattern was still observable after the surface was covered by a rather thick layer, e.g. 25 Å, as described above.

In the coverage range from 2.5 Å to 16 Å, the $AS - d$ plots of the substrate signals may be described by a simple model proposed initially by Zhu et al. [8]. The model is called the quasi-isotropic growth model on the basis of a number of assumptions, and its detail and the formulas for the simulation will appear elsewhere [9]. The solid curves a and b in fig. 4 are simulated by means of the isotropic growth model and an anisotropic growth model, respectively. It is seen that only the isotropic growth model predicts the correct behaviour of Ni growth upon (1×1) TiO$_2$(110). It may be expected that, after completion of a hexagonal close-packed Ni monolayer with the sixfold symmetry, the anisotropic features of the TiO$_2$(110) surfaces may be weakened very much, leading to an isotropic growth of Ni islands. The density of the Ni islands in the coverage range from 2.5 Å to 16 Å, before coalescence occurs, is estimated on the basis of the isotropic growth model to be 2.9 × 10^{13} cm^{-2}.

The present work was supported by the Danish Center for Surface Reactivity.

Appendix

Given below are the scheme and formulas for simulation of the positions of LEED beams (i, j) from an overlayer that is inclined to a substrate plane by an angle α. Let the substrate have a rectangular lattice, and the overlayer a hexagonal lattice. The unit-cell vectors of the former are $\mathbf{a}_1$ and $\mathbf{b}_1$, and the corresponding reciprocal vectors $\mathbf{a}_1^*$ and $\mathbf{b}_1^*$. The vectors of the latter are $\mathbf{a}_2$ and $\mathbf{b}_2$, and the corresponding reciprocal vectors $\mathbf{a}_2^*$ and $\mathbf{b}_2^*$. The direction of $\mathbf{b}_1$ is parallel to $\mathbf{b}_2$, both lying along the axis of the tilt. Two sets of reciprocal rods from the substrate and the overlayer, together with the Ewald sphere, are depicted in a side view in fig. 5, where the real and reciprocal lattices from the substrate and the overlayer (top views), respectively, are also inserted.

LEED patterns are normally viewed along the incident beam direction perpendicular to the substrate plane. The coordinates of the diffracted beam $(x_{i,j}, y_{i,j})$ in the plane perpendicular to the beam direc-

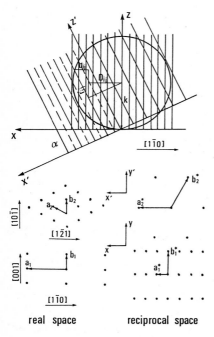

real space   reciprocal space

Figure 5   Scheme of two sets of reciprocal rods from the substrate and the overlayer together with the Ewald sphere (a side view). The real and reciprocal lattices from the substrate and the overlayer (top views), respectively, are inserted.

tion are of interest for comparison to the LEED pattern photographs. The x-coordinate is oriented in the direction of the tilt. These coordinates can be obtained with reference to the Ewald construction. For the given case

$$x_{i,j} = (D_{i,j} + d_{i,j})/\mid k \mid, \quad y_{i,j} = j \mid b_2^* \mid \sin 60^0/ \mid k \mid$$

where i, j are indexes of LEED diffracted beams from the overlayer $(i, j = 0, \pm 1, \pm 2, \cdots)$. $D_{i,j}$ is the distance between the center of the Ewald sphere and the center of the circle in which the reciprocal lattice plane i intersects the Ewald sphere, projected onto the plane perpendicular to the incident beam (see fig. 5). $d_{i,j}$ is the difference in x coordinate of the jth beam in the plane i with reference to that of the center of the circle of intersecting plane i. The radius of the circle of intersection i is $r_{i,j}$

$$r_{i,j} = [\mid k \mid^2 - (\mid k \mid \sin \alpha + i \mid a_2^* \mid + j \mid b_2^* \mid \cos 60^0)^2]^{1/2}$$

and

$$D_{i,j} = (\mid k \mid \sin \alpha + i \mid a_2^* \mid + j \mid b_2^* \mid \cos 60^0) \cos \alpha$$

$$d_{i,j} = [r_{i,j}^2 - (j \mid b_2^* \mid \sin 60^0)^2]^{1/2} \sin \alpha$$

where $\mid k \mid = \sqrt{2mE}/\hbar = 0.512\sqrt{E}$, with $\mid k \mid$ in Å^{-1} units and E (the beam energy) in eV units, $\mid a_2^* \mid = 2\pi/(\mid a_2 \mid \cos 30^0)$ and $\mid b_2^* \mid = 2\pi/(\mid b_2 \mid \cos 30^0)$.

References

[1] H. Onish, T. Aruga, C. Egawa and Y. Iwasawa, Surf. Sci. **199**, 54 (1988).
[2] (a) K. Tamura, M. Owari and Y. Nihei, Bull. Chem. Soc. Jpn. **61**, 1539 (1988). (b) K. Tamura, U. Bardi and Y. Nihei, Surf. Sci. **216**, 209 (1989).
[3] H. R. Sadehi, D. E. Resasco, V. E. Henrich and G. L. Haller, J. Catal. **104**, 252 (1987).
[4] J.-W. He and P. J. Møller, Surf. Sci. **178**, 934 (1986).
[5] P. J. Møller and M.-C. Wu, Surf. Sci. **224**, 265 (1989).
[6] P. J. Møller and J.-W. He, J. Vacuum Sci. Technol. **A5**, 996 (1987).
[7] M.-C. Wu and P. J. Møller, Surf. Sci. **224**, 250 (1989).
[8] Q.-G. Zhu, A.-D. Zhang, E. D. Williams and R. L. Park, Surf. Sci. **172**, 433 (1986).
[9] M.-C. Wu and P. J. Møller, to be submitted.

Characterization of (100) and (111) Surfaces of MgO by Reflection Electron Microscopy

M. Gajdardziska-Josifovska, P.A. Crozier, and J.M. Cowley

Department of Physics, Arizona State University, Tempe AZ 85287, USA

Abstract. We have employed reflection electron microscopy to study the topography and the crystal structure of (100) and (111) surfaces of MgO prepared by annealing in oxygen. We find that the (111) surface is reconstructed with a $(\sqrt{3}\times\sqrt{3})R30°$ structure which is composed mainly of oxygen.

1. Introduction

MgO is an important material for understanding the basic properties of ionic-crystal surfaces. It has a rocksalt structure and the ionicity of its bonds is very close to 2. It is a suitable model surface for insulating oxides because it is relatively stable under electron and ion bombardment. From the technological viewpoint, the characterization of the (100) surfaces of MgO has been motivated by its numerous uses as a substrate for epitaxial growth of metal films [1], high T_C superconducting films [2] and catalysis test systems [3].

Two of the surfaces of MgO are of special interest because of their very different fundamental properties. The (100) surface has the lowest surface energy and it is the natural cleavage plane of the crystal. Due to the ease of preparation this surface has been studied extensively with many different surface techniques under both UHV and non-UHV conditions [4]. The square array of Mg and O ions at the surface was shown to be very close to a bulk termination (the rumpling and/or surface relaxation appears to be less than 3% [5]). Both the net surface charge and the dipole moment normal to the surface are zero which is the theoretical requirement for good stability of ionic surfaces [6]. On the contrary, the surface charge of a bulk terminated (111) surface is infinite and there exist a dipole moment normal to the surface because in the <111> direction the crystal consists of alternating close packed planes of Mg^{2+} cations and O^{2-} anions The electrostatic arguments indicate that the surface energy of the (111) surface would be infinite (or very large for small crystals) and that the surface could exist only through substantial reconstruction or adsorption of charged species. Since cleaving along the (111) plane is very difficult, the number of studies of the (111) surface has been limited. Henrich [7] and Onishi et al. [8] have investigated the effect of Ar bombardment and UHV annealing of the mechanically and chemically polished (111) surfaces using low energy electron diffraction (LEED). They showed that the resulting surfaces were highly corrugated into sets of {100} planes.

We have studied the effect of oxygen annealing on the (100) and (111) surfaces of MgO. We use the technique of reflection high energy electron diffraction (RHEED) and reflection electron microscopy (REM) to characterise the surface crystal structure and topography. Reflection electron energy-loss spectroscopy (REELS) with 100keV electrons is used to assess the elemental composition of the surface. Although these studies are performed under non-UHV conditions it is important to note that, for relatively inert surfaces (such as noble metals and oxides), the surface crystal

Springer Series in Surface Sciences, Vol. 24   **The Structure of Surfaces III**
Editors: S.Y. Tong · M.A. Van Hove · K. Takayanagi · X.D. Xie
© Springer-Verlag Berlin, Heidelberg 1991

structure is not destroyed by thin amorphous contamination layers [9]. Furthermore, because reflection microscopy relies on Bragg scattering of the high energy electrons from the crystal surface, accurate information about these "burried interfaces" can be obtained.

2. Experimental Conditions

Fresh (100) surfaces were produced by cleaving MgO crystals in air before each experiment and (111) surfaces were prepared by cutting and mechanical polishing. High temperature annealing of the samples was performed in an alumina-muffle-tube furnace under a steady flow of research-grade oxygen. Two annealing temperatures were used (1550°C and 1700°C) in combination with two annealing times (22h and 50h). Additional details of sample preparation procedures are given elsewhere [10].

All REM images and RHEED patterns were recorded under surface resonance condition [11] in a JEOL 2000FX transmission electron microscope (TEM) operated at 100keV. Larger incident angles were used to decrease the foreshortening in the images: 800 specular spot in the <001> azimuth was used to image the (100) surfaces; 666 specular spot in the <211> azimuth was used for the (111) surfaces. REEL spectra were acquired at a range of incident angles in a Philips 400T TEM operated at 100keV. This microscope is fitted with a Gatan 607 magnetic sectored spectrometer coupled to a serial recording system capable of single electron pulse counting.

3. Results and Discussion

(100) Surface

An REM image and accompanying RHEED pattern from a freshly cleaved (100) surface is shown in fig.1. The surface does not cleave smoothly, and the area of the terraces is very limited. The steps run predominantly in the <010> and <001> directions and the finer steps merge to form coarse steps which can be many tens of atomic layers in height.

The effect of the high temperature annealing on the cleaved (100) surfaces is shown in fig.2. The curved appearance of the steps indicates that there has been a considerable motion of the atoms on the surface during annealing. The lower temperature annealing (fig.2a) results in large terraces with ragged ledges. The fact that the steps on the surface do not branch or merge suggests that the terraces are a single atomic layer in height. Annealing at higher temperatures gives more

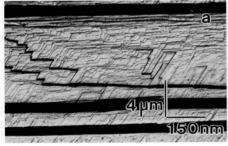

Fig.1 REM image (a) and RHEED pattern (b) of freshly cleaved (100) surface of MgO formed with the (800) specular reflection at surface resonance condition.

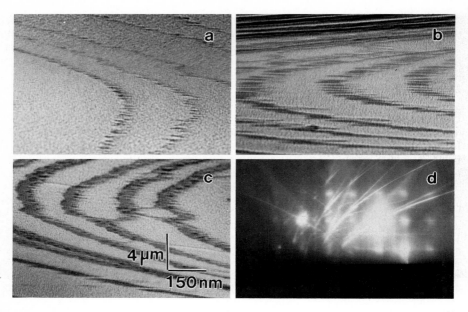

Fig.2 REM images of (100) surfaces which have been annealed in oxygen at: (a) 1550°C for 22h, (b) 1700°C for 22h, (c) 1700°C for 50h. Typical RHEED pattern from annealed surfaces is shown in (d).

pronounced raggedness of the steps (fig.2b) which increases with annealing time (fig.2c). Calcium precipitation from the bulk of the crystal is known to occur at high temperatures [12] and it could be the source of the observed step structure.

(111) Surface

As polished (111) surfaces give rise to diffuse RHEED patterns which are due to the damaged surface region. High temperature oxygen annealing was observed to restore the crystallinity of the surface and produced regions with extended terraces (fig.3). The surface steps do not exhibit decoration and some of them appear to branch. At lower temperature (fig.3a) a mottled contrast is prominent on the surface which may be due to residues from the polishing. This surface feature is removed by the higher temperature annealing (fig.3b).

The most interesting property of the oxygen annealed (111) surface is observed in the RHEED pattern in fig.3c. It consists of a ring of superlattice spots whose spacing parallel to the shadow edge is one third of the spacing of the bulk reflections. These spots indicate that a surface reconstruction has taken place. They were observed in the diffraction patterns from any part of the (111) surfaces annealed in four different runs and in two different furnaces. The geometry of the surface diffraction pattern was not changed by prolonged irradiation with the electron beam nor by exposure to air. The superlattice spots could still be observed from the surfaces which were kept in a desiccator for two months indicating that the annealed (111) surfaces are extremely stable. The study of the RHEED patterns at different azimuths shows that the reconstruction is a ($\sqrt{3}$x$\sqrt{3}$)R30° structure (in Wood's notation). The unit cell has dimensions a=b=5.16Å, α=60°.

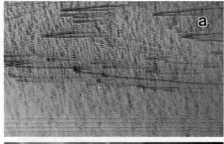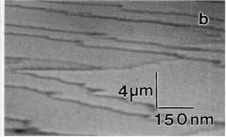

Fig.3 REM images of (111) surface of MgO after annealing in oxygen at: (a) 1550°C for 22h and (b) 1700°C for 50h. Images formed with 666 specular reflection at condition shown in (c). RHEED pattern shows superlattice spots indicating surface reconstruction.

Floquet and Dufour [13] have observed the same reconstruction on the oxygen annealed (111) surfaces of NiO. Their Auger spectra show a presence of Si atoms on the surface which was attributed to the diffusion of the impurity atoms from the bulk to the surface. We employed REELS to identify the atomic species associated with the reconstruction on the (111) surface of MgO [14]. We used spectra from the cleaved (100) surface of MgO as a standard to compare with the spectra from annealed (111) surface. Spectra were recorded from the specular reflection under diffracting geometries which satisfied the surface resonance condition. Since the incident angles for surface resonance are different for the two surfaces, a number of diffracting conditions were studied to facilitate the comparison of the results (table 1). For each surface and constant diffraction condition, the ratio between the O and Mg K ionisation edges was found to vary by less than 15% as a function of the position of the probe on the surface. The average ratios are given in table 1 and it can be seen that the variation with the diffracting condition is almost 40% for the (100) surface.

Table 1 Mg/O signal ratios from (100) and (111) surfaces under different diffracting conditions.

Surface	Specular Reflection	Incident Angle (mrad)	Mg/O Signal Ratio
(100)	(400)	17.5	0.134±0.005
	(600)	26	0.148±0.006
	(800)	35	0.107±0.006
(111)	(444)	30.4	0.05 ±0.02
	(666)	45.6	0.04 ±0.02

However, the most prominent difference in the ratio occurs between the two surfaces, the Mg/O ratio from the (111) surface being 2-3 times smaller than the ratio from the (100) surface. This difference, in conjunction with the absence of other peaks in the REEL spectra from the (111) surface, indicates that the reconstructed surface is composed of oxygen atoms or molecules.

4. Conclusion

We have studied the topography and the crystal structure of cleaved (100) surfaces of MgO and polished (111) surfaces of MgO which have been subject to high temperature annealing in an oxygen rich atmosphere. Wide single atom high terraces with ragged steps were observed on the (100) surfaces annealed at 1550°C. Step decoration was observed for the samples annealed at 1700°C, the density of which increased for longer annealing time. A $(\sqrt{3} \times \sqrt{3})R30°$ reconstruction was discovered on the annealed (111) surface. The surface structure was found to be stable in air and the reflection energy-loss spectroscopy indicates that the surface is terminated with oxygen atoms or molecules.

Acknowledgement

This work was supported by U.S. DOE grant DE-FG02-90ER45228 and was performed at the ASU HREM facility and the materials science laboratory.

References

[1] J.-W. He and P. J. Moller, Surf. Sci. **178**, 934(1986).
[2] C. B. Carter, S. R. Summerfelt, L. A. Tietz, M. G. Norton and D. W. Susnitzky, Inst. Phys. Conf. Ser. No.98, Ch.9, 415(1989).
[3] Y. C. Lee, P. Tong and P. A. Montano, Surf. Sci. **181**, 559(1987).
[4] V. E. Henrich, Rep. Prog. Phys. **48**, 1481(1985); H. Nakamatsu, A. Sudo and S. Kawai, Surf. Sci. **223**, 193(1989); X. D. Peng and M. A. Barteau, Surf. Sci. **224**, 327(1989).
[5] C. Duriez, C. Chapon, C. R. Henry and J. M. Rickard, Surf. Sci. (1990) in press.
[6] P. W. Tasker, J. Phys. C **12**, 4977(1979).
[7] V. E. Henrich, Surf. Sci. **57**, 385(1976).
[8] H. Onichi, C. Egawa, T. Aruga and Y. Iwasawa, Surf. Sci. **191**, 479(1987).
[9] T. Hsu, Ultramicroscopy **11**, 167(1983).
[10] P. A. Crozier, M. Gajdardziska-Josifovska and J. M. Cowley, J. Electron Microscopy Technique (1990) submitted.
[11] A. Ichimiya, K. Kambe and G. Lehmpfuhl, J. Phys. Soc. Jap. **49**, 684(1980).
[12] R. C. McClune and P. Wynblatt, J. Am. Cer. Soc. **66**, 111(1983).
[13] N. Floquet and L.-C. Dufour, Surf. Sci. **126**, 543(1983).
[14] Z. L. Wang and J. M. Cowley, Surf. Sci. **193**, 510(1988).

MgO(100) Structural Investigation Using EELFS and EXFAS Techniques

M. De Crescenzi[1], N. Motta[1], F. Patella[1], A. Sgarlata[1], F. Arciprete[1], A. Balzarotti[1], M. Benfatto[2], and C.R. Natoli[2]

[1]Dipartimento di Fisica, Università di Roma II, "Tor Vergata", I-00173 Roma, Italy
[2]INFN, Laboratori Nazionali di Frascati, I-00044 Frascati, Italy

The use of low energy electrons (E_p=1-3 KeV) in the reflection scattering geometry has greatly contributed to the knowledge of a wide series of structural parameters on atomic scale because of their low penetration depth and their strong coupling with the elementary excitations of the investigated system. We present recent experimental and theoretical results obtained by means of two electron techniques, namely EELFS (Extended Energy Loss Fine Structure) and EXFAS (Extended Fine Auger Structure) which relate to the most commonly used surface techniques to study the short range order in terms of radial distribution function, coordination numbers, thermal effects, etc. The EELFS analysis obtained above the oxygen and magnesium ionization K edges of a MgO(100) surface suggests that the surface lattice parameters are identical to those of the bulk so that, within our experimental resolution, we exclude any inward relaxation of the topmost layer, as suggested by previous measurements.

1. Introduction

The determination of surface atomic arrangements is still a difficult task because most of the structural techniques are not capable of providing direct information on the atomic geometry of the surface unit cell [1]. Moreover , a full structural determination requires complex electron scattering models or analysis of photoexcited electron spectra, all of which rely on crystal potentials and atomic interactions which are not precisely known. Thus it is necessary to develop a structural probe which combines the use of a simple experimental apparatus with a straightforward method to extract the local order information around the investigated atom. All these requirements hold for the EELFS (Extended Energy Loss Fine Structure) spectroscopy [2] that measures the fine structures which extend for few hundreds of eV above the ionization edges of a surface atom. The surface sensitivity of the technique is ensured by the use of electrons of 2 keV with a mean free path of about 10 Å so that few atomic surface layers are involved. The major attraction of the technique stems from the simplicity of the data analysis which is close to that of the EXAFS spectroscopy [3]. This has been demonstrated by many experimental and theoretical results [4,5,6] and it is based on the dominance of the dipolar term over all other inelastic channels (monopole , quadrupole etc.) in the matrix element of the inelastic scattering cross section above a core edge. In this paper we demonstrate the complete tunability on the lattice site of the EELFS technique in the case of MgO(100) by analyzing the radial distributions F(R) in the neighbourhood of the magnesium and oxygen atom , respectively. Through a fitting procedure using the EXAFS formula [7] and the phase shifts of McKale et al.[8], we are able to investigate the two F(R) up to the third nearest neighbours and to recognize the nature of the different scatterers. Our structural result is reliable within 0.02 Å and this allows us to exclude in a straightforward way any surface relaxation of the first few atomic layers which has been proposed recently [9]. Furthermore we present a complete analysis of the features observed in the Auger spectrum above the $L_{2,3}VV$ Mg transition. The extended fine structure (EXFAS) is observed as a modulation of the secondary electron distribution at higher kinetic energies with respect to the Auger line [10].

The proposed mechanism [10,11] involves direct recombination of a continuum core ionized electron with its core hole. The released energy in this autoionization process is supplied by the system to a valence band electron which escapes from the sample with a kinetic energy greater than that of the corresponding CVV Auger transition. Within this interpretation scheme the extended features originate from the final-state interference experienced by the core ionized electron wave function. Thus the analysis of the EXFAS structure should give the same structural information as the EELFS technique with,in addition, a greater surface sensitivity because the collected EXFAS electrons have a smaller escape depth due to their low kinetic energies (50-400 eV).

2. Experiment and Results

The experiments were performed in an UHV system with a base pressure of 10^{-10} Torr. The system was equipped with a Riber single pass cylindrical mirror analyzer (CMA) to obtain the electron energy losses and fine structure measurements in addition to the Auger analysis. The electron beam was at normal incidence to the sample surface. Cleaned MgO(100) single crystal surfaces were obtained after repeated heating of the sample at 400 °C in 10^{-6} Torr oxygen partial pressure. The

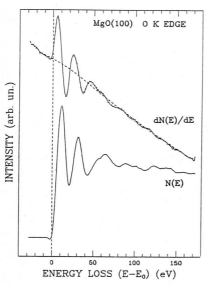

Fig.1- (upper curve) Extended energy loss fine structure above the oxygen K edge of a MgO(100) surface.The spectrum is collected as the first derivative of the electron yield N(E) and a modulation of 8 V_{p-p}.

Auger spectra performed after prolonged electron beam exposure revealed no compositional modification and contaminants within the Auger sensitivity.

The EELFS data have been collected as a derivative of the electron yield using a lock in amplifier to obtain the extended feature as far as possible from the core edge onset.

We report in the upper part of Fig.1 the first derivative energy loss spectrum detected above the oxygen K edge. Fig.2 reports the energy loss spectrum performed above the Mg K edge. To obtain a closer correspondence with the EXAFS modulation function $\chi(K)$ the raw data have been numerically integrated and the extended fine structure has been isolated from the smooth atomic scattering contribution by means of a background subtraction. In Fig.2 we report a one to one comparison between the EELFS features and the fine structure measured in the electron energy loss in transmission mode using an electron microscope [12].

The Fourier transform of $\chi(K)$, where K is the wave vector measured from the onset, is shown in Fig.3 for the oxygen K edge to visualize the different shells surrounding the oxygen absorbing atom. We observe two prominent peaks at 1.9

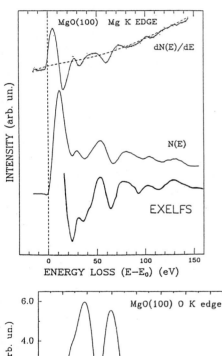

Fig.2 - (upper curve) EELFS features above the magnesium K edge of MgO(100) surface. The lower curve shows the numerical integration of the EELFS spectrum.The EXELFS spectrum [12] above the same edge obtained in transmission with an electron microscope is also shown for comparison.

Fig.3- Magnitude of the Fourier transform of the integrated EELFS above the oxygen K edge. The various peaks corresponding to the different neighbours of the MgO cage are not corrected for the proper EXAFS-like phase shift.

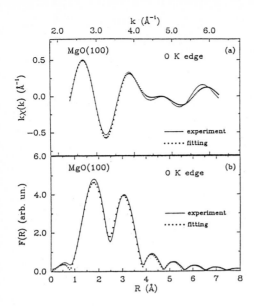

Fig.4- (a) solid line: EELFS signal obtained after backfourier inversion of the F(R) shown in Fig.3 between 1.0 and 3.8 Å ; dotted line : fitting curve obtained using the EXAFS formula [7] and the bulk MgO lattice distances for the first two O-Mg and O-O shells. The theoretical phase shifts of McKale et al. [8] have been used. (b) Fourier transforms of the curves shown in (a).

and 2.95 Å which may be assigned to the O-Mg and O-O bond lengths respectively. The results are in qualitative agreement with previous EELFS investigations [13,14]. To obtain a complete structural characterization (distance R_J coordination number N_J, and Debye-Waller factor σ^2) we have isolated, by means of a backfourier procedure, the two prominent peaks of Fig.3 and the oscillating features have been fitted using the EXAFS formula [7]. The backscattering amplitudes and the phase shifts used are those of McKale et al. [8] computed for the K edge X-ray absorption spectra in the spherical wave approximation.

The fitting curves, shown as dotted line in Fig.4 (a), for the oscillating features and in Fig.4(b) for the corresponding F(R), are in good agreement with the experimental spectra. The structural values obtained both for the O-Mg and O-O bond lengths and their coordination numbers are in excellent agreement with the crystallographic data for bulk MgO.

Recent EELFS investigations on the MgO(100) surface have suggested [9] an inward relaxation of the topmost layer by 15± 3 % with respect to the distance in the bulk. The evidence for inward relaxation is claimed because of the disagreement between experiment and calculation. Such disagreement, however, can alternatively depend upon the use of phase shift data [8] which have been computed with an improved method .

In order to exclude definitively the presence of surface relaxation we have analyzed the local environment at the Mg site. In Fig. 5(a) and 5(b) we show the results of the EELFS analysis above the Mg K edge of Fig.2, following the procedure described for the oxygen K edge. Also in this case we obtain a good agreement (both in radial position and in the coordination numbers) with the theoretical computation using the phase shifts of McKale et al. [8] and the bond lengths of the bulk MgO.

Fig.6(a) shows the extended feature measured in the high kinetic energy background above the $L_{2,3}VV$ Mg Auger transition. In the past these features have

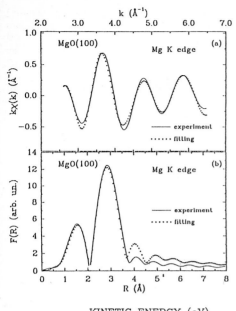

Fig.5- (a) solid line :EELFS signal obtained after backfourier inversion of the F(R) around the Mg K edge between 1.0 and 3.8 Å. The dotted line represents the fitting curve obtained using the EXAFS formula and the bulk MgO bond lengths for the first two Mg-O and Mg-Mg shells. (b) Fourier transform of the curves shown in (a).

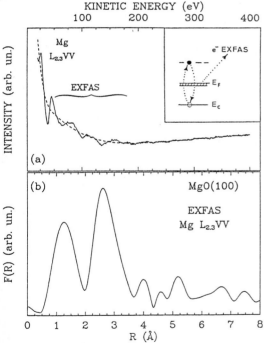

Fig.6- (a) Auger signal for a clean MgO(100) surface above the $L_{2,3}VV$ Mg transition showing the location of the EXFAS structure. The inset shows schematically the electron energy levels involved in the EXFAS process [10,11]. (b) Fourier transform of the EXFAS features of (a).

been interpreted as a diffraction effect [15]. We have recently demonstrated the EXAFS-like nature of these features originated by an autoionization process [10,11].

Sizeable EXFAS signals has been measured in all d-metals displaying narrow and high density of valence states, while they are absent for Si and Al which have low s-p DOS and large valence band width . In the case of MgO a narrow oxygen 2p band is superimposed on to the broad and much less intense Mg s band.[13] This explains the appearance of the EXFAS features also for a system without d valence states. The Fourier transform shows a radial distribution function similar to that obtained with the EELFS technique around the Mg site, shown in Fig.5 (b). A better agreement of the distances obtained with the two techniques would require a more precise knowledge of the phase shifts which are not available at present for the EXFAS analysis.

References

1- G.Ertl and J.Küppers,"Low Energy Electrons and Surface Chemistry " 2nd ed.VCH Publishers, Weinheim,West Germany 1985.

2- M.De Crescenzi,"Critical Reviews in Solid State and Material Sciences" 15,279 (1989) and references therein.

3- P.A.Lee, P.M.Citrin, P.Eisenberger, and B.M.Kincaid, Rev.Mod.Phys.53,769 (1981).

4- M.De Crescenzi,Surf.Sci.162,838(1985).

5- M.De Crescenzi, L.Lozzi, P.Picozzi, S.Santucci, M.Benfatto, and C.R.Natoli, Phys.Rev. B39,8409 (1989).

6-Y.U.Idzerda,E.D.Williams,T.L.Einstein,and R.L.Park,Surf.Sci.160,75 (1985).

7- E.A.Stern,D.E.Sayers,and F.W Lytle, Phys.Rev.B11,4836 (1975).

8-A.G.McKale,B.W.Veal,A.P.Paulinkas,S.-K.Chan,and G.S.Knapp,J.Am.Chem.Soc.110, 376 (1988).

9- A.Santoni,D.B.Tran Thoai,and J.Urban, Solid State Commun.68,1039 (1988).

10- M.De Crescenzi, E.Châinet, and J.Derrien,Solid State Commun.57,487 (1986).

11- M.De Crescenzi,A.P.Hitchcock, and T.Tyliszczak,Phys.Rev.B39,9839 (1989).

12- T.Manoubi, and C.Colliex,J.Microsc.Spectrosco.Electron.10,441 (1985).

13- Y.C.Lee,P.Tong,and P.A.Montano,Surf.Sci.181,559 (1987).

14- C.R.Henry,and H.Poppa,Thin Solid Films 189, 303 (1990).

15- A.P.Janssen.R.C.Schoonmaker,A.Chambers, and M.Prutton,Surf.Sci.45,45 (1974).

LEED Studies of Krypton and Nitrogen Monolayers on MgO Single Crystals

T. Angot and J. Suzanne

CRMC2-CNRS, Faculté des Sciences de Luminy, Département de Physique, Case 901, F-13288 Marseille Cedex 9, France

Abstract. Low energy electron diffraction (LEED) studies of krypton and nitrogen monolayers adsorbed on the ultra-high vacuum cleaved MgO(100) single crystal surface are reported. LEED isotherms of krypton feature vertical steps, unlike nitrogen. They give isosteric heats of adsorption at half coverage of 2.7 and 2.6 kcal/mol for Kr and N_2 respectively. Kr gives a hexagonal incommensurate structure with a well defined epitaxial orientation in the (010) direction. For nitrogen LEED patterns at submonolayer coverage show a poorly ordered phase becoming a more well ordered (2xn) commensurate 2D solid at monolayer completion, with $n \geq 3$, indicating a uniaxial compression along the [011] direction.

1. Introduction

Monolayers of simple Van der Waals molecules adsorbed on MgO (100) surfaces give interesting surface systems where frustration symmetry may occur. Depending on the size of the molecule compared to the surface corrugation of the substrate, we expect a wide range of behaviors owing to the competition between the natural tendency to natural packing of the adsorbate and the surface field of the substrate which will tend to impose its own square symmetry.

Experiments have been performed on very uniform MgO powders : volumetric adsorption isotherm measurements [1,2], neutron diffraction on CD_4 [2,3] and argon [2], X-ray diffraction on krypton and xenon [4] and He-scattering on CH_4 [5]. They have shown that CD_4 (CH_4) and Ar monolayers have commensurate structures and that Kr and Xe present a hexagonal incommensurate packing. However, in the case of these two latter adsorbates, information about orientational epitaxy is lacking.

We report here Low Energy Electron Diffraction (LEED) studies of monolayers of krypton and nitrogen adsorbed on MgO (100) single crystals, cleaved in situ, under ultra-high vacuum (UHV) conditions. They are the continuation of work which has been undertaken previously on crystals cleaved outside the UHV chamber [6,7,8]. The results presented here show a considerable improvement in the quality of the surface as deduced from the much more well defined phase transitions and structure of the adsorbed phases. The nitrogen molecule is particularly interesting because of its quadrupole moment which interacts strongly with the electric field of the ionic surface. We can expect that this will increase the frustration phenomena due to the competition between the molecule-molecule quadrupole and the molecule-substrate interactions.

2. Experimental set-up and procedure

The experimental set-up has been described elsewhere [7,8]. We have added an in-situ cleaver built in our laboratory. It is small enough to be mounted on the cryogenic sample holder. The cooling stage has been modified. We use a closed-cycle refrigerator (Displex) that we have adapted to our UHV system. The cold head is thermally linked to the movable sample holder via two copper braids. It allows to work within the 30 K-300 K temperature range [9]. The temperature is measured with a platinum resistor (100 Ω at 0°C). A temperature controller drives a heater on the sample holder and allows a temperature regulation at ± 0.02 K. The absolute accuracy is ± 1 K. MgO single crystals are 3N (Kr) and 4N (N_2) grade from W&C Spicer Limited.

The overlayer structure is determined from LEED spectra and the thermodynamics are measured from LEED adsorption isotherms [7,8]. Patterns are recorded using a video camera and grabbed on a Macintosh IIcx computer. Before any adsorption experiment, the surface cleanliness is checked using a CMA Auger spectrometer. Within the 10-1200 eV energy range, the spectra show no visible surface contamination, even after numerous experiments for many weeks. Auger analysis of MgO samples cleaved outside the UHV chamber, in the air or under inert atmosphere (same conditions as in ref. 6,7,8) shows clearly carbon contamination with a ratio of the carbon (270 eV) to oxygen (502 eV) peaks equal to 0.2 - 0.3.

3. Results

3.1 Krypton

Eight adsorption isotherms have been measured for 40 K ≤ T ≤ 48 K. We show in figure 1 two isotherms : one obtained in the previous study on a crystal cleaved outside, under inert atmosphere [7] and another one obtained in the present study, that is on a MgO crystal cleaved in-situ under UHV conditions. The sharpness of the phase transition (step in the isotherm) of the latter isotherm indicates a very uniform defect free surface and a great improvement with respect to previous experiments [7,8]. The same improvement has been observed in the case of argon and xenon [9].

The heat of condensation we have obtained is Q_c (Kr) = 2.7 ± 0.2 kcal/mol in very good agreement with the value obtained on powders at higher T, for the transition gas 2D - liquid 2D [1].

The diffraction pattern is shown in figure 2a. The twelve spots around the specular spot at the center of the pattern are characteristic of two hexagonal domains rotated 90° apart. These domains are incommensurate. One direction of the hexagonal cell is along the [010] direction. The other spots are due to double diffraction as shown in figure 2b. From the diffraction pattern of figure 2a, we can calculate the nearest neighbor distance between Kr atoms which is 4.14 ± 0.04 Å at 42.6 K, in agreement with the X-ray experiments [4].

Upon adsorption of two and more layers, the thick film keeps the same epitaxial orientation. The LEED results show that the growth of thick films is probably of the Stranski-Krastanov type with two uniform layers plus crystallites.

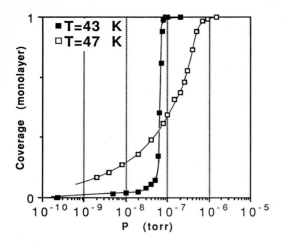

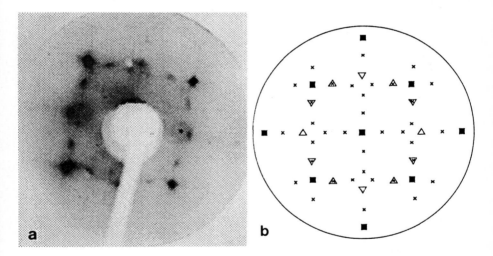

Figure 1 : LEED adsorption isotherms of krypton on MgO (100).
■   UHV cleaved MgO crystal (this work).
□   MgO crystal cleaved outside (Ref. 7,8).

Figure 2 : a) LEED pattern of a krypton monolayer on MgO (100) at T=40 K and P=2 10^{-7} torr. Electron energy 113.8 eV. b) Schematic diagram of figure 2a) showing the diffraction from the two krypton hexagonal domains rotated 90° apart (△ , ▽), the double diffraction (x) and the MgO diffraction spots (■).

3.2   Nitrogen

Figure 3 shows a set of adsorption isotherms. Unlike krypton or the other gases (Ar, Xe, CH_4) [9], the step is no longer vertical but presents a slope instead. This is the signature of a continuous transition from a dilute to a dense state. The value of the

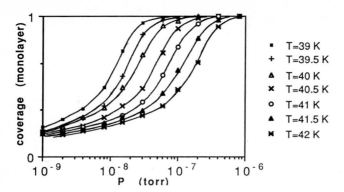

Figure 3 : LEED adsorption isotherms of nitrogen on MgO (100) obtained by video-camera (average of the substrate spots).

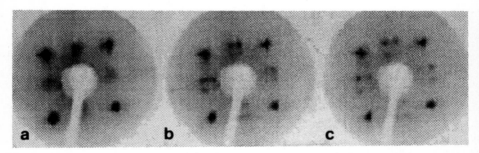

Figure 4 : LEED pattern of nitrogen/MgO(100) at T= 30 K and electron energy = 111.8 eV. a) Half monolayer coverage, small domains close to the ($\sqrt{2}$x$\sqrt{2}$) R°45 structure. b) almost complete monolayer (P=10^{-9} torr) and c) full monolayer (P=1.5 10^{-6} torr), 2xn commensurate structure.

isosteric heat of adsorption at half monolayer coverage, 2.6 ± 0.2 kcal/mol, agrees with adsorption volumetric experiments [2,11].

Figure 4 shows a set of LEED patterns from half coverage to full monolayer. At half coverage, the LEED pattern presents a poorly ordered phase, as shown by broad spots, compatible with small domains close to the ($\sqrt{2}$x$\sqrt{2}$) R°45 commensurate structure (figure 4a) . The diffraction pattern becomes much more well defined when coverage increases toward monolayer completion. Figure 4 b&c corresponds to a (2xn) structure with n≥3 indicating a uniaxial compression along the [011] direction.

4. Discussion - Conclusion

Our LEED study clearly shows the influence of the surface corrugation potential of MgO on the structure of the two adsorbates.

Krypton has no commensurability at least at short distances. This arrangement is determined, in a first approximation, by the van der Waals diameter of the atom. The epitaxial orientation of the krypton monolayer is at 45° from that of the argon film [6,9]. This difference is not well understood. It is likely that the krypton incommensurate (floating) solid layer is more sensitive to defects such as steps. Indeed, steps are known to orient along the [010] direction [10]. In the case of argon, the registry stabilization energy may be strong enough to impose the orientation to the argon crystallites.

Nitrogen monolayers are adsorbed at very low undersaturation ($P_1/P_0 \simeq 10^{-6}$) compared to krypton ($P_1/P_0 \simeq 10^{-2}$), showing the strong interaction of the substrtate field with the quadrupole moment of the molecule. But the fact that isotherms are bent at such low temperatures leads one to think that the substrate imposes a molecular orientation leading to a molecule-molecule interaction which is weakly attractive or repulsive.

The value of the heat of adsorption for krypton monolayers is in good agreement with theoretical calculations using pair-wise potentials [7]. The case of nitrogen is more complicated and only single molecule adsorption energy calculations have been performed [12]. They give a value between 2 and 3 kcal/mol depending on the value taken for the effective surface ionic charge.

References

[1] J. P. Coulomb, T. S. Sullivan and O. E. Vilches, Phys. Rev. B 30 (1984) 4753.

[2] K. Madih, Thesis, Université d'Aix-Marseille II, France 1986.

[3] J.P. Coulomb, K. Madih, B. Croset and H.J. Lauter, Phys. Rev. Lett. 54 (1985) 1536.

[4] D. Degenhardt, H. J. Lauter and R. Frahm, Surf. Sci. 215 (1989) 535.

[5] D.R. Jung, Jinhe Cui, D.R. Frankl, G. Ihm, H.Y. Kim and M.W. Cole,
Phys. Rev. B 40 (1989) 11893.

[6] T. Meichel, J. Suzanne et J.M. Gay, Compt. Rend. (Paris) t.303, Ser. II, 11 (1986) 989.

[7] T. Meichel, J. Suzanne, C. Girard and C. Girardet, Phys. Rev. B 38 (1988) 3781.

[8] T. Meichel, Thesis, Université d'Aix-Marseille II, France 1987.

[9] T. Angot, Thesis, Université d'Aix-Marseille II, France (1990).

[10] C. Duriez, C. Chapon, C.R. Henry and J.M. Rickard, Surf. Sci. (1990), in press.

[11] M. Trabelsi, K. Madih and J.P. Coulomb, Phase Transition, in press.

[12] A. Lakhlifi and C. Girardet, Surf. Sci. to be published.

Structural Analysis of a Thin CaO Layer Formed by Electron Bombardment Heating on CaF$_2$(111) by Means of Chemical-State Discriminated XPED

C. Akita[1], T. Tomioka[2], M. Owari[3], A. Mizuike[2], and Y. Nihei[3]

[1]Central Research Laboratory, Taiyo Yuden Co., Ltd.,
  562 Hongo-tsukanaka, Haruna-machi, Gunma 370-33, Japan
[2]Department of Industrial Chemistry, Science University of Tokyo,
  1-5 Kagurazaka, Shinjuku-ku, Tokyo 162, Japan
[3]Institute of Industrial Science, University of Tokyo,
  7-22-1 Roppongi, Minato-ku, Tokyo 106, Japan

Abstract. CaF$_2$ in the surface layer with a thickness of a few nanometers converted to CaO by electron bombardment heating above 300°C, but CaO was not formed by the lamp heating. This conversion is considered to occur by electron bombardment of the sample surface in heating. Furthermore it was found that CaO grew epitaxially on CaF$_2$(111) from chemical-state discriminated X-ray photoelectron diffraction (XPED) measurements. The crystallographic orientation of the CaO epitaxial layer was directly determined from the analysis of XPED patterns.

1. Introduction

X-ray photoelectron diffraction (XPED) is one of the useful methods giving information on the structural arrangement of the surface[1,2]. The angular distribution of the X-ray photoelectron intensities from a single-crystal sample exhibits many fine structures due to XPED from the regular arrangement of atoms. XPED is a method of measuring the X-ray photoelectron intensities for each angle and its advantages are similar to those of X-ray photoelectron spectroscopy (XPS). XPED is sensitive to the surface layers of a few nanometers thickness. In XPED measurement only X-rays are employed and destruction is much smaller in comparison with electron beam irradiation. In recent years, many investigations of XPED have been reported on various samples including pure crystals, adsorbate-substrate systems and surface altered layers. Most of the authors who have reported XPED patterns regarded the XPED measurement as a means of performing geometric analysis of atomic sites. The basic concept in these works is that the XPED pattern is the fingerprint of the emitter site.

    CaF$_2$ (fluorite) is particularly susceptible to electron beam damage[3,4]. As a result of the exposure of the CaF$_2$ surface to the electron beam, incident electrons induce the desorption of fluorine atoms in the surface region at first and further electron doses cause desorption of fluorine atoms in several surface layers of CaF$_2$ and aggregation of Ca atoms in the surface region. When the damaged CaF$_2$ surface is exposed to oxygen gas, oxygen occupies the surface fluorine vacant site for the lightly damaged surface, while oxygen atoms react with

Springer Series in Surface Sciences, Vol. 24 **The Structure of Surfaces III**
Editors: S.Y. Tong · M.A. Van Hove · K. Takayanagi · X.D. Xie
© Springer-Verlag Berlin, Heidelberg 1991

the Ca atoms to form CaO in the surface region for the heavily damaged surface. However, a study of the structure of CaO formed on CaF_2(111) has not been reported.

In this study, CaF_2(111) was damaged by electron bombardment heating and CaO was formed in the surface layers by the reaction with residual H_2O gas in the instrument. We report the effect of electron bombardment heating on CaF_2 by means of XPS and discuss the structure of CaO formed in the surface layers of CaF_2(111) from XPED patterns. In successive steps of the experiment from native CaF_2 to CaO formed by electron bombardment, Ca or O atoms may exist in two or more different chemical states. Therefore chemical-state discrimination is essential for the study of these systems. Because XPED patterns are the angular distributions of XPS peak intensity, by XPED measurements for chemically shifted XPS peaks, the structure can be analyzed for each phase which contains specific chemical state atoms[5]. In this study, structural analysis by chemical-state discriminated XPED was made for the first time.

2. Experimental

The X-ray photoelectron spectrometer used for the measurement has been described elsewhere[1]. In the XPED measurements, the polar angle of photoelectron taken-off was scanned by rotating the sample by 1^O steps about an axis lying on the sample surface and the azimuthal scan was carried out by rotation by 1.5^O steps about surface normal. The data acquisition time for F1s, Ca2p and O1s was 3, 5 and 10 minutes, respectively. XPED patterns were obtained by plotting XPS peak areas as functions of the angle. A 10mmx10mmx0.5mm piece of the polished CaF_2 (111) was loaded into a vacuum chamber as received. No impurity was detected except small amounts of carbon and oxygen.

Argon ion bombardment was performed until the C1s peak could no longer be detected. The pressure of argon gas was maintained at around $2x10^{-4}$ Torr during ion bombardment. The ion acceleration voltage was 400V. Then, photoelectron spectra and XPED patterns for F1s, O1s and Ca2p were measured by AlKα excitation.

After the above procedures, the sample was heated by electron bombardment of the biased mounting plate of the sample at 300 to 400^OC for 30 minutes. The potential difference between the mounting plate and the filament was about 60V. XPED measurement was performed subsequently. To compare the influence of the heating method, the sample was also subjected to the lamp heating after argon ion bombardment.

3. Results and discussion

3.1 Photoelectron spectra

After argon ion bombardment, a small O1s peak was observed at a maximum of 530.4eV. The O1s peak could never be removed by further argon ion bombardment. This O1s peak is considered to result from H_2O adsorbed on the surface of CaF_2, because residual gas analysis showed that the partial pressure of H_2O was about $2x10^{-8}$ Torr and the O1s peak disappeared by lamp heating at 300^OC.

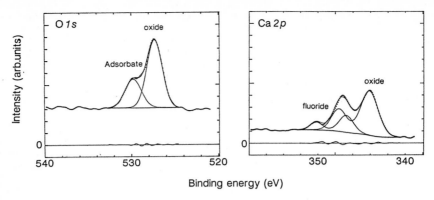

Fig.1. X-ray photoelectron spectra of O1s and Ca2p after electron bombardment heating at 300°C.

Figure 1 shows photoelectron spectra of O1s and Ca2p after
electron bombardment heating at 300°C. The O1s peak consisted
of two components and the Ca2p spectrum had three maxima.
These photoelectron spectra were deconvoluted to two sets of
components as shown in Fig.1. The O1s peak at a maximum of
ca. 530eV was assigned to adsorbate oxygen and one of ca.
528eV was assigned to oxide oxygen. The two Ca2p components
were assigned as follows: The Ca2p peak at $2p_{3/2}$ maximum of
about 347eV was calcium fluoride and the one at 344 to 345eV
was calcium oxide because of the necessity of having a counter
cation of oxide oxygen. These spectra indicate that CaO was
formed on the surface of $CaF_2(111)$ after electron bombardment
heating at 300°C. The thickness of the CaO layer was estimated
to be 2-3nm from the mean free path of photoelectrons. When
the sample was further heated at 400°C by electron bombardment,
the surface region within the depth of information of photoelectrons was almost converted to CaO.

3.2 XPED

XPED patterns from the sample heated at 300°C by electron bombardment are shown in Fig.2. For O1s or Ca2p, the intensity
(area) of each component together with total intensity were
plotted. The XPED pattern of F1s and fluoride Ca2p were essentially the same as those from clean $CaF_2(111)$ surface. Therefore bulk CaF_2 lay beneath the CaO surface layer. The XPED
patterns of O1s from CaO showed significant modulation in
contrast to O1s of the adsorbate. For the Ca2p XPED patterns
both oxide and fluoride components showed significant modulation. The modulation pattern for oxide Ca2p was different from
that of fluoride Ca2p. This indicates that the atomic arrangement around the oxide Ca is completely different from the
arrangement around the fluoride Ca.
At this stage, the sample consists of three phases: CaF_2
substrate, CaO overlayer and H_2O adsorbate. In this system,
calcium atoms exist in the overlayer and the substrate. Ca2p
photoelectrons excited in the CaO overlayer and those excited
in the CaF_2 substrate can be discriminated through the chemical
shift of XPS peaks. From separately plotting the two photo-

678

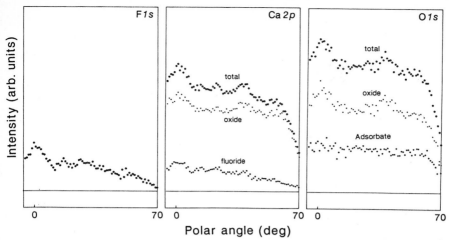

Fig.2. X-ray photoelectron diffraction (XPED) patterns of F1s, Ca2p and O1s after electron bombardment heating at 300°C.

electron intensities evaluated by XPS peak deconvolution, two XPED patterns were obtained, each of which reflected the structure of the individual phase. Therefore the procedure described above - chemical-state discriminated XPED - enables the separate discussion of the atomic arrangement around specific atoms of the same element but in a different phase. This particular advantage is best realized by the combination of chemical state discrimination by XPS and structural analysis of XPED. This study is the first example of chemical-state discriminated XPED.

After heating at 400°C, XPED patterns for CaO became more clear. The modulation amplitude of Ca2p XPED pattern from CaO was comparable to that from bulk CaF_2. This result indicates that thick CaO layer grew epitaxially on $CaF_2(111)$[6].

The crystallographic orientation of this epitaxial CaO layer can be determined as follows: (1)CaO crystallizes in the NaCl-type structure. (2)A polar angle scan XPED pattern of Ca2p or O1s from the CaO layer showed the largest peak at 0°. This indicates that one of the low index axes of the CaO epitaxial layer was parallel to the surface normal[7]. (3)An azimuthal scan XPED pattern of Ca2p or O1s at a polar angle of 36° showed three-fold symmetry. From this fact together with (1) and (2), it is obvious that the <111> axis is parallel to the surface normal. (4)In the O1s and Ca2p polar angle scan XPED patterns, a common large peak except 0° appears at 35°. Therefore one of the low index axes lies at 35° on the plane of the polar angle scan. From a simple consideration of the NaCl-type structure, this axis appears to be a <110> axis. The above discussion leads to the model for the orientational relation between CaF_2 substrate and CaO surface layer shown in Fig.3. The atomic arrangement in the boundary is thought to be more complex than shown in Fig.3, because the lattice mismatch between CaF_2 and CaO is 13%. Apart from detailed atomic arrangement in the boundary, this model is consistent with the dominant peaks on XPED patterns from the $CaO/CaF_2(111)$ system.

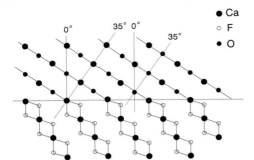

Fig.3. A model of orientational relation between CaO surface layer and CaF$_2$(111) substrate.

4. Conclusion

CaF$_2$ in the surface layer with a thickness of a few nanometers converted to CaO by electron bombardment heating at above 300°C. CaF$_2$ was damaged to desorb fluorine atoms by electron bombardment to the surface. The Ca atoms remaining in the surface immediately reacted with residual H$_2$O gas in the instrument to form CaO surface layer.

In this system, Ca or O atoms existed in two different chemical states and two components of the same element can be discriminated only through the chemical shift of XPS peaks. Accordingly separate XPED patterns were obtained on the basis of chemical-state discrimination. These XPED patterns indicated that CaO(111) grew epitaxially on CaF$_2$(111). Furthermore XPED patterns also clarified the orientational relation between CaF$_2$ substrate and CaO surface layer. This is the first time a structure based on XPED patterns has been obtained by chemical-state discrimination. This study clearly shows that the chemical-state discriminated XPED is a powerful tool for structural analysis of epitaxial thin films.

Acknowledgements

In the peak deconvolution procedure, the algorithm developed by Dr. A. E. Hughes of CSIRO Division of Material Science and Technology was used with small modifications.

References

1. M. Owari, M. Kudo, Y. Nihei and H.Kamada: J. Electron Spectrosc. Relat. Phenom. 22 (1981) 131.
2. U. Bardi, K. Tamura, M. Owari and Y. Nihei: Appl. Surf. Sci. 32 (1988) 352.
3. C. L. Strecker, W. E. Moddeman and J. T. Grant: J. Appl. Phys. 52 (1981) 6921.
4. K. Saiki, Y. Sato and A. Koma: Jpn. J. Appl. Phys. 28 (1989) L134.
5. Y. Nihei: Bunseki (1984) p.628 [in Japanese].
6. K. Tamura, M. Owari and Y. Nihei: Bull. Chem. Soc. Jpn. 61 (1988) 1539.
7. S. Y. Tong, H. C. Poon and D. R. Snider: Phys, Rev. B32 (1985) 2096.

Index of Contributors